8th Annual International Conference on Science and Engineering (8th AIC 2018)

Connecting the World through Innovation and Sustainable Development

IOP Conference Series: Materials Science and Engineering
Volume 523

Aceh, Indonesia
12 - 14 September 2018

Editor:

Sugiarto

ISBN: 978-1-5108-9083-1
ISSN: 1757-8981

Printed from e-media with permission by:

Curran Associates, Inc.
57 Morehouse Lane
Red Hook, NY 12571

Some format issues inherent in the e-media version may also appear in this print version.

Printed by Curran Associates, Inc. (2019)

For permission requests, please contact the Institute of Physics
at the address below.

Institute of Physics
Dirac House, Temple Back
Bristol BS1 6BE UK

Phone: 44 1 17 929 7481
Fax: 44 1 17 920 0979

techtracking@iop.org

Additional copies of this publication are available from:

Curran Associates, Inc.
57 Morehouse Lane
Red Hook, NY 12571 USA
Phone: 845-758-0400
Fax: 845-758-2633
Email: curran@proceedings.com
Web: www.proceedings.com

Table of contents

Volume 523

The 8th Annual International Conference (AIC) 2018 on Science and Engineering

12–14 September 2018, Aceh, Indonesia

Accepted papers received: 27 March 2019
Published online: 8 July 2019

Preface

Papers

Mathematical Sciences & Physical Sciences And Engineering

Civil and Structural Engineering

Electrical, Computer Engineering & Information System

Architecture, Urban and Rural Planning

Geology, Mining, Petroleum, Mechanical And Industrial Engineering

IOP Conf. Series: Materials Science and Engineering **523** (2019) 011001 doi:10.1088/1757-899X/523/1/011001

ANNUAL INTERNATIONAL CONFERENCE AIC 2018
UNIVERSITAS SYIAH KUALA

Preface the 8th AIC 2018

The 8th AIC 2018 International Conference is is hosted by Syiah Kuala University (recognizably abbreviated as UNSYIAH), the largest and the oldest national university in Aceh. The AIC is an annual international conference carried out regularly by Syiah Kuala University since 2011. The 8th AIC 2018 International Conference held on September 12-14, 2018, in Banda Aceh, Aceh Province, Indonesia.

Under the theme of the conference " Connecting the World through innovation and suistainable development ", this conference features a rich program, including Syiah Kuala University Innovation Expo 2018, a keynote speech delivered by Prof. Dr. Ahmad Fauzi Ismail, Universitas Teknologi Malaysia, Malaysia, and two invited speeches by Prof. Dr. Azman Bin Abd Ghani, University of Malaya, and Prof. Marwan, Syiah Kuala University. The 8th AIC on Science and Engineering 2018 offers a special opportunity to bring together professors, researchers and scholars around the globe, and serves as a platform to deliver innovative research results and latest trends and development in the fields of sciences and engineering.

The scientific papers published in these proceeding have been revised and approved by the scientific committee of the 8th AIC on Science and Engineering 2018. All the papers exhibit clear, concise, and precise expositions that appeal to a broad international readership interested in Science and Engineering. To all members of the organizing committees, the international scientific committee, the reviewers, and the collaboration partners, we would like to thank all of them for their tremendous efforts to organize this conference successfully.

We want to thank the organizing committee for having a successful conference, the UNSYIAH and Sponsors supporting the conference and we hope that all the attendees benefit from this conference.

Best Regards,

Dr. Heru Fahlevi
General Chair of AIC-SE 2018

IOP Conf. Series: Materials Science and Engineering **523** (2019) 011002 doi:10.1088/1757-899X/523/1/011002

ANNUAL INTERNATIONAL CONFERENCE
AIC 2018
UNIVERSITAS SYIAH KUALA

PROCEEEDING

The 8th AIC on Sciences and Engineering

The Annual International Conference 2018
Syiah Kuala University

"Connecting the World through Innovation and Sustainable Development"

Banda Aceh, Aceh, Indonesia
September 12-14, 2018

IOP Conf. Series: Materials Science and Engineering **523** (2019) 011002 doi:10.1088/1757-899X/523/1/011002

Advisory Board

IOP Conf. Series: Materials Science and Engineering **523** (2019) 011002 doi:10.1088/1757-899X/523/1/011002

Editorial Board

IOP Conf. Series: Materials Science and Engineering **523** (2019) 011002 doi:10.1088/1757-899X/523/1/011002

Organizing Committee

General Chair
Dr. Heru Fahlevi, Syiah Kuala University, Indonesia

General Co-Chair
Wahyu Rinaldi, Syiah Kuala University, Indonesia

Treasurer and Finance Chair
Dr. Ira Devi Sara, Syiah Kuala University, Indonesia

Conference Chair
Dr. Eng. Sugiarto, M.Eng, Syiah Kuala University, Indonesia

Editor in-Chief
Dr. Saiful, Syiah Kuala University, Indonesia

Website Developer
Muhammad Chandra Gunawan, Syiah Kuala University, Indonesia

IOP Conf. Series: Materials Science and Engineering **523** (2019) 011002 doi:10.1088/1757-899X/523/1/011002

International Scientific Committee

Scientific Committee Chair

Dr. Eng. Sugiarto, ST., M.Eng (Syiah Kuala University, Indonesia)

Scientific Committee Members

Prof. Ahmad Fauzi Ismail (Universiti Teknologi Malaysia)
Prof. Azman A Ghani (University of Malaya Kuala Lumpur)
Dr.-Ing. Joewono Prasetijo (Universiti Tun Hussein Onn Malaysia)
Dr. Cristian Tosa (Universitatea Tehnica Cluj-Napoca, Rumania)
Dr. Tien Dung Chu (University of Transport & Communications, Vietnam)
Dr. Satria Antonie (King Abdulaziz University, Saudi Arabia)
Dr. Billy G. Adhiperdana (Padjadjaran University, Indonesia)
Dr. Yusuf wibisono (Brawijaya University, Indonesia)
Dr. Abu Maskur (Sebelas Maret University, Indonesia)
Prof. Dr. Ir. Marwan (Syiah Kuala University, Indonesia)
Prof. Dr. Ir. Sri Aprilia, M.T (Syiah Kuala University, Indonesia)
Dr. Nasrul AR, MT (Syiah Kuala University, Indonesia)
Dr. Fitri Arnia, S.T., M.Eng.Sc (Syiah Kuala University, Indonesia)
Dr. Ira Devi Sara, ST., M. Eng.Sc (Syiah Kuala University, Indonesia)
Dr. Dra. Siti Rusdiana, M.Eng (Syiah Kuala University, Indonesia)
Dr. Febriani, S.Si, M.Si (Syiah Kuala University, Indonesia)
Muhammad Bahi, S.Si.,M.Sc.,Ph.D (Syiah Kuala University, Indonesia)
Dr. Muhammad Dani Supardan (Syiah Kuala University, Indonesia)
Dr. Sri Mulyati, ST, MT (Syiah Kuala University, Indonesia)
Dr. Husni, ST, M.Sc., M. Eng (Syiah Kuala University, Indonesia)
Dr.-Ing. Teuku Edisah Putra (Syiah Kuala University, Indonesia)
Dr. Ir. Mohd. Iqbal, M.T (Syiah Kuala University, Indonesia)
Dr. Muhammad Rizal, S.T, M.Sc (Syiah Kuala University, Indonesia)
Dr. Ir. Mochammad Afifuddin, M.Eng (Syiah Kuala University, Indonesia)
Dr. Azmeri, ST, MT (Syiah Kuala University, Indonesia)
Dr. Munira Sungkar, S.T., M.T (Syiah Kuala University, Indonesia)
Dr. Nasaruddin, S.T.,M.Eng (Syiah Kuala University, Indonesia)
Dr. Ramzi Adriman, S.T, M. Sc (Syiah Kuala University, Indonesia)

IOP Conf. Series: Materials Science and Engineering **523** (2019) 011003 doi:10.1088/1757-899X/523/1/011003

Welcome Speech from Rector of Syiah Kuala University

In the name of Allah, the Most Beneficent and the Most Merciful. May peace, mercy, and blessings of Allah be upon you.

Dear colleagues, professors, lecturers, researchers, ladies and gentlemen. On behalf of Syiah Kuala University, I would like to express my sincere gratitude and welcome you to the 8th Annual International Conference (AIC) on Science and Engineering 2018. Moreover, I honorably welcome our keynote and invited speakers Prof. Dr. Ahmad Fauzi Ismail, Universitas Teknologi Malaysia, Malaysia, Prof. Dr. Azman Bin Abd Ghani, University of Malaya, and Prof. Marwan, Syiah Kuala University.

It is my hope that the 8th AIC on Science and Engineering 2018 would be able to achieve its objective in providing an effective forum for academician, researchers, and practitioners to advancing knowledge, research, and technology for humanity. There has always been a gap between peoples and communities who can make effective use of technology and those who cannot. 'Humanity' (or being 'humane') can also in one usage refer to civil rights and social causes, or in other words to people treating each other with care, compassion, and dignity; respecting the common 'humanity' in the other person. Therefore, despite what we may have been falsely led to believe, matters of 'human' rights have no relation to politics or any part of the so-called political spectrum: they do not belong to any political party or faction, nor to any individual or organization; they are part of our common nature, part of the collective 'humanity' spoken for by our very existence. For that matter, all academicians, researchers, and practitioners should consider humanity things to make a peaceful world. With humanity, we change to the better environment and a better life. We believe knowledge, research and technology for all people, for minorities, people with disabilities, and to help anyone. No matter how much we can accomplish by ourselves, whether it be research or development, it is never sufficient in this world of knowledge. Therefore, the focal drive of this conference is to exchange ideas, and by participating in this exchange, it is hoped that all parties who may benefit from the conference can apply it in managing activities in their areas. It is pleasing to note that the agenda of this conference covers a wide range of interesting topics related to all theoretical and practical aspects, but not limited to Mathematical Sciences, Physics Sciences and Engineering, Chemical Sciences and Engineering, Geology, Mining and Petroleum Engineering, Mechanical and Industrial

IOP Conf. Series: Materials Science and Engineering **523** (2019) 011003 doi:10.1088/1757-899X/523/1/011003

Engineering, Civil and Structural Engineering, Architecture and Planning, Electrical and Computer Engineering, Computer Sciences and Information Systems.

Last but not least, my deepest gratitude goes to the Advisory Board, Organizing Committee, International Scientific Committee, institutions, companies, and volunteer who have directly and indirectly supported the success of this seminar. The committee has organized a vibrant scientific program and is working hard to present highly respected and internationally notorious speakers to lead it. Although we try our finest to be professional, on behalf of Syiah Kuala University, please accept our sincere apologies should there be inconveniences that occur before, during, or after the event. I wish you a very productive conference with exciting and encouraging discussions and exchange of knowledge so that together we can anticipate a future of groundbreaking knowledge, research, and technology for humanities.

May God bless us all with good health to make this event a successful and enjoyable one!

Best Regards,

Prof. Dr. Ir. Samsul Rizal, M.Eng
Rector of Syiah Kuala University

ANNUAL INTERNATIONAL CONFERENCE
AIC 2018
UNIVERSITAS SYIAH KUALA

Welcome Speech from General Chair

Assalamualaikum Wa Rahmatullahi Wa Barakatuh,

On behalf of the organizing committees, I would like to welcome all of you to Banda Aceh, Indonesia for the 8th Annual International Conference (AIC) on Science and Engineering 2018 as a part of the 2018 Annual International Conference (AIC) Syiah Kuala University. The AIC is an annual international conference carried out regularly by Syiah Kuala University since 2011.

Under the theme of the conference " Connecting the World through innovation and suistainable development ", this conference features a rich program, including Syiah Kuala University Innovation Expo 2018, a keynote speech delivered by Prof. Dr. Ahmad Fauzi Ismail, Universitas Teknologi Malaysia, Malaysia, and two invited speeches by Prof. Dr. Azman Bin Abd Ghani, University of Malaya, and Prof. Marwan, Syiah Kuala University. The 8th AIC on Science and Engineering 2018 offers a special opportunity to bring together professors, researchers and scholars around the globe, and serves as a platform to deliver innovative research results and latest trends and development in the fields of sciences and engineering.

All accepted and presented papers in this conference will have the opportunities to be published in IOP Material Science and Engineering Proceedings and indexed by Sci-Scopus, Thomson Reuters ISI, and EI Compendex. The conference has received 110 submitted papers, whereby 80 papers have been accepted by the committees for presentation and to be included in the proceedings. These papers on various topics are divided into 10 parallel sessions in the conference. To all members of the organizing committees, the international scientific committee, the reviewers, and the collaboration partners, we would like to thank all of them for their tremendous efforts to organize this conference successfully.

We look forward to having a successful conference, and we hope that all the attendees enjoy and benefit from this conference.

Best Regards,

Dr. Heru Fahlevi
General Chair

IOP Conf. Series: Materials Science and Engineering **523** (2019) 011004 doi:10.1088/1757-899X/523/1/011004

UNIVERSITAS SYIAH KUALA

Keynote Speaker

Nano-enabled Membrane Technology for Water and Wastewater Treatment

Ahmad Fauzi Ismail*, Zulhairun Abdul Karim, Goh Pei Sean

Advanced Membrane Technology Research Centre (AMTEC), Universiti Teknologi Malaysia, 81310 UTM Skudai, Johor Darul Ta'zim, Malaysia.
School of Chemical and Energy Engineering, Faculty of Engineering, Universiti Teknologi Malaysia, 81310 UTM Skudai, Johor Darul Ta'zim, Malaysia.
*Corresponding author. Tel.: +607 553 0360, Fax: +607 557 1700
Email: afauzi@utm.my / fauzi.ismail@gmail.com

Abstract

Recent developments in nanotechnology have resulted a breakthrough in water and wastewater treatment by membrane technology. Nano-enabled membranes have emerged as a significant innovation for providing safe and clean water as environmental discharge or municipal consumption though various membrane processes such as reverse osmosis, forward osmosis, adsorption, photocatalytic, nano- and ultrafiltration. The use of nanomaterials with amazing properties renders extraordinary capability of the newly developed membranes to treat all kinds of contaminated, nonpotable water containing salts (seawater), heavy metals, oil, solvent, and other organic pollutants. Wide ranges of nanoparticles, ranging from carbon allotropes (graphene oxide, carbon nanotubes), metal oxides, and clay minerals have been used to improve the performance of the membranes in terms of water permeability, selectivity, anti-fouling, anti-bacterial, adsorption capacity, and mechanical stability. Considerable attention is placed on tailoring the microstructure and surface chemistries of the nanoparticles involving optimization of the synthesis procedures and chemical modifications. Nano-enabled membranes with unique morphologies and configurations can be attained by tuning the fabrication techniques or using special devices (such as multiple orifice spinnerets). In essence, different membrane processes require different

IOP Conf. Series: Materials Science and Engineering **523** (2019) 011004 doi:10.1088/1757-899X/523/1/011004

innovative approaches to battle the current challenges for possible industrial deployment. The future of membrane technology for vast application in water and wastewater treatment relies on the advancement of nanotechnology. Intensive effort should be placed to expand the laboratory-scale research towards commercial level while taking into consideration the feasibility of the innovation to ensure cost effective technology can be proposed.

ANNUAL INTERNATIONAL CONFERENCE
AIC 2018
UNIVERSITAS SYIAH KUALA

Invited Speaker 1

Classification and Geochemistry of the Granitic rocks from Sumatera : an overview

Azman A Ghani
Department of Geology
University of Malaya Kuala Lumpur
Malaysia

Abstract

The Sumatera Mainland tectonically can be divided into three units: from southwest to northeast, they are the Woyla terrane, West Sumatra block, and East Sumatra block. The granitic rocks in Sumatera exposed mainly in the Woyla and East Sumatera block. The granites of Sumatera can be divided into two groups (1) the tin bearing and (2) the volcanic arc groups. The tin bearing group occurs mainly in the islands off eastern coast of Sumatera (Bangka, Belitung) and the volcanic arc suites which confines to the Barisan Range and western coastal area of Sumatera island. Geochemically the granitic rocks of volcanic arc group is more diverse ranging from granite to gabbro whereas the tin bearing group is mainly syenogranite and monzogranite. The tin bearing group is comparable to the Main Range granite (Peninsular Malaysia and Southern Thailand) in term of the composition and their age (200- 220Ma). The volcanic arc suite is younger in age ranging from 50 to 150 Ma. Th /Yb vs Nb/Yb diagram indicate that magma from both granitic groups originated from continental as their source rock. It is however the granitic magma from the volcanic arc group is clay poor whereas the tin bearing granites have more clay rich material.

IOP Conf. Series: Materials Science and Engineering **523** (2019) 011004 doi:10.1088/1757-899X/523/1/011004

Invited Speaker 2

Microwave heating to accelerate solid catalyzed chemical reactions

Prof. Dr. Marwan

Chemical Engineering Department, Syiah Kuala University
Banda Aceh – Indonesia
E-mail: marwan@unsyiah.ac.id

Abstract

Microwave has been utilized to enhance many chemical reactions due its effect on intermolecular interaction among the reactants. The system also offers simple, clean, efficient, economic, and more environmentally friendly technique. As the result, the microwave irradiation has emerged as new tool in organic synthesis. In this opportunity, some progress on the catalytic reactions involving microwave heating is briefly reported. The biodiesel production from palm oil using hydrated calcined seashell catalyst was carried out under microwave irradiation. The reaction was accomplished in a very short time at much milder conditions than in the conventional reactor with a maximum yield of 96.0% in 10 min reaction. Recently, triacetin, an alternative biodiesel additive, was prepared by esterification of glycerol using activated natural zeolite catalyst. The conversion value of glycerol was more than 95% after 60 min and the reaction took place at much milder condition compared with those reported elsewhere with non-microwave heating. These findings suggest the microwave effect could overcome mass transfer barrier that often limit various solid catalytic reactions.

Peer review statement

All papers published in this volume of *IOP Conference Series: Materials Science and Engineering* have been peer reviewed through processes administered by the proceedings Editors. Reviews were conducted by expert referees to the professional and scientific standards expected of a proceedings journal published by IOP Publishing.

IOP Conf. Series: Materials Science and Engineering **523** (2019) 012001 doi:10.1088/1757-899X/523/1/012001

Risk factor for mortality of children among victims in Southern Thailand

Z M Kesuma[1,*], L Rahayu[2], M Eso[3], Rasudin[4], R Saputra[5]
[1,2,5] Statistics Department, Faculty of Mathematics and Natural Science, Syiah Kuala University, Banda Aceh, Indonesia
[3]Department of Mathematics and Computer Science, Faculty of Science and Technology, Prince of Songkla University, Pattani, Thailand
[4] Informatics Department, Faculty of Mathematics and Natural Science, Syiah Kuala University, Banda Aceh, Indonesia

Email:kesumaku@yahoo.com

Abstract. Conflict can be defined as any type of controversy interactions between two or more parties which occur because of the differences in status, aim, value or perception. The consequence of the conflict in a region was the incidence of fatalities, both the injured and died. Children are the society at greater risk to be the victims because of the conflict. In Southern Thailand, conflicts occur in 4 provinces which the Muslim majority, those are Pattani, Yala, Narathiwat, and Songkhla. The aims of this research have identified the factors that influence the risk of child deaths during the conflict in Southern Thailand and determined the models by influence factors. The used method was the binary logistic regression method. This method was chosen because the response variable that used is dichotomous. The dichotomous response variable was the risk of children died or not during the conflict in Southern Thailand. The data used were the number of children under 18 years old who were victims of the conflict from 2004 until 2014 in Southern Thailand. The results showed that the influence risks of child death during conflict in Southern Thailand are gender (odds ratio = 1.99; CI 95% = 1.38-2.88), type of guns that used was bomb (odds ratio = 0.16; CI 95% = 0.1-0.26) and the others (odds ratio = 2.46; CI 95% = 1.55-3.9), and the place of the hardness occurred was open field (odds ratio = 2.9; CI 95% = 1.54-5.48). Generally, children there were in the open field and female have the greater risk of death to be victimized by bombs, rifles or any other types of guns during conflict in Southern Thailand.

1. Introduction

Conflict can be defined as one part of social interaction in the dissociative form that occurs in the seizure of purpose by involving a conflicting person or group (Littlejohn and Domenici, 2007). Basically the initial causes of conflict in a region are individuals or small groups in society.

*Corresponding author: kesumaku@yahoo.com

According to a report from the National Counterterrorism Center (2011), there have been war conflicts in 72 different countries resulting in 70% of injuries and 30% of deaths. Most of the victims of war conflict are the Muslim group, and the most vulnerable job target of becoming victims are the police or other military groups.

The Asian continent is one of the continents with the most conflict events, spreading from the Middle East to South Asia. Vulnerable conflicts in the Asian region are created by the historical relics of colonialism from their former colonies that combine two or more ethnic opposites in one territory (Bastari, 2013).

Conflict in Southeast Asia is also common. Religious issues are central to the emergence of conflicts in the region. Most of the conflicts in Southeast Asia involved minority Muslim groups who were victims. Countries that often conflict in Southeast Asia are Southern Thailand and Myanmar. Southern Thailand is a region in Thailand bordering the Malaysian peninsula. This place consists of 14 regions namely Narathiwat, Pattani, Yala, Songkhla, Krabi, Phuket, Nakhon Si Thammarat, Phang Nga, Phatthalung, Ranong, Surat Thani, Trang and Satun. The majority of the population is Buddhist, except in 5 provinces of Pattani, Yala, Narathiwat, Songkhla and Satun which are predominantly Muslim. Of the five provinces in the southern part of the Islamic Republic, there are four areas that suffered considerable losses due to the conflict, namely Pattani, Yala, Narathiwat and Songkhla.

One of the earliest causes of the conflict in southern Thailand was when a group of people attacked an army base in Narathiwat province, seized a large number of weapons and killed four soldiers. Over the next few days the conflict continues with several schools being burned, houses and government buildings attacked unending bombings and killings.

Most of the victims were civilians, women and children. They suffered injuries and some were killed by gunfire, bombs or other types of weapons. Children are people who are at greater risk to become victims of conflicts that occur in a region. This is because they feel or see directly the violence that occurred.

The purpose of this study was to understand how much the risk of death of children during the war conflict that occurred in Southern Thailand. The method of analysis used is binary logistic regression. This method is chosen because the response variable used is a dichotomy. The dichotomous response variable in this study was the risk of children being killed or not during the conflict in Southern Thailand. While the predictor variables in this study were gender, religion, age group, type of weapon used, province, and place of violence. Binary logistic regression method can illustrate the relationship of several predictor variables with the dichotomous response variable.

2. Literature review

2.1. Conflict

Conflict is one part of social interaction in the form of dissociative that takes place by involving people or groups that challenge each other with the threat of violence (Budiyono, 2009). The potential for conflict in an area is very vulnerable, this is because the conflict is inherent. Inherent is a potential conflict that will always exist and happen in every space and time, anywhere and anytime.

According to Dahrendrof's theory, conflict will lead to change and development. In conflict situations, the groups involved take actions to make changes in the social structure. And if the conflict happens so great there will be a radical change. These changes are included in cultural change (Dahrendorf, 1959).

Most victims of conflict in an area according to Peleg et al. (2004) are women and children. In armed conflict, usually innocent and uninvolved children are often targeted by victims of violence, murder, or torture. The conflicts that occur can have a direct impact on the growth and development of mental health of children. This is because they feel or see directly the violence that occurred.

Based on the above explanation, in this study the variables chosen as predictor variables to see the factors that cause children to be victims of the conflict are gender, religion, age group, type of weapon used, province, and place of violence. The definitions of each predictor variable are as follows:

- *Gender*

Sex factors are particularly vulnerable to the victims of conflict. Genders are more likely to be victims when the conflict is women. While children of both men and women have the same risk in conflict areas, because their ability to survive is still very weak.

- *Religion*

Religion is a belief shared by a person r group of people who live in a certain area. According to Busaban et al. (2014), most of the victims of the war conflict that occurred in Southern Thailand were a group of Muslims.

- *Age group child.*

According to MOH (2009), the age-sharing of children according to school age consisted of children under five (0 to <5 years), pre-school children (5 to <6 years) and school-aged children (6 to ≤18 years).

- *Weapon type*

Several types of weapons used by military and non-military personnel involved in the conflict in Southern Thailand are rifles and explosives such as bombs and cannons.

- *Province*

The most vulnerable areas of conflict in Southern Thailand include four provinces: Pattani, Yala, Narathiwat and Songkhla (Busaban, 2014).

- *Place of violence*

Many places were targeted during the conflict. Among the places often targeted during the conflict were schools, residents' residences, offices, highways, military posts and other public places where there were many gathered communities.

2.2. Binary logistic regression

Logistic regression is a regression analysis used to see the relationship between response / bound variable with some predictor / free variable, either numerical or categorical. Logistic regression allows one to predict the discrete results of dichotomous-shaped response variables with predictor variables having different types such as discrete, continuous, and dichotomous (Hosmer dan Lemeshow (2000).

Binary logistic regression is a regression method that can describe the relationship between a dichotomous/binary response variable with several predictor variables. It says dichotomous because it consists of two categories namely "1" if successful and "0" if it fails. Therefore, the response variable following the Bernoulli distribution with the probability function is:

$$f(y) = \pi^y (1-\pi)^{1-y} \; ; y = 0,1 \tag{1}$$

Where if $y = 0$ then $f(y) = 1-\pi$ and if $y = 1$ then $f(y) = \pi$.

The logistic regression function can be written as:

$$f(z) = \frac{1}{1+e^{-z}} \text{ equivalent } f(z) = \frac{e^z}{1+e^z} \tag{2}$$

Where :

$z = \beta_0 + \beta_1 x_1 + \cdots + \beta_k x_k$; k = the number of independent variables.

2.3. Parameter estimation

The most commonly used method of parameter estimation in logistic regression is the Maximum Likelihood Estimation (MLE) method for obtaining maximum completion of the likelihood function. Each observation for the logistic regression model is a random variable from the Bernoulli distribution. The principle of Maximum Likelihood method is to estimate β to maximize likelihood function. Suppose that n sampled vector of random variables is taken $Y_1, Y_2, \ldots, Y_n$ with $Y_i = [y_1 y_2 \ldots y_n]^T$ and probability of h-category, $p_h(x_i)$ then likelihood function is:

$$L(\beta_0, \beta_1, \ldots, \beta_i) = \prod_{i=1}^{n} [\pi_i^{y_i}(1-\pi_i)^{1-y_i}] \quad (3)$$

Where : y_i = observations on i th variables

π_i = Probability for i th predictor variabel

2.4. Parameter significance testing

After the parameter is predicted, the next step is to test the significance of the parameters. The test used to test the significant β coefficients of the model can be tested partially or simultaneously.

2.4.1. Simultaneous testing. Performed as an attempt to examine the role of predictor variables in the model simultaneously.

Hipotesis:

H_0: $\beta_1 = \beta_2 = \ldots\ldots = \beta_k = 0$, (simultaneously the predictor variable has no significant effect on the response variable)

H_1 : There is at least one $\beta_i \neq 0$ (i = 1, 2, 3..., k), (at least one predictor variable has a significant effect on the response variable)

Test Statistics used is *G likelihood ratio test:*

$$G = -2\ln\left[\frac{L_0}{L_1}\right]; G = -2\ln\left[\frac{\left(\frac{n_1}{n}\right)^{n_1}\left(\frac{n_0}{n}\right)^{n_0}}{\prod_{j=1}^{n} p_j^{y_j}(1-p_j)^{1-y_j}}\right] \quad (4)$$

$$G = 2[A - B] \quad (5)$$

where,

$A = \sum_{j=1}^{n}(y_j\, ln(1-p_j) + (1-y_j)\, ln(1-p_j))\; B = [n_1\, ln(n_1) + n_0\, ln(n_0) - n\, ln(n)]$

n_1 = number of categorical observations 1

n_0 = number of categorical observations 0

n = number of observation

y_j = response variable on observation jth, with j=1, 2, 3...n.

p_j = probability of observation jth

The test statistic G follows the chi-square distribution with the rejection criterion $G > \chi^2_{(db,\alpha)}$ (Hosmer and Lemeshow, 2000).

2.4.2. Partial Test. Partial test is done to know the significance of parameters to the dependent variable. Partial / individual test results will show whether an independent variable is feasible to enter the model or not (Agresti, 1990), with the following hypothesis:

H_0: Predictor variable i^{th} is not significant to the dependent variable

$\beta_i = 0$, for i = 1,2,..., k

H_1: Predictor variable i^{th} is significant to the dependent variable

$\beta_i \neq 0$, for i = 1,2,..., k

where:

i = number of predictor variables

Test Statistics:

$$\text{Wald } (W) = \frac{\beta_i}{SE(\beta_i)} \tag{6}$$

The ratio generated from the test statistic, under the hypothesis will follow the normal distribution (Hosmer dan Lemeshow, 2000). So the decision is obtained by comparing test statistic with standard normal distribution *(Z)*. Rejected criteria for H_0, if $W > Z_{\alpha/2}$.

2.5. Goodness of Fit Test

Goodness of fit is a test used to see the suitability of a model. If the Hosmer and Lemeshow Test is satisfied then the model is able to predict the observed value or the model is appropriate.

Hipotesis:

H_0: Model is appropriate

H_1: Model is not appropriate

Statistical test *Hosmer and Lemeshow* $(\hat{C})$ which is calculated based on the value $y = 1$ is formulated as:

$$\hat{C} = \sum_{r=1}^{g} \frac{(o_r - n_r\bar{p}_{1r})}{n_r\bar{p}_{1r}(1-\bar{p}_{1r})} \tag{7}$$

Where:

$\bar{p}_{1r}$ = average estimated probability of group success r^{th}

o_r = number of samples of successful events r^{th} group.

n_r = total sample r^{th} group, with $r = 1, 2, \ldots, g$.

Test statistics $\hat{C}$ is approximately *Chi-Square* distribution, with degree of freedom g-2, (Hosmer dan Lemeshow, 2000). Rejects H_0 if $\hat{C} > \chi^2_{\alpha:g-2}$.

2.6. Interpretation of Logistics Regression Parameter

In logistic regression modeling, parameter interpretation aims to find out the meaning of parameter estimation value on predictor variable. There are two types of predictor variables, which are categorical variables and continuous variables. The way used to interpret logistic regression parameters from categorical variables is by odds ratio.

Odds is the proportion of probability of a successful event with an unsuccessful event in a category. The odds ratio is the ratio of the odds value for category x = 1 to the odds for the category x = 0, in the same predictor variable assuming the other predictor variable is constant. The odds ratio is expressed by ψ and is written as:

$$\psi = \frac{\left(\frac{\pi(1)}{1-\pi(1)}\right)}{\left(\frac{\pi(0)}{1-\pi(0)}\right)} \tag{8}$$

where :

$\pi(1)$ = Probability of success event

$\pi(0)$ = Probability of failure event

To interpret logistic regression parameters of continuous variables it is assumed linear logit function to independent variable. Suppose that the independent variable is continuous, and functions $g(x) = \beta_0 + \beta_1 x_1$ has the same properties as the parameters in linear regression. Every increase of one unit x, the value of $g(x)$ will increase β_1, or $\beta_1 = g(x+1) - g(x)$ for every value (Hosmer dan Lemeshow, 2000).

3. Methods

The data for this study was obtained from *Deep South Co-ordination Center* (DSCC), Prince of Songkla University (PSU), which was compiled by police, military and local government reports. It comprises of the number of children under 18 who were victims of the conflict from 2004 to 2014 in Southern Thailand. The total sample used is 1,475 data in 4 provinces in Southern Thailand, namely Pattani, Yala, Narathiwat and Songkhla.

Table 1. Data of child victims of South Thailand conflict

Province	Sample Size
Narathiwat	530
Pattani	406
Yala	413
Songkhla	126
Total	1.475

Software used for data processing in this research is R, library Epicalc. The method used to analyze the data is binary logistic regression method. This method can identify any variables that affect the risk factors of child mortality during conflict in Southern Thailand.

Table 2. Definition of research variables

Respon Variable	Category
Child mortality risk (Y)	1 = not die
	2 = die
Predictor Variabel	**Category**
Sex (X_1)	1 = Female
	2 = Male
Religion (X_2)	1 = Non-Moeslem
	2 = Moeslem
Child Age Group (X_3)	1 = 0 to <5 tahun
	2 = 5 to <6 tahun
	3 = 6 to ≤ 18 tahun
Type of weapon used (X_4)	1 = Bomb
	2 = Gun
	3 = Other
Province(X_5)	1 = Narathiwat
	2 = Yala
	3 = Pattani
	4 = Songkhla
Place of violence (X_6)	1 = Main road
	2 = Private House
	3 = Worship place
	4 = School
	5 = Offices
	6 = Market
	7 = Official residence
	8 = Military Base
	9 = Field

The data analysis procedure used in this research are: Inferencing analysis using binary logistic regression method. The steps of data analysis using binary logistic regression are as follows:

- *Perform parameter estimation by using Maximum Likelihood Estimation* (MLE).

- *Perform testing of parameter significance*

There are two tests used to test the significance of parameters, namely partial test and simultaneous test. Partial test using Wald (W) test while simultaneous test using Test G or Likelihood Ratio as test statistic. Both of these tests were performed to answer the previously formulated hypothesis.

- *Goodness of Fit Test*

Is used to determine whether the model is able to predict the value of its observations or whether the model is appropriate. The test statistic used is Hosmer Test and Lemeshow. Rejection criteria H0 if $\hat{C} > \chi^2_{a:g-2}$

- *Model Interpretation*

The way to interpret logistic regression parameters from categorical variables is by using odds ratios. Increasing one unit x then odds will rise by e^{β} times. In addition, model interpretation can also be done by analyzing the probability of influential variables in the model.

4. Result

4.1. Descriptive Analysis

Descriptive analysis of child victims of conflict data in Southern Thailand aims to describe the characteristics of the research variables.

Table 3. Characteristics of research data (n=1.475)

Variable	Category	n(%)
Child mortality risk (Y)	1 = not die	1.247(84.5%)
	2 = die	228(15.5%)
Sex (X_1)	0 = Female	953(64.6%)
	1 = Male	522(35.4%)
Religion (X_2)	0 = Non-Mouslem	523(35.5%)
	1 = Mouslem	952(65.5%)
Child Age Group (X_3)	1 = 0 to <5 tahun	200(13.6%)
	2 = 5 to <6 tahun	45(3.1%)
	3 = 6 to ≤18 tahun	1.230(83.4%)
Type of weapon used (X_4)	1 = Bomb	593(40.2%)
	2 = Gun	749(50.8%)
	3 = Other	133(9.0%)
Province (X_5)	1 = Narathiwat	530(35.9%)
	2 = Yala	413(28.0%)
	3 = Pattani	406(27.5%)
	4 = Songkhla	126(8.5%)
Place of violence (X_6)	1 = Main road	772(52.3%)
	2 = Private House	185(12.5%)
	3 = Worship place	15(1.0%)
	4 = School	23(1.6%)
	5 = Offices	45(3.1%)
	6 = Market	258(17.5%)
	7 = Official residence	80(5.4%)
	8 = Military Base	20(1.4%)
	9 = Field	77(5.2%)

Table 3 Shows that the respondents who died during the South Thai conflict amounted to 228 children, and the rest were injured. A total of 953 girls became victims during the conflict in Southern Thailand. Most of them are mouslem. Children aged 6 to ≤18 years were found to be more victims during the conflict. This means that most of the victims are children of school age. There are 749 children

who have been affected by gun and the rest are hit by bombs or other types of weapons. In total it can be said that 503 victims were from Narathiwat province, 413 victims were from Yala province, 406 were from Pattani province and the rest were from Songkhla province. The scene of the most violence during the South Thai conflict was on the highway with 772 occurrences.

4.2. Cross Tabulation

Cross tabulation is used to calculate cases that have a combination of values of two or more variables.

Table 4. Frequency Distribution of Child mortality risk

Variable	Not Die		Die		Total	
	n	%	n	%	n	%
Sex *(X_1)*						
Female	770	61,7	183	80,3	953	64,6
Male	477	38,3	45	19,7	522	35,4
Total	**1247**	**100**	**228**	**100**	**1475**	**100**
Religion *(X_2)*						
Non-Mouslem	452	36,2	71	31,1	523	35,4
Mouslem	795	63,8	157	68,9	952	64,6
Total	**1247**	**100**	**228**	**100**	**1475**	**100**
Child Age Group *(X_3)*						
0 to < 5 tahun	169	13,6	31	13,6	200	200
5 to < 6 tahun	37	3	8	3,5	45	45
6 to ≤ 18 tahun	1041	83,4	189	82,9	1230	83,4
Total	**1247**	**100**	**228**	**100**	**1475**	**100**
Type of weapon use *(X_4)*						
Bomb	570	45,7	23	10,1	593	40,2
Gun	591	47,4	158	69,3	749	50,8
Other	86	6,9	47	20,6	133	9
Total	**1247**	**100**	**228**	**100**	**1475**	**100**
Province *(X_5)*						
Narathiwat	456	36,6	74	32,4	530	35,9
Yala	333	26,7	80	35,1	413	28
Pattani	339	27,2	67	29,4	406	27,5
Songkhla	119	9,5	7	3,1	126	8,6
Total	**1247**	**100**	**228**	**100**	**1475**	**100**
Place of violence *(X_6)*						
Main road	654	52,4	118	51,8	772	52,3
Private House	140	11,2	45	19,7	185	12,5
Worship place	13	1	2	0,9	15	1
School	18	1,4	5	2,2	23	1,6
Offices	38	3	7	3,1	45	3,1
Market	232	18,6	26	11,4	258	17,5
Official residence	79	6,3	1	0,4	80	5,4
Military Base	16	1,3	4	1,8	20	1,4
Field	57	4,6	20	8,8	77	5,2
Total	**1247**	**100**	**228**	**100**	**1475**	**100**

Girls have a higher rate as victims (died/injured) compared to boys, during the South Thai conflict (64.6%). The majority of the victims were Muslim (64.6%), with ages 6 to <18 years (83.4%). The percentage of victims affected by gunfire was also very high (50.8%). Roads are the most vulnerable place for children to be affected by the conflict (52.3%). (See Table 4).

4.3 Parameter Significance Testing

Parameter significance test of the predictor variable is done to determine whether the parameters obtained have a significant effect on the model or not.

Table 5. The relationship between risk factors and child mortality

Variable	OR [95% CI]	*Wald's Test* (*p-value*)	L-R *Test* (*p-value*)
Sex Female (ref = Male)	1,99 [1,38; 2,88]	< 0,001	< 0,001
Religion Mouslem (ref = Non-mouslem)	0,88 [0,63; 1,23]	0,456	0,458
Child Age Group (ref= 6 to ≤ 18 thn)			0,548
0 to < 5 thn	1,09 [0,7; 1,71]	0,69	
5 to < 6 thn	1,6 [0,69; 3,7]	0,275	
Type of weapon use (ref= gun)			< 0,001
Bomb	0,16 [0,1; 0,26]	< 0,001	
Other	2,46 [1,55; 3,91]	< 0,001	
Province (ref= Narathiwat)			0,067
Yala	1,43 [0,98; 2,08]	0,065	
Pattani	0,94 [0,64; 1,39]	0,762	
Songkhla	0,6 [0,24; 1,52]	0,279	
Place of violence (ref= Main road)			0,009
Private House	1,28 [0,85; 1,92]	0,242	
Worship place	1,06 [0,22; 5,21]	0,941	
School	2,3 [0,75; 7,01]	0,143	
Official residence	0,39 [0,15; 1,02]	0,054	
Military Base	1,23 [0,75; 2,03]	0,418	
Field	0,44 [0,05; 3,57]	0,441	
Official residence	0,76 [0,24; 2,41]	0,643	
Military Base	2,9 [1,54; 5,48]	0,001	

The determination of the reference category used is based on the interest on the category to be seen or based on the highest number of victims. Based on likelihood ratio test value of P (LR-test) presented in table 5, it appears that there is at least one predictor variable has a significant effect on risk factor of child mortality. (*P value*$<\alpha$ or 0,001 < 0.05). On the other hand, based on the value of the partial test P(*Wald's-test*) which is presented in table 5, it can be seen that there are 3 variables that significantly influence the risk of death, they were sex (X1), type of weapon used (X4) and place of violence (X6).

4.4. Goodness of Fit Test

Goodness of Fit Test was done by Hosmer dan Lemeshow test.

Based on Hosmer and Lemeshow test results, we found Chi-Square value = 4.796, df = 8. We conclude that it fails to reject H0, and the model fits the data.

4.5. Interpretation of Binary Logistic Regression Model

After testing the significance of the parameters and fit model test, then the next step is to interpret the model. One way to interpret binary logistic regression models is the odds ratio. Here is an interpretation of these variables: *Sex* (X_1) Girls are more likely to be at risk of being killed in conflict in Southern

Thailand by 1.99 times than boys with 95% of confidence intervals (1.38, 2.88); *Types of weapons used (X4)* Victims affected by the bomb were at risk of dying by 0.16 times compared with the ones who was exposed to a gun with a 95% confidence interval (0.1, 0.26); As for those affected by weapons other than bombs and rifles have a risk of dying 2.46 times compared to those who were victims of a rifle with 95% confidence interval (1,55; 3,91); *Place of violence (X6)* Children in the field had a 2.9 times higher risk of death than those ones on the highway with a 95% confidence interval (1.54; 5.48).

5. Discussion

Study aimed to examine the risks of child death during conflict in Southern Thailand. The sample encompassed more girls than boys due to the situation in the field. To get better insight on child risk, the sample size and comparison between male and female samples should be balanced. That way comparison between different types of child age group could be done, giving insight in how specific factors affecting risk of child death during the conflict in Southern Thailand.

Finally, even though the study gives observational data, it confirmed early findings of gender, type of weapon (bomb and other) and place of violence (field) are risk factors that should be taken into account in seeing the relationship between the conflict area and the risk of children becoming victims.

6. References

[1] Agresti, A. 1990. *Categorical Data Analysis*. John Wiley Sons. Inc, New York.

[2] Barenbaum, J., Ruchkin, V., and Schwab-Stone, M,. 2004. *Journal of Child Psychology and Psychiatry*. **45**(1): 41–62.

[3] Bastari, G.R. 2013. *Conflict in South Asia.*

[4] Budiyono. 2009. *Sosiologi Untuk SMA dan MA kelas XI.* Departemen Pendidikan Nasional, Jakarta.

[5] Busaban, C., Kuning, M., McNeil, N. 2014. *Kasetsart J.* **35**: 84-91.

[6] Casella, G., and Berger, R.L. 2002. *Statistical Inference*. Duxbury, USA.

[7] Dahrendorf, R. 1959. *Class and Class Conflict in Industrial Society*. Stanford University Press, California.

[8] Departemen Kesehatan RI. 2009. *Kondisi Pencapaian Program Kesehatan Anak Indonesia*. Depkes RI, Jakarta.

[9] Hosmer, D. W and Lemeshow, S. 2000. *Apllied Logistic Regression.* John Wiley dan Sons, Canada.

[10] Marohabout, P., Choonpradub, C., and Kuning, M. 2009. *Songklanakarin Journal of Social Sciences and Humanities*. **15**(6): 883-895.

[11] McCloskey, L.A., Figueredo, A.J., Koss, M.P. 1995. *Child Development*. **66**:1239-1261.

[12] National Counterterrorism Center. 2011. *2010 Report On Terrorism*

[13] Peleg, K., Daniel, L., and Stein, M. 2004. *Annals of Surgery*. **239**(3): 311-318.

[14] Stoner, J.A.F, and Wankel, C. 1986. *Management.* Prentice-Hall International, Inc. New Jersey.

[15] Thnaprarnsing, P. 2009. *Solving the Conflict in Southern Thailand.* US Army War College. Pennsylvania.

IOP Conf. Series: Materials Science and Engineering **523** (2019) 012002 doi:10.1088/1757-899X/523/1/012002

Generalized additive models fitting with autocorrelation for sea surface temperature anomaly data

S Ananda[1,*], Miftahuddin[1]
[1]Department of Statistics, Faculty of Mathematics and Sciences, Syiah Kuala University, Banda Aceh 23111, Indonesia

E-mail: shafia.ananda@students.stat.unsyiah.ac.id

Abstract. Climatic conditions in Sumatra Island are affected by Sea Surface Temperature Anomaly (SSTA) in the Indian Ocean or more commonly referred to as Indian Ocean Dipole (IOD). Extreme climate events also closely related to SSTA. Several climate features that affect SSTA such as air temperature, precipitation rain, relative humidity, wind speed, and solar radiation. SSTA is an increase or decrease in the mean of Sea Surface Temperature so required the analysis to assess extreme climatic events to the risk due to the occurrence of anomalies. Generalized Additive Models (GAM) with autocorrelation can be used to model this phenomenon. GAM method accommodates the nonlinear influence between response variables and predictor variables. The data used in this research is the time series data, daily data from 2006-2017 and there are gaps in it, where there is an autocorrelation value. The purpose of this research is to get a representative model and to know the factors that influence to SSTA. The results show that GAM's best model with autocorrelation is GAM model by including month and year variables with the monthly autocorrelation structure. Factors that affecting SSTA are the air temperature, month and year as time covariates.

1. Introduction

The Earth's climate system is affected by many parameters. Sea Surface Temperature (SST) is one of those parameters. The SST parameter is very useful in obtaining indications of Earth's climate and variability, such as tropical climate variability. SST prediction data is used as an indicator to detect many marine phenomena, such as the Indian Ocean Dipole (IOD), monsoon, and El Nino Southern Oscillation (ENSO) consisting of El Nino and La Nina phenomena [1]. 71% of the Earth's surface is an ocean, and 97% of all water in the earth with a total volume of more than 1 billion km^3 is sea water, so it is inevitable that the oceans greatly affect the movement and circulation of the atmosphere and weather in any region of the earth [2] .

Indonesia's climate conditions are strongly influenced by Indonesia's position and its surrounding atmospheric and surface status, such as the Sea Surface Temperatures Anomaly (SSTA) in the Indian Ocean region. SSTA in the Indian Ocean region is often referred to as Indian Ocean Dipole (IOD) or Dipole Mode Index (DMI) [3]. The SSTA is also closely related to extreme climatic events. The province of Aceh is a province located on the westernmost part of the island of Sumatra and faces directly with the Indian Ocean so that it has a considerable impact due to climate change caused by the incident SSTA. Key weather and climate features and also affect SSTA such as air temperature, rainfall, humidity, wind, and solar radiation [4].

*Corresponding author: shafia.ananda@students.stat.unsyiah.ac.id

An analysis that examines extreme climatic events is needed to minimize the adverse effects of the SSTA event. One method for estimating models in extreme studies is Generalized Additive Models (GAM) [5]. GAM methods can accommodate well with non-linear influence between response variables and predictor variables. The modelling started from model fitting using GAM whereas previously it has been shown that linear regression model as fundamental fitting for SST data [6]. The data used in this research comes from the National Oceanic and Atmospheric Administration (NOAA) with 8N90E position, which is a point close to Aceh Province.

2. Literature Review

2.1. Sea Surface Temperature Anomaly (SSTA)

SSTA in western Sumatra and eastern Africa is often referred to as the Indian Ocean Dipole (IOD) or Dipole Mode Index (DMI), an sea surface temperature anomaly in the Indian Ocean characterized by a zonal pattern. SSTA is an increase or decrease in the mean of Sea Surface Temperature (SST). SSTA is calculated based on the result of the reduction between the actual SST value and the average SST value of the place in question [7].

If the actual SST value is higher than the average value, then the SSTA value will be positive. Otherwise, if the actual SST value is lower than the average value, then the SSTA value will be negative. Indian Ocean Dipole (IOD) has its own classification. The classification of IOD is distinguished:

1) Normal IOD, if the IOD value is in the range of -0.4 to 0.4.
2) IOD is strong, if the IOD value is in less than -0.4 and more than 0.4.

If the IOD value is < -0.4 then there is an increase in water vapor which in turn increases rainfall. Conversely, if the value of IOD is > 0.4 then there is a decrease in the number and incidence of rain so causing the incidence of drought [8].

2.2. Generalized Additive Models (GAM)

Generalized Additive Models (GAM) is the extended of Generalized Linear Models (GLM) by replacing linear functions $\sum_{j=1}^{p} \beta_j X_j$ with additive functions $\sum_{j=1}^{p} f_j(X_j)$ [9]. GAM generalizes the additive model into the form of exponential family distribution. The general form of the GAM model can be formulated as follows:

$$g(\mu) = \beta_0 + \sum_{j=1}^{p} f_j(X_j) \tag{1}$$

where:

$g(\mu)$ = *link function*
β_0 = constant coefficient
f_j = *smoothing function* of the predictor variable
X_j = predictor variable to j

2.3. Autocorrelation

GAM with autocorrelation structure refers to the Autoregressive Moving Average (ARMA) model. ARMA is a family of models for analyzing time series. The notation ARMA(p,q) refers to a model with p autoregressive terms and q moving-average terms [10]. The GAM model that has been formed is then formulated into a GAM model with autocorrelation in which the residual model has an autocorrelation value. The basic idea is to modelling year and month with autocorrelation structure with different order into GAM model [11].

3. Result and discussion

3.1. Testing Distribution and Link Function Determination

Based on the tests performed, SSTA histogram forms a normal distribution. But to prove it needed Kolmogorov Smirnov test and obtained P-value is 0.4325 so it can be concluded that the SSTA data is normal distribution. There is one link function that is suitable for normal distribution data that is identity. Therefore the link function used in this research is link function identity.

3.2. Modeling

3.2.1. Linear Model. The first modeling done before forming a GAM model in this research is to make annual and seasonal pattern using Linear Model to know the pattern of SSTA occurrence based on annual effect and seasonal effect. The results obtained can be seen in Figure 1.a and 1.b below:

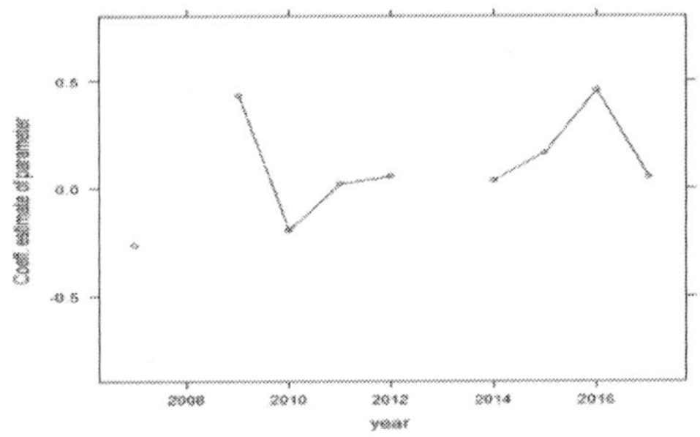

Figure 1.a. Annual pattern from annual effects

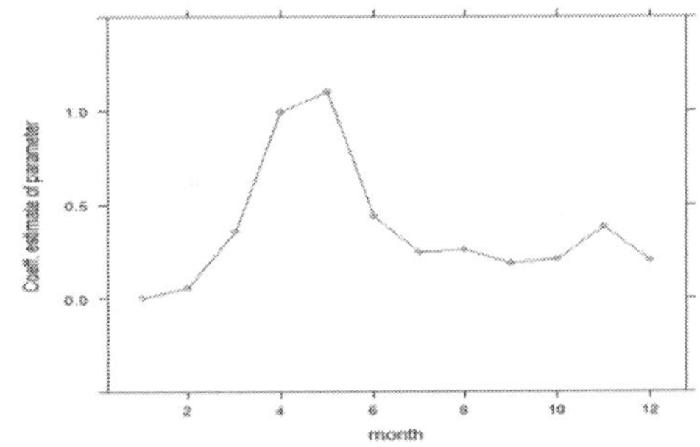

Figure 1.b. Seasonal Pattern from seasonal effects

Figure 1.a for the annual effect shows a strong SSTA decline from 2009 to 2010. While for 2011-2016 the annual effect increased strongly and declined again in 2017. Figure 1.b for seasonal effects indicates that an increase of SSTA from January to February, and a strong increase from February to May (peak season), as [12] has done for SST data. Further declines from May to October and an increase again in November.

3.2.2. Outlier Examination

a) *Studentized Residuals*

An observation is said to be the outlier data if the absolute value of studentized residuals is greater than 2 [13]. Figure 2 show a visualization of studentized residuals:

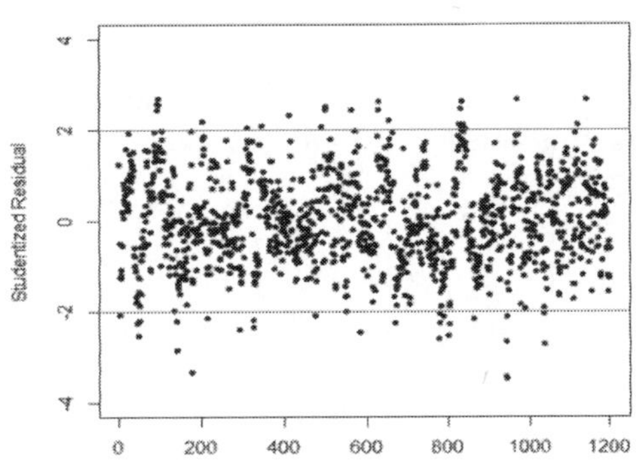

Figure 4. Studentized Residual plot

Based on figure 4 above, it can be seen that there is an observations that an outlier because has an absolute value of studentized residuals greater than 2. The picture above also shows there is 53 outlier data

3.2.3. Influential Observation Analysis

a) *Cook's Distance* (Di)

An observation is said the influential observation if the value of Di is greater than 4/(n-k-1) [13]. Table 1 show the value Di based of data from studentized residual value:

Table 1. *Cook's Distance* (Di)

Obs	*Studentized Residual*	Di	Obs	*Studentized Residual*	Di	Obs	*Studentized Residual*	Di	Obs	*Studentized Residual*	Di
2	-2.0725	0.0065	293	-2.3872	0.0030	627	2.6205	0.0053	834	2.6124	0.0066
46	-2.2668	0.0063	312	2.0576	0.0022	628	2.4447	0.0042	835	2.0047	0.0036
47	-2.5327	0.0084	326	-2.1998	0.0045	656	2.2091	0.0033	838	2.1075	0.0045
51	-2.2168	0.0058	327	-2.3427	0.0048	669	-2.2566	0.0043	844	2.0306	0.0036
93	2.4372	0.0040	347	2.0901	0.0032	778	-2.5990	0.0066	865	-2.1666	0.0033
94	2.5839	0.0049	414	2.3140	0.0065	779	-2.2610	0.0050	942	-2.6719	0.0055
95	2.5541	0.0044	474	-2.0879	0.0025	780	-2.0555	0.0043	943	-3.4377	0.0066
96	2.6898	0.0065	489	2.0648	0.0031	789	-2.1205	0.0036	944	-3.4843	0.0082
141	-2.2050	0.0052	498	2.5010	0.0036	802	-2.5304	0.0054	946	-2.1184	0.0024
142	-2.8581	0.0092	500	2.4487	0.0043	803	-2.2764	0.0041	968	2.6747	0.0049
176	-3.3452	0.3951	549	-2.0011	0.0028	828	2.1064	0.0037	1037	-2.0497	0.0033
204	2.1761	0.0043	562	2.4340	0.0050	831	2.4649	0.0060	1038	-2.7172	0.0052
214	-2.1419	0.0052	583	-2.4631	0.0152	833	2.0622	0.0045	1120	2.1232	0.0030
									1141	2.6702	0.0049

By using the value 4/(nk-1) = 4/(1202-5-1) = 0.00334, then based on table 1 it can be seen that observations 293, 312, 347, 474, 489, 549, 656, 865, 946, 1037, and 1120 are the outlier data and also have the value of Cook's Distance < 0,00334. So that it can be said that the-11 data is data the outlier data that is not the influential observations. So the data can be excluded from observation. The next model is a model by excluded the outlier data that is not the influential observations.

3.2.4. Generalized Additive Models (GAM). Before built a GAM model with autocorrelation, the first GAM model that built in this research is a GAM model with 5 predictor variables (climate features). The ANOVA table of the GAM model is as follows:

Table 2. GAM Model

Source of Variation	Edf	Ref.df	F	P-value
Air Temperature	7,766	8,557	541,72	2×10^{-16}
Precipitation Rain	8,565	8,933	22,40	2×10^{-16}
Relative Humidity	3,574	4,561	12,46	$9,78\times10^{-11}$
Wind Speed	8,881	8,995	93,29	2×10^{-16}
Shortwave Solar Radiation	4,355	5,404	11,01	$8,74\times10^{-11}$

Table 2 shows that all predictor variables are significant because they have P-value < 0.05. The visualization between actual data with fitted GAM model results for SSTA is show in figure 5:

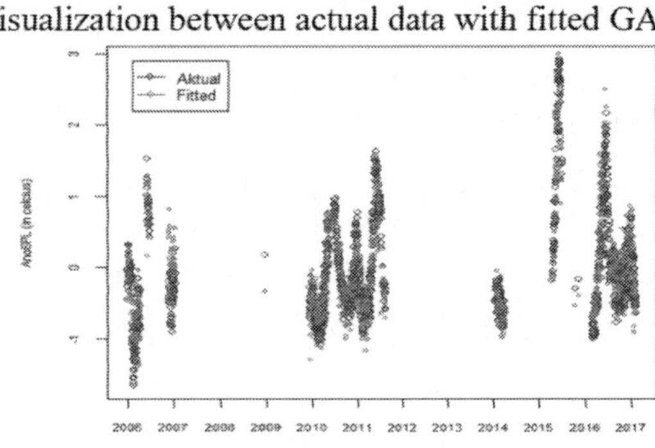

Figure 5. GAM model visualization

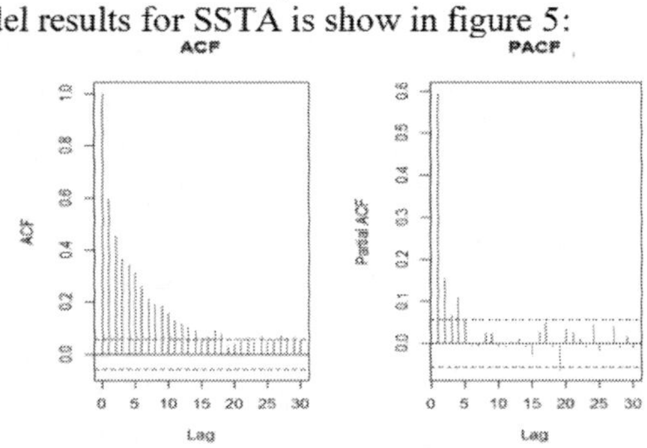

Figure 6. ACF and PACF plot GAM model

Figure 5 shows that the model is not compatible with the data because the fitted model has not approached the actual data and is still spreading. In addition there is a gap on visualization because data containing NA is excluded from the study. After that, checking of GAM's residual model was performed using the Autocorrelation Function (ACF) and Partial Autocorrelation Function (PACF) plots to build GAM models with autocorrelation. Figure 6 shows that the plot of the ACF decreases exponentially and the PACF plot drops dramatically after the first lag so the GAM model with autocorrelation that suitable is the AR (1) model.

3.2.5. GAM with Autocorrelation. After build a GAM model with 5 predictors variable, then reseracher built several GAM models with autocorrelation that show in tabel 3 below:

Table 3. Summary of GAM Models with Autocorrelation Structures

Model	Time Variable	Autocorrelation	R^2 (adjusted)	AIC
GAM_1	Month and Year	Month	73,40%	-1022,501
GAM_2	Month and Year	Year	60,70%	-1512,852
GAM_3	Month	Month	72,10%	-1014,341
GAM_4	Month	Year	60,70%	-1514,852
GAM_5	Year	Month	50,80%	-1006,123
GAM_6	Year	Year	37,90%	-1413,718

Based on table 3 it can be concluded that the best model is GAM_1 model because it has the highest R^2 (adjusted) value and low AIC value.

3.2.6. Visualization of the Best Model. Visualization between actual data with fitted GAM_1 model can be seen in figure 7 below:

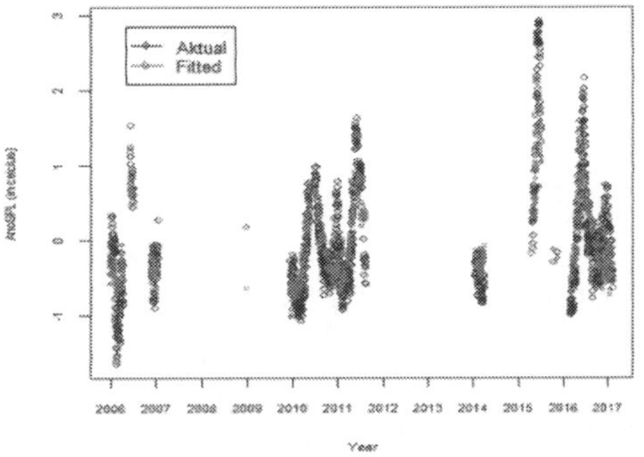

Figure 7. Visualization the best model (GAM_1 model)

Based on the visualization in figure 7 can be seen that the model has been compatible with the data because the fitted model has approached the actual data and not spread. The picture also shows that IOD turn up to be maximum around 2015 and 2016 with SSTA $\geq$ 2°C. While the negative IOD's occurred around 2007, 2010, and 2014.

3.3. Parameter Significance Testing

The parameter significance test is performed to find out whether the predictor variable used in this research has significant effect on the response variable. The significance testing of the best GAM model show in table 4:

Table 4. Parameter Significance Testing

Coefficient	Estimation	P-value	Coefficient	Estimation	P-value
Intercept	-0,0386	0,4709	Wind Speed	-0,0056	0,6956
Air Temperature	-0,1154	0,0138*	Shortwave Solar Radiation	$7{,}95\times10^{-10}$	0,9999
Precipitation Rain	-0,1004	0,2111	Month	0,6147	$5{,}02\times10^{-9}$***
Relative Humidity	0,0226	0,4336	Year	-0,1298	$1{,}68\times10^{-4}$***

Descriptiom: Significant on α 0,1% (***), 1% (**), 5% (*) and 10% (.)

Based on Table 4 can be seen that the variables that significantly affect SSTA is air temperature, month, and year with a significance level of 5%.

IOP Conf. Series: Materials Science and Engineering **523** (2019) 012002 doi:10.1088/1757-899X/523/1/012002

4. Conclusion

The best model in this research is GAM_1 model, that is GAM model which includes month and year factor with month autocorrelation structure. Measurements of R^2 adjusted and AIC GAM model are 73.40% and -1022,501, respectively. The GAM_1 model has also been compatible with the data because the result of fitted GAM_1 model approached the actual SSTA data. Factors that affect the SSTA is air temperature, month, and year, for period 2006-2017 at 8N90E.

5. Acknowledge

The authors thank the mentor, Dr. Miftahuddin for the help so that this research can be done.

6. References

[1] Miftahuddin. 2016. *Semirata PTN Barat Bidang Ilmu MIPA*. Sriwijaya, 0:732-741.

[2] Reid, C., Marshall, J. Logan, D., Kleine, D. 2009. *Coral Reefs and Climate Change*. The University of Queensland, Australia.

[3] Pramudia, A., Estiningtyas, W., Susanti, E., Sucianti. 2015. *Fenomena dan Perubahan Iklim Indonesia serta Pemanfaatan Informasi Iklim untuk Kalender Tanam*. Agency for Agricultural Research and Development, Ministry of Agriculture.

[4] Tjasjono, B. 1999. *Klimatologi Umum*. ITB, Bandung.

[5] Handayani, L. 2014. *Statistical Downscaling* dengan Model Aditif Terampat untuk Pendugaan Curah Hujan Ekstrim. *Tesis*. Bogor Agricultural Institute, Bogor.

[6] Miftahuddin. 2016. Fundamental fitting of the SST Data using Linear Regression Models. Program Book 12th ICSMA. ISBN 978-1-5090-3385-0.

[7] Swarinoto, Y. Makmur E.E.S. 2009. *Bulletin of Meteorology Climatology and Geophysics*, **5**(3): 55-56.

[8] BMKG. 2018. *Analisis Dinamika Atmosfer*.

[9] Hastie, T. dan Tibshirani, R. 1986. Statistical Science, **1**(3):297-318.

[10] Hamilton J.D. 1994. *Time Series Analysis*. Princeton University Press, Princeton.

[11] Wood, S. N. 2006. *Generalized Additive Models: An Introduction with R.* Chapman & Hall/CRC Press, London.

[12] Miftahuddin. 2017. Assessment of Sea Surface Temperature in the Indian Ocean using Generalized Additive Models. *Proceeding SEMIRATA*. Jambi, 0:225-237

[13] Fox, John. 2009. *Regression Diagnostics*. McMaster University, Canada.

IOP Conf. Series: Materials Science and Engineering **523** (2019) 012003 doi:10.1088/1757-899X/523/1/012003

Application of ARCH model on nutmeg price forecasting in South Aceh district

A Rusyana[1], **R Ferdiana**[1], **M E Putri**[1,*]
Department of Statistics, Faculty of Mathematics and Natural Sciences, Syiah Kuala University, Banda Aceh 23111, Indonesia

E-mail: meiditaekaputri71@gmail.com

Abstract. The price of nutmeg which is changing and unstable in South Aceh results to volatility. Volatility is a variance pattern of time series, especially the time series of finances caused by variance not constant. This results the possibility of heteroscedastic data, so that it needs to be made a model of a particular approach to measure residual volatility problems. The data used in this research is the price of nutmeg in South Aceh district from January 2012 to December 2016 which has volatility. ARCH (Autoregressive Conditionals Heteroscedastic) is a model used to resolve the residuals variance which is not constant in financial time series. Then this model was developed into a Generalized Autoregressive Conditional Heteroskedastic (GARCH) to avoid too high orders on ARCH models and make the variance is always positive. The purpose of this research is to get the best ARCH/GARCH model for nutmeg price in South Aceh district and get the nutmeg price forecast for January 2016 until December 2017. The results of this research show that price of nutmeg in South Aceh district is stationary to the mean after the second differencing and the result of the ARCH Lagrange Multiplier test shows the problem in the 4 lag so the data are modelled with ARCH. The best models for forecasting price of nutmeg in South Aceh district is using model of ARIMA (2,2,0)-ARCH (4). While the results of the verification of the model are obtained the value of the MAPE of 6.12 percent.

1. Introduction

In the financial sector generally has a very high level of volatility. The high Volatility is shown by a phase where relatively high fluctuation and then followed the low fluctuation and high returns, with the other words, the mean and the variance are not constant. The presence of high volatility causes difficulty of the estimation and prediction of the movement of these variables. The model which assumes that the residual variance is not constant in time series data was developed by Engle (1982) in autoregressive model called conditionals heteroskedastic (ARCH).

Nutmeg is one of export commodities originating from South Aceh. According to BPS Aceh in year 2016, Aceh has nutmeg production of 6,620 Tonnes where the South Aceh Regency has the highest i.e. 5,747 tonnes. Changes of rise and fall of the price of nutmeg in South Aceh Regency cause occurrence of volatility. Since model of the ARCH was published by Engel in 1982, the model has been widely used to describe the behavior of a time series data volatility. Therefore, this research was

[*]Corresponding author: meiditaekaputri71@gmail.com

conducted to obtain the best ARCH model in forecasting the price of nutmeg in South Aceh district from January 2016 until December 2017.

2. Literature Review

2.1 Stationary Data

There are two types of stationary, namely stationary in the mean and stationary in variance. Stationary in the mean is data fluctuations around the mean value of a constant. Augmanted Dickey Fuller (ADF) test is usually used to test stationery in mean [1]. Differencing is used to convert a time series data of nonstationary becomes stationary in the mean [2].

A time series data are said to be stationary in the variance if structure of the data have constant data fluctuations from time to time [3]. If the data are not stationary in the variance then the data transformation is used to stabilize or get a constant variance. This transformation is called the Box-Cox transformation [4].

2,2 Model of Autoregressive Integrated Moving Average (ARIMA)

A model with stationary data through differencing process is called ARIMA. If the data are stationary on the process of differencing d times and apply the ARMA (p,q) then these models are called ARIMA (p, d, q). The form of the equation for model of ARIMA is [3]:

$$\phi_p(B)(1-B)^d Z_t = \theta_0 + \theta_q(B)a_t \quad (1)$$

2.3 Identification model of ARIMA

Model of ARIMA can be identified from pattern of the ACF and PACF plot if data is already stationary in the mean and variance, see table 1.

Table 1. ACF and PACF criteria on the model of ARIMA

Proses	*Autocorrelation Function* (ACF)	*Partial Autocorrelation Function* (PACF)
AR(p)	Exponentially down trend	Cut off after lag p
MA(q)	Cut off after lag q	Exponentially down trend
ARMA(p,q)	Cut off after lag (q-p)	Cut off after lag (p-q)

2.4 Model of Autoregressive Conditional Heterokedastic (ARCH) and the effects test of ARCH

Volatility is a measure of the uncertainty of time series data indicated by the presence of fluctuations. These fluctuations cause the residual variance not constant and heteroskedastis. ARCH is a model that can be used to overcome the residual variance not constant in financial time series data. Where the equation of the model can be seen as follows [5]:

$$\sigma_t^2 = \alpha_0 + \alpha_1\varepsilon_{t-1}^2 + \alpha_2\varepsilon_{t-2}^2 + \dots + \alpha_q\varepsilon_{t-q}^2 \quad (2)$$

where:

σ_t^2 : variance of the residual at time t
α_0 : component constant
α_q : parameter of ARCH with order q
ε_{t-q}^2 : residual squared at time *t-q*

LM ARCH test is used to find out the heteroscedasticity element in the data time series with hypothesis [5]:

$H_0 : \alpha_1 = \alpha_2 = \dots = \alpha_q = 0$ (There is no effect of heteroscedasticity)
$H_1 : \alpha_i \neq 0, i = 1,2,\dots q$ (There is an effect of heteroscedasticity)

2.5 Akaike Information Criterion (AIC)
The selection of the best model is made based on the value of AIC of the smallest [6].

$$\text{AIC} = \ln\hat{\sigma}^2 + \frac{2}{n}r \quad (3)$$

where:
n = number of error
r = the number of parameters in the model of ARIMA
$\hat{\sigma}^2$ = estimator of error variance

2.6 Mean Absolute Percentage Error (MAPE)
MAPE is a measurement of error calculated by formulae :

$$MAPE = \frac{1}{n}\sum_{t=1}^{n}\frac{|Z_t - \hat{Z}_t|}{Z_t} \text{ x } 100 \quad (4)$$

dimana:
Z_t : actual value at time period *t*.
$\hat{Z}_t$: forecast value for time period *t*.
n : number of observations.

3. Method
This study uses data of price of nutmeg in a South Aceh District from 2012 until 2016 obtained from the Department of trade, industry, cooperatives and UKM in South Aceh District. The number of data used is 60 from January 2012 to December 2016. In this study, data are divided into two parts: the data from January 2012 to December 2015 are as training data while the data from January until December 2016 are data for testing.

4. Results and discussion

4.1 Time Series Plot
Based on figure 1, the highest nutmeg prices in South Aceh district occurred in August to September 2013 Rp 62.000 per Kg. Then the nutmeg price changes for the following months decreased monthly to reach the lowest price in June and July of 2015 Rp 35.000 per Kg.

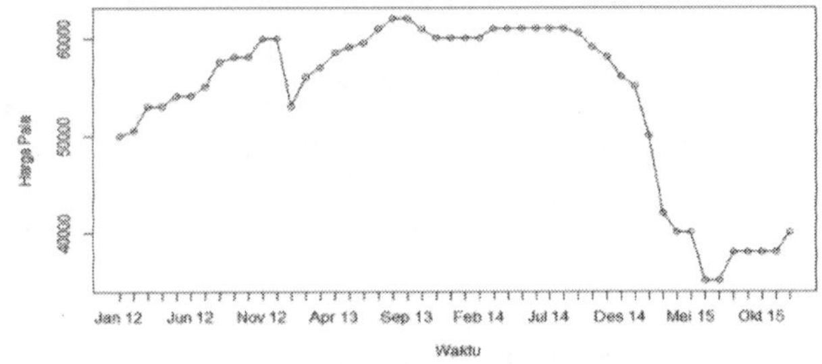

Figure 1. Plot of nutmeg price data in South Aceh District from January 2012 untill December 2015.

4.2 Stationarity test
The result of the Augmanted Dickey Fuller (ADF) test gives the value of the P-value of 0.561. This shows that its decision rejects H_0 with $\alpha = 5\%$ which it means data are not stationary, so the data need for differencing. After the first differencing is done then got P-value 0.549 from ADF test. This indicates that the data is still not stationary in mean, so data need to be the second differencing. On the second differencing, P-value is 0.015, where the result indicates that data have been stationary in mean, because P-value < α.

IOP Conf. Series: Materials Science and Engineering **523** (2019) 012003 doi:10.1088/1757-899X/523/1/012003

Box-Cox test acquired value of λ=1.999924. This shows that the data have been stationary in variance and do not require transformation because the value of λ is obtained already past 1.

4.3 Identification of ARIMA Model

Model identification uses ACF and PACF plots before and after transformation in figure 2 and figure 3.

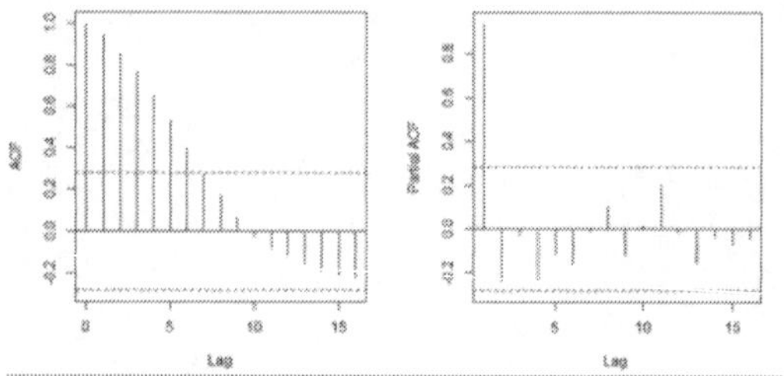

Figure 2. Plot of ACF and PACF before differencing level 2.

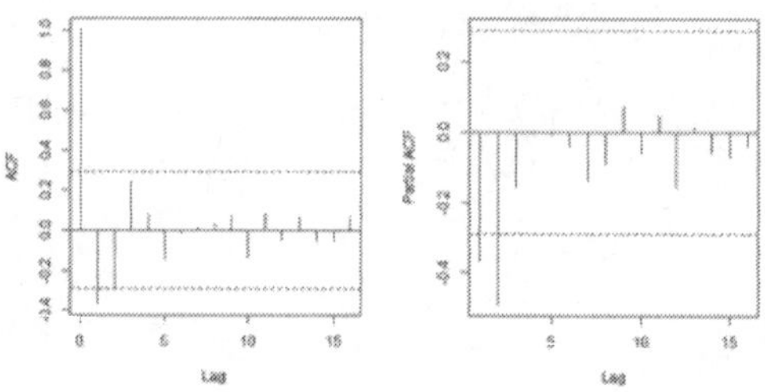

Figure 3. Plot of ACF and PACF after differencing level 2.

Figure 2 shows the plot of ACF Downtrend exponentially and plot the PACF cut off after the first lag. While the Fig. 3 showed that ACF and PACF plots have cut off and can be seen there are still some lag line. Based on figure 2 and 3, Five models in table 2 are selected.

Table 2. Tentative model of ARIMA.

Order ARIMA	Parameter Model						AIC
	AR(1)	AR(2)	AR(3)	MA(1)	MA(2)	MA(3)	
(1,2,0)	**-0,369****	-	-	-	-	-	853,796
(2,2,0)	**-0,549****	**-0,488****	-	-	-	-	**843,159**
(3,2,0)	-0,632**	-0,576**	-0,158	-	-	-	843,972
(2,2,3)	**0,133 ****	**-0,947****	-	**-0,884****	**1,170****	**-0,599****	845,191
(3,2,2)	0,394	-0,088	0,333**	-0,119**	0,119	-	845,700

Significant at the α=10% (**)

Based on Table 2, there are three significant coefficient of the whole model, namely, model of ARIMA(1,2,0), ARIMA(2,2,0), and ARIMA(2,2,3). To get the best of the three models then the selection is based by the smallest AIC value. The model of ARIMA (2,2,0) has the smallest AIC value.

4.4 Testing of ARCH Effect

Effect of ARCH test is used to check that there is a problem of heteroscedasticity of ARIMA (2,2,0) model residual. From testing of ARCH-LM to the residual of the model of ARIMA (2,2,0), p-value is obtained 0.00121 on lag 4. Then the decision is rejection of H_0 with α = 0.05 so there is effect of ARCH or there is heteroscedasticity problem. The existence of the element of the ARCH can also be seen from the plot of ACF and PACF residual square.

ACF and PACF graph in Figure 4 shows that there is a lag that still over of barlet line, it also shows that the nutmeg data price in South Aceh district has problem of heteroscedasticity, so ARCH model is good to be applied.

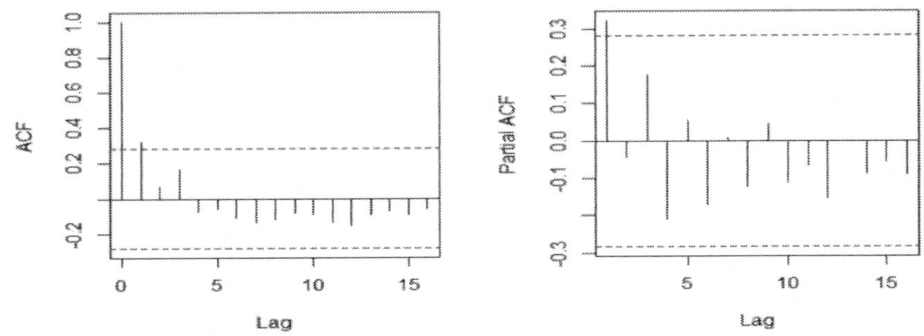

Figure 4. The plot of ACF and PACF residual square of the ARIMA(2,2,0) Model.

4.5 Identification of ARCH Model

Testing of ARCH-LM previously can be inferred, there is the effect of the forth lag in ARCH so that the model of the ARCH is suitable for forecasting nutmeg price in South Aceh District. Then the model can also be identified with ACF and PACF residual square shown in Fig. 4, where there are some early lag over the line of bartlett.

Table 3. Tentative model of ARCH.

Order ARCH	Parameter Model					AIC
	α_0	α_1	α_2	α_3	α_4	
ARCH(1)	4,297 x10^6**	4,222 x 10^{-1}	-	-	-	856,8146
ARCH(2)	4,071 x10^6**	2,434 x 10^{-1}	1,144 x 10^{-1}	-	-	839,8755
ARCH(3)	3,845 x10^6**	2,044 x 10^{-1}	5,091 x10^{-9}	7,729 x 10^{-2}	-	824,1773
ARCH(4)	3,619 x10^6**	1,480 x 10^{-1}	4,307 x 10^{-2}	6,197 x 10^{-2}	9,601 x 10^{-8}	**808,9155**

Keterangan: Signifikan pada α=5% (**)

Table 3 shows that the model of the ARCH (4) has the smallest AIC if it is compared three of the other models. So the ARCH (4) model is suitable to be used in forecasting the nutmeg price in South Aceh district to 24 period. The ARCH (4) model is

$$\sigma_t^2 = 3{,}7557 \times 10^6 + 1{,}6440 \times 10^{-1}(\varepsilon_{t-1}^2) + 3{,}3014 \times 10^{-2}(\varepsilon_{t-2}^2) + 2{,}9528 \times 10^{-2}(\varepsilon_{t-3}^2) + 1{,}1344 \times 10^{-8}(\varepsilon_{t-4}^2) \quad (5)$$

4.6 Forecasting of ARIMA-ARCH

Forecasting is done with combining of the ARIMA and ARCH model, they are ARIMA (2,2, 0)-ARCH (4) which can be written as:

$$Z_t = 0{,}96313\, Z_{t-1} - 1{,}07374 Z_{t-2} + 1{,}03687\, Z_{t-3} + 3{,}7557 \times 10^6 + 1{,}6440 \times 10^{-1}(\varepsilon_{t-1}^2) + 3{,}3014 \times 10^{-2}(\varepsilon_{t-2}^2) + 2{,}9528 \times 10^{-2}(\varepsilon_{t-3}^2) + 1{,}1344 \times 10^{-8}(\varepsilon_{t-4}^2) \quad (6)$$

After getting the best models of ARIMA(2,2,0) - ARCH(4), forecasting is conducted where forecasting is divided two, namely training and testing forecasting. Training forecasting is the nutmeg price from January until December 2016, while testing forecasting is from January until December 2017. The result of the training and testing forecasting can be seen in figure 4.

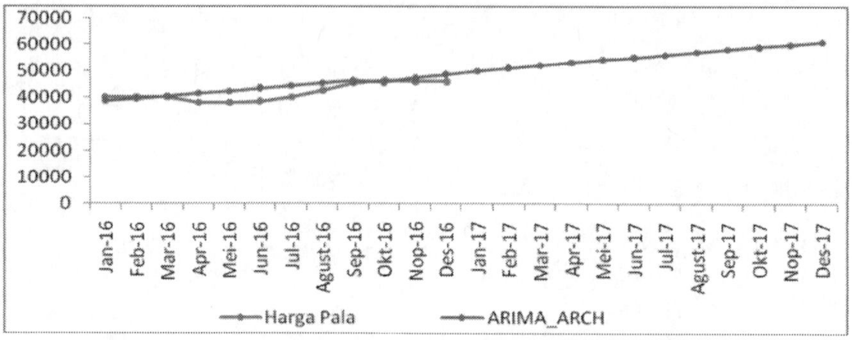

Figure 5. Comparison graph testing and forecasting.

The value of the MAPE obtained from actual and forecasting data from January to December 2016 is 6.12 percent. It shows the average of forecasting error rate is 6.12 percent. Based on the MAPE, the forecasting with ARIMA(2,2,0)-ARCH (4) is categorized well to forecast the price of nutmeg in South Aceh Regency.

5. Conclusions

1. The best model obtained to predict the price of nutmeg in South Aceh Regency is a ARIMA (2, 2, 0)-ARCH (4) which can be written as.

$$Z_t = 0,963Z_{t-1} - 1,074Z_{t-2} + 1,037Z_{t-3} + 3,756\times10^{6} + 1,644\times10^{-1}(\varepsilon_{t-1}^{2})$$
$$3,301\times10^{-2}(\varepsilon_{t-2}^{2}) + 2,953\times10^{-2}(\varepsilon_{t-3}^{2}) + 1,134\times10^{-8}(\varepsilon_{t-4}^{2})$$

2. The result of nutmeg price forecasting in South Aceh district in January 2016 until December 2017 tends to increase each month about Rp 976,- and average error level is relatively low 6.12 %.

6. References

[1] Gujarati and Porter 2009 *Dasar-Dasar Ekonometrika* (Jakarta: Salemba Empat)
[2] Rusyana A, Nurhasanah, Marzuki and Flancia M 2016 SARIMA Model for Forecasting Foreign Tourists at the Kualanamu International Airport *Proc. Int. Conf. Mathematics, Statistics, and their Applications (ICMSA)* 155
[3] Wei W W S 2006 *Time Series Analysis Univariate and Multivariate Methods* 2nd ed (USA: Addision-Wesley Publishing Company)
[4] Montgomery D C, Jennings C L, and Kulahci M 2008 *Introduction to Time Series Analysis and Forecasting* (New Jersey: John Wiley & Sons)
[5] Juanda B and Junaidi 2012 *Ekonometrika Deret Waktu* (Bogor: IPB Press)
[6] Brooks C 2008 *Introductory Econometrics for Finance.* 2nd ed (New York: Cambridge University Press)

IOP Conf. Series: Materials Science and Engineering **523** (2019) 012004 doi:10.1088/1757-899X/523/1/012004

Application of generalized additive models with P-Spline basis for infant mortality rate data

D Ahya[1,*], **Miftahuddin**[1]
[1]Department of Statistics, Faculty of Mathematics and Sciences, Syiah Kuala University, Banda Aceh 23111, Indonesia

E-mail: desfira.ahya@students.stat.unsyiah.ac.id

Abstract. Infant Mortality Rate (IMR) is an indicator of health that reflects the state of the health status of a population. The infant mortality rate in Aceh province period of 2014 has the highest position compared to the previous 4 years that amounted to 15/1000 live births. Generalized Additive Models (GAM) is a method that can handle the condition of the data which response variable does not have to be normally distributed and related with the predictor variable also does not have to be linear. Fitting GAM models with P-spline base can produce a smoother curve and avoid the occurrence of misfitting models. Application of GAM P-spline bases in this research is used data number of IMR in Aceh province period of 2012-2015. The purpose of this research is to get the best fitting model of IMR with GAM models P-spline bases. The results show that the best model of GAM P-spline which can be explain IMR in Aceh Province period of 2012-2015 with knots 7 and GCV is 0,0723.

1. Introduction

Infant mortality rate (IMR) is one of the health indicators that determine the level of public health. Reidpath and Allotey [1] stated that IMR is an essential indicator for the health of the entire population, reflecting that structural factors such as economic development, living conditions, social welfare, and environmental quality affecting the health of the entire population have an impact on the IMR. It is because the IMR is sensitive to changes in the level of health and welfare of the community. During 2012-2014, IMR in Aceh Province continues to increase. In 2014, IMR is 15/1000 live birth [2]. The increased of IMR indicates that the more babies who died. It cannot be left alone, because the child's survival determines the quality of human resources in the future.

A research in analyzing of the infant mortality rate in Aceh Province show that factors affecting of infant mortality were percentage of poor households who use clean water, the number of community health centers and health facilities, the latest birth attendants with the help of TBAs and family, and nutritional coverage of toddlers receiving care [3]. In addition, research [4] analyzing the risk factors for low birth weight associated with IMR using the logistic regression analysis method.

Based on the description above, then in this research analyzing IMR using Generalized Additive Models (GAM) method. The GAM model can be more flexible than the additive model because it can accommodate complex data issues, such as nonnormality errors, nonlinear relationships, and autocorrelation variables [5]. A smoothing method in the GAM model that can give better results is a smoothing spline [6]. Smoothing spline approaches are various, among others original spline, spline

*Corresponding author: desfira.ahya@students.stat.unsyiah.ac.id

type M, spline relaxed, weighted spline, B-spline, P-spline, spline truncated and others. In this research used one spline bases that is P-spline. P-spline is one of the most popular smoothing approaches due to its simplicity and flexibility [7]. The purpose of this research is to get the best fitting model of IMR in Aceh Province period 2012-2015 period with GAM models P-spline bases.

2. Literature Review

2.1. Infant Mortality Rate

Infant Mortality Rate (IMR) is the number of infant deaths under one year of age per 1000 live births in a given year [8]. Based on the cause, infant mortality is distinguished by endogenous and exogenous factors. Endogenous infant mortality (neonatal mortality) is the incidence of death occurring in the first month since the baby was born, generally caused by a factor brought by birth, inherited by the parents at the time of conception or obtained from his mother during pregnancy. While exogenous mortality (neonatal post-mortem) is infant mortality occurring between the age of one month or until one year, caused by factors related to environmental influences. The following formula can calculate the calculation of infant mortality.

$$IMR = \frac{number\ of\ infant\ mortality\ below\ one\ year}{the\ number\ of\ live\ births\ in\ a\ given\ year} \times 1.000 \qquad (1)$$

2.2. Generalized Additive Models

Generalized Additive Models (GAM) is the development of Generalized Linear Models (GLM) by replacing linear functions $\sum_{j=1}^{p} \beta_j X_j$ with additive functions $\sum_{j=1}^{p} f_j (X_j)$ [10]. GAM generalizes the additive model into the form of exponential family distribution. The general form of the GAM model can be formulated as follows:

$$g(\mu) = \beta_0 + \sum_{j=1}^{p} f_j (X_j) \qquad (2)$$

where: $g(\mu)$ = link function; f_j = smoothing function of the predictor variable
β_0 = constant coefficient; X_j = predictor variable - j

2.3. Penalized Spline (P-spline)

One of the unique aspects of the GAM model is the nonparametric function estimated by the scatter diagram smoothing function. Hastie & Tibshirani (1990) used smoothing spline to produce the best model. Then Eilers and Marx combine spline approach and spline regression into Penalized Spline (P-Spline). P-Spline is the regression obtained based on the least squared with a penalty of roughness. P-spline has many similarities with smoothing spline, but the type of penalty used on P-spline is more general than smoothing spline [11]. The general form of the P-spline can be formulated as follows:

$$f(x_j) = \beta_0 + \beta_1 x_j + \cdots + \beta_p x_j^p + \sum_{k=1}^{K} (x_j - k_K)_+ \qquad (3)$$

where: β_0 = constant of *P-spline*; p = degree *spline*
β_1 = coefficient of *P-spline -p*; K = number of knots
x_j = predictor variable to-*j*; k_K = value of knots to-K

From the formula the P-spline function indicates that the spline is a broken off polynomial model, but is still continuous on the knots. Knots can be interpreted as a focal point in the spline function so that the curve formed is segmented at that point. The number of knots is the number of points where functional behavior changes at different intervals.

Knot selection is very important, because it affects the model to be selected. The selection of knots on the P-spline model used Generalized Cross Validation (GCV) criteria. If compared with other methods such as Cross Validation (CV), the GCV method has asymptotically optimal properties [12]. The optimum knot point is marked with a minimum of GCV value. GCV values are defined as follows:

$$GCV(K) = \frac{MSE(K)}{[n^{-1}tr(I - A(K))]^2} \tag{4}$$

where: n = number of data; tr = trace; I = identity matrix; A = $\boldsymbol{X}(\boldsymbol{X}^T\boldsymbol{X})^{-1}\boldsymbol{X}^T$

3. Method

This research uses data from the Aceh Health Profile book 2012-2015 period and Aceh Dalam Angka book 2013-2016 period. The observation unit is the district/city in Aceh Province in 2012-2015. The data used consisted of 1 response variable and 9 predictor variables. The amount of data used is 92 observations. The variables are defined as in Table 3.1.

Table 3.1. Variables in this research

Variable	Explanation	Type of Data
Y	Number of infant mortality (person)	Numeric
X_1	Childbirth assisted by health personnel (%)	
X_2	ASI Exclusive (%)	
X_3	Low birth weight baby (person)	
X_4	Number of health personnel (person)	
X_5	Healthy house (%)	
X_6	The poor population (%)	
X_7	Percentage of married women under the age of 15 years (%)	
X_8	Year	Category
X_9	District/City	

4. Results and discussion

4.1. Descriptive Analysis

The descriptive analysis aims to the representation of the data that has been collected. Here is presented a descriptive analysis like scatter plot matrix from data IMR in Aceh Province 2012 to 2015 in Figure 4.1. Scatter plot matrix is useful for knowing the shape of the histogram, values of Pearson correlation, and regression lines of data. Figure 4.1 show that the histogram shape of response variable Y (IMR) is not symmetrical, and the predictor variable also has not symmetrical histogram shape. The response variable Y (IMR) has a histogram shape extending to the right. The highest Pearson correlation values of the between Y (IMR) and the predictor variables were 0.60, 0.36, 0.14, -0.14, 0.11 and 0.053. The highest linear correlation is Y (IMR) with X_3 (LBWB) which is indicates a strong related of both variables. While the lowest correlation is in Y (IMR) with X_7 (percentage of married women under the age of 15 years) that is equal to 0.053, that indicates is both of variables has weakly related.

IOP Conf. Series: Materials Science and Engineering **523** (2019) 012004 doi:10.1088/1757-899X/523/1/012004

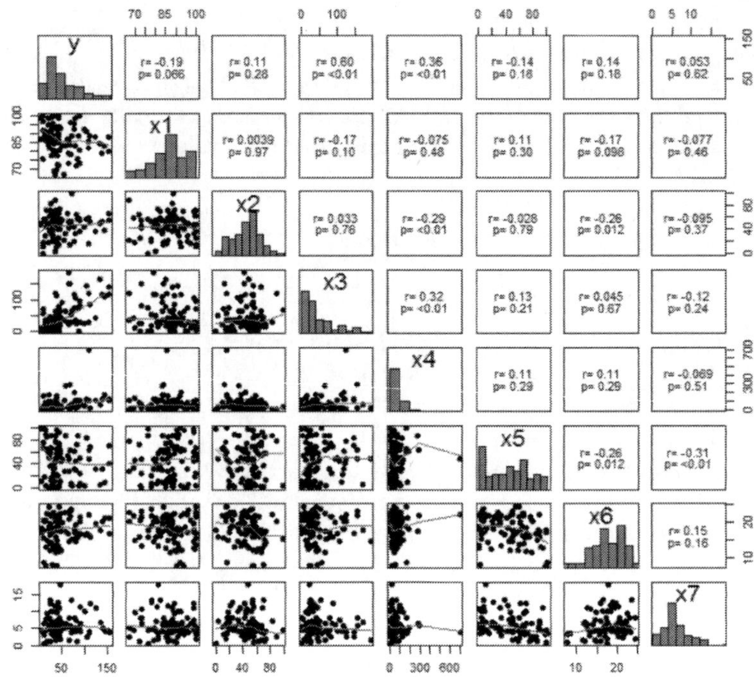

Figure 4.1. *Scatterplot matrix* of IMR in Aceh Province in 2012-2015 period

4.2. GAM of *P-spline bases*

GAM model to be formed consists of 9 variables and 2 of them with categorical types of data, that is year and district/city. Both of variables are incorporated into the model as dummy variables and the comparison factors in each variable are year of 2012 and Aceh Barat District. GAM model established using Gamma family and link function log.

The best fitting GAM model with P-spline bases selected based on the optimal knot point. The location of different knot points will result in different models. The optimal knot point is obtained based on minimum GCV criteria. The following table shows the optimal knot points with minimum GCV values.

Table 4.1. Value of GCV for each knot

Knot	GCV
5	0,0767
6	0,0726
7	**0,0723**
8	0,0740
9	0,0743
10	0,0746

Table 1 show that the optimal knot is to use point knot 7 with minimum of GCV value is 0.0723. So the best model fitting of GAM based on P-spline using knot 7. The complete model of GAM based P-spline from Infant Mortality Rate in Aceh Province is as follows:

IOP Conf. Series: Materials Science and Engineering **523** (2019) 012004 doi:10.1088/1757-899X/523/1/012004

$$
\begin{aligned}
g(\mu) &= \beta_0 + \sum_{j=1}^{7} f(x_j) + factor(x_8) + factor(x_9) \\
&= 3{,}9188 + f(x_1) + f(x_2) + f(x_3) + f(x_4) + f(x_5) + f(x_6) + f(x_7) + 0{,}2304x_8(2013) + \\
&\quad 0{,}367x_8(2014) + 0{,}2131x_8(2015) - 0{,}3304x_9(2) + 0{,}4253x_9(3) - 0{,}7750x_9(4) - \\
&\quad 0{,}2488x_9(5) - 0{,}8382x_9(6) + 0{,}6581x_9(7) - 0{,}3099x_9(8) - 0{,}8441x_9(9) + 0{,}6157x_9(10) + \\
&\quad 0{,}3214x_9(11) - 1{,}0858x_9(12) - 0{,}3075x_9(13) + 0{,}9105x_9(14) - 0{,}91x_9(15) - \\
&\quad 0{,}4449x_9(16) - 0{,}6540x_9(17) - 0{,}4713x_9(18) + 0{,}8232x_9(19) - 0{,}8005x_9(20) - \\
&\quad 2{,}0693x_9(21) - 0{,}4679x_9(22) - 1{,}5131x_9(23)
\end{aligned}
$$

4.3. Visualization

Based on the model that has been established, the visualization of GAM based on P-spline model is as follows.

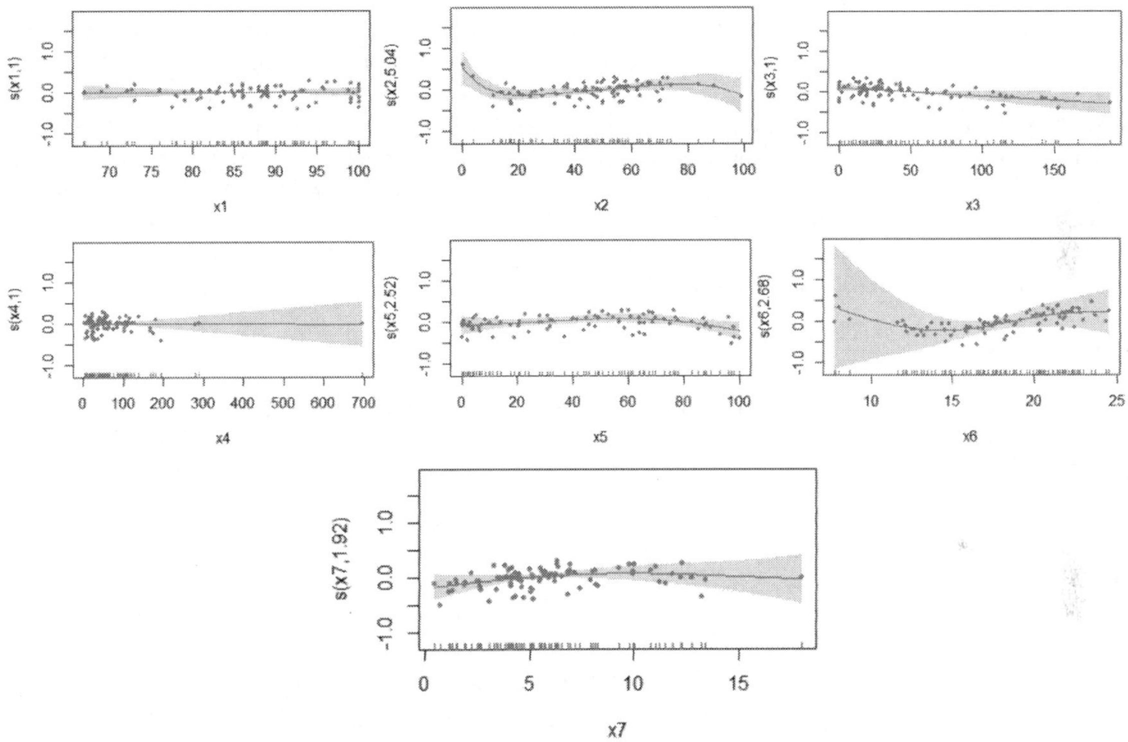

Figure 4.2. Visualization of GAM based on P-spline

Figure 4.2 is a plot of the best GAM model based on P-spline bases using knots 7. The shaded area of gray is a 95% confidence interval. Then the y-axis of each plot is a smooth component which is useful for view residual spread and the general shape of the curve. From the seven plots smoothing, the curve of smoothing are different, where the smoother plots are X_2 (ASI exclusive), X_5 (healthy house), X_6 (the poor population), and X_7 (percentage of married women under 15). In the X_4 plot (number of health workers) there are a number of health workers far different from others, that indicates there is one district/city with a large number of health workers. While other districts/cities have the number of health workers with an adjacent number. Similarly, plot X_6 there are some poor people with a small amount. While plots X_1, X_2, X_3, X_5, and X_7 have the same criteria, which has a value that is not much different from the others. While the visualization of variables with categorical data types are as follows.

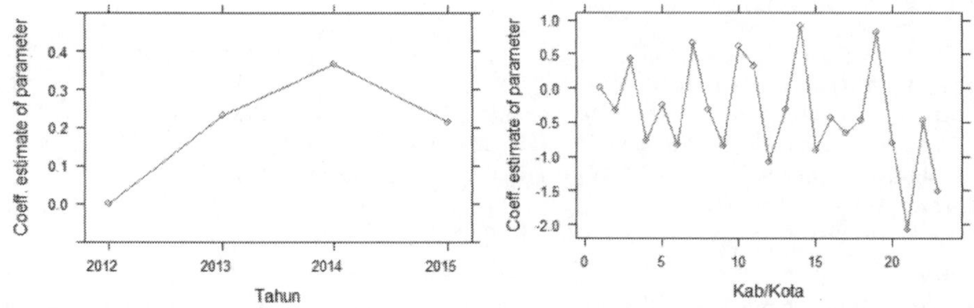

(a) (b)

Figure 4.3. The curve of IMR based on (a) year; (b) district/city

Figure 4.3 (a) shows the Infant Mortality Rate continues to increase from year of 2012 to 2014, but in 2015 Infant Mortality Rate is declines. It can be seen also that the highest Infant Mortality Rate in 2014 with the year used as a comparison is 2012. And year of 2013 and 2015 have the value of Infant Mortality is almost the same. Figure 4.3 (b) shows that the estimated infant mortality rate in each district varies. By using the Aceh Barat as comparison factor, the highest infant mortality rate was in the district-14 (Bireuen City) and the lowest was the district-21 (Sabang). While the infant mortality rate approaching to the comparison factor of Aceh Barat District is the 2nd, 5th, 8th, 13th, 16th, 17th, 18th and 22th districts. Each of these districts are Aceh Barat Daya, Aceh Selatan, Aceh Tengah, Bener Meriah, Langsa, Lhokseumawe, Nagan Raya, and Simeulue.

5. Conclusions

1. The best-fitting model of GAM P-spline for Infant Mortality Rate data in Aceh Province period of 2012-2015 using a knot point 7 with a minimum GCV is 0.0723.
2. GAM model on P-spline bases with comparative factor year of 2012 and Aceh Barat Dsitrict is:

$$g(\mu)=\beta_0+\sum_{j=1}^{7} f(x_j)+factor(x_8)+factor(x_9)$$

6. References

[1] Reidpath, D. D., and Allotey, P. 2003. *Journal of Epidemiology and Community Health*, **57**(5), 344-346.

[2] Dinas Kesehatan Aceh. 2015. *Profil Kesehatan Provinsi Aceh Tahun 2015.*

[3] Harahap, A.R. 2014. Pemodelan Angka Kematian Bayi di Provinsi Aceh Dengan Menggunakan Generalized Poisson Regression. *Skripsi.* Universitas Syiah Kuala, Banda Aceh.

[4] Jackson, dkk. 2008. Journal of the Georgia Public Health Association, Vol. 1 No. 1.

[5] Miftahuddin. 2017. *Assessment of Sea Surface Temperature in the Indian Ocean using Generalized Additive Models*. Prosiding SEMIRATA 2017.

[6] Hastie, T. and Tibshirani, R. 1990. *Generalized Additive Models 4th ed.* Chapman and Hall.

[7] Hall, P. and Opsomer, J. D. 2005. **92**:1-105.

[8] WHO. 2012. World Health Organization, Amerika.

[9] FEUI. 2010. *Dasar-Dasar Demografi*. Salemba Empat, Jakarta.

[10] Hastie, T. and Tibshirani, R. 1986. *Statistical Science*, **1**(3):297-318.

[11] Ruppert, D., Wand, M.P., and Carrol, R.J. 2003. *Semiparametric Regression*. Cambridge

[12] Wahba, G. 1990. *Spline Models for Observational Data*. SIAM Pensylvania.

IOP Conf. Series: Materials Science and Engineering **523** (2019) 012005 doi:10.1088/1757-899X/523/1/012005

Grouping the districts in Indonesia based on value of science subjects of National Exam using K-Means clustering method

R Ferdhiana[1,*], **T FAbidin**[2], **A Mardhiah**[3]
[1] Department of Statistics, Syiah Kuala University, Banda Aceh
[2] Department of Informatics, Syiah Kuala University, Banda Aceh
[3] Department of Statistics, Syiah Kuala University, Banda Aceh

E-mail: ridha.ferdhiana@unsyiah.ac.id

Abstract. Implementation of the National Examination (NE) for senior high schools for each district in Indonesia is done as one way to mapping the success of education in each the district. Mapping can be done by grouping districts based on attribute values obtained on the implementation of the NE, which in this study using the K-means clustering method. The attribute of NE values used are the values for the science majors, which consist of 6 subjects, where from 514 districts only 510 districts own it. The purpose of the study in addition to grouping districts, will also predict the group of 4 districts that have no NE values. Prediction is done by entering the value of the district Human Development Index (HDI) that has no UN value into 95% confidence interval of HDI from groups formed from k-means clustering. Once known where the district is in a group based on the HDI group then the value of its UN will follow the confidence interval of the UN value of the group. The results of this study indicate that the number of clusters (k) selected to classify districts in Indonesia as many as 5 clusters. Group prediction results show that one district belongs to the V cluster, while the other three districts are predicted to be members of the I cluster.

1. Introduction

The National Exam (NR) is the final evaluation to measure the student learning outcomes and the quality of education in Indonesia and been implemented since 1965 [1]. Regulation of the Ministry of Education and Culture of the Republic of Indonesia No. 5 of 2015, article 1, paragraph 5, stated that the National Examination, hereinafter referred to as the NE, is an activity of measuring and assessing the achievement of national graduate competencies on certain subjects [2]. The NE has undergone several changes and the development of government policies from the beginning set. By 2015, the government sets the NE as one of the minimum passing standards that students must achieve and become a reference point for continuing education to the next level.

High school education in Indonesia has not yet become a compulsory education. The 2015/2016 educational dataset [3] shows that the number of new high school students, both those who attend public high school and vocational schools only amounts to 3,175,407 people, while the new junior high school (SMP) students are 3,435,984 people, difference of about 260,000 people. In fact, the high school level is the level of education that must be passed in order to continue studies to the higher school level, i.e., college or university.

*Corresponding author: ridha.ferdhiana@unsyiah.ac.id

The high school level is divided into 4 majors: Natural Science (IPA), Social Sciences (IPS), Language, and Religion, but the dominant majors are the science and social science. Data from the Education Assessment Centre (Puspendik) of the Ministry of Education and Culture (Kemendikbud) shows that the science and social sciences majors are majors with the interest of more than two other majors. Department of science is also more desirable than IPS because the selection of study programs in universities more freely done by students majoring in science.

Data from Puspendik stated that the average value of UN in Indonesia in the academic year 2014/2015 for high school level is 61.29 with an average score of UN for science majors of 61.84 out of 100.00, thus the average UN for IPS slightly below IPA. The subjects included in the science are Bahasa Indonesia, English, Mathematics, Physics, Chemistry, and Biology. Furthermore, the average value of the UN in 2015 is grouped in four categories, which is very good if the average of UN ≥ 85, good if ≥ 70, enough if ≥ 55, and less if <55.

UN value, in addition to being used as a reference for continuing study to the next level, is also used as a mapping of the success of teaching and learning activities [2]. This mapping can then be used to improve learning and decision-making activities. This mapping, if done using only the average reference value of the NE which are very good, good, enough, and less like the guidance of puspendik, will give a misleading mapping result. The misleading is due to the NE score of 6 subjects, one school could be very strong in mathematics-based sciences but lacking in language and biology, or strong on biology and chemistry but lacking in mathematics and physics.

Economics development in one region certainly has a significant effect on the success of the education in that region. Indonesia is an unequal country for economic development [4], so there is a lack of success in education in all its regions. There are 514 districts in Indonesia spread over 34 provinces and in the 5 largest islands. The islands of Java and Bali are the ones with higher economic levels than the other islands, thus the districts within them have more success in their educational system. If the mapping only sees the average NE score, then only districts in Java and Bali and some cities outside them are included in good category. For that, another way of mapping is needed by looking at the all value of NE subjects. One way to do that is by using the clustering method.

Out of 514 districts in Indonesia only 510 districts have the value of NE majoring in science in the academic year 2014/2015, thus the grouping will be done for the 510 districts.. Four districts that do not have the value of NE majoring in Science are Administrasi Kepulauan Seribu, Nduga, Puncak, and Intan Jaya. The absence of NE value data might be caused by the absence of science majors in the area, or there is data input error, or indeed no data. Because education closely related to the Human Development Index (HDI), those 4 districts NE will be predicted based on the value of HDI.

2. Proposed Method

2.1 Data and Variable

The data used in this study was taken from the website of the national exam results report by the Center of Education Assessment (Puspendik) and consist of the average NE values from each district in Indonesia. The HDI variable, which taken from Indonesia Bureau of Statistics (BPS), will be used to predict the value of NE for the four districts that have missing NE value. The variables used are the average values of Bahasa Indonesia (X_1), English (X_2), Mathematics (X_3), Physics (X_4), Chemistry (X_5), Biology (X_6), and the average value of standard deviation of Bahasa Indonesia (X_7), English (X_8), Mathematics (X9), Physics (X_{10}), Chemistry (X_{11}), Biology (X_{12}), and HDI (X_{13}).

2.2 Data Analysis Procedure

The procedure of data analysis in this study has the following stages:

1. Grouping of districts/cities based on the value of NE majoring in Science using K-means method. The stages in performing clustering analysis using K-means method are as follows:
   a) Determine the number of clusters (k) using the Elbow method.

b) Measure the quality of cluster number (k) using clustering quality measurement method: R-square (R^2), Average Proportion of Non-overlap (APN), Average Distance (AD), Average Distance between Means (ADM), and Figure of Merit (ADM).
c) Collect the districts groups according to the best number of clusters (k) selected.

2. Grouping the value of HDI for each districts based on the result of cluster in step 1a.
3. Predict the clusters of four districts that do not have the NE value based on the group of their HDI.

3. Result and Discussion

3.1 Determining the Best Number of Cluster (k)
The number of clusters (k) will be determined based on the Sum of Squares Error (SSE) value or Total Within Sum of Squares (TWSS). The TWSS value shows the cluster variance, the smaller the result of TWSS, the better-formed cluster [5]. The best candidates of the number of cluster are chosen when the graph of TWSS shown a sharp turn, which is similar to elbow, thus called elbow method [6]. The graph of elbow method is shown in Figure 1 and it is clear that the number of clusters (k) to be selected for clustering lies between 3 up to 5 clusters.

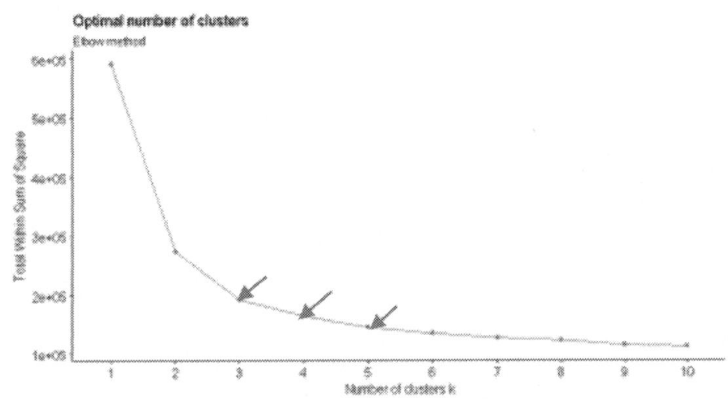

Figure 1. Graph of Elbow method

3.2. Measuring the quality of cluster number (k)
To determine the best number of cluster, the R-square (R^2) and the qualities of measurements are used. R^2 is a measurement to check the height of the quality of clustering results by looking at its ability to distinguish existing data on a cluster with existing data on other clusters. The value of R^2 ranges from 0 to 1, and the best R^2 value is when it is close to 1, which shows the maximum differences between the clusters [7].

Quality of measurements consists of three values. First is the Average Proportion of Non-overlap (APN), which performs calculations on the average proportion of observations that are not placed within the same cluster, with the grouping based on complete data and grouping by data with a single column deleted. The value of APN ranges from 0 to 1 [0.1], the smaller the resulted APN value can show highly consistent grouping results [8]. Second, Average Distance (AD) is a cluster validation calculation based on the average distance between observations placed on the same cluster, with the groupings based on the complete data and grouping by data with one column deleted [8]. The value of AD ranges from 0 to infinity (∞) [0, ∞], the smaller the resulted AD value can show very consistent grouping results. The third is Average Distance Between Means (ADM) which its calculation based on the average distance between cluster centres with the observations on the same cluster, with groupings based on complete data and groupings based on data with one column deleted. The ADM value ranges from 0 to infinity (∞) [0, ∞], the smaller the resulted ADM value can show very consistent grouping results.

The quality measurements result is shown in Table 1. And the bolded number shows the optimal number of the measurements, the 3 clusters has two optimal measurements while 5 clusters has three, thus it is decided to use 5 clusters.

Table 1. Results of clustering quality measurement based on number of clusters (k)

Clustering Quality Measurement	The Number of Cluster(k)		
	3	4	5
R-square(R^2) (%)	67,640	72,264	**75,600**
Average Proportion of Non-overlap (APN)	**0,0527**	0,1014	0,2120
Average Distance (AD)	26,521	24,987	**24,533**
Average Distance between Means (ADM)	**2,5643**	4,0911	7,0979
Figure of Merit (ADM)	5,7873	5,4181	**5,3170**

3.3. Cluster of Districts Based on NE Value

Based on the clustering quality measurement, the K-Means algorithm is run to get five clusters of district based on NE value with the results of the number of districts in each cluster are are 141, 169, 77, 86, and 177 respectively. The summary for the number of each cluster of districts per-province is shown in Table 2.

Figure 2 shows the boxplots of each subject for each cluster. The purpose of the boxplot is to make comparison of the subject average value by cluster. Based on the Figure, it seen that the IV cluster is the cluster with the best ranking, followed by the II cluster, then the V cluster, the III cluster, while the I cluster is cluster with the lowest rank for every subjects. This finding is match the preliminary epiphany that the district with higher economic status will have a better education system, as it is clearly shown that most districts in Java and Bali Island are laid in the IV, II, and V clusters. While, most districts in Papua, Nusa Tenggara Barat, and Nusa Tenggara Timur mostly laid in the I and III cluster.

Table 2. Cluster of districts/cities per province based on the NE value (1/2)

No.	Province	Cluster										Total
		I		II		III		IV		V		
		n	%	n	%	n	%	n	%	N	%	
1	DKI Jakarta	-	-	3	60.00	-	-	2	40.00	-	-	5
2	West Java	-	-	11	40.74	2	7.41	5	18.52	9	33.33	27
3	Central Java	-	-	8	22.86	-	-			27	77.14	35
4	DI Yogyakarta	-	-	-	-	-	-	1	20.00	4	80.00	5
5	East Java	-	-	20	52.63	-	-	14	36.84	4	10.53	38
6	Aceh	-	-	13	56.52	1	4.35	8	34.78	1	4.35	23
7	North Sumatera	1	3.03	10	30.30	3	9.09	17	51.52	2	6.06	33
8	West Sumatera	1	5.26	11	57.89	-	-	2	10.53	5	26.32	19
9	Riau	-	-	9	75.00	-	-	3	25.00	-	-	12
10	Jambi	1	9.09	3	27.27	3	27.27	-	-	4	36.36	11
11	South Sumatera	-	-	7	41.18	1	5.88	5	29.41	4	23.53	17
12	Lampung	1	6.67	2	13.33	4	26.67	-	-	8	53.33	15
13	West Kalimantan	-	-	4	28.57	3	21.43	-	-	7	50.00	14
14	Central Kalimantan	1	7.14	-	-	5	35.71	-	-	8	57.14	14
15	South Kalimantan	-	-	7	53.85	1	7.69	1	7.69	4	30.77	13
16	East Kalimantan	-	-	1	10.00	4	40.00	-	-	5	50.00	10
17	North Sulawesi	-	-	7	46.67	1	6.67	2	13.33	5	33.33	15
18	Central Sulawesi	2	15.38	1	7.69	8	61.54	-	-	2	15.38	13
19	South Sulawesi	-	-	14	58.33	-	-	9	37.50	1	4.17	24

Table 2. Cluster of districts/cities per province based on the NE value (2/2).

No.	Province	Cluster										Total
		I		II		III		IV		V		
		n	%	n	%	n	%	n	%	N	%	
20	Southeast Sulawesi	-	-	11	64.71	-	-	3	17.65	3	17.65	17
21	Maluku	5	45.45	3	27.27	-	-	-	-	3	27.27	11
22	Bali	-	-	-	-	-	-	9	100	-	-	9
23	West Nusa Tenggara	-	-	3	30.00	2	20.00	1	10.00	4	40.00	10
24	East Nusa Tenggara	12	54.55	1	4.55	7	31.82	-	-	2	9.09	22
25	Papua	10	38.46	7	26.92	4	15.38	1	3.85	4	15.38	26
26	Bengkulu	-	-	1	10.00	5	50.00	-	-	4	40.00	10
27	North Maluku	2	20.00	2	20.00	3	30.00	1	10.00	2	20.00	10
28	Bangka Belitung	-	-	-	-	7	100	-	-	-	-	7
29	Gorontalo	-	-	2	33.33	2	33.33	1	16.67	1	16.67	6
30	Banten	-	-	2	25.00	2	25.00	-	-	4	50.00	8
31	Riau Islands	1	14.29	-	-	2	28.57	-	-	4	57.14	7
32	West Sulawesi	-	-	1	16.67	3	50.00	-	-	2	33.33	6
33	West Papua	3	23.08	5	38.46	-	-	1	7.69	4	30.77	13
34	North Kalimantan	1	20.00	-	-	4	80.00	-	-	-	-	5

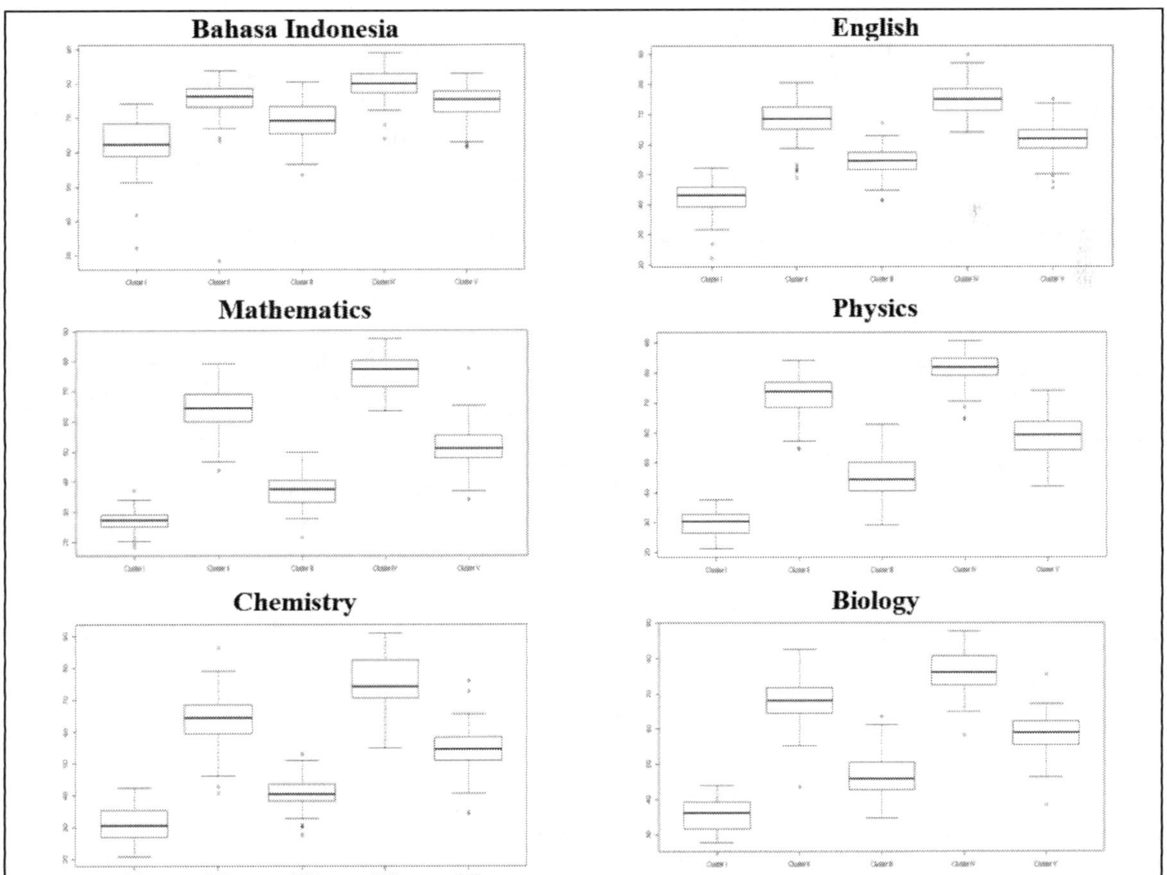

Figure 2. Boxplot cluster rating by subjects

3.4. HDI Score of Cluster of Districts Based on NE Value

Based on the cluster obtained above, then the HDI value of each district in the cluster is grouped. This grouping aims to see whether educational success is comparable to overall human development. Statistical measurements of the grouped HDI values are then calculated as shown in Table 3. Based on the mean HDI value for each cluster, it can be seen that the HDI rating is the same as the result rating of the NE value. Thus, we can conclude that education is an important factor in Human Development Index.

Table 3. Summary statistics of HDI Score for each cluster

Summary Statistics	Cluster				
	I	II	III	IV	V
Minimum	52,78	43,55	44,32	53,73	40,91
Maximum	70,89	80,96	79,63	83,37	81,20
Mean	61,82	68,13	64,97	70,57	66,90
Standard Deviation	4,53	6,72	5,41	6,14	5,68
Median	60,96	67,45	64,36	70,36	67,12
95% Confidence Interval					
Lower Limit	60,39	69,64	64,69	69,25	67,43
Upper Limit	63,25	72,18	67,56	73,02	70,15

3.5. The Prediction of Cluster of Districts that has no NE Value

The HDI grouping aims to predict the NE value for districts that does not have it, which are district of Administrasi Kepulauan Seribu, Nduga, Puncak and Intan Jaya. The 95% Confidence Interval (CI) shown in Table 3 is then be used to predict the district cluster according to their HDI values. The cluster where those districts belong is given in Table 5. Thus, the Administrasi Kepulauan Seribu is belong to the V cluster , while those other 3 belong to the I cluster though their HDI values are far below the 95% CI of HDI in the I cluster. Table 4 shows the prediction values of each subjects for those 4 districts. Figure 3 give overall mapping of cluster rating of NE values in Indonesia. The mapping shows that most Java and Bali Islands are in highest rating while Papua and Sumbawa and Kalimantan islands are in the lowest rating.

Table 4. The Prediction of NE values per subjects of districts that has no NE values

Districts/cities	Cluster	Subjects	Confidence Interval 95%	
			Lower Limit	Upper Limit
Administrasi Kepulauan Seribu	V	Indonesian	73,39	75,15
		English	60, 68	62,44
		Mathematics	50,57	52,67
		Physics	57,57	59,90
		Chemistry	53,48	55,63
		Biology	57,44	59,16
Nduga Puncak Intan Jaya	I	Indonesian	59,07	64,54
		English	40,19	44,01
		Mathematics	25,86	28,32
		Physics	28,49	31,19
		Chemistry	29,43	32,74
		Biology	34,24	37,22

Table 5. The clusters of districts that has no NE value

Districts	HDI	Cluster	Confidence Interval 95%	
			Lower Limit	Upper Limit
Administrasi Kepulauan Seribu	68,84	V	67,43	70,15
Nduga District	25,47	I	60,39	63,25
Puncak District	39,41	I	60,40	63,26
Intan Jaya District	44,35	I	60,41	63,27

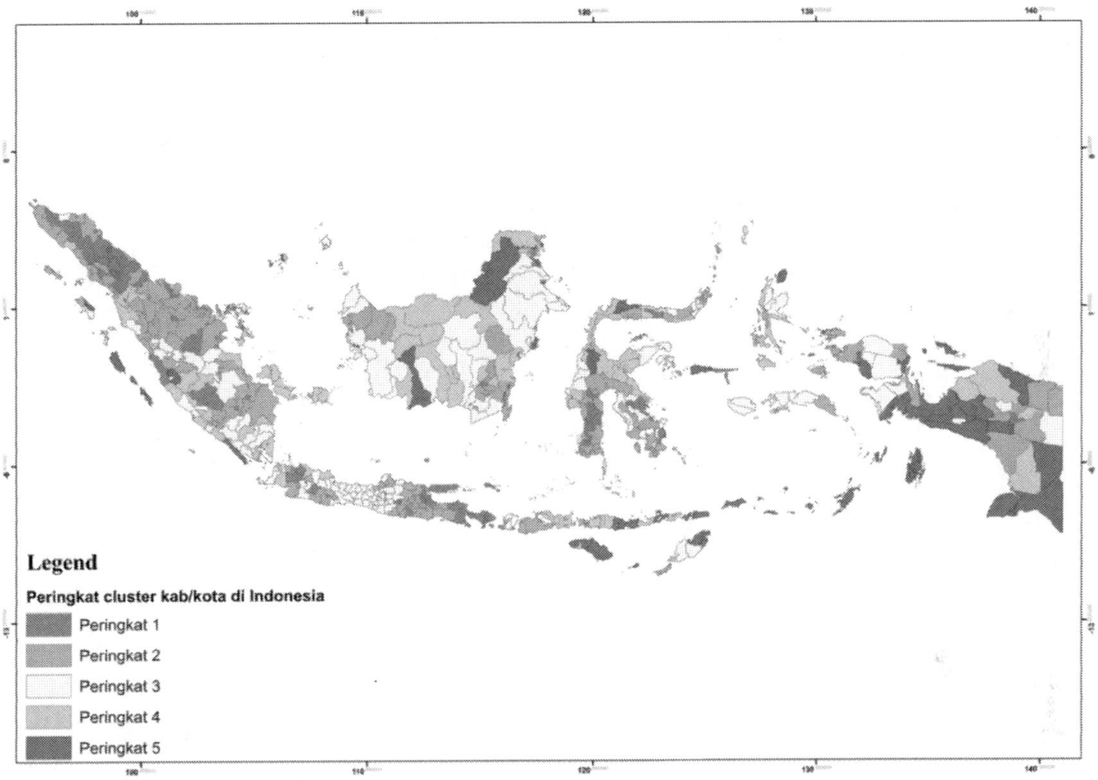

Figure 3. Map of cluster districts ranking by island in Indonesia

4. Conclusion

The conclusions of cluster districts analysis based on NE value and HDI score using K-means clustering method can be revised as follows:

1. The best number of cluster (k) to cluster districts based on UN value is 5
2. The order of the rating of clusters is the IV, II, V, III and last the I cluster.
3. The cluster of those districts that has no NE values are Administrasi Kepulauan Seribu City belong to the V cluster, while Nduga, Puncak and Intan Jaya belong to the I cluster.

5. Acknowledgments

This research was supported by Lembaga Penelitian dan Pengabdian kepada Masyarakat Syiah Kuala University through Penelitian Lektor Scheme. We thank our colleagues from Pusat Penilaian Pendidikan, Kementerian Pendidikan Dan Kebudayaan who provided data. We also thank our colleagues in Statistics Department of Syiah Kuala University for the support and discussions.

6. References

[1] Setneg RI (Sekretariat Negara Republik Indonesia) 1966 *Penetapan Presiden Republik Indonesia Nomor 19 Tahun 1965 tentang Pokok-Pokok Sistem Pendidikan Nasional Pancasila* (Jakarta : Setneg RI)

[2] Mendikbud RI (Menteri Pendidikan dan Kebudayaan Republik Indonesia) 2015 *Peraturan Menteri Pendidikan dan Kebudayaan Nomor 5 Tahun 2015 tentang Kriteria Kelulusan Peserta Didik, Penyelenggaraan Ujian Nasional, dan Penyelenggaraan Ujian Sekolah/Madrasah/ Pendidikan Kesetaraan pada SMP/MTs atau yang Sederajat dan SMA/MA/SMK atau yang Sederajat* (Jakarta : Mendikbud RI).

[3] Kementerian Pendidikan Dan Kebudayaan ,Sekretariat Jenderal Pusat Data Dan Statistik Pendidikan Dan Kebudayaan. 2016 *Statistik Sekolah Menengah Atas (SMA) 2015/2016*.

[4] Andi Irawan 2015 *Bulletin of Indonesian Economic Studies,* 51:1, 148-149,

[5] Calinski, Tadeusz, and Harabasz, Joachim. 1974 *Communications in Statistics*. 3(1): 1-27.

[6] Kodinariya, Truptil M. and Maknawa, Prashant R. 2013 *International Journal of Advance Research in Computer Science and Management Studies*. 1(6): 90-95.

[7] Gudono 2011 *Analisis Data Multivaria (*Yogyakarta: BPFE)

[8] Datta, Susmitta, and Datta, Somnath. 2003 *Bioinformatics* 19(4): 459–466.

IOP Conf. Series: Materials Science and Engineering **523** (2019) 012006 doi:10.1088/1757-899X/523/1/012006

Comparison of cox models in detecting factors affecting healing rate of dengue hemorrhagic fever

Miftahuddin[1,*], **I Salsabila**[1], **A Gul**[2]

[1] Department of Statistics, Faculty of Mathematics and Sciences, Syiah Kuala University, Jln. Teuku Nyak Arief, Darussalam, Banda Aceh, Aceh, 23111, Indonesia
[2]Department of Statistics, Shaheed Benazir Bhutto Women University Peshawar, Pakistan

E-mail: miftah@unsyiah.ac.id

Abstract. Dengue Hemorrhagic Fever (DHF) is an epidemic of disease that usually becomes a benchmark of good or bad condition of the environment and health facilities of a region. When the disease is outbreak and does not get serious treatment, it causes death. One of the first steps that can be done to deal with this disease is to know the factors that affect the healing rate of DHF patients. This study aims to detect factors affecting the healing rate of DHF patients by using Cox regression analysis based on Cox model family, such as Cox Proportional Hazard (PH) model, extended Cox model with one and two heaviside function, and stratified Cox model and get best model for this study. The data used is secondary data consisting of 107 in-patients DHF period January-December 2016 at Regional General Hospital dr. Zainoel Abidin Banda Aceh, Indonesia. Based on Cox Proportional Hazard model, the variable that significantly affect the patient's healing rate is Age. Based on the Extended Cox model with one heaviside function, the variables that significantly affect the patient's healing rate are Age, Sex, Number of Platelets, Clinical Degrees III and Number of Leukocytes multiplied by the time function. Based on the Extended Cox model with two heaviside functions, the variables that significantly affect the patient's healing rate are Age, Sex, Number of Platelets, Clinical Degrees III and Number of Leucocytes multiplied by second time function. Based on the stratified Cox model, there are no variables that significantly affect the patient's healing rate. The best model based on Akaike Information Criterion value is the Stratified Cox model.

1. Introduction

Dengue Hemorrhagic Fever (DHF) is a disease caused by dengue virus that is transmitted by mosquito aedesaegypti. This disease often appears as an Extraordinary evidence with a relatively high number of morbidity and mortality. The number of dengue fever cases in Banda Aceh during 2016 were 152 (men 77 and women 75 cases). Number of DHF cases, according to Comunity Health Center, can be seen in Figure 1 below (Health Office of Banda Aceh, 2016). DHF is a disease that is often prevalent in the community and needs to get serious handling. For this purpose, identification of the factors that affect the healing rate of DHF patients are of crucial importance.

*Corresponding author: miftah@unsyiah.ac.id

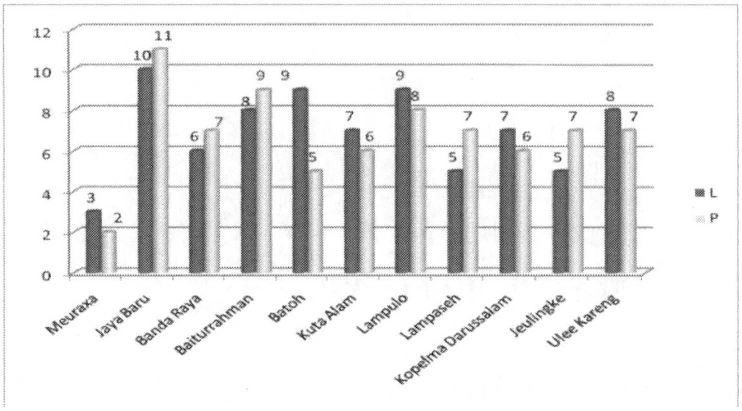

Figure 1. Number of DHF cases according to Community Health Center in Banda Aceh 2016.

Generally, the statistical method applied to analyze the relationship between dependent and independent variables is regression analysis. If the dependent variable is considered as the time until going out of remission, the analysis used is related to survival analysis.In epidemiological studies, one of the most commonly used regression analyzes for survival analysisis is the Cox regression model. This model makes it possible to isolate variables that have small effect on survival. In addition, the model allows to estimate the risk or danger of death for an individual based on the prognostic variables. This study investigated the factors that influence the healing rate of DHF patients. Cox regression analysis technique is implemented, where comparisons are made among the Cox model family, such as Cox Proportional Hazard, extended Cox with one and two heaviside functions, and stratified Cox model to get best model. Based on these models, there are different results about the factors that affect the healing rate of DHF patients, with identification a set of factors.

2. Materials and Methods

This study used secondary data consisting of 107 Dengue Hemorrhagic Fever (DHF) patients in the period of January-December 2016 at Regional General Hospital dr. Zainoel Abidin Banda Aceh. The variables used are detailed in Table 1.

Table 1. Research Variables.

Symbol	Variable Name	Variable Type	Unit
T	Duration of patient treated (survival time)	Numerical	Days
d	Status 0 = censored, 1 = uncensored		
X_1	Age	Numerical	Years
X_2	Sex 0 = female, 1 = male	Categorical	
X_3	Hemoglobin level	Numerical	gr/dl
X_4	Hematocrit level	Numerical	%
X_5	Number of Platelets	Numerical	Thousand/mm^3
X_6	Number of Leukocytes	Numerical	Thousand/mm^3
X_7	Degree DHF 0 = none 1 = Degree I 2 = Degree II 3 = Degree III 4 = Degree IV	Categorical	

2.1 Dengue Hemorrhagic Fever (DHF)

Dengue Fever is an acute febrile viral disease frequently presenting headaches, bone or joint and muscular pain, rashes and leukopenia as symptoms. Dengue Hemorrhagic Fever (DHF) is characterized by four major clinical manifestations: high fever, haemorrhagic phenomena, often with hepatomegaly and, in severe cases, signs of circulatory failure.Such patients may develop hypovolemic shock resulting from plasma leakage.This is called Dengue Shock Syndrome (DSS) and can be fatal [1]. According to [2], several tests are commonly used to determine whether antibodies to dengue virus are present in the body by counting the number of antibodies to dengue virus and complete blood count (hemoglobin, leucocytes, hematocrit and platelets).

Hemoglobin (Hb) is a complex protein containing iron and red in the erythrocytes. Normal Hb level for women is11.4 to 15.1 g/dl while for males is 13.4 to 17.7 g/dl. Leukocytes or white blood cells are blood cells containing the nucleus. The number of leukocytes within normal limits for men and women is 4300-11300/mm^3. Platelets are the smallest part of cellular elements in the bone marrow and are important in clotting and hemostasis process. Under normal conditions, the number of Platelets for women and men is 150-400 thousand/mm^3. Hematocrit (Ht) is a number that indicates the percentage of solids in blood to blood fluid. For normal Ht levels for women 38-42% while for men 40-47% [3].

2.2 Survival Analysis

In survival analysis, is focused to time-related data, ranging from time origin or start point to the occurrence of a special event or end point [4]. The subjects are called censored when the follow up time ends before the event occurs. In such cases the actual duration of time to the event is not known or 'censored' by the study. The outcome variable for each subject is therefore composed of 'time' and the 'status' at the end. Mathematically, the status is 1 if the event takes place and 0 otherwise [5]. The causes of censored data include:

a. Loss to follow up, occurs when objects move, dies or refuses to participate

b. Drop out, occurs when the treatment is stopped for some reason.

c. Termination of the study, occurs when the study period ends while the observed object has not reached a failure event [6].

2.3 Cox Proportional Hazard Model

Cox Proportional Hazard regression model was applied to find the correlation of predictor variables to survival time. The hazard function of different individuals is assumed to be proportional every time [7]. In general, the Cox Proportional Hazard regression model can be expressed as follows.

$$h(t,\mathbf{X}) = h_0(t)\exp\left[\sum_{i}^{p}\beta_i X_i\right] \quad (1)$$

A key assumption of the Cox regression model is proportional hazards. The proportional hazards assumption means that the hazard ratio is constant over time, or that the hazard for an individual is proportional to the hazard for any other individual [8]. There are three methods commonly used to assess the PH assumption: (1) graphical, using, say, log–log survival curves; (2) using an extended Cox model; and (3) using a goodness-of-fit (GOF) test. There are two options to consider if the PH assumption is not satisfied for one or more of the predictors in the model. We suggest the application of a stratified Cox (SC) model, which stratifies on the predictor(s) not satisfying the PH assumption, while keeping in the model those predictors that satisfy the PH assumption. In addition the option, which involves using time-dependent variables is also considered [6].

2.4 Extended Cox Model

In the Cox regression model, there can be variables which involve t. Such variables are called time-dependent variables. A time dependent variable is defined as any variable whose value for a given subject may differ over time(t). If there are time-dependent variables in the model, the Cox regression model can be used but can no longer satisfy the proportional hazards assumption. Therefore, extended Cox regression model should be used instead [9]. The general model of this method is as follows.

$$h(t,\mathbf{X}(t)) = h_0(t)\exp\left[\sum_{i=1}^{p1}\beta_i X_i + \sum_{j=1}^{p2}\beta_j X_j(t)\right] \quad (2)$$

The formation of extended Cox model to overcome non-proportional hazard is done by adding time function, i.e. g(t) = t on variable that does not satisfy the PH assumption. One Heaviside function is of the form g(t), which takes on the value 1 if t is greater than or equal to some specified value of t, called t, and takes on the value 0 if t is less than t. An extended Cox model which contains a single Heaviside function is shown here, where g(t) = 1 if $t \geq t_0$ and 0 if $t < t_0$ then

$h(t, X(t)) = h_0(t) \exp [\beta E + \delta E\, g(t))]$, where for $t \geq t_0$: $g(t) = 1 \rightarrow E \times g(t) = E$
so $h(t, X(t)) = h_0(t) \exp [(\beta + \delta)E]$, and $\widehat{HR} = \exp[\hat{\beta} + \hat{\delta}]$
whereas for $t < t_0$: $g(t) = 0 \rightarrow E \times g(t) = 0$ so $h(t, X(t)) = h_0(t) \exp [(\beta E]$, and $\widehat{HR} = \exp[\hat{\beta}]$.

There is actually an equivalent way to write this model that uses two Heaviside functions in the same model. This alternative model is shown here. The two Heaviside functions are called $g_1(t)$ and $g_2(t)$. Alternative model with two Heaviside functions is shown here.

$g_1(t) = 1$ if $t \geq t_0$ and 0 if $t < t_0$ and $g_2(t) = 1$ if $t < t_0$ and 0 if $t \geq t_0$

If $t \geq t_0$: $g_1(t) = 1$, and $g_2(t) = 0$ then

$h(t, X) = h_0(t) \exp [\delta_1(E \times 1) + \delta_2(E \times 0)] = h_0(t) \exp [\delta_1 E]$, so that $\widehat{HR} = \exp[\hat{\delta}_1]$

If $t \geq t_0$: $g_1(t) = 0$, and $g_2(t) = 1$ then

$h(t, X) = h_0(t) \exp [\delta_1(E \times 0) + \delta_2(E \times 1)] = h_0(t) \exp [\delta_2 E]$, so that $\widehat{HR} = \exp[\hat{\delta}_2]$

2.5 Stratified Cox Model

In comparison to the extended Cox model, a Stratified Cox models utilized. The method used is grouping of variables that do not satisfy the PH assumption into several strata. The general model of this method is as follows.

$$h_g(t, \mathbf{X}) = h_{0_g}(t) \exp [\beta_1 X_1 + \beta_2 X_2 + \dots + \beta_p X_p] \quad g = 1,2, \dots, k^* \tag{3}$$

The stratified Cox regression model is a modification of the Cox regression model by the stratification of a covariate that does not satisfy the proportional hazards assumption. Covariates that are assumed to satisfy the proportional hazards assumption are included in the model, whereas the predictor being stratified is not included [10]. According to [11], a stratified Cox model can be used that allows separate baseline hazards but common coefficients between strata.

2.6 Testing the Significance of Parameters

The test used to test the significance of the β coefficient of the model may use partial test and simultaneous test.

2.6.1 Partial Test

Partial test is done to know the significance of parameters to the dependent variable. Testing the significance of parameters using Wald [12] with the following hypothesis. $H_0 : \beta_j = 0$ and $H_1 : \beta_j \neq 0$, with statistics test:

$$w = \left(\frac{\hat{\beta}_j}{SE(\hat{\beta}_j)}\right)^2$$

The W test statistic also referred to as Wald's test statistic with $SE(\hat{\beta}_j)$ is a standard estimate of parameter error. The rejection zone is if $|W| > Z_{\alpha/2}$ or $W^2 > \chi^2_{(v,\alpha)}$ with degrees of freedom v.

2.6.2 Simultaneous Test

Tests simultaneously performed to examine the significance of the overall coefficient β with the following hypothesis. $H_0 : \beta_1 = \beta_2 = \dots = \beta_p = 0$ and H_1 : at least one $\beta_j \neq 0$; $j = 1, 2, ..., p$

Test statistics: $G = -2\,\{L(\hat{\beta}) - Lp(0)\}$

where, $L(\hat{\beta})$ = Partial log likelihood for models with predictor variables

$Lp(0)$ = Partial log likelihood for models without predictor variables

The test statistic G follows the Chi-square distribution. The rejection criterion H_0 if the value of G > $\chi^2_{(v,\alpha)}$ with df (degree of freedom) is degrees of freedom and a is the error rate used.

2.7 Model Selection

Akaike Information Criterion (AIC) is used for selection of the optimal model. The AIC is a popular method for comparing the adequacy of multiple, possibly non-nested models. Current practice in cognitive psychology is to accept a single model on the basis of only the "raw" AIC values. It makes

IOP Conf. Series: Materials Science and Engineering **523** (2019) 012006 doi:10.1088/1757-899X/523/1/012006

difficult to un-ambiguously interpret the observed AIC differences in terms of a continuous measure such as probability [13]. The model with the smallest AIC value is the best model that can explain the prediction data with the real data. AIC is defined as:

$$\text{AIC} = e^{\frac{2k}{n}}\left(\frac{\sum_{i=1}^{n} u_i^2}{n}\right)$$

Where k is the number of parameters in the model including the intercept, n is the number of observations (samples) and u is the residual. The smallest standard value of the AIC depends on the number of parameters and the number of observations present in the model.

3. Results and Discussion

Characteristics of patient data on Dengue Hemorrhagic Fever (DHF) can be seen from the descriptive statistics presented in tables 2 and 3 below.

Table 2. Descriptive statistics of DHF data.

Variable	Minimum	Maximum	Mean	Deviation Standard
Duration of patient treated (days)	2	21	4663	2656
Age (years)	1	64	14.77	10.75
Hemoglobin (Hb) (gr/dl)	6	19	11392	2338
Hematocrit (%)	20	57	36317	6535
Platelet (thousand/mm^3)	12	344	622055	51301
Leukocyte (thousand/mm^3)	1	20	5.616	3982

Table 3. Distribution of DHF patients by sex, clinical degree and status

Variable	Category	Quantity	Percentage
Sex	Female	35	33
	Male	72	67
Clinical Degree	Degree I	34	32
	Degree II	63	59
	Degree III	8	7
	Degree IV	0	0
	None	2	2
Status	Censored	2	1.87
	Uncensored	105	98.13

Table 2, the average hospitalization of patients treated was 4 days with a standard deviation of 2 days. The minimum time of patients treated is 2 days and maximum is 21 days. The average patient age is 14 years old with the youngest of 1 year old and at most 64 years. Hemoglobin, hematocrit, platelets and leukocytes patients had an average of 11.3392 g/dl, 36317%, 62205 thousand/mm^3 and 5616 thousand/mm^3. The average platelet of patients was in abnormal condition, but the maximum value is still in normal condition. While the average leukosit still in normal condition, but its maximum value is in abnormal condition. Whereas table 3, the percentage of DBD male patients is (67%) and women is (33%). Based on Clinical Degree, DHF patients are more classified in DHF II, ie 59%, compared to Degree I, III, and no degree. Based on the status, DHF patients are more in the uncensored status (stated by the expert doctor repatriated because it has improved) that is 98.13% compared to censored status (run from the hospital and died) that is 1.87%.

3.1 Cox Proportional Hazard Model

The formation of Cox Proportional Hazard (PH) model was performed to find out the relationship between survival time and the variables suspected to affect survival time. The variables are age, sex, Hb level, hematocrit level, Number of Platelets, Number of Leukocytes, and dengue degrees. The results of Cox PH model formation can be seen from the following table 4. The Cox Proportional Hazard model equation obtained is as follows.

IOP Conf. Series: Materials Science and Engineering **523** (2019) 012006 doi:10.1088/1757-899X/523/1/012006

$$h(t, X) = h_0(t) \exp[-0{,}02445X_1 + 0{,}43895X_2 + 0{,}12534X_3 + 0{,}05925X_4 - 0{,}00472X_5 - 0{,}03806X_6 - 0{,}38981X_7(1) - 0{,}60638X_7(2) - 1{,}24453X_7(3)] \quad (4)$$

Table 4. Summary of Cox PH model.

	Coef	P	Likelihood ratio test	P-value
Age (X_1)	-0.02445	0.014	24.67	0.003
Sex (X_2)	0.43895	0.063		
Hemoglobin level (X_3)	0.12534	0.129		
Hematocrit level (X_4)	-0.03806	0.188		
Number of Platelets (X_5)	-0.00472	0.056		
Number of Leukocytes (X_6)	0.05925	0.081		
Degree I ($X_7(1)$)	-0.38981	0.602		
Degree II ($X_7(2)$)	-0.60638	0.411		
Degree III ($X_7(3)$)	-1.24453	0.134		

After obtaining parameter estimation for Cox PH model formation, all parameters must be tested for significance. The significance test can be done with two stages, i.e. simultaneous test using Likelihood Ratio Test, and partial test using Wald Test. Based on Table 4, with a significance level of 5% (α = 0.05), it can be done simultaneous significance test so it is known that there is at least one independent variable that affects the model. It is concluded pursuant to p-value (0.003) which is smaller than α (0.05). To determine which variables are influential, then it can be done partial test. Based on table 4, the p-values for the partial test (p) of all independent variables are different. The P-value for the Age(X_1) is smaller than α, whereas the p-values for the Sex(X_2), Hb level(X_3), Hematocrit level(X_4), Number of Platelets(X_5), Number of Leukocytes(X_6), Degree I ($X_7(1)$), II ($X_7(2)$) and III($X_7(3)$) is greater than α. Therefore, it can be concluded that the Age variable has significant effect on the model, while the Hb level, Hematocrit level, Number of Platelets, Number of Leukocytes, Degree I, II and III have no significant effect on the model. The next step is to test the PH assumption to find out whether the PH assumption for the model has been met. The PH assumption can be tested using Goodness-of-Fit (GoF) test. The following table presents the results using GoF test.

Table 5. The Result of PH assumption test using Goodness-of-Fit test

	Rho	Chisq	P
Age(X_1)	0.0946	0.913	0.33929
Sex(X_2)	0.1097	1.493	0.22176
Hemoglobin level(X_3)	-0.0417	0.269	0.60429
Hematocrit level(X_4)	-0.0815	0.95	0.32983
Number of Platelets(X_5)	0.112	1.866	0.17191
Number of Leukocytes(X_6)	-0.2221	7.588	0.00588
Degree I ($X_7(1)$)	-0.0932	0.954	0.32858
Degree II ($X_7(2)$)	-0.0495	0.263	0.60830
Degree III ($X_7(3)$)	-0.0829	0.786	0.37533
Global	NA	12.201	0.20220

Based on table 5, with H_0 the data satisfy the PH assumption and the 5% significance level, the decision to reject H_0 for the Number of Leukocytes(X_6), and receive H_0 for the Age(X_1), Sex(X_2), Hb level(X_3), Hematocrit level(X_4), Number of Platelets(X_5), Degree I ($X_7(1)$), II ($X_7(2)$) and III ($X_7(3)$). This means that the assumption of PH is satisfy for the Age, Sex, Hb level, Hematocrit level, Number of Platelets, Degree I, II and III variables and that are not satisfy for the Number of Leukocytes variable. Because there are variables that do not satisfy the PH assumption, another alternative is needed for this data modeling in order to avoid non-proportional hazard conditions.

3.2 Extended Cox Model

The formation of extended Cox model to overcome non-proportional hazard is done by adding time function, i.e. $g(t) = t$ on variable that does not satisfy the PH assumption. Note that the Heaviside function is the form g(t), which takes the value of 1 if t is greater than or equal to some certain t value, called t_0, and takes the value 0 if t is less than t_0.

For this case, suppose we want to assess the assumption of PH using the Heaviside function (HF) to produce a constant hazard ratio for less than 10 days, and a constant hazard ratio of 10 days or more of follow-up. There are two possible ways to model the involvement of one HF depending on whether the experiment uses one or two Heaviside functions.

Model 1 (with one Heaviside function) :

$$h(t, X(t)) = h_0(t) \exp [\beta_1X_1 + \beta_2X_2 + \beta_3X_3 + \beta_4X_4 + \beta_5X_5 + \beta_6X_7(1) + \beta_7X_7(2) + \beta_8X_7(3) + \delta_1(X_6 \text{ x } g(t))]$$

where $g(t) = 1$ if $0 \leq t < 10$ days and 0 if $t \geq 10$ days.

Model 2 (with two Heaviside functions) :

$$h(t, X(t)) = h_0(t) \exp [\beta_1X_1 + \beta_2X_2 + \beta_3X_3 + \beta_4X_4 + \beta_5X_5 + \beta_6X_7(1) + \beta_7X_7(2) + \beta_8X_7(3) + \delta_1(X_6 \text{ x } g_1(t)) + \delta_2(X_6 \text{ x } g_2(t))]$$

where $g_1(t)=1$ if $0 \leq t<10$ days and 0 if $t \geq 10$ days, $g_2(t)=1$ if $t \geq 10$ days and 0 if $0 \leq t<10$ days.

If we choose to use model 1, then the estimation of the extended Cox model parameter is given by the following table 6. The equation of the extended Cox model with one heaviside function is obtained as follows.

$$h(t, X(t)) = h_0(t) \exp[-0.02034X_1 + 0.38485X_2 + 0.10662X_3 - 0.02945X_4 - 0.00436X_5 - 0.17576X_7(1) - 0.40253X_7(2) - 0.95231X_7(3) + 0.05917(X_6 \text{ x } g(t))] \quad (5)$$

Table 6. Summary of Extended Cox model with one heaviside function.

	Coef	P	Likelihood ratio test	P-value
Age(X_1)	-0.02034	0.0036	18.63	0.03
Sex(X_2)	0.38485	0.0256		
Hemoglobin level(X_3)	0.10662	0.1150		
Hematocrit level(X_4)	-0.02945	0.1735		
Number of Platelets(X_5)	-0.00436	0.0172		
Degree I ($X_7(1)$)	-0.17576	0.5476		
Degree II ($X_7(2)$)	-0.40253	0.1194		
Degree III ($X_7(3)$)	-0.95231	0.0383		
Number of Leukocytes(X_6) x $g_1(t)$	0.05917	0.0364		

Based on table 6, with a significance level of 5% ($\alpha = 0.05$), p-value (0.03) smaller than α (0.05) can conclude that there is at least one independent variable that significantly affects the model. To determine which variables are influential, then it can be done partial test. Based on Table 6, the p-values for the Age(X_1), Sex(X_2), Number of Platelets(X_5), Degree III ($X_7(3)$) and Number of Leukocytes(X_6) x $g_1(t)$ smaller than α, while p-value for Hb level(X_3), Hematocrit level(X_4), Degree I ($X_7(1)$) and II ($X_7(2)$) greater thanα. Therefore, it can be concluded that the variables Age, Sex, Number of Platelets, Degree III and Number of Leukocytes x time function have a significant effect on the model, while the Hb level, Hematocrit level, Degree I and II have no significant effect on the model. Then to find out whether the condition of non-proportional hazard has been resolved, it can be tested the PH assumption using GoF test. The following table presents the test results using GoF test.

Based on table 7, with Null Hypothesis (H_0) the data satisfy the PH assumption and the 5% significance level ($\alpha = 0.05$), the decision to receive H_0 for the Age(X_1), Sex(X_2), Hb level(X_3), Hematocrit level(X_4), Number of Platelets(X_5), Degree I ($X_7(1)$), II ($X_7(2)$), III ($X_7(3)$) and Number of Leukocytes(X_6) x $g_1(t)$. This means that the PH assumption issatisfy for all variables, i.e. Age, Sex, Hb level, Hematocrit level, Number of Platelets, Degree I, II, III, Number of Leukocytes x time function. Since all variables satisfy the PH assumption, the non-proportional hazard condition is resolved.

Table 7. The Result of PH assumption test using Goodness-of-Fit test.

	Rho	Chisq	P
Age(X_1)	0.0746	0.2770	0.599
Sex(X_2)	0.0786	0.4020	0.526
Hemoglobin level(X_3)	-0.0334	0.1219	0.727
Hematocrit level(X_4)	-0.0530	0.2509	0.616
Number of Platelets(X_5)	0.0767	0.4902	0.484
Degree I ($X_7(1)$)	-0.1282	0.7958	0.372
Degree II ($X_7(2)$)	-0.522	0.0672	0.795
Degree III ($X_7(3)$)	-0.0877	0.7164	0.397
Number of Leukocytes(X_6) x $g_1(t)$	-0.1204	1.7553	0.185
Global	NA	4.1598	0.901

Furthermore, if we choose to use model 2, then the estimation of extended Cox model parameters is given by the following table.

Table 8. Summary of Extended Cox model with two heaviside function.

	Coef	P	Likelihood ratio test	P-value
Age(X_1)	-0.01860	0.0076	19.62	0.03
Sex(X_2)	0.39466	0.0245		
Hemoglobin level(X_3)	0.10956	0.1065		
Hematocrit level(X_4)	-0.03092	0.1560		
Number of Platelets(X_5)	-0.00406	0.0228		
Degree I ($X_7(1)$)	-0.17615	0.5519		
Degree II ($X_7(2)$)	-0.40552	0.1239		
Degree III ($X_7(3)$)	-0.97773	0.0409		
Number of Leukocytes(X_6) x $g_1(t)$	0.05658	0.0522		
Number of Leukocytes(X_6) x $g_2(t)$	-0.25119	0.0371		

The equation of the extended Cox model with two Heaviside function is obtained as follows.

$$h(t, X(t)) = h_0(t)\exp[-0.01860X_1 + 0.39466X_2 + 0.10956X_3 - 0.03092X_4 - 0.00406X_5 - 0.17615X_7(1) - 0.40552X_7(2) - 0.97773X_7(3) + 0.05658(X_6 \text{ x } g_1(t)) - 0.25119(X_6 \text{ x } g_2(t))] \quad (6)$$

Table 8, with a significance level of 5% ($\alpha = 0.05$), p-value (0.03) smaller than α (0.05) can conclude that there is at least one independent variable that significantly affects the model. To determine which variables are influential, then it can be done partial test. Based on table 8, the p-values for the Age(X_1), Sex(X_2), Number of Platelets(X_5), Degree III ($X_7(3)$) and Number of Leukocytes(X_6) x $g_2(t)$ smaller than α, whilep-value for the Hb level(X_3), Hematocrit level(X_4), Degree I ($X_7(1)$), II ($X_7(2)$) and Number of Leukocytes(X_6) x $g_1(t)$ greather thanα. Therefore, it can be concluded that the Age, Sex, Number of Platelets, Degree III andNumber of Leukocytes x $g_2(t)$ have significant effect on the model, while Hb level, Hematocrit level, Degree I, II and Number of Leukocytes x $g_1(t)$ have no significant effect on the model. Furthermore, the following table presents the PH assumption test results using Goodness-of-Fit test.

Table 9. The Result of PH assumption test using Goodness-of-Fit test.

	Rho	Chisq	P
Age(X_1)	0.1224	0.7461	0.388
Sex(X_2)	0.0998	0.6935	0.405
Hemoglobin level(X_3)	-0.0259	0.0738	0.786
Hematocrit level(X_4)	-0.0687	0.4272	0.513
Number of Platelets(X_5)	0.1087	0.9569	0.328
Degree I ($X_7(1)$)	-0.1352	0.8673	0.352
Degree II ($X_7(2)$)	-0.0505	0.0623	0.803
Degree III ($X_7(3)$)	-0.1085	1.1865	0.276
Number of Leukocytes(X_6) x $g_1(t)$	-0.1344	2.2268	0.136
Number of Leukocytes(X_6) x $g_2(t)$	-0.1969	0.7281	0.393
Global	NA	5.6424	0.844

Table 9, with Null Hypothesis (H_0) the data satisfy the PH assumption and the 5% significance level ($\alpha = 0.05$), the decision to receive H_0 for the Age(X_1), Sex(X_2), Hb level(X_3), Hematocrit level(X_4), Number of Platelets(X_5), Degree I ($X_7(1)$), II ($X_7(2)$), III ($X_7(3)$), Number of Leukocytes(X_6) x $g_1(t)$ andNumber of Leukocytes(X_6) x $g_2(t)$. This means that the PH assumption issatisfy for all variables, i.e. Age, Sex, Hb level, Hematocrit level, Number of Platelets, Degree I, II, III, Number of Leukocytes x time function (both of $g_1(t)$ and $g_2(t)$). Since all variables satisfy the PH assumption, the non-proportional hazard condition is resolved.

3.3 Stratified Cox Model

How to resolve non-proportional hazard using the time-dependent variable method or extended Cox model is done. In comparison of these results, we can create a stratified Cox model. Below is the estimation of model parameters using stratified Cox model.

Table 10. Summary of Stratified Cox Model.

	Coef	P	Likelihood ratio test	P-value
Age (X_1)	-0.02640	0.33	7.54	0.5
Sex (X_2)	0.52183	0.21		
Hemoglobin level (X_3)	0.11796	0.47		
Hematocrit level (X_4)	-0.01067	0.11		
Number of Platelets (X_5)	-0.00588	0.90		
Degree I ($X_7(1)$)	-0.34255	0.82		
Degree II ($X_7(2)$)	-0.27620	0.86		
Degree III ($X_7(3)$)	-1.05038	0.55		

Based on table 10, with a significance level of 5% ($\alpha = 0.05$), we can test the significance of the parameters simultaneously. Based on p-value (0.5) greater than α (0.05) it can be concluded that all independent variables have no significant effect on the model so no partial test is required. The equation of the stratified Cox model is obtained as follows.

$$h_g(t, X) = h_0(t) \exp[-0.02640X_1 + 0.52183X_2 + 0.11796X_3 - 0.01067X_4 - 0.00588X_5 - 0.34255X_7(1) - 0.27260X_7(2) - 1.05038X_7(3)] \quad (7)$$

Based on table 11, with Null Hypothesis (H_0) the data satisfy the PH assumption and the 5% significance level ($\alpha = 0.05$), the decision to receive H_0 for the Age(X_1), Sex(X_2), Hb level(X_3), Hematocrit level(X_4), Number of Platelets(X_5), Degree I ($X_7(1)$), II ($X_7(2)$) and III ($X_7(3)$). This means that the PH assumption is satisfy for all variables, ie Age, Sex, Hb level, Hematocrit level, Number of Platelets, Degree I, II and III. Because all variables satisfy the PH assumption, the non-proportional hazard condition is resolved. The following table presents the PH assumption test results using GoF.

Table 11. The Result of PH assumption test using Goodness-of-Fit test.

	Rho	Chisq	P
Age(X_1)	-0.02736	0.08708	0.768
Sex(X_2)	-0.00219	0.00055	0.981
Hemoglobin level(X_3)	0.02153	0.06023	0.806
Hematocrit level(X_4)	0.06403	0.37673	0.539
Number of Platelets(X_5)	-0.01598	0.04203	0.838
Degree I ($X_7(1)$)	-0.03221	0.05858	0.309
Degree II ($X_7(2)$)	-0.02026	0.02335	0.879
Degree III ($X_7(3)$)	-0.02374	0.04298	0.836
Global	NA	0.72754	0.999

After getting some Cox models, then the best Cox model is selected by comparing the AIC values for each model. Here is a comparison table of each model.

Table 12. The comparison of Cox PH, Extended Cox, and Stratified Cox Model

Model	AIC
Cox PH	777.162
Extended Cox with one Heaviside	822.6516
Extended Cox with two Heaviside	823.6601
Stratified Cox	121.0142

Based on table 12, the AIC value for the stratified Cox model (121.0142) is lower than the Cox PH model (777.162), extended Cox model with one Heavisidee (822.6516), and two Heaviside function (823.6601). According to AIC criteria, the model with the lowest AIC value is the best model that can explain the data. Thus, the Stratified Cox model is the best model based on the results of this study.

4. Conclusions

Based on Cox Proportional Hazard model, the variable that significantly affect the DHF patient's healing rate is Age. Whereas the Extended Cox model with one Heaviside function, the variables that significantly affect the DHF patient's healing rate are Age, Sex, Number of Platelets, Clinical Degrees III and Number of Leukocytes multiplied by the time function.

For the Extended Cox model with two Heaviside functions, the variables that significantly affect the DHF patient's healing rate are Age, Sex, Number of Platelets, Clinical Degrees III and Number of Leucocytes multiplied by second time function. Whereas based on the stratified Cox model, there are no variables that significantly affect the DHF patient's healing rate. By using AIC criteria, the best model that can explain the data of the DHF patients is stratified Cox model because it has minimum AIC value compared to Cox PH, extended Cox model with one and two Heaviside function.

5. Acknowledges

The authors would like to thank the Department of Statistics, Faculty of Mathematics and Sciences, and the Institute of Research and Public Services of Syiah Kuala University. We also would like to thank the anonymous reviewers for their valuable suggestions.

6. References

[1] World Health Organization 1997 *Dengue Haemorrhagic Fever: Diagnosis, Treatment, Prevention and Control* (WHO Library Cataloguing in Publication Data).

[2] Chuansumrit, A. Tagnararatcchakit K, Hongeng S 2003 *Dengue Hemorrhagic Fever in Children with Underlying Hematologic-oncologic diseases*. (Thai J Hematol Transf Med).

[3] Hadinegoro S.R. and Sattari, H.I 2004 *Tatalaksana Demam Berdarah Dengue di Indonesia* (Departemen Kesehatan RI dan Direktorat Jenderal Pemberantasan Penyakit Menular dan PenyehatanLingkungan, Jakarta).

[4] Collet, D 1994 *Modeling Survival Data in Medical Research* (New York)

[5] Chongsuvivatwong, Virasakdi 2012 *Analysis of Epidemiological Data Using R and Epicalc. Epidemiology Unit* (Prince of Songkla University).

[6] Kleinbaum, D.G. and Klein, M 2005 *Survival Analysis a Self-Learning Text* (Springer, New York)

[7] Hutahaean, L.P., M.A. Mukidand T. Wuryandari 2014 *Jurnal Gaus*sian 3(2): 173 – 181.

[8] Therneau, T. M. and Grambsch, P. M 2000 *Modelling Survival Data: Extending the Cox Model* (Springer, New York).

[9] Fisher, L. D. ve Lin D. Y 1999 *Annual Review of Public Health* 20, 145 – 157.

[10] Klein, John P. and Moeschberger, M. L. *Survival Analysis Techniques for Censored and Truncated Data*, (Springer, New York, 1997).

[11] Everitt, Brian. and Rabe-Hesketh, Sophia 2001 *Analyzing Medical Data Using S-PLUS.* (Springer, New York).

[12] Hosmer, D.W. and Lemeshow, S 2000 *Applied Survival Analysis Regression Modeling of Time to Event Data* (John Wiley and Sons, Inc, New York).

[13] Wagenmakers, Eric-Jan. and Farrel, Simon 2004 *Psychonomic Bulletin & Review*, 11 (1), 192-196.

IOP Conf. Series: Materials Science and Engineering **523** (2019) 012007 doi:10.1088/1757-899X/523/1/012007

Kovats Retention Index analysis of flavor and fragrance compound using Biplot Statistical method in gas chromatography systems

R Idroes[1,2], A F Japnur[1], R Suhendra[3], A Rusyana[4,*]

[1]Department of Chemistry, Faculty of Mathematics and Natural Sciences, Syiah Kuala University, Kopelma Darussalam, Banda Aceh 23111, Indonesia
[2]Department of Pharmacy, Faculty of Mathematics and Natural Sciences, Syiah Kuala University, Kopelma Darussalam, Banda Aceh 23111, Indonesia
[3]Department of Informatics, Faculty of Mathematics and Natural Sciences, Syiah Kuala University, Kopelma Darussalam, Banda Aceh 23111, Indonesia
[4]Department of Statistics, Faculty of Mathematics and Natural Sciences, Syiah Kuala University, Kopelma Darussalam, Banda Aceh 23111, Indonesia

E-mail: asep.rusyana@unsyiah.ac.id

Abstract. A lot of retention parameters have been introduced and used in chromatographic instrumentation analysis, one of which is the retention index. Retention index is a secondary parameter which in its calculation requires data retention time and dead time. This study uses data from flavor and fragrance retention indices with a variation of columns, boiling points and log Kow which aims to identify the level of similarity between compounds based on retention index using biplot statistical method. Biplot analysis is used to view object relationships based on the variables. The results showed that biplot could describe the relationship of a compound as an object and its retention index as a variable of 99.45%. Based on the results, the compounds can be classified into three groups. The first group has a specific retention index that is lower than other groups. The second group has a high Log Kow whereas the third group is dominated by the influence of DB-Wax column compared to other columns. Columns DB-1, OV-101, and DB-5 have a positive correlation with the correlation level close to 1.00.

1. Introduction

Various parameters have been introduced and are used in the chromatographic analysis which can explain the analytical aspects of the data. The most commonly used parameter for explaining the data of the analysis results is retention factor. It can be used for both qualitative and quantitative analysis [1]. There are many retention parameters of chromatography, but only the retention time and dead time parameters as the primary parameters generated by the chromatography instrument, while the rest are secondary parameters [2].

The retention time is relatively sensitive to slight changes in experimental conditions such as column changes, column length, column diameter, carrier gas, flow rate, pre-column, and more, while

*Corresponding author : asep.rusyana@unsyiah.ac.id

the retention index Kovats is more reliable and reproducible. The Kovats retention index is a concept introduced in gas chromatography to convert retention time into a more reliable and reproducible system. This retention index was first introduced by E. Kovats in 1958 on a gas chromatographic system experiment. The retention index of a compound is calculated relatively to two standard compounds so that the effect of the instrument can be ignored and the precision can be increased [3]. Thus in this study, the data from retention index of flavor and fragrance compounds with a variation of the column is used to identify the level of similarity among compounds based on the pattern of Kovats retention index data using biplot statistic method.

2. Literature Review

Biplot, a two-dimensional mapping of the principal component analysis (PCA), was firstly introduced by Gabriel (1971), thus it is often called as Gabriel's biplot or PCA Biplot. Biplot is a descriptive statistical technique that is useful for simultaneously presenting observational objects and p variants in flat space, so that the characteristics of the variables, observation objects and relative positions between observed objects and variables are able to be analyzed visually. Biplot refers to two types of information contained in the data matrix. The information in the row corresponds to the sample and in the column to the variable [4].

The information obtained from Biplot is the degree of relative similarity among observation objects, the relationship among variables, the value of the variables in an object, and the variability of the variables. The relationship between the variables is that if the angle of two variables < 90° then the correlation is positive, if the angle of two variables > 90° the correlation is negative, and if the smaller the angle is, the stronger the correlation will become. An object has the property of a variable if the object is adjacent to that variable. The closer the position of the object to the variable, the more dominant character of the variable on the object [5].

Biplot was developed based on Singular Value Decomposition (SVD). Suppose an X data matrix of size n x p containing n observations and p variables corrected to its mean and r-value can be written as:

$$X = ULA' \tag{1}$$

where: the matrix U dan A each sized (nxr) and (pxr) so that $U'U = A'A = I_r$. L is a diagonal matrix size (rxr) with its diagonal elements is the square root of eigen value of $X'X$ or XX' so:

$$\sqrt{\lambda_1} \geq \sqrt{\lambda_2} \geq \cdots \geq \sqrt{\lambda_r} \tag{2}$$

Column matrix A is an eigen vector corresponding to the eigen value λ from matrix $X'X$ or XX'. Matrix columns U can be calculated through:

$$U_i = \frac{1}{\sqrt{\lambda_i}} \times \alpha_i \tag{3}$$

with λ_i is eigen value i-th from matrix $X'X$ or XX', and α_i is column i-th matrix A.

$$X = UL^{\lambda}L^{1-\lambda}A' \tag{4}$$

if $G = UL^{\lambda}$ and $H' = L^{1-\lambda}A'$, so:

$$X = GH' \tag{5}$$

factor (i, j)-th matrix X can be written:

$$X_{ij} = g_i'h_j' \tag{6}$$

where: $i = 1,2,3, \dots, n$ and $j = 1,2,3, \dots, p$ with g_i' and h_j' each is a matrix row G and H'.
If $r(X) = 2$ so g_i' and h_j' depicted in a 2-dimensional space.
The percentage of diversity displayed by biplot can be defined as follows:

$$\rho_2 = (\lambda_1 + \lambda_2) \Big/ \sum_{i=1}^{r} \lambda_i \tag{7}$$

where:
λ_1 = the first largest eigen value
λ_2 = the second largest eigen value
λ_i = the largest eigen value i-th from $X'X$, $i = 1,2,3, \ldots, r.$, [4]

3. Methods

3.1 Data Collection

In this study, the researchers took data on retention index of homologous n-alkane and flavor and fragrance compounds as chromatographed using OV-101, DB-1, DB-5, and DB-Wax columns, as well as Boiling Point (BP) and Log Kow data from the literature of Goodner K.L. 2008, *Practical retention index models of OV-101, DB-1, DB-5, and DB-Wax for flavor and fragrance compounds, LWT - Food Sci. Technol.*, vol. 41, no. 6, pp. 951–958 [6].

The columns used in this study were OV-101 column with dimethylpolysiloxane, DB-1 with 100% -dimethylpolysiloxane, DB-5 with (5%-Phenyl)-methylpolysiloxane and DB-Wax with polyethylene glycol as stationary phase respectively. The column information used is obtained from https://www.agilent.com/en/products/gas-chromatography/gc-columns.

3.2 Retention index determination

Retention index determination can be performed with a simple interpolation between the logarithm of the n-alkane compounds before and after being eluted solutes, at each temperature is defined as 100 times n_c (carbon number) [7,8].

3.3 Data Analysis

The data which have been collected, are initially converted or transferred into XLSTAT software, then the data outliers are removed. The next stage of preprocessing performed on the data to eliminate variations in data that are not related to analytical information. After preprocessing, the data analysis using biplot method is used to obtain data classification model.

4. Result and Discussion

Data on the retention index of n-alkane, flavor, and fragrance compounds with variations of OV-101, DB-1, DB-5, and DB-Wax columns, as well as Boiling Point and Log Kow data were used to identify similarity levels based on the pattern of Kovats retention index data using the Biplot statistics method.

Biplot statistical method was used to identify the similarity level up to 99.45%, PC1 87.79% and PC2 11.66% (figure 1). The result of Biplot which is based on the pattern of the compound retention index, can be grouped into three groups.

The first group (I) consists of relatively shorter carbon n-alkane compounds (pentane, hexane, heptane, octane, nonana, and decane). The n-alkane compounds or homologous series of n-alkanes are often used in the determination of kovats retention index because they have robust retention index results. This is based on the retention index calculation process performed by simple interpolation logarithms between the two n-alkana, before and after eluted solutes, at any temperature defined as 100 times nc (carbon number), such as Retention Index for Ethane = 200, Butane = 400, Hexane = 600, and so on [9,10].

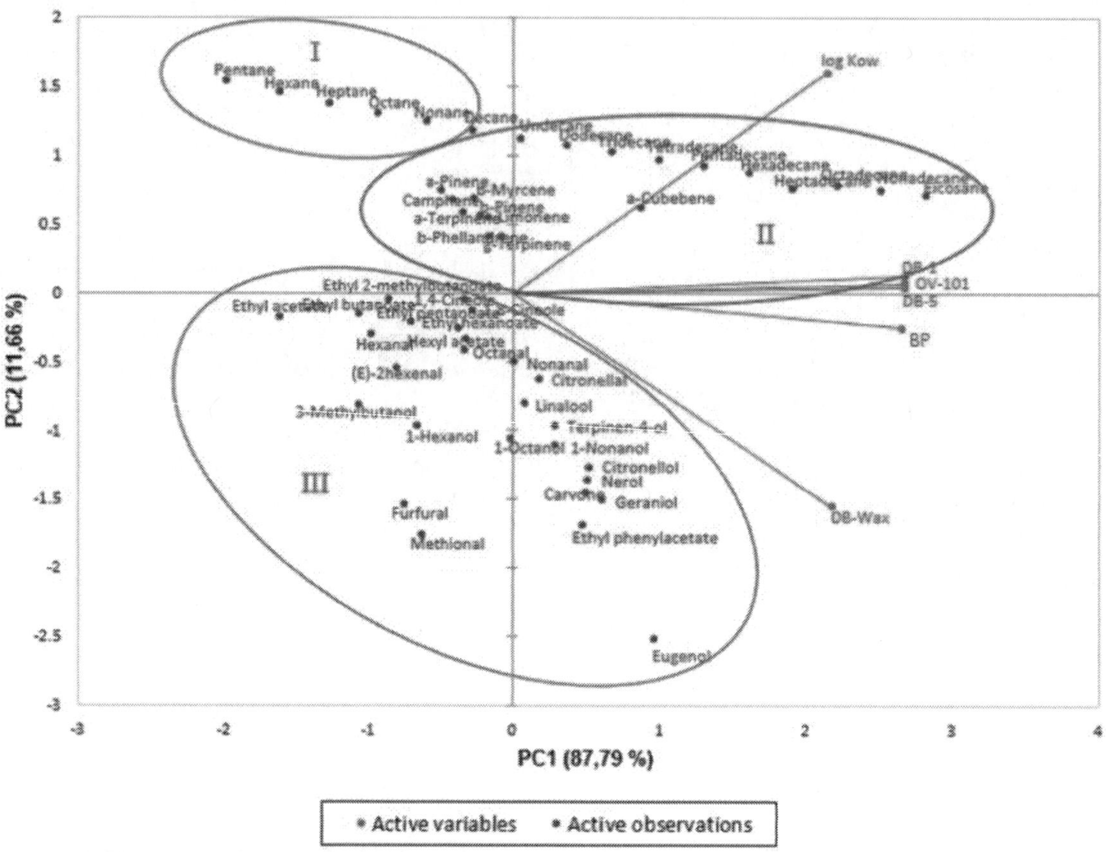

Figure 1. Biplot analysis results of retention index data of flavor and fragrance compounds

The second group (II) has a higher Log Kow value than the others. Based on the retention index pattern, the compounds in this group consist of an n-alkane compound with a higher carbon chain (undecane, dodecane, tridecane, tetradecane, pentadecane, hexadecane, heptadecane, octececane, nonadecane, and eicosane) as the result, these compounds have higher retention index values as opposed to the first group. The terpenoid compounds comprising (a-pinen, b-mycenen, b-pinen, camphen, a-terpenen, limonene, b-phellandene, g-terpenen, and a-cubebene) are considered to have a degree of polarity that is close to the n-alkane homologous series with a higher carbon chain.

The third group (III) comprises alcohol group compounds (linalool, 3-methylbutanol, 1-hexanol, 1-octanol, 4-ol, 1-nonanol, citronellol, nerol, geraniol, and eugenol), carboxylic acids (ethyl butanoate, ethyl 2- methylbutanoate, ethyl pentanoate, ethyl hexanoate), esters (ethyl acetate, hexyl acetate, ethyl phenylacetate), and aldehydes (hexanal, octanal, furfural, methional, nonanal). This group has the dominant similarity to the pattern of retention index data generated from the DB-Wax column. This is because groupings of these compound groups have a higher retention index than other compounds in the DB-Wax column.

Table 1. Correlation between variables

Variables	OV-101	DB-1	DB-5	DB-Wax	BP	log Kow
OV-101	1	0,998	0,999	0,794	0,983	0,811
DB-1	0,998	1	0,999	0,782	0,981	0,821
DB-5	0,999	0,999	1	0,801	0,986	0,803
DB-Wax	0,794	0,782	0,801	1	0,855	0,312
BP	0,983	0,981	0,986	0,855	1	0,734
log Kow	0,811	0,821	0,803	0,312	0,734	1

Table 2. The standard deviation of retention index compound using OV-101, DB-1, DB-5, and DB-Wax columns

SDV Retention Index					
Compound	SDV[1]	SDV[2]	Compound	SDV[1]	SDV[2]
(E)-2hexenal	149,67	21,20	1,8-Cineole	102,26	11,37
Ethyl acetate	147,72	9,85	Limonene	93,43	8,50
Ethyl butanoate	142,34	54,00	g-Terpinene	101,38	25,79
3-Methylbutanol	238,47	9,87	1-Octanol	248,63	10,00
Hexanal	148,89	10,07	Citronellal	174,79	12,29
Ethyl 2-methylbutanoate	108,44	5,86	Terpinen-4-ol	216,23	12,17
Ethyl pentanoate	126,39	12,12	1-Nonanol	247,35	3,79
a-Pinene	52,13	6,81	Citronellol	269,20	5,03
1-Hexanol	252,96	9,61	Ethyl phenylacetate	276,88	17,69
Camphene	62,52	2,00	Carvone	267,69	20,53
Furfural	308,68	30,62	Geraniol	299,23	14,42
b-Pinene	67,70	2,52	a-Cubebene	64,69	16,17
Methional	294,45	26,31	Eugenol	403,11	23,03
b-Myrcene	83,83	0,58	1,4-Cineole	92,50	11,79
Ethyl hexanoate	121,35	11,27	Linalool	224,93	8,14
Hexyl acetate	131,43	10,12	Nerol	283,49	11,72
Octanal	150,10	10,97	Nonanal	151,39	10,02
a-Terpinene	91,92	12,66	b-Phellandrene	115,00	31,09

SDV1 Standard deviation from the column retention index of OV-101, DB-1, DB-5, and DB-Wax
SDV2 Standard deviation from the column retention index of OV-101, DB-1, and DB-5

DB-1, OV-101, and DB-5 columns have a positive correlation among them because they are close to each other (see Figure 1). The data of Table 1 shows the column correlation is close to 1. The DB-1, OV-101, and DB-5 columns are relatively non-polar so they have a higher similarity level, whereas DB-Wax column has different polarity. Based on the data Table 1, DB-Wax column has a lower correlation level against other columns.

From the Table 2, we could identify that the standard deviation of retention index compound using DB-Wax column is greater than without using it, because the data of retention index compound using DB-Wax column is strongly influenced by the column broadening effect.

5. Conclusion

Based on kovats retention index data, flavor and fragrance compounds used and OV-101, DB-1, DB-5, and DB-Wax columns, as well as boiling point and Kow log can be grouped into three groups using the Biplot statistics method. The first group consists of the n-alkanes group with lower carbon numbers and a lower retention index. The second group consists of the n-alkanes group with higher carbon numbers and terpenoid group compounds that have a higher Kow log values. The third group consists of alcohols, aldehydes, esters and carboxylic acids which are dominantly the retention index pattern in the DB-Wax column. Columns DB-1, OV-101, and DB-5 have a positive correlation close to 1.00.

6. References

[1] Domínguez J G and Diez-Masa J 2001 *Pure Appl. Chem.* **73** 969–92
[2] Furr H C 1989 *J. Chromatogr. Sci.* **27** 216–20
[3] Mahmudi, Saiful, khairan and Idroes R. 2015 *J. Nat.* **15** 1–9
[4] Johnson R A and Wichern D W 2007 *Applied Multivariate Statistical Analysis*
[5] Gabriel K R 1971 *Biometrika* **58** 453–67
[6] Goodner K L 2008 *Food Sci. Technol.* **41** 951–8
[7] Idroes R 2009 *Maj. Farm. Indonesia.* **20** 133–40
[8] Pratiwi S U T, Lagendijk E L, de Weert S, Idroes R, Hertiani T and Van den Hondel C 2015 *Int. J. Appl. Res. Nat. Prod.* **8** 1–13
[9] Idroes R 2005 *Malaysian Journal of Analytical Science.* **9** 224–2
[10] Idroes R 2010 *Indonesian Journal of Chemistry.* **1** 58–10

IOP Conf. Series: Materials Science and Engineering **523** (2019) 012008 doi:10.1088/1757-899X/523/1/012008

Activities inhibition methanol extract Laban Leaf (*Vitex pinnata*) on growth of bacteria *S. mutans* Atcc 31987

C A Nuraskin[1,2,*], Marlina[3], R Idroes[3,4], C Soraya[5], Djufri[6]

[1] Graduate School of Mathematics and Applied Sciences, Syiah Kuala University, Kopelma Darussalam, Banda Aceh 23111, Indonesia

[2] Poltekkes Kemenkes Aceh Jl. Soekarno Hatta, Tingkeum, Darul Imarah, Lheu Blang, Banda Aceh, Kabupaten Aceh Besar, Aceh, 23231, Indonesia

[3] Chemistry Department, Faculty of Mathematics and Natural Sciences, Syiah Kuala University, Kopelma Darussalam, Banda Aceh 23111, Indonesia

[4] Pharmacy Department, Faculty of Mathematics and Natural Sciences, Syiah Kuala University, Kopelma Darussalam, Banda Aceh 23111, Indonesia

[5] Conservative Dentistry Department, Faculty of Dentistry, Syiah Kuala University., Kopelma Darussalam, Banda Aceh 23111, Indonesia

[6] Biology Education Department, Faculty of Teacher Training and Education, Syiah Kuala University, Kopelma Darussalam, Banda Aceh 23111, Indonesia

E-mail: cutaja82@yahoo.co.id

Abstract. People in Aceh, especially in the area of *Ie Seu'um* using the surrounding plants to treat diseases. One of the medical plants is laban (*Vitex pinnata*). Laban leaves are used to treat fever, hypertension, and toothache. Various bacteria found in the oral cavity, but only a few bacteria cause dental caries, including *S. mutans*. Based on the laban plant ethnobotany information, an evaluation of antibacterial activity on bacterial bioindicator of *S. mutans* and phytochemical screening had been performed. Laban plants contain flavonoids, saponins, and tannins. This study aims to determine the inhibition activity of laban leaf methanol extract (*Vitex pinnata*) on the growth of *S. mutans* bacteria in vitro. The antimicrobacterial activity was tested by using the Kirby-Bauer method with the extract concentration of 30%, 40%, 50% and 60%. The results showed that there was an inhibition activity with the significance of 0.000 < 0.05, therefore the hypothesis in this research can be accepted that Laban leaf methanol extract (*Vitex pinnata*) has inhibition activity, the higher the concentration the higher inhibition power.

1. Introduction

The use of traditional medicines to treat diseases is quite attractive to most of Indonesian people, because of its mild side effects [1]. People in Aceh, especially in the area of *Ie Seu'um* use the surrounding plants to treat diseases. One of the plants used as alternative medicine is laban plants (*Vitex pinnata*), known under various names, Aceh (*bak mane*) [2], Indonesia (*laban*), *Millettia pinnata* (L.) *Panigrahi*, *Pongamia glabra Vent*, *Derris indica*, *keranja* (Hindi, Bengali, Sanskrit), *ki pahang laut*, *sea-nut*, and *pongam* oil tree (Malay) [3].

*Corresponding author : cutaja82@yahoo.co.id

Laban plants contain the components of flavonoids, saponins, and tannins. Laban leaves (*Vitex pinnata*) can be used for the treatment of diseases such as fever and hypertension [4]. The Punjabi (Pakistan) community uses laban (*Vitex pinnata*) plants to treat oral diseases [3]. Phytochemical test results of laban leaves shows the content of alkaloids, flavonoids, saponins, sterpenoids, tannins,[2]. Therefore, laban plant (*Vitex pinnata*) is potential as an antibacterials [5].

Various bacteria are found in the oral cavity, but only a few bacteria cause dental caries, one of which is *S. mutans*. These bacteria can be easily mount the tooth surfaces [6]. *S. mutans* is a bacteria that can grow well in an acidic atmosphere and can produce acid as a result of carbohydrate fermentation. The acid produced by this bacterium can cause tooth demineralization [7]. Therefore, the activity of inhibition of methanol leaf extract of Laban leaf (*Vitex pinnata*) in *Ie Seu'um* to *S. mutans* bacteria is needed, and Laban leaf methanol extract (*Vitex pinnata*) can be used as a raw material of anti-bacterial drugs that can prevent dental caries. The area of *Ie Seu'um* is one of the Manifestations of the existing geothermal system in Aceh Besar. This area is located in the geothermal outflow zone of Mount Agam, Aceh Besar, located 20 km from the top of the mountain and 35 km from the city of Banda Aceh [8]. Associated with the inhibitory activity of methanolic extract of laban leaves (*Vitex pinnata*) against *S. mutans* bacteria in vitro until now has never been published. In Particular laban plant (*Vitex pinnata*) from the area of *Ie Seu'um* Aceh Besar.

2. Materials and Methods

2.1. Sampling of Plants

Laban leaf (*Vitex pinnata*) collections were collected on February 3rd, 2018, from *Ie Seu'um* Mesjid Raya, Aceh Besar District, Aceh Province with purposive sampling method. Sampling was done at four locations with following coordiantes: sample point 1 at coordinates (5°32'48.9" N 95°32'55.1" E), sample point 2 (5°32'51.4" N 95°32'54.3" E), sample point 3 (5°32'53.7" N 95°32'54.6" E), and the sample point 4 (5°32'45.7" N 95°32'57.8" E). Plant identification was done in Biology Department Faculty of Mathematics and Natural Sciences of Syiah Kuala University with number 558 / UN11.1.28.1 / DT / 2018. Coordinate *Ie Seu'um* and Laban (*Vitex pinnata*) can be seen in Figure 1 and Figure 2 respectively.

Figure 1. Sampling coordinates in *Ie Seu'um* area, Aceh Besar.

Figure 2. Laban (*Vitex pinnata*) in *Ie Seu'um* Area

2.2. Instruments and Materials

Instruments used included an analytical balance sheet, hot plate, media bottle, ose needle, shaker, tweezers, spiritus burner, autoclave, incubator (ecocell), colony counter, and sliding range. Materials used were Laban leaf (*Vitex pinnata*), *S. mutans* ATCC 31987 from Faculty of Veterinary Medicine Unsyiah, Nutrient agar (NA), Nutrient Broth (NB), Muller Hinton Agar (MHA), Disk Blank, Alcohol 70% , Aquades, Plastic wrap, and Aluminum foil, sulfoxide) [9].

2.3. Procedure

2.3.1. Maceration.

1 kg of dried leaves sample were mashed [10]. The obtained simplicia was then weighed [11], macerated with methanol solvent for 3 × 24 hours, evaporated with a rotary evaporator to obtain the methanol extract [12]. The extract was divided into several concentrations, 30%, 40%, 50% and 60%. Sterile aquadest was used as the negative control, while the positive control was *Chlorhexidine* 0.2%.

2.3.2. S. mutans bacterial culture.

Cultures of *S. mutans* was grown in NA agar medium, subsequently was incubated at 37° C for 2x24 hours in an incubator. One ose needle cultured *S. mutans* in NA medium was taken and inserted into a liquid tube of NB. The tube was inserted into the anaerobic jar, incubated at 37° C for 24 hours in the incubator. The turbidity was compared with one of McFarland 0.5. If the turbidity of *S. mutans* in liquid medium NB equal to the turbidity of McFarland 0.5 then the number of *S. mutans* is estimated to be 1.5 x 108 CFU/ml.The next was performing gram staining procedure. Each disc was dipped into a petri dish containing Laban leaf methanol extract (*Vitex pinnata*) with concentrations of 30%, 40%, 50% and 60%. A sterile cotton bud was dipped into a *S. mutans* culture in suspension of NB medium, then was smeared evenly on the surface of MHA media, the disc was placed on the surface of MHA media.The disk paper was previously dipped into *Chlorhexidine* 0.2% before being placed on the MHA media surface. Then it was incubated at 37°C for 24 hours [13]. The procedure was conducted with three repetitions, then the inhibition zone was measured [14].

3. Result and Discussion

The results of the inhibitory activity of Laban leaf methanol extract (*Vitex pinnata*) can be seen in Table 1.

Table 1. Analysis and statistic of confidence level of the inhibition activity of Laban leaf methanol extract (*Vitex pinnata*) against *S. mutans* ATCC 31987

Experimental treatment	Average (mm)	Confidence Interval 95%		Standard Deviation*
		Lower Limit	Upper Limit	
30%	6.567	6.423	6.71	0.06
40%	6.867	6.487	7.246	0.15
50%	7.2	6.952	7.448	0.1
60%	10.433	9.916	10.95	0.21
Chlorhexidine 0.2 % Control (+)	19.6	15.696	23.504	1.57
Aquades Control (-)	6	6	6	0

* Three repetitions

Table 1. shows the number of treatment measures for each three times repetition, with different mean values with concentrations of 30%, 40%, 50% and 60%. This shows an increase in inhibition activity of laban leaf methanol extract (*Vitex pinnata*) against *S. mutans* bacteria. All the experimental treatment showed relative low standard deviations.

Table 2. ANOVA table to see the effect of laban concentration on inhibition power of *S. mutans*.

Source of Variation	Sum of Squares	Degree of Freedom	Mean Square	F	P-value
Treatment	407.824	5	81.565	191.917	0.000
Error	5.100	12	0.425		
Total	412.924	17			

Table 2 is the result of ANOVA test showing that there is an inhibition activity with significance of 0.000 < 0.05, therefore the hypothesis in this research is acceptable in which the extract of laban leaf methanol (*Vitex pinnata*) has inhibitory activity against *S. mutans* bacteria.

Table 3. Duncan's advanced test

Treatment	Subset for alpha = 0.05
30%	6.567
40%	6.867
50%	7.200
60%	10.433
Chlorhexidine 0.2% Control (+)	19.600
Aquades Control (-)	6.000

Table 3. exhibit the mean values of treatments at concentrations of 30%, 40%, 50% and 60% were not significantly different. However the mean values of treatments 30%, 40%, 50% and 60% in comparison to positive control and negative control treatments were significantly different. Meanwhile the mean value of positive control treatment was significantly different from the negative control treatment.

The test results showed an increase in inhibitory activity of laban leaf methanol extract (Vitex pinnata) against S.mutans bacteria. Each experimental treatment has a different standard deviation value. Positive methanol extract contained alkaloids, steroids, terpenoids, saponins, flavonoids and phenolics [2]. In line with S. Thenmozhi's research, laban plants are potential antibacterial [5]. The result of ANOVA test confirmed that there was an inhibitory activity to S.mutans bacteria with significance of 0.000 <0.05. This is due to the greater concentration of laban extract, the more inhibitory zone to S. mutans. This difference was caused by several factors, including the amount of inoculum, incubation time, extract concentration, and antibacterial type. The larger the inoculum the smaller the inhibit zone is formed. The inhibitory zone formed in Chlorhexidine 0.2% (the positive control) was greater than the inhibitory zone of laban leaf methanol extract. Meanwhile, the sterile aquadest (the negative control) did not form a zone of inhibition. The antibacterial activity of laban leaf methanol extract was in accordance with the results of the study that extracts with 60% concentration had a larger inhibitory zone compared with concentrations of 30%, 40% and 50%.

4. Conclusion

Based on the results of a test study of inhibitory activity laban leaves methanol extract (*Vitex pinnata*) against *S. mutans* bacteria in vitro, it can be concluded that laban leaves methanol extract (*Vitex pinnata*) has an activity in inhibiting the growth of bacteria *S. mutans* in vitro. Based on the results of in this study, it can be concluded that the higher the concentration, the higher the inhibition power.

5. References

[1] Lawal, Nwogwugwu, J. O., Faloye, B., Olatunji, B. P., Adesoga, A.C. 2010 *African Journal of Pharmacy and Pharmacology*, *4*(1), **1–7.**

[2] Idroes R, Khairan, Fakri F, Zulfendi, 2016 *Monographs of screening for plant activity that have the potential as antimicrobial material in the area of Ie Seu'um* Aceh Besar, (Syiah Kuala Univ Press.)

[3] Muqarrabun L M R, Ahmat N, Ruzaina S A S, Ismail N H, Sahidin I 2013 *Ethnopharmacology* 150(2) **395–420**

[4] Goh, M. P. Y., Basri, A. M., Yasin, H., Taha, H., Ahmad, N. 2017 *Asian Pacific Journal of Tropical Biomedicine*, *7*(2), **173–180.**

[5] Thenmozhi1 and U. Subasini 2016 *Journal of Biomedical AND Pharmaceutical Sciences*, *3*(1),

191–201.
[6] Forssten SD, Bjorklund, M, Ouwehand AC. 2010 *Nutrien Dent J.* :2 (1): **290 - 289.**
[7] Kidd E 2005 *Essential Of Dental Caries*, (New York : Oxford University Press)
[8] Idroes R, Yusuf M, Alatas M, Subhan, Lala A, Saiful, Suhendra R, Idroes G M and Marwan 2018 *IOP Conf. Ser.: Mater. Sci. Eng.* **334**
[9] Mahmudah F L, Atun S 2017 *S. mutans* 22 (1) **59–66**
[10] Biswas SK, Cowdhury, karmakar UK, Shill MC, R. 2011 *Applied Pharmaceutical Science.*, 01(06): **112- 114.**
[11] Florensius Rinaldi F, Arsyik Ibrahim, Jaka Fadraersada, L. R. 2016 *Identification of Secondary Matabolit and Toxicity Test of Laban Bark Methanol Extract (Vitex pinnata) by the Brybene Shrimp Lethaltty Test* (BSLT) method, **20–21.**
[12] Hermansah A, Harlia, Zahara T A 2015 *Jkk* 4 (2) **67–71**
[13] Wan AKL, Seow WK, Walsh LJ, and B. P. 2002 *Australia Dental Journal*, 47(1*)*, **21–26.**
[14] Khafagi IK, D. A. 2000 *Journal of Ethnopharmacology*, 71: **365–376.**

IOP Conf. Series: Materials Science and Engineering **523** (2019) 012009 doi:10.1088/1757-899X/523/1/012009

The potential of five therapeutic medicinal herbs for dental treatment : A review

D S Ningsih[1,2,*], R Idroes[3,5] , B M Bachtiar[4], Khairan[5]

[1] Graduate School of Mathematics and Applied Sciences, Syiah Kuala University, Kopelma Darussalam, Banda Aceh 23111, Indonesia
[2] Department of Dental Materials, Faculty of Dentistry, Syiah Kuala University,Banda Aceh 23111, Indonesia
[3] Department of Chemistry, Faculty of Mathematics and Natural Sciences, Syiah Kuala University, Kopelma Darussalam, Banda Aceh 23111, Indonesia
[4] Oral Biology Department, Faculty of Dentistry, University of Indonesia, Jakarta, 10430, Indonesia
[5] Department of Pharmacy, Faculty of Mathematics and Natural Sciences, Syiah Kuala University, Kopelma Darussalam, Banda Aceh 23111, Indonesia

E-mail: dee_aceh@yahoo.co.id

Abstract. Indonesia, especially in Aceh Province, has a huge biodiversity of natural resources in herbal medicine. *Calotrophis gigantea (L), Acrostichum aureum (L), Plucheaindica (L), Cissus adnata* and *Abutilon indicum (L)* are considered to have secondary metabolite that are potential for dental therapeutic treatments because the compounds have anti-inflammatory, antibacterial, antifungal, antioxidant, antiseptic and wound healing ability. The utilization of medicinal herbs in dentistry is still limited, thus in this study we provide valuable information of each five medicinal herbs for dental therapeutic treatments. The results indicated that all the five medicinal herbs have different abilities and potentials in treating diseases / abnormalities in the oral cavity. The selection and the use of appropriate medicinal herbs can give an effective therapeutic treatment to improve the health of oral cavity. The qualities of the five herbs are directly proportional to their ability as an alternative therapeutics in dentistry. To produce a good herbal medicines in dentistry,a lot of supporting researches are needed.[1]

1. Introduction

Herbs are plants that can be used to cure diseases. The use of herbs is greatly improved from year to year because the raw material has a high availability, inexpensive price and low side effects compared to commercial chemical drugs [1]. The raw materials for herbal medicine in Indonesia, especially in Aceh, is readily available thus can be used afor alternative therapies. Some herbs are grown in the province include: Calotrophic gigantea (L), *Acrostichum aureum* (L), *Pluchea indica* (L), *Cissus adnata* and *Abutilon indicum* (L).

Those plant grow in the wild shrubs and in a tropical climate. The plants are commonly used by the public as antioxidants, painkillers, antiviral/bacterial, anti-inflammatory, antidiabetic, wound

*Corresponding author : dee_aceh@yahoo.co.id

IOP Conf. Series: Materials Science and Engineering **523** (2019) 012009 doi:10.1088/1757-899X/523/1/012009

medications, cardioprotection, anti-cancer, and anti-ageing. In the field of dentistry, the use of the five is still very limited. Some researches suggests that the use of these herbs are still limited to as antimicrobial, anti-inflammatory and wound medications [1,2].

The antibacterial activity studies of these plants are still limited especially on the bacteria that cause caries. Sharma et al., (2015) and Fatimatuzzahra *et al.* (2016) reported that *Calotropis gigantea* could inhibit the growth of *Streptococcus mutant*, *Lactobacillus casei Enterococcus fecalis* and could heal wounds [3-5]. Sharma et al., (2015) also reported that the secondary metabolites from *Calotropis gigantea* can be widely used as an antiseptic, antibacterial, antimicrobial, antifungal, antioxidant, antiviral, anticancer, anti-tumour analgesic, antioxidant, antiviral, anticancer, anti-tumour and analgesic [3]. This is agreed by Agbor (2015), stated that this plant is potential to be used as a pain reliever (analgesic), gingivitis, periodontitis, ulcerative stomatitis corner, ulcers, and oral thrush in dental treatment, and also some other malignant diseases such as cancers and tumours [6].

The potential of these plants are correlated to their secondary metabolites such as phenolic, terpenoids, polysaccharides, and related phenolic compounds such as flavonoids, phenyl propanoid, rosmarinik acid, catechins, tannins, and polyketides. These secondary metabolites are believed to have antiinflammatory, estrogenic, inhibition of enzymes, antimicrobial activity, antiallergic activity, antioxidant, antitumor, and has a vascular activity. The terpenoids that have been isolated from these plants such as mono and sesquiterpenes, iridoids, and saponins are believed to have anticancer, antimalarial, anti-nflammatory, antibacterial, and antivirus activity. These plants are also known to contain alkaloids, glucosides, cardiac glycosides or esters [7-10]. Based on the reports above, in this short review, we would like to present the potential use of these five therapeutic medicinal herbs in dental treatments. It is expected that this review would provide valuable information pertaining the five therapeutic medicinal herbs such as *Calotropis gigantea* (L), *Acrostichum aureum* (L), *Pluchea indica* (L), *Cissus adnata* and *Abutilon indicum* (L) in the field of dentistry.

2. Review on Therapeutic Medicinal Herbs

Calotropis gigantea (L)

Calotropis gigantea (L), also known as Biduri, has a large availability in Indonesia, especially in Aceh Province. All parts of this plant can be used to treat diseases because of the secondary metabolites contents such as alkaloids, steroids, heart glycosides, and terpenoids [3,11]. Some researchers have identified that there are 21 compounds in the leaves and fruits parts of this plant, 43 compounds in the flowers, and some compounds are identified in latex [12]. The common compounds found in leaves, fruits, and flowers are *3*-Hexene-1-ol, benzaldehyde, benzyl alcohol, ± linalol, oct-*3*-en-2-ol, phenethyl alcohol, α-terpineol, *2,4*-dimethyl -acetophenone, *4*-vinyl guaiacol and *n*-tetradecane, while the latex contains lignanglycosides, pinoresinol 4-*O*-[6 ''-*O*-vanilloyl]-β-D-glucopyranoside and other phenolic compounds such as 6'-*O*-vanilloyltachioside, 6'-*O* –vanilloylisotachioside, and cardenoline such as uskarin (Figure 1). Stems and seeds of this plant also have a potential use for medicinal treatment because of the containing halocellulose, cellulose, lignins and alkali-soluble substances [13,14].

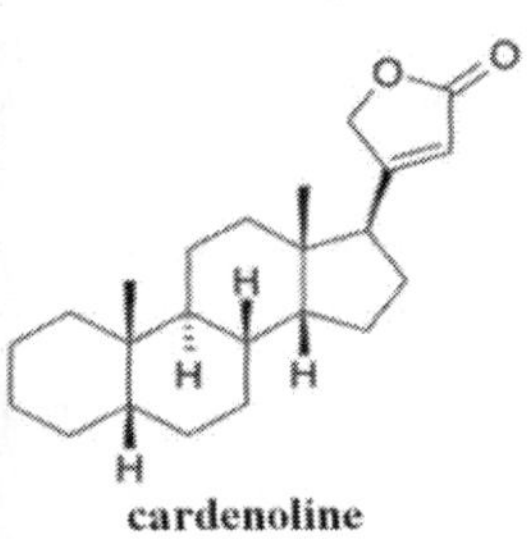

Figure 1. The cardenolide compound isolated from *Calotropis gigantea* (L) [14]

The pharmacological benefit of the secondary metabolites from this plant are usually used as an analgesic, antiasthma, sedative, antiinflammatory, antidiarrhea, hepatoprotective, and anticancer. Another study also mentioned that this plant is a potential antibacterial, antioxidant and reinforcing material composites [3,13-15]. The potential evaluation of this plant in dentistry is still limited, the recent reports mentioned that this plant is potentially used as a anticarcinogenic medicine that inhibits *Streptococcus mutants* and *Lactobacillus sp.,* where these bacteria are known to have a role in cavities formation [3].

***Acrostichum aureum* L**

Acrostichum aureum L or also known as sea fern, is normally grow in brackish water area (mangrove area) [16]. Sea fern is very easy to grow in tropical regions in countries like Brazil, Ecuador, Paraguay, India, Sri Lanka, Bangladesh and Indonesia. Ethnobotanically, this plant is often used traditionally as analgesic (pain killer), antifibrinolytic and antitumor [16-18] Hossaini et al (2011) reported that *Acrostichum aureum* contains glycosides, saponins, flavonoids, steroids, fatty acids and long-chain hydrocarbon compounds. [19]

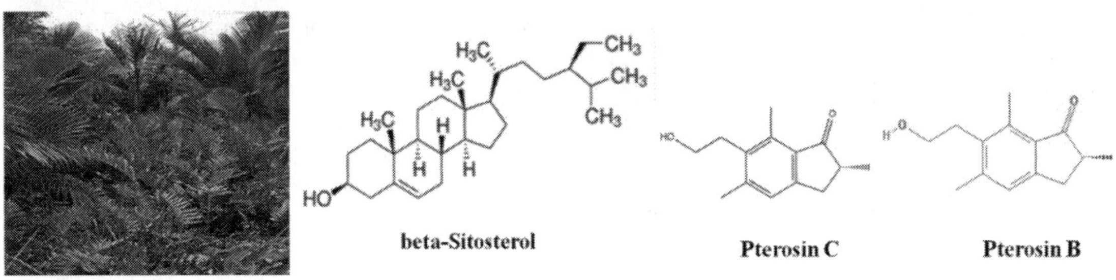

Figure 2. Sea fern (*Acrostichum aureum* L) and its compounds such as β-sitosterol, Ptreosin C, and Pterosin.

Two new sesquiterpenes (Figure 2) compounds such as (2*R*, 3*S*)-dulfated pterosin C and (2*S*, 3*S*)-sulfated pterosin C, or also known as, (2*R*, 3*S*)-pterosin C and (2*R*)-pterosin B and β-sitosterol are reported to have been succesfully isolated from aerial parts of this plant. Pterosins C is reported to have citotoxicity effect on AGS, HT-29, MDA-MB-231, MCF-7, and NIH3T3 cell lines in mice with the MIC_{50} value ranged from 23.9 to 68.8 pM [17]. The activity of this plant in dentistry treatment has been not yet reported, but ethnomedicine studies reported that this plant can be used as an alternative therapeutic in dentistry [16-18].

***Pluchea indica* (L)**

Pluchea indica (L) is widely available in Asia and Australia, in Indonesia this plant also known as Bluntas (Figure 3). *Pluchea indica* is often used as antipyretic, antiinflammatory, muscle relaxant, hepatoproteksi, astringent, wound medication and ulcers. Pharmacologically, this plant has a high antioxidant activity. Other references mentioned that this plant also has antibacterial, antimalarial, antifungal, antitumor activity and wound healing properties [4,20,21]. The activities of this plant is associated with the secondary metabolites it contained, such as tannins, flavonoids and essential oils. The leaves of *Pluchea indica* have been reported to contain flavonoids, saponins, phenolic hydroquinone, alkaloids, sterols, tannins, and reducing sugars [21-24].

Figure 3. A. *Pluchea indica* (L) and B. *Cissus adnata* (L).

In dentistry, this plant has been used as an antibacterial activity against *Streptococcus mutans*, Enterococcus fecalis, and *Lactobacillus* and *Fusobacterium nucleatum*. This plant also has been used as wound healing on gingival enlargement and as astrigen on mouthwash [4,5]. Also, this plant has been reported to act as anticarcinogenic and antibacterial root-canal in endodontic treatment failure[20].

Cissus adnata

Cissus adnata is well known as *Cissus pallida*, this plant is classified as vines herbs (Figure 3). The use of this plant is not as popular as other *Cissus* family such as *Cissus quadrangularis*. However, this plant is reported to contain alkaloids, carbohydrates, flavonoids, phenols, terpenoids, saponins, and tannins. This plant also contains antioxidant compounds such as polyphenols which are known for the antioxidant, antibacterial, and antihelmintic activities [24]. The mostly used parts of this plant for therapeutic treatments are leaves, stems, and roots. The leaves of this plant can be used as therapeutic in treating wounds, abdominal pain, peptic ulcer disease, syphilis, ulcers, broken bones, urolithiasis, arthritis pain, and reducing pain during menstruation. The stalks are also often used as therapy for jaundice and paralysis. The roots of this plant are very useful as an anti-diuretic, blood purifier and elephantiasis. Also, this plant is helpful to treat cancer, hemorrhoids, dysentery, epilepsy, fever, asthma, malaria, and food poisoning.[25-26]

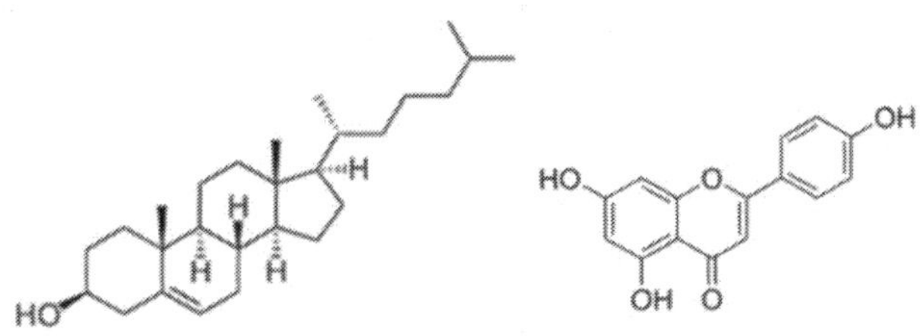

triterpenoid **apigenin**

Figure 4. Structure triterpenoid and apigenin isolated in *Cissus adnata*

Phytochemical screening showed this plant contains α- and β-amyrins, β-sitosterol, ketosteroid, carotene, vitamin C, tannins, and phenols. While, the leaves contain triterpenoids, flavonoids, and apigenin (Figure 4). The pharmacological studies showed that this plant has an antioxidant activity and also can be used as therapeutic treatment in dentistry [24-26]. However, the literature of *Cissus* as therapeutic agents in dentistry is very limited and still required an effective research for the development of this plant as therapeutics in dentistry.

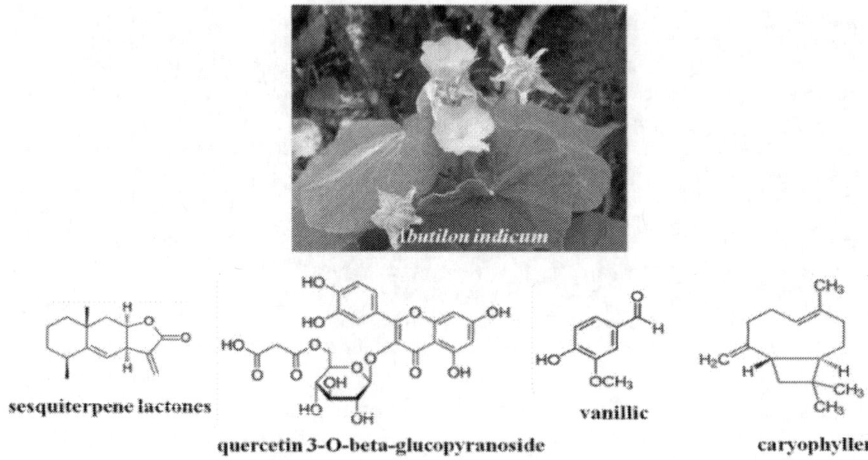

Figure 5. The compounds that isolated from *Abutilon indicum*: sesquiterpene lactones, quercetin *3-O*-beta-glucopyranoside, vanillic, and caryophyllene.

Abutilon indicum (L)

Abutilon indicum (L) is classified as herbal plant from the family of *Malvaceae*. This plant widely grow in the Aceh Province. In Indonesia, this plant has several names, Jeukeupa (Aceh), Belalang Sumpah (South of Sumatera), Cemplok (Java) and Kembang Sore (Moluccas). All Parts of this plant such as leaves, seeds, flowers and roots are widely used as a medical therapeutics.

This plant has been reported to contain various secondary metabolites such as saponins, flavonoids, alkaloids, hexoses, *n*-alkane mixtures (C_{22}-C_{34}), alkanols, β-sitosterol, vanillic, fumaric acid, *p*-coumaric, caffeic acid, sesquiterpene lactones and amino acids. In addition, this plant also contains essential oils such as α-pinene, caryophyllene, caryophyllene oxide, endesmol, farnesol, borenol, geraniol, geranyl acetate, elements and 1:8-cineole. Seven flavonoids compounds have been isolated from the flowers of *Abutilon indicum,* the compounds are; luteolin, chrysoeriol, luteolin 7-*O*-beta-glucopyranoside, chrysoeriol 7-*O*-beta-glucopyranoside, apigenin 7-*O*-beta-glucopyranoside, quercetin 3-*O*-beta-glucopyranoside ,quercetin 3-*O*-alpha-rhamnopyranosyl (1-6)-beta-glucopyranoside. Other compounds such as guanethidine, eugenol (4-allyl-2-methoxyphenyl), vitamin E also isolated from *Abutilon indicum*. [27-29]. The activities of those flavonoids compounds are presented in Table 1.

Table 1. The compound and its biological activity of plant Abutilon indicum [29]

Name of the compounds	Biological activity
Benzofuran, 2,3-Dihydro-	Neuropathic pain, analgesic, anti-inflammatory
4H-Pyrans-4-one 2,3-dihydro-3,5-dihydroxy-6- methyl-	Antimicrobial, antioxidant, anti-inflammatory
α-D-Glucopyranose, 4-OD-galactopyranosyl-	Preservative
Hexadecanoic acid, ethyl ester	Antioxidant, nematicide, pesticides, hypocholesterolemic, hemolytic, reductase inhibitors, anti-sandrogenic flavor
Phenol, 3-methyl-5- (1-methylethyl) -, Methylcarbamate	Antifungal, anti-cancer
9,12,15-Octadecatrienoic acid, 2,3-bis [(trimethyl) oxy] propyl ester	Anti-inflammatory and CNS depressant activity
Phytol	Antimicrobial, anticancer, anti-inflammatory, diuretic
vitamin E	Antiageing, analgesic, anti-diabatic anti-inflammatory, antioxidant, anti-dermatitic, anti-leukemic, anti-tumor, anti-cancer, hepatoprotective, hypocholesterolemic, anti-ulcerogenic, vasodilator, anti-spasmodic, anti-

IOP Conf. Series: Materials Science and Engineering **523** (2019) 012009 doi:10.1088/1757-899X/523/1/012009

	bronchitic, anticoronary
Guanethidine	antihypertensive drug
trans-isoeugenol Antioxidant	Anticancer, antioxidant
γ-Tocopherol	Antioxidant, anticancer
Stigmasterol	Anti-inflammatory anti-osteoarthritic
β-sitosterol	Anti-inflammatory, anti-pyretic, anti-arthritic, anti-ulcer,
Campesterol	Anti-inflammatory

Pharmacologically, this plant is widely used as an anti-inflammatory, antifibrinolytic, antioxidant, antibacterial, antiulser, antifungal and wound medication. Abutilon indicum capability is already being developed in dentistry as antiinflamsi analgesic, anti-bacterial are also caused by the content of eugenol from A. indicum [28-30].

3. Potential Herbal Medicines in Dental Treatment

The oral cavity is the gateway contamination in the human body. Due to high contamination, many bacteria can be detected in the oral cavity. Dewhirst et. al., (2010) indicates at least 619 of taxon are identified in the oral cavity [31]. Due to the frqeuent occurrence of the infections in the oral cavity, thus preventive treatment of the diseases is needed. One of the prevention can be done by exploiting theses five herbal medicine as therapeutic in dental treatment.

Based on several studies mentioned aboved the five plants of *Calotrophic gigantea* (L), *Acrostichum aureum* (L), *Pluchea indica* (L), *Cissus adnata* and *Abutilon indicum* (L) are considered to have polyphenolic compounds such as flavonoids. Flavonoids are polyphenolic compounds play an important role for an antidote to catch free radicals (free radical scavenger) as antioxidant and antimicrobial. Shoibe and You (2012) reported that the flavonoids from *Calotrophic gigantea* (L) and *Cissus adnata* (L) play an important role in preventing cancer, cardiovascular, macular degeneration, cataracts and asthma diseases, and can improve immune function [13,24,32]. Another flavonnoids compounds such as isolaflavans, gallic acid, and quarcetin are potentials to develop in dental teurapheutic. *Calotrophic gigantea* reported contains gallic acid and flavonoids as much as 19.86 ± 0:02 mg/gl and quersetin as much as 5.10 ± 0.60 mg/gm, this causes *Calotropis gigantea* to have a capability to inhibit dental caries causing bacteria with the inhibition zone of 16 mm [33]. While, *Abutilon indicum* is reported to contain flavonoids of 64.6 ± 1.20 mg/gl and quersetin of 85.3 ± 0.8 mg/gm [30]. Meanwhile, *Pluchea indica* has been reported to contain flavonoids as much as 0.831 ± 0.129 mg/g [23]

Several studies have not been able to explain the mechanism of these flavonoids compounds against the bacterial. However, based on the literature, the mechanism of the flavonoids on the bacterial inhibition is by damage the extracellular membrane of the bacteria, the other machanism is by forming a complex with soluble extracellular proteins of the bacteria, the other possible mechanism is by interacting with the cell wall of the bacteria. However, Singh (2006), reported that the flavonoids compounds from these plants are potential to be used for wound-healing treatment.[8] Rahman et al., (2014) also reported that flavonoids were able to stimulate red blood cells (eritrosites) in proliferating and differentiation [5]. The cardenoline and terpenoids compounds from these plants are able to prevent of tumors and cancers cells, these compounds are reported to be able to induce the apoptosis in tumors cell and also able to regulate cancers cells through multiple signaling pathways [33]. The steroids from these plants are also reported to have inhibition capability against the development of tumors and cancers cells. The terpenoids compounds from these plants are reported to be capable of stimulating the signaling cells through the NF-κB pathway through different mechanisms, particularly through the IκB phosphorylation, DNA binding, translocation of p65, and others [34]. The eugenol has been used traditionally in Asian countries as antiseptic, analgesic and antibacterial, eugenol also used as refresher in cosmetics and food products. Meanwhile, in dentistry eugenols is used as cement restorations, painkillers and medicamen intracanal [35].

4. Conclusion

These five plants could potentially be used for dental therapy in dentistry because of secondary metabolites, were very supportive. These compounds cause the five plants to have a function as antimicrobial, anti-inflammatory, anti-cariogenic, anti-cancer, anti-tumor and wound healing medicine.

5. Reference

[1] Kumar G, Jalaluddin Md, Rout P, Mohanti R, Dileep CL. 2013 *Journal of Clinical and Diagnostic Research* **7(8)**1827-9
[2] Kunwar RM, Mahat L, Acharya RP and Bussmann RW. 2013 *Journal of Ethnobiology and Ethnomedicine* **9(1)**: 9-24
[3] Sharma M, Tandon S, Aggarwal V, Bhat K, Kappadi D, Chandrashekhar P, and Dorwal R. 2015 *J. Conserv. Dent* **18(6)**:457-60
[4] Fatimatuzzahra N, Ning NS, Feny F, Darsono A, Salasia SIO. 2016 *Sain J. Vet* **34(2)**: 182-7.
[5] Ab Rahman MR, FA Razak, Bakri MM. 2014 *Evid. Based. Complement. Alternat. Med* 1-9
[6] Agbor AM, Naidor S., 2015 *Evidence-Based Complement Altern Med.* 1-10.
[7] Orhan DD, Ozcelik B, Özgen S, and Ergun F. 2010 *Microbiol. Res.* **165(6)**: 496-504.
[8] Singh MR, Govindarajan R, Nath V, Rawat AKS, and Mehrotra S. 2006 *J.Ethnopharmacol* **107(1)**: 67-72.
[9] Singh R, and Mendhulkar VD. 2015 *J. Chem. Pharm. Res* **7(6)**: 205-11.
[10] Thoppil RJ and Bishayee A. 2011 *World J. Hepatol* **3**: 228-49.
[11] Kanchan T and Atreya A. 2016. *Calotropis gigantea. Wilderness Environ. Med*. **27(2)**: 350-1
[12] Singh M and Javed K.. 2015 *Eur. Chem. Bull* **4(10)**, 477-80
[13] You H, Lei M, Song W, Chen H, Meng Y, Guo D, Liu X, and Hu L. 2013 *Steroids* **78**(10): 1029-34.
[14] Ashori A and Bahreini Z. 2009 *Journal of Composite Materials* **43(11)**: 1297-1304
[15] Kumar NS and Balamurugan V. 2015 *Res. J. Phytochem* **9(3)**:137-43
[16] Yamamoto T, Tsuda Y, Mori GM, Cruz MV, Shinmura Y, Wee AKS, Takayama K, Asakawa T, Yamakawa T, Suleiman M, Núñez-Farfán J, Webb EL, Watano Y, and Kajita T. 2016. *Appl. Plant Sci* **4(9)**:1-5
[17] Uddin SJ, Jason TLH, Beattie KD, Grice ID, and Tiralongo E. 2011 *J. Nat. Prod* **74**: 2010-13.
[18] Sureshkumar J, Silambarasan R, Bharati KA , Krupa J, Amalraj S, Ayyanar M. *Journal of Ethnopharmacology **219** 269-87*
[19] Hossain H, Jahan AI, Nimmi I, Hossain A, Kawsar H. 2011 *Bangladesh Pharmaceutical Journal* **14(2)**: 107-9
[20] Pargaputri AF, Munadziroh E, and Indrawati R. 2016.. *Dent. J (Magazine of doctor. Dental)* **49(2)**: 93-98
[21] Sen T, Chaudhuri AKN. **1991** *Journal of Ethnopharmacology* **33(1-2)** 135-41
[22] Widyawati PS, Wijaya CH, Hardjosworo PS, and Sajuthi D. *Hayati J. Biosci*. **20(3)**: 117-26.
[23] Andarwulan N, Batari N, Sandrasari DA, Bolling B, Wijaya H. 2010 *Food Chem* **121(4)**: 1231-5.
[25] Laitonjam WS, Yumnam, RS, Kongbrailatpam B D. 2011 *International Research Journal of Pure & Applied Chemistry* **1(1)**: 1-13
[24] Shoibe M, Chy M, Alam M, Adam M, Islam M, Nihar S, Rahman N, Suez E. 2017 *Biomedicines* **5(63)**:1-18
[26] Rashid R, Towsif FN, Bushra FA and Tahia F. 2016 *Dhaka Univ. J. Pharm. Sci.* **15(1)**: 69-71
[27] Singh R, and Mendhulkar VD. 2015 *J. Chem. Pharm. Res*. **7(6)**: 205-11.
[28] Irena M and Maria S. 2002 *Acta Pol Pharm* **59(3)**:227-9
[29] Radhakrish K, Mohan A, Chandra SM, and Velavan S. 2016 *Res. J. Phytochem* **11(1)**: 11-9.

[30] Srividya AR, Dhanabal SP, Jeevitha S, Vishnu VV, RR Kumar. 2012 *Indian J. Pharm. Sci* **74 (2)**: 163-7.
[31] Dewhirst FE, Chen T, Izard J, Paster BJ, Tanner, ACR., Yu WH, Lakshmanan A, Wade WG *Journal of Bacteriology* **192(19)**: 5002-17
[32] Wen S, Chen Y, Lu Y, Wang Y, Ding L, and Jiang M. 2016 *Fitoterapia* **112**: 74-84.
[33] Kumar NS and Balamurugan V. 2015 *Res. J. Phytochem* **9(3)**:137-43
[34] Jain HN, Dhingra N, Narsinghani T, and Sharma R. 2016 *Experimental Oncology* **38**: 158-68
[35] Lee YY, Hung SL, Pai SF, Lee YH, That SF. 2007 *J. Endod.* **33(6)**: 698-702.

IOP Conf. Series: Materials Science and Engineering **523** (2019) 012010 doi:10.1088/1757-899X/523/1/012010

Geochemistry of warm springs in the *Ie Brôuk* hydrothermal areas at *Aceh Besar* district

R Idroes[1,*], M Yusuf[1], M Alatas[1], Subhan[1], A Lala[1], Muslem[1], R Suhendra[2], G M Idroes[3], Suhendrayatna[3], Marwan[4,5] M Riza[3]

[1] Department of Chemistry, Faculty of Mathematics and Natural Sciences, Syiah Kuala University, Kopelma Darussalam, Banda Aceh 23111, Indonesia

[2] Department of Informatics, Faculty of Mathematics and Natural Sciences, Syiah Kuala University, Kopelma Darussalam, Banda Aceh 23111, Indonesia

[3] Department of Chemical Engineering, Faculty of Engineering, Syiah Kuala University, Kopelma Darussalam, Banda Aceh 23111, Indonesia

[4] Graduate School of Mathematics and Applied Sciences, Syiah Kuala University, Kopelma Darussalam, Banda Aceh 23111, Indonesia

[5] Department of Geophysics Engineering, Faculty of Engineering, Syiah Kuala University, Kopelma Darussalam, Banda Aceh 23111, Indonesia

E-mail: rinaldi.idroes@unsyiah.ac.id

Abstract. Indonesia is a country with the largest geothermal potential resource in the world, estimated of 28,617 MW or equivalent to 40% of the world's geothermal potential. Geothermal energy is environmentally friendly and quite economical compared to fossil energy. *Seulawah Agam* is one of Indonesia's geothermal fields that has several manifestation zones, including *Ie Brôuk* (BH1 and BH2) manifestations. The analysis for this study was conducted by acid-base titration, UV-Vis Spectrophotometry and Atomic Absorption Spectroscopy (AAS). The data analysis of cations and anions were plotted in liquid chemistry worksheet plotting spreadsheet version 3 Powell geoscience September 2012 by Powell and Cumming. The type of manifestation fluid is bicarbonate, obtained from the geoindicator Cl-SO_4-HCO_3, while the water condition is immature based on Na-K-Mg ternary diagram. The *Ie Brôuk* reservoir temperature was 289 °C and 291 °C obtained from Na/K geothermometer Tonani (1980) and Na/K Giggenbach (1988) respectively. Therefore, the *Ie Brôuk* manifestation classified as a high-temperature geothermal system.

1. Introduction

Indonesia's geothermal energy potential reaches 28,617 megawatts (MW) [1] or about 40% of the total geothermal energy potential in the world [2] scattered in several regions in Indonesia such as Sumatra, Java, and parts of eastern Indonesia [3]. Geothermal energy is environmentally friendly and economical enough [4] when compared to fossil fuels. Indonesia has started the exploration of geothermal energy since 1970 [5]. Aceh is one of the provinces that has some active volcanoes, i.e *Jaboi*, *Seulawah Agam*, *Geuredong*, and *Burni Telong* which suggest to have the potential of geothermal energy [6], one of which is Mount *Seulawah Agam* located in *Aceh Besar* district [7]. Ie-

*Corresponding author : rinaldi.idroes@unsyiah.ac.id

Seu'um manifestation around *Seulawah Agam* Mountain is one manifestation of which the geothermal exploration has been carried out, where the reservoir temperature is 188.7±9.3 °C and the water type is chloride [8]. In this study, the determination of *Ie Brôuk* (BH1 and BH2) characteristic was determined using geoindicator equation of Cl-SO_4-HCO_3 and ternary diagram of Na-K-Mg, while the temperature interval of *Ie Brôuk* reservoir measured using the equation of Geothermometer Na/K [9, 10, 11, 12, 13, 14] and Geothermometer Na/K/Ca [15].

2. Methods and Materials

2.1. Site Sampling and Field Data Measurement

Ie Brôuk geothermal manifestations are located in the village of *Lam Apeng*, *Lamteuba*, *Seulimum* sub-district, *Aceh Besar*, that is in the up-flow zone of *Seulawah Agam*. The characteristics of *Ie Brôuk* manifestations were shown by the slightly turbid water color, the warmth, the amount of air bubbles and small debit. The manifestations were surrounded by bushes, mossy rocks, leeches and many trees. The maps of sampling location was observed through satellite image (google earth), geological map, temperature map, and lineament density map in figures 1, 2, 3 and 4 respectively. Before the water sample was taken, the warm spring surrounding area was cleaned up. Initial measurements were carried out which included sampling point coordinates, elevation, temperature, pH, conductivity and total dissolved solids (TDS) of surface water. The geothermal surface water samples were filtrated using Whatman filter paper with a 0.45μm pore size, then the filtrate was put into the polyethylene (PE) bottle and acidified by adding 3N HNO_3 solution for pH adjustment for cations measurements, while for anions measurement, the filtrate was not acidified.

2.2 Materials and Instruments

The materials used in this study were Whatman filter paper with a pore size of 0.45 μm, demineralized water, NaCl solution, HNO_3 solution, HCl solution, $AgNO_3$ solution, orange methyl indicator, phenolphthalein indicator, potassium chromate indicator, sulfaver powder, potassium standard solution (1000 mg/L), sodium (1000 mg/L), calcium (1000 mg/L) and magnesium (1000 mg/L). Instruments used in this study included Portable pH meter, TDS meter, Conductometer (Schott Instrument), Portable Temperature (Fisher Scientific Traceable), Atomic Absorption Spectroscopy (AAS) (Shimadzu 6200), UV-Vis Spectrophotometer (DR/2010), GPS (Garmin 62S), analytical balance, polyethylene bottles, universal pH indicator, gloves, plastic bucket, titration apparatus, spatulas, hot plate and cooler boxes.

2.3. Measurements and Data Processing

Analysis of anion content of bicarbonate (HCO_3), Chloride (Cl) and sulphate (SO_4) was respectively carried out by acidimetry, argentometry (mohr) and UV-Vis spectrophotometry methods. Analysis of the cation content of Potassium (K), Magnesium (Mg), calcium (Ca) and sodium (Na) of warm water samples were measured using Atomic Absorption Spectroscopy (AAS). The uncertainty determination of cations concentrations is based on S_c equation (standard deviation of concentration). The Standard deviation in concentration can be found from the standard error of the estimate, also called as the standard deviation of the regression [16].

The data content of cations and anions then were plotted in liquid chemistry worksheet plotting spreadsheet version 3 Powell geoscience September 2012 by Powell and Cumming [17]. The output of the geothermal fluid type and geothermal fluid chemical equilibrium were exhibited in Cl-SO_4-HCO_3 and Na-K-Mg ternary diagram respectively. The depth temperature estimate was calculated using geothermometer Na-K and Na-K-Ca equations from several references which can be seen in table 1.

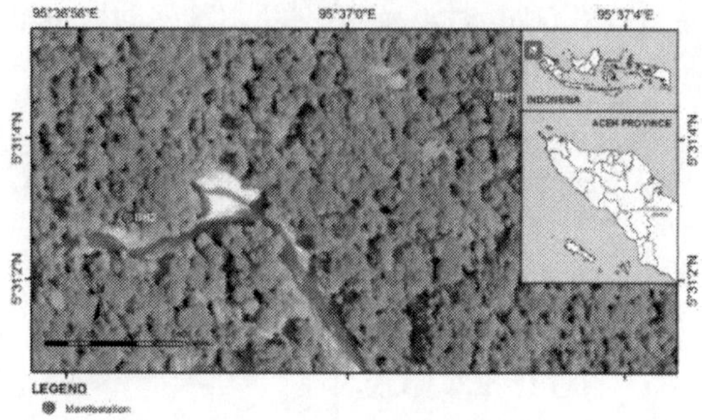

Figure 1. Satellite imagery map location manifestation *Ie Brôuk* (BH1 and BH2)

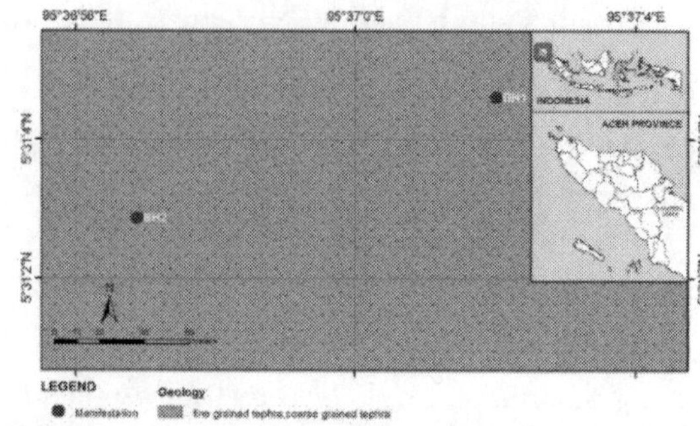

Figure 2. Geological map location manifestation *Ie Brôuk* (BH1 and BH2)

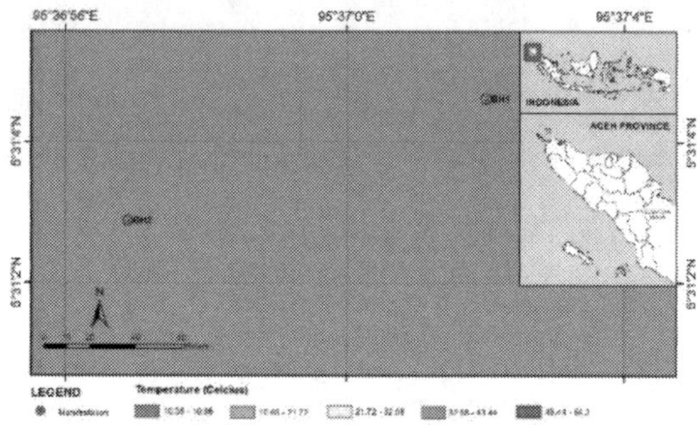

Figure 3. Temperature map *Ie Brôuk* (BH1 and BH2)

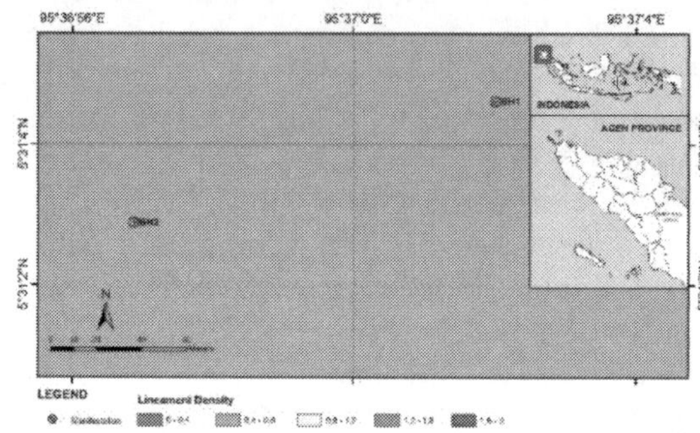

Figure 4. Lineament density map *Ie Brôuk* (BH1 and BH2)

Table 1. Temperature equations (°C) for geothermometers

Geothermometer	Equations
Na – K Truesdell (1976) [13]	T = [855.6 / (0.857 + log(Na/K))] – 273.15
Na – K Tonani (1980) [14]	T = [833 / (0.780 + log(Na/K))] – 273.15
Na – K Arnorsson (1983) [11]	T = [1319 / (1.699 + log(Na/K))] – 273.15
Na – K Fournier (1979) [9]	T = [1217 / (1.483 + log(Na/K))] – 273.15
Na – K Nieva & Nieva (1987) [12]	T = [1178 / (1.470 + log(Na/K))] – 273.15
Na – K Giggenbach (1988) [10]	T = [1390 / (1.750 + log(Na/K))] – 273.15
Na-K-Ca Fournier and Truesdell (1973) [15]	$T = \frac{1647}{\log\frac{Na}{K} + \beta\left[\log\left(\frac{Ca^{1/2}}{Na}\right) + 2{,}06\right]} - 273$ for : β = 4/3, if T < 100°C; β = 1/3, if T > 100°C

3. Result and Discussion

3.1. Location and Surface Characteristics of Ie Brôuk Manifestation

Measurements of surface water characteristics and warm water sampling were conducted on November 20th, 2017 at two sampling locations BH1 and BH2. The measurements were carried out on-site with five repetitions and the uncertainty measurement can be seen in table 2.

Table 2. *Ie Brôuk* manifestation surface characteristics

Sampling Points	Coordinate N	Coordinate E	Elevations (m)	T_{water} ±*Sd* (°C)	pH±*Sd*	Conductivity ±*Sd* (mV)	TDS±*Sd* (mg/L)
BH1	5^0 31.077	95^0 37.034	210	40.2 ±0.05	7.13±0.03	602.4±1.14	298.4±1.14
BH2	5^0 31.048	95^0 36.948	197	49.5 ±0.09	7.04±0.03	553.2±0.84	125.64±0.93

The temperature at sampling point BH2 (49,5±0,09 °C) is higher than BH1 (40.2±0.05 °C). *Ie Brôuk* manifestation has neutral pH and conductivity along with TDS is relatively high due to organic material dissolved in sedimentary rocks around the manifestation [19]. The sample of warm waters from BH1 and BH2 were analyzed in the laboratory to determine the anion and cation contents.

3.2. The Fluid Type of Ie Brôuk Manifestation

The fluid type (sulphate, bicarbonate or chloride types) of *Ie Brôuk* manifestation (BH1 and BH2) were determined based on the dominant value of the anion content that can be seen in Table 3.

Table 3. The concentration of the Bicarbonate, Sulphate and Chloride anions of warm spring samples at *Ie Brôuk* Manifestation

Sampling Points	HCO_3^-±*Sd* (mg/L)	SO_4^{2-}±*Sd* (mg/L)	Cl^-±Sd (mg/L)
BH1	292.6±0.26	33.87±0.85	12.98±0.01
BH2	284.03±0.03	12.15±0.01	21.19±0.04

Ie Brôuk manifestation has a higher content of bicarbonate anion compared to sulphate and chloride. The anion data was used to plot Cl-SO_4-HCO_3 ternary diagram on determining the fluid type of manifestation. Based on figure 5, BH1 and BH2 are indicated as bicarbonate fluid type. This observation signifies the effect of high CO_2 content from the gas or steam condensation process at geothermal springs [19].

3.3. Ternary Diagram Na–K–Mg

The Na-K-Mg ternary diagram is used to explain geothermal fluids at equilibrium in lithology based on cation content. The results of the laboratory measurement of the calcium, sodium, potassium and magnesium cation content for each sampling point are described in table 4.

Table 4. The concentration of the Potassium, Sodium, Calcium and Magnesium cations of warm spring samples at *Ie Brôuk* manifestation

Sampling Point	K^+±*Sc* (mg/L)	Na^+±*Sc* (mg/L)	Ca^{2+}±*Sc* (mg/L)	Mg^{2+}±*Sc* (mg/L)
BH1	18.87±0.20	116.30±0.23	3.60±0.16	32.75±0.01
BH2	19.61±0.20	101.62±0.23	4.23±0.18	60.38±0.01

The cation data were plotted on the ternary diagram of Na-K-Mg to identify the condition of water, whether or not it has reached the equilibrium in lithology. The warm spring samples of *Ie Brôuk* manifestation (BH1 and BH2) indicated as immature water (figure 6). This condition showed that the concentration of dissolved cations in geothermal water had been mixed with the surface water, so it has not reached chemical equilibrium yet [18]. Cation data is also used to estimate reservoir depth temperatures based on geothermometer equations.

IOP Conf. Series: Materials Science and Engineering **523** (2019) 012010 doi:10.1088/1757-899X/523/1/012010

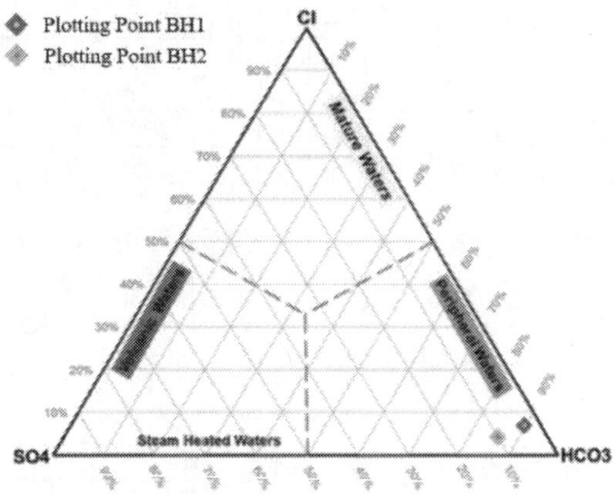

Figure 5. Ternary Diagram $Cl-SO_4-HCO_3$ at *Ie Brôuk* geothermal manifestation

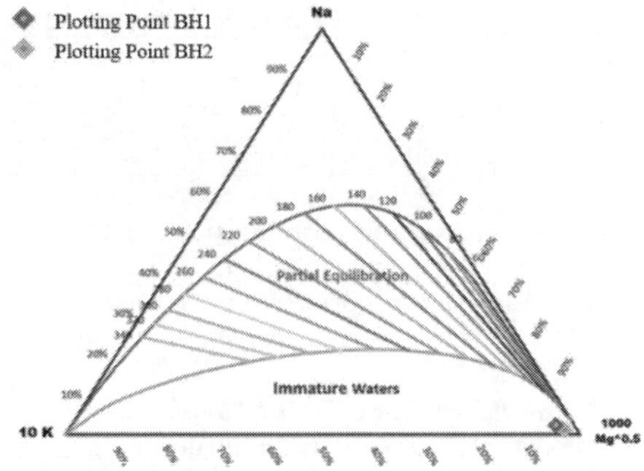

Figure 6. Ternary Diagram Na-K-Mg at *Ie Brôuk* geothermal manifestation

3.4. Determination of Reservoir Temperature at Ie Brôuk Geothermal Manifestations

Geothermometer is an equation that is used to estimate the temperature of the subsurface (reservoir). Chemical equilibrium temperature is highly influenced by cation concentration that is used as temperature control of the reservoir [19]. The result of the calculation from several geothermometers on *Ie Brôuk* geothermal manifestation can be observed in Table 5.

Table 5. Depth temperature calculation using several geothermometers for *Ie Brôuk* geothermal manifestations

Sampling Points	Na-K-Ca Fournier and Truesdell [15]	Na/K Fournier [9]	Na/K Truesdell [13]	Na/K Giggenbach [10]	Na/K Tonani [14]	Na/K Nieva & Nieva [12]	Na/K Arnorsson [11]
BH1	181 °C	262 °C	246 °C	274 °C	289 °C	248 °C	250 °C
BH2	175 °C	281 °C	271 °C	291 °C	318 °C	266 °C	273 °C

The estimation of the reservoir depth temperature of the *Ie Brôuk* manifestation BH1 is 289 °C (Na/K Tonani, 1980) and BH2 is 291 °C (Na/K Giggenbach, 1988) (Table 5). The depth temperature is classified into three geothermal systems: low temperature (<125 °C), medium temperature (125-225 °C) and high temperature (>225 °C) [18]. Based on this classification, the manifestation of *Ie Brôuk* is the high temperatures geothermal system. According to calculated equations on the depth temperature estimation, all geothermometers Na/K delivered the same result in an average of 180 to 300 °C and the depth temperature is good if the result is close to 300 °C [19]. In this study, only the geothermometer Na/K Tonani (1980) and Na/K Giggenbach (1988) showed the good result.

4. Conclusion

The geothermal fluid of *Ie Brôuk* manifestation is bicarbonate water type based on a $Cl-SO_4-HCO_3$ ternary diagram with neutral pH. The manifestation of *Ie Brôuk* based on the Na-K-Mg ternary diagram indicates the manifestation to be immature water. This condition showed that the concentration of dissolved cations in geothermal water had been mixed with surface water, so it has not reached chemical equilibrium yet. In this study, the estimation of the reservoir depth temperature of the *Ie Brôuk* manifestation BH1 is 289 °C (Na/K Tonani, 1980) and BH2 is 291 °C (Na/K Giggenbach, 1988). The results of this calculation explained that the manifestation of *Ie Brôuk* is classified as a high temperature geothermal system.

5. References

[1] Nasruddin, Alhamid I, Daud M, Surachman Y, Sugiyono A, Aditya H B and Mahlia T M I 2016 *Renewable and Sustainable Energy Reviews*, *53*, 733–740

[2] Mulyana R 2015 *Unlocking Indonesia's Geothermal Potential, Renewable Energy and Energy Conservation Ministry of Energy and Mineral Resources ISBN 978-92-9254-901-5*

[3] Darma S, Harsoprayitno S and Setiawan B 2010 *Proceedings World Geothermal Congress 2010 Bali, Indonesia*

[4] Mwangi M N 2010 *Bulletin*, 24–35

[5] Hochstein M P and Sudarman S 2008 *Geothermics* vol. 37 no. 3 pp. 220–266

[6] Jonan I 2017 *Menteri Energi dan Sumber Daya Mineral: Potensi Panas Bumi Indonesia*.

[7] Mathieu T, Antonaria, Rustandi D, Armstrong J, Brophy P, Sikumbang I, Nugroho H, Primana J R, Marizi M, Nugroho H T, Simamora W P, Sunandar, Sinaga D V C R, Widyastuti N L, Hendra Wijaya P 2014 *Geothermal Handbook for Indonesia Directorate for Energy Resources, Mineral and Mining*

[8] Idroes R, Yusuf M, Alatas M, Subhan, Lala A, Saiful, Suhendra R, Idroes G M and Marwan 2018 *IOP Conf. Ser.: Mater. Sci. Eng*. 334 012002

[9] Fournier R O 1979 *Geoth. Res. Council Trans*. 3, 221-224

[10] Giggenbach W F 1988 *Geochim. Cosmochim. Acta*, 52, 2749-2765

[11] Arnorsson S 1983 *Geothermics*, 12, 119-128

[12] Nieva D and Nieva R 1987 *Developments in geothermal energy in Mexico part 12 - A cationic composition geothermometer for prospection of geothermal resources Heat recovery systems and CHP*, 7, 243-258

[13] Truesdell A H and Fournier R O 1976 *Proceedings 2nd UN Symposium on the development and use of geothermal resources, San Francisco*, 1, 837-844

[14] Tonani F 1980 *In: Proc. Adv. Eur. Geoth. Res., Second symposium, Strasboug*, 428-443

[15] Fournier R O, Truesdell A H 1973 *Geochim. Cosmochim.* Acta, 37, 1543–1550.

[16] Skoog D A, Holler F J and Crouch S R 2007 *Principles of Instrumental Ana*lysis, *Sixth Edition, (*Brooks Cole*)* vol. 6, 12-19

[17] Powel T and Cumming w 2010 *Geotherm. Reserve. Eng* vol February, no. 1-3, p.10

[18] Hochstein M P 1990 *Classification and Assessment of Geothermal Resources In: Dikson MH and Fanelli M (eds) Sma ll geothermal resources*, (NITAEWNDP Centre for Small Energy Resources, Roma-Italy) 31-59

[19] Nicholson K 1993 *Geothermal Fluids Chemistry and Exploration Technique*. (*Springer-Verlag*, Inc.: Berlin Heidelberg New York)

Antibacterial cream formulation of ethanolic *Pliek U* extracts and ethanolic residue hexane *Pliek U* extracts against *Staphylococcus aureus*

R. Rahmad[1], N Earlia[2,*], C Nabila[3],, I Inayati[3], M Amin[4], C R S Prakoeswa[5], K Khairan[2], R Idroes[2,3]

[1]Department of Chemistry, Faculty Mathematics and Natural Sciences, Universitas Syiah Kuala, Banda Aceh, Indonesia, 23111
[2]Graduate School of Mathematics and Applied Sciences, Universitas Syiah Kuala, Banda Aceh, Indonesia, 23111
[3]Department of Pharmacy, Faculty Mathematics and Natural Sciences, Universitas Syiah Kuala, Banda Aceh, Indonesia, 23111
[4]Department of Biology, Faculty of Mathematics and Natural Science, Universitas Negeri Malang, Malang, Indonesia, 65145
[5]Department of Dermatology and Venereology, Faculty of Medicine, Universitas Airlangga, Surabaya, Indonesia, 60132

Email: nanda.lia@mhs.unsyiah.ac.id

Abstract. Indonesia as the largest archipelago country in the world is also known as a country with plenty of natural resources that can be used as traditional medicinal plants. *Pliek U* is a fermented crude coconut meat has been used as an ingredient in Aceh traditional food and also used as a traditional topical medicine in Aceh. *Pliek U* is potential as a source of free fatty acids (FFAs) such as lauric acid and monolauric acid which have antimicrobial properties. The aim of this study is to determine the activity ethanolic *Pliek U* extracts (EPUE) and ethanolic of residue hexane of *Pliek U* extracts (ERHPUE) and also the activities of their cream formulation against *Staphylococcus aureus*. The antibacterial activity was determined by disc difussion method with clindamycin as a positive control. The antibacterial activity results showed that EPUE has antibacterial activity against *Staphylococcus aureus* with diameter inhibition zone of 8.66±0.57 mm, while ERHPUE and their cream have no activity against *Staphylococcus aureus*.

1. Introduction

The Acehnese people have traditionally processed the coconut into traditional coconut oil for consumption. The fermented coconut meat is then dried and processed into coconut products such as *Pliek U* [1]. Coconut is rich in fatty acids such as lauric acid which has an antibacterial effect against *Propionibacterium acnes*, *Staphylococcus aureus* and *Staphylococcus epidermidis* [2].

The research that has been done by Nurliana *et al.* in 2009 concluded that ethanolic *Pliek U* extract and ethanolic of residue hexane of *Pliek U* extracts able to inhibit the growth of *Staphylococcus aureus* with MIC_{50} values of 19.33 ± 0.47 mm and 18.33 ± 0.47 mm respectively [3].

* Corresponding author : nanda.lia@mhs.unsyiah.ac.id

Staphylococcus aureus is known to be a major bacterial in the complication of atopic dermatitis. Up to now, for the management of atopic dermatitis, there is no scientific evidence that shows *Pliek U* extract is effective moisturizer for adult patients with atopic dermatitis.

The Development of *Pliek U* as a therapeutic substance candidate should be continued by formulating this substance in pharmaceutical form. One of the appropriate forms is cream, because it can be smeared easily on the skin and can be absorbed easily. Based on the description above, the aim of this study is to analyze the activity ethanolic *Pliek U* extracts (EPUE) and ethanolic of residue hexane of *Pliek U* extracts (ERHPUE) and also the activities of their cream formulation against *Staphylococcus aureus*.

2. Methods

2.1 Samples collection

Samples were obtained from the household production located in Gampong Paya Bieng, Bireuen, Aceh Province, Indonesia. The sample obtained is *Pliek U*.

2.2 Processing of ethanolic *Pliek U* extracts (EPUE)

The *Pliek U* extraction was performed by macerating 250 g of *Pliek U* with 1500 mL ethanol 96% and being shaked at a speed of 130 rpm at 28 °C for 48 hours. It is then filtered using a fritted glass filter connected to a vacuum pump. The *Pliek U* residue was re-extracted in a separating funnel twice using 300 mL of ethanol 96%. The obtained filtrate was concentrated by using a rotary evaporator at 50 ° C to obtain ethanolic *Pliek U* extracts (EPUE)

2.3 Processing of ethanolic of residue hexane of *Pliek U* extracts (ERHPUE)

Extraction of *Pliek U* was done by macerating 250 g of *Pliek U* with 1500 mL hexane solvent with a ratio of 1: 6, then shaked with a speed of 130 rpm at 28°C for 48 hours. It is then filtered using a fritted glass filter connected to a vacuum pump. Filtering results obtained by macerate and residue. The residue was re-extracted using 300 mL of hexane solvent in the separation funnel twice to separat hexane layer and the residue. The residue is then re-extracted using 1500 mL ethanol 96% in the same procedure. The ethanol macerate was concentrated using a rotary evaporator at a temperature of 50 ° C, to obtain the ethanolic of residue hexane of *Pliek U* extracts (ERHPUE).

2.4 Formulation and processing of cream

Formulations of cream are made using vanishing cream as a base with an ingredients namely stearic acid, cera alba, vaseline album, triethanolamine, propylenglycol, and aquadest. The cream is made of 2 formulations and each formulation is made up to 3 preparations with a weight in each preparation is 100 g. The cream formulation is made ie F_0 which does not contain extract and F_1 contains 4%.

Table 1. The cream formulation

Materials	Concentration (%)	
	F_0	F_1
EPUE/ERHPUE	-	4
Stearic acid	14.2	14.2
Cera alba	2	2
Vaseline album	8	8
Trietanolamine	1.5	1.5
Propylengycol	8	8
Aquadest	ad 100	ad 100

2.5 Evaluation of EPUE/ERHPUE cream

Evaluation of cream includes cycling test, organoleptic, homogeneity, pH, dispersion, adhesion and viscosity.

2.5.1 Cycling test. The cream is put into a plastic pot and covered with aluminum foil. The cream is placed in the oven at 45 ± 2°C for 24 hours, then placed in the freezer at 5 ± 2°C for 24 hours (1 cycle). This cycle is done as much as 6 cycles for 12 days [4].

2.5.2 Organoleptic test. Observations made visually on the cream include the shape, color, odor, and the presence or absence of separation between the cream base and *EPUE/ERHPUE* [5].

2.5.3 *Homogeneity test.* 0,5 g cream was applied to object glass and then the particle size was observed [6].

2.5.4 pH test. The cream is weighed as much as 1 g and put into the beaker, then added 10 mL of distilled water and stirred. pH meters that have been calibrated with aquades (pH 7.0) are fed into the solution and pH was read on the monitor.

2.5.5 Dispersion. A total of 1 g of cream is placed between two round glasses. A load of 5 g is placed on the glass and left for 1 minute. The diameter of the cream distribution is measured. The other load continuously added after 1 minute with a weight of 5 g to obtain a constant diameter of cream distribution. The recommendation of dispersion is 5 - 7 cm [5].

2.5.6 Viscosity. 100 g cream is put in a beaker and placed under the International Rheology Viscometer. Then the spindle is inserted into the cream and the viscosity of the cream is measured. The recommendation of viscosity is 5000 - 20000 mPa [7].

2.6 Culture of Staphylococcus aureus. The identification process of *Staphylococcus aureus*, the manufacture of *Staphylococcus aureus* media and regeneration was performed at the Microbiology Laboratory, Faculty of Medicine, Syiah Kuala University, Darussalam, Banda Aceh, Indonesia.

2.7 Antibacterial activity test. The antibacterial activity test of the cream was done by disc diffusion method. The advantages of the disk diffusion method are easy performance and less requirement of special equipment [8].

2.8 Processing of EPUE 4% and ERHPUE 4% and their cream formulation. EPUE 4% and ERHPUE 4% were prepared by dissolving 0.4 g of EPUE and 0.4 g ERHPUE in 10 ml ethanol, then soaked the disc for 30 minutes. Each 1 g of EPUE 4% cream and ERHPUE 4% was soaked in 1 mL sterile distilled water and then soaked in a blank disc for 30 minutes.

2.9 Antibacterial activity test of EPUE 4%, ERHPUE 4% and their cream formulation. The antibacterial activity test was performed with seven types of test samples: EPUE 4%, ERHPUE 4%, F_1(cream of EPUE 4% and cream of ERHPUE 4%), clindamycin (positive control), ethanol 96% (negative control) and cream F_0 / base (negative control). The antibacterial activity test method was performed using the disc method. The discs are then immersed in six sample types: EPUE 4%, ERHPPUE 4%, F_0, F_1 and 96% ethanol. The bacterial inoculum is scratched on MHA media using a sterile cotton bud. Next, put the disc paper soaked in EEPU 4%, EERHPU 4%, F_0, F_1 and ethanol 96% above the surface of MHA media. The positive control used is clindamycin disc

3. Result and Discussion

3.1 Sample extraction

The organoleptic characteristics of ethanolic *Pliek U* extracts (EPUE) obtained are blackish brown color, typical smell of *Pliek U* and its consistency is thick. The amount of EPUE obtained was 60 g with a rendemen percentage of 24%. The EPUE obtained then was tested for its activity against Staphylococcus aureus, and also was used to prepare its cream.

The ethanolic of residue hexane of *Pliek U* extracts (ERHPUE) obtained was 52.2 g with an extraction rate up to 20%. Its organoleptic features are dark brown, typical smell of *Pliek U* and thick.

3.2 Evaluation of cream

Based on Table 2 it can be seen that in F_0 and F_1 creams there is no change either the color, odor, shape, or separation. The F_0 and F_1 creams remain homogeneous from prior to the cycling test up until the 6th cycle. The pH value of the F_0 and F_1 is above 6.5, with the highest pH reached 8.1 (cream base) when cycling test 0 and 1, the lowest pH 6.7 (ERHPUE cream) was reached at the 6th cycle. Spreading well at no load, 50 g and the highest 100 grams when cycling test 0 with a value of 6.2; 6.6 and 7.0 (cream base) respectively, and lowest dissipation is 3.1; 3.2 and 3.5 (ERHPUE cream) when 5th and 6th cycle No change shows that the creams of F_0 and F_1 are stable during storage at two different temperatures and the cream is homogeneous if the distribution of particles is evenly distributed and there is no breakdown in the cream.

Table 2. Organoleptic, homogeneity, pH and dispersion

Cycle			Organoleptic				Homogeneity	pH	Dispersion (cm)		
			Color	Odor	Shape	Separation					
0	EPUE	F_0	-	-	-	-	Homogeneous	8.1	6.2	6.6	7.0
		F_1	-	-	-	-	Homogeneous	7.1	4.1	4.5	4.9
	ERHPUE	F_0	-	-	-	-	Homogeneous	8.1	5.5	6.2	6.6
		F_1	-	-	-	-	Homogeneous	6.8	4.1	4.2	4.3
1	EPUE	F_0	-	-	-	-	Homogeneous	8.1	6.0	6.5	6.9
		F_1	-	-	-	-	Homogeneous	7.0	4.0	4.4	4.7
	ERHPUE	F_0	-	-	-	-	Homogeneous	8.1	5.4	5.8	6.0
		F_1	-	-	-	-	Homogeneous	6.9	4.0	4.1	4.2
2	EPUE	F_0	-	-	-	-	Homogeneous	8.0	5.9	6.4	6.8
		F_1	-	-	-	-	Homogeneous	7.0	3.9	4.2	4.6
	ERHPUE	F_0	-	-	-	-	Homogeneous	8.0	5.2	5.7	6.1
		F_1	-	-	-	-	Homogeneous	6.9	3.8	4.0	4.3
3	EPUE	F_0	-	-	-	-	Homogeneous	8.0	5.8	6.2	6.6
		F_1	-	-	-	-	Homogeneous	7.0	3.9	4.2	4.5
	ERHPUE	F_0	-	-	-	-	Homogeneous	8.0	5.0	5.4	5.7
		F_1	-	-	-	-	Homogeneous	7.1	3.6	4.0	4.2
4	EPUE	F_0	-	-	-	-	Homogeneous	8.0	5.7	6.1	6.4
		F_1	-	-	-	-	Homogeneous	7.0	3.8	4.1	4.4
	ERHPUE	F_0	-	-	-	-	Homogeneous	8.0	5.0	5.4	5.6
		F_1	-	-	-	-	Homogeneous	7.0	3.4	3.6	3.7
5	EPUE	F_0	-	-	-	-	Homogeneous	7.9	5.7	6.0	6.4
		F_1	-	-	-	-	Homogeneous	7.0	3.7	4.0	4.3
	ERHPUE	F_0	-	-	-	-	Homogeneous	8.0	5.0	5.3	5.6
		F_1	-	-	-	-	Homogeneous	6.9	3.1	3.2	3.5
6	EPUE	F_0	-	-	-	-	Homogeneous	7.9	5.6	6.0	6.4
		F_1	-	-	-	-	Homogeneous	7.0	3.6	3.9	4.2
	ERHPUE	F_0	-	-	-	-	Homogeneous	7.9	5.0	5.1	5.4
		F_1	-	-	-	-	Homogeneous	6.7	3.1	3.2	3.5

- : no change.
F_0 : cream base.
F_1 : cream formulation.

The acidity of the cream affects the acceptance of the cream onto the skin. The pH of the cream should be in accordance with the pH of the skin which is between 4.5-6.5. Too-alkaline pH preparations cause scaly skin, whereas too acid pH will cause skin irritation [9]. The pH of F_0 and F_1 does not satisfy the expected pH value due to the pH value of the cream F_0 and F_1 exceeds the maximal threshold. The pH value of F_1 is lower than the pH value of F_0. The difference in pH values between F_0 and F_1 can be due to the pH of EPUE is 4.0 (acid) and ERHPUE is 4.1 (acid). The pH of the F_0 and F_1 cream decreased during storage caused by hydrolysis of the acidic compound due to the temperature change during the cycling test [10]. Spreading test is done to know the cream ability to be evenly spreaded on skin. Recommended spread of cream is 5 - 7 cm. Cream that has a spreading power with a range of 5 - 7 cm will be easily applied and spread on the skin [5].

F_1 distribution does not meet the requirement where the spreading power of F_1 is 3.1 - 4.9 cm (EPUE) and 3.1 - 4.3 (ERHPUE), meanwhile the spreading energy of F_0 is qualified with the diameter of cream distribution in the range of 5.6 - 7.0 cm. Factors that affecting the power of spread is the addition of extracts. The addition of extract to the cream causes the volume of water used to be less so that the consistency of the cream becomes tough, resulting in the spread of cream oil becomes decreased [10]. Therefore, the dispersion energy of the F_1 cream (containing EPUE and ERHPUE) is smaller than the spreading power of F_0. The distribution of F_0 and F_1 creams is smaller after cycling test which can be affected by viscosity, where the viscosity of F_0 and F_1 is greater after cycling test.

The viscosity value is related with whether or not the sample is easy to be removed from the container. When the sample is easily removed from the containe, the viscosity value meets the packaging requirements, and it can be easily used [5]. The viscosity test results can be seen in Table 3. Recommended viscosity values is between 5000-20000 mPa [7]. The viscosity values of F_0 and F_1 are qualified to be considered as a good viscosity for cream.

Table 3. Viscosity test

	Cycle	Cream	Viscositys (mPa)
EPUE	Before cycling	F_0	5182,44
	After cycling	F_0	5803,64
	Before cycling	F_1	7366,22
	After cycling	F_1	7982,17
ERHPUE	Before cycling	F_0	7562,94
	After cycling	F_0	9134,37
	Before cycling	F_1	7730,04
	After cycling	F_1	9458,49

F_0 : cream base.
F_1 : cream formulation.
Recommeded viscosity value : 500-20000 mPa.

3.3 Antibacterial activity test

Inhibitory zone of EPUE 4%, ERHPUE 4%, F_1(cream of EPUE 4% and ERHPUE 4%), ethanol and F_0 (negative control), and clindamycin (positive control) is exhibited in table 4.

Table 4. Antibacterial test

Samples	Diameters of the inhibitory zone (mm)			D_{mean}	Classification (Morales *et al*)
	D_1	D_2	D_3		
F_1 (cream of EPUE 4%)	-	-	-	0	
EEPU 4%	-	-	-	0	
F_1 (cream of ERHPUE 4%)	-	-	-	0	
ERHPUE 4%	9.0	8.0	9.0	8.66±0,57	Moderate
Control + (clindamycin)	15	15	14.0	14.66±0,57	Strong
Control – (ethanol 96%)	-	-	-	0	
Control – (F_0/base cream)	-	-	-	0	

The mean inhibitory zone diameter of EPUE 4% is 8.66 ± 0.57 mm which, according to Morales *et al.*(2003), the inhibitory zone includes moderate antibacterial activity [11]. The absence of 4% ERHPUE activity against *Staphylococcus aureus* indicates that suspected fatty acid compounds that have anti-bacteria activity in *Pliek U* have been extracted into hexane solvent. Fatty acid compounds such as lauric acid, capric acid, myristic acid, palmitic acid, oleic acid and linoleic acid are nonpolar and the hexane solvent is also nonpolar which attracts nonpolar compounds, so there is no inhibitory zone in ERHPUE 4%. In contrary with Nurliana *et al.* (2009)

which suggested that the inhibitory zone diameters of EPUE and ERHPUE (crude extract) against *Staphylococcus aureus* are 19.33 ± 0.47 and 18.33 ± 0.47 mm respectively and were classified as a strong antibacterial activity [3] . The difference between this study and Nurliana's research were due to different concentration of the samples tested. EPUE 4% cream and ERHPUE 4% cream also do not have an inhibitory zone formed because the cream can not diffuse well into the disc so that no inhibition zone is formed. Positive control (clindamycin) resulted in an inhibitory zone of 14.66 ± 0.57 mm which was a strong activity. The negative control used was 96% ethanol as a 4% EEPU counter and base cream (F_0) as a comparison of 4% EPUE cream and 4% ERHPUE (F_1). Negative control (ethanol and base cream) did not show any antibacterial activity characterized by there is no inhibition zone. These results suggest that ethanol solvent and cream base (F_0) does not affect the antibacterial test results of the extract and their cream.

4. Conclusion

Based on the results of the study it can be concluded that EPUE 4% has antibacterial activity against *Staphylococcus aureus*, while ERHPUE 4% and their cream do not have antibacterial activity against *Staphylococcus aureus*.

5. Reference

[1] Arpi N 2013 *Jurnal Teknik Industri Pangan* **5** 17-22
[2] Vaughn AR, Clark AK, Sivamani RK and Shi VY 2018 *Am J Clin Dermatol.* **19** 103-17
[3] Nurliana M S, Sudirman LI and Sanjaya AW 2009 *Forum Pascasarjana.* **32** 1-10
[4] Maulina ID *2011 Uji Stabilitas Fisik dan Aktivitas Antioksidan Sediaan Krim yang Mengandung Ekstrak Umbi Wortel (Daucus carota L.) (Jakarta :* Universitas Indonesia Press)
[5] Pramitasari M 2011 *Formulasi dan Uji Aktivitas Antijamur Krim Minyak Sereh (Cymbopogon citratus (DC) Stapf.) dengan Basis Vanishing Cream Terhadap Candida albicans dengan Metode Sumuran* (Jember :Fakultas Farmasi Universitas Jember)
[6] Anonim 2005 *Cosmetic products stability Guide 1st ed* (Brazil : National health Surveillance Agency Press)
[7] Lachman L, Lieberman H A and Kanig JL 1994 *Teori dan Praktik Farmasi Industri. (Jakarta :* Universitas Indonesia Press)
[8] Prayoga SE 2013 *Perbandingan Efek Ekstrak Daun Sirih Hijau (Piper betle L.) Dengan Metode Difusi Disk Dan Sumuran Terhadap Pertumbuhan Bakteri Staphylococcus Aures* (Jakarta : Universitas Islam Negeri Syarif Hidyatullah)
[9] Iswindari D 2014 *Formulasi dan Uji Aktivitas Antioksidan Krim Rice Bean Oil* (Jakarta : Universitas Islam Negeri Syarif Hidyatullah)
[10] Suhery WN, Fernando A 2016 *Jurnal Farmasi Indonesia* **13** 112
[11] Morales G, Sierra P, Mancilla, Paredes A, Loyola LA, Gallardo O 2005 *Journal of the chilean chemical society* **48** 35-41

Geochemistry of Sulphate spring in the *Ie Jue* geothermal areas at Aceh Besar district, Indonesia

R Idroes[1,*], M Yusuf [1], M Alatas[1], Subhan[1], A Lala[1], Muhammad[2], R Suhendra[3], G M Idroes[4], Marwan[5,6]

[1] Department of Chemistry, Faculty of Mathematics and Natural Sciences, Syiah Kuala University, Kopelma Darussalam, Banda Aceh 23111, Indonesia
[2] Faculty of Teacher Training and Education, Abulyatama University, Lampoh Keudee, Aceh Besar 23372, Indonesia
[3] Department of Informatics, Faculty of Mathematics and Natural Sciences, Syiah Kuala University, Kopelma Darussalam, Banda Aceh 23111, Indonesia
[4] Department of Chemical Engineering, Faculty of Engineering, Syiah Kuala University, Kopelma Darussalam, Banda Aceh 23111, Indonesia
[5] Graduate School of Mathematics and Applied Sciences, Syiah Kuala University, Kopelma Darussalam, Banda Aceh 23111, Indonesia
[6] Department of Geophysics Engineering, Faculty of Engineering, Syiah Kuala University, Kopelma Darussalam, Banda Aceh 23111, Indonesia

E-mail: rinaldi.idroes@unsyiah.ac.id

Abstract. Geochemical studies of geothermal manifestations in the Northern Zone of Mt. *Seulawah Agam*, one of which is *Ie Jue* manifestation, has been conducted. In this study, water geochemical analysis of cations and anions content was performed to observe the surface characteristics of manifestations, the type of water manifestation, the water balance of reservoir depth and to estimate depth temperature using geothermometer equations. Cations and anions measurements were conducted using atomic absorption spectroscopy, UV-Vis spectrophotometer and acid-base titration. The data analysis was determined using spreadsheet version 3 Powell Geoscience Ltd.3 September 2012 by Powell & Cumming. The geothermal water *Ie Jue* manifestation was acidic pH, water sulphate type was obtained from the geoindicator Cl-SO_4-HCO_3 and the immature water was based on plot Na-K-Mg ternary diagram. Its reservoir temperatures ranged from 690 °C to 761 °C in accordance with Na/K geothermometer equations based on results of Na/K Giggenbach and Fournier. The *Ie Jue* manifestations indicated a high-temperature geothermal system (high enthalpy) that could be characterized when the average temperature reached higher than 225 °C and suitable for power plant development.

1. Introduction

Indonesia is located in the area of "The Ring of Fire", which is a belt that has the most volcano up to more than 200 volcanoes, where 129 active volcanoes spread over from Sumatra, Java, Sulawesi to eastern parts of Indonesia [1]. There is a high amount of geothermal energy potential, so that research

*Corresponding author : rinaldi.idroes@unsyiah.ac.id

on the potential or prospect of geothermal in various regions of Indonesia is important in order to obtain a qualitative and quantitative picture of the geothermal energy potential [2].
Potential geothermal resources in Indonesia have been exploited such as Kamojang, Darajat, Salak Mountain, Dieng, Sibayak, Wayang Windu, Lahendong, Ulu Belu, Mataloko and Ulumbu [3]. Therefore, it is also necessary to explore and exploit other geothermal resources especially in Aceh province.

Aceh province is one of the areas in Sumatra that has geothermal energy potential, especially Mount *Seulawah Agam*. Based on the previous exploration, there are found many locations of manifestations such as *Ie Seu'um, Ie Jue, Ie Brôuk, Alue Ie Masam, Alue Ie Suum, Alue Utuen Pineung* and others. However there are only a few research to discover and to discuss the geothermal prospect of Mount *Seulawah Agam*. Among several potential geothermal fields in Aceh, only 2 (two) are ready to be developed, namely *Jaboi* geothermal field in *Sabang* and *Seulawah Agam* in *Aceh Besar* [4].

Exploration of geothermal energy of sulphate springs has been widespread throughout the world such as the hot springs of South Solfatara area of Saint Lucia, Lesser Antilles island [5], Gonghe geothermal waters, Northwestern China [6], Groundwaters of Mt. Vulture volcano, southern Italy [7], Thermo-mineral springs of the Cerna Valley, Romania [8], Los Humeros geothermal system, Mexico [9] and Urban sandstone aquifer, Birmingham, UK. In relation to the geothermal field of Mount *Seulawah Agam* in *Aceh Besar*-Indonesia regency, the determination of reservoir depth temperatures and geothermal fluid types in *Ie Seu'um* manifestation area has been conducted[11], meanwhile in the *Ie Jue* manifestations, the determination has not been conducted. One of the first stages of the exploration of geothermal sources is to estimate the depth temperature and reservoir fluid type based on the geochemical composition. This research will cover the determination of geothermal surface fluid characteristic, geothermal chemical fluid equilibrium based on Na-K-Mg triangle diagram [12] and determining geothermal fluid type based on triangle diagram of Cl-SO_4-HCO_3 [12], whether the fluid type is chloride, sulphate or bicarbonate. The depth temperature estimation of the reservoir is based on chemical composition using Na/K geothermometer equation [12,13,14,15,16].

2. Methods and Materials

2.1. Water Sampling and Field Data Measurement

The location of the geothermal manifestation of *Ie Jue* is found in *Lamteuba Droe* Village, *Seulimum* Sub-district, *Aceh Besar* District, which is an upflow zone of Mount Seulawah Agam. Manifestations are found in the form of hot sulphate springs with hot water output and hot steam smells of sulfur. The color of the hot water produced is cloudy due to mixing with acidic clay water. There are no plants and few trees that grow outside the area of manifestation (stress vegetation). For sampling location map via satellite image (google earth) and temperature map can be seen respectively in figures 1 and 2. Before the water sample is taken, the initial measurement as field data covering point of the coordinate of sampling point, elevation, surface water temperature, degree acidity (pH), conductivity and total dissolved solids (TDS). The geothermal water is taken at large and high-temperature water discharge. Hot water samples were filtered using Whatman filter paper with a pore size of 0.45 μ m, incorporated in a polyethylene bottle with the addition of 3N HNO_3 solution to pH = 2 (acidified), acid addition is not carried out for the cation and anion measurement.

2.2. Materials and Instruments

The materials used were demineralized water, 0.45 μ m filter paper, NaCl solution, HNO_3 solution, methyl orange indicator, phenolphthalein indicator, chromate potassium indicator, sulfaver (powder kit), HCl solution, $AgNO_3$ solution, standard solution of potassium (1000 mg/L), sodium (1000 mg/L), calcium (1000 mg/L) and magnesium (1000 mg/L).

The instruments used are Portable pH meter and Conductometer (Schott Instrument), Portable Temperature (Fisher Scientific Traceable), Atomic Absorption Spectroscopy (AAS) instruments (Shimadzu 6200), UV-Vis Spectrophotometer (DR/2010), GPS (Garmin 62S), analytical balance, polyethylene bottle, universal pH indicator, glove, plastic bucket, a set of titrations, spatula, hot plate and cool boxes.

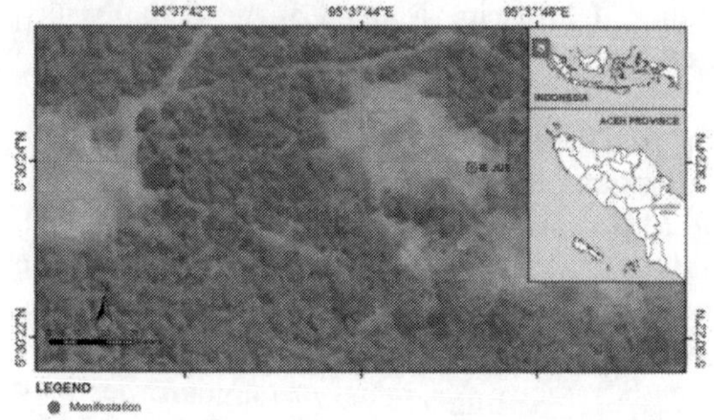

Figure 1. Satellite Imagery (google earth) of sulphate spring in the *Ie Jue* manifestation and sampling point location at the geothermal area.

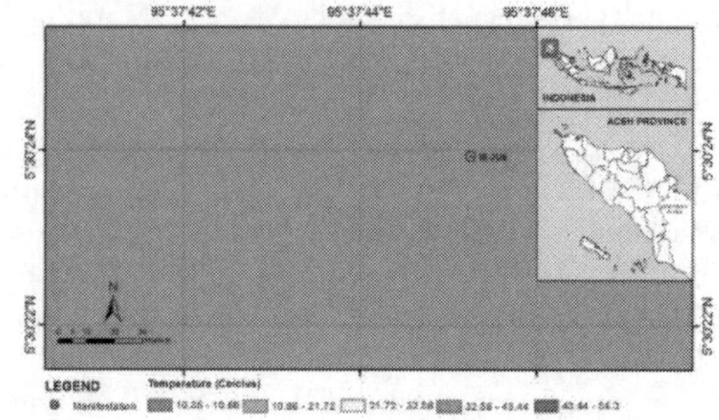

Figure 2. Temperature Map of sulphate spring in the *Ie Jue* manifestation at the geothermal area.

2.3. Measurements and Data Processing

An analytical method for cation content (K, Mg, Ca, Na) in a geothermal water sample is carried out using AAS instrument. For the analysis of anion content of bicarbonate, chloride and sulphate, the titration method of acidimetry, argentometry (mohr) and UV-Vis spectrophotometry respectively are carried out. The content of cations and anions is input in spreadsheet version 3 Powell Geoscience Ltd.3 September 2012 by Powell & Cumming [17]. The determination of geothermal water type is carried out by the triangle diagram of Cl-SO_4-HCO_3 [12], meanwhile the determination of geothermal chemical equilibrium is carried out by Na-K-Mg triangle diagram [12]. The depth temperature estimation uses the Na/K geothermometer formula with various references such as table 1. For the uncertainty measurement of cation concentration is calculated according to formula (1) [18].

$$S_C = \frac{S_r}{m}\sqrt{\frac{1}{M} + \frac{1}{N} + \frac{(Y_c - Y_{ave})^2}{m^2 S_{xx}}} \quad (1)$$

Where S_c is standard deviation of concentration, S_r is standard deviation of regression, m is slope, M is the number of replicated results, N is the number of point in the calibration curve (number of standards), Y_c is the mean response for the unknown. Y_{ave} is the mean value of y for calibration results and the quantity S_{xx} is the sum of the squares of the deviation of x values from the mean [18]

Table 1. Geothermometers equations for an estimate of the reservoir temperature.

No	Geothermometers	Equations	References
1	Na – K	T = [1390 / (1.750 + log(Na/K))] – 273.15	Giggenbach, 1988 [12]
2	Na – K	T = [1217 / (1.483 + log(Na/K))] – 273.15	Fournier, 1979 [13]
3	Na – K	T = [855.6 / (0.857 + log(Na/K))] – 273.15	Truesdell, 1976 [14]
4	Na – K	T = [833 / (0.780 + log(Na/K))] – 273.15	Tonani, 1980 [15]
5	Na – K	T = [1319 / (1.699 + log(Na/K))] – 273.15	Arnorsson, 1983 [16]

3. Result and Discussion

As a preliminary data, the determination of the surface geothermal properties of *Ie Jue* manifestations was carried out. Measurements were done in-situ with five repetitions, to obtain the average value and the value of uncertainty measurement as shown in Table 2.

Table 2. On-site characteristics of water sample for *Ie Jue* manifestation

Coordinate		Elevation	T_{water} (°C)	pH	Conductivity	TDS
N	E	(m)			(mV)	(mg/L)
5° 30.414	95° 37.736	265	93.55 ± 0.64	5.77 ± 0.04	85.42 ± 1.36	886.1 ± 3.94

Ie Jue manifestation was in the form of sulfuric hot springs with acidic water output (pH = 5.77 ± 0.04) and accompanied by a sulfur-containing hot vapor. The surface water temperature reached 93.55 ± 0.64 ° C and the conductivity value and the TDS value were relatively high. The color of hot water that was cloudy due to mixed acidic clay. There was no vegetation surrounding the manifestation (vegetation stress).

Determination of geothermal fluid type in *Ie Jue* manifestation was done by measuring anion content in the geothermal water sample. The measured anion content was sulphate, bicarbonate and chloride. According to the result of anion measurement, sulphate concentration had the highest concentration level that was 480.0 ± 0.36 mg/L compared with the concentration of bicarbonate and chloride that was respectively equal to 36.0 ± 0.26 mg/L and 4.99 ± 0.10 mg/L.

Based on the plot results in the triangular diagram of Cl-SO_4-HCO_3 geothermal fluid manifestation *Ie Jue* had a water sulphate fluid type (Figure 3). This type of fluid could be interpreted as steam-heated water which was characterized by an acidic pH and makes its ability to dissolve the surrounding rocks [19].

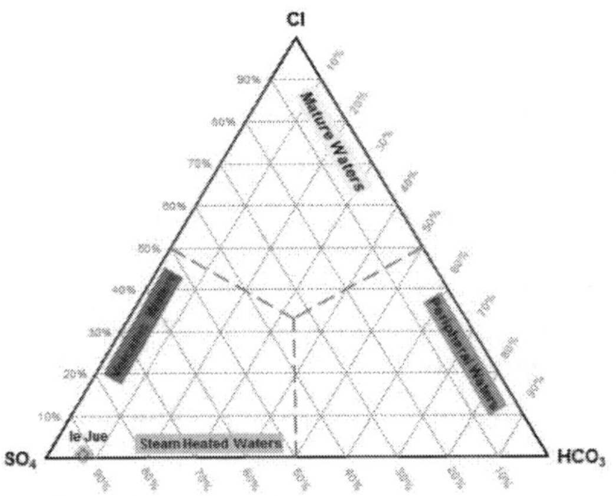

Figure 3. Cl-SO_4-HCO_3 ternary diagram of the geothermal water for *Ie Jue* manifestation.

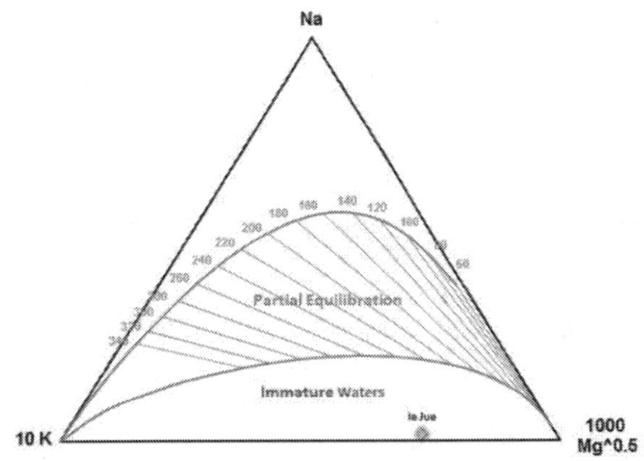

Figure 4. Na-K-Mg ternary diagram of the geothermal water for *Ie Jue* manifestation.

The Na-K-Mg triangle diagram is an indicator to explain the fluid equilibrium at a depth of the reservoir. The result of the measurement of cation concentration from a geothermal water sample of *Ie Jue* manifestation is exhibited by Table 3.

Table 3. The chemical concentration (cations) of a water sample for *Ie Jue* manifestation.

No	Cations	Concentrations (mg/L) ± S_c
1	Sodium	54.82 ± 0.26
2	Potassium	111.04 ± 0.11
3	Calcium	9.33 ± 0.11
4	Magnesium	8.62 ± 0.01

^{Sc} standard deviation of concentration

The result of plot of Na-K-Mg triangle diagram based on cation content can be seen in Figure 4. The manifestation of *Ie Jue* is in immature water state. This condition indicated that the influence of surface water mixed with the fluid during the formation of hot springs. In addition, hot water is also affected by the interaction between the water and the rocks in the heat condition [20].

Geothermometer is a form of equation used to estimate the subsurface temperature of the earth (reservoir). Chemical equilibrium to temperature is strongly influenced by the concentration of cations used as temperature prediction of a reservoir [19]. So far geothermometer equations have been proposed by several different authors. The result of calculation from some geothermometer to the geothermal manifestation of *Ie Jue* can be seen in Table 4.

Although it had been calculated by many geothermometer equations, all the results gave the same result where the temperature is above 300 °C. Among those equations, only Fournier (1979) and Giggenbach (1988) geothermometers are best used for high-temperature reservoirs [19]. In this study, the estimated temperature of the reservoir depth of *Ie Jue* manifestation is 690-761 °C using the geothermometer equation Na/K Giggenbach (1988) and Fournier (1979).

Table 4. Calculated reservoir temperature with geothermometers equation of the geothermal water for *Ie Jue* manifestation.

No	Geothermometers	Temperatures Estimation (°C)
1	Na/K Giggenbach, 1988 [12]	690
2	Na/K Fournier, 1979 [13]	761
3	Na/K Truesdell, 1976 [14]	1280
4	Na/K Tonani, 1980 [15]	1592
5	Na/K Arnorsson, 1983 [16]	1086

Based on reservoir depth temperatures obtained through geothermometer calculations, it is known that *Ie Jue* manifestations including high-temperature geothermal system (high enthalpy) are characterized by temperature gain > 225 °C, thus suitable for power plant development [21]. Based on this survey, it is estimated that the geothermometer temperature of *Ie Jue* manifestation ranged from 690-761 °C. The accuracy of this method of hot water geothermometer needs to be verified further with the gas geothermometer because the area of the *Ie Jue* manifestation also found a steam source of fumarole.

4. Conclusion

Geothermal fluid in *Ie Jue* manifestation has sulphate spring type based on triangle Cl-SO_4-HCO_3 and acidic diagram (pH = 5.77 ± 0.04). Manifestation of the *Ie Jue* is in immature water state based on Na-K-Mg triangle diagram, this condition indicates that the influence of surface water mixed with fluid at the time of hot spring formation. Estimated temperature of the reservoir depth of *Ie Jue* manifestation is 690-761°C. Based on geothermometer calculations, *Ie Jue* manifestations can be identified as high-temperature geothermal (high enthalpy) systems, making them suitable for power plant development.

5. References

[1] Mansoer W R and A Idral 2015, *Proceedings World Geothermal Co,gress 2015 Melbourne, Australia*, 19-25 April 2015
[2] Hochstein M P and Sudarman S 2008 *Geothermics* vol. 37 no. 3 pp. 220–266
[3] Bertani R *Proc. World Geotherm. Congr.* 19-25 April 2015 no. 19–25 April 2015 pp. 1–19
[4] Ismail N and Ramadhan S 2013 *Prosiding Semirata* FMIPA Universitas Lampung
[5] Joseph E P, Fournier N, Lindsay J M, Robertson R and Beckles D M 2013 *J. Volcanol. Geotherm. Res.* vol. 254, pp. 23–36
[6] Liu M. *et all* 2017 *Procedia Earth and Planetary Science* 17, 157–160
[7] Paternoster M, Parisi S and Caracausi A 2010 *Geochemical Journal*; 44(2): 125-135
[8] Wynn J G, Sumrall J B and Onac B P 2010 *Chemical Geology*; 271(1-2): 31-43

[9] Serrano R G M, Jacquier B and Arnold M 1996 *Journal of Volcanology and Geothermal Research*; 73(1): 99-118
[10] Bottrell S, Tellam J and Bartlett R 2008 *Applied Geochemistry*; 23(8): 2382-2394
[11] Idroes R, Yusuf M, Alatas M, Subhan, Lala A, Saiful, Suhendra R, Idroes G M and Marwan 2018 *IOP Conf. Ser.: Mater. Sci. Eng.* 334 012002
[12] Giggenbach W F 1988 *Geochim. Cosmochim. Acta*, 52, 2749-2765
[13] Fournier R O 1979 *Geoth. Res. Council Trans.* 3, 221-224
[14] Truesdell A H and Fournier R O 1976 *Proceedings 2nd UN Symposium on the development and use of geothermal resources, San Francisco*, 1975, 1, 837-844
[15] Tonani F 1980 *In: Proc. Adv. Eur. Geoth. Res., Second symposium, Strasboug*, 428-443
[16] Arnorsson S 1983 *Geothermics*, 12, 119-128
[17] Powell T and Cumming W 2010. *Geotherm. Reserv. Eng.*, vol. February, no. 1–3, p. 10
[18] Skoog D A, Holler F J and Crouch S R 2007 *Principles of Instrumental Analysis, Sixth Edition, (Brooks Cole)* vol. 6, 12-19
[19] Nicholson K 1993 *Geothermal Fluids Chemistry and Exploration Technique.* (Springer-Verlag, Inc.: Berlin Heidelberg New York)
[20] Giggenbach W F 1991 *Chemical techniques in geothermal exploration in UNITAR/UNDP Guidebook: Application of geochemistry in resources development*, 119-144
[21] Hochstein M P 1990 *Classification and Assessment of Geothermal Resources In: Dikson MH and Fanelli M (eds) Small geothermal resources,* (NITAEWNDP Centre for Small Energy Resources, Roma-Italy), 31-59

IOP Conf. Series: Materials Science and Engineering **523** (2019) 012013 doi:10.1088/1757-899X/523/1/012013

Isolation and identification of Endophytic Fungus *Fusarium sp* from Agarwood (*Aquilaria sp*) population originated from the forest of Aceh Tamiang district, Indonesia

Zulfendi[1,2], Rinaldi Idroes[3,4], Khairan[3,*]

[1] Graduate School of Mathematics and Applied Sciences, Syiah Kuala University, Kopelma Darussalam, Banda Aceh 23111, Indonesia
[2]Human Resources Development Agency, Banda Aceh, Aceh Province
[3]Department of Chemistry, Faculty of Mathematics and Natural Sciences, Syiah Kuala University, Kopelma Darussalam, Banda Aceh 23111, Indonesia
[4] Department of Pharmacy, Faculty of Mathematics and Natural Sciences, Syiah Kuala University, Kopelma Darussalam, Banda Aceh 23111, Indonesia

E-mail: khairankhairan@unsyiah.ac.id

Abstract. *Aquilaria sp* is one of higher plant types that can produce agarwood. Agarwood is commonly stimulated by the biological response of endophytic fungal infection of *Fusarium sp*. This research aims to isolate and identify endophytic fungus *Fusarium sp* from *Aquilaria sp* populations originated from the forest of Aceh Tamiang District. The pure isolates of Fusarium *sp* derived from *Aquilaria sp* that reinfected on *Aquilaria sp* populations to stimulate agarwood formation. The endophytic fungal isolates of *Fusarium sp* were growth on a selective media of Sabouraud Dextrose Agar (SDA) containing antibiotics *Streptomycin* 10 µg/dl and *Chloramphenicol* 25µg/dl. Ten percent of infected *Aquilaria sp* from the population was taken aseptically. Isolation and identification of endophytic fungus Fusarium *sp* were conducted from June to August 2016. The results showed that within 10 infected *Aquilaria sp*, several microbe species were found such as three *Fusarium sp*, six *Aspergillus sp*, two *Rhizopus sp*, two *Penicillium sp*, one *Bacillus sp*, one *Actinomyces sp* and one *Streptomyces sp* species. The result of identification towards three species of endophytic fungus *Fusarium sp* that has been purified were srongly suspected to be *Fusarium solani, Fusarium oxysporum,* and *Fusarium monyliformae.*

1. Introduction

Aquilaria sp is one of the *Thymelaeaceae* family plant. This plant is also known as aloe plant because its ability to stimulate the formation of resin or aloe resin (agarwood). *Aquilaria* has several species, but species that grow a lot in Sumatra, Borneo, and Malaysia are *Aquilaria microcarpa, Aquilaria ardesiacus, Aquilaria hirta, Aquilaria beccariana* and *Aquilaria ardesiacus* [1]. The province of Aceh, particularly in Aceh Tamiang Regency, has a large biodiversity of *Aquilaria sp*. This plant has high non-wood forest economic value. Up to this point, this potential did not get much attention from both

*Corresponding author : khairankhairan@unsyiah.ac.id

educational and government institutions because of the lack of knowledge and methods of aloe cultivation, especially in Aceh Tamiang district.

Fusarium sp is one of stimulant (probiotic) for agarwood formation on *Aguilaria spp* stem [2]. several injection and infusion methods of *Fusarium sp* has been commercialized to stimulate agarwood formation on *Aquilaria sp plant*, but this method has not showed satisfying results. It may be caused by several factors like biological factor. For that reason, it is important to conduct *in situ* endophytic fungal isolation of *Fusarium sp* from *Aquilaria sp* stem from the forest area population for reinfection using *Fusarium sp*. Reinfection is essential to stimulate the in situ formation of endophytic fungus *Fusarium sp* so that the secondary metabolites produced by endophytic fungus is effective against the pathogen bacteria, thus the quality of the agarwood is improved. Some endophytic fungal species which had been commercialized as probiotics for reinoculum in *Aquilaria sp* cultivation *are Torula sp., Cladosporium sp., Aspergillus tamarii, Pencillium citrinum, Botryodiplodia theobromae, Cytosphaera mangiferae, Epicoccum granulatum, Fusarium cylindriscorpu, Fusarium oxysporum, Fusarium solani* and *Chaetonium globosum.*

Agarwood can be found in *xylem* stem of *Aguilaria sp* (Figure 1), this is formed by biological response to *Fusarium sp* either naturally or artificially with endophytic fungal injection method. Agarwood is an export commodity which has a high economic value as a raw material for perfume and pharmaceutical industries or medicines. Agarwood produces fragrant aroma when burned because it contains *agarospiral* and *jinkohol-eramol.*

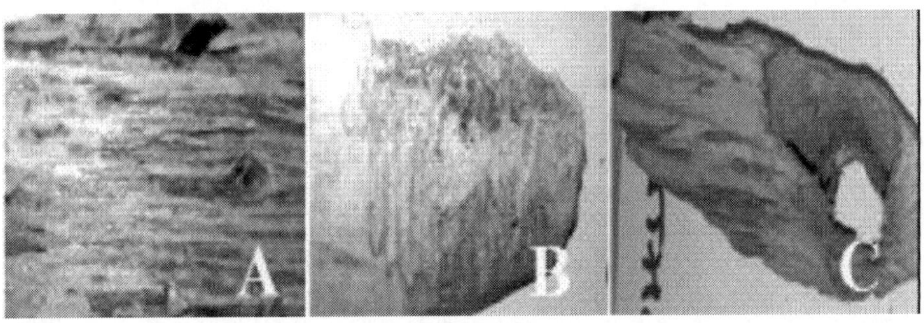

Figure 1. A; resin or aloes resin (Agarwood) in xylem stem of *Aguilaria sp* occured naturally, B; resin or aloes resin (Agarwood) formed by the endophytic fungal injection method, C; resin or aloes resin (Agarwood) of Kemedang type.

Agarwood that formed, is a product from fungal infection process that is opportunistic on xylem infected *Aguilaria sp.* It suspected that the formed resin is the resistance of phytoalexins in cambium producing aromatic oils. For a long time, the resin will be formed in endophytic infected xylem. The resin generally is light brown and later turns black. Besides containing *agarospiral* and *jinkohol-eramol,* the resin also contains other chemical compounds such as *ethyl methoxyphenyl chromate* (27%) and *phenylethyl chromate* (15 %).

Endophytic fungus on agarwood is opportunistic and symbiotic microorganisms that live in the plant cell and stimulate secondary metabolite formation helping plant as a self defense against the predator or other pathogen bacteria. Endophytic microorganisms can be either bacteria or fungus in *xylem* or *floem tissue*. Sometimes their presence does not cause a disturbance on plants[3]. Endophytic fungus *Fusarium sp* is the most predominant fungi found on the *Aquilaria malaccencis Lamk.,* stem. Agarwood is generally formed on the top, middle and bottom parts of the stem [4]. Some studies indicate that the endophytic fungus *Fusarium sp* has an important role as a stimulant (probiotics) for *Aquilaria sp* in the agarwood formation from response to fungal infection (endophyte) in xylem tissue. This research aims to map the plant population of *Aquilaria sp.* in Aceh, particularly in Aceh Tamiang Regency, followed by isolation and identification of endophytic fungus *Fusarium sp.* This study also aims to find out the type of endophytic fungus that presents in *Aquilaria sp.* There are four selected

IOP Conf. Series: Materials Science and Engineering **523** (2019) 012013 doi:10.1088/1757-899X/523/1/012013

populations of *Aquilaria sp* such as a population area of forest Sekrak, Seruwai, Kampung Pipa and Selele, Aceh Tamiang regency. In this research, isolation and identification methods of endophytic fungus *Fusarium sp* were performed on selective media *Sabouraud Dekstrosa Agar* (SDA) containing *Streptomycin* [10µg/dl] and *Chloramphenicol* [25µg/dl]. To prevent the occurrence of contamination and to ensure the purity of endophytic fungal isolates of *Fusarium sp*, macroscopic and microscopic inspection were conducted to see the fungus morphology.

2. Experimental

2.1. Materials

Saboraud Dekstrosa Agar (SDA) media, *Nutrient Agar* (NA) media (Oxoid), *chloramphenicol, Nystatin,* aquadest, and sodium hypochlorite 5.3%. The materials and Tools Used for sterilization are alcohol 70%, sodium hiphoklorit (klorox) 0.5%, the sterile water and sterile wipes to dry samples. The media used is MA 2% (Malt Agar) containing 20 g of extract malt, 20 g of agar, 1000 mL of tap water and 100 mg of Streptomycin to suppress the growth of bacteria. The equipment used were a test tube, , measuring cup, thermometer, Erlenmeyer (size 1000 ml, 500 ml and 250 ml), beaker glass, aluminium foil, laminar air flow cabinet, tea strainer, tweezer, scissors, Petri dish, bag, cork borer, Oose needle and measuring cup [5]. All glassware is sterilized with the dry heat sterilization techniques using oven at 170 °C for 40 minutes.

2.2 Methods

2.2.1 Sampling of Endophytic Fungal *Fusarium sp* on *Aquilaria sp.* Population. The survey of *Aquilaria sp* population was done in Aceh Tamiang forest area starting at early June 2017 until June 2018. The *Aquilaria sp* population selected on four locations of forest area which is the natural habitat of *Aquilaria sp* i.e. Sekrak, Seuwai, and Selele and Kampung Pipa. Sampling of endophytic fungus *Fusarium sp* is done by observing the wounded limb, stem or branch that have colour range from brown to black, which are allegedly contained resin. The wounded bark was cleaned, slashed with knives, and tested the aroma with burned test. For the endophytic fungal isolation of *Fusarium sp* on plant parts containing resin was done with sterile sculpture of xylem 10 times. Samples are put into adhesive plastic aseptically and taken to the laboratory to do the isolation and identification of fungal species.

2.2.2 Preparation of Media Saboraud Dekstrosa Agar (SDA). 6.5 g SDA powder was dissolved in 975 mL aquadest in an Erlenmeyer, the flask was then heated until SDA dissolves homogeneously. The solution was sterilized by autoclave at 121 °C for 15 minutes. After that, the solution was cooled at room temperature until the temperature dropped to 40-50 °C. 25 mL of the SDA media solutions were taken and put in the erlenmeyer 100 mL followed by added 100 mg antibiotic of Streptomycin and then was dissolved homogeneously (Solution A). 25 mL SDA media solution was taken and put in other erlenmeyer followed by the addition of 250 mg antibiotic of chloramphenicol and then was dissolved homogeneously (Solution B). Solution A and Solution B were mixed homogeneously until the temperature dropped to 40 °C. The mixture was poured into petridish (9 mm diameter), each of petridish was 18-20 mL. The mixture was cooled until it was solidified. This SDA media was used to isolate endophyte fungung *Fusarium sp*. For isolate culture rejuvenation media of *Fusarium sp*, 5-8 mL SDA solutions were taken and put into sterile tubes and then was placed on the 30-45 °slope position until it was solidified [6].

2.2.3 Preparation of Media Nutrient Agar (NA). 28 g of NA powders were dissolved with 1000 mL sterile aquadest in an erlenmeyer 1 L. The solution was heated and dissolved homogeneously then sterilized by autoclave at 121 °C for 15 minutes. The NA solution was cooled at room temperature until the temperature dropped to 40-60 °C. 20 mL NA solution were introduced into an erlenmeyer

100 mL followed by the addition of 100 mg antibiotic of Nystatin, dissolved homogeneously and cooled until the temperature dropped to 40 °C. The solution was poured into petridish (9 mm diameter), each of petridish was 18-20 mL. The solution was cooled until it was solidified. This media was a selective media for bacterial growth [6].

2.2.4. Inoculation of endophytic fungal isolate. Inoculation was done by cutting the wood fiber containing agarwood into 0.5-1.0 cm long bars. The wood fiber was sterilized using sodium hypochlorite 5.3% for 1 minute followed by several times rinsing with sterile aquadest and alcohol 70% for 15 seconds, respectively. The wood fiber was drained above sterile filter paper for several minutes at 30 °C in an incubator. In aseptic circumstances, the pieces of wood fiber were placed above the petridish containing SDA media and antibiotic. The petridish was incubated at room temperature in the septic cabinet for 48-72 hours. Fungus that growth was identification both in macroscopic and microscopic. The identification of edophytic fungal isolate was performed using NA contained media with the same previous method.

3. Results

Aquilaria sp population selected on four different locations in the forest that were the natural habitat of *Aquilaria sp* namely. *Sekrak*, *Seuwai*, *Selele* and *KampungPipa* can be observed in Table 1.

Table 1. The endophyte fungal sampling location on some *Aquilaria sp* populations on four different locations in Aceh Tamiang Regency.

No	Sampling Location	Number stem samples of *Aquilaria sp*	number endophytic fungal samples
1.	*Sekrak*	3	3
2.	*Selele*	2	2
3.	*Kampung Pipa*	2	2
4.	*Seruwai*	3	3
	Total	10	10

Sampling of wood fiber containing agarwood for isolation of the endophytic fungus in ten different sampling locations can be observed in Figure 2.

Figure 2. The sampling of endophytic fungus *Aquilaria sp*. a; samping of endophytic fungus with three month incubation commercial inoculation, b; sampling of endophytic fungus from xylem tissue of *aquilaria sp*., c; sampling of endophytic fungus from natural infected fungus

Table 2. Endophytic fungal and bacteria isolates isolated from ten stems of *Aquilaria sp* from four different forest in *AcehTamiang*.

No.	Sample Code	endophytic fungal isolates	endophytic bacteria isolates
1	01	*Aspergillus sp* and *Fusarium oxysporum,*	Gram positive rods branch (*Actinomices/ Streptomyces*)
2	02	*Fusarium oxysporum, Penicilium sp.* and *Rhizopus sp.,*	Gram positive rods branch (*Actinomices/ Streptomyces*), dan *Bacillus* sp
3	03	*Asfergillus sp., Curvularia sp., Fusarium solani,* and *Fusarium oxysporum,*	Gram positive stem bark (*Actinomices/ Streptomyces*), and *Bacillus* sp
4	04	*Aspergillus sp.,* and *Fusarium moniliformae*	Gram positive stem bark (*Actinomices/ Streptomyces*), and *Bacillus* sp
5	05	*Fusarium solani, Fusarium oxysporum,* and *Trichoderma sp. (kapang)*	Gram positive stem bark ((*Actinomices/ Streptomyces*), and *Bacillus* sp
6	06	*Fusarium oxysporum,* and *Penicillium sp.,*	Gram positive stem bark *(Actinomices/ Streptomyces)*
7	07	*Aspergillus sp.,* and *Fusarium solani,*	Gram positive stem bark (*Actinomices/ Streptomyces*), and *Bacillus* sp
8	08	*Fusarium oxysporum, Trichoderma sp., (kapang),* and *Fusarium solani,,*	Gram positive stem bark (*Actinomices/ Streptomyces*),
9	09	*Aspergillus sp., Rhizopus nigricans, Trematostroma sp,. (kapang),* and *Trichoderma (kapang.*	Gram positive stem bark (*Actinomices/ Streptomyces*)
10	10	*Aspergillus sp., Fusarium moniliformae, Fusarium oxysporum,* and *Trichoderma (kapang)*	Gram positive stem bark ((*Actinomices/Streptomyces*), and *Bacillus sp*

Sample of wood fiber was put into sterile adhesive plastic aseptically (can be seen in Figure 3) and taken to laboratory for isolatation and indentification of fungal and bacterial species that produced from agarwood. *Lacto phenol cotton blue (LPCB)* was used to determine the morphology of endophytic fungus and bacteria such as conidiospore, conidia, hifa and septa that can be seen in Figure 4.

The observation of conidium structures of fungus that grows on the media was conducted under the microscope with 400x magnification. Based on the surface structure of the colony, texture, color, aerial hyphae vegetative, and the morphology of pseudohypha, septahypha,conidia form (microconidia and macroconidia), conidiospore and the mycellium of endopyhtic fungus *Fusarium sp*, we concluded that the endopyhtic fungal isolates of *Fusarium sp* were *Fusarium solani, Fusarium oxyforum* and *Fusarium monyliformae* (Figure 4).

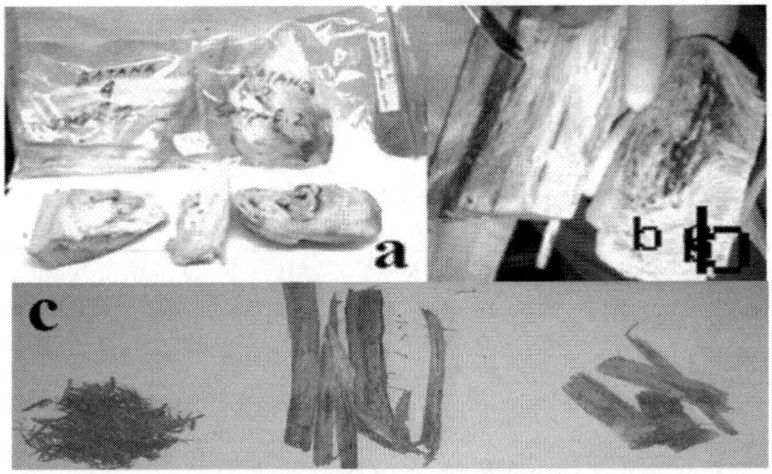

Figure 3. The wood fibre containing agarwood from sampling techniques, a; sample of wood fibre containing agarwood in sterile adhesive plastic, b; endophytic fungus from xylem, c; endophytic fungus from wood fibre of *Aquilaria sp*.

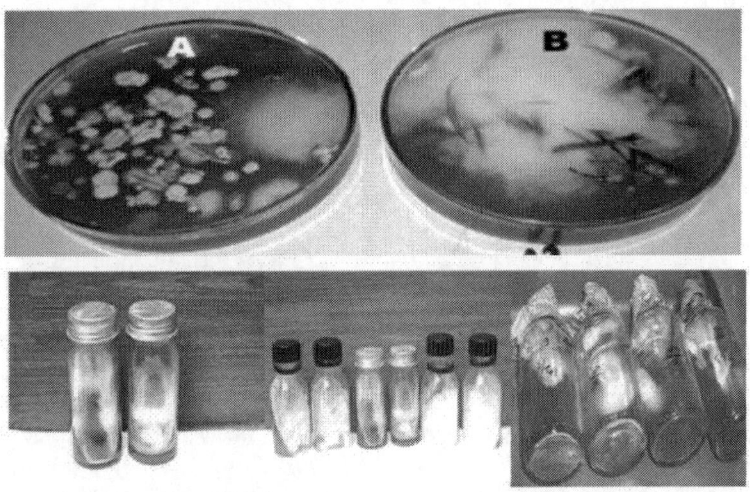

Figure 4. The endopyhtic fungal isolates of *Fusarium solani, Fusarium oxyforum* and *Fusarium monyliformae* grown on SDA media, A; *Fusarium oxyforum* grown on SDA media, incubation at room temperature, day 5th incubation, B; *Fusarium oxyforum* grown on SDA media, incubation at room temperature, on day 7th of incubation.

To determine the type of bacteria, Gram-positive or Gram-negative, from endophytic bacterial isolates of agarwood, resin staining test was conducted using a solution of Gram Staining microscopically at 1000x magnifications followed by biochemistry fermentation test. This observation showed that the bacterial isolates commonly in basil form. The bacterial are Gram-positive (+) if the color of the bacterial cells change from violet to purple and Gram-negatif (-) if the cells were red in color.

Research results of *Aquilaria sp* population from the four different locations in Aceh Tamiang with macroscopic, microscopic and biochemistry test, we found thirteen (13) endophytic fungal isolates and three (3) endophytic bacteria isolates. The thirteen endophytic fungal isolates consisted of three (3) *Fusarium sp*, six (6) *Aspergillus sp*, two (2) *Rhizopus sp* and two (2) 1s *Penicillium sp* species. The three (3) species of endophytic bacterial isolates was Gram-positive bacteria, namely *Bacillus sp, Actinomyces sp and Streptomyces sp.* Results identification of endophytic fungal and endophytic bacterial isolates can be seen in Table 2. The results of the identification of three (3) endophytic fungal isolates of *Fusarium sp*, we found three species namely *Fusarium solani*, the *Fusarium oxyforum* and *Fusarium monyliformae*.

4. Discussion

The population of *Aquilaria sp* from four isolation locations in Aceh Tamiang Regency mostly found in Sekrak and Seuwai areas reached 0.6-1 stem per hectare while the endophytic sample containing agarwood mostly found in Kampong Pipa behind the palm plantation border of PTP Aceh Tamiang. This area can be used as a location of agarwood agro-industry reinfection research. In addition, the forest is still natural and the local people ever get natural agarwood.

Fusarium sp generally breed with conidiophore which produces macroconidia and microconidia on each hypha area of mycelium, conidia is a black- that are proliferated spores antigen for tissue *fitoaleksin* compounds in cambium of *Aquilaria sp*. Microconidia is considered as foreign matter by plants. The *fitoaleksin* compound can be a resin mixed with black conidia, so the resin can be brown and fragrant. The resin will be accumulated on the *xylem* and *floem*. This vessels prevent the wound to spread to other tissue.

From the three species found from the isolation, only *Fusarium solani* and *Fusarium oxyforum* that were reported produced the natural agarwood. Both of this *Fusarium sp* fungal spesies allegedly act in the agarwood formation or in trials of infected birds on the stem of *Aquilaria sp*. This is an opportunity to develop agarwood agro-industry reinfected in Aceh Tamiang. Beside on infecting, *Fusarium oxysporum* caused withered decease on potato and banana [7]. Both of this fungus will produce fermentation of secondary products [8]. The secondary product will be further analyzed in order to determine the content of metabolites contained in both species of *Fusarium sp* [9].

5. Conclusion

The population of *Aquilaria sp* in Aceh Tamiang mostly found in Sekrak, Seuruwai, Selele, and Kampung Pipa. In this study, we found that endophytic fungus containing agarwood producing an aromatic fragrant are commonly derived from Selele and Kampung Pipa. Based on the identification towards the endophytic fungus and bacteria that isolated from the infected *Aquilaria sp*, we have found thirteen (13) endophytic fungus and bacteria isolates. In this study, we concluded that the endophytic fungal isolates of *Fusarium sp* are *Fusarium solani, Fusarium oxyforum* and *Fusarium monyliformae*. These two spesies of *Fusarium solani and Fusarium oxyforum* are expected to be responsible in resin and secondary metabolite products formation in *Aquilaria sp*. Both of these possible *Fusarium sp* will produce secondary fermentation products [8], secondary product will be further analyzed to determine the content of metabolites in both species of *Fusarium sp* [9].

6. References

[1] Wayne V 2013 *Isolasi And Caharateristization Of Indophytes Isolat From Akar Gaharu* (Department Of Molecular Biotecnology Faculty Of Resource Science And Tecnology University Malasia Serawak, Malaysia)
[2] Rangasamy O, Raoelison G, Fransisco E, Rakoloiriana K, Chenk, Suzane U R, Jullie O L, Ammenah G F, Anwar H S 2007 *Journal of Etnopharmacology* 1092 **331-337**
[3] Hallman J, Quadt-Hallman A, Mahafee WF, Kloepper J W 1997 *J Microbiol* 43**895-914**
[4] Lisdayani 2014 *Reisolasi and identification of Fungi on plants Agarwood (Aquilaria Malaccencis Lamk.) that had been Inoculated* (Faculty of Agriculture University of North Sumatra Medan)
[5] Sunariasih N P, Suada I K, Sutini N W, 2014 *E-Jurna lAgroekoteknologi Tropika* 3 **2**
[6] WHO 2008 *Clinical and Laboratory Standards Institute (CLSI). Document M31-A3. Performance Standards for Antimicrobial Disk and Dilution Susceptibility Tests for*

Bacteria Isolated from Animals Approved Standard, Third Edition (CLSI Wayne Pennsylvania)
[7] Suryanti I A PRamona Y Meitini W *Plants Jurnal Biologi* 28 2**37-41**
[8] Jones W P and Kinghorn A D 2012 *Methods Mol. Biol.* 20 **323–351**
[9] Margino S 2008 *Indonesia Pharmaceutical Magazine Indonesia* 19(2)**86–94**

Isolation of Pectin from coffee pulp Arabica Gayo for the development of matrices membrane

U Hasanah[1,2], M Setyowati [3], Edwarsyah[2], R Efendi[4], E Safitri[4,*], R Idroes [4,5], L Y Heng[6,7], N D Sani[8]

[1] Graduate School of Mathematics and Applied Sciences, Syiah Kuala University, Kopelma Darussalam, Banda Aceh 23111, Indonesia
[2] Department of Fishery, Faculty of Fishery and Ocean Science, Teuku Umar University, 23615 West Aceh, Indonesia
[3] Department of Agronomy, Faculty of Agriculture, Teuku Umar University, 23615 West Aceh, Indonesia
[4] Department of Chemistry, Faculty of Mathematics and Natural Sciences, University of Syiah Kuala, Banda Aceh 23111, Indonesia
[5] Department of Pharmacy, Faculty of Mathematics and Natural Sciences, University of Syiah Kuala, Banda Aceh 23111, Indonesia
[6] School of Chemical Sciences and Food Technology, Faculty of Science and Technology, Universiti Kebangsaan Malaysia, 43600 UKM Bangi, Selangor Darul Ehsan, Malaysia.
[7] Southeast Asia Disaster Prevention Research Initiative (SEADPRI-UKM), LESTARI, Universiti Kebangsaan Malaysia, 43600 UKM Bangi, Selangor Darul Ehsan, Malaysia.
[8] Sanichem resources Sdn. Bhd. No 7 & 7A Jalan Timur 6/1A Mercato Enstek, Bandar Estek 71060, Negeri Sembilan, Malaysia.

E-mail: e.safitri@unsyiah.ac.id

Abstract. Pectin from Arabica Gayo coffee pulp has been successfully isolated using citric acid as a solvent. The pectin extracted from the coffee pulp is determined by the extraction conditions. This study aims to produce pectin from dried coffee extract using citric acid using two factors, namely the difference factor of citric acid (1: 5 and 1: 20 b/v) and extraction time 75 and 150 minutes. The optimum ratio of Arabica Gayo coffee pulp with citric acid was 1:20 b/v with rendemen 7,8% with 125 minute extraction time and at the temperature of 80 °C and pH 4. The methoxyl content of isolated pectin was 12.71% and has been determined as high methoxyl pectin (HMP). The result of analysis with FTIR is known that the extract produced is pectin with comparison with pectin standard. The pectin has been used as a matrix membrane and showed its smooth surface gel form and smaller diameter from scanning electron microscopy (SEM).[1]

*Corresponding author : e.safitri@unsyiah.ac.id

IOP Conf. Series: Materials Science and Engineering **523** (2019) 012014 doi:10.1088/1757-899X/523/1/012014

1. Introduction

Coffee is one of the most important food commodities in the world and ranks second after crude oil. These plants can usually grow in tropical and subtropical regions [1]. A Number of countries are known as coffee producers including Indonesia which is the fourth largest global coffee producer with an annual production of 686,763 tons (in 2007). Arabica and Robusta coffee (Coffee canephora) are the most famous coffee produced in Indonesia. both coffee possess distincttaste and caffeine content and are cultivated in different areas. Arabica coffee, the most popular coffee are usually grown in the highlands. This coffee is well known for its low caffeine content and favourable non acidic taste In Indonesia, this coffee is known as Gayo Arabica Coffee and are grown central Aceh. Similarly, Arabica coffee is also grown in Central America like Mexico, Guatemala, Costa Rica, Dominica, Jamaica.

It is common knowledge that the seeds from the coffee plant are processed into coffee beans which can be consumed directly afterwards. During this process, 55% to 60% useless material are produced from the pulp (part mesocarp), skin (eksokarp part), mucilage and parchment (endocarp part) [2]. Commonly, it is used for animal feeding or starting material for organic fertilizer. Carbohydrate forms the largest composition of coffee pulp which consists of pectin (35%), protein (5.2%), fiber (30.8%) and minerals (10.7%) while the mucilage portion contains water (84.2%), protein (8.9%), sugar (4.1%) and ash (0.7%) [2]. Therefore, based on these composition, coffee pulp becomes a potential material for pectin recourse.

Pectin is a natural anionic polysaccharide which is located in plant cell walls and are highly biodegradable [5]. It is a high value functional food which is widely used in gel and polymer formation [6]. With the availability of carboxylic groups, it was predicted that pectin can be utilized as matrices for biomolecule attachment. Pectin that is commonly found in the industry is a type of high methoxyl pectin (HMP). The HMP which minimally contains 7% of methoxyl and low methoxyl pectin (LMP) which maximally contains 7%, of methoxyl. It has The ability to form gel with sugar and acid. otherwise it will be able to form gel in the presence of calcium ions [3]. The formation of gel from pectin with high degree of methylation is also affected by pectin concentration, percentage of sugar and pH. The greater the concentration of pectin, the easy the gel will be formed. The formation of high methoxyl pectin gel takes place through hydrogen bonds between free carboxyl groups and hydroxyl groups [4].

This work is aimed to isolate pectin from Gayo Arabica coffee using organic acid solvent (citric acid). onto the best of our knowledge, there is no previous report about the use of citric acid as a solvent to extract pectin. citric acid is a more environmentally friendly option than strong acid [7] It is a weak acid that has lower dissociation than strong acid. In addition, the use of weak acid can prevent depolymerization of protons from pectin [8]. In order to conform the formation of pectin, the isolated pectin will be characterized using Fourier Transform Infra Red Spectroscopy 400FT-IR/FT-NIR while the morphology gel matrix membrane will be evaluated using scanning electron microscopy

2. Materials and Methods

2.1 Material and Instrument

The main material used in this work was Gayo Arabica coffee pulp obtained from smallholder farmers in Takengon Central of Aceh, Indonesia and was dried in a green house for one month. 96% Ethanol was obtained from Emsure and citric acid, phenopthaline were purchased from Fisons. Sodium hydroxide, hydrochloric acid (HCl) were obtained from Merck. The isolated pectin was characaterized using 400 FT-IR/FT-NIR Spectrophotometer and Scanning Electron Microscopy (SEM) Fesem Merlin.

2.2 Methodology

Gayo Arabica Coffee was obtained from Takengon, Central Aceh of Indonesia. The first layer of coffee seed was then dried for one month in a green house. Dried coffee peel was crushed using a blender followed by pectin extraction using various citric acid concentration (w / v) (1: 5, 1:10, 1:15, 1:20) In order to find the optimum yield of pectin, different length of extractions (75, 100, 125, 150 minutes) were investigated at temperature 85° C [9].

The extracts were filtered and cooled. Next, these filtrates were precipitated using 96% ethanol. The precipitate was washed with 30 mL of 96% ethanol for three times through filtration process and dried at 600C. The yield of extracted pectin was determined. In addition, the degree of methoxylation (DM) was calculated using acid base titration method and specific functional groups of pectin were evaluated based on FTIR spectra.

3. Result and Discussion

3.1 Effect of Pulp and Solvent Ratio on Yield Percentage

The yield of pectin obtained from extraction depends on the type of solvent and method used. Figure 1 presents a profile of rendement percentage on ratio of solvent and time of extraction.

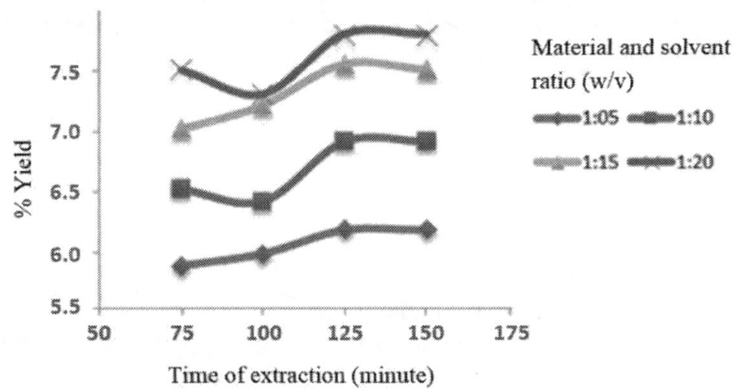

Figure 1. Effect of coffee pulp to solvent ratio at various extraction times on rendemen percentage.

Percentage of pectin rendemen was influenced by ratio of solvent and extraction time. Figure 1 shows that the percentage of rendement has increased after 100 minutes of extraction time and the optimum rendement was reached at 125 minutes extraction time for all pulp to solvent ratio. The highest result, 7.8%, was obtained when the extraction was conducted at ratio 1: 20 (w/V) pulp to solvent. This is due to the greater volume of solvents used. After 150 minutes of extraction time, the rendemen becomes stable because the ability of acid to hydrolize protopectin to pectin has reached its maximum capacity When the coffee pulp to solvent ratio is increased against a longer extraction time a spread distribution of particles and expansion of surface contact was observed.. previous studies using different starting materials such as banana peels, apples, and lemon byproducts reported that pulp to solvent ratio and a longer time extraction time will affect the spread distribution of particles and expand the surface contact of particles and solvents. This will increase the production yield of pectin [10,11]. In this experiment the best ratio obtained was 1:20 after 125 minutes extraction time, with the percentage of rendements becoming stable. Therefore it can be concluded that the optimun extraction time was reached at 125 minutes.

Higher pectin content is caused by the decomposition of protopectin (cell wall) into smaller pectin particles due to partial acid hydrolysis during extraction, increased temperature and time. Pectin consists of α (1-4) related units of galactic acid or methyl ester [12]. Glycosidic bond is one type of ether bond, so it can pass the hydrolysis reaction under the right conditions. In this case, it is assumed that the higher polymer hydrolysis of the pectin molecule to the lower polymer causes increased solubility in water and allows for easier extraction. At low temperatures, it may not be sufficient to allow the hydrolysis of protopectin by acid, this results in lower pectin yields[13].

3.2 Metoxyl Content Analysis
Metoxyl content in pectin is used to classify typesof pectin like low metoxyl or high metoxyl pectin. In this work, the metoxyl content is in the range of 7.75% to 12.71%. The optimum result was found at pulp to solvent ratio of 1:20 and at 125 minutes extraction time. The lowest metoxyl content of 7.75% occurred at the lowest experimental condition of ratio 1:5 with 75 minutes extraction time. Figure 2 shows that metoxyl content was not affected by extraction time but is highly influenced by the pulp to solvent ratio. the metoxyl content also affects the solubility of pectin in water and will determine its application. the isolated pectin in this work was indicated to be high metoxyl pectin. Based on the FOOD Chemical Codex (1996) standard, pectin methoxyl content > 7% is classified as high and <7% is considered as low.

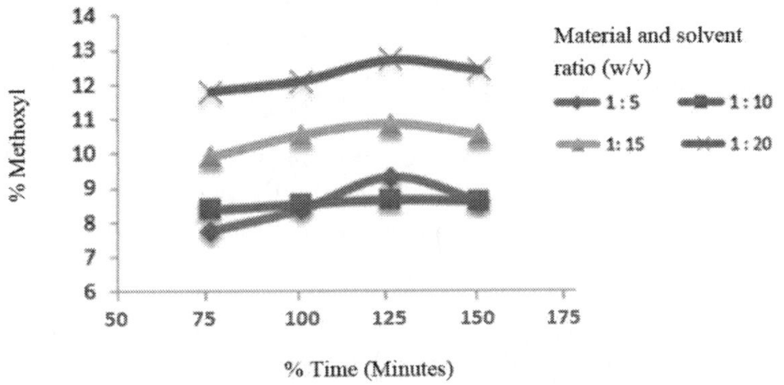

Figure 2. The Effect of coffee pulp to solvent ratio and extraction time on the percentage of methoxyl.

3.3 FTIR Analysis and SEM
Isolated pectin from Gayo Arabica coffee pulp was analyzed using FTIR and the morphology of its surface was characterized by SEM. Figures 3 and 4 depict the characteristics of pectin functional group. The FTIR spectrum in Figure 3 shows the wave mumber range of 950 and 1200 cm^{-1} as the 'finger print' area which identifies polysaccharides as carbohydrates. It has been confirmed by the similarity of finger print pectin standard as reported by Černá et al. 2003[14]. Degree of esterification (DE) of pectin can be determined by the relationship between peak area of the free carboxyl functional group (1650 cm^{-1}) and the ester group (1750 cm^{-1})[15].

IOP Conf. Series: Materials Science and Engineering **523** (2019) 012014 doi:10.1088/1757-899X/523/1/012014

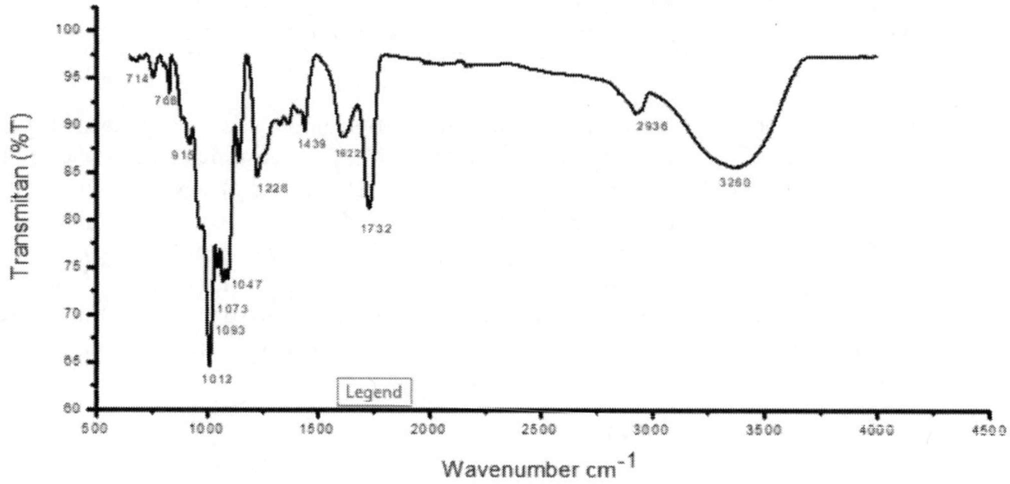

Figure 3. FTIR spectrum of extracted pectin Arabica Gayo coffee.

The morphology of isolated pectin surface was characterized using a Scanning Electron Spectroscopy (SEM). The pectin surface image was taken at 2 μ m magnification. As seen in Figure , pectin in gel form has a smooth area and homogenous surface . This characteristic indicates that pectin can beapplied as a matrix membrane .

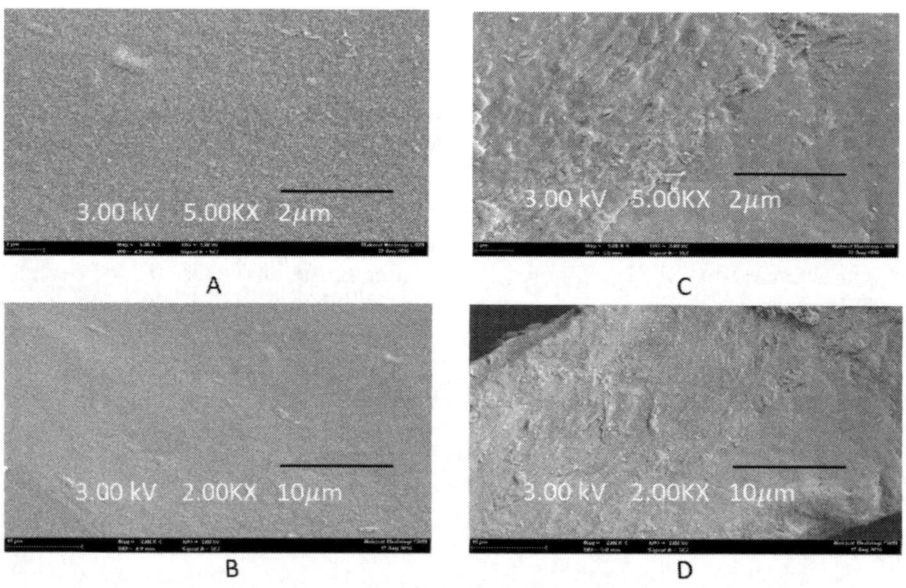

Figure 4. SEM analysis of Pectin: A: 2 μm enlargement gel , B: 2 μm enlargement powder, C: 10 μm magnification gel, D: 10 μm magnification powder.

4. Conclusion

Pectin from Arabica Gayo coffee pulp has been successfully extracted using acid solvent and has been categorized as high metoxyl content. FTIR analysis has confirmed the functional group as pectin. In addition, the isolated pectin can be formed into gel and has homogeneous surface area that appropriate used as matrix membrane.

IOP Conf. Series: Materials Science and Engineering **523** (2019) 012014 doi:10.1088/1757-899X/523/1/012014

5. References

[1] Esquivel P, and Jiménez V M 2012 *Food Research International* **46** 488-495.
[2] Ballesteros L F, Teixeira J A, and Mussatto, S I 2014 *Food and Bioprocess Technology* **7** 3493-3503.
[3] Glicksman 1969 Gum Technology in The Food Industry. Academic Press. NewYork.
[4] Guichard, E S, ssanchou A, Descovieres, and Etievant P 1991 *J. Food Science* **56** :1621
[5] Devasenathipathy, R. et al 2014 *Electrochimica Acta*. **135** 260–269.
[6] Batista, K A. et al 2014 *Journal of Molecular Catalysis B: Enzymatic* **102** pp.25–32.
[7] Chan S Y, and Choo W 2013 *Food Chemistry* **141** 3752–3758.
[8] Kermani Z J, Shpigelman A, Kyomugasho C, Van Buggenhout S, Ramezani M, Van Loey A, and Hendrickx ME 2014 *Food Chemistry* **161** 199–207.
[9] Paulo H F P, Tulio I S O, Morsyleide F R, Fabio L C, Graham K M, Nikolaus W, Keith W W, and Henriette M C A 2016 *International Journal of Biological Macromolecules* **88** 373-379.
[10] Happi E T, Ronkart S N, Robert C, Wathelet B, and Paquot M 2008 *Food Chemistry* **108** 463-471.
[11] Garna H, Mabon N, Robert C, Cornet C, Nott K, Legros H, Wathelet B, and Paquot M 2007 *J. of Food Science* **72** C1-C9.
[12] Chan S Y, and Choo W S 2013 *Food Chemistry* **141** 3752-3758.
[13] Lucia C V, Reinaldo F T, and Carmen L de O P 2012 *Food Science and Technology* **49** 108-116.
[14] Marcela C, Antonio S B, Alexandra N, Silvia M R, Ivone D, Jana C, Manuel A C 2003 *Carbohydrate Polymers* **51**(4):383–89.
[15] Gnanasambandam, Ravin and Proctor A 2000 *Food Chemistry* **68** (3):327–32.

IOP Conf. Series: Materials Science and Engineering **523** (2019) 012015 doi:10.1088/1757-899X/523/1/012015

Development of bioplastic from wheat Janeng starch for food packaging

Saiful[1,*], H Helwati[1], S Saleha[1], T M Iqbalsyah[1]
Chemistry Department, Faculty of Mathematics and Natural Science, Syiah Kuala University Banda Aceh, Indonesia 23111

E-mail: saiful@unsyiah.ac.id

Abstract. Agricultural and industrial products need a high performance of packaging that can preserve its properties in a long period. A high-performance bioplastic was developed in this study from wheat Janeng starch. The bioplastics were prepared by casting on the glass plate and formed by phase inversion method. Bioplastics are made from a mixture of starch and glycerol provides transparent films with high tensile strength and elongation value. The optimum bioplastics compositions were 12 % (w/w) of starch and 5 % (w/w) of glycerol as plasticizer. Glycerol was used as a plasticizer to modify its mechanical properties. The addition of glycerol with the concentration of 5 % (w/w) produced a bioplastic with a high tensile strength of 20.95 kgf/mm^2 and 42.69 % of elongation. The bioplastics are made from janeng starch can be used for packaging apples and tomatoes slice with a high protective barrier.

1. Introduction

The high desire of consumers to obtain high-quality groceries and fresh durable has attracted the interest of researchers to develop a packaging system that is a multi-function[1]. The packaging system is based on the development of bioplastics by using biopolymers are biodegradable such as polysaccharides, proteins, fat, or a combination of these compounds. The biopolymer-based plastics have been widely applied as a protective or barrier for the transfer of fat, moisture, air, and smell for fresh fruits and vegetables, frozen foods, meat, and sweets [2,3].

Bioplastics biopolymers as base material attracted the interest of many researchers and industry because it is environmental friendly and is a substitute packaging material derived from petroleum raw materials such as polystyrene and polyethylene which is not decomposed material [4]. This non-biodegradable plastics are uses in a short time, after which discharged into the environment as waste. Waste from petroleum-based plastic waste is one of the factors that cause environmental pollution. Processing or recycling process generally requires a high cost. Besides, recycled plastic products are lack of thermal and mechanical properties as original polymer plastic.

Therefore, the developments of biodegradable bioplastics are focused by researcher for food packaging materials. One of the organic compounds that can be a source of biodegradable plastic is a polysaccharide [5] which is widely available. Janeng starch is an alternative source for the provision of polysaccharides in which this material is easily made, easily processed, and easily obtained with low production costs of *Dioscorea spp* [6]. In Aceh, *D. hispida* is known as the Janeng. Utilization of

* Corresponding author: saiful@unsyiah.ac.id

carbohydrate from janeng as food is minimal due to toxic compounds containing in janeng bulb such as dioscorine, diosgenine and a precursor cyanide compound causes neurological disorders. However, these all chemical compounds are polar compounds that it is easily removed by washing with water.

Janeng also has advantages compared to other root crops, which can be grown in all types of soil, either latosol, alluvial, and podzolic [7]. Another advantage janeng is a high carbohydrate content, which is about 56-78% of the dry weight of janeng starch [6]. The janeng starch is an interesting source of carbohydrate-based; base on its the physical and chemical properties has the potential to use as bioplastics. In this study, bioplastic-manufacturing processes have been studied from raw material janeng starch. The bioplastic quality is influenced by starch concentration and glycerol as plasticizer. The addition of plasticizers can reduce stiffness and improve the mechanical properties and thermal resistance of the bioplastic. The final bioplastics are expected to be applied for packaging of tomato and apple slice.

2. Materials and Methods

Janeng (*Dioscorea hispida*) samples were collected in Aceh Besar and used for starch production. chemicals are purchased from supplier company used in experiments in pro analysis grade. Tomato and apple are collected from a local market. The equipment used in this study include knives, blender, hot plate, micrometer, analytical balance, magnetic stirrer, a stopwatch, a pH meter, vacuum pump, UV-VIS spectrophotometer, and infrared spectrophotometer.

Isolation of starch janeng

A total of 2 kg janeng tuber peeled and cleaned. Furthermore, tubers cut into small pieces and rinsed with distilled water. Then the tuber pieces are blended at low speed with a sodium bisulfite solution (1.12 g / L) for 5 minutes. The homogenate was rinsed with distilled water and squeezed using a soft porous fabric. Extortion is repeated several times until the juice is clear. Juice left in place for about a day and night is obtained the precipitate. The precipitate is washed with distilled water and filtered with a Buchner filter to remove dioscorine, and HCN contained in the precipitate. The qualitative test is done by using a solution of 0.5 M $AgNO_3$. Furthermore, the white precipitate obtained is dried in an oven of 70 ° C for 24 hours. Finally, the dried white powder is crushed and sifted with a 100 mesh sieve and referred to as wheat janeng flour.

Preparation bioplastics

A total of 5 g starch janeng plus 95 g of distilled water and stir until blended. The mixture was heated on a hotplate at a temperature of 70-75 ^{0}C stirrer until gelatinization conditions are reached (this state is achieved about 10-15 minutes). Then the starch paste is cooled until it reaches 45-50 °C and then added plasticizer 2.5 g of glycerol and stir until homogeneous mixtures. Starch paste is cooled to room temperature and then casting was done on glass plates with a thickness of ± 0.3 mm. The thin layer is drying in the oven at the temperature of 30-35 ^{0}C for 24 hours. Finally, the films are carefully removed for further characterized and application. A variation of janeng starch concentration and glycerol concentration are investigated in the manufacturing of bioplastic. The various janeng starch concentration are 5, 7.5, 10, and 12% (w/w), and various concentrations of glycerol are 0, 2.5, 5, and 7.5% (w/w).

Characterization of bioplastics

Bioplastics obtained were characterized and tested for thickness, tensile strength, elongation, and water absorption and solubility. FTIR analysis was carried out to characterize the functional groups of bioplastics.

Application of bioplastics for fruits and vegetable packaging

Application of bioplastics for fruits packaging was performed by wrapping tomatoes and apple slices with bioplastics during storage for seven days and compared with the fruit without wrapping with bioplastic. Ten selected panelists have observed the color change in the organoleptic test. The change of vitamin C concentration in tomatoes and apples before and after wrapping with bioplastic was determined using the spectrophotometric method. The fruits sample of 10 gram was homogenized in 25 ml of 4% solution $H_2C_2O_4$, and the mixture was filtered by vacuum filter. The obtained filtrate was diluted to 100 ml with a solution of 4% $H_2C_2O_4$. Then 1 ml of $CuSO_4$ solution pH of 6 is added to aliquots of the sample, and the absorbance read at 249 nm wavelength using a UV-Vis Spectrophotometer. A calibration curve was prepared with a standard solution of ascorbic acid in solution $H_2C_2O_4$ 4%.

3. Results and Discussion

Starch Janeng Isolation

D. hispida tubers contain a high dioscorine and HCN levels and should be removed during the isolation process and purification of starch. In previous studies reported that the levels of HCN in *D. hispida* reach 700 mg/kg of tubers [6]. Janeng starches were isolated utilizing depositing slurry of starch. By this way, the starch will be present in the filtrate, while the other non-starch component will be retained by refining filter. The obtained filtrate was deposited, and finally, the precipitate was washed repeatedly by aquadest. The washing step was performed repeatedly to remove all the content dioscorine and HCN. Janeng starch free of HCN is indicated by a negative qualitative reaction using $AgNO_3$ reagent. The obtained starch was reduced particles size by using a blender and sieved with a 100 mesh sieve. *D. hispida* tubers contain 15% (w/w) of dry starch.

Bioplastics preparation and characterization

Preparation of starch-based bioplastics is via phase inversion method. The casting solution is prepared by heating a suspension of starch reaches the gelatinization temperature of about 70-75 °C. The addition of other components is done at cooler temperatures and stirring until a homogeneous mixture. The casting solution is printed on glass plate and bioplastics will be formed after a solidification process by evaporation of the solvent. The quality of bioplastics from starch janeng influenced by the addition of plasticizers. Glycerol has been added in casting solution to improve physical and chemical properties of bioplastic. The use of glycerol as plasticizers was affected the elasticity of the film and also increase in water vapor and oxygen permeability.

Tensile strength is one of the mechanical properties to measure the strength of the film is used as a packaging material. Bioplastic tensile strength values of pure starch obtained ranged between 0 -5 kgf/mm^2. The highest tensile strength values were obtained in pure starch bioplastic with a concentration of 12% (w/w). The addition of plasticizer influences the value of the tensile strength and elongation of the film. The optimum tensile strength of 20.95 kgf/mm^2 is obtained at 5 % (w/w) glycerol concentration as the plasticizer. The elongation increased from 10.83 % to 42.9 % by adding 5% (w/w) glycerol in the casting solution. The characteristics of bioplastics against water absorption and solubility are one of the essential indicators to show the level of loss of bioplastics in contact with water or water vapor. The water absorption of bioplastic was obtained 61.35 %, and the bioplastic solubility was 17.29 %. The amount of water that can be absorbed by bioplastics and the amount of bioplastic dissolved in the water is done by immersion test for 24 hours. The characteristics of bioplastics for water absorption and solubility of bioplastics are one of the important indicators to show the level of loss of bioplastic in contact with water or water vapor. The high amount of water absorbed is presumably because the nature of water and plasticizers are very polar so they tend to absorb more water. Water absorption capacity can also be linked to the chemical structure of materials that have a functional group (OH) that can absorb water [10] . Based on solubility, the solubility value of glycerol is near complete, which can dissolve as a whole in water.

The ability of bioplastics to regulate oxygen transport is largely determined by the basic structure and chemical properties of polymers as constituents of these bioplastics. The ability to pass gas particles is very dependent on the porosity of the film and the swelling degree of the film. Oxygen permeability in packaging films is useful for estimating the shelf life of packaged products. The value of oxygen permeability of bioplastic starch janeng modified by the addition of glycerol obtained an average value of 5,883 x 10^{-9} (cm^3.cm / cm^2.s.cmHg). This lower oxygen permeability due to bioplastic janeng starch-glycerol has a denser structure and has a good interconnection in structure because the starch and glycerol have hydrogen bonds [11].

FTIR analysis

The results of measurements of the FTIR spectra of pure starch bioplastic and added plasticizer glycerol in Figure 1. Based on the spectra in Figure 1 can be informed some common functional groups contained in the starch molecule. The infrared spectrum of starch showed a significant absorption at wavenumbers 3000 to 3600 cm^{-1} indicates the stretching vibration of OH groups. Broad peak and a high intensity indicate the hydrogen bonds in the starch molecule. Peaks that appear in the region below 1450 cm^{-1} shows the bending vibration of the methylene group. -CH stretching vibration of alkanes indicated by the peak at 2884 cm^{-1} region. The existence of ether and alcohol C-O bond can be seen in the peak that appeared at the 1316 and 1024 cm^{-1}. FTIR spectra showed a band at 1639 cm^{-1}, which shows the carbonyl glucose from starch. FTIR spectra of the bioplastic that have been added plasticizer mixture show the difference in the width and intensity of the resulting peaks associated with pure compound and mixing ratio. Glycerol is a compound containing an OH group and the CH-alkanes. The spectra revealed that there is some specific interaction between the hydroxyl groups of starch and the hydrogen atom of the hydroxyl group from glycerol that cause the increase in the peak intensity and width of the stretching vibrations of the -OH group of bioplastic.

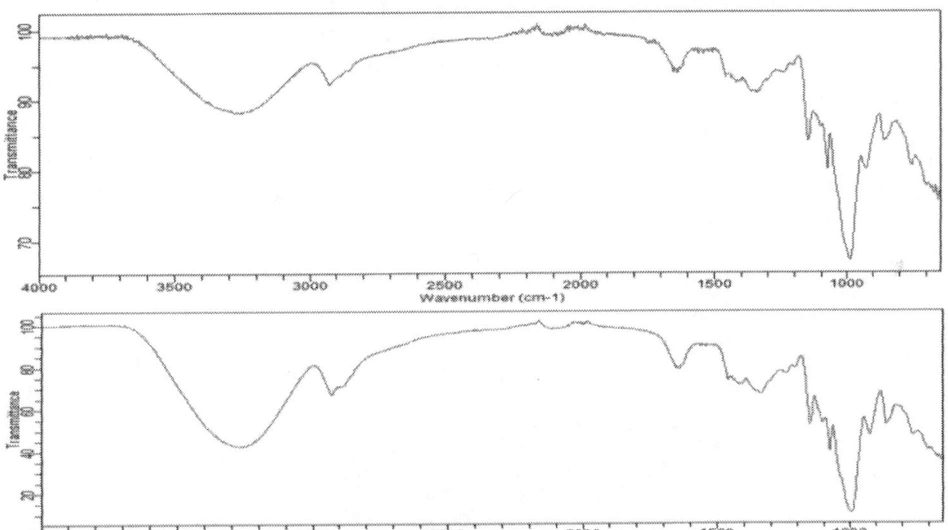

Figure 1 FTIR spectra of starch bioplastic and the addition of glycerol

Application of bioplastics for fruits and vegetable packaging

Application of bioplastics for fruits packaging was performed by wrapping tomatoes and apple slices by looking at the color change with the bioplastics wrapped pieces of fruit without wrapping. Pure starch bioplastics, glycerol modified starch bioplastics, and commercial wrapping plastic were illustrated their ability in protecting tomatoes and apple slices. The protecting barrier ability of plastics was observed for storage period seven days. Packaging is done by wrapping and storage is done for seven days at room temperature (27 ° C - 29 ° C), as shown in Figure 2. Tests were also conducted on the plastic wrapping commercial as a comparison (control).

.

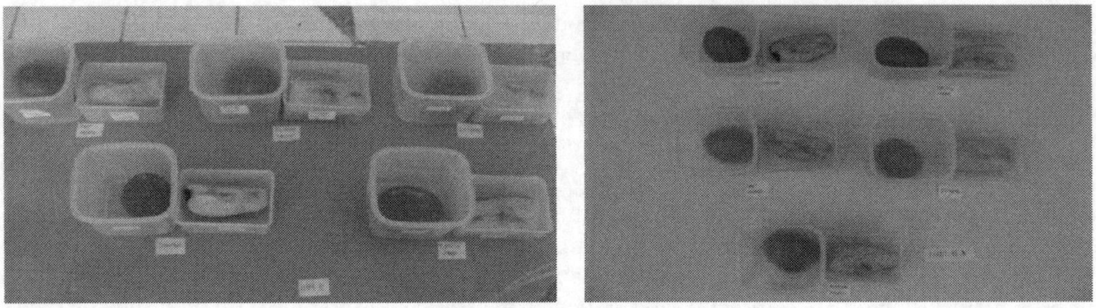

Figure 2. Visual illustration of the testing process as bioplastic packaging pieces of apples and tomatoes (Left: first day and Right: after seven days)

Ten selected panelists have observed the color change in the organoleptic test. From the results of data analysis, it was found that the storage of tomatoes and slices of apple on the seventh day showed the color of the fruit wrapped in bioplastics was significantly different from the fruit that was not wrapped at a 95% confidence level. The fruit that is not wrapped in bioplastics has turned dark on day three while the fruits wrapped in bioplastic starch, especially those modified with glycerol, have not been significantly different after seven days. From these results, indicate that bioplastics as packaging can protect the fruit.

Vitamin C content in tomatoes and apple slices wrapped starch bioplastics, commercial plastic wrap and tomatoes not wrapped (control) are shown in Figure 3. Figure 3. confirmed that the vitamin C content of apples wrapped in bioplastics higher than the apple that is not wrapped. This agrees with the literature, the content of vitamin C in apples obtained by 5 mg in 100 g of fruit. The tomatoes wrapped with bioplastic was also proved that the vitamin C content maintained and higher that unwrapped tomatoes. In 100 g fresh tomatoes vitamin C levels obtained by 9 mg. From the results show that vitamin C in slices apple and tomatoes wrapped with bioplastic are comparable and more protected level content of Vitamin C.

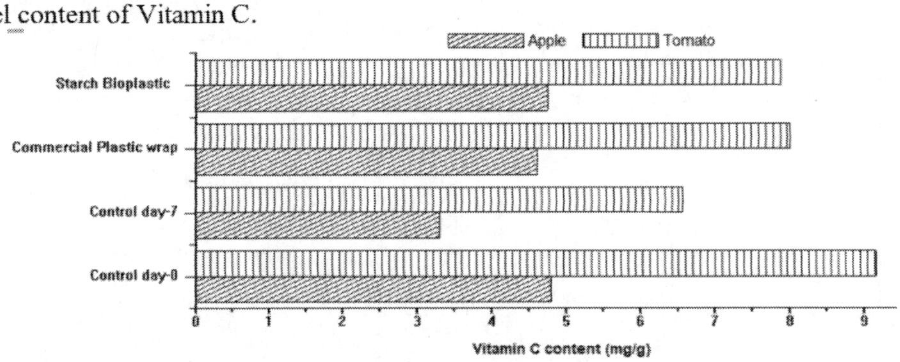

Figure 3. The test results of vitamin C on apple slices and tomatoes

4. Conclusions

Janeng starch is obtained from the isolation of tuber janeng has been used for the manufacture of bioplastics. Bioplastics can be made using the phase inversion method by drying and evaporation of solvent printing solution at room temperature. Bioplastics are made from a mixture of starch and glycerol provide transparent film, the value of tensile strength and high elongation. The bioplastics have a high value of the tensile test is equal to 20.95 kgf / mm^2 and the elongation of 42.69%. Bioplastics are made from starch janeng can be used for packaging apples and tomatoes.

5. Acknowledgments
This research was funded by the Syiah Kuala University Competitive Research Grant.

6. References
[1] Ortega-Toro R., Rodrigo Ortega-Toro, Jeannine Bonilla Pau Talens, Amparo Chiralt 2017 *Future of Starch-Based Materials in Food Packaging* (Elsevier)
[2] Mihai A., Grumezescu, and Alina Maria Holban 2018 *Food Packaging and Preservation A volume in Handbook of Food Bioengineering* (Academic Press)
[3] Rizzo, V. and G. Muratore 2009 *Journal of Food Engineering* **90(1)** 124
[4] Castillo Luciana A., Sonia Farenzena, Esteban Pintos, María Susana Rodríguez, Olivia V. López, 2017 *Food Packaging and Shelf Life*, **14 Part B** . 128-136
[5] Siracusa, V., Pietro Rocculi, Santina Romani and Marco Dalla Rosa 2008 *Trends in Food Science & Technology* **19(12)** 634
[6] Saleha S., Nurdin Saidi, Saiful, Murniana, Saida Rasnovi, Teuku M. Iqbalsyah 2018 *Jurnal Natural* **18(1)** 1-6
[7] Sukarsa, E., 2010 *Tanaman Gadung*, (Lembang : Balai Besar Pelatihan Pertanian (BBPP))
[8] Iqbalsyah, T. M., Oesman, F. dan Alghifari, A. F 2010 *Proceeding of Semirata BKS PTN Wilayah Barat MIPA* **24** 487-499.
[9] Iqbalsyah, T. M., Oesman, F. dan Nuryadi, F 2011 *Proceeding of Seminar Himpunan Kimia Indonesia* 2011 127-13
[10] Cao, N., X. Yang and Y. Fu 2009 *Food Hydrocolloids* **23(3)** 729
[11] Chen, C.-H. and L.-S. Lai 2008 *Food Hydrocolloids* **22(8)** 1584.

IOP Conf. Series: Materials Science and Engineering **523** (2019) 012016 doi:10.1088/1757-899X/523/1/012016

The study on composites formation from HDPE and sawdust/rice husk as raw materials

F Mulana[1,2,*], M Mariana[1], M N Hadman[2], S H Hanipah[2]
[1] Department of Chemical Engineering, Faculty of Engineering, Syiah Kuala University, Banda Aceh, Indonesia
[2] Faculty of Chemical Engineering, Universiti Teknologi MARA, Shah Alam, Malaysia

E-mail: seulanga20@gmail.com

Abstract. In this study, composites were prepared using compression molding method from cheap raw materials of high-density polyethylene (HDPE) as matrix and mixture of rice husk and sawdust as filler. The small amount of coupling agent was added during preparation of composites and for comparison composite without addition of coupling agent was also prepared. The coupling agent used in this study was maleic anhydride (MA). The aim of this study is to form composites with and without addition of coupling agent and to know the effect of coupling agent addition and different ratio of matrix and filler toward mechanical properties of produced composites. Composites that contained different ratio of matrix and filler were pressed and heated at temperature of 170°C. The composites were tested its mechanical properties (bending strength and tensile strength) and water absorption capacity. The results indicated that the bending strength of the composite increased with addition of 4 wt% of maleic anhydride and with increased the percentage of filler until 50%. As for the tensile strength, the addition of 4 wt% of maleic anhydride as coupling agent increased the tensile strength significantly. The highest bending strength value of 12.8 MPa was obtained from composite prepared from raw materials contained same amount of matrix and filler and addition of 4 wt% of maleic anhydride. While the highest tensile strength value of 6.1 MPa was obtained from composite prepared from raw materials contained ratio percentage of matrix and filler (40:60 of wt%) and addition of 4 wt% of maleic anhydride.

1. Introduction

Composite board or also known as wood-plastics composite are a product which is widely used since last 40 years [1]. Due to its wide usage, it is constantly changing in term of usage and application. It can be said that composite board has been used as a replacement for regular wood or board. They are used in wide variation, whether used for structural or non-structural application. It can be used indoor as well as outdoor, for example it can be used as construction materials, garden and yard components. It even is used inside the automotive (engine and interior), household item and consumers good. However, the most popular usage of composite are the construction industry and automotive application [1].

* Corresponding author: seulanga20@gmail.com

IOP Conf. Series: Materials Science and Engineering **523** (2019) 012016 doi:10.1088/1757-899X/523/1/012016

Composite board usually made up of filler which come from wood fiber or organic material and matrix which also known as thermoset or thermoplastic [2]. Thermoset is basically a type of plastic that once cured, cannot be melted by heating such as epoxies and phenolic. While, thermoplastic is basically a type of plastic that can be repeatedly melted. For the matrix, thermoplastics are more preferable due to its ability to be melted multiple times allowing it to be shaped and recycled. The composite have characteristics such as low density, low manufacturing cost, renewable and recyclable [3].

Composite board can be derived from used item. For example, the organic material can be come from used wood particle or any plant waste, such as the sawdust, rice husk and many more. The matrix is not necessarily comes from virgin thermoplastics. Used or recycled thermoplastics can also be used as the matrix [4]. These properties allow the composite to become more popular. Now, waste from organic materials can be easily found. Sawdust is usually left behind at the sawmill. The rice husk and rice stalk from the rice field usually being burned and causing air pollutions. With formation of composites with these used materials, waste from those industries can be minimized, and pollution can be reduced. Those biomass materials of sawdust and rice husk have several advantages such as low density, low cost, renewable and recyclable and biodegradability properties [5, 6].

Despite its big popularity, the composite board is usually not strong enough compared to regular board, due to its low adhesion between the matrix and the filler. Most of the matrix, especially thermoplastics, is non-polar (hydrophobic) substances, which are not compatible with polar (hydrophilic) wood fibers and, therefore, resulting the poor adhesion between polymer and wood fiber. In order to improve the adhesion between the wood fibers and polymer, addition of coupling agent is needed [6]. Coupling agent is basically a chemical that is added to the mixture in small quantities, so it allow bonding to occur between its surface with the other surface such as the polymer and the wood fibers. The coupling agent will encourage the spreading of particle that will increase the adhesion between the polymer and wood fibers. So, the addition of coupling agent will increase the strength of the composites that will allow the composite to be used even more frequently in many applications.

There are several studies have been conducted by researchers regarding the formation of composite by using the sawdust and rice husk separately [4, 5, 6]. But there is no research has been done for the formation of composite by using mixture of sawdust and rice husk. So, due to that, mixture of sawdust and rice husk is selected as the fillers. As for coupling agent, the most popular coupling agent of maleic anhydride (MA) is selected. 4 wt% of maleic anhydride was added into mixture of matrix and filler because according to [6, 7], the most optimum value in term of mechanical properties was obtained when coupling agent was added at 4 wt% of the matrix and fillers weight. Different ratio of matrix and filler was performed, because according to [8], similar ratio of matrix and filler was the most suitable for the melt-blending formation.

The aim of this study is to form composites from high-density polyethylene as matrix and mixture of rice husk and sawdust as the filler with and without addition of maleic anhydride as coupling agent and to know the effect of addition of coupling agent and different ratio of matrix and filler toward mechanical properties and water absorption capacity (%) of produced composites.

2. Methodology

2.1. Equipments and Materials

The equipment used for this study included stirred reactor consisting of a beaker (Pyrex), motor with stirrer (Fisher Scientific, max. speed of 250 rpm), oil bath (silicone oil), hot press (Boy22A), grinder and +80-200 mesh size of sieve with shaker, oven, digital scale, and thermometer (0-300°C). The materials used in this study were rice husk and sawdust as the filler, high density polyethylene (Sigma-Aldrich) as matrix, xylene as solvent and maleic anhydride (Sigma-Aldrich) as the coupling agent. Rice husk was obtained from BERNAS Sekinchan and sawdust was obtained from a workshop in Faculty of Civil Engineering, UiTM Shah Alam.

2.2. Formation of Composites

2.2.1. Preparation of Matrix and Filler

In order to eliminate the extractive substance that may still be contained inside the filler of rice husk and sawdust, it was soaked with hot water at temperature of 100°C for about 2 hours. During soaking, the filler was stirred at the medium speed. After that, the filler was dried up in the oven for 24 hours. Later, the filler was grinded and sieved with +80-200 mesh size sieves [7]. Then, the the filler was dried again by using oven at temperature of 110°C for 24 hours in order to remove the excess moisture content to be less than 8% by weight [9]. After dried for 24 hours, the filler was stored in the desiccator to prevent any moisture to absorb into the filler.

2.2.2. Formation of Composites

Composites were formed by using compression molding method. Initially, 35 g of virgin HDPE was placed inside the flask, and 100 mL of xylene was added into the beaker slowly. The flask was immensed into the oil bath that was heated at the temperature of 145°C. After the matrix was fully melted, 15 g of filler comprised of 7.5 g of dried sawdust and 7.5 g of dried rice husk, was added to the flask and stirred using stirrer until both matrix and filler were fully homogenized for about 20 minutes. Maleic anhydride with amount of 4% by weight was added to the mixture inside the flask. The ratio of matrix and filler were varied at mass ratio of 70:30, 60:40, 50:50 and 40:60. After the mixture was fully homogenized, the mixture was removed from the flask and allowed to cool down naturally until room temperature and until all the solvent evaporated. For further compression, it was compressed by using hotpress at temperature of 170°C for 30 minutes. The composites then cooled naturally and tested. Another composite with the same variation of ratio of matrix and filler was also prepared but without addition of maleic anhydride for comparison.

2.2.3. Testing and Analysis of Prepared Composites

Mechanical properties of composites that include tensile and bending strength were tested using Universal Testing Machine Model SHIMADZU in Concrete Lab, Faculty of Civil Engineering, UiTM Shah Alam. Water absorption capacity test was performed as procedure written in SNI 03-2105-2006 [10]. The results of bending strength, tensile strength and water absorption capacity value from this study were later compared to the standard of SNI 03-2105-2006 (the quality of particle board). According to SNI 03-2105-2006, the minimum standard of bending strength and the tensile strength is 9.81 MPa and 0.15 MPa, respectively and the water absorption capacity value was less than 12 %.

3. Results and Discussion

3.1. Mechanical Properties of Prepared Composites

Maleic anhydride is a bridge that connecting the polar of wood fiber with the nonpolar of polymer which are not compatible with polar (hydrophilic) wood fibers and, therefore, resulting the better adhesion between polymer and wood fiber. Theoretically, mechanical properties of composites are associated with ability of composite to resist external forces that acting on the composite.

Figure 1 shows results of tensile strength for all composites with and without addition of maleic anhydride at different ratio of matrix of HDPE and filler of rice husk and sawdust. According to figure 1, the increasing of percentage of filler in the mixture of matrix and filler resulted in increasing of tensile strength of the prepared composites. The highest tensile strength of 6.1 MPa was obtained when the percentage of filler was at 60% with the addition of maleic anhydride. The lowest tensile strength of 1.3 MPa was obtained on composite contained 30 wt% filler without addition of the maleic anhydride.

From the above results, it indicated that tensile strength of all composites by using HDPE met the standard of SNI 03-2105-2006. The highest tensile strenght of 6.1 MPa in this study was a bit higher compared to similar research using raw material of only rice husk as filler with tensile strenght of 5.75 MPa in our previous paper [7] and with tensile strenght of 5.69 MPa in [6]. The highest tensile strenght of composite in both previous papers was also found on samples after addition of 4 wt% of coupling agent.

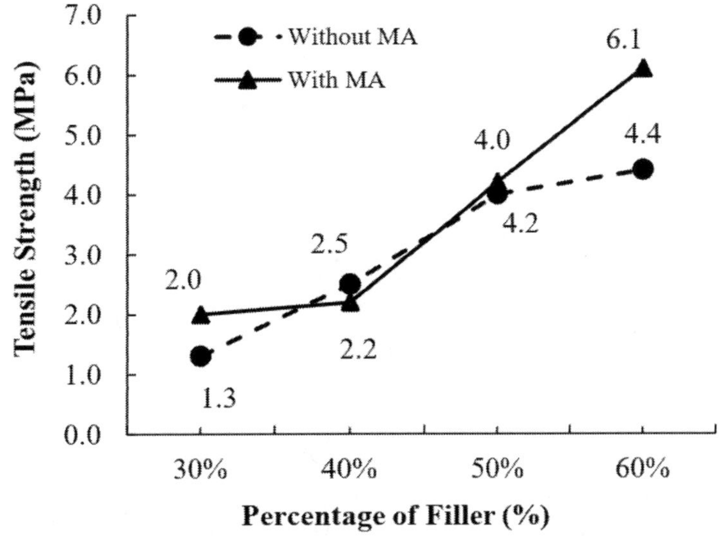

Figure 1. The effect of percentage of filler (%) toward tensile strength value of prepared composites with addition and without addition of maleic anhydride (MA) coupling agent.

Figure 1 also showed that the prepared composite with addition of 4 wt % of maleic anhydride was stronger than the prepared composite without addition of maleic anhydride. The presence of small amount of maleic anhydride induced the stickness between rice husk/sawdust that has properties as hydrofilic and HDPE as hydrophobic material was improved. The maleic anhydride acted as bridge link between both matrix and filler to increase adhesion properties inside the composite and finally increased the tensile strenght of composite. The similar structure and formation mechanism of this prepared composite were studied previously by another researcher [11]. However, there was slight inconsistency the result with 40% percentage of filler, where the tensile strength was higher at sample without addition of maleic anhydride. This inconsistency may occur due to the presence of air bubble inside the prepared composite. Slight inconsistency in the thickness of the composite may also affect the results. This inconsistency of thickness may due to the method of compression molding; where it does not really control the thickness compared to the injection molding method [1, 12].

Figure 2 showed results of bending strength for all composites with and without addition of maleic anhydride at different ratio of matrix of HDPE and combined filler of rice husk and sawdust. Figure 2 showed that the presence of maleic anhydride content in the mixture resulted in increasing of bending strength of the prepared composites almost twice. The highest bending strength was obtained when the percentage of filler was 50 wt% with the addition of 4 wt % maleic anhydride and the value was 12.8 MPa. This bending strength value was higher than that of a minimum standard of 9.81 MPa as stated in SNI 03-2105-2006. However, all bending strength value of the prepared composites without addition of maleic anhydride did not meet SNI 03-2105-2006.

As in tensile strength, for the bending strength value there was slight inconsistency in the result. At 40% filler, the bending strength was lower on a sample with addition of maleic anhydride. This inconsistency may occur due to the presence of air bubble inside the prepared composite and also due to the thickness of the composite as explained previously. To understand the resistance of composite when contacting to moisture or wet environment, so all prepared composites were performed the water absorption testing by immersing composites in water. The testing results of water absorption were shown in figure 3. Composite board is a substance that has the hygroscopis properties so that it is able to absorb moisture from the surrounding environment to achieve equilibrium and likewise when composite board was immersed in water it would expand. Water absorption capacity was determined based on the weight of the specimens after being immersed in water in accordance with the method requirement in SNI 03-2105-2006.

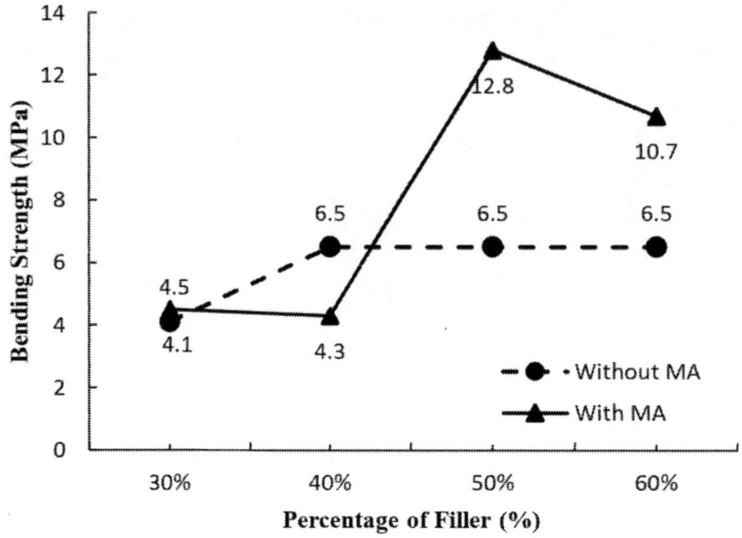

Figure 2. The effect of percentage of filler (%) toward bending strength value of prepared composites with addition and without addition of maleic anhydride (MA) coupling agent.

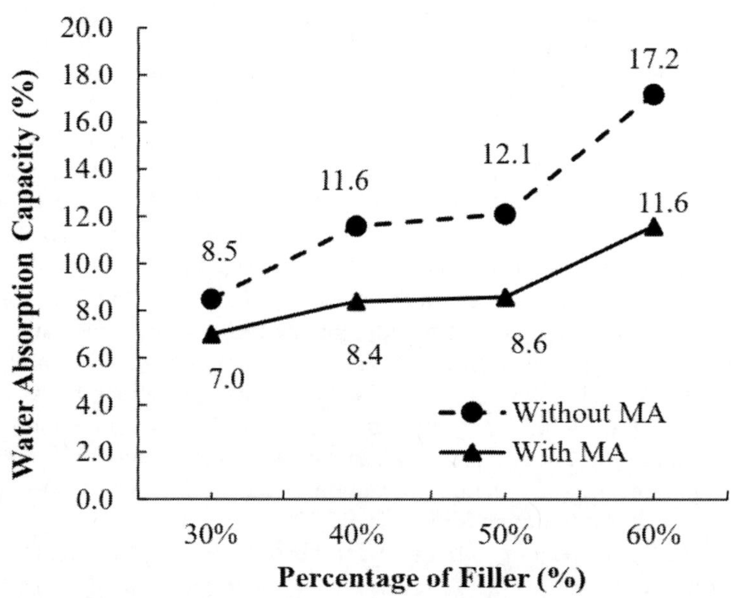

Figure 3. The effect of percentage of filler (%) toward water absorption capacity (%) value of prepared composites with addition and without addition of maleic anhydride (MA) coupling agent.

The results showed that the water absorption capacity from the prepared composites with addition of 4 wt % of maleic anhydride on average range from 7.0 % - 11.6%, which these results indicated that the water absorption capacity in all prepared composite board with addition of coupling agent met SNI 03-2105-2006. On the other hand, the water absorption capacity of composites without addition of maleic anhydride only met the standard on samples with percentage of filler 30% and 40%. Therefore, it can be seen clearly that the composites with addition of maleic anhydride had a better properties of water absorption resistance compared to the composites without addition of maleic anhydride [13].

4. Conclusion

The bending strength and the tensile strength of the composite increased with addition of maleic anhydride coupling agent and also increased as the percentage of filler increased. The addition of 4 wt % of maleic anhydride coupling agent affected significantly the mechanical properties of prepared composite. The presence of maleic anhydride coupling agent in composites resulted in increasing adhesion properties between rice husk/sawdust and HDPE thereby increasing the mechanical physical properties of composites.

Tensile strength, bending strength and water absorption capacity values of prepared composites with addition of 4 wt % of maleic anhydride met the minimum standards of SNI 03-2105-2006. The highest tensile strength and bending strength of composites with the addition of 4 wt % maleic anhydride was 6.1 MPa and 12.8 MPa, repectively. In addition, the water absorption capacity of composites with addition of 4 wt % of maleic anhydride was in the range of 7.0 % – 11.6 %.

References

[1] Schwarzkopf M J and Burnard M D 2016 *Wood-Plastic Composites—Performance and Environmental Impacts*, in *Environmental Impacts of Traditional and Innovative Forest-based Bioproducts*, Springer 19-43.
[2] Clemons C 2002 *Forest Products Journal* **52** (6) 10.
[3] Hung K C, Yeh H, Yang T C, Wu T L Xu J W and Wu J H 2017 *Polymers* **9** 726
[4] Mulana F, Hisbullah dan Iskandar 2011 *Jurnal Rekayasa Kimia dan Lingkungan* **8** (1) 17-22
[5] Moreno D D P and Saron C 2017 *Composite Structures* **176** 1152-1157
[6] Mulana F, 2012 Composites prepared from rice husk and recycled/ virgin HDPE with addition of coupling agent *The Proc. of 2nd AIC Unsyiah & 8th IMT-GT Uninet Biosciences Conference,* 22-24 November 2012, Banda Aceh, Indonesia
[7] Mulana F, 2014 The effect of MA/MAPE combination as coupling agent in formation of rice husk and recycled HDPE-based composites *Proc. of the 4th AIC Unsyiah in conjunction with 9th AIWEST-DR,* October 22-24, 2014, Banda Aceh, Indonesia
[8] Lu J Z, Wu Q and McNabb H S 2007 *Wood and Fiber Science* **32** (1) 88-104.
[9] Schmidt E 1993 *Methods for preparation of composite materials* US Patent 5,217,655
[10] Badan Standarisasi Indonesia (BSN) 2006 Papan Partikel SNI 03-2105-2006
[11] Yang H S, Kim H J, Park H J, Lee B J and Hwang T S 2007 *Composite Structure* **77** 45 - 55
[12] Yam R C M and Mak D M T 2014 *Journal of Cleaner Production* **67** 277-284
[13] Gilormini P and Verdu J 2018 *Polymer* **142** 164 - 169

IOP Conf. Series: Materials Science and Engineering **523** (2019) 012017 doi:10.1088/1757-899X/523/1/012017

Deposition of polydopamine on the surface of Polyvinylidene Fluoride (PVDF) membrane as A UV-Shielding layer

S Muchtar[1], M Wahab[1], S Mulyati[2,*], M Riza[1,2], N Arahman[2]
[1]Doctoral School of Engineering Science, Syiah Kuala University, Banda Aceh, Indonesia 23111
[2]Department of Chemical Engineering, Syiah Kuala University, Banda Aceh, Indonesia, 23111

E-mail: sri.mulyati@unsyiah.ac.id

Abstract. This paper discusses about the influence of polydopamine layer in enhancing UV resistance of PVDF membrane for membrane photocatalytic reactor application. The PVDF membrane was prepared by common NIPS method using DMAc as a solvent and PEG as a pore-forming additive. The polydopamine layer was deposited on the membrane surface by the facile dip-coating method through Tris-buffered dopamine polymerization. The UV-shielding effect of PDA layer was studied by comparing the changes in pure and PDA coated membrane coating before and after UV irradiation. The studied effects are in terms of changes in chemical, morphological structure and mechanical properties which observed by means of ATR-FTIR, FESEM instrumental analysis, and tensile measurement, respectively. The IR analysis showed that after PDA coating, the membrane surface is rich of catecholamine groups which greatly contributed as free radical scavengers. The change in chemical structure was seen on pure membrane which attributed to the rearrangement of polymer chemical structure caused by UV-induced photodegradation. FESEM imaging results showed that with PDA coating the membrane surface showed minimal damage in comparison to that of non-coated PVDF membrane. These results altogether confirmed, that the PDA layer can protect the membrane surface from UV-initiated free radicals attack.

1. Introduction

In general, it is a well-known fact that most of the polymeric materials are sensitive to environmental influences such as temperature, humidity, and light [1-3]. Same case goes to the polymer-based membranes, if it is continuously exposed to light (either UV or visible light), for instances photocatalytic membrane, as the time passes the membrane will be damaged because the quantum energy of UV light is sufficient to break the major polymer chains and disrupt the molecular structure of a polymer [3]. This phenomenon is commonly known as photodegradation. Photodegradation will result in morphological damage to the membrane, as well as a decrease in mechanical strength that culminates in the short lifespan of the membrane [2, 4].

Photodegradation occurs in three major stages: initiation, propagation, and termination. Initiation is a stage that occurs when energy from UV light is absorbed by the membrane which can break down the chemical chains of the polymer and produces free radicals. At the propagation stage, free radicals

* Corresponding author : sri.mulyati@unsyiah.ac.id

will react with oxygen and produce peroxy radicals. These radicals will steal the H atoms from the polymer chain and react to form new hydrocarbon radicals such as hydroperoxides etc. this will result in the backbone cleavage of O-O bonds which will be followed by the occurrence of bond scission. Termination is when at least two free radicals combine with each other and result in the formation of new components which eventually lead to changes in the morphological structure and mechanical failure [4, 5].

One of the solutions to prevent or at least reduce the occurrence of photodegradation of polymeric membranes is by protecting it using a scavenger material which capable of being a membrane shielding from UV attack. One of such materials is polydopamine (PDA) [6-9]. In addition to being frequently used as a membrane modifier due to its hydrophilic characteristics [10], PDA is also reported to have high resistance to UV. The UV shielding activity of polydopamine is attributed to the rich content of the catecholamine functional groups (OH and NH) which capable of catching the UV-initiated free radicals before they get to attack the membrane [11-13]. In this study, the shielding ability of the PDA layer was studied. The membrane used was Polyvinylidene Fluoride (PVDF), PDA was incorporated on the surface of the membrane by dip-coating technique. The observed effects are in terms of changes in chemical and morphological structure as well as mechanical properties.

2. Experimental

2.1. Materials

The membrane employed in this study was based on PVDF (Solef 6012, Solvay Specialty Polymer, USA), with DMAc as a solvent and PEG 1000 Da as a pore-forming agent. Both of those materials were purchased from Wako Pure Chemical Industries, Japan. Dopamine hydrochloride (Sigma Aldrich, Germany) and Tris (2-amino-2-hydroxymethyl-1,3-propanediol) (Wako Pure Chemical Industries, Japan) were used to prepare dopamine coating solution.

2.2. Preparation and Modification of PES Membrane

The membrane was prepared by following the procedure reported in our previous work [14]. In detail, PVDF, DMAc, and PEG were mixed with the composition ratio of 15:85:10 wt% then stirred with heating at 60°C to form a homogeneous solution. This solution was then cast on a glass plate using a casting knife with a controlled thickness of 2 mm and then dipped in a water-filled coagulation bath until the membrane film was detached from the glass plate. The finished membrane was then dipped in a 15mM dopamine solution (Tris buffer: pH 8.8) with dopamine concentration of 2 g/L for 2 hours. After the dip-coating was completed, the membrane was then rinsed using distilled water and isopropanol alternately to wash the weakly-bonded PDA on the membrane surface.

2.3 UV Radiation experiment

UV exposure to the membrane was performed using a UV-Lamp instrument (22W, SUV-16 254nm, AS ONE, Japan) which placed in a tightly sealed lab-made box. The membrane sheet was placed at a distance of 5 cm below the UV lamp. The experiment was conducted for 5 days, the sample was taken on day 3 and day 5 for characterizations.

2.4 Membrane Characterization

Characterization on surface morphology, chemical composition and mechanical properties of the membrane before and after UV radiation was performed using the Field-Emission Scanning Electron Microscopy (FE-SEM, JSF-7500F, Jeol Co. Ltd., Japan), the Fourier-Transform Infrared Spectroscopy Attenuated Total Reflectance instrument (FTIR-ATR) (Thermo Scientific iD5 ATR- Nicolet iS5 FTIR Spectrophotometer, Japan) and a tensile test instrument (Autograph AGS-J, Shimadzu Co., Japan), respectively. Before analysis, all samples were dried using freeze drier (FD-1000, Eyela, Japan).

3. Results and Discussion

3.1. Membrane Chemical Composition

Using Fourier Transform Infrared Spectroscopy-Attenuated total reflection (FTIR-AR), analysis of the membrane surface chemical structure was performed to confirm the successful deposition of the PDA layer. Figure 1 shows that after coating with the PDA, there is a noticeable change in the chemical composition of the membrane marked by the emergence of the major components of the PDA, i.e, amine and hydroxyl groups [8, 12]. For detail, in the wavenumber range of 1500-1711 cm^{-1} appeared a new peak due to the bending and scissoring vibration of the N-H group, in addition, at wavenumber ranging from 3300 to 3600 cm^{-1} marks the presence of OH and NH groups of catecholamines contained in the PDA [8, 12, 15]. These are the specific groups which cause PDA to have a superhydrophilic and high free radical scavenger traits.

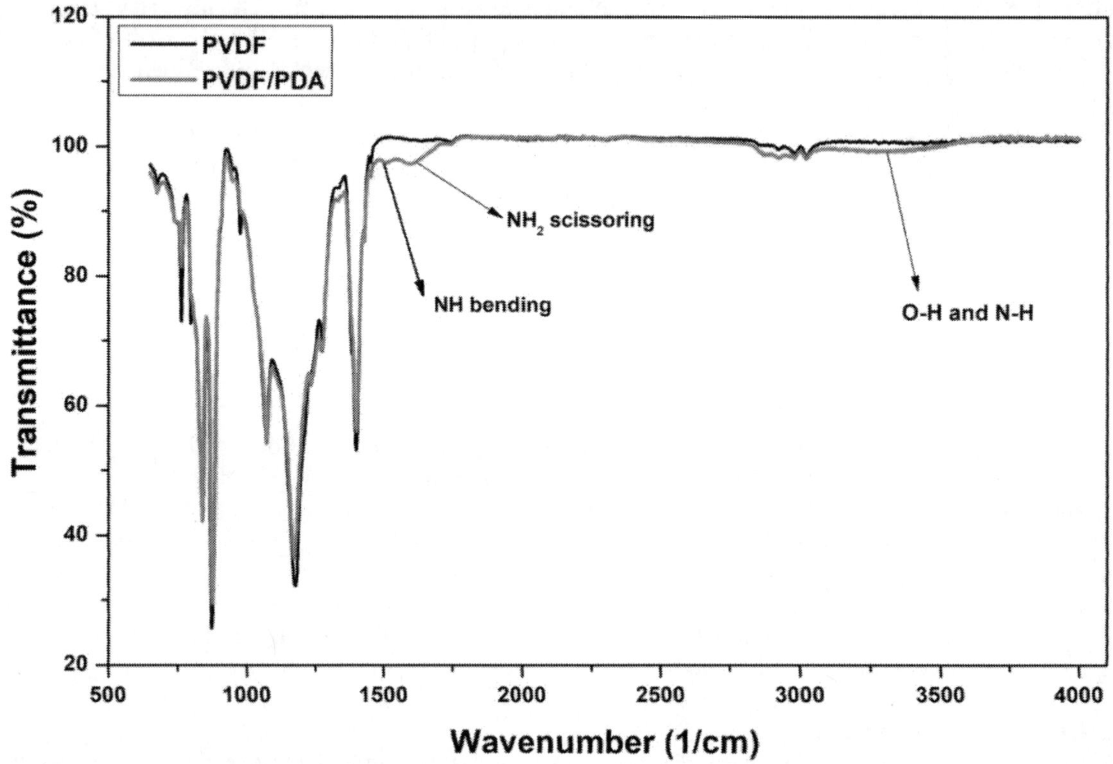

Figure 1. IR Spectra results of PVDF membranes with and without PDA coating

To further confirm the ability of PDA layer as a UV shielding against the membrane, FTIR analysis was also done on the membrane after UV radiation. It is shown in Fig. 2 that, for a pure PVDF membrane, after 5 days of UV exposure, a new peak occurs in a wavenumber 1717 cm^{-1} which denotes a C=O group of ketone [16]. The sudden presence of ketone and aldehyde on an irradiated membrane is attributed to the effects of photodegradation. When UV is irradiated on a polymeric material, it will result in the formation of free radicals which will disrupt the main chemical structure of the polymer, which through several stages of the reaction will produce new compounds, which are generally reported as ketones [5]. As for the PDA coated membrane, there is almost no change in IR spectra results either before or after UV exposure. This is due to the presence of functional groups of catecholamine (OH and NH) of PDA that have high radical scavenging activity which operates by deactivating UV-induced free radicals through the electron donor process [11, 17, 18].

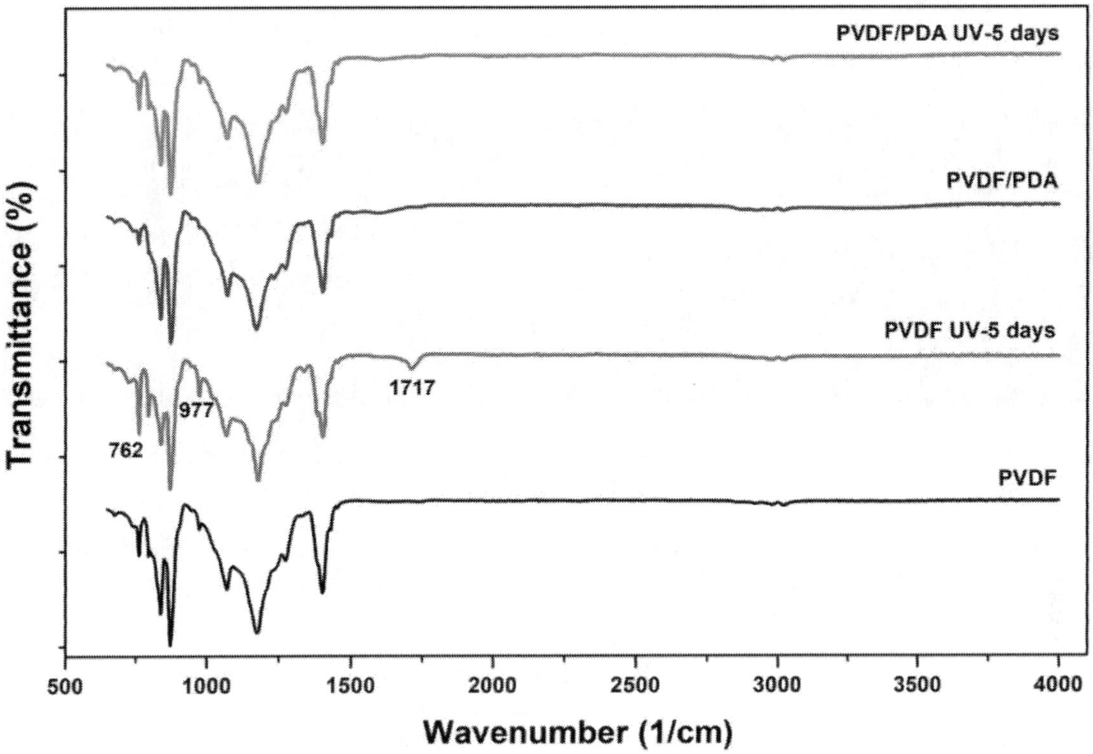

Figure 2. IR Spectra results of PVDF membranes with and without PDA coating before and after UV irradiation

3.2. Morphological Structure

To further explore the effect of PDA coating conditions as UV shielding layer for the membrane, FESEM observation was conducted on pure and PDA coated PVDF membranes of different dopamine concentration and coating time before and after 5 days of UV illumination. From Fig. 3, it is clear that the pure PVDF membrane is severely damaged after UV exposure. Even with the exposure duration of three days only, the deformation that occurs is already significant. Deformation is seen in the form of cracks on the surface also the enlargement of pore size.

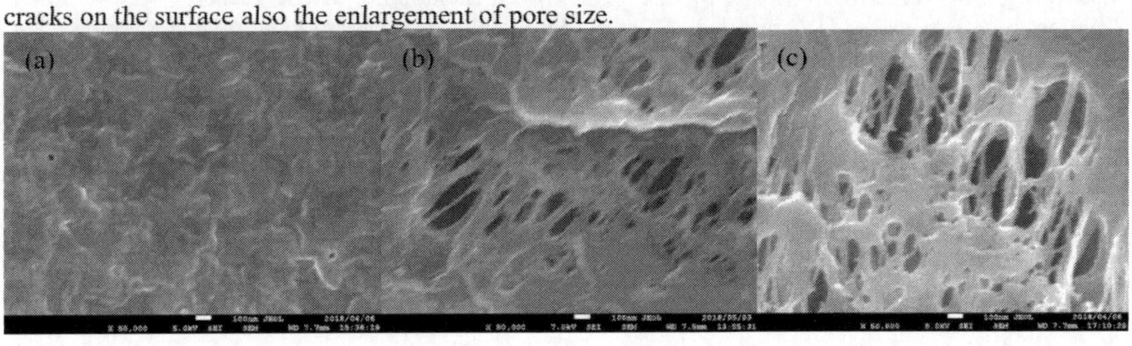

Figure 3. FE-SEM imaging of PVDF membrane (a) before, (b) after 3 days, and (c) 5 days of UV exposure

IOP Conf. Series: Materials Science and Engineering **523** (2019) 012017 doi:10.1088/1757-899X/523/1/012017

On the irradiated PDA modified membrane (Fig. 4), a similar type of damage is seen but with way more minimal scale. This once again confirms that the PDA coating is successful in blocking the occurrence of photodegradation of the membrane.

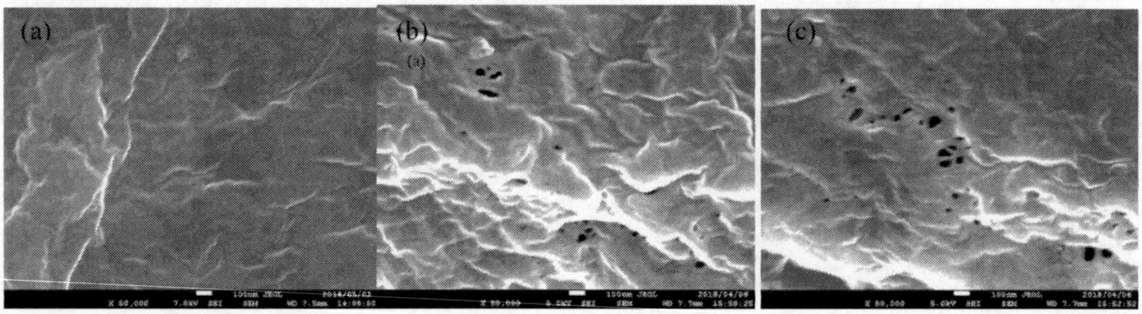

Figure 4. FE-SEM imaging of PDA coated PVDF membrane (a) before, (b) after 3 days, and (c) 5 days of UV exposure

As an additional information, minimal damage is still visible on the membrane coated with PDA presumably because the coating time is still relatively short i.e. only 2 hours. As reported by Zhu and co-workers [19] on their extensive study on characteristics of polydopamine layer on PVDF membrane, in short coating time the polymerization that occurs is not long enough to create a thicker deposited PDA layer on the membrane surface which means free radical scavenger functional groups such as N-H and OH are not sufficient to completely protect the membrane from UV attack. Similar results and reasoning also affirmed by some other relevant literature [20, 21].

3.3. Mechanical Properties

To better explain the effect of PDA coating parameters on UV resistance, mechanical strength data of all the prepared membranes before and after 5 days irradiation are presented in the form of tensile and elongation at break retention as shown in Fig. 5 and 6, respectively. Retention of mechanical properties is obtained by comparing the value of tensile or elongation at break of the sample after UV radiation and that of before UV radiation. As seen from Fig. 5, after only 3 days of irradiation, the tensile at break of original PVDF membrane already decreased significantly by almost 40%. After 5 days of irradiation, the tensile continued to fall dramatically by 60% from its original (before irradiation) tensile value. Meanwhile, for the PDA-coated membrane, the decrease is very minimal even after being exposed to UV for 5 days. This finding is closely correlated with the surface damage caused by UV-induced photodegradation [3] which is greatly supported by the FE-SEM imaging results in Section 3.2. Damage formed on the membrane makes the membrane weakens and breaks easily during tensile testing. In the PDA coated membrane, reducement of tensile is only by 7% on the third day of exposure and almost stable even on the fifth day. This is because the damage formed on modified membrane is not as severe as that of pure PVDF membrane.

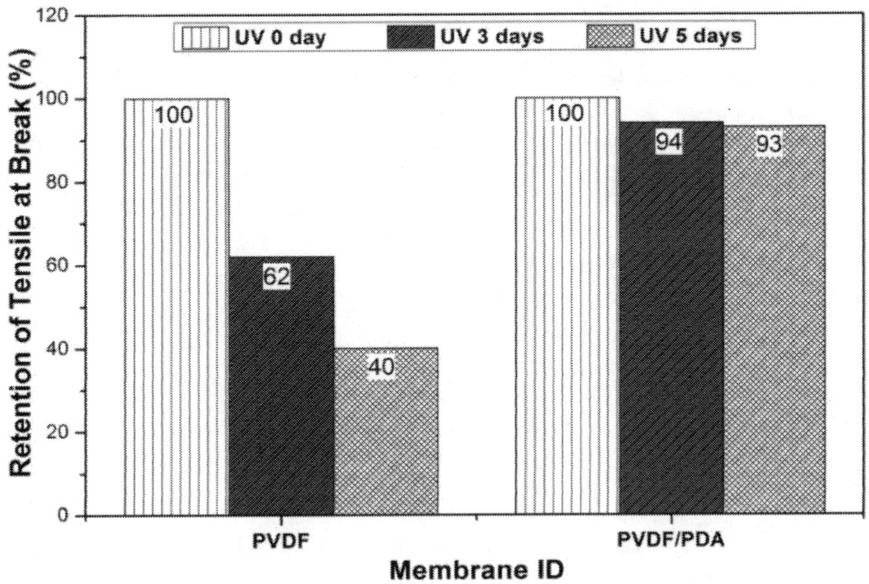

Figure 5. Tensile at break retention of membranes before and after UV exposure

A similar tendency is also observed from elongation at break results. As seen in Figure 6, the ductility of membrane decreased significantly with increasing duration of UV exposure. With only 3 days of exposure, the elongation at break of pure PVDF dropped by 75% (retention value of 25%). As for membrane which shielded by PDA layer, the decline also occurs but not as severe. From the results in Fig. 6, it can be seen that after 5 days exposure membrane is still able to retain about 50% of its ductility. The weakening of the membrane is attributed to the occurrence of embrittlement caused by photodegradation [1, 22].

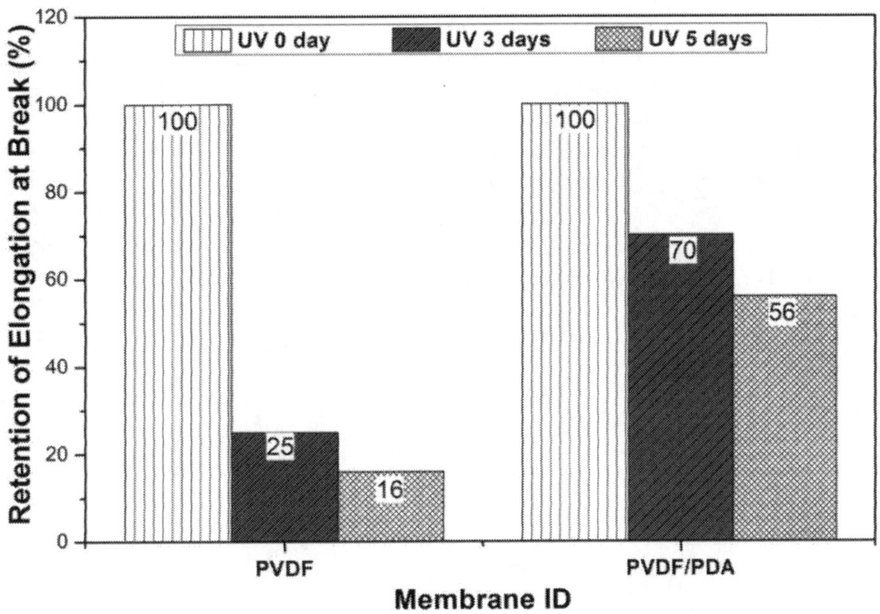

Figure 6. Elongation at break retention of membranes before and after UV exposure

4. Conclusions

Polydopamine layer has been deposited on the surface of the PVDF membrane to improve its resistance to UV radiation. The effectivity of PDA as a membrane UV shielding layer was studied from the membrane changes in terms of chemical composition, morphological structure, and mechanical properties. From the results of the characterization, it was clearly confirmed that PDA layer successfully minimized the effects of UV-induced photodegradation on the membrane.

5. Acknowledgment

This research was funded by the Ministry of Research, Technology and Higher Education of the Republic of Indonesia under 'PMDSU' scholarship.

6. References

[1] Cui H, Hanus R and Kessler M R 2013 *Polym. Degrad. Stab.* **98** 2357-65
[2] Rivaton A and Gardette J L 1999 *Polym. Degrad. Stab.* **66** 385-403
[3] Lee Mei J, Ong Chi S, Lau Woei J, Ng Be C, Ismail Ahmad F and Lai Soon O 2016 *J. Polym. Eng.* **36** 261
[4] Rånby B 1989 *J. Anal. Appl. Pyrolysis* **15** 237-47
[5] Rabek J F 1994 Polymer Photodegradation: Mechanisms and experimental Methods. (Sweden: Springer Science & Business Media)
[6] Feng K, Hou L, Tang B and Wu P 2015 *J. Membr. Sci.* **490** 120-8
[7] Wu H, Liu Y, Mao L, Jiang C, Ang J and Lu X 2017 *J. Membr. Sci.* **532** 20-9
[8] Liu Z, Hu J, Sun Q, Chen L, Feng X and Zhao Y 2017 *Macromol. Res.* **25** 431-8
[9] Phua S L, Yang L, Toh C L, Guoqiang D, Lau S K, Dasari A and Lu X 2013 *ACS Appl. Mater. Interfaces* **5** 1302-9
[10] Syawaliah, Mulyati S, Muzaitun and Mulyasari R 2018 *IOP Conf. Ser.: Mater. Sci. Eng.* **352** 012052
[11] Yen G C and Hsieh C L 1997 *Biosci. Biotechnol. Biochem.* **61** 1646-9
[12] Liebscher J, Mrówczyński R, Scheidt H A, Filip C, Hădade N D, Turcu R, Bende A and Beck S 2013 *Langmuir* **29** 10539-48
[13] Wypych G 2015 Handbook of UV Degradation and Stabilization (Second Edition): ChemTec Publishing) pp 37-65
[14] Muchtar S, Wahab M Y, Fang L-F, Jeon S, Rajabzadeh S, Takagi R, Mulyati S, Arahman N, Riza M and Matsuyama H 2018 *J. Appl. Polym. Sci.* **136** 47312
[15] Liu Y, Ai K and Lu L 2014 *Chem. Rev.* **114** 5057-115
[16] Lambert J B 1987 Introduction to Organic Spectroscopy (New York: Macmillan)
[17] Nimse S B, and Pal, D 2015 *RSC Adv.* **5** 27986-8006
[18] Ju K-Y, Lee Y, Lee S, Park S B and Lee J-K 2011 *Biomacromolecules* **12** 625-32
[19] Juang R-S, Huang C and Hsieh C-L 2014 *J. Taiwan Inst. Chem. Eng.* **45** 2176-86
[20] Li B, Liu W, Jiang Z, Dong X, Wang B and Zhong Y 2009 *Langmuir* **25** 7368-74
[21] Xi Z-Y, Xu Y-Y, Zhu L-P, Wang Y and Zhu B-K 2009 *J. Membr. Sci.* **327** 244-53
[22] Rabello M S and White J R 1997 *Polym. Degrad. Stab.* **56** 55-73

Phytochemical screening and *In vitro* Cytotoxic activity of Hexane extract of Temurui (*Murraya koenigii* (Linn.) Spreng) leaves against Human Cervical Cancer (HeLa) cell line

U Amna[1],*, H Halimatussakdiah[1], P Wahyuningsih[1], N Saidi[2], R Nasution[2], S Y Astryna[2]

[1] Chemistry Department, Faculty of Engineering, Universitas Samudra, 24416 Kota Langsa, Aceh, Indonesia

[2] Chemistry Department, Faculty of Mathematics and Natural Sciences, Universitas Syiah Kuala, 23111 Banda Aceh, Aceh, Indonesia

E-mail: ulil_amna@unsam.ac.id

Abstract. Cancer is a chronic disease caused by the growth of abnormal cells in body tissues and includes the second deadly disease in the world where the number of sufferers increases every year. Some chemotherapy prevention agents using synthetic drugs have been used to treat cancer, but it is relatively expensive and cause poisoning that limits their use. The aim of this study was to evaluate the phytochemical and develop natural anticancer drugs from hexane extract of Temurui (*M. koenigii* (Linn.) Spreng) leaves. The phytochemical analysis showed the presence of terpenoids and steroids. Then, the hexane extract of Temurui leaves was screened for *in vitro* cytotoxic activity against human cervical cancer (HeLa) cell line by using the MTT assay. The result showed a very strong cytotoxic activity effect with CD_{50} values less than 1 μg/ml. This indicated as a potent cytotoxic activity agent for HeLa cancer cells. Therefore, it is expected to conduct further research for cytotoxic test of other cancer cell lines so that it could be developed as raw materials for the manufacture of new drugs.

1. Introduction

Cancer is a chronic disease caused by the growth of abnormal cells in the body's tissues that destroy the normal cells. Cervical cancer is the most common genital cancer and one of the leading causes of death among female population. It accounts for approximately 12% of all cancers in women and it is the second most common cancers in women worldwide most especially in developing country, including Indonesia [1, 2, 3]. WHO reports that cervical cancer is the second most dominant cancer after breast cancer in Indonesian women, with prevalence in woman aged 15 to 44 years [1]. Among the gynaecological cancers, cervix cancer is the most common cancer, followed by cancer of the ovary, the uterus, the vulva, the vagina, and fallopian tube cancer. Indonesia is the 4th ranking country in South-East Asia, with the highest cervical cancer incidence after Cambodia, Myanmar, and Thailand. A comprehensive statistic estimated in 2012, reports that about 20,928 new cervical cancer cases are diagnosed annually in Indonesia. The incidence rate of cervical cancer is 17 per 100,000 women per year [3].

* Corresponding author : ulil_amna@unsam.ac.id

The increasing cases of deaths caused by cancer encouraged researchers to conduct research to find potential anticancer drugs. Some chemotherapy prevention agents using synthetic drugs have been used to treat cancers, but it is relatively expensive and cause poisoning that limits their use. Nowadays, research about finding anticancer agent from plants is widely developing. The present review presents that most of secondary metabolites isolated from large number of plant families showed specific emphases on their potential development as anticancer agents [4, 5]. WHO noted that 65-80% of diseases from the human body can be treated with drugs from nature [6]. The use of medicines from natural product was increased and demanded relatively caused of cheaper compared to synthetic drugs and believed to be safer and minor side effects [7, 8].

Research on medicinal plants has been based on the development of ethnobotany and the chemotaxonomy of plants and also emphasized by the researchers on the availability of local natural products. This research focused on the development of anticancer agent from Aceh local plant, temurui. Temurui is a local name of *M. koenigii* (L.) Spreng and widely spread in the province of Aceh. The majority of Acehnese people use this plant as spices. Traditionally this plant has also been used as herbs, condiments and also used to treat various types of ailments in Indian traditional system [9]. Research on *M. koenigii* (L.) Spreng as a bioactivity has been widely studied and reportedly active as antitumor, antioxidant, antimutagen, anti-inflammatory, antidiabetic, stimulant and antibacterial [9]. Research of this plant as anticancer has also been widely reported in several countries, including HT-29 intestinal cancer [10], HL-60 blood cancer and HeLa cervix cancer [11], HTB-37 colon cancer, liver HB-8065 [12] and breast cancer MBA-NB-231 [13].

In Indonesia, especially in Aceh, research on the potential of *M. koenigii* (L.) as anticancer agent has never been reported before. Refers to chemotaxonomy review, *M. koenigii* (L.) Spreng can be potentially active as anticancer. Based on the increasing the number of cervical cancer patients in Indonesia, the researchers focus to develop of *M. koenigii* (L.) Spreng leaves as a natural product for cervical cancer drugs. The results of this study are expected to contribute in the medical to develop *M. koenigii* (L.) Spreng as a natural source for anticancer drug and can be widely used as a safe anticancer drug.

2. Methodology

2.1 Plant Material and Bioindicator

M. koenigii leaves were collected from Langsa, Aceh (Indonesia) in February 2018. The bioindicator used ini this research is human cervical cancer (HeLa) cell line.

2.2 Extraction

The air-dried leaves (1 Kg) of plant materials were ground and extracted with n-hexane by maceration method for 3 x 24 hours, the maceration was repeated until the filtrate is clear. The extract solution was filtered and evaporated by rotary evaporator to give hexane extract.

2.3 Phytochemical Screening

2.3.1 Alkaloid.

About 2 g of plant materials were crushed then added 1 mL of ammonia. Furthermore, 10 mL of chloroform was added, then crushed and filtered. The filtrate was added 10 mL of sulfuric acid 2N, shaken vigorously, left for a minute until the sulfuric acid solution and chloroform separated. The sulfuric acid layer is taken and divided into three test tubes and each test tube is tested by Meyer, Dragendorff, and Wagner reagents to determine the presence of alkaloids. The addition of Meyer reagent established white precipitate, Dragendorff' reagent caused reddish precipitate, and Wagner reagent raised yellow precipitate. Those results indicate the presence of alkaloids.

2.3.2 Terpenoid, Steroid, and Saponin.

Ten grams of plant materials were finely ground, then extracted with hot methanol. The obtained filtrate was concentrated with rotary evaporator to yield methanol extract. The methanol extract was partitioned

IOP Conf. Series: Materials Science and Engineering **523** (2019) 012018 doi:10.1088/1757-899X/523/1/012018

with hexane. The soluble extract in hexane was tested with the Liebermann-Burchard reagent. The blue or green color exhibits the presence of steroids and red color for terpenoids. The insoluble residue in hexane is added water and shaken vigorously. The presence of the stable foam for 30 minutes indicates the existence of saponins, if positive for saponins.

2.3.3 Flavonoid.
Plant materials (10 g) was extracted with methanol and concentrated. The concentrated methanol extract was partitioned with hexane. The residue was extracted with 10 mL of 80% ethanol, subsequently added 0.5 mg of magnesium and HCl 0.5 M. The pink or purple color shows the presence of flavonoids.

2.3.4 Phenol.
Plant materials was tested by Ferric Chloride. Add 3 – 4 drops of $FeCl_3$ solution into extract, the formation of bluish black color exhibits the phenol compound.

2.3.5 Tannin.
About 0.5 g of plant materials were boiled in 10 ml of water in the test tube and then filtered. Add a few drops of $FeCl_3$ 0.1%. Forming of a brownish green or bluish black colour indicates tannins.

2.4 MTT Assay
Cytotoxic activity in this study was treated against breast cancer (MCF-7) cell line. The cell was recognized from the American Type Cell Collection (ATCC). Medium without compound was used as negative control. The cell was cultured using Roswell Park Memorial Institute Medium (RPMI) 1640, Dulbecco's Modified Eagle's Medium (D-MEM), Fetal Bovine Serum (FBS) 5% and Penicillin 100 U/mL, Streptomycin 100 U/mL, maintained at 37°C in 5% CO_2 atmosphere and counted using hemocytometer. The MTT assay was carried out in the 96-wells plate. Briefly, a volume of 100.0 μl of complete growth medium was added into each well of 96-wells flat bottom microtiter plate (Nunclon, USA). The hexane extract was varied with concentration of 1000, 500, 100, 50, 20, 10, 5, dan 1 μg/ml, aliquoted into wells in triplicate and serially diluted. A volume of 100.0 μl of $1x10^5$ cells/mL MCF-7 cells were seeded into 96-wells flat microtiter plates and incubated for 24 hours in CO_2 incubator. After 24 hours incubation, a volume of 100.0 μl of MTT solution was added into each well and incubated for 4 hours. The culture medium was removed and the SDS 10% in 0.1 N HCl solution was added to each well to solubilise the formazan formed. The plate was red using the plate reader at 595nm wavelength (Infinite M200, Tecan, Switzerland) [14].

3. Result and Discussion

3.1 Phytochemical Screening
Phytochemical screening aim to identify the compound groups contained in the sample. The phytochemical screening was carried out on leaves and hexane extract of *M. koenigii* (Linn.) Spreng using various phytochemical reagents. Examination on leaves showed the active phytochemical classes as alkaloids, terpenoids, flavonoids, phenols, and tannins, while hexane extract just presence the terpenoids as showed in Table 1.

Phytochemical screening of Hexane Extract of *M. koenigii* (Linn.) Spreng Leaves showed the presence of terpenoids, a nonpolar secondary metabolite. Hexane is a nonpolar solvent that caused the secondary metabolites extracted by hexane should be a nonpolar. Alkaloids, flavonoids, Saponins, Phenols and Tannins are polar secondary metabolite. So, they should not presence in the extract of hexane. Major classes of anticancer compounds include alkaloids, terpenoids, flavonoids and lignans [15]. The presence of terpenoid from *M. koenigii* (Linn.) Spreng leaves also reported from previous studies [16, 17, 18].

IOP Conf. Series: Materials Science and Engineering **523** (2019) 012018 doi:10.1088/1757-899X/523/1/012018

Table 1. Phytochemical Screening of *M. koenigii* (Linn.) Spreng

Secondary Metabolites	Leaves of *Murayya koenigii* (L.) Spreng	Hexane Extract of *Murayya koenigii* (L.) Spreng Leaves
Alkaloid	+	-
Terpenoid	+	+
Steroid	-	-
Saponin	-	-
Flavonoid	+	-
Phenol	+	-
Tannin	+	-

Terpenoids composed of "isoprenoid" units constitute one of the largest groups of natural products accounting for more than 40 000 individual compounds, with several new compounds being discovered every year. Terpenoids are synthesized from two five-carbon building blocks. Based on the number of building blocks, terpenoids are classified into several classes, such as monoterpenes, diterpenes, triterpenes and tetraterpenes. The diverse array of terpenoid structures and functions has provoked increased interest in their commercial use. Terpenoids have been found to be useful in the prevention and therapy of several diseases, including cancer, and also use as antimicrobial, antifungal, antiparasitic, antiviral, anti-allergenic, antispasmodic, antihyperglycemic, anti inflammatory, and immunomodulatory properties [19].

3.2 Cytotoxic Activity

In this study, the hexane extract from *M. koenigii* (Linn.) Spreng leaves was evaluated for cytotoxic activity against HeLa cell line. The cytotoxicity of the extract was assayed at various concentrations of 1000, 500, 100, 50, 20, 10, 5, and 1 µg/ml under continuous exposure for 72 h, are expressed in CD_{50} values and are summarized in Table 2. Then, viability data is plotted with concentration doses to determine CD_{50} (Figure 1). Results showed as CD_{50} represent the extract concentration doses that reduced the mean absorbance at 595 nm to 50% of those in the untreated control wells. The CD_{50} value was obtained from the plot of the concentrations of extract versus percent of cell viability. The value was used to describe the degree of cytotoxicity of the extract towards cell lines [14].

Table 2. Cytotoxic Activity of Hexane Extract of *M. koenigii* (Linn.) Spreng Against Servical Cancer (HeLa) Cell Line

Hexane Extract (µg/ml)	I	II	III	Average	%Viability
1000	0.016	0.029	0.031	0.025	2.484
500	0.016	0.028	0.033	0.026	2.517
100	0.414	0.337	0.119	0.290	28.441
50	0.034	0.365	0.083	0.161	15.757
20	0.149	0.095	0.050	0.098	9.611
10	0.162	0.202	0.047	0.137	13.436
5	0.054	0.068	0.076	0.066	6.473
1	0.058	0.064	0.081	0.068	6.636
Cell control	1.004	1.012	1.043	1.020	100.000

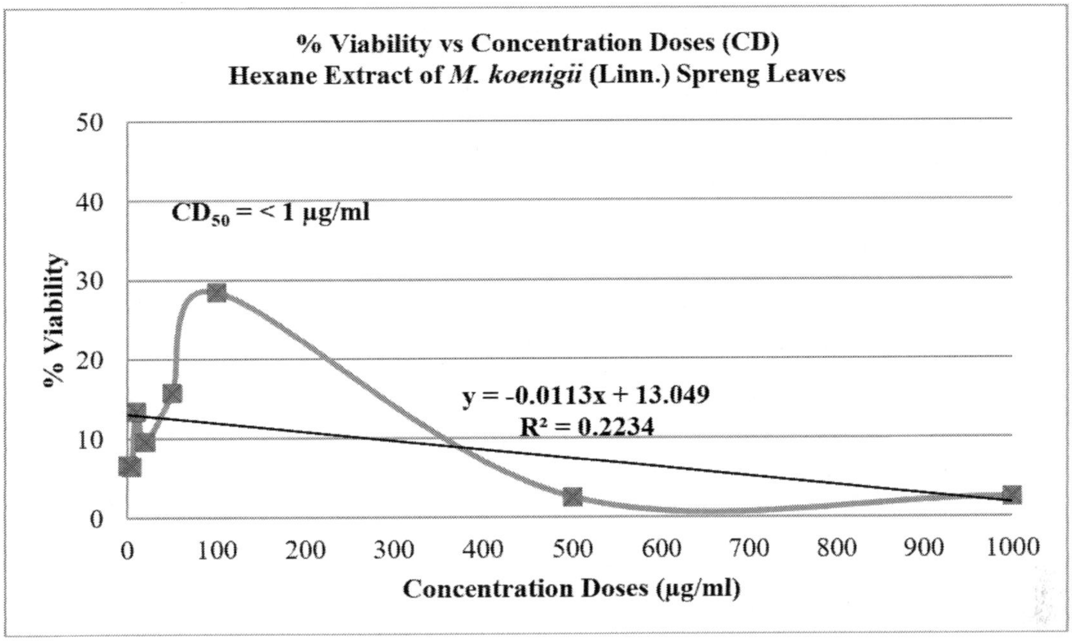

Figure 1. Viability (%) vs Concentration Doses (µg/ml) of Hexane Extract from *M.koenigii* (L) Spreng Leaves

Based on the result of cytotoxicity against HeLa cell line, hexane extract of *M.koenigii* leaves showed cytotoxic activity with CD_{50} value less than <1 µg/ml. Previous studies, vincristine, a conventional drug of cancer, was use as positive control against HeLa cell line and showed CD_{50} value of 0,4 µg/ml [20]. Compare with the positive control of vincristine, hexane extract of *M.koenigii* leaves gave a very good cytotoxic activity against HeLa cell line. Plant extract with CD_{50} less than 1000 µg/ml identified as active against cancer cell line. This cytotoxic activity of hexane extract from *M.koenigii* leaves is contributed by secondary metabolites contained in the plant that can kill or inhibit cancer cell growth. This result showed a potential natural product of *M. koenigii* and could be developed as anticancer agent.

4. Conclusion

The phytochemical screening performed on the hexane extract of *M. koenigii* (Linn.) Spreng showed the presence of terpenoids. This extract showed a very strong cytotoxic activity effect against HeLa cell line with CD_{50} values less than 1 µg/mL. It indicated as a potent cytotoxic activity agent for HeLa cancer cell. Therefore, it is expected to conduct further research for cytotoxic test of other cancer cell lines so that it could be developed as raw materials for the manufacture of new drugs.

5. Acknowledgment

The authors would like to thanks to The Ministry of Research, Technology and Higher Education of Republic of Indonesia (KEMENRISTEKDIKTI RI) who have fully funded this research through the grant of Partnership Research of Higher Education (PKPT). Acknowledgments are also addressed to Universitas Samudra and Universitas Syiah Kuala for all technical supports.

6. References

[1] Ali C I, Makina N E and Ezenduka P O 2016 *Gynecol Obstet (Sunnyvale)* **6** 378 doi:10.4172/2161-0932.1000378

[2] Balogun M R, Odukoya S C, Oyediran M A and Ojomu P I *African Journal of Reproductive Health* **16** 75-82

[3] Nurcahyanti A D R 2016 *Journal of Cancer Biology and Research* **4** 1078
[4] Jain R and Jain S K 2011 *Asian Pacific Journal of Tropical Biomedicine* **1** S147-S150
[5] Bhanot A, Sharma R and Noolvi M N 2011 *International Journal of Phytomedicine* **3** 09-26
[6] Belayachi L, Luquero C A, Merghuob N, Bakri Y, Mattos S F, Amzazi S and Villalonga P 2013 *European Journal of Medicinal Plants* **3** 310-332
[7] Moglad E H O, Abdalla O M, Algadir H A, Koko W S and Saadabi A M 2014 *International Journal of Medicine and Medical Sciences* **1** 32-37
[8] Hassan R, Hussein F, Hawraa M, Akram H, Ahmad K, Ahmad D and Bassam B 2013 *International Research Journal of Pharmacy* **4** 132-136
[9] Handral H K, Pandith A and Shruthi S D 2012 *Asian Journal of Pharmacetical and Clinical Research* **5** 5-14
[10] Sumalatha D and Nithya T G 2014 *Asian Journal of Pharmacetical Clinical Research* **7** 83-86
[11] Tan S P, Ali A A, Nafiah M A, Awang K and Ahmad K 2015 *Tetrahedron* doi:10.1016/j.tet.2015.04.037
[12] Patterson J and Verghese M 2015 *Journal of Pharmacology and Toxicology* **10** 49-59
[13] Yeap S K, Abu N, Mohamad N E, Beh B K, Ho W Y, Ebrahimi S, Yusof H M, Ky H, Tan, S W and Alitheen N B 2015 *BMC Complementary and Alternative Medicine* **15** 306 doi:10.1186/s12906-015-0832-z
[14] Amna U, Hasnan M H H, Ahmad K, Ali A A, Awang K and Nafiah M A 2015 *International Journal of Pharmaceutical Sciences Review and Research* **32** 15-20
[15] Prakash O, Kumar A, Kumar P and Ajeet 2013 *American Journal of Pharmacological Sciences* **1** 104-115
[16] Tan S P, Ali A A, Nafiah M A, Amna U, Ramli S A and Ahmad K 2017 *Chemistry of Natural Compounds* 980-981 doi:10.1007/s10600-017-2177-y
[17] Sharmila S, Jeyanthi R L, Merina P D, Saduzzaman M and Sashi B 2013 *Research Journal of Pharmaceutical, Biological and Chemical Sciences* **4** 1104-1109
[18] Jain M, Gilhotra R, Sigh R P and Mittal J 2017 *Biology and Med*icine **2** 236–256
[19] Thoppil R J and Bishayee A 2011 *World Journal of Hepatology* **3** 228-249
[20] Amna U, Nafiah M A and Ahmad K 2017 *Malaysian Journal of Chemistry* **19** 67 – 74

Synthesis of Kieserite fertilizer by using natural magnesite Ore as raw material

F Mulana[1,*], Mustanir[2], Z Fuadi[3]

[1] Department of Chemical Engineering, Faculty of Engineering, Syiah Kuala University, Banda Aceh, Indonesia

[2] Department of Chemistry, Syiah Kuala University, Banda Aceh, Indonesia

[3] Department of Mechanical Engineering, Syiah Kuala University, Banda Aceh, Indonesia

E-mail : seulanga20@gmail.com

Abstract. Magnesite ore is one kind of rare minerals that can be found in area of Aceh Besar District Province of Aceh, NTT Province and Southeast Sulawesi Province. The further treatment of this rare magnesite mineral containing mainly magnesium to form kieserite fertilizer product with composition magnesium and sulphur would be more benefit. The purpose of this study is to synthesize kieserite fertilizer from magnesite ore as raw material with addition of sulphur and to see whether the composition of prepared product meet the specification of kieserite fertilizer as stated in SNI 02-2807-1992. The synthesis was carried out by mixing the treated natural magnesite ore and the solid sulphur in a ball mill. The product of kieserite fertilizer was then analyzed according to SNI 02-2807-1992. The results showed that kieserite fertilizer can be produced from raw material of magnesite ore and a product met the SNI 02-2807-1992. The prospective product of kieserite fertilizer was obtained after mixing magnesite ore with addition 20 wt% of solid sulphur. The composition of prepared kieserite fertilizer was in range of 26.72-30.20 wt% for MgO, 20.29-23.93 wt% for S and 0.40-0.76 wt% for free water content with the prospective product composition was 30.20 wt% for MgO, 23.93 wt% for S and 0.45 wt% for free water content.

1. Introduction

Inorganic fertilizer is one type of fertilizers that commonly used in palm plantations. Kieserite fertilizer as inorganic fertilizer must contain a minimum of 25.5 wt% of magnesium in the form of MgO and 21 wt% of sulphur in the form of S as standardized in SNI 02-2807-1992 [1]. Magnesium is a macronutrient that assist in the formation of chlorophyll and assist in transport of phosphate in plants that produce more fresh fruit bunches (FFB) and high oil content [2]. Whereas, sulphur is responsible for formation of amino acids and growth of shoots, seedlings and helps the formation of root nodules and serve to increase the resistance of plants to fungi [3].

The synthesis of kieserite fertilizer can be formed from natural magnesite ore. Deposit of magnesite ore in Indonesia is only found in three provinces, namely in NTT Province, Southeast Sulawesi Province and Aceh Province. The deposit of magnesite ore in Aceh Province is found in Aceh Besar District with deposit around 84.45 million cubic meters [4].

* Corresponding author: seulanga20@gmail.com

Nowadays, community groups or local euntepreneurs have mined magnesite ore by direct mining it from deposit location, then packed with a particular brand and sold it as a magnesium fertilizer with a high Mg content but with very low S content [5,6]. Kieserite as a natural mineral with chemical formula of $MgSO4.H2O$ in the form of a salt deposit was first discovered in Germany [7]. Considering a huge magnesite ore deposits located in Aceh Besar District, Aceh Province and a large amount of kieserite fertilizer necessity to support the productivity of oil palm plantations, therefore this study was conducted to assess the process of kieserite fertilizer formation from magnesite ore and solid sulphur so that these produced kieserite fertilizers could conform to SNI 02-2807-1992 [1] to increase products added value. Some previous studies have been done on the process of kieserite fertilizer formation with various materials and methods [8,9]. Researchers generally use variety sources of magnesium and sulfur to form kieserite fertilizer. Several other researchers used sulphuric acid and solid sulphur [10]. The results of previous study showed that the use of different sources of sulphur would also produce different product composition and by-products generated. Therefore, in this study we use of natural resources of magnesite ore from Aceh Province as raw material to form kieserite fertilizer. This study is an innovative effort in order to increase the added value of the raw mineral of magnesite ore.

2. Methods

2.1. Materials and Equipments

Materials used in this research were the natural magnesite ore of Aceh Besar District and a solid sulphur mineral purchased from supplier in the market. The chemicals were purchased from commercial available supplier. Equipments used included mortar, ceramic bowl, milling tools, sieves (Test-Sieve BS 410-69), vibration apparatus (Fritsch), erlenmeyer, thermometers, digital scales (Acis), mixer, ball mill, oven and furnace.

2.2. Research Procedure

The formation process of kieserite fertilizer was as following phases: firstly, raw materials of magnesite ore was collected from area of Kuta Cot Glie Sub-District in Aceh Besar District. Next, raw material of magnesite ore was dried under sunlight for a couple of days in order to reduce the water content. Then, the dried magnesite ore was crushed by using a mortar in a ceramic bowl to reduce size of material before conducting milling in a ball mill apparatus and finally performed a sieving process to obtain the same size of +80-100 Mesh. The grain size of magnesite ore is then performed a calcination process at a temperature of 700 °C [11]. Finally, the certain amount of calcined magnesite ore was mixed and milled with solid sulphur in the variety of weight (KF-01 for 20 wt% of S; KF-02 for 25 wt% of S; KF-03 for 30 wt% of S and KF-04 for 35 wt% of S samples) to form kieserite fertilizer in a ball mill at milling speed of 300 rpm for around 30 minute. The produced kieserite fertilizer was then stored in dessicator for further analysis.

2.3. Analysis Procedure

Analysis procedures of raw material and final products of kieserite fertilizer was performed following the procedure in the SNI 02-2807-1992. The analysis included content analysis of magnesium as MgO, content analysis of sulfur as S and content analysis of free water. All of analysis results were subsequently processed and displayed in the form of tables and figures.

3. Results and Discussion

Recently, communities who live in surrounding of magnetite ore deposit location and or entrepreneurs had been mining the mineral by using a simple methods and equippments. After mining and cleaning from some impurities, they generally sold minerals in bulk or put in bags of 40 or 50 kg without any other treatment. Some of them named their product as kieserite fertilizer not as magnesium fertilizer [5,6]. Those products without any treatment still have composition of natural magnesite ore which is lacking of sulphur content and therefore it is not appropriate if named as kieserite fertilizer when we refers to SNI 02-2807-1992 because according to SNI "kieserite fertilizer is a mineral material containing nutrients of magnesium at least 25.5% by weight and sulfur at least 21% by weight". The detail of any composition in kieserite fertilizer referring to SNI 02-2807-1992 can be seen in Table 1.

Table 1. Requirements of kieserite fertilizer based on SNI 02-2807-1992

No	Parameter	Requirements
1	Magnesium content as MgO, %	Min. 25.5
2	Sulphur content as S, %	Min. 21.0
3	Free water content, %	Max 0.5

Based on the facts and activity in the mining site and refer to above SNI 02-2807-1992, fertilizers that mined and sold by communities/entrepreneurs are more accurately named as magnesium fertilizer rather than kieserite fertilizer. The natural magnesite ore mineral that mined by communities/entrepreneurs without any treatment was dominated by the content of MgO as much as 32.72 wt % and contained only 7.46 wt% of S as can be seen in Table 2. The composition of natural magnesite ore and other products were tested by using titrimetry and gravitymetry analysis in a known and accredited laboratory of PT Sucofindo, Medan, Indonesia. Because of the lacking of sulphur content in natural magnesite ore so that we attempted to synthesize kieserite fertilizer in accordance with the Indonesian National Standard (SNI) by utilizing natural magnesite ore minerals from Aceh Besar District, Aceh Province as a raw material.

Table 2. Composition of natural and calcined magnesite ore and composition of prepared kieserite fertilizers

No	Parameter	Natural magnesite ore	Calcined magnesite ore	Produced kieserite fertilizer			
				KF-01	KF-02	KF-03	KF-04
1	Magnesium content as MgO, %	32.72	37.8	30.20	26.74	28.08	26.72
2	Sulfur content as S, %	7.46	0.32	23.9	20.29	21.96	21.14
3	Free water content, %	na	3.5	0.45	0.76	0.47	0.40

Note: KF = kieserite fertilizers; na = not available; calcined at 700 °C

The process used to synthesize kieserite fertilizer was initiated with the size reduction process of

IOP Conf. Series: Materials Science and Engineering **523** (2019) 012019 doi:10.1088/1757-899X/523/1/012019

both raw materials namely the natural magnesite ore that contains of approximately 32.72 wt% of MgO and solid sulfur that contains of approximately 99.9 wt% of S. After milling and sieving at size of +80-100 Mesh, the natural magnesite ore was calcined at 700 °C under air atmosphere. The product of calcination was measured its composition as can be seen in Table 2.

The calcined magnesite ore contained 37.8 % of MgO and 0.32 % of sulphur. The contain of sulphur was similar to sulphur content in our previous research [12]. The calcination process has increased the content of MgO from 32.72 % of natural magnesite ore to 37.8 % of magnesite ore. During calcination process at high temperature of 700 °C under air atmosphere, some impurities in natural magnesite ore was evaporated and resulted in also increasing the content of MgO. The calcination process also decreased the content of sulphur in mineral because sulphur was changed to sulphur dioxide dan evaporated to air.

Theoretically, mixing both the calcined magnesite ore as raw material and solid sulphur would produce a product with high content of MgO and S that can be determined precisely and accurately. However in practice there are several factors that can change the content of MgO and S in the product of kieserite fertilizer. One of factors is the homogeneous mixing of raw materials and the precision of the methods of analysis of the product. The composition of all prepared kieserite fertilizer can be seen in Table 2.

Table 2 showed that magnesium content as MgO from four products of kieserite ranged from 26.72 – 30.2%, sulphur content as S ranged from 20.29 – 23.9 % and free water content ranged from 0.40 – 0.76. From these results, it can be seen that product labeled with KF-02 has fulfilled in SNI 02-2807-1992 in magnesium content only but not yet met the criteria of sulfur content and free water content.

According to SNI 02-2807-1992 the content of magnesium as MgO must be at least 25.5 wt%, content of sulphur as S must be at least 21 wt% and free water content must be a maximum of 0.5 wt%. Other products labeled as KF-01, KF-03 and KF-04 have fulfilled all requirements as stated in SNI 02-2807-1992. However, considering the price of solid sulphur is higher than the price of mineral magnesite ore then the more economical product is the product that used the lowest solid sulphur amount and product must meet all requirements of SNI 02-2807-1992. Therefore, based on the result of laboratory analysis, the prepared product labeled with KF-01 was considered as prospective product in this study.

A B C

Figure 1. Photographs of natural magnesite ore (A), after sieving of 80-100 mesh (B) and after milling with sulphur (C)

Figure 1 showed photographs of natural magnesite ore, after sieving and calcination process and after milling with solid sulphur. The figure showed clearly that the size of raw material of natural magnesite ore was still unorder with the white greyish color due to the high content of magnesium (Figure 1 (A)). The size became similar after sieving with size of +80-100 mesh and the color became brighter due to calcination at high temperature (Figure 1 (B)).

The prepared kieserite fertilizer as can be seen in Figure 1 (C) had a powdered apparent with the color of kieserite fertilizer product tend to be grey yellowish due to addition some amount of solid sulphur.

4. Conclusion

Natural mineral of magnesite ore from Aceh has a huge potential to be used as raw material for formation of kieserite fertilizer. Kieserite fertilizer can be made from milling of calcined magnesite ore and solid sulphur and the product met the Indonesian National Standard (SNI) 02-2807-1992.

The prospective composition of produced kieserite fertilizer was 30.2 wt% for MgO content, 23.9 wt% for S content and 0.45 wt% for free water content. This kieserite fertilizer product was obtained from milling of calcined magnesite ore and 20 wt% of solid sulphur product.

5. References

[1] Standar Nasional Indonesia 1992 *Pupuk Kiserit* SNI Number 02 - 2807- 1992
[2] Pahan I 2008 *Panduan Lengkap Kelapa Sawit* ISBN 979489995X Penerbit Penebar Swadaya
[3] Nurman I 2018, http://ceritanurmanadi.wordpress.com/2012/01/26/fungsi-sulfur-s-atau-belerang-bagi-tanaman//, explored on 7 May 2018
[4] Dinas PE, Aceh Besar 2013 *Perkiraan Potensi Serpentenit Menurut Luas dan Cadangan* Dinas Pertambangan dan Energi, Kabupaten Aceh Besar
[5] Anonimous, 2018a, *Kieserite Bola Dunia,* http://tanraju.blogspot.com/p/kieseriete-bola-dunia.html explored on 11 June 2018
[6] Anonimous, 2018b, *Magnesium kieserit green Sumatera* https://agrounited.wordpress.com/magnesium-kieserit-green-sumatera/ explored on 11 June 2018
[7] Benckiser R 2013 *Preparation of Magnesium Sulphate* Advancing of Chemical Science
[8] Jamieson C S 2011 *The variable spectrum of kieseritee: Grain size and temperature effects* EPSC Abstracts **6**
[9] Hong X L, De W Z and Yan Y 2013 *Advanced Materials Research*, **807 – 809** 2788-2792
[10] Messick D L and Brey C D 2012 *"Sulphur Fertilizers – New Products Add to Conventional Sources to Offer a Widge Range of Options"* The Sulphur Institute, Washington, USA
[11] Gence N 2011 *Enrichment of Magnesit Ore* Eng. & Arch. Fac. Osmangazi University **14**
[12] Ismayanda M H dan Mulana F 2014 *Jurnal Rekayasa Kimia dan Lingkungan,* **10** (2) 78-84

IOP Conf. Series: Materials Science and Engineering **523** (2019) 012020 doi:10.1088/1757-899X/523/1/012020

The influence of Fe_3O_4 on Magnetic Chitosan composite preparation for Methylene blue removal from water

I Ishmaturrahmi[1], R Rahmi[1,*], I Mustafa[1]
Chemistry Department, Syiah Kuala University, Banda Aceh, Indonesia, 23111

Email:rahmi@fmipa.unsyiah.ac.id

Abstract. The influence of Fe_3O_4 on magnetic chitosan composite preparation for methylene blue removal from water had been studied. Fe_3O_4 was loaded on chitosan with various content. The prepared magnetic chitosan composites were characterized by Fourier Transform Infrared (FT-IR), Scanning Electron Microscopy (SEM) and X-Ray Diffraction (XRD) analysis. Results showed Fe_3O_4 loading on magnetic chitosan composites preparation improved methylene blue adsorption capacity of chitosan. The best ratio of chitosan/Fe_3O_4 obtained in this work was 0.35g/0.5g. The methylene blue adsorption capacity of the magnetic chitosan composite was higher than pure chitosan. Magnetic composite can be used as potential adsorbent for methylene blue removal from water.

1. Introduction

Waste is an increasingly crucial environmental problem as technology develops and changes in the times. One of the wastes that become environmental problems is dye. Staining products in various industries such as textiles, varnishes, leather and plastics generally produce toxic waste andhighly carcinogenic. They give mutagenic effects on humans and aquatic animals [1]. Methylene blue is a harmful dyestuff used by the textile industry. Methylene blue can cause irritation of the digestive tract, cyanosis and skin irritation. This compound is quite stable so it is difficult to degrade in nature and harmful to the environment. In high concentrations, methylene blue can increase COD (Chemical Oxygen Demand) and can damage the balance of environmental ecosystems characterized by the death of aquatic organisms around the waste disposal site [2]. Therefore, further treatment for dyestuff waste before disposal to the environment is necessary.

Treatments for wastewater have been widely used, one of which is adsorption [3]. The adsorption process is the best alternative because it is able to eliminate the polluting dye effectively by using adsorbents that have high absorption capacity. Currently, the development of waste-based adsorbents such as chitosan derived from shrimp shells waste with high adsorption capacity and biodegradation [4]. The ability of chitosan in adsorption is still limited due to its weakness as easily coagulated and easily soluble in acid and difficult to separate from the solution. Therefore, chitosan was often combined with other materials [5] such as the use of Fe_3O_4 to facilitate chitosan separation from solution after adsorption [1].

* Corresponding author : rahmi@fmipa.unsyiah.ac.id

In this work, chitosan was combined with Fe_3O_4 obtained from coastal sand of Syiah Kuala, Aceh. Magnetic material is an appropriate solution for the combination of chitosan where insoluble and paramagnetic properties allow separation to be more efficient [6].

2. Experimental

2.1. Materials
Chitosan was purchased from Tokyo Chemical Industry Co., Ltd. Japan (obtained from shrimp shell where deacetylation degree was 75.0-85.0%). Iron sand were collected from coastal sand of Syiah Kuala, Aceh. HCl, NH_4OH, NaOH, acetic acid and methylene blue.

2.2. Isolation of Fe_3O_4 particles
Fe_3O_4 particles were isolated by coprecipitation method. The iron sand was dissolved into 35 mL HCl at a temperature of 70^{0}C and stirred for 30 minutes using a magnetic stirrer. The solution was filtered with filter paper. After 30 minutes the NH_4OH was added to the solution while stirring with a magnetic stirrer and heated at a temperature of 70^{0}C for 30 minutes. The obtained precipitate was washed with distilled water until neutral pH was reached. Then the precipitate was dried in the oven at 70^{0}C for 2 hours [7,8].

2.3. Preparation of magnetic chitosan composite
Magnetic chitosan was prepared by mixing 0.35 g of chitosan into 20 mL of 2% acetic acid and stirred using a magnetic stirrer for 2 h. Fe_3O_4 particles were then added to the chitosan solution with several compositions (0.1, 0.2, 0.3, 0.4, 0.5, 0.6 and 0.7 grams) and stirred manually for 1 hour. Afterwards the dope solution was dropped into 3M NaOH solution to form the microsphere and then washed with distilled water until the pH neutral reached and dried. Each magnetic chitosan composites obtained were examined their adsorption capacity on methylene blue adsorption. The magnetic chitosan composite with highest adsorption capacity of methylene blue was then characterized by using Fourier Transform Infra Red (FTIR), X-Ray Diffraction (XRD) and Scanning Electron Microscopy (SEM).

2.4. Adsorption studies
Methylene blue adsorption onto the composite was performed in batch experiments. 0,1 g of the composite was placed into erlenmeyer flasks containing 10 mL methylene blue solution. The mixture was shaken at room temperature, 150 rpm for 25 minutes. The mixture was separated by using permanent magnet and the absorbance of methylene blue in the solution was determined by using UV-Visible spectrometer. Methylene blue adsorption capacity (q, mg/g) and removal efficiency (R,%) were calculated using the following equations:

$$q = \frac{(Co-Ct)V}{m} \quad (1)$$

$$R(\%) = \frac{Co-Ct}{Co} \times 100 \quad (2)$$

Where C_0 (mg/g) and C_t(mg/L) are the initial methylene blue concentration and final concentration of methylene blue after contacting with the composite for certain time, t (min). V (L) is the volume of methylene blue and m (g) is the mass of the composite. UV-Visible spectrometry analysis was carried out at 655 nm. Calibration curve was obtained by using methylene blue concentrations of 1, 2, 3, 4, 5, 6 and 7 ppm.

IOP Conf. Series: Materials Science and Engineering **523** (2019) 012020 doi:10.1088/1757-899X/523/1/012020

3. Result and discussion

3.1. Characterization

The chemical structures of the raw materials and magnetic chitosan composite were studied by FTIR spectroscopy. FTIR spectra of chitosan, and magnetic chitosan composite are shown in Fig. 1.

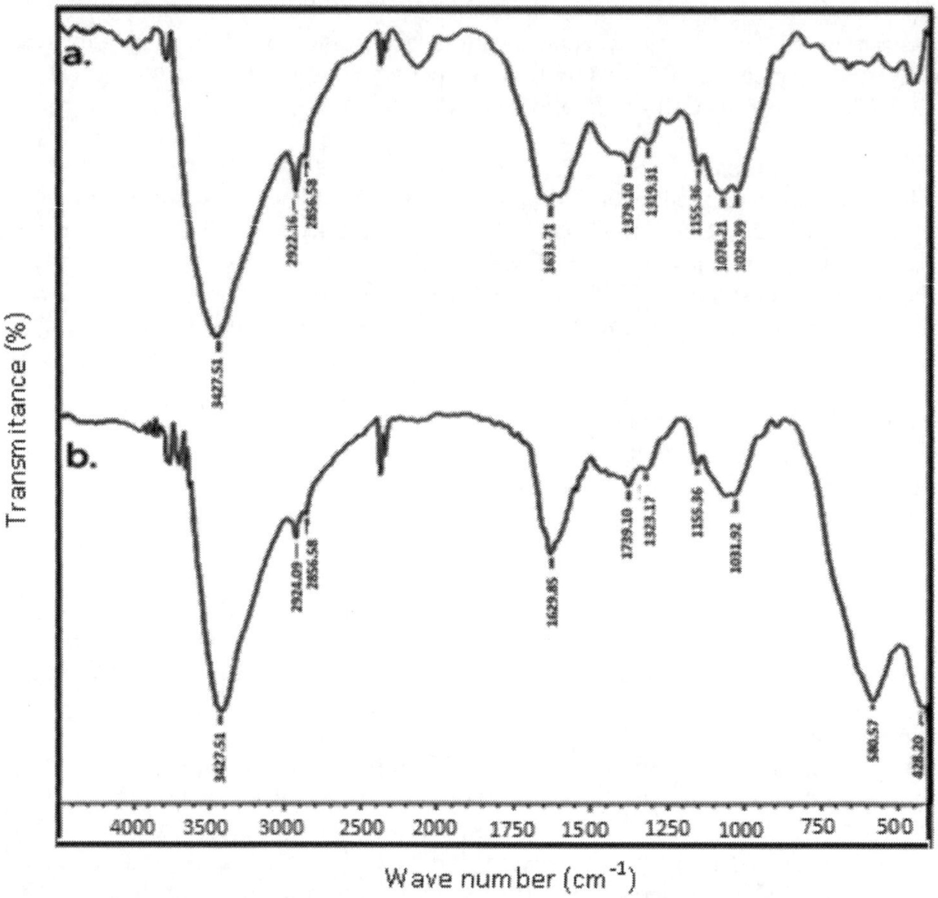

Figure 1. Spectra of chitosan (a) and magnetic chitosan composite (b)

Figure 1a exhibits typical FTIR spectrum of chitosan, where $-NH_2$ stretching vibration band appears at wave number 3427.51 cm^{-1} and amide stretching vibration at wave number 1633.71 cm^{-1}. Compared with FTIR spectrum of chitosan, FTIR spectrum of magnetic chitosan composite shows some differences. Band at wave number 2922,16 cm^{-1} of chitosan spectrum shifts to wave number 2924,09 cm^{-1} on magnetic chitosan composite spectrum. Band at wave number 1029.99 cm^{-1} shifts to wave number 1031,92 cm-^{1}. New bands, typical bands of FeO, are found at wave numbers 580.57 cm^{-1} and 428.20 cm^{-1} on magnetic chitosan composite spectrum. These results show interaction between chitosan and Fe_3O_4 in the composite.

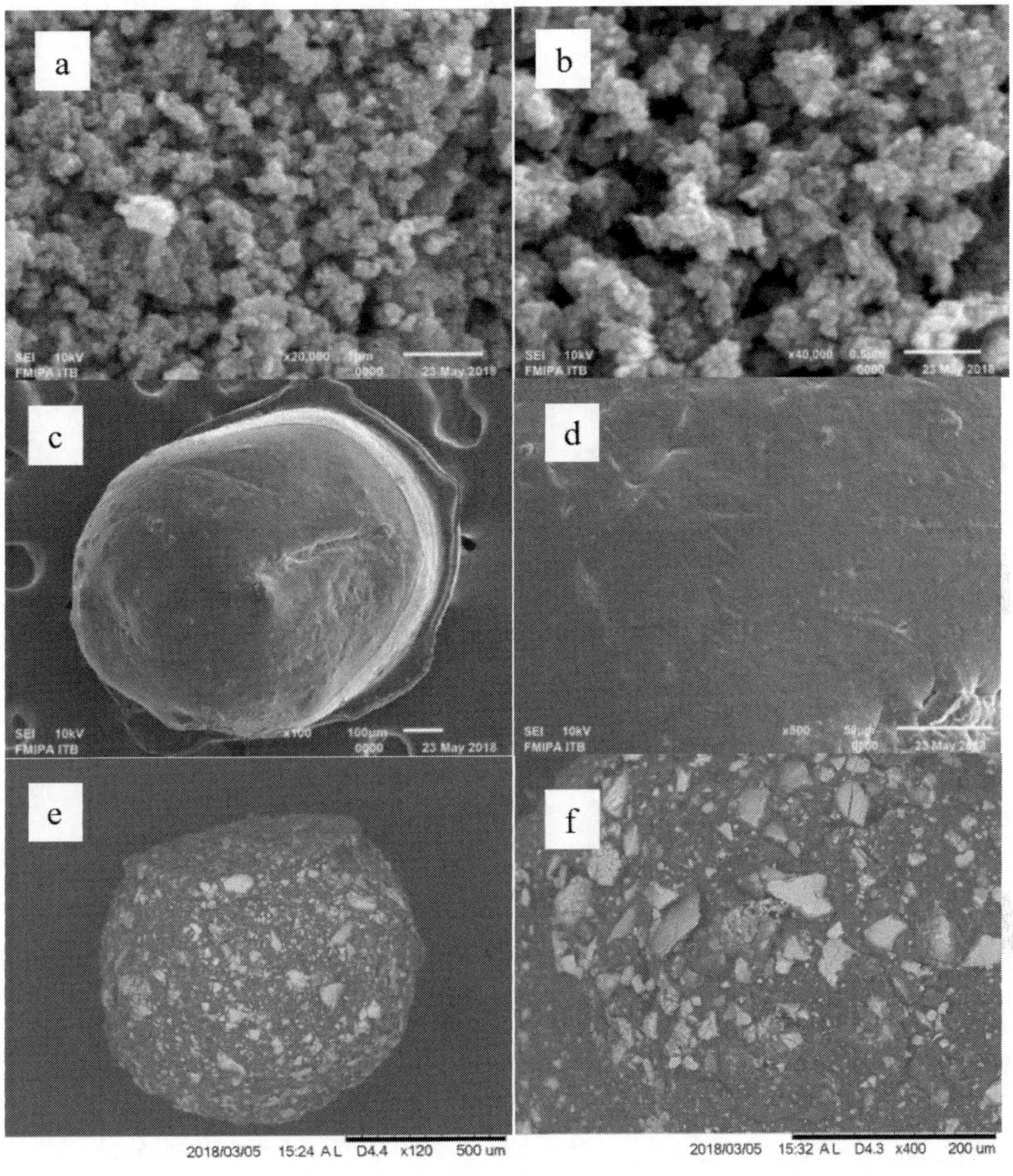

Figure 2. SEM images of Fe_3O_4 (a,b), bead chitosan (c,d) and magnetic chitosan composite (e,f).

Particle size and morphology of materials are studied by SEM. The SEM images of Fe_3O_4, chitosan bead and magnetic chitosan composite bead are shown in Figure 2. Figure 2a and 2b are SEM images of Fe_3O_4 particles with different magnification. Figure 2c and 2d are SEM images of chitosan bead. The surface of chitosan bead is smoother than the surface of magnetic chitosan composite bead (Figure 2e and Figure 2f). The difference of surface morphology of chitosan bead and magnetic chitosan composite bead is due to the Fe_3O_4 distribution on the surface magnetic chitosan composite bead.

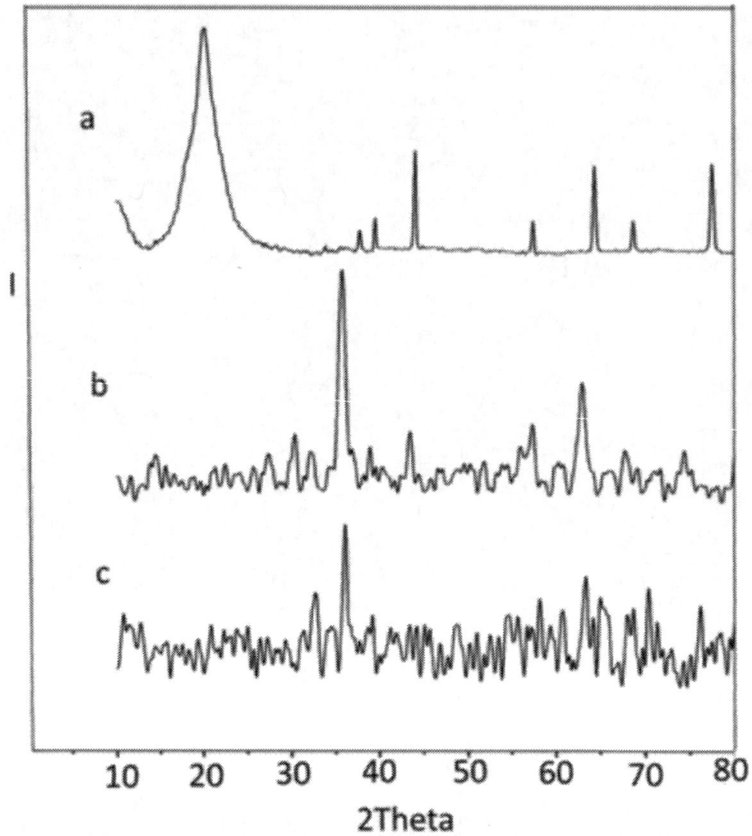

Figure 3. XRD patterns of chitosan (a), Fe_3O_4 (b) and magnetic chitosan composite (c).

Figure 3 shows the XRD pattern of chitosan (Figure3a), Fe_3O_4 (Figure3b) and magnetic chitosan composite (Figure 3c). Chitosan peak (Figure 3a) in this study shows 2θ at 19.5°. XRD pattern of chitosan exhibits 2θ=19.5°. Karimi [9] and Ge [10] reported that chitosan has typical peak at 2θ=19°-20°. Figure. 3b shows the typical XRD pattern of Fe_3O_4 at 2θ=30,10°, 35,44° and 62,59°. XRD pattern of Fe_3O_4 (Figure.3b) shows that Fe_3O_4 has nano size particle with size 0.0587 nm. It was obtained by Scherrer formula, the crystallite size D is given by [11].

$$D = \frac{0{,}9\lambda}{B(\cos\frac{2\vartheta}{2})} \quad (3)$$

Where 0.9 is a correction factor used for the determination of the full width half maximum (FWHM), the wavelength λ is 0.5418 Å, B is the FWHM, and $\cos\frac{2\vartheta}{2}$ is the cosine of the angle.

The XRD pattern of magnetic chitosan composite (Figure 3c) shows the loss of chitosan crystallinity at 2θ=19,5°. Associated with that, there is a low Fe_3O_4 diffraction pattern at 2θ 35.42°, 43.05° and 62.5°. The combination of chitosan and Fe_3O_4 (magnetic chitosan composite) forms amorphous material.

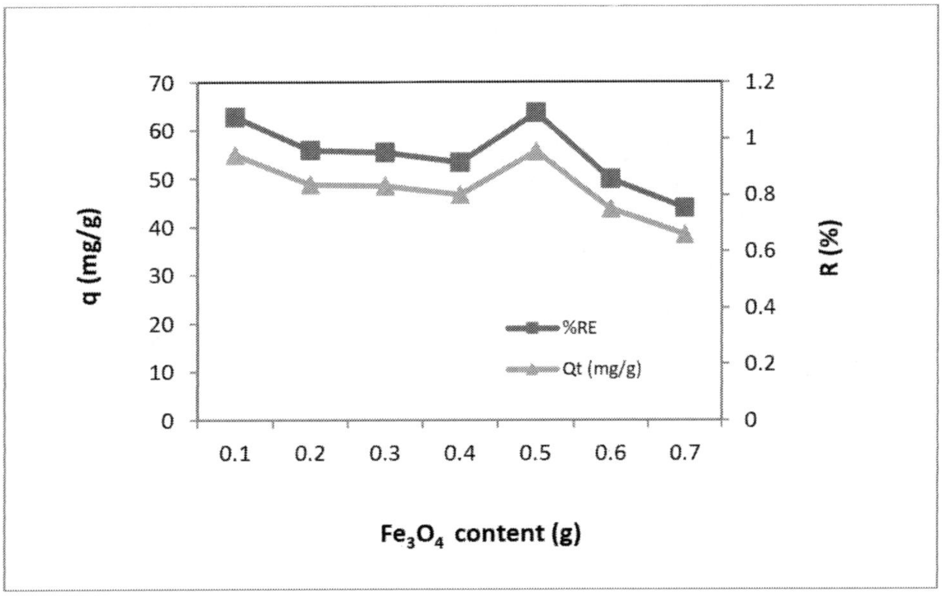

Figure 4. Adsorption capacity of methylene blue onto magnetic chitosan composites with different content of Fe_3O_4 particles

Figure 4 shows methylene blue adsorption capacity of magnetic chitosan composites with different content of Fe_3O_4 particles. Adsorption capacity of magnetic chitosan composites that contain Fe_3O_4 particles more and less than 0.5 gram are lower than adsorption capacity of magnetic chitosan composite that contain 0.5 gram of Fe_3O_4 particles. In low content of Fe_3O_4 particles the main component of adsorben is chitosan where Fe_3O_4 particles were covered by polymer chains of chitosan. By increasing Fe_3O_4 particles content, the adsorption capacity decrease. However, when Fe_3O_4 particles content is 0.5 gram, chitosan can not totally cover the Fe_3O_4 particles that lead interaction between methylene blue and Fe_3O_4 particles resulting in increasing in adsorption capacity. In high content of Fe_3O_4 particles, chitosan content decrease and its ability to cover Fe_3O_4 particles also decrease. The main component of adsorbent is Fe_3O_4 particles that has lower adsortion capacity than chitosan. Therefore, the best magnetic chitosan composite is obtained at Fe_3O_4 particles content of 0.5 gram where it shows the high adsorption capacity (62,7%). The magnetic chitosan composite at Fe_3O_4 particles content of 0.1 gram also shows high adsorption capacity (63,6%), however it was not easily separated from solution by magnet permanent due to low content of Fe_3O_4.

4. Conclusions

Adsorption capacity of chitosan can be improved by Fe_3O_4 loading on magnetic chitosan composites preparation. The ratio of magnetic chitosan composites preparation was obtained 0.35g/0.5g (chitosan/Fe_3O_4) where it showed the highest adsorption capacity (63,6%). This result showed the methylene blue adsorption capacity of the magnetic chitosan was higher than pure chitosan.

5. References

[1] Yang D, Qiu L, and Yang Yaling 2016 *Journal of chemical and engineering data*. **63** 147-158.
[2] Hassan W, Farooq U, Ahmad M, Athar M, and Khan M A 2017 *Arabian Journal of Chemistry*. **10** 1512-1522.
[3] Pavan F A A C Mazzocato and Y Gushikem 2008. *Bioresource Technology*. **99** 3162–3165.
[4] M W Wan, C C Kan, B D Rogel, M L P Dalida 2010 *Carbohydr. Polym*. **80** 891–899.
[5] Nguyen V C, and Pho Q H 2014 *The Scientific World Journa*. **2014** 1-9.
[6] Zhou J, Dong Z, Yang H, Shi Z, Zhou X, and Li R 2013 *Applied Surface Science*, **279** 360–366.

IOP Conf. Series: Materials Science and Engineering **523** (2019) 012020 doi:10.1088/1757-899X/523/1/012020

[7] Hong M K, Park B J, and Choi H J 2007 *Aplications and Material Science*. **204** 4182-4185.
[8] Fahlepy M R, Tiwon VA, Subaer 2017 *Journal of Physics*. Serries **997** 012036
[9] Karimi M H, Mahdavinia G R, Massoumi B, Baghban A, and Sarae M 2017 *Journal Biological Macromolecules*. **113** 361-375.
[10] Ge H Hua T and Chen X 2016 *Journal Carbohydrate Polimers*. **141** 188–196.
[11] Zepeda A M, Gonzales D, Heredia L G, Marquez K, Perez C, Pena E 2018 *Microchemical Journal* **141** 188–196.

IOP Conf. Series: Materials Science and Engineering **523** (2019) 012021 doi:10.1088/1757-899X/523/1/012021

Factorial experimental design for superabsorbent carbonaceous polymer through inverse suspension polymerization method

S N Ainda Mazlan[1,*], S S Jamari[1]
[1]Faculty of Chemical and Natural Resources Engineering, Universiti Malaysia Pahang, 26300 Gambang, Kuantan, Pahang, Malaysia

E-mail: norainda91@gmail.com

Abstract. This paper studies on the effect of synthesis parameters towards the performance of superabsorbent carbonaceous polymer (SPC) via inverse suspension polymerization method. The SPC consists of acrylic acid, acrylamide, and carbon filler. In this work, four independent factors (synthesis parameters) i.e. the content of carbon filler, the content of initiator, the content of crosslinker and reaction temperature that affecting the water absorbency of SPC was investigated. A 2^4 full factorial design was used to investigate the effect of independent factors as well as the interaction factors on the water absorbency of SPC. Apart from reaction temperature, other factors were shown to have the significant effect on the water absorbency of SPC. The results showed the order of significance: initiator's content > crosslinker's content > carbon filler's content > reaction temperature. Meanwhile, interaction factor of carbon filler's and crosslinker's content had the strongest effect on the water absorbency amongst the other interactions. The highest water absorbency of the SPC was achieved at the conditions (filler's content (A): 0.05wt%; initiator's content (B): 0.10wt%; crosslinker's content (C): 0.001wt%, and reaction temperature: 70°C).

1. Introduction

Superabsorbent polymers (SAP) are three-dimensionally cross-linked hydrophilic polymers with the ability to absorb, swell and retain large volumes of water, saline water or aqueous solutions up to hundreds of times their own weight without dissolving [1,2]. High-performance superabsorbent materials conventionally are synthesized from the polymerization process of acrylic esters, polyacrylamide and other unsaturated monomers [3,4]. Inverse suspension polymerization technique is an attractive method to synthesis SAPs in powder or microspheres (beads) form [5].

In this work, SAPs have been incorporated with carbon fiber (filler) due to the value added of the carbon element in soil. The incorporation of these fillers ensures a good balance between high swelling and adequate mechanical properties and even can create new ones (e.g. biocompatibility, biodegradability, antimicrobial, etc.) [6].

Back in the time, one-factor-at-a-time (OFAT) approach is used for running experiments. By using OFAT, the estimation of the factor effects is less precise and interaction between factors cannot be

* Corresponding authors: norainda91@gmail.com

estimated because there is no information and this can mislead the optimal conditions of the process. On the other hand, Design of Experiments (DOE) is an alternative answer to the above challenges. DOE is one of the most useful statistical techniques as it offers a number of advantages. It able to cover a wide range of experimental statistics, obtain unambiguous results with a minimum number of experiments and discover the presence of interaction between the factors of the process. When the interaction effect of several factors on the response need to be study, factorial designs are widely applied particularly 2^k factorial design in the beginning stages of experimental work. 2-level factorial designs have been introduced in screening the factors in which to determine the degree of effect from the factors [7,8].

There has not been any literature reported on the application of full factorial design to determine the effects of factors on the water absorbency of superabsorbent carbonaceous fiber polymer (SPC) via inverse suspension polymerization method. In the present study, the SPC will be synthesized by inverse suspension polymerization with acrylic acid (AA), acrylamide (AM) and carbonaceous fiber as raw material. The effects of carbonaceous fiber filler, N,N'-methylenebisacrylamide (NNMBA) crosslinking agent, ammonium persulfate (APS) initiator and reaction temperature on water absorption are explored. This preparation of SPC using inverse suspension polymerization will become the major challenge due to the properties of the materials used and it is still new. Therefore, 2^4 level factorial design was used to design the experiment, generate a model and optimize the process variable. The overall objective is to maximize the water holding capacity as well as improving the properties of SPC.

2. Experimental

2.1. Materials

Chemicals used for synthesizing the SPC were acrylic acid (AA), acrylamide (AM), N,N'-methylenebis(acrylamide) (NNMBA), ammonium persulfate (APS), sodium hydroxide (NaOH), sorbite anhydride monostearic acid ester (Span-80), cyclohexane and methanol supplied by Merck. Carbonaceous fiber (filler) was obtained by processing Kenaf via hydrothermal carbonization process (HTC).

2.2. Methods

2.2.1. Production of superabsorbent carbonaceous fiber polymer (SPC) via inverse suspension polymerization method.

A continuous phase was prepared by mixing cyclohexane and span-80 in a five-neck round bottom flask equipped with a magnetic stirrer, reflux condenser, nitrogen line and immersed in a water bath. Then, the specific amount of carbonaceous fiber (0.01-0.05wt%) and NNMBA (0.001-0.004wt%) were added to the continuous phase. The mixture was then stirred with 300rpm agitation speed until it was well-mixed. Next, a dispersed phase was prepared by adding partially neutralized acrylic acid, sodium hydroxide, acrylamide and the specific amount of APS (0.10-0.25wt%) in a beaker. After that, the dispersed phase was added slowly into the flask that contains the continuous phase and purged with nitrogen gas for 15 minutes. It was then stirred and the reaction was held for 3 hours at the desired temperature (50-70°C). The resulted polymers were then filtered, washed with methanol and dried at 60°C in an oven for 24 hours.

2.2.2. Experimental design.

In the present study, 4 factors e.g. content of filler, initiator, crosslinker, and reaction temperature were taken into account to investigate their effects on the water absorbency of SPC using a 2^4 full factorial design. Table 1 shows the designed factors and levels to be employed for the experiments. A total of 16 experiments runs were conducted. Factor level was coded as -1 (low level) and +1 (high level) where low level indicates the lowest range of the factors and high level indicates the highest range of the factors. The experiments were carried in 3 replicates. The order in which the experiments were run randomized to avoid systematic errors. The outputs of the experimental design were analyzed with Design Expert (DX7) software to evaluate the effects as well as the statistical parameters. The response was analyzed using ANOVA based on the p-value with 95% of confidence level.

Table 1. 2^4 full factorial design matrix and the results.

Standard Order	Run Order	Factor 1 A: Filler wt%	Factor 2 B: Initiator wt%	Factor 3 C: Crosslinker wt%	Factor 4 D: Temperature ^0C	Response Water Absorbency g/g	
						Standard Deviation	Average
1	2	0.01 (-1)	0.10 (-1)	0.001 (-1)	50 (-1)	11.218	105.684
2	4	0.05 (+1)	0.10 (-1)	0.001 (-1)	50 (-1)	15.609	160.292
3	5	0.01 (-1)	0.25 (+1)	0.001 (-1)	50 (-1)	16.547	66.311
4	15	0.05 (+1)	0.25 (+1)	0.001 (-1)	50 (-1)	1.591	84.877
5	13	0.01 (-1)	0.10 (-1)	0.004 (+1)	50 (-1)	0.967	105.675
6	1	0.05 (+1)	0.10 (-1)	0.004 (+1)	50 (-1)	2.948	81.873
7	12	0.01 (-1)	0.25 (+1)	0.004 (+1)	50 (-1)	0.613	70.519
8	7	0.05 (+1)	0.25 (+1)	0.004 (+1)	50 (-1)	1.285	64.785
9	11	0.01 (-1)	0.10 (-1)	0.001 (-1)	70 (+1)	14.808	128.644
10	6	0.05 (+1)	0.10 (-1)	0.001 (-1)	70 (+1)	8.130	178.247
11	14	0.01 (-1)	0.25 (+1)	0.001 (-1)	70 (+1)	2.448	64.338
12	3	0.05 (+1)	0.25 (+1)	0.001 (-1)	70 (+1)	8.106	73.260
13	9	0.01 (-1)	0.10 (-1)	0.004 (+1)	70 (+1)	24.271	108.924
14	10	0.05 (+1)	0.10 (-1)	0.004 (+1)	70 (+1)	1.752	101.011
15	16	0.01 (-1)	0.25 (+1)	0.004 (+1)	70 (+1)	3.347	54.644
16	8	0.05 (+1)	0.25 (+1)	0.004 (+1)	70 (+1)	0.853	52.885

Factor	Level		Units	Type
	Low (-1)	High (+1)		
Filler	0.01	0.05	wt%	Numeric
Initiator	0.1	0.25	wt%	Numeric
Crosslinker	0.001	0.004	wt%	Numeric
Temperature	50	70	^0C	Numeric

2.2.3. Water absorbency test.

Tea bag method was used to measure the water absorbency of the synthesized SPC. A specific amount of SPC was put into the tea bag and then immersed in a beaker of distilled water. After 24 hours, the tea bag was allowed to drain for 10 minutes to remove the excess water and then been weighed. Water absorbency was calculated using below equation;

$$\textbf{Water Absorbency, } Q\left(\frac{g(H_2O)}{g\ sample}\right) = \frac{m_2 - m_1}{m_1} \quad (1)$$

Where m_2 is the weight of tea bag after reach equilibrium and m_1 is the initial weight of tea bag before immersed in distilled water [10].

3. Result and discussion

3.1. Screening of factors affecting the water absorbency of SPC

A total of 16 experimental runs were performed to screen significant process parameters as per 2^4 full factorial design (Table 1). The four parameters were studied simultaneously at a low and high level to determine the degree of the effect of the factors on the response. Independent factor e.g. filler's content (0.01 to 0.05wt %), initiator's content (0.1 to 0.25wt %), crosslinker's content (0.001 to 0.004wt %), and reaction temperature (50 to 70°C) have been taken into account to investigate their effect on the water absorbency. The water absorbency was ranged from 52.885 g water/g sample to 178.247 g water/g sample. The highest water absorbency of the SPC was achieved at the conditions (filler's content (A): 0.05wt%; initiator's content (B): 0.10wt%; crosslinker's content (C): 0.001wt%, and reaction temperature: 70°C).

3.2. Statistical analysis on water absorbency

The significance of the interaction between the factors on water absorbency was evaluated through an effects list of % contribution more than 1 as shown in Table 2. It is noted that the interaction of B*C had the strongest effect on the water absorbency of SPC. The order of significance of main factors: B > C > A > D.

Table 2. Effects list of standardized effects for water absorbency.

Term	Standardized Effects	Sum of Squares	% Contribution
A-filler	11.56	534.66	2.65
B-Initiator	-54.84	12030.31	59.59
C-Crosslinker	-27.67	3061.88	15.17
D-Temperature	2.74	30.08	0.15
AB	-6.56	172.27	0.85
AC	-21.36	1825.58	9.04
AD	0.65	1.70	8.420E-003
BC	16.18	1047.02	5.19
BD	-13.08	684.70	3.39
CD	-4.09	66.88	0.33
ABC	12.62	636.87	3.15
ABD	-2.07	17.13	0.085
ACD	4.31	74.45	0.37
BCD	0.54	1.18	5.839E-003
ABCD	-0.91	3.31	0.016

The relative importance of main effects and their interactions can be seen by Pareto chart as shown in Figure 1. The bonferroni limit line in the chart indicates the maximum statistically significant effect magnitude meanwhile t-value limit line indicates the minimum statistically significant effect magnitude. Any effect that extends beyond this two lines is potentially important. According to the chart in which the bar lengths are proportional to the absolute value of the estimated effects with 95% confidence level, it was possible to verify that the B demonstrates the most significant effect on the water absorbency. The effect of C on water absorbency was half of the effect of B. Even though, the interaction of A*C was lower in comparison with main effects of B and C, but it was also statistically significant at 95% confidence level. The interaction of A*A was followed by the interaction of B*C. Although the main effect of A, the interaction of A*B*C and the interaction of B*D have significant effects on water absorbency, their levels were only a small amount compared to others. Meanwhile, the linear effect of D had been verified as not statistically significant on the water absorbency.

The blue colour bar shows a negative effect of a given factor. A negative effect of a given factor (i.e., B and C) indicates that the respond (water absorbency of SPC) decreases with the increase in the value of that factor. While the orange colour bar shows a positive effect of a given factor. A positive effect of a given factor (i.e., A) indicates that the response increases with the increase in the value of that factor [9]. This was consistent with the findings of [2] and [4] as the increase of carbon filler will give greater values of water absorbency due to the carbonaceous fiber has cross-linked with the polymer chain.

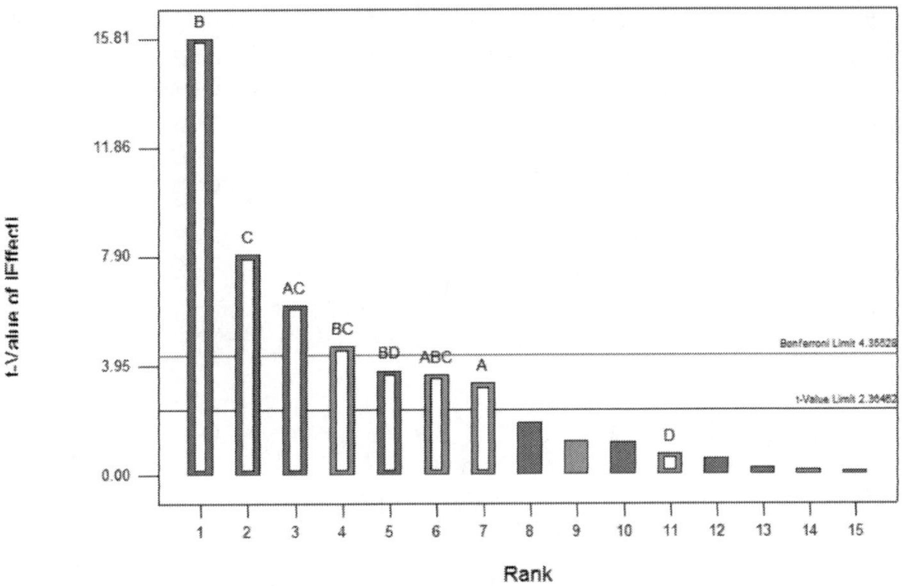

Figure 1. Pareto chart of the standardized effects.

Figure 2 shows the ANOVA analysis for SPC. The p-value of main effects of A, B, and C were less than 0.05 implies the significance of the independent factors. It was necessary to examine any interactions that were important in the experimental design analysis; the interactions of A*C, B*C, B*D, and A*B*C show their significance at p-value smaller than 0.05. The main effects of B and C showed a remarkable model ($p<0.0001$), thus they were the most statistically significant effects from all.

Source	Sum of Squares	df	Mean Square	F Value	p-value Prob > F	
Model	19851.09	8	2481.39	51.56	< 0.0001	significant
A-Filler	*534.66*	*1*	*534.66*	*11.11*	*0.0125*	
B-Initiator	*12030.31*	*1*	*12030.31*	*249.95*	*< 0.0001*	
C-Crosslinker	*3061.88*	*1*	*3061.88*	*63.62*	*< 0.0001*	
D-Temperature	*30.08*	*1*	*30.08*	*0.62*	*0.4552*	
AC	*1825.58*	*1*	*1825.58*	*37.93*	*0.0005*	
BC	*1047.02*	*1*	*1047.02*	*21.75*	*0.0023*	
BD	*684.70*	*1*	*684.70*	*14.23*	*0.0070*	
ABC	*636.87*	*1*	*636.87*	*13.23*	*0.0083*	
Residual	336.91	7	48.13			
Cor Total	20188.00	15				

Figure 2. ANOVA analysis.

Two model terms; B and C were selected for model building with values of Prob>F less than 0.0001 which indicate both of the model terms were the most statistically significant. In the valid range from -1 to +1 for the experimental conditions used in this study, the determination coefficient of the proposed mathematical model can explain about 98.33% of the variance (R^2).

4. Conclusion

The effects of four factors e.g. the content of carbon filler (A), the content of initiator (B), the content of crosslinker (C) and reaction temperature (D) on water absorbency of SPC were studied using 2^4 full factorial design. All independent factors except the reaction temperature were found to have a significant effect on the water absorbency. The sequence of the significant main effects with respect to decreasing of effect was found to be: initiator's content > crosslinker's content > carbon filler's content > reaction temperature. Meanwhile, interaction factor of carbon filler's and crosslinker's content had the strongest effect on the water absorbency amongst the other interactions. The highest water absorbency of the SPC was achieved at the conditions (filler's content (A): 0.05wt%; initiator's content (B): 0.10wt%; crosslinker's content (C): 0.001wt%, and reaction temperature: 70°C). The results showed that full factorial design is convenient in investigating the effect of a large number of factors with a minimum number of experiments instead of OFAT.

5. Acknowledgement

The authors acknowledge the support of Universiti Malaysia Pahang for providing laboratory facilities and financial assistance under Post Graduate Research Grant Scheme (PGRS1803102).

6. References

[1] Chang C, Duan B, Cai J and Zhang L 2010 J. *Eur. Polym.* **46** 92–100
[2] Tuan Zakaria M E , Jamari S S , Yeong Y L and Ghazali S 2017 *IOP Conf. Ser.: Mater. Sci. Eng.* **204** 012013
[3] Doraji S S, Golchin A and Ahmadi S H 2010 *Journal of Soil and Water (Agricultural Sci. and Industries)* **24** 306-16
[4] Jamari S S, Ghazali S and Wan Yaacob W S N 2015 *J. Adv. Agric. Tech.* **2** 156-9
[5] Ahmed E M 2015 *J. Adv. Res.* **6** 105–21
[6] Marcos R G, Fauze A A, André R F, Alessandro F M, Alexandre T P, Magali F T D, Adley F R and Edvani C M 2015 *Eur. Polym. J.* **72** 365-85
[7] Khai E L, Morad N, Tjoon T T and Beng T P 2012 *APCBEE Procedia* (Hong Kong) vol 1 (Elsevier B.V.) p 59 – 65
[8] Matt T 2015 Available from : https://www.qualitydigest.com/inside/quality-insider-article/understanding-design-experiments.html
[9] Deniz B , Dursun S and Dilek S Ö 2015 *Arabian Journal for Science and Engineering* **40** 109-16
[10] Wan Yaacob W S N, Jamari S S and Ghazali S 2015 *Springer Science+Business Media Singapore* 11-7.

IOP Conf. Series: Materials Science and Engineering **523** (2019) 012022 doi:10.1088/1757-899X/523/1/012022

Characterization of adsorbent derived from Coconut Husk and Silica (SiO_2)

M Mariana[1,*], F Mulana[1], S Sofyana[1], N P Dian [2], M R Lubis[1]

[1] Department of Chemical Engineering, Faculty of Engineering, Syiah Kuala University, Darussalam, Banda Aceh, Indonesia
[2] Undergraduate Student of Department of Chemical Engineering, Faculty of Engineering, Syiah Kuala University, Darussalam, Banda Aceh, Indonesia

E-mail: mariana_hasyim@yahoo.com

Abstract. Activated carbon is one of the most commonly used adsorbents because of its high adsorption effectiveness. However, an adsorbent with one type of activated carbon causes a lacking of adsorption efficiency. Therefore, in this study we prepared activated carbon from two types of raw materials. The type of raw material is a combination of coconut husk and silica with several products of activated carbon that contained coconut husk and silica with mass ratio of 70:30, coconut husk with and without activation and pure silica (SiO_2). In this study, raw materials were physically heated at 800 °C for 2 hours and chemically activated with H_2SO_4 solution for 2 hours. FTIR analysis identified a chemical group of Si-O in adsorbent at 898,83 – 1442,74 cm^{-1}. XRD analysis showed that the prepared activated carbon of combination coconut husk:silica; coconut husk with activation; coconut husk without activation and pure silica had silica (SiO_2) composition of 83%; 75%; 62.4% and 63.2%, respectively. The water content in activated carbon of combination coconut:silica; coconut husk with activation; coconut husk without activation and pure silica (SiO_2) was 4.41%; 3.87%; 3.21% and 2.71%, respectively and the ash content was 3.9%, 2.04%; 1.86% and 0.32% respectively. Both water content and ash content met Indonesian National Standard (SNI) No. 06-3730-1995 with the maximum of ash and water content was 10% and 15%.

1. Introduction

Environmental pollution occurred so far either because human activities or natural processes can change the state of the environment. These changes result in decreasing quality of the environment and cannot function properly. Most environmental pollution is caused by industrial waste. With the increase of industrial growth, the amount of waste generated either solid waste, liquid or gas also increased significantly. Liquid waste of tofu industry and batik industry contains various organic compounds such as ammonia and synthetic dyes. Methylene blue is one of the synthetic dyes contained in batik industrial liquid waste. The content of methylene blue in water that exceeds the permitted quality standards can cause human health problems in the form of nausea, vomiting, abdominal and chest pain, headaches, excessive sweating and hypertension, but it can also cause the death of aquatic biota [1,2].

* Corresponding author : mariana_hasyim@yahoo.com.

The nitrogen content in ammonia can cause toxic effect if the compound is in the form of nitrites [3,4]. Nitrogen from nitrite can produce methemoglobin that is a process of inhibiting the transport of oxygen in the bloodstream. According to Government Regulation No. 82 year 2001 on Water Quality Management and Water Pollution Control, the permitted ammonia content is 0.5 mg NH_3/L and Regulation of Environmental Ministry of Indonesia No. 14 year 2013 also states that the amount of methylene blue allowed to be discharged into the environment is only 0.5 mg/L. Therefore, both types of waste before disposal into the environment need to be treated first. Various liquid wastes processing process has been applied, such as adsorption process using adsorbent. The advantages of adsorption process using adsorbent are cheap, easy in operation and environmental friendly [5,6].

Coconut husk as raw material combined with silica was used in this research to produce activated carbon. Coconut husk, an agricultural solid waste that has not been used optimally, contains lignin, cellulose and porous structures and through controlled combustion process coconut husk can be converted to ash which is a source of amorphous silica so it is suitable for use as an adsorbent. The ability of adsorption increases with the increase of silica content in the adsorbent. Silica (SiO_2) is inert and has good adsorption and ion exchange properties, easily modified with certain chemical compounds [7]. To increase the silica content in the activated carbon product, some pure silica needs to be added in the preparation of the adsorbent to produce adsorbents with good adsorption capability and ion exchange. Various studies related to the addition of silica in a particular adsorbent have been made to increase the adsorbent reactivity. Silica and bentonite were utilized to produce adsorbents for adsorption of Fe metal ion in batik liquid waste [8]. In this study, we attempted to produce activated carbon by using two types of raw materials that are derived from the biomass of coconut husk and pure silica.

The purpose of this research is to produce the activated carbon from the combination of coconut husk and silica with certain mass ratio. In this initial study, the produced activated carbon was characterized before being used in the adsorption process of ammonia and methylene blue. The characterization was necessary to know the reactivity of the adsorbent by knowing the type of functional groups, the amount of silica, water content and ash content in the prepared activated carbon.

2. Methodology

2.1. Materials and tools

The materials used in this study were coconut husk and pure powder silica (SiO_2) (Merck), H_2SO_4 98 v% (Merck) and aquades. While the tools used were desiccator, muffle furnace, pH meter, +80-100 mesh mechanical sieve, analytical balance, filter paper and electric oven.

2.2. Adsorbent activation and characterization

Coconut husk weighed as much as 10 g then was carried out the carbonization process using muffle furnace to form carbon [9]. The pure silica was then dissolved by mixing in water with a mass ratio of 10:1, then the activated charcoal from coconut husk and pure silica were mixed and stirred using a magnetic stirrer for 2 hours. After the stirring and filtration process, the wet activated charcoal was heated in the oven for 1 day to remove water content.

Dry activated charcoal was then immersed in 500 ml of sulfuric acid solution for 24 hours. After that it was filtered and rinsed with distilled water until reaching a neutral pH and followed by calcination process using muffle furnace at 800 °C for 2 hours [10]. After the carbonization process, the activated carbon was sieved with a size of +80-100 mesh and the fine activated carbon was stored in a desiccator. The characterization of samples was performed by XRD D-6000 (Shimadzu) to identified composition of adsorben, SEM typed JSM 6510 LA((JEOL) to know morfology of adsorben, and FTIR Prestise -21 (Shimadzu) to identified functional groups.

3. Results and Discussion

3.1. Morphology of adsorbents

Figure 1(a) and 1(b) is SEM image of activated carbon from coconut husk with activation process and without activation process, respectively. While, figure 1(c) and 1(d) is SEM image of activated carbon activated carbon from coconut husk:silica with mass ratio of 70:30 and pure silica (SiO_2), respectively. SEM results show that the activated carbon from combination of coconut husk:silica with mass ratio of 70:30 has larger pore than that of from coconut husk only and pure silica. The SEM results also showed the presence of impurities particles attached to activated carbon without activation process as seen in figure 1(a).

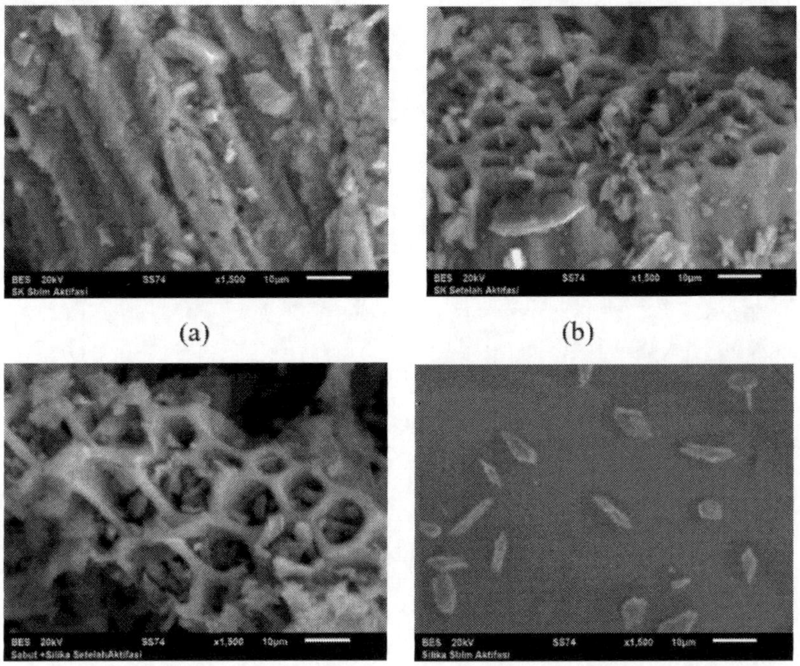

(a) (b)

(c) (d)

Figure 1. SEM image of activated carbon from coconut husk without activation process (a) coconut husk with activation process (b), coconut husk:silica with mass ratio of 70:30 (c), and pure silica (d). the highest magnification was 1500 X.

3.2. Composition analysis using X-Ray Diffraction (XRD)

The XRD analysis results of four types of adsorbents are showed in figure 2 to 5.

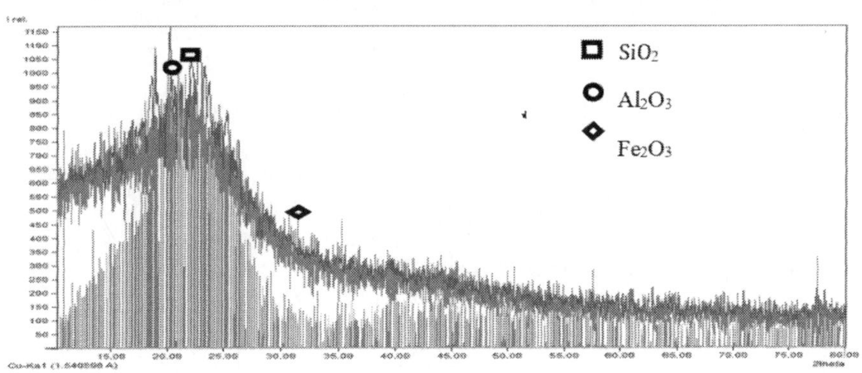

IOP Conf. Series: Materials Science and Engineering **523** (2019) 012022 doi:10.1088/1757-899X/523/1/012022

Figure 2. XRD profile of activated carbon from coconut husk:silica with mass ratio of 70:30

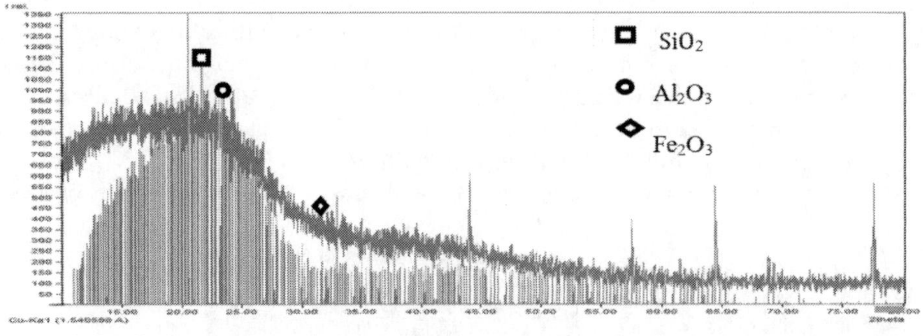

Figure 3. XRD profile of activated carbon from coconut husk without activation process

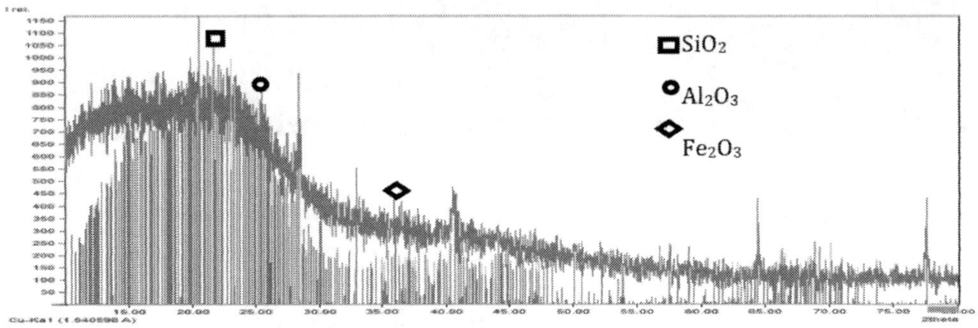

Figure 4. XRD profile of activated carbon from coconut husk with activation process

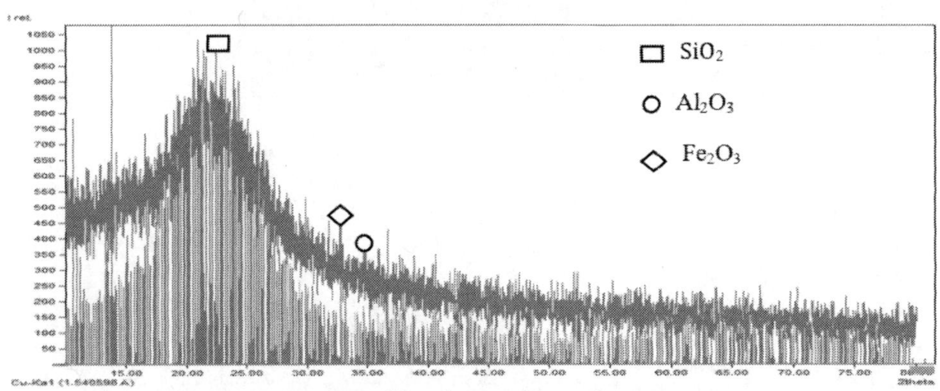

Figure 5. XRD profile of pure silica adsorbent

Figures 2 to 5 showed the XRD diffractogram of activated carbon from coconut husk and silica. SiO_2 peaks of XRD diffractogram of the activated carbon from coconut husk:silica with mass ratio of 70:30, activated carbon from coconut husk without activation process, activated carbon from coconut husk with activation process, pure silica adsorbent were identified at 2θ around 22.54°, 21.98°, 21.96°, 22.24°, respectively. The XRD analysis results shown in figures 2 to 5 interpreted the composition of SiO_2, Al_2O_3 and Fe_2O_3 of all samples as summaried in Table 1.

Table 1. Chemical composition in each adsorbents

Type of adsorbent	Composition (weight %)		
	SiO_2	Al_2O_3	Fe_2O_3
coconut husk : silica (70:30)	83	14.7	2.3
coconut husk without activation process	62	21.9	15.7
coconut husk with activation process	75	16.2	6.8
pure silica without activation process	63.2	16.7	20.1

Table 1 showed that composition of SiO_2 in the activated carbon from coconut husk:silica with mass ratio of 70:30 was higher than composition of SiO_2 in pure silica adsorbent and composition of SiO_2 in activated carbon from coconut husk with activation process. The higher of SiO_2 content in the sample induced the reactivity of the activated carbon from coconut husk:silica with mass ratio of 70:30 was better than the reactivity of pure silica adsorbent and activated carbon from coconut husk with activation process [11]. Silica plays an important role in the adsorption process because it has a good adsorption and ion exchange capability [12].

The content of SiO_2 from activated carbon from coconut husk with activation process was higher than that of the activated carbon from coconut husk without activation process. The activation of adsorbent with sulfuric acid affected the release of impurities on the adsorbent surface easily so that the active site pores inside the sorbent became broader [13].

3.3. Functional groups analysis of prepared activated carbon

Table 2 and figure 6 showed that the activated carbon from coconut husk:silica with mass ratio of 70:30, activated carbon from coconut husk without activation process, activated carbon from coconut husk with activation process, pure silica adsorbent has a Si-O functional group at peak of 1175 cm^{-1}, 1202 cm^{-1}, 1162 cm^{-1}, 1178 cm^{-1}, respectively. These results confirmed that the spectra pattern and wavenumber located at 1175 cm^{-1} of the activated carbon from coconut husk:silica with mass ratio of 70:30, is the closest to the standard wavenumber reference of 1022.27 cm^{-1} for Si-O functional groups [14]. Figure 6 shows FTIR spectra of 4 types of adsorbents.

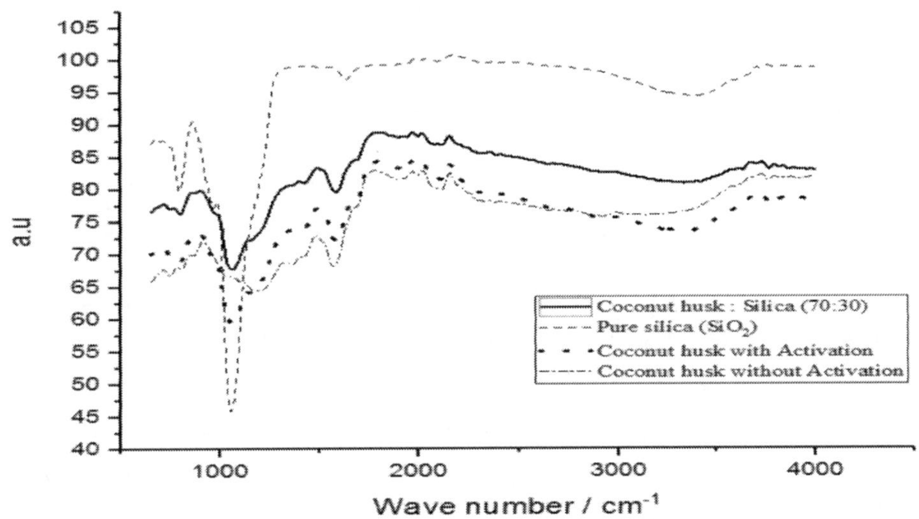

Figure 6. FTIR spectra of 4 types of adsorbents

Table 2. FTIR functional groups of each adsorbents

Type of Adsorbents	Functional Groups	Wavenumber / cm^{-1}	
		Measured	References [14]
coconut husk : silica (70:30)	Si-O-Si	1175	898,83-1442,75
coconut husk with activation process	Si-O-Si	1162	898,83-1442,75
coconut husk without activation process	Si-O-Si	1202	898,83-1442,75
pure silica without activation process	Si-O-Si	1178	898,83-1442,75

3.4. Water content & ash content

Characterization of adsorbents is used to determine the physical and chemical properties of the adsorbent that later would be useful to be used in the adsorption process. In this study, we characterized the water content and ash content of the four types of adsorbents as shown in table 3.

Table 3. The water content and ash content of four adsorbents

Parameter	Indonesian National Standard (SNI) No. 06-3730-1995	Sorbents	Results (%)
Water content	Maximum 15%	coconut husk : silica (70:30)	4.41
		coconut husk without activation process	3.21
		coconut husk with activation process	3.87
		pure silica without activation process	2.71
Ash content	Maximum 10%	coconut husk : silica (70:30)	3.90
		coconut husk without activation process	1.86
		coconut husk with activation process	2.04
		pure silica without activation process	0.32

The water content can affect the adsorption capability of adsorbent. The higher water content causes the smaller capability of adsorbent to adsorb adsorbate. In this study, the water content of each samples was below 15% that meets the Indonesia national standard. These results indicated that the characterization result of the water content of all adsorbents was much lower than the standard one so that all adsorbents has potential to be used in adsorption process of ammonia and methylene blue of liquid wastewater.

The characterization of ash content is carried out to determine the amount of metal oxides or mineral salts and impurities contained in the adsorbent. The ash content obtained from each adsorbent was below 10%, which means that the adsorbent used in the research is feasible to use and meet the standard. The highest ash content was found in the activated carbon of the combination type of coconut husk: silica (70:30) which was 3.90%, followed by the activated carbon of coconut husk with activation process, the activated carbon of coconut husk without activation process and pure silica was 2.04%, 1.86 % and 0.32%, respectively. The high ash content found in the activated carbon of the combination type of coconut husk: silica (70:30) and the activated carbon of coconut husk with activation process due to the use of chemicals as activating agents often results in contamination of the produced activated carbon. Generally activators leave undesired materials, such as oxides which are not soluble in water at the time of washing process [15].

3.5 Surface area
Characterization of activated carbon coconut husk and silica was done by BET method to determine the surface area. The result show that the specific surface area of activated carbon coconut husk : silica (70:30) was 52.924 m^2/g with pore volume of 0.0948 cm^3/g and pore size of 71.712 nm. The specific surface area of activated carbon from coconut husk with activation was 3.0438 m^2/g with pore volume of 0.00751 cm^3/g and pore size of 98.6295 nm, The specific surface area of activated carbon without activation was 2.8119 m^2/g with pore volume of 0.00753 cm^3/g and pore size of 107.1203 nm, and the specific surface area of pure silica (SiO_2) was 23.0385 m^2/g with pore volume of 0.1688 cm^3/g and pore size of 293.24 nm. Those BET data also showed that the activated carbon of coconut husk : silica (70:30) can be classified as mesoporous size. According to IUPAC, mesoporous diameter lies in the range of 2-50 nm. Activated carbon of coconut husk : silica (70:30) had a higher ratio of surface area to pore volume compared to pure silica (SiO_2) and coconut husk only. This makes the activated carbon of coconut husk : silica (70:30) more reactives.

4. Conclusion
1. The XRD analysis confirmed that the highest silica composition of 83 % was observed in the activated carbon from coconut husk: silica (70:30). The silica composition in the activated carbon of coconut husk with activation process, the activated carbon of coconut husk without activation process and pure silica was 75 %, 62.4 % and 63.2 %, respectively.
2 The activated carbon of coconut husk with activation process had higher silica composition compared to the activated carbon of coconut husk without activation process. FTIR spectra identified the presence of Si-O fungsional group in all type of prepared adsorbent. SEM image showed that the activated carbon from coconut husk:silica (70:30) had larger and broader pores compared to the activated carbon of coconut husk with and without activation process.
3. The specific surface area of the activated carbon from coconut husk:silica with mass ratio of 70:30 was higher than the specific surface area of pure silica (SiO_2) and coconut husk only.

5. References
[1] Firiani, Dyah, Dwita O and Lusiana 2015 *Jurnal Gradien* **11** 2 1091-1092
[2] Alouani M E, Alehyen, Achouri M E and Taibi M 2018 *Journal of Materials and Environmental Sciences* **9** 1 32–46
[3] Irmanto and Suyata 2009 *Jurnal Molekul* **4** 2 105–114
[4] Lesbani A, Yusuf S and Yeni I D 2013 *Jurnal Penelitian Sains* **16** 1 10–13
[5] Luo X, Yan Q, Wang C, Luo C, Zhou N and Chensheng J 2015 *Int. J. Environ. Res. Public Health* **12**
[6] Foo K Y and Hameed B H 2010 *Chemical Engineering Journal* **156** 2–10.
[7] Hardyanti, Ika S, Isni N, Dyan S H HP, Evalisa A and Emas, A P W 2017 *Jurnal Sains Terapan* **3** 2 37-41
[8] Ika S H, Isni N, Dyan S H, Evalissa A and Emas A W P 2017 *Jurnal Sains Terapan* **3** 2 34–41
[9] Abdullah A, Saleh A and Novianty I 2013 *Jurnal Al Kimia* **1** 2 32–43
[10] Mariana and Oobuchi Y 2004 *Journal of the Chinese Institute of Chemical Engineers* **35** 4 491–494

[11] Sarengat N, Setyorini I and Prayitno 2015 Pengaruh Penggunaan Adsorben Terhadap Kandungan Amonia (NH3-N) Pada Limbah Air Industri Karet RSS in *Seminar Nasional Kulit, Karet, dan Plastik Ke-4 Yogyakarta* 75–84
[12] Sinta, Ida N, Outu S and Sri R S 2015 *Jurnal Kimia* **9** 2 217-225.
[13] Linda T, Sri S and Eti R 2015 *Jurnal Sains dan Teknologi Lingkungan* **7** 2 70
[14] Rosalina, Tun T E R and Sri S 2016 *Jurnal Biopropal Industri* **7** 1 41

IOP Conf. Series: Materials Science and Engineering **523** (2019) 012023 doi:10.1088/1757-899X/523/1/012023

The use of bentonite of Bener Meriah Aceh to improve the mechanical properties of Polypropylene-Montmorillonite Nanocomposite

J Julinawati[1,*], S Gea[2], E Eddiyanto[3], B Wirjosentono[2], I Ichwana[4]
[1]Chemistry Department, Faculty of Mathematics and Natural Science, Universitas Syiah Kuala, Jalan Tgk. Tanoh Abe Nomor 3, Kopelma Darussalam, 23111 Banda Aceh, Indonesia
[2]Chemistry Department, Faculty of Mathematics and Natural Science, Universitas Sumatera Utara, Jalan Bioteknologi No.1 Kampus USU, 20155 Medan, Indonesia
[3]Chemistry Department, Faculty of Mathematics and natural science, Universitas Negeri Medan, Jl. Willem Iskandar Psr. V Medan Estate, 20222 Sumatera Utara, Indonesia
[4]Department of Agricultural Engineering-Faculty of Agriculture, UniversitasSyiah Kuala. Jl. Tgk. Hasan KruengKalee No. 3 Kopelma Darussalam, 23111 Banda Aceh, Indonesia

E-mail: julinawati@unsyiah.ac.id

Abstract. The research on application of bentonite Bener Meriah of Aceh to improve the mechanical properties of polypropylene-montmorillonite nanocomposite has been conducted. Bentonite was isolated into the nano-sized montmorillonite and was used as a filler of polypropylene-montmorillonite nanocomposites with the addition of PP-g-MA as a compatibilizer and octadecylamine as a modifier of MMT. The results showed that bentonite of Bener Meriah Aceh contained montmorillonite with 70.6% content. Based on the result of mechanical properties test, it was ddiscovered that montmorillonite isolated from Bentonite of Bener Meriah could improve the mechanical properties of PP-MMT nanocomposite in composition ratio of PP/PP-g-MA/MMT is 85/10/ 5.

1. Introduction

Bentonite is an abundant natural resource in Indonesia, such as in Java, Sumatra, and Sulawesi, which contain more than 380 million tons but has not been optimally utilized. In Aceh, natural bentonite can be found in the districts of North Aceh, Bener Meriah, Sabang, Central Aceh, and Simeulue [1,2,3] which reaches the amount of 2,618,224,030.20 tons [4]. To date, research on Aceh bentonite has been conducted only on several studies [5] and the bentonite that has been used only the one that found in North Aceh. Therefore, bentonite processing from other districts in Aceh is required. Bentonite that will be used in this research is bentonite from Bener Meriah, which is one of the areas in the Province of Aceh which has bentonite with a thickness of 1 m width of 20 Ha and its content reaches 520,000 tons [4] which has not been utilized.

*Corresponding authors julinawati@unsyiah.ac.id

Bentonite is natural clay whose main component is the mineral montmorillonite (85%), with the chemical formula of $M_x(Al_{4-x}Mg_x)Si_8O_{20}(OH)_4.nH_2O$. Montmorillonite (MMT) is a filo-silicate mineral that has the ability to expand and can be intercalated and exfoliated, thus it is widely used as filler of nanocomposite to enhance the properties of the nanocomposite [6]. When an exfoliation occurs, the mechanical and rheological properties of the nanocomposite increase dramatically when compared with the pure polymer [7]. Several researches on the addition of MMT in polypropylene (PP) nanocomposites has been conducted and indicated that MMT can improve some of these nanocomposite properties such as mechanical properties [8,9,10,11,12], thermal properties [13], fire retardancy properties [14], and increased degrees of degradation [16]. It is expected that MMT from Bener Meriah can also be used to improve the mechanical properties of nanocomposites.

Nanocomposites can be obtained by mixing the silicate layers of MMT with PP by melting intercalation method. Mixing of silicate layers from MMT in PP can be increased by using functional oligomers as compatibilizers. Several studies have reported using polypropylene-graft-maleic anhydride (PP-g-MA) as a compatibilizer [15].MMT is also modified using a long organic alkyl chain, which is called a modified organo-silicate layer (OMLS) or organo-clay.The organic clay will change the hydrophilic into hydrophobicproperties of MMT, which is allowing the MMT interface to interact with several different polymer matric.The Organic compounds commonly used to modify MMT are alkynammonium. The research was conducted several stages are characterizing of bentonite, isolationof the MMT nanoparticle, preparation and characterize of PP-MMT nanocomposites.

2. Materials and Methods

2.1. Materials

Polypropylene was obtained from Aldrich (density 0.896 gr/cm^3, melting point 176 °C), PP-g-MA is polypropylene which has been grafted with maleic anhydride, density 0,934 g/cm^3, melting point 156 °C, Mn 3,900 (GPC), Mw 9100 (GPC), oktadecylamine, ($C_{18}H_{39}N$, 90%) was obtained from Aldrich, CH_3Br (density 2,10 gr/ml), alcohol, montmorillonite K10 (standart) was obtained from Aldrich, montmorillonite was isolated from bentonite of Pantanlah, Bener Meriah of Aceh.

2.2. Methods

2.2.1. Isolation nano montmorillonite from natural bentonite Bener Meriah.

Montmorillonite nanoparticles can be obtained by preparing 1 kg of bentonite sample and filtered with a 100 mesh sieves, then it was dried in oven at the temperature of 105 °C for 4 hours. Subsequently, the sample was fractionated. Fractionation was done with sedimentation by weighing 40 grams of 100 mesh bentonite and added with 2L aquades to form the suspension. Bentonite suspension was given ultrasonic waves for 15 minutes at 750 watts at room temperature. Furthermore, the suspension is left in a flat place and kept away from all vibrations. Precipitation that occur within 15 minutes are taken by pouring the suspension into another container and leaving the filtrate again.The precipitate formed in the next 3 days is filtered back and taken its filtrate. The floating fraction in the filtrate is stirred again, then the filtrate is left for a week and collected of precipitate is formed. This precipitate was dried in an oven at 105 °C for 3 hours, then crushed and sieved using 200 mesh sieve. This fraction is stored in a desiccator. The identification of the fraction (montmorillonite) was carried out using FT-IR, X-RD diffraction and SEM [16].

2.2.2. Preparation of polypropylene-montmorillonite nanocomposite (PP-MMT).

Variation of the composition materials of PP, PP-g-MA, and montmorillonite for preparing PP-MMT nanocomposite can be seen in Table 1. Materials of various compositions of PP, PP-g-MA and montmorillonite were coumpounded respectively in a Haake Rheomix 3000 internal laboratory mixer with high rotor intensity, operating at 180 °C at 65 rpm for 10 minutes. Samples for mechanical tests were made from the nanocomposite produced.

Table 1. Composition of material of polypropylene-montmorillonite nanocomposite

Code of sample	PP (% wt)	PP-g-MA (% wt)	Montmorillonite (% wt)	
			Bener Meriah	Commercial
A	100	0	0	
B	95	5	0	
C	95	0	5	
D	90	5	5	
E	85	5	10	
F	80	5	15	
G	85	10	5	
H	80	15	5	
I	85	10		5

2.3. Characterization

Samples of bentonite and nano montmorillonite were characterized by X-Ray Flourencense (XRF) PANalytical, Axios Advance. The percentage of montmorillonite be calculated using the Meyer equation [17]. The functional group of bentonite, nano montmorillonite and nanocomposite speciment were analyzed by using FTIR of Perkin Elmer type spectrum 100 at wave numbers 400 to 4000 cm^{-1}. The X-Ray Diffraction scattering patterns of bentonite, nano montmorillonite and nanocomposite were recorded using XRD Shimadzu 6000, operating at Cu Kα radiation generated at 40kV and 30 mA. The sample were scanned in range of 3° to 50° with scan rate 5°(2θ)/minute. Surfaces morfologies of nano montmorillonite and nanocomposite was analyzed by Scanning Electron Microscopy (SEM) Merck Zeiss type EVO MA 10 at 20 kV. Mechanical tests was done using the equipment a Lloyd LR/10KN Universal, operation at room temperature and a speed of 50 mm min^{-1}, according to ASTM D638 and ASTM D256 standard. Nanoparticle was analyzed by Particle Size Analyzer (PSA) of LS 100 Q Coulter.

3. Result and discussion

3.1. XRF analysis

XRF analysis was chosen asa method for the assessment of the quantitative results. The result of characterization of bentonite of Bener Meriah Aceh by XRF can be seen in Table 2. Table 2 shows the chemical composition of bentonite from Bener Meriah, the amount of Na_2O is greater than CaO, this indicates that the bentonite from Bener Meriah is Na-Bentonite as described in [18].

Table 2. Percentage of chemical composition of bentonite Bener Meriah

Chemical composition of bentonite Bener Meriah	Weight (%)
SiO_2	65.39
Al_2O_3	16.6
Fe_2O_3	2.69
CaO	0.09
MgO	0.84
Na_2O	0.76
K_2O	0.84

3.2.Meyer test

The content of montmorillonite in Bener Meriah bentonite was analyzed using the Meyer equation, (1972) and the result was 70.6%. The percentage of montmorillonite is different for each region, it is also suspected to be influenced by the process of bentonite formation refer to [18]. Meyer equation :

$$\text{Percentage of montmorillonite} = \frac{\text{dry weight of montmorillonite}}{\text{dry weight of bentonite}} x\ 100 \tag{1}$$

3.3. FTIR analysis

FTIR analysis was performed to determine the functional group of a sample. The FTIR spectra for bentonite of Bener Meriah are shown in Figure 1. The presence of absorption bands at 3620.42 cm^{-1} which can be attributed of Al-OH or Si-OH stretching, an absorption band at 3444,41 cm^{-1} and 1637,54 cm^{-1}, attributed to OH stretching and OH bending respectively. Characteristic absorption of Si-O stretching, Al-O, Si-O bending and Si-O-Si can be observed at 1031,20 cm^{-1}, 538.15 cm^{-1} and 467,10 cm^{-1}respectively. The FTIR spectra of bentonite Bener Meriah showed similar absorption bands to those found in [19].

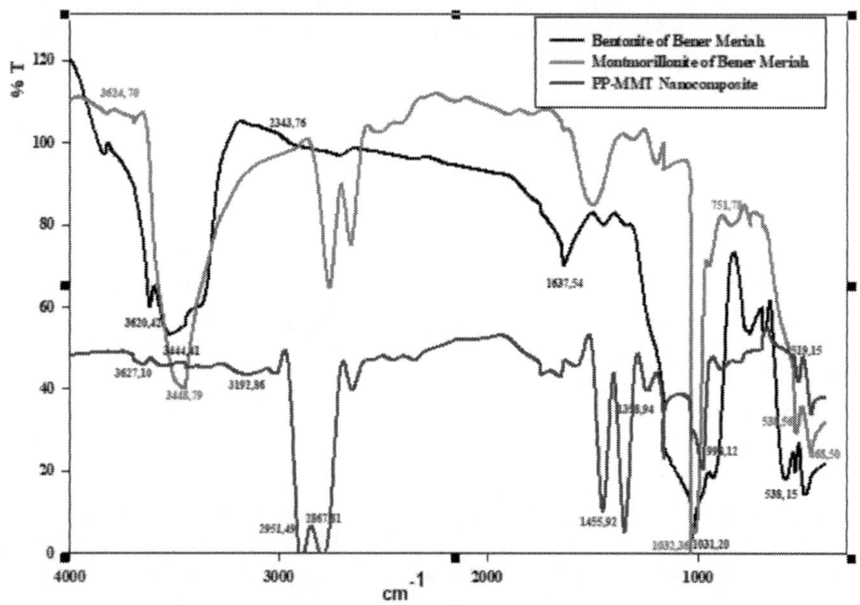

Figure 1. FTIR spectrum of bentonite Bener Meriah, montmorillonite of Bener Meriah and PP-MMT nanocomposite (PP/PP-g-MA/MMT: 85/10/5)

The FTIR spectra for montmorillonite of Bener Meriah in Figure 1 showed absorption bands characteristic of montmorillonite in 3624,70, 3448,79, 1736,87, 1032,36, 751,78, 538,56, and 468,50 cm^{-1}, these spectra are similar to the standard montmorillonite [20]. Figure 1 also showed of FTIR spectra for PP-MMT nanocomposite (PP/ PP-g-MA/MMT: 85/10/5). Absorption bands at 3627,10 - 3192,86 cm^{-1} at FTIR spectra of PP-MMT nanocomposite can be attributed N-H of octadecylamine. Octadecylamine was used on the preparation of nanocomposites by mixing modified polypropylene with PP-g-MA and montmorillonite which have modified with octadecylamine. Absorption bands at 2951,49 -2867,81 cm^{-1} correspond to alkane of PP and oktadecylamine, while absorption bands at 1455,92-1358,99 cm^{-1} and 998,12-519,15 cm^{-1} correspond to C-H of alkane of PP. Characteristic absorption of Si-O-Si from MMT can be observed at 1166,99-1041,88 cm^{-1}. This indicates that MMT modified with octadecylamine has been intercalated into the galleries of PP.

IOP Conf. Series: Materials Science and Engineering **523** (2019) 012023 doi:10.1088/1757-899X/523/1/012023

3.4. XRD analysis

The characterization of sample by X-Ray Diffraction (XRD) aimed to verify the existence minerals of Bentonite Bener Meriah and about exfoliation and intercalation of PP-MMT nanocomposite. The XRD patterns of bentonite Bener Meriah can be seen in Figure 2.

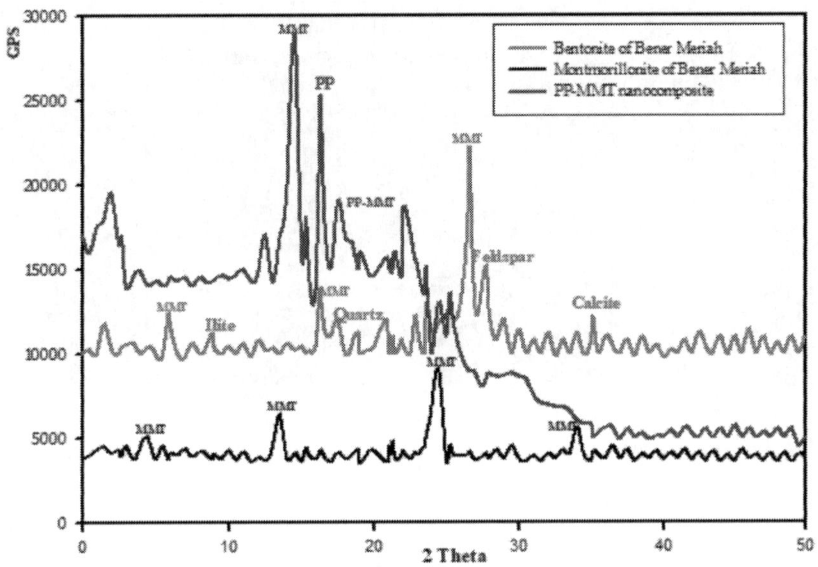

Figure 2. XRD spectra of bentonite, montmorillonite of Bener Meriah and PP-MMT nanocomposite

XRD spectra of bentonite Bener Meriah in Figure 2 shows the presence of montmorillonite (2θ : 5.86°, 16.29°, 21.93°, 23.65°, 26.64°, 27.83°, 35.14°) as majority phases. The patterns of diffraction peaks corresponding to planes (001) and (020), confirming the presence of montmorillonite in bentonite of Bener Meriah. The data also showed the presence of impurities such quartz (2θ : 20.87°), ilite (2θ : 8.9°), fieldspar (2θ : 25.12°) and calcite (2θ : 29.78°) [16,21]. In addition, the peak of 2θ at 5.86° and 16.29° in the Figure 2 showed that bentonite of Bener Meriah is Na-bentonite [22]. The XRD patterns of montmorillonite show only peak of 2θ : 5.47°; 14.54°; 25.25°, 35.04°, which is the result of isolation from bentonite of Bener Meriah [23]. The pattern XRD of the PP-MMT nanocomposite in the composition ratio of PP/ PP-g-MA/ MMT is 85/ 10/ 5 showed peak of 2θ at 2.4 to 2.7°, that indicates that PP chains insert into the interlayers of MMT and hence the polymer intercalated or exfoliated structure is achieved [24].

In addition, angular shifts from the top of MMT also explain the occurrence of intercalation and exfoliation in the silicate layer. The new peak are detected at 2θ : 15,3° and 17.6° corresponding to plane (110) and (040) of polypropylene(PP). Peak of 2θ at 18,9° and 22,8° shows the presence PP containing MMT modified. The reduced XRD peaks detected in PP-MMT nanocomposites can be ascribed to the irregular state of modified MMT distributed in the PP matrix [17,23,24].

3.5. SEM analysis

The Scanning Electron Microscopic (SEM) technique was used to explore the surface morphologies of the sample. Figure 3(A) shows the surface morphologies of montmorillonite which isolated from bentonite Bener Meriah.The figure showed that structure of MMT Bener Meriah has layered pores which randomly distributed with different sizes [25,26]. The existence of these layered pores causes MMT to be used as a nanocomposite filler to enhance the properties of the nanocomposite. Figure 3 (B) shows the surface morphologies of PP-MMT nanocomposite. The figure showed that montmorillonite Bener Meriah modified dispersed uniformly and homogenous in the PP matrix which is considered an indication of the successful preparation of the PP-MMT nanocomposite. The result indicates that exfoliation and intercalation of PP in MMT may occur so as to produce compatible nanocomposites [27].

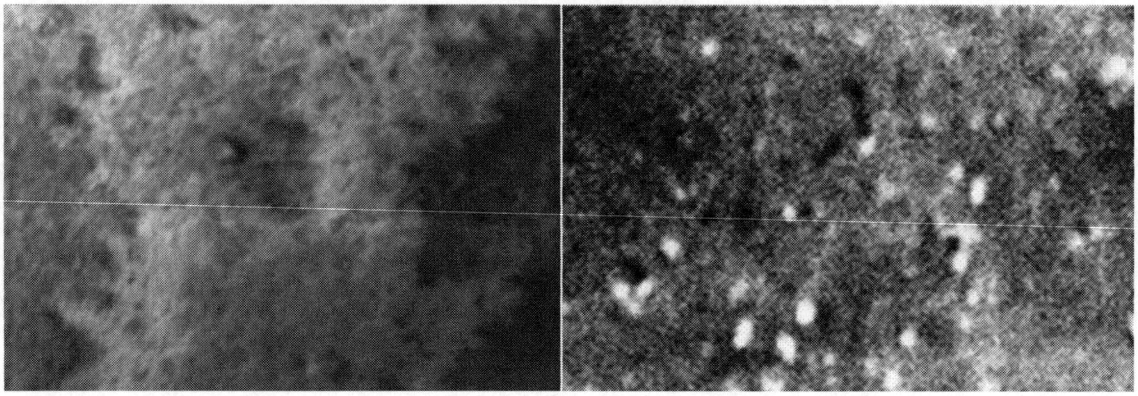

(A) (B)

Figure 3. SEM of montmorillonite(A) and PP-MMT nanocomposite of (PP/PP-g-MA/MMT: 85/10/5) (B)

3.6. PSA analysis

Montmorillonite which has isolated from bentonite was processed into nanoparticles by precipitation [16]. To analyze particle size of montmorillonite was used Particle Size Analyzer. The average result of particle size of montmorillonite Bener Meriah was 67,8 nm, can be seen in Figure 4. Then, this montmorillonite was used to preparation of nanocomposite.

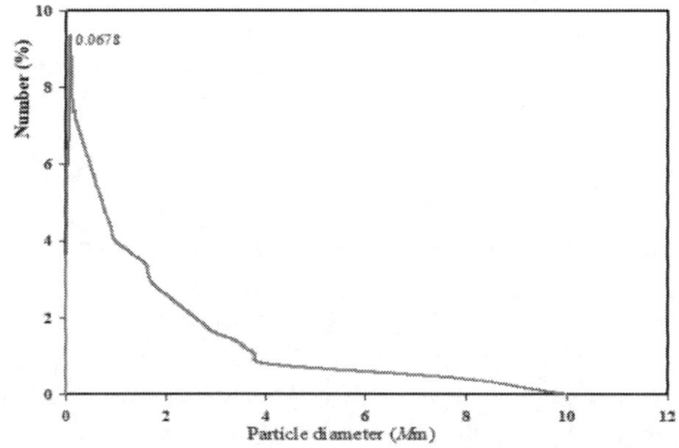

Figure 4. Particle size of montmorillonite Bener Meriah

3.7. Mechanical properties of PP-MMT nanocomposite

Figure 5, 6 and 7 shows the mechanical properties of PP-MT nanocomposite, the figure showed that the addition of MMT Bener Meriah which has modified significantly increased the tensile strength, elastic modulus and elongation of PP-MMT nanocomposite [11,20,21].The optimum of tensile strength, elastic modulus and elongation of PP/PP-g-MA/ MMT occurs when the ratio of composition of PP/PP-g-MA/MMT is 85/10/5. These values peak at 31.75 MPa for tensile strength, 374,41Mpa for elastic modulus and 8.48% for elongation. The value of mechanical properties of PP-MT nanocomposite which the addition of MMT Bener Meriah modified showed similar with mechanical properties of PP-MT nanocomposite which the addition of MMT commercial modified.

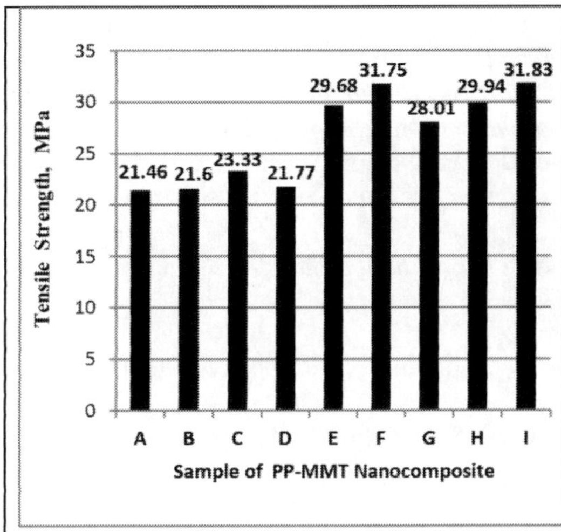

Figure 5. Graph of the relationship between PP/PP-g-MA/MMT composition with tensile strength of PP-MMT nanocomposite

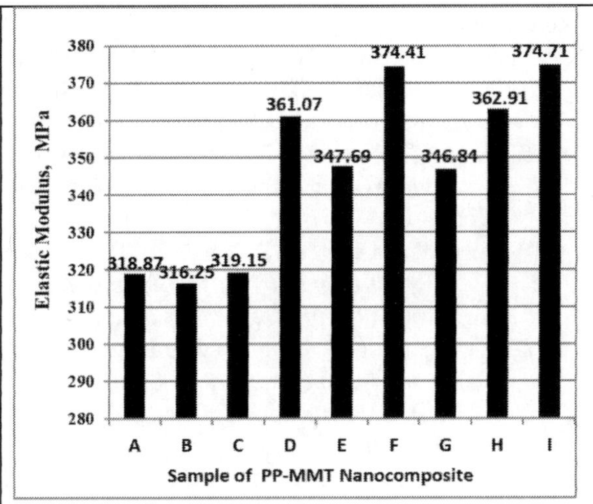

Figure 6. Graph of the relationship between PP/PP-g-MA/MMT composition with elastic modulus of PP-MMT nanocomposite

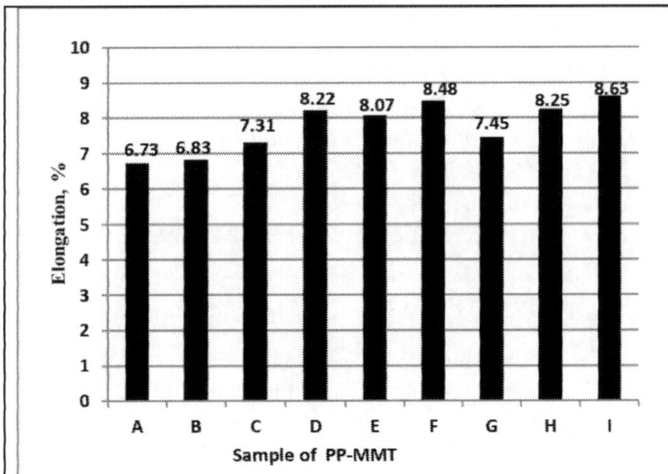

Figure 7. Graph of the relationship between PP/PP-g-MA/MMT composition with elongation of PP-MMT nanocomposite

4. Conclusion

It can be concluded that bentonite is Na-bentonit and has percentage of montmorillonite of 70.6%. Bentonite Bener Meriah was isolated into nano-sized montmorillonite and used as a filler of polypropylene-montmorillonite nanocomposites by the addition of PP-g-MA as a compatibilizer and octadecylamine as a modifier of MMT. The result indicates that exfoliation and intercalation of PP in MMT may occur to produce compatible nanocomposites. Based on mechanical properties test, it showed that the addition of MMT Bener Meriah which has modified significantly increased the tensile strength, elastic modulus and elongation of PP-MMT nanocomposite in the composition ratio of PP/PP-g-MA/MMT is 85/10/ 5.

5. References

[1] Syuhada, Rachmat W, Jayatin and Saeful R 2009 *Jurnal Nanosains & Nanoteknologi* **2** 8-51
[2] Zulfikar, Adrian Z, Djadja T and Irwan M 2006 *Proceeding Pemaparan Hasil-Hasil Kegiatan Lapangan dan Non Lapangan Tahun 2006 Pusat Sumber Daya Geologi* **1** 1
[3] Kaelani M S, Iwan A H, Irwan M, Asep S, Jubbel B, Abdul F and Yeni A 2007 *Proceeding Pemaparan Hasil Kegiatan Lapangan dan Non Lapangan Tahun 2007 Pusat Sumber Daya Geologi* **1** 1
[4] Wastoni C P, and Martua R P 2009 *Proceeding Pemaparan Hasil-Hasil Kegiatan Lapangan dan Non Lapangan tahun 2009 Pusat Sumber Daya Geologi*, **1** 1
[5] Lubis S 2007 *Jurnal Rekayasa Kimia dan Lingkungan,* **6** 77
[6] Qina H, Shimin Z, Chungui Z, Meng F, Mingshu Y, Zhongjun S, Shousheng Y 2004 *Polymer Degradation and Stability* **85** 807
[7] Cipriano B H, Kashiwagi T, Zhang X, Raghavan S R 2009 *ACS Applied Material and Interface* **1** 130
[8] Ding C, Demin J, Hui H, Baochun G, Haoqun H 2005 *Polymer Testing* **24** 94
[9] Castel C D, Pelegrini T Jr, Barbosa R V, Liberman S A, Mauler R S 2010 *Composites Part A* **41** 185
[10] Kord B 2012 *Turk. J. Agric.* **36** 510
[11] Barleany D R, Rudi H and Santoso 2011 *Prosiding Seminar Nasional Teknik Kimia* "Kejuangan" **1** DO3-1- DO3-6
[12] Leszczy´nska A, Njuguna J, Pielichowski K, Banerjee J R 2007 *Thermochimica Acta* **453** 75
[13] Wang Z M, Nakajima H, Manias E and Chung T C 2003 *Macromolecules* **36** 8919
[14] Shi D, Robert K Y L, Wei Y, Jinghua L 2007 *European Polymer Journal* **43** 3250
[15] Liu, Lim S H, Zhao J H, Juay M S, Yong and Lu X H 2005 *SIMTech Technical Reports* **6** 33
[16] Fisli A, 2008 *Indonesian Journal of Materials Science* **10** 12
[17] Meyer W A 1972 *AFS Cast Metals Res. J.* **8** 161
[18] Tekmira 2008 *Bentonit http://www.tekmira.esdm.go.id/*
[19] Rahardjo A K, Maria J J S, Alfin K, Nani I, Suryadi I 2011 *Journal of Hazardous Materials* **190** 1001
[20] Ulfah F, Irwan N 2014 *Molekul* **9** 155
[21] Lumingkewas S 2006 *Eugenia* **12** 242
[22] Manias E, Touny A, Wu L, Strawhecker K, Lu B and Chung T C 2001 *Chem. Mater.* **13** 3516
[23] Zhihong Y, Minghu W, Gang C, Shengjuan L, Panpan P and Qiaolian Z 2016 *Polymers & Polymer Composites* **24** 331
[24] Nejad S J, Ahmadi S, Abolghasemi H, Mohaddespour 2007 *e-Polymer* **7** 1
[25] Lee S Y, Kang I A, Doh G H, Kim W J, Kim J S, Yoon H G, and Wu Q 2008 *eXPRESS Polymer Letters* **2** 78
[26] Bharadwaj R K, Mehrabi A R, Hamilton C, Trujillo C, Murga M, Fan R, Chavira A and Thompson A K 2002 *Avery Research Center* **43** 3699
[27] Bahrami S H, Mirzaie Z 2011 *World Applied Sciences Journal* **13** 493

IOP Conf. Series: Materials Science and Engineering **523** (2019) 012024 doi:10.1088/1757-899X/523/1/012024

Comparison between maceration and microwave extraction techniques of strawberry fruit (*fragaria sp*) and antioxidant activity test

N Fitri[1,*], Halimatussa'diah[1], D Fitriastuti[1]
Department of Chemistry, Universitas Islam Indonesia
Jl. Kaliurang KM 14.5, Besi, Sleman, Yogyakarta, Indonesia.

Email: noor.fitri@uii.ac.id

Abstract. Extraction of strawberry fruit (*Fragaria sp*) by maceration and microwave and antioxidant activity test was studied. The objectives of this research are (1) to compare the maceration and microwave extraction techniques, and (2) to determine the antioxidant activity of strawberry fruit extract. The research steps consist of sample preparation, maceration and microwave extraction with 96% ethanol solution, phytochemical screening test and antioxidant activity test with DPPH method. The yield of extraction of 5.77%, 2.12%, 1.55% and 2.58% is achieved at 24 hours maceration and 3, 5 and 7 min microwaves, respectively. The phytochemical screening result shows that the strawberry fruit ethanol extract contains tannins, flavanoids, alkaloids, and saponins compounds. The identification result of flavanoids compounds by UV-Vis spectophotometer reveals that the strawberry fruit ethanol extract is interpreted to contain isoflavones compounds. The FTIR spectra displays the existence of specific function groups of flavanoids compound such as OH, C-O alcohol, C=C aromatic, C-H aromatic, C-H alifatic, C=O and C-O ether. Antioxidant activity test by DPPH method reveals that strawberry fruit ethanol extract at 24 h maceration and 3, 5 and 7 min microwave containing IC_{50} of 50.61 ppm and 67.97, 118.45 and 61.42 ppm, respectively. Moreover, LC-MS-MS analysis indicates the presence of isoflavones compounds peak i.e. formononetin and daidzin.

1. Introduction

Indonesia is known as a country with biodiversity abundance. Indonesia has a tropical climate that support to be host of the development of medicinal plants cultivation. One of the many crops grown in Indonesia is strawberry, especially in West Java. Strawberry fruit is one of sources of bioactive compounds, ascorbic acid, anthocyanin, phenol compounds and has antioxidant property [1,2].

Strawberry (*Fragaria sp*) is a fruit plant with high economic value with sour taste, sweet and fresh. Strawberry fruit contains vitamin C, rich in fiber, folic acid, potassium and antioxidants. The content of strawberry can be used as an alternative to improve heart health, reduce the risk of cancer and beneficial to health [3,4]. Strawberry is rich in color pigments. The red color of strawberry is caused by anthocyanin and has antioxidant properties. Strawberry have a high antioxidant content compared to other fruits and vegetables, so strawberry can be used to overcome the problem of free radical diseases such as cancer, stroke and the aging process. Strawberry also contain low sugar so it is suitable for diabetic diet and arthritis. Due to its high antioxidant, this fruit can also be used to smooth and brighten the skin [5]. The antioxidant substances in strawberry can serve to neutralize free radical compounds and inhibit oxidative processes [6]. Various extraction techniques with various solvents have been performed to extract the antioxidant active substances in strawberry. In this study, we compared the extraction techniques i.e. between maceration and microwave techniques to extract the

*Corresponding author: noor.fitri@uii.ac.id

active substances and to test the antioxidant activity of the strawberry fruit extract. Microwave technique showed advantages in terms of high extraction efficiency and antioxidant activity of extract within shortest extraction time [7]. The solvent used in the extraction process is ethanol. The antioxidant activity test method applied in this study is DPPH method *(2,2- Diphenyl-1-Picryhydrazyl)*. The DPPH method is easy to use, fast, thorough enough and sensitive to evaluate the antioxidant activity of natural material compounds [8].

2. Methods

2.1 Instruments and materials

The experiment was performed by means of a set of glassware, blender, cuvet, rotary evaporator, microwave, UV-Vis Spectrophotometry, FTIR and LC-MS/MS. The materials include strawberries, ethanol, methanol, $FeCl_3$ 1%, concentrated HCl, Mg powders, Mayer reagent, Wagner reagen and Dragendorff reagent, aquadest, NaOH, $AlCl_3$, CH_3COONa, H_3BO_3 and DPPH *(2,2- Diphenyl-1-Picryhydrazyl)*.

2.2 Sample Preparation

Strawberries were obtained directly from Inggit Strawberry orchards in Banyuroto, Sawangan, Magelang in Central Java. The samples were washed with water, dried and blended until they become a refined powder, which were ready for the extraction process.

2.3 Maceration Extraction Method

Strawberry powder was macerated for 2 days (M extract). Sample observation was done every 6 hours. The filtrat was then filtered, collected and evaporated with a vacuum rotary evaporator at 37°C until a viscous extract obtained.

2.4 Microwave Extraction Method

Sample A was treated by microwave with power *100* watts with time variation 3 (WA), 5 (WB) and 7 (WC) minutes. Samples of WA, WB and WC were filtered and evaporated with vacum rotary evaporator at 37°C until obtaining viscous extract.

2.5 Testing of strawberry fruit extract includes

Phytochemical screening tests include tannins, flavanoids, alkaloids, and saponins tests. Identification of flavonoid compounds used shift reagents and UV-Vis Spectrophotometry [9]. Identification of compounds of strawberry extract was tested using FTIR. Antioxidant activity test was performed by means of DPPH method. Identification group of compound structure of strawberry extract was done by using LC-MS-MS.

3. Results and discussion

This study was aimed to compare the maceration and microwave techniques for the extraction of strawberries and antioxidant test of strawberry extract. The yield of extraction of 5.77%, 2.12%, 1.55% and 2.58% is achieved at 24 hours maceration and 3, 5 and 7 min microwaves, respectively. It means that, the yield of the maceration technique is greater (5.77%) than that of the microwave extraction technique (2.58%) using 96% ethanol. The effectiveness of an extraction process is influenced by the type of solvent used as a filter, the particle size of the sample, the technique and duration of extraction. Maceration technique performed for 2 days so that the longer the extraction time, the more active substances that can be extracted to achieve the balance of diffusion between solvent and solute. The advantages of microwave extraction process has a relatively short time only 3-7 minutes, while the maceration extraction takes 2 days [7].

IOP Conf. Series: Materials Science and Engineering **523** (2019) 012024 doi:10.1088/1757-899X/523/1/012024

3.1. Phytochemical Screening

Phytochemical screening is a qualitative test of chemical compounds in plants. Phytochemical screening was performed to determine the class of active compounds or secondary metabolite compounds contained in extract of strawberry fruit. It is determined by looking at the color change reaction that occurred after the addition of a reagent used. The results of phytochemical tests shown that the extract contained tannin, flavanoids, alkaloids, and saponin (Table 1).

Table 1. Results of phytochemical screening test of strawberry fruit extract

No.	Compounds group	Reagent	Color	Result
1.	Tannin	$FeCl_3$ 1%	Blackish green	+
2.	Flavonoids	Mg + HCl solid	Red Blood	+
		Dragendrof	Orange	+
3.	Alkaloids	Mayer	White yellowish	+
		Wagner	Reddish brown	+
4.	Saponins	Aquadest	Shaped foam	+

3.2. Identification of flavanoid compounds by UV-Vis Spectrophotometers

Table 2 shows the UV VIS wavelength before and after reactions with Shift reagent. Based on the Table 2, band I and band II of all samples were interpreted to contain different flavonoid compounds. This results indicated to the presence of methylation and glycosylation which is particularly present in C-3, C-5, C-7 and C-4 'hydroxyl, which results in a lower shift [9].

3.2.1. Determination of flavanoid compound structure from M sample

Based on Table 2, after addition of the 2M NaOH solution, the M solution was turned to yellow and underwent a leftward hypochromic shift due to the ionization of the alkaline-sensitive hydroxyl group. The shift was occurred in band I of 10 nm and band II of 20 nm. It can be seen that band I and band II shift but does not show the presence of hydroxy group at flavonoid core.

After addition of 5% $AlCl_3$ solution, the M solution color changed to yellow and causes a 10 nm shift of batochromic to the band II indicating that it is likely to be suspected in the C-5 position of the OH group present in the type of flavonoid group of isoflavones, whereas in the absence of the band I shift which means there is no possibility of a hydroxy group at the flavonoid core [10].

Table 2. UV VIS wavelength of the extract solution before and after reaction with shift reagent.

Sample	Wavelength	Extract + CH_3OH	Extract + CH_3OH + NaOH	Extract + CH_3OH + $AlCl_3$	Extract + CH_3OH + CH_3COONa	Extract + CH_3OH + CH_3COONa + H_3BO_3
M	Band I	363	353	*n.d*	368	378
	Band II	263	243	273	263	270
WA	Band I	362	343	*n.d*	380	381
	Band II	260	279	273	270	263
WB	Band I	363	343	*n.d*	*n.d*	382
	Band II	274	245	273	273	268
WC	Band I	363	343	384	*n.d*	384
	Band II	260	283	272	277	263

n.d = not detected

Table 3. Interpretation of wavelength changes of M extract solution with the addition of Shift reagents.

Treatment	Wavelength (nm)		Shift (nm)		Substitution
	Band I	Band II	Band I	Band II	
Extract + CH_3OH	363	263	-	-	-
Extract + CH_3OH + NaOH	353	243	-10	-20	-
Extract + CH_3OH + $AlCl_3$	*n.d*	273	-	+10	5-OH (isoflavones)
Extract + CH_3OH + CH_3COONa	368	263	+5	Tetap	7-OH (isoflavones)
Extract + CH_3OH + CH_3COONa + H_3BO_3	378	270	+15	+7	*o*-diOH in ring B, *o*-diOH in ring A (6,7 atau 7,8)

Based on data interpretation of maximum wavelength change by using addition of shift reagent, the extract of strawberry fruit possibly contains isoflavones. It can be proven by the formed spectrum which exhibits the characteristics of isoflavones with the possible presence of hydroxy groups in C-5, C-7 and C-8. The structure of the isoflavone trihydroxy compound is shown in Figure 1.

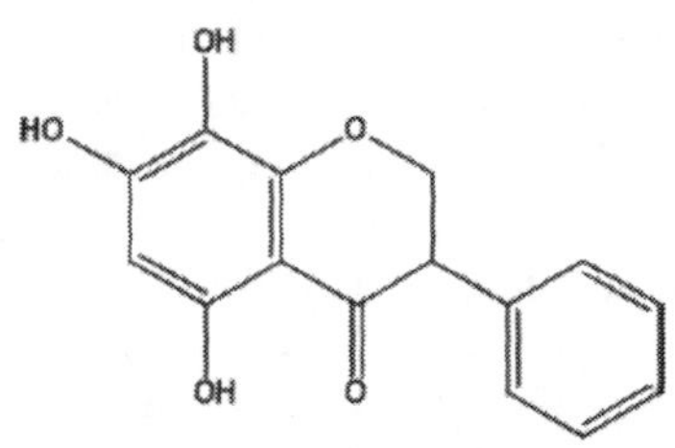

Figure 1. Structure of the 5,7,8-Trihydroxy Isoflavone

3.2.2. Determination of flavanoid compound structure from WA sample

The result of UV-Vis spectra of WA sample with addition of 2M NaOH solution are hypochromic shift in band I of 19 nm and batochromic shift in band II of 19 nm. It can be seen that in band I and band II has shift but does not show the presence of hydroxyl group at flavonoid core. Addition of $AlCl_3$ 5% solution results a 13 nm shift in batochromic of band II indicates that C-5 is suspected to be a 5-OH group. Whereas in band I does not experience a shift which implies that it does not show the presence of a hydroxyl group at the flavonoid core. Addition of the CH_3COONa solution causes a batochromic shift in band I of 18 nm and band II of 10 nm indicating that it is likely to be suspected in position C-7 of the OH group present in the type of flavonoid class of isoflavones.

Addition of shift reagents CH_3COONa and H_3BO_3 results a shift in the band I of 19 nm indicating that the possibility of o-dihydroxyl in ring B, while the shift in the band II of 3 nm, although shifted but did not show the presence of hydroxy group at the flavonoid core (Table 4).

Based on interpretation of data in Table 4, the compound contained in WA extract possibly is isoflavones. It can be seen from the formed spectrum is characteristic of isoflavones with the presence of hydroxy group on C-5 and C-7. The structure of the isoflavone 5.7-dihydroxy compound is shown in Figure 2.

Table 4. Interpretation of wavelength changes of WA extract by the addition of shift reagents.

Treatment	Wavelength (nm)		Shift (nm)		Alleged Substitution
	Band I	Band II	Band I	Band II	
Extract + CH_3OH	362	260	-	-	-
Extract + CH_3OH + NaOH	343	279	-19	+19	-
Extract + CH_3OH + $AlCl_3$	*n.d*	273	-	+13	5-OH (isoflavones)
Extract + CH_3OH + CH_3COONa	380	270	+18	+10	7-OH (isoflavones)
Extract + CH_3OH + CH_3COONa + H_3BO_3	381	263	+19	+3	o-dihydroxyl in ring B

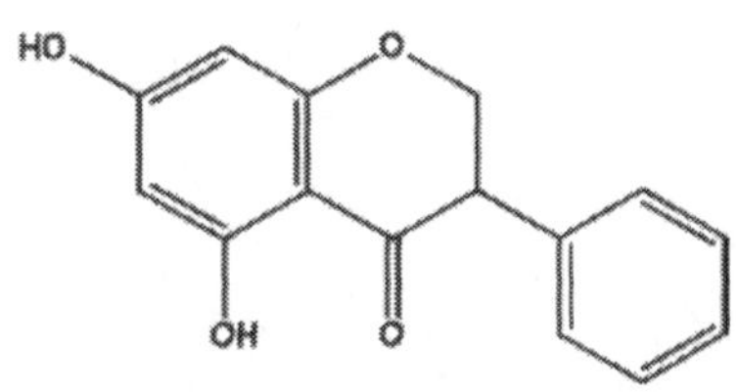

Figure 2. Structure of the Isoflavone 5.7-dihydroxy

3.2.3. Determination of flavanoid compound structure of WB extract

Table 5 shows the wavelength change of WB extract solution before and after reaction with Shift reagent. Reaction with 2M NaOH results a hypochromic shift in band I of 20 nm and a band of II of 29 nm. After addition of 5% $AlCl_3$ solution and CH_3COONa solution, there are no significant shifting of wavelength. it indicates that posibbly there no hydroxyl group in the flavonoid core. Reaction with CH_3COONa and H_3BO_3 solution results a batochromic shifting in band I of 19 nm. It is suspected that the o-dihydroxyl group in ring B, where as in band II, there is no significant shifting. Based on interpretation of the wavelength change, the WB extract possibly contains isoflavones. The isoflavone structure is shown in Figure 3.

3.2.4. Determination of flavanoid compound structure of WC extract

Table 6 shows the wavelength change of WC extract solution before and after reaction with Shift reagent. Reaction with 2M NaOH results a hypochromic shift in band I of 20 nm and a batochromic band II of 23 nm. Reaction of WC extract with of 5% $AlCl_3$ solution results a shift in band I of 21 nm indicating the possibility of suspected o-dioxide in ring A. However, in band II shifts 12 nm, reveals the presence of OH group at position C-5. Reaction with CH_3COONa and H_3BO_3 solution results a batochromic shifting in band I of 17 nm indicating the presence of OH group at position C-7. Addition of CH_3COONa and H_3BO_3 solution to the WC extract solution causes a shift of batochromic in band I of 21 nm indicating that the presence of o-diOH group in ring B. However, there is no significant shifting on band II.

Table 5. Interpretation of wavelength changes of WB extract by the addition of Shift reagents.

Treatment	Wavelength (nm)		Shift (nm)		Alleged Substitution
	Band I	Band II	Band I	Pita II	
Extract + CH_3OH	363	274	-	-	-
Extract + CH_3OH + NaOH	343	245	-20	-29	-
Extract + CH_3OH + $AlCl_3$	*n.d*	273	-	-1	-
Extract + CH_3OH + CH_3COONa	*n.d*	273	-	-1	-
Extract + CH_3OH + CH_3COONa + H_3BO_3	382	268	+19	-6	o-diOH in ring B

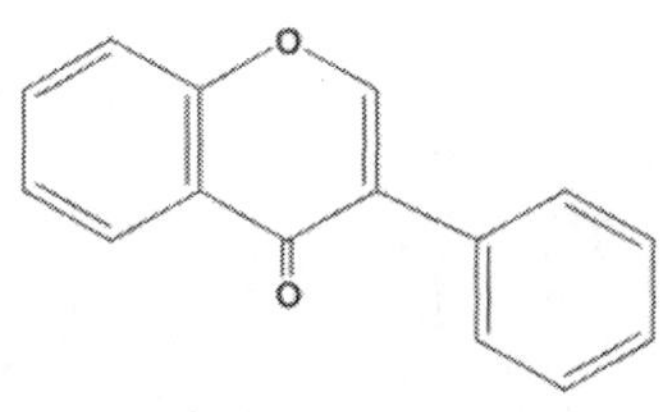

Figure 3. Structure of the Isoflavone

Table 6. Interpretation of wavelength changes of WC extract by the addition of Shift reagents.

Treatment	Wavelength (nm)		Shift (nm)		Alleged Substitution
	Band I	Band II	Band I	Pita II	
Extract + CH_3OH	363	260	-	-	-
Extract + CH_3OH + NaOH	343	283	-20	+23	
Extract + CH_3OH + $AlCl_3$	384	272	+21	+12	o-diOH in ring A, 5-OH (isoflavones)
Extract + CH_3OH + CH_3COONa	*n.d*	277	-	+17	7-OH (isoflavones)
Extract + CH_3OH + CH_3COONa + H_3BO_3	384	263	+21	+3	o-diOH in ring B

Based on interpretation of wavelength change using shift reagent, the WC extract possibly contains isoflavones. It can be seen from the formed spectrum is characteristic of isoflavones with the presence of dihydroxy group on C-5 and C-7 (Figure 4).

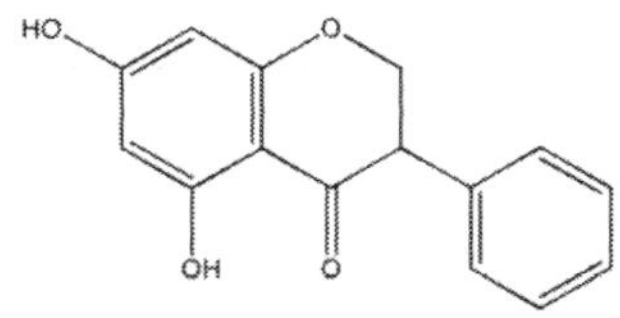

Figure 4. Structure of the Isoflavone 5,7-dihydroxy.

3.3. IR spectrum of Strawberry extract

The four extracts of the strawberries were then analyzed using FTIR which aimed to identify the functional group. Identification of compounds using FTIR to further strengthen the presence of flavonoid compounds. Based on infrared spectrum of strawberries extracts in Figure 5, it shows that there are several functional groups in the crude extract. Absorption at 3271.48 cm^{-1} indicates the presence of -OH groups in aliphatic and aromatic groups. It is a result of vibration of intramolecular hydrogen bonds which is supported by the vibration of CO alcohol bending on 1051.09 cm^{-1}. Both of these absorption peaks indicate that there is an OH group of alcohols attached to the carbon atoms. This assumption is reinforced by the presence of C = C aromatic absorption at 1620.67 cm^{-1}. The strong absorption peaks of the aromatic C - H bend are present at the 816.81 cm^{-1}. The 2932.52 cm^{-1} wave number is suspected the existence of alpha C-H. A strong absorption band of 1714.71 cm^{-1} indicates the presence of a carbonyl group (C = O), suspected as a common feature of the flavonoid group compound. The absorption peak of the vibration of the C-O ether group (1101.56 cm^{-1}). Based on the assumption reinforced by the presence of functional groups such as OH, CO alcohol, C = C aromatic, CH aromatic, CH aliphatic, C = O and CO ether functional groups possessed by flavonoid compounds [11,12]. Based on data of UV-Vis and infrared spectrum, the strawberries extract possibly contains isoflavone compounds with the possibility of hydroxy group at C-5, C-7 and C-8 on the A ring.

3.4. Antioxidant Activity of Strawberry Extract

Antioxidant activity of crude strawberry extract was assessed using DPPH method [13]. The DPPH method uses 2,2 diphenyl-1-picrylhydrasyl as a stable free radical compound. This method is the commonly used antioxidant assay for plant extract. Four crude strawberries extract, M, WA, WB, and WC extract were assessed to determine its antioxidant activity. Based on IC_{50} value, M extract has a very strong activity (50.6 ppm). Where as WA (68.0 ppm) and WC (61.4 ppm) extracts have strong activities, and WB has a medium antioxidant activity (118.5 ppm) (Table 7). This result reveals that the maceration technique is the suitable technique to extract antioxidant substance from strawberry fruit compare to microwave technique. The maceration technique was conducted in room temperature whereas using the microwave technique resulted a higher temperature in extraction process. Possibility, a higher temperature could be damage the active compounds.

3.5 Identification of antioxidant activity compound structure using LC-MS-MS [14]

The quadrupole MS / MS Triple Q mass spectrometer with an electrospray (ESI) mode and quantification of the Selected Reaction Monitoring (SRM) method were employed for identification of flavonoids in Fragaria sp. The SRM mode can provide high specifity, sensitivity and by the structures of unknown of flavonoids may be acceessed based on the m/z of both ion and fragment ion obtained through mass transition [15,16]. Moreover, the absorption spectral data were compared with that reported in the literature. Fragment ions $[M-H]^+$ was identified at m/z 253-223, the parent ion $[M^+]$ at m/z 267 and absorption maximum at 263 nm obtained due to there are of hidroxyl groups from isoflavon that revealing the compound to be **formononetin**. The presence of two hexose moieties linked to a luteolin aglycone in the structure for peak 3 was identified by the fragment ions at m/z [M-H]$^{+}$ at m/z 294.5-295.5. The parent ion [M^{+}] at m/z 415 and λ max at (368, 353 nm bond 1 and 263, 243 nm bond 2) obtained was tentatively identified as a derivate of **daidzin** by comparison with the literature [17].

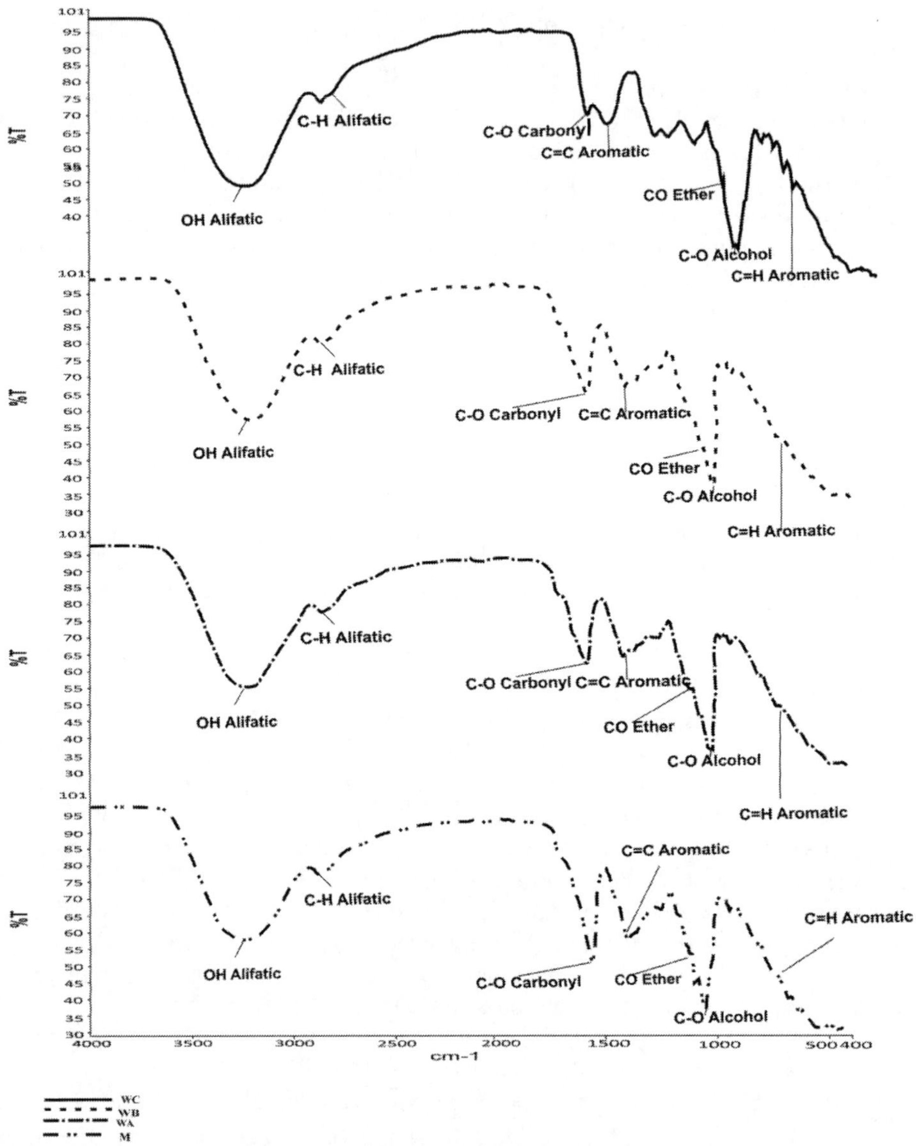

Figure 5. Infrared Spectrum of strawberry extract.

Table 7. The strength of antioxidants based on IC_{50} value

Method	IC_{50} (ppm)	Very strong (< 50 ppm)	Strong (50-100 ppm)	Medium (100-150 ppm)	Weak (150-200 ppm)
M	50.6	√			
WA	68.0		√		
WB	118.5			√	
WC	61.4		√		

4. Conclusions

Phytochemical screening test indicates the presence of compounds tannin, flavonoids, alkaloids and saponins in crude strawberry extract. The identification result using LC-MS-MS shows the existence of isoflavones compound peak i.e. formononetin and daidzin. The extraction techniques comparation result of strawberry fruit ethanol extract shows the highest yield and the best antioxidant activity with maceration techniques. This shows that the antioxidant activity of strawberry fruit extract with maceration technique is more active than the microwave technique. The maceration technique is the suitable extraction technique to achieve the highest antioxidant activity of Strawberry extract.

5. References

[1] Vierkötter_A and Krutman J 2009 *Dermatoendocrinol* **4**(1), 227-231.

[2] Panico A M, Garufi F, Nitto S, Mauro R D, Longhitano R C, Magri G, Catalfo A, Serrentino M E, and Guidi G D 2009 *Pharmaceutical Biology* **47**(3) 203–208.

[3] Ehsani B, Moghaddam and Khanizadeh S (2009) *An Overview in Antioxidant Properties of Strawberry* **in** *New Plant Physiology Research* (New York), pp 203-216.

[4] GolKhoo S, Ahmadi AR, Hanachi P, Barantalab F, Vaziri M, *Pak. J. Biol. Sci.,* 2008, **11(18)**, 2254-2258.

[5] Kumalaningsih S 2006. Antioksidan Alami, *Penangkal Radikal Bebas "Sumber, Manfaat, Cara Penyajian dan Pengolahan".* Trubus Agrisarana, Surabaya.

[6] Selawa W, Revolta M, Runtuwene J, and Citraningtyas G 2013 *Pharmacon* **2**(1): 18-22.

[7] Hayat K, Hussain S, Abbas S, Farooq U, Ding B, Xia S, Jia C, Zhang X, Xia W 2009 Separation and Purification Technology **70** (1) 63-70.

[8] Klejdus B, Vacek J, Benesová L, Kopecky J, Lapcík O, Kubán V, *Anal. Bioanal. Chem.*, 2007, **389**(7-8), 2277-2285.

[9] Markham K R 1981 *Cara Mengindentifikasi Flavonoid* (Penerbit ITB: Bandung).

[10] Silverstein R M 1986 *Penyedikan Spektrometrik Senyawa Organik* Edisi Ke-4, a.b. A J Hartomo and Anny Victor Purba (Erlangga: Jakarta).

[11] Svehla G 1990 *Buku Teks Analisis Anorganik Kualitatif Makro dan Semimikro* Edisi Kelima. translated by Setiono L and A H Pudjaatmaka (PT Kalman Pusaka: Jakarta).

[12] Sastroadmidjojo H, 1996 *Sintesis Bahan Alam* Cetakan Ke-1 (Liberty: Yogyakarta).

[13] Uwa, L M 2017 *J. Curent Trends Biomedical Engineering & Biosciences* **7** 001-003.

[14] Sulaiman T C and Indira B 2016 *J. Basic and Applied Science* **5** 231-235.

[15] Mantovani D, Cardozo L F, Santos C L, de Souza L V, Watanabe S 2009 *J. Chromatogr. Sci.,* **47**(9), 766-769.

[16] Dinelli G, Carretero A S, Silvestro R D, Marotti I, Fu S, Benedettelli S, Ghiselli L, Gutiérrez AF, 2009 *J. Chromatogr. A.*, **1216** (43), 7229-724.

[17] Zhang Q, Yu J, Wang Y, and Su W 2016. *Moleculs* **21** 989.

Effects of temperature and duration of drying on the quality of powdered *Asam sunti*

S Mulyati[1,*] F A Pramesthy[2], F Meutia[2], A Rinaldi[2], S M Siregar[2], S Muchtar[3]
[1]Department of Chemical Engineering, Syiah Kuala University, Banda Aceh, Indonesia, 23111
[2]Undergraduate Student of Chemical Engineering, Syiah Kuala University, Banda Aceh, Indonesia, 23111
[3]Sudent at Doctoral School of Engineering Science, Syiah Kuala University, Indonesia

E-mail: sri.mulyati@unsyiah.ac.id

Abstract. Generally, Acehnese use *asam sunti* as a spice in cooking. *Asam sunti* is made from dried *belimbing wuluh (Averrhoa bilimbi Linn)*, it has a flat shape with brownish colour. To increase the aesthetic value, ease of storage, and maintain its quality, *asam sunti* is produced in powder form. The quality of *asam sunti* is determined from several parameters such as water content, ash content and oxalic acid content. The powdered *asam sunti* was prepared by using Tray and Spray Dryer. The effects of temperature and duration of drying on the quality of *asam sunti* have been studied. Aesthetically, it is seen that the powdered *asam sunti* has better appearance than that of the original form. The water, ash and oxalic acid contents in the powdered *asam sunti* dried using a spray dryer was lower than that of dried by tray dryer. The optimum temperature was obtained at 80 °C. The optimum drying time were 45 minutes and 3 hours for spray and tray dryer, respectively. Water, ash and oxalic acid contents in optimum condition with tray dryer were 49.7%, 26.3% and 10.4%, consecutively. Meanwhile, by using spray dryer, the product has higher quality in terms of water content (4.7%), ash content (14.25%) and oxalic acid level (9.48%).

1. Introduction

Drying is a process of removing moisture from the material to improve the quality and shelf life of the product. Drying process has been applied to food preservation since centuries ago. The aim of drying a food product is to reduce the water content in the food to a safe limit to the increase shelf life of the product and to provide microbiological stability and reduce food damage caused by chemical reactions [1]. Moreover, by the drying process, the products are expected to have similar level of moisture [2].

In the food and vegetable industries, powdered product provides advantages such as reducing packaging expenditure, facilitating storage and transporting, and more stable than liquid products [3]. Some researchers have applied drying process to obtain powdered end-products, for example, *sinteractive thoria* powder which obtained by using freeze drying method and microwave drying [4], powdered milk by using drum dryer [5], and kimchi powder by spray dryer [6].

Belimbing wuluh (Averrhoa bilimbi Linn) is well known in Indonesia because it has sour taste and a distinctive aroma, it is often used in cooking [7-8]. In Aceh, people use or cultivate bilimbi into semi-

* Corresponding author : sri.mulyati@unsyiah.ac.id

IOP Conf. Series: Materials Science and Engineering **523** (2019) 012025 doi:10.1088/1757-899X/523/1/012025

dried salty bilimbi or commonly known as *asam sunti*. *Asam sunti* is widely used by the people of Aceh as one of cooking spices as it provides distinct sour flavor, color and increase thickness to cuisine. Prior to usage as a spice, *asam sunti* has to be milled because it is still in a semi-dried form. The resulting *asam sunti* is brown with elastic texture and flat-shaped form as shown in Figure 1.

As seen in Fig.1, the shape and appearence of *asam sunti* seems less attractive. Increasing the aesthetics of *asam sunti* can be done by changing its texture into powder by drying. Besides increasing the aesthetic value, drying can also prolong the shelf life of *asam sunti*, enhance portability, as well as achieve uniform structure and quality of the product. In addition, the economical value of powder *asam sunti* can be increased and can be recognized by all Indonesian people outside Aceh. In this study, the powdering process of *asam sunti* was done by drying technique using Tray Dryer and Spray Dryer. The influence of operation conditions such as drying time and drying temperature to the quality of *asam sunti* i.e. water content, ash content and oxalic acid content are studie. Besides, this research also aims to find the optimal condition of the research variables performed on both methods of drying.

2. Methods

2.1. Materials

The materials used were fresh *belimbing wuluh* which obtained in the vicinity of Banda Aceh city. Kitchen salt was used as *asam sunti* preservative. NaOH 0.1 N was used as a standard solution for the analysis of oxalic acid level, Phenolphtalein indicator was used as an indicator solution in titration process. Maltodextrin was used as a filler on drying with Spray Dryer. All of those three compounds were purchased from Merck. Whatman filter paper was used as a solution filter. The tool used were blender, analytical scales, oven, Tray Dryer (DSR-115, ISUZU SOYOKAZE), Spray Dryer (Leybold Didactic, Germany), desiccator, blast furnace, electric bath, titration equipment, petri dish, porcelain cup, clipping cup, volume pipette, glassware, filter paper, plastic and bucket.

2.2. Preparation of asam sunti

A total of 10 kg fresh *belimbing wuluh* was washed and weighed. *Belimbing* wuluh was blanched for 5 minutes in boiling water, then it was dried on the floor with plastic on its surface for ± 8 hours, allowed overnight then re-dried for another ±8 hours. Following that, the dried belimbing wuluh was put in a plastic container and sprinkled with salt. It was then settled for 1 night, and dried again for ± 8 hours. *Asam sunti* was obtained when the dried belimbing wuluh turned into brown color.

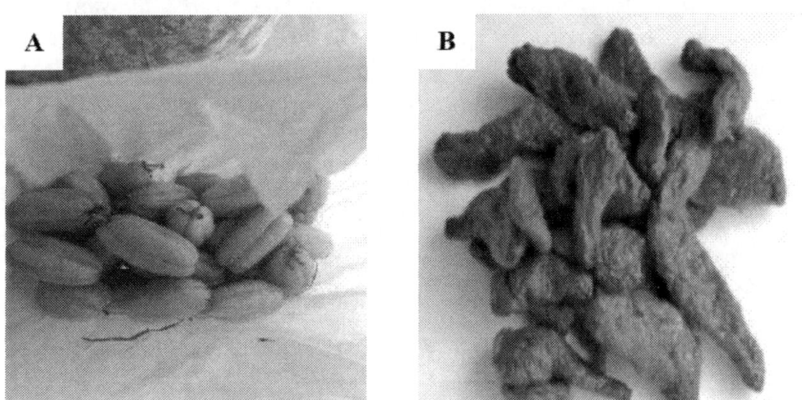

Figure 1. Belimbing wuluh (A) and *Asam sunti* Aceh (B)

2.3. Fabrication of powdered asam sunti

Asam sunti which was produced from 10 kg of belimbing wuluh was cut into small pieces, and blended to form an *asam sunti* paste. 100 gram of generated *asam sunti* paste was weighed then dried with two different drying methods i.e. using Tray Dryer and Spray Dryer. The quality of powdered *asam sunti* was analyzed in terms of water, ash, and acid content

2.4. Quality analysis of product

The water, ash and oxalic acid content were analyzed to evaluate the quality of the resulting *asam sunti* powder. Water content was analyzed by following SNI 01-2891-1992 point 5.1. Meanwhile, the analysis done for ash content was using SNI 923.03 AOAC 1998. As for the oxalic acid, the analysis was done by using acid base titration method.

3. Results and Discussion

The success of a drying process can be confirmed from the reduction of the water content to the nonstick grain level, or in other words, all the feed (liquid or paste) can be converted into powder. Figure 2 shows the *asam sunti* powder dried with tray and spray dryers. In general, the powdered *asam sunti* which produced with both drying methods are slightly brownish. The size of the powder produced by Spray Dryer is seemed to be more uniform and finer compared to that of with the tray dryer.

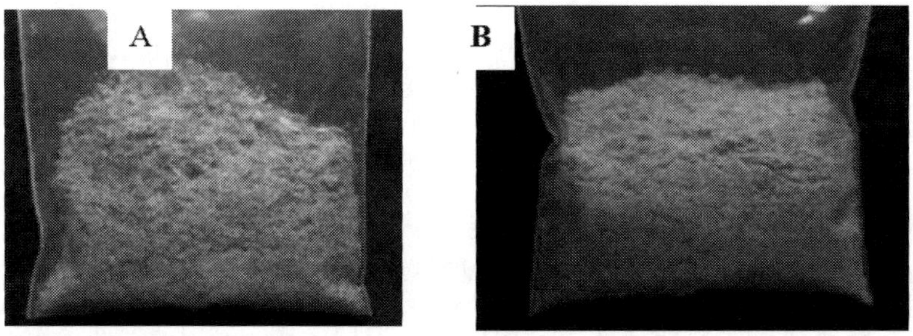

Figure 2. Powdered *Asam sunti* dried by (A) : Tray Dryer, (B) : Spray Dryer

3.1. Effect of temperature and drying time on water content in powdered asam sunti

Water content is one of the chemical properties which indicates the amount of water in the food. It has effect on enzyme and microbe activities which cause organoleptic change and nutritional value [9]. Water content is very influential on the quality of food, therefore, during food processing, water is usually issued or reduced by evaporating or thickening and drying. Reduced moisture level can increase resistance to damage and can preserve food [10]. In addition, drying make material become more durable, as well as reducing volume of foodstuffs to facilitate storage and packing space.

The drying process of *Asam sunti* involves heat and mass transfer. Heat transfer occurs from the heating medium (air) to the *asam sunti* paste. Simultaneously, mass transfer of the paste into the air also occured. The consequence of mass transfer event is the decrease of water content of *asam sunti* paste thus it will become *asam sunti* in the powder form. The reducement of water content due to drying is shown in Figure 3. The water content of *asam sunti* paste is originally 70.59%, and reduced to 30.85% after drying using tray dryer, and down to 4.29% by spray dryer. The water content decreases with the increase of drying air temperature (from 50-90 °C). This is due to the increased amount of water evaporated into the air.

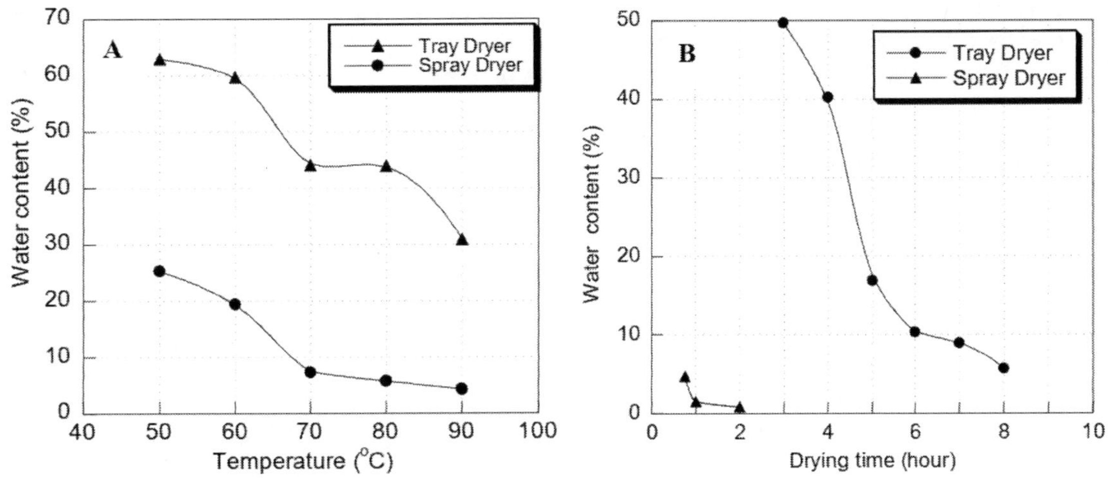

Figure 3. Effect of temperature (A) and drying time (B) on water content in powdered *asam sunti*

3.2. Effect of temperature and drying time on ash content in powdered asam sunti

Ash is a residual inorganic substance resulted from combusting an organic material. The ash content is related to the minerals of a material consisting of two types of salt, namely organic salts (i.e. salt of malic acid, oxalate, acetate, pectate) and inorganic salts (i.e. salts in the form of chloride, phosphate, carbonate, sulfate and nitrate) [12]. Ash content is a property that shows the total amount of minerals in a food product [9]. The ash content of *asam sunti* paste before dried is 14.91%. Increasing the drying temperature increased ash content in *asam sunti* powder. As shown in Figure 4., the drying results show that the lowest ash content in *asam sunti* powder sample is 14.82% for that dried by tray dryer, while the ash content of the spray-dried sunti sample is lower than that of original sample content, namely 6.33%.

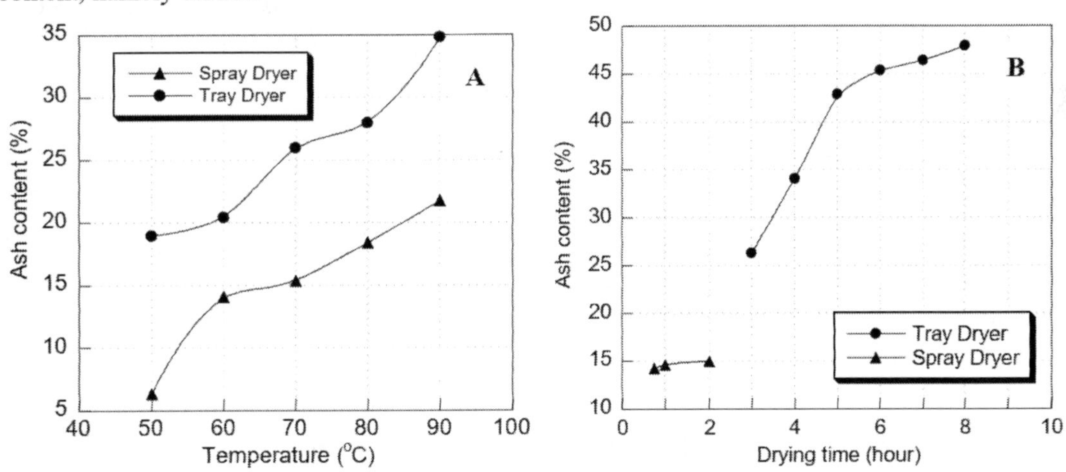

Figure 4. Effect of temperature (A) and drying time (B) on ash content in powdered asam sunti

3.3 Effect of Temperature and Drying Time on Oxytic Acid Level in Powdered Asam sunti

Oxalic acid is a natural constituent of plants which usually ingested by animals and human [13]. When plants or foods containing oxalic acid are consumed by human, it will induce bad effects on health because oxalate has the capability to bind calcium and other minerals. Excessive consumption of oxalic acid can lead to stone formation in the urinary tract when acid is excreted in urine. That is why

Asam sunti with high level content can cause kidney stones in the human body [14]. Therefore, the oxalic acid level present in *asam sunti* is expected to be as low as possible. The relationship between temperature and duration of drying and oxalic acid level in powdered *asam sunti* can be seen in Figure 5.

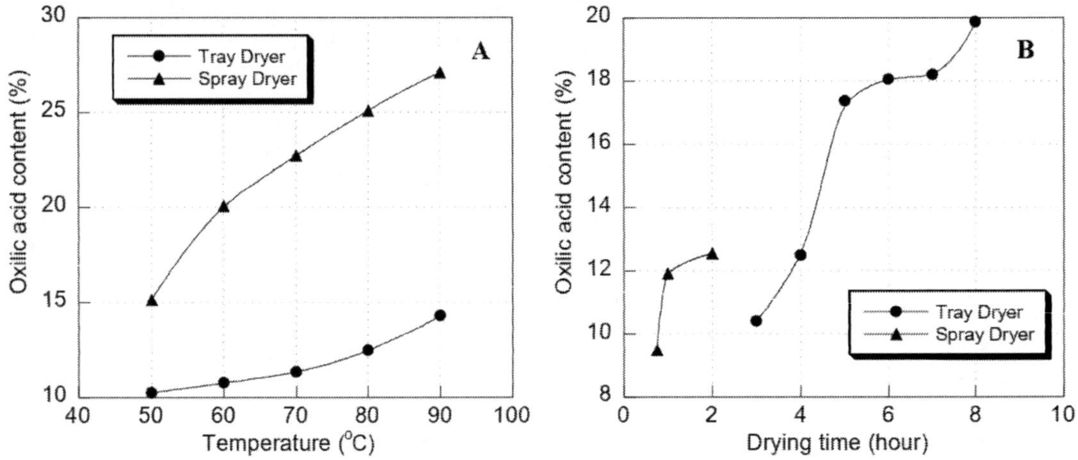

Figure 5. Effect of temperature (A) and long drying (B) on oxalic acid levels on powdered *asam sunti*

Figure 5 shows the increment tendency of oxalic acid level with increasing temperature and duration of drying. This is caused by the hardening of the outside skin of dried fruit hence organic acids are trapped in it [15]. Moreover, the higher the drying temperature and the longer the drying time will decrease water content and will increase the mineral content in a material [11]. Drying with Tray Dryer gives a lower oxalic acid than drying with Spray Dryer. The results show that the lowest oxalic acid level in the *asam sunti* powder is 10.41% for the sample dried by Tray Dryer, whereas the oxalic acid content in *asam sunti* dried by Spray Dryer is much lower than the original sample content level, namely 9.48% with drying time of 0.75 hours or 45 minutes.

4. Conclusion

Asam sunti is one of main cooking ingredients for the people of Aceh. To facilitate the storage and enhance the aesthetic of *asam sunti*, which commonly flat, chewy, and brown, it had been made into powder by using a Tray Dryer and Spray Dryer drying methods. It has been confirmed that, the temperature and duration of drying affect the quality of *asam sunti* obtained. Drying with Tray Dryer gives the lowest water content at 30,85% while Spray Dryer can decrease water up to 4,29%. For ash content, drying with Spray Dryer produces lower ash content than Tray Dryer. The oxalic acid content of the lowest powdered *asam sunti* is obtained at 9.48%. This value is obtained with operating conditions of 80 °C and 45 minutes long.

5. Acknowledgement

The authors would like to thank the Ministry of Research, Technology and Higher Education Directorate of Student Affairs through the Student Creativity Program (PKM-P) which has funded this research. Thank you also to Syiah Kuala University for facilitating the implementation of this research

6. References

[1] Welsh, Z C Kumar, and Karim A 2017 Energy Procedia **110** 465
[2] Margaris D P and Ghiaus A.-G 2006 *J. Food Eng.* **75** 542
[3] Esteve M J, Frígola A, Rodrigo C and Rodrigo D 2005 *Food Chem. Toxicol.* **43** 1413
[4] Annie D, Chandramouli V, Anthonysamy S, Ghosh C and Divakar R 2017 *J. Nucl. Mater.* **484** 51

[5] Aalaei K, Rayner M and Sjöholm I 2016 *LWT Food Sci. Technol.* **73**, 675
[6] Park H-J, Lee Y and Eun J-B 2016 *Biocatalysis and Agricultural Biotechnology* **5** 193
[7] Wijayakusuma H M H, Dalimartha S and Wirian A S 1993 *Tanaman Berkhasiat Obat di Indonesia* (Pustaka Kartini)
[8] Tampubolon O T 1995 *Tumbuhan Obat* (Bharantara : Jakarta)
[9] Riansyah A, Supriadi A and Nopianti R 2013 *Jurnal Fishtech* **2** 53
[10] Nielsen S S 2010 *Determination of Moisture Content*.(US :Springer)
[12] Fitriani S, Akhyar A and Widiasuti *Sagu* **12** 1
[13] Sudarmadji S, Haryono B and Suhardi 1997 *Prosedur Analisis untuk Bahan Makanan dan Pertanian* (Liberty:Yogyakarta)
[14] García-Fernández A J, Espín S, Gómez-Ramírez P and Martínez-López E 2014 *Oxalates A2 - Wexler, Philip. Encyclopedia of Toxicology* (Oxford: Academic Press.)
[15] Irhami 2012 *Kajian Penangan Pasca Panen Belimbing Wuluh(Averrhoa bilimbi.L) Sebagai Bahan Baku Pembuatan Bubuk Asam sunti Menggunakan Penggeringan Kabut(Spray Dryer)* (Institut Pertanian Bogor:Bogor)

IOP Conf. Series: Materials Science and Engineering **523** (2019) 012026 doi:10.1088/1757-899X/523/1/012026

Sensory and physical quality assessment of *Sie reuboh* (Cooked Meat- an Acehnese traditional cuisine)

D Hasni[1,*], N M Erfiza[1], C Nilda[1], M Faiz[1]
[1]Agricultural Product Technology Department, Agriculture Faculty, Syiah Kuala University, Darussalam - Banda Aceh 23111, Indonesia

E-mail: hasni_dian@unsyiah.ac.id

Abstract. *Sie reuboh* is an Acehnese traditional cuisine which uses red-meat as the main protein and is prepared specifically with addition of beef tallow, traditional vinegar and blend of spices. The dish was extremely popular, has a long shelf life, exquisite and complex taste. However due to the changing in life styles and modernization, nowadays the dish hardly prepared at home and has varied of recipes. Based on previous studies, the updated recipe was formulated by varying the amount of beef tallow and vinegar as independent variables. Furthermore, this study aims to assess the physical quality (pH and cooking weight loss) and sensory attributes (color, aroma, flavor, texture). The sensory attributes were divided onto ten specific attributes, which were assessed with Qualitative Descriptive Analysis (QDA) method. The analysis of variance (ANOVA) showed that addition of beef tallow significantly affected ($P \leq 0.01$) the umami flavor of *sie reuboh* , whereas addition of vinegar significantly ($P \leq 0.01$) affected pH, cooking weight loss, color, meat flavor, sour flavor, tenderness and chewiness of produced *sie reuboh* descriptively. The sensory profiles of each treatment were also described, where the increasing amount of vinegar tend to decrease pH, prevent cooking weight loss and elevate color, tenderness and sour flavor of the product. [1]

1. Introduction

A traditional such as *sie reuboh,* a cuisine from Aceh Province-Indonesia, is considered as one of indigenous knowledge which carries on the local cultural wisdom. According to [1], *sie reuboh* is red-meat boiled with palm vinegar, dried and then fried or served as preferred. Its appearance is quite similar with dendeng, the thin slices of dry cooked red-meat, but it is moister and has a complex acid flavor. In ancient times, this cuisine was commonly used as a packed meal during war, since it has a long shelf-life. Its long shelf-life is manifested by a combination of palm vinegar and beef tallow added during preparation [2,3]. Palm (*Arenga pinnata*) vinegar is known as natural food's preservative, which commonly used in South East Asean cuisines. Palm vinegar contains acetic acids as the product of vinegar fermentation. The addition of vinegar has been reported to decrease pH, where foods with lower pH (higher acidity) is considered to be more resistance towards microbial growth [4]. An acetate acid also helps to sustain the food quality hedonically [5]. On the other hand, the addition of beef tallow was reported to increase the shelf-life of *sie reuboh* in room temperature. Beef tallow which has solid form in room temperature, is covering the food surfaces, this condition successfully inhibits the microbes to interact with food at room temperature during storage [3].

*Corresponding author : hasni_dian@unsyiah.ac.id

However due to the changing in life styles and the consumer health concerns, the preference towards *sie reuboh* is also rapidly change. Modern life styles nowadays prefer to convenience and ready to eat foods, as well as healthier foods. Many reports mentioned that the re-heated processing of *sie reuboh* established higher peroxide value since it consisted of many amount of beef tallow [3], [6]. The mentioned factors above have brought *sie reuboh* as feast cuisine, where the dish is eaten for occasional moments only [7]. Moreover, as the dish is not required to be preserved for a long time storage anymore, the recipe is changed, especially for the amount of palm vinegar and beef tallow being used. Previous research reported by [5] mentioned that consumers prefer to use up to 5% of palm vinegar and up to 50% of beef tallow for *sie reuboh* recipes, which is less than previous formulation. Therefore, this research aims to observe the effect of the developed recipe conducted by [5] towards the physical properties and sensory assessment of the product, both descriptively and hedonically.

2. Materials and Methods

2.1. *Materials*

This study used the meat from thigh (beef round) and beef tallow, of 3 year old Aceh local female cow which was freshly bought at Lambaro Kaphe traditional market, Aceh Besar district, palm vinegar (pH 3.6) from 3 month fermentation process, was bought at Ulee Kareng traditional market, Banda Aceh. The palm vinegar was then pasteurized before used, whilst the red-meat and beef tallow was cut onto square bites (4 cm x 6 cm x 2 cm) with weight 90-100 g per pieces. The recipe percentage was in the basis of the used red-meat [5]. The cooking utensils used were digital scales, blender, basins, gas stove, clay pot and wood stirrer.

2.2. *Statistical Methods and Analytical Procedures*

2.2.1 Experimental design.

This study used a Randomized Block Design with two independent variables. The first variable was percentage of palm vinegar being used (C1=2.5% and C2=5%), whilst the second variables was percentage of beef tallow being used (L1=0%, L2=25% and L3=50%). Then all the treatment combinations were done in three repetitions, resulting the total experiments units of 18 samples. Analysis of Variance (ANOVA) and Duncan Multiple Range Test (DMRT) was used for statistical analysis.

2.2.2. Sie reuboh preparation.

The *sie reuboh* cooking process was done refers to [5]. The tallow was firstly trimmed and removed from the fresh meat and the trimmed meat was cut into a block shape with the average weight of 90-100 g per piece, then the meat pieces was washed and drained for 5-10 minutes. The red-meat, salt and lime extract and tallow were placed in *beulanga* (a traditional cooking clay pot) and well-mixed. All spices such as dried ground red chilies (5%), cayenne pepper (1%), garlic (1%), galangal (5%), ginger (1%), turmeric (0.3%), slat (1%), lime juices (2%) and water (100%), were coarsely blended with spoon for 1-2 minutes, mixed with meat and added drinking water. Then all ingredients were cooked for 60 minutes without cover, then vinegar was added based on the treatment level and then continued to boil for ± 30 minutes. After the cooking process was done, *sie reuboh* was allowed to cool down to room temperature, then was placed into vacuum plastic bags and stored over night at 5°C. Prior to analysis, *sie reuboh* was re-heated for 15 minutes and drained, then for each experimental unit, the meat was cut into 1 cm thickness without the curry, followed how the dish was normally eaten. The samples were randomly served to 8 trained panelists based on SNI 2346-2006 sensory evaluation procedure and [8], [9]. Figure 1 shows finished *sie reuboh* after the cooking process is done.

Figure 1. *Sie reuboh*

2.2.3. Physical property determination.
Acidity (pH) was determined by a pH meter (LaMotte pH plus Direct) based on [10], where about 10 grams of beef sample were minced using food processor, and then the sample was mixed with 100 ml of water. The homogenous and liquefied sample was placed in a beaker glass to measure its pH by pH meter, each sample was repeated three times. For cooking weight loss determination, the raw beef was weighted (W1) and then was cooked as *sie reuboh* cooking procedure. The cooked beef was weighted (W2) and cooking weight loss was calculated as below equation (1):

$$Cooking\ loss\ (\%) = \frac{W1-W2}{W1} x\ 100\% \quad (1)$$

2.2.4. Sensory analysis.
The sensory quality was descriptively assessed with a Qualitative Descriptive Analysis (QDA) Method [8]. The QDA used 8 trained panelists who were used to made and consumed *sie reuboh* on daily basis. The panelist were selected from several districts in Aceh Besar regency. They received training three times before the real test in order to habituate them with the test procedures. This QDA assessed the intensity of color, aroma (sour, spicy and meaty), flavor (sour, spicy, salty and umami) and texture (tenderness and chewiness) of *sie reuboh* . The intensity was measured using 5-point descriptive scales (1= very weak; 2=weak; 3=just right; 4= strong and 5=very strong) (Setyaningsih et al., 2010). The obtained given score then calculated and normalized based on [9] with a 95% confidence interval for the standard distribution. The mean data of each treatment combination used for ANOVA. The correlation between sensory quality based on QDA and the consumer acceptance towards *sie reuboh* reported by [5] was also conducted.

3. Results and Discussion

3.1. Physical quality
Sie reuboh in this study had the cooking weight loss values about 40.47–44.52% and average 41.82%. The result of ANOVA showed that addition of vinegar has significantly influenced ($P \leq 0{,}05$) on the cooking weight loss of *sie reuboh* as showed in Figure 2. $DMRT_{0.05}$ indicated that the cooking weight loss value of *sie reuboh* which was added with 2,5% palm vinegar had a significant difference to *sie reuboh* added with 5% palm vinegar, where 2,5% vinegar had lower cooking weight loss than the other one (Figure 2). This result agrees with [12] who showed that when beef was immersed in 40 ml of vinegar, it had higher cooking weight loss value than the beef immersed in 20 ml of vinegar. Increasing the percentage of vinegar could increase the cooking weight loss of beef. Acidic properties in vinegar can extract meat protein due to its capability to weaken the miofibril protein bonds that result in a lot free water that comes out of the meat [13]. Later, it also stated that vinegar could affect coking loss value because acetic acid in vinegar can hydrolyse beef protein so that the nutrition of the beef decline [13], as well as the moisture contents [14]. The higher cooking weight loss value was the

lower meat quality due to loss of nutrition [15]. According to [16], increasing cooking weight loss value can be caused by rising temperature of cooking and heating duration during meat processing. Figure 3 showed the effect of palm vinegar addition towards pH of produced *sie reuboh*. As palm vinegar given increased (5%), the pH of *sie reuboh* decreased. The addition of vinegar has been reported to decrease pH [4], as reported as well by [3], [13]. In this research, higher acid treatments tend to increase the color quality of product as can be explained in color attributes.

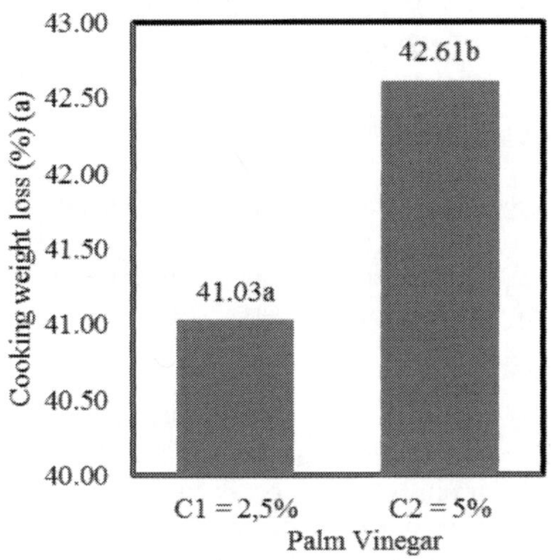

Figure 2. Effect addition of vinegar to cooking weight loss of *sie reuboh* (coefficient of variant was 0,55%)/ (The following values mark with different letter indicated the significantly different)

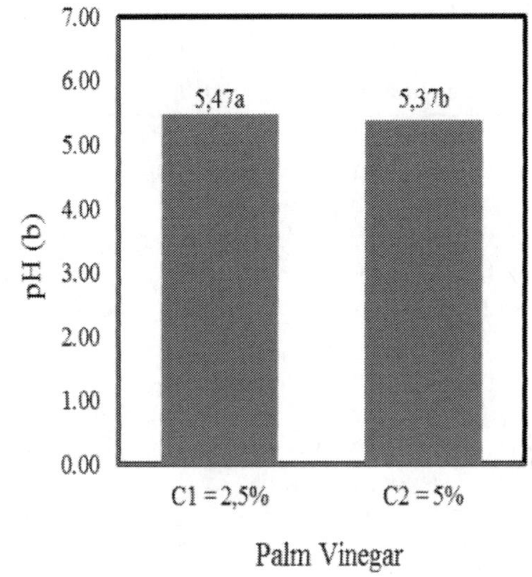

Figure 3. Effect addition of vinegar to pH *sie reuboh* (coefficient of variant was 0,25%). (The following values mark with different letter indicated the significantly different)

3.2 Color attributes

Brightness or color *sie reuboh* descriptively measured in range between 2.50 to 3.93 (pale to bright) with average value of 3.13 (just right). ANOVA shows that the amount of palm vinegar added has significantly influenced ($P \leq 0.01$) the color of *sie reuboh* descriptively, whereas the addition of tallow has showed any influence ($P > 0.05$). Furthermore DMRT test as seen in Table 1 shows that both of treatment levels of palm vinegar have different performed for the brightness of *sie reuboh* , where the trend shows a linear relation between the percentage of added palm vinegar and the increasing brightness level of *sie reuboh* descriptively. Even the spices being used in this product also affects the color of *sie reuboh* such as turmeric which is contributed to the yellow color. But since the amount for all treatments are similar, palm vinegar could be consider as main cause for the brightness. Five percent of palm vinegar produced *sie reuboh* with brighter color than the 2.5%. This might be occurred due to the acidity level of used palm vinegar (pH 3.6). Lower pH of palm vinegar could prevent the meat oxidation during cooking, which at the end could lead to myoglobin denaturation. Therefore by prevention of this oxidation reaction, the brightness or color of meat could be stable during cooking. Moreover in lower pH condition, muscle protein in red meat will be fractioned and enable meat to reflect the light and perform the brighter color [17].

Table 1 Effect of palm vinegar and tallow towards color and aroma of *sie reuboh* descriptively

Factors/Level Treatments		Colour/Brightnes[b]		Sour Aroma[b]		Spicy Aroma[a]		Meaty Aroma[b]	
Palm Vinegar (%)	C1 (2.5%)	2.58	±0.02a	2.58	±0.07a	3.08	±0.08	3.54	±0.03b
	C2 (5.0%)	3.69	±0.07b	3.84	±0.13b	3.27	±0.14	3.04	±0.12a
Tallow (%) (ns[a])	L1 (0%)	3.08	±0.61	3.50	±0.71	3.40	±0.05	3.17	±0.65
	L2 (25%)	3.22	±0.66	3.20	±0.66	3.43	±0.09	3.13	±0.63
	L3 (50%)	3.10	±0.62	2.93	±0.59	3.78	±0.01	3.38	±0.65

[a] Factors and treatments are not significant towards measured parameters.
[b] Factors and treatments are significant towards measured parameters
Sensory attributes are measured by its intensity descriptively with range scale 1 to 5 (1= very weak; 2 = weak; 3 = just right, 4 = strong and 5 = very strong).
a and b the notation as symbol the statistic differentiation within treatment levels, where numbers followed by same alphabetic symbol in similar row show any differentiation exists

3.3. Aroma attributes

Aroma attributes assessed in this research consist of sour aroma, spicy aroma and meaty aroma. The ANOVA shows that only percentage of added palm vinegar has significant influence towards the aroma of *sie reuboh* descriptively, namely for sour and meaty aroma. On the other hand the percentage of beef tallow added during cooking showed any influences for three descriptive attributes. Moreover underlining the spicy aroma of *sie reuboh*, which is not influenced by both factors, this condition could be occurred due to the amount of spices using in this research were similar for all treatment, since it is not the dependent factor.

Sour aroma of all samples of *sie reuboh* is measured in range of 2.43-4.20 (weak to strong) with average score 3.21 (just right) and meaty aroma is in range 2.63-3.63 (just right to strong) with mean value is 3.29 (just right). From Table 1, it could be seen that the addition of palm vinegar influenced the sour and meaty aroma of *sie reuboh* , where palm vinegar showed direct proportion with sour aroma but it performed inversed proportion with meaty aroma. Sour aroma is an aroma produced by organic acid added or contained in the product, respectively acetic acid inside the palm vinegar. Therefore as the percentage of palm vinegar added increased the sour aroma intensity also increased respectively [13].

On the other hand, the reaction of palm vinegar factor towards meaty aroma was reversed. Meaty aroma defines as aroma which is originally came from fresh meat, descriptively the meaty aroma is combination of bloody and fishy of meat. Several researchers have mentioned that lipid oxidation as the essential cause of meaty aroma [18-20]. So it is potential the meaty aroma increases as the tallow given the cooking process of *sie reuboh* . However, the intensity of meaty aroma decreased as the percentage of palm vinegar increased. In addition, a certain prediction stated by [21] that acetic acid is diffused in meat during cooking process and led to protein denaturation, which finally produced the volatile compounds during the maillard reaction.

3.4 Flavour attributes

In this research, flavor intensity is measured based on its hot spiciness, sourness, saltiness and umami descriptively. ANOVA statistically mentioned that palm vinegar factor only has significant effect ($P \leq 0.05$) on sour flavor, whilst beef tallow factor only influenced ($P \leq 0.05$) the umami flavor. For measured parameters which are unaffected ($P > 0,05$) by dependent factors, namely as hot-spicy and salty flavor could be caused by the amount of chilly, salt and other spices in this research are similar for all unit experiments.

Table 2 shows the DMRT0.05 effect of dependent factors towards flavor of *sie reuboh* , where sour flavor intensity showed the raising trend as the percentage of palm vinegar increased, and has significantly differentiated for both treatment levels. This might be occurred due to the acetic acid compounds in palm vinegar, which is produced during the fermentation process. Acetic acid considers to have specific flavor, which is commonly used in South East Asian cuisine as synthetic flavor agents. In detail in Table 2, it is clearly showed that the increasing trends are also performed by others three flavor attributes, but without any statistical differentiation between the treatment levels followed. Umami flavor of all *sie reuboh* samples descriptively measured in range of 2.20 to 3.93 (weak to strong) with average score is 3.15 (just right). Moreover, The DMRT0.05 statistically mentioned that the addition of beef tallow up to 50% is successfully elevated the umami flavor which reached the score 3.87 (strong) that is significantly different than other lower levels. This increasing intensity could be caused by the presence of primer product of lipid oxidation as the given percentage of beef tallow higher [22] which is also interacted with formation with umami flavor in maillard reaction during cooking process [21]. Later several reports also added that lipid such as tallow derived volatiles in larger amounts [22]; and plays multiple roles in flavor development, in terms of producing distinctive flavor during the lipid oxidation during the thermal cooking processing [22],[23],[20], as occurred in *sie reuboh* .

Table 2 Effect of palm vinegar and tallow towards flavor of *sie reuboh* descriptively

Factors/Level Treatments		Measured Parameters							
		Hot-spicy Flavor [a]		Sour Flavor [b]		Salty Flavor [b]		Umami[b]	
Palm Vinegar	C1 (2.5%)	3.30	±0.04	2.74	±0.07a	3.08	±0.07	3.03	±0.02
	C2 (5.0%)	3.52	±0.13	3.56	±0.16b	3.19	±0.10	3.27	±0.24
Beef Tallow	L1 (0%)	3.67	±0.16	3.50	±0.38	3.37	±0.07	2.33	±0.06a
	L2 (25%)	3.37	±0.12	3.10	±0.31	3.17	±0.05	3.25	±0.13b
	L3 (50%)	3.20	±0.05	2.85	±0.18	2.87	±0.00	3.87	±0.03c

Table 3 Effect of palm vinegar and tallow towards texture of *sie reuboh* descriptively

Factors/Level Treatments		Measured Parameters			
		Tenderness[b]		Chewiness[b]	
Palm Vinegar (%)[b]	C1 (2.5%)	2.77	±0.08a	3.30	±0.16b
	C2 (5.0%)	3.60	±0.09b	2.49	±0.13a
Beef Tallow (%)[a]	L1 (0%)	3.20	±0.47	3.18	±0.48
	L2 (25%)	3.32	±0.22	2.55	±0.32
	L3 (50%)	3.03	±0.19	2.95	±0.06

[a] Factors and treatments are not significant towards measured parameters
[b] Factors and treatments are significant towards measured parameters
Sensory attributes are measured by its intensity descriptively with range scale 1 to 5 (1= very weak; 2 = weak; 3 = just right, 4 = strong and 5 = very strong).
a,b and c are the notations used as symbol the statistic differentiation within treatment levels, where numbers followed by same alphabetic symbol in similar row shows any differentiation exists

3.5 Texture attributes

As meat products, texture of *sie reuboh* is descriptively assessed based on the tenderness and chewiness of meat during gustation and chewing process. Based on ANOVA, only palm vinegar factor has significant influence ($P \leq 0.05$) towards the tender and chewy texture of *sie reuboh*. All samples of this cuisine is reported have tenderness scale in range within 2.57 to 3.87 (just right to strong) with average score 3.18 (just right) descriptively. Based on Table 3, it is clearly seen that the

IOP Conf. Series: Materials Science and Engineering **523** (2019) 012026 doi:10.1088/1757-899X/523/1/012026

increasing level of palm vinegar given in formulation product, could significantly elevated the tenderness of *sie reuboh* , as 5% of palm vinegar produced meat with higher tenderness level (3.60 – strong). This might be occurred since the palm vinegar has ability to hydrolyze the protein muscles which impact the activity of binding water and loose the interval spaces within the muscles. On the other hand, chewiness emphasizes to the hard feeling occurred during meat chewing process in the mouth, where the score of all samples is varied from 2.10 to 3.87 (weak to strong) and average score is 2.89 (just right). $DMRT_{0.05}$ showed that when less percentage palm vinegar given, the chewiness score is higher. This phenomenon shows that chewiness is revised version of tenderness as presented in Table 3 below.

3.6 Correlation between the measured parameters with sie reuboh consumer acceptance scores

Table 4 shows the correlation between the physical/sensory quality of *sie reuboh* and consumer acceptance. Firstly discussed is the correlation within pH and color which assessed hedonically. The coefficient is -0.48 which refers to negative relationship within both parameters. This negative correlation is also proven by the consumer acceptance towards color and flavor of *sie reuboh* elevated as the pH level decreased due to the higher percentage of palm vinegar being used. On the other hand cooking weight loss statistically proven to have positive correlation (+0.40) with consumer acceptance towards texture of *sie reuboh* . It is can be explained by as the tendency of higher acceptance of texture increases aligned with higher value of cooking weight loss, which led to the tenderness of meat.

Table 4 Correlation between the physical quality of *sie reuboh* and consumer acceptance

Physical Measurement	Consumer Acceptance							
	Color		Aroma		Flavor		Texture	
	Corr	R^2	Corr	R^2	Corr	R^2	Corr	R^2
Ph	--	-0.48	Ns	0,16	+	0,55	ns	-0.38
Cooking weight loss	ns	0,10	Ns	-0.07	ns	0,00	+	0,41
Sensory Descriptive Measurement	**Consumer Acceptance**							
	Color		Aroma		Flavor		Texture	
	Corr	R^2	Corr	R^2	Corr	R^2	Corr	R^2
Color/Brightness	+++	0,81	+	0,40	+	0,45	ns	0,30
Sour Aroma	++	0,60	Ns	0,21	Ns	0,30	++	0,52
Spicy Aroma	ns	0,04	-	-0,43	Ns	-0,23	+	0,47
Meaty Aroma	--	-0,50	Ns	-0,09	Ns	-0,13	-	-0,47
Hot-spicy Flavor	ns	-0,10	Ns	-0,26	Ns	-0,01	ns	0,29
Sour Flavor	+	0,45	Ns	0,19	Ns	0,17	+	0,45
Salty Flavor	ns	-0,14	Ns	-0,28	Ns	-0,21	+	0,48
Umami	-	0,44	++	0,61	++	0,67	ns	-0,12
Tenderness Texture	+	0,61	Ns	0,32	+	0,47	++	0,66
Chewiness Texture	-	-0,53	Ns	-0,31	-	-0,46	ns	-0,32

+++/++/+ positive correlation exists within the parameters with very strong/strong/slightly strong
---/--/- negative correlation exists within the parameters with very strong/strong/slightly strong
ns not significantly correlated

Table 4 also presents the correlation within the sensory quality and consumer acceptance. *Sie reuboh* which descriptively has brighter color tend to have higher consumer acceptance, as proven by very strong positive correlation (+0.82). The similar trend also appears for correlation between tenderness, sour aroma and sour flavor towards consumer acceptance for texture attributes. All above parameters are linked with palm vinegar factor, as a cause of intensity of sour and flavor aroma, as well as tenderness in *sie reuboh* increases.

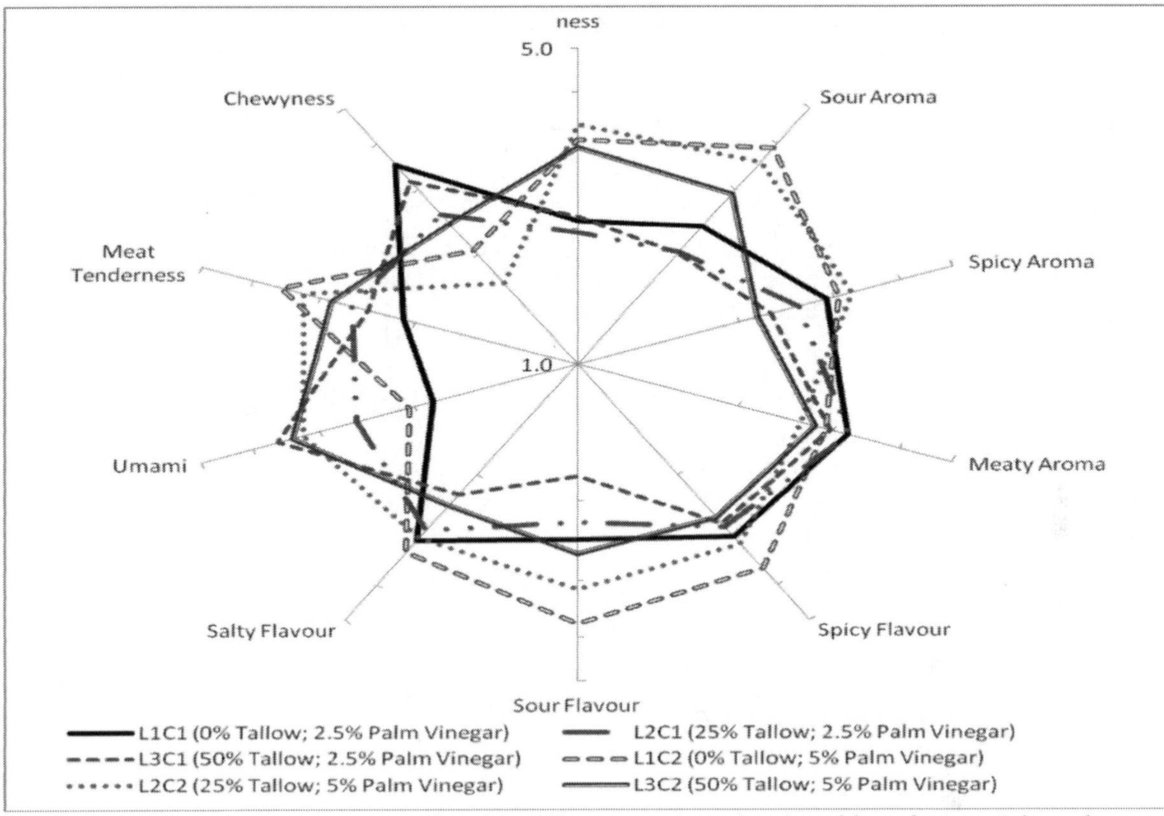

Figure 4. Sensory profile of *sie reuboh* with different treatment levels, with scale 1 to 5 (very low intensity to very strong intensity).

3.7. Sensory profile of 'Sie rebouh'

Sensory profile of *sie reuboh* could be seen in Figure 3, where all ten descriptive attributes assessed in this research are presented in spider web diagram, performs each different sensory profile for all six unit experiments. The Figure uses five scale intensity as performed during data collection where 1 refers to very weak intensity and 5 refers to very strong intensity. Unit experiment which is combination of absence of beef tallow (L1=0%) and highest percentage of palm vinegar added (C2=5%) produced *sie reuboh* with high intensity of sour aroma and flavor, brightness and tenderness of *sie reuboh* rather than the units which used less percentage of palm vinegar. However, the above attributes will have less intensity as the percentage of beef tallow increased (L2= 25% and L3=50%). Therefore it can be clearly said that palm vinegar factor is substantially forming up the sensory profile by determination its color, sour aroma and sour flavor, whilst the beef tallow factor has major contribution to formulate the meaty aroma and umami flavor of *sie reuboh* . After detail investigation of *sie reuboh* sensory profile, it can be said the best unit experiment with optimum physical and optimum quality is combination of 25% beef tallow and 5% palm vinegar (L3C2).

4. Conclusions

Taking everything into consideration it can be concluded that there is a considerable changes in formulation of *sie reuboh* recipes and the changes also has major impact on physical quality and sensory quality of the cooked meat product. As percentage of palm vinegar used increases, pH of product decreases and the sensory quality especially the brightness color, the sour aroma and flavor as well as tenderness of product tend to elevate onto higher intensity, whilst the contradiction occurred towards meaty aroma and chewiness attributes. Palm vinegar factor also built up positive correlation between brightness, sourness and tenderness of product and whole consumer acceptance hedonically. On the other hand, factor of beef tallow has considerable influenced towards umami flavor. Lastly mentioned that the best combination treatment which produced optimum sensory quality of *sie reuboh* is combination of 25% beef tallow and 5% palm vinegar (L3C2) with this sensory quality descriptively (3.40 color; 3.60 sour aroma, 3.20 spicy aroma, 2.60 meaty aroma, 3.10 spicy flavor, 3.10 sour flavor, 3.0 salty flavor, 3.2 umami, 3.2 tenderness and 1.70 chewiness).

5. Acknowledgements

All the involved researchers pleasantly acknowledge Ministry of Research and Higher Education of Indonesia, in the name of Syiah Kuala University, who funded this research through 'Hibah Laboratorium Uji Sensorik Tahun 2018.

6. References

[1] Hidayah, Z. 2015. *Ensiklopedi Suku Bangsa di Indonesia*. 2nd Edition (Jakarta :Yayasan Pustaka Obor Indonesia)
[2] Fitri, C. A. 2002. *Thesis*. Progam Pascasarjana, Institut Pertanian Bogor.
[3] Suhairi, L. 2007. *Skripsi*. Retrieved from Institute Pertanian Bogor Repository, Bogor.
[4] Cahyadi W, 2009. *Analisis dan Aspek Kesehatan Bahan Tambahan Pangan*, Edisi Kedua. (Jakarta : Bumi Aksara)
[5] Hasni, D. Erfiza, N.M. Faiz, M dan Syahrina. U. 2017 *Proceeding of 2nd International Conference on Science and Technology* (Malaysia : University of Mataram and University of Malaya)
[6] Lawrie, R. A. 2003. *Ilmu Daging*. Edisi Kelima (Jakarta :UI-Press)
[7] Nur A, Nelly M, dan Nur R 2015 *SEL*. 2(2): 72-76.
[8] Meilgaard MC, Civille GV, and Carr BT 2009 *Sensory Evaluation Techniques*. 2nd edition. (CRC Press, Boca Raton)
[9] Association Meat Science of America (AMSA) 2015 *Research guidelines for cookery, sensory evaluation and instrumental tenderness measurement of meat*. 2nd edition. (AMSA. Illinois)
[10] Association of Official Analytical Chemists 1995 *Official methods of analysis*. 16th edition. (Association of Official Analytical Chemists. Washington DC, USA)
[11] Setyaningsih, D, Anton A, dan Maya P. S. 2010 Analisis Sensori untuk Industri Pangan dan Agro. (IPB Press, Bogor)
[12] Tarigan, A. 2004 *Jurnal Seminar Nasional Teknologi Peternakan dan Veteriner*.
[13] Masyitah I, Arief I, Suryati T 2016 *Jurnal Ilmu Produksi dan Teknologi Hasil Peternakan*. 4(1).
[14] Erfiza, NM, Hasni, D and Syahrina, U. 2018 *Jurnal Teknologi Industri Pertanian Indonesia* 10(1): **28-35**.
[15] Jamhari E, and Rusman 2007 *Buletin Peternakan*. 31(2): 94-100. ISSN 0126-4400.
[16] Soeparno. 2005. *Komposisi Karkas dan Teknologi Daging*. (Fakultas Peternakan. Pascasarjana UGM. Yogyakarta)
[17] Purnamasari, E., M. Zulfahmi dan I. Mirdhayati. 2012. *Jurnal Peternakan*. 9(1): 1-8.
[18] Gray JI., Gomoa EA, Buckley DJ. 1996, *Meat Science*, 43: 111–123.
[19] Komariah Sri R, and Sarjito 2009 *Buletin Peternakan*. 33(3): 183-189.
[20] Khan MI, Cheorun Jo, Tariq MR 2016 *Meat Science* 110: 278-284
[21] Purba, M. 2014 *Jurnal WARTAZOA*. 24(3): 109-118.
[22] Sartika, R. A. D. 2008 *Jurnal Kesehatan Masyarakat Nasional*. 2(4): 154-160
[23] Mottram, D. S., & Edwards, R. A. 1983 *Journal of the Science of Food and Agriculture*, 34, 517–522.

IOP Conf. Series: Materials Science and Engineering **523** (2019) 012027 doi:10.1088/1757-899X/523/1/012027

Effect of varieties and processing practices on the physical and sensory characteristics of Gayo Arabica specialty coffee

Y Abubakar [1,*], D Hasni[1], M Muzaifa[1], Sulaiman[2], Mahdi[3], H P Widayat
[1]Department of Agricultural Product Technology, Agriculture Faculty, Syiah Kuala University, Banda Aceh 23111, Indonesia.
[2] Department of Economic Management, Economy and Business Faculty, Syiah Kuala University, Banda Aceh 23111, Indonesia
[3]Gayo Cupper Team, Simpang Mongal, Takengon, Aceh, Indonesia

E-mail: yusya.abubakar@unsyiah.ac.id

Abstract. Nowadays, sensory characteristics are considered as basic requirement for coffee quality, which determined its future commercial value. Gayo Arabica coffee is one of specialty coffees known by international buyers due to its unique taste and complex aroma. However, the sensory characteristics of gayo arabica coffee vary based on type of local varieties and the processing technique practiced. The present study aims to determine the effect of variety and processing methods towards physical and sensory properties of Gayo Arabica coffee. Four Gayo local varieties (Tim-tim, Borbor, Ateng Super and Multi-varieties) and three processing practices (full-wash, semi-wash and honey natural process) were investigated. The measured parameters were moisture contents and cupping quality based on Specialty Coffee Association of America (SCAA) procedure. The statistical analysis showed that local varieties and processing attributes has significant influenced ($P \leq 0.01$) on the certain sensory attributes. The interaction of both factors showed significant effects ($P \leq 0.01$) towards flavour, acidity and balance attributes. Ateng Super variety showed a high score of overall cupping attributes whilst among processing practices it showed no differences. In conclusion, all experiments produced specialty coffee based on SCAA cupping test. Ateng Super local varieties showed the highest score (85.36 ± 0.14) and full-wash processing produced the highest cupping quality compared to others (84.79 ± 0.03). The Q-grader also remarked that Gayo Arabica coffee has distinguished flavours such as chocolate, dark-chocolate, caramel, nutty and spicy in all experiments. However, several unpleasant remarks also appeared in honey natural process such as sour, slightly aged and over fermented taste.

1. Introduction

World coffee consumption has grown steadily in the last few years and rises up to 158,901,000 bags (60 kg per bags) in 2017. This figure is depicting an increase of 4.7% in coffee consumption, compared to that of 2014. As coffee consumption grows, number of consumer who requires a good quality coffee also increases, especially consumer who wants a specialty grade coffee, [1, 2]. It has also reported that export market of Indonesian coffee (both robusta and arabica) increases by 25% in 2017 [3,4]. Therefore, Indonesia, as the third world coffee producer who produced 10,902,000 bags of

* Corresponding author: yusya.abubakar@unsyiah.ac.id

IOP Conf. Series: Materials Science and Engineering **523** (2019) 012027 doi:10.1088/1757-899X/523/1/012027

green bean in 2017, became an important supplier of good quality coffee. Therefore, Indonesian farmer should improve their farming practice, as well as maintain their coffee bean quality.

Consumer drinks coffee mainly because of its magnificent taste and the health benefit. A good quality coffee should have a high cup score and provide a great taste demanded by coffee drinkers. Coffee taste (cup quality) relates strongly with its green bean quality, including variety, geographical location, farm altitude, harvesting practice, postharvest handling, and storage [5], continued with roasting process [6]. Flament and Thomas [7] mentioned that, due to heat application, bean roasting has triggered chemical reaction, named as Maillard reactions, which is leading to the formation of unique chemical compound which will impart specific flavour characteristics and taste of coffee. Type and amount of compound formed highly depend on the composition of precursors available in the green bean. One type of coffee may produce different compositions of volatile and non-volatile compound resulting in a quiet different coffee flavour, taste and aroma [8].

One of the main region producing arabica coffee in Indonesia is Gayo Highland (in Aceh Province), which contribute to about 60-70% of Indonesian arabica coffee production. As demand and price of Arabica coffee increases, as well as the shifting trend towards specialty and single origin specialty coffee. Farmers or coffee producers are forced to improve their practices and coffee quality, in order to win the competition in world market. For the case of arabica Gayo coffee, it has known in the world market as one of Indonesian specialty coffee with a premium quality, but improper postharvest handling and processing practices actually block out this chance.

Farmers in the Gayo Highland area carry out several processing methods such as full wash, semi wash, honey natural and few others. However, the quality needs to be improved [9]. Abubakar et al. [10] reported that processing methods and coffee variety influenced sensory characteristic of gayo arabica coffee. Furthermore, they reported that superior local variety such as Tim-Tim, Borbor, and Ateng Super produced different flavour characteristic when harvested from farms at different altitude. Farmers or collectors who plan to produce high quality coffee should consider these factors. Unfortunately, the relationship among these factors has not explored widely yet, especially the influence between superior varieties and processing methods, which was designed to be objective of this study. The results of this research is expected to be used by farmers to optimize their processing technique to obtain a high quality coffee, which in turn will increase the coffee price and their income accordingly.

2. Material and methodology

This research used randomized complete block design with two factors, the first factor was Gayo local varieties (V1: Tim-tim; V2: Borbor 2; V3: Ateng Super or Dwarf Tree and V4: Multi-varieties) and the second factor was green bean processing (P1= full-wash, P2= semi-wash and P3= honey natural process). Each treatment was repeated three times, which made the total number of samples to be 36. The data then analyzed by ANOVA and Duncan Multi Range Test with SPSS version 20.0.

The green bean processing was conducted in the farm plantation areas, whilst the moisture contents analyzed in Food Analysis and Agricultural Product Laboratory- Department of Agricultural Product Technology – Syiah Kuala University. The cupping test and physical appearance were conducted by Gayo Cupper Team (GCT) as Q-Grader in Takengon, Central Aceh- Indonesia.

Materials used in this research was Arabica coffee (*Coffea arabica,* L), specifically three local varieties which were mostly planted in Gayo Highland, Aceh Tengah, Indonesia. Tim-tim (V1) was collected from coffee farmers in Bies with elevation 1400 meter above sea level (m.asl). Borbor was collected from Pilar Jaya, Aceh Tengah with elevation 1600-1700 m.asl, then Ateng Super was gathered from Bies Penantan farmers, which has elevation around 1500 m.asl. For Multi-varieties, which refers to mixed variety in Gayo Highland, was collected from Bies and Bies Penantan farmers. The recommendation of elevation level for each variety refers to [10].

The research procedures divided onto three stages, which were collection of coffee cherries in the location mentioned above, green bean processing based on treatment levels and the analytical procedures. Green bean processing stages were divided onto three categories, full wash, semi wash and honey natural. Full wash (P1) and semi wash (P2) were started with pulping the coffee cherries. Then for full wash (P1) the beans were rinsed and soaked onto water overnight, then it was washed up until clean. Later, the coffee was dried up with sun drying until the moisture content reached approximately 10-12%, then the grain was hulled to obtain the full-wash green bean. For semi wash (P2) the wet beans were rinsed and fermented in container (without water) overnight, then it was washed until clean and dried up under the sun until it reached about 35-40% moisture contents. Then they were hulled to remove the coffee parchment (to produce "kopi labu"; a local term for the half wet bean) and then was continued with sun-drying until moisture contents reached 10-12%. For honey natural process (P3), it refers to pulped dry process, where after pulping, the grains were directly sun-dried until their moisture reached 10-12%. Then the dry coffee grain was hulled to obtain the green bean.

GCT who were professional team recognized by local authority and accessed by coffee exporter did cupping test. The team has three members holding Q-grader certificates that recognized by international buyers. Cupping test was an assessment of sensory quality by descriptive method based on Specialty Coffee Association of America [12]. The coffee preparation started by roasting the coffee bean at maximum one day prior to analysis. The roasting was done in medium roasting degree with colour associated with 55 Agtron scale. Then the coffee bean is ground up to 20 meshes. The brewing techniques started with poured 8.25 gr of grounded coffee with 150 ml hot water (93°C). During the test, both ground and brewed coffee were used.

The cup test objectively measured coffee quality based on ten attributes (fragrance/aroma, flavour, acidity, aftertaste, body, balance, uniformity, clean cup, sweetness and overall). All of the sensory attributes were assessed sequentially. Fragrance and aroma firstly examined during 3-5 min after brewing. Fragrance note was also done at first step, by noising the ground coffee to detect the characteristics of the coffee. The second step was involved flavour, acidity, aftertaste, body and balance, where the temperature of coffee drink was around 71°C, which is commonly occurred roughly 8-10 min from the infusion. The last step was assessment of uniformity, clean cup, sweetness and overall. Finally, when the cupping test were done, all the notes and scores were marked on the cupping form. The score using numeric values from 6 to 9 is shown in table 1. The validity of each score is based on the flavor judgement experiences of the trained assessor. Later the total score was summed up, as accumulation of all individual score of all attributes constitutes. In order to obtain final scores, the defect value was subtracted from the total score. Final score represents the overall quality of the coffee. Scoring key has proven to be a meaningful way to describe the range of coffee quality for the final score.

3. Results and discussion

3.1 Sensory Profile of Gayo Arabica Coffee Single Origin Based on Processing Methods

Sensory profile of gayo arabica coffee single origin based on different processing methods are shown in Table 1. All of local varieties were classified as specialty coffee, even when it was treated in different processing methods. The Ateng Super local varieties obtained the highest score amongst others (85.36±0.14) and full-wash processing produced the highest cupping quality compared to others (84.79±0.03). It could be observed that for Tim-tim (V1) and Borbor (V2) varieties, full-wash processing method produced coffee with highest cupping score compared to that of other processes. However, Ateng Super or Dwarf Tree (V3) variety processed with honey natural method produced coffee with higher cupping score, while Multi-varieties (V4) obtained high final score if processed with semi-wash method. Bhumiratna et al. [13] mentioned that fragrance of coffee characterized by the origin of coffee varieties. However, it is not considered as the main cause, because coffee quality is influenced by numerous factors and holistic coffee processing practice [14]. As for full-wash and honey natural processing, full-wash tended to produce more aromatic fragrance such as fruity, floral and acidic attributes (sour) as shown in table 3 for both Tim-tim and Borbor varieties. This finding is supported also by Duarte et al. [15] and Leloup et al. [16] who observed that the full wash method

IOP Conf. Series: Materials Science and Engineering **523** (2019) 012027 doi:10.1088/1757-899X/523/1/012027

produced more acidic attributes and less bitter, burnt and woody notes, as an impact of the mucilage removal during the wet or full wash process. Mucilage removal is considered as a cause of undesirable fragrance such as soury, herby or greenish which usually found in green bean coffee [13] [17].

In contrast, semi wash process which is a common coffee processing practice originated from Gayo Highland, ultimately can be considered as the main caused of pleasant nutty, choco or chocolate like notes where these notes are regularly found as characters of semi wash process, as presented in table 3 and also highlighted by [10]. Nutty flavor can be defined as a light, brown, slightly musty aromatic flavor associated with nuts where Choco-like or chocolate like refers to sweet, dusty often bitter aromatics associated with cocoa beans or cacao derived product [13].

Table 1. Total score and fragrance remarks of all treatment combinations

	Treatments	Final Score	Fragrance Remarks
	V1P1	84.76±0.24	Spicy, floral, nutty, grassy, citrusy, Choco-powder, salty, tobacco.
	V1P2	84.17±0.38	2/5 sour, over ripe, 1/5 fermented, floral, syrupy, fruity, brown tea.
	V1P3	83.67±0.52	Dark Choco, nutty, floral, caramel, citrusy, syrupy, greenish, tobacco.
	V2P1	84.75±1.09	Spicy, sugar palm, creamy, toasty, grassy, Choco, fruity, chocolate,.
	V2P2	84.00±1.00	Floral, tea like, grassy, malty, syrupy, dry fruit, pineapple, chocolate.
	V2P3	83.75±.1.00	Choco, dark caramel, nutty, fruity, malty, spicy, herby, bold, tobacco.
	V3P1	85.00±0.87	Bright Choco powder, nutty, fruity, grassy, sugar brown, citrusy.
	V3P2	85.25±0.25	Fruity, black tea, floral, over ripe, juicy, chocolate, herby, tobacco.
	V3P3	85.83±0.58	Spicy, floral, bright nutty, grassy, citrusy, Choco-powder, tobacco.
	V4P1	84.67±1.01	2/5 sour, over ripe, 1/5 fermented, banana, floral, syrupy, fruit tea like.
	V4P2	84.92±0.80	Dark Choco, nutty, floral, caramel, citrusy, syrupy, greenish, biscuit.
	V4P3	84.33±0.76	Spicy, sugar palm, creamy, toasty, grassy, greenish, Choco, fruity.
Average		84.59	

Table 2. Influence of local varieties, processing method and their interaction towards parameters

Measured Parameters	Local Varieties (V)		Processing Methods (P)		Interaction	
	F-Table	F Value	F-Table	F Value	F-Table	F Value
Moistures^{ns}		0.41		0.02		1.18
Fragrance**		8.51		0.02		0.89
Flavour**		6.03		0.68		3.68
Aftertaste^{ns}		2.92		0.14		2.50
Acidity*	4.72	2.21	5.61	1.25	2.51	2.92
Body**		7.92		12.80		1.05
Balance*		2.17		3.65		3.59
Overall**		7.92		7.44		1.35
Total**		6.46		1.10		1.58

^{ns} Factors and its interaction do not have any significant influence on the measured parameter.
*/** Certain factor has significant/very significant influence towards the measured parameter.

3.2. Fragrance attribute

Fragrance attribute of Gayo single origin coffee were varied from 7.60 to 8.16 (very good to excellent quality scale) with a mean value of 7.98±0.04 (excellent) as shown in Figure 1. As stated in Table 2, only local varieties had significant (P≤0.0) influenced on this attributes, where all treated varieties were significantly different from each other. Ateng Super yielded the highest score of fragrance (8.16) amongst others, whereas Tim-tim placed as the lowest while Borbor and Multi-varieties were in between. These observations are supported by Hulupi et al. [18] who stated that in term of fragrance, which measured quality of ground coffee aroma, Tim-tim has lower score than Borbor. As the local varieties used here were considered as of optimum quality based on its farm elevation referred to [10], the results were satisfactory since all of varieties were yielded very good up to outstanding qualification, according to SCAA standard.

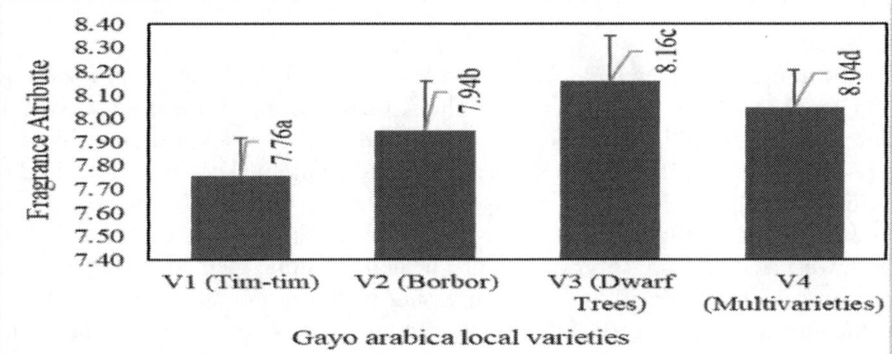

Figure 1. Influence of local varieties (P≤0.01) on fragrance attributes of Gayo Arabica coffee, $DMRT_{0.05}$= 0.179, 0.189 and 0.195 (numbers which followed by different alphabet showed significant differences).

3.3. Flavor attribute

Table 3. Effect of interaction factors on flavor, acidity and balance attributes

Treatments	Flavor	Acidity	Balance
V1P1	7.93±0.12b	7.50±0.00a	8.00±0.00b
V2P1	7.93±0.12b	7.93±0.17b	7.87±0.12ab
V3P1	7.87±0.12b	7.60±0.17ab	7.60±0.17a
V4P1	8.00±0.00c	7.53±0.25a	7.60±0.17a
V1P2	7.80±0.00b	7.50±0.00a	8.00±0.00b
V2P2	7.93±0.12b	7.93±0.29b	7.70±0.17ab
V3P2	7.93±0.12b	7.80±0.00b	7.83±0.29ab
V4P2	7.93±0.12b	7.70±0.17ab	7.88±0.12ab
V1P3	7.60±0.17a	7.87±0.12b	7.60±0.17a
V2P3	7.90±0.10b	7.90±0.29b	7.67±0.15a
V3P3	8.10±0.17c	7.80±0.00b	7.87±0.12ab
V4P3	7.93±0.12b	7.60±0.17ab	7.60±0.17a
DMRT0.05	0.180-0.210	0.266-0.312	0.249-0.298

numbers which followed by different alphabet showed significant differences).

Flavor of Gayo single origin coffee were varied from 7.60 to 8.10 (very good to excellent quality scale) with an average score of 7.91±0.05 (excellent) based on SCAA standards. Variability of flavors among local varieties processed by semi wash and full wash were very small, except flavor of multi-varieties processed by full wash that differed significantly from others. However, variability of flavors of local coffee processed by honey natural method were very significant. Within honey natural process, flavor of Ateng Super which was the highest, differed significantly from the other three varieties, whereas flavor of Tim-tim being the lowest and those of Borbor and Multi-varieties were in between (Table 3). Tim-tim and Borbor has been suggested to be planted in farm with elevation above 1200 m.asl for diseases and cupping quality reasons [18]. Moreover, Abubakar et al. [10] also found that Tim-tim harvested from farm at 1200 m.asl and Borbor from 1600-1700 m.asl yielded a higher cupping scores than those from lower or higher farm elevation level. Ateng Super, as a hybrid of Katai varieties, showed good cupping quality on flavor and fragrance.

3.4. Acidity attribute

Acidity attributes from all coffee samples were varied between 7.47 and 7.87 (very good scale quality) with an average score of 7.63±0.07 (very good). The ANOVA assumed that only interaction between the two factors that had significant influenced to the acidity of brewed coffee. Table 3 shows that Borbor variety yielded the highest acidity scores in all processing methods examined in this research, where the acidity of each variety were comparable across all processing method. However, it is extremely surprising that Tim-tim variety processed by full wash and semi wash methods yielded the bean with the lowest acidity score compared to that of honey natural method. Ateng Super and Multi-varieties have similar range of acidity in all processing methods. For the multi-varieties, this could occur due to the absence of characteristics notes, as multi-varieties were coming from or a mix of all varieties planted in the region. Some researchers also proposed that acidity could be influenced by climate condition as well as farm elevation [20]. In this research, Borbor were manually harvested from the areas, which was higher than that of other varieties (1600-1700 m.asl).

It can be seen in Figure 3 that wet processing, for both full wash and semi wash, produced coffee with higher acidity. The acidity from wet process was significantly different from the one that from honey natural, which was considered as dry process. This finding is also supported by Mazzafera and Purcino [21] who mentioned that wet processing has superior coffee with higher acidity and more aroma than the coffee obtained from dry process. The differences were due to the impact of pulping and fermentation process [22].

3.5 Balance attribute

Balance score was varied from 7.60 to 8.00 in all samples (very good to excellent quality) with a mean value of 7.77±0.04 (very good). ANOVA showed that processing method and interaction from both factors had significant influences on balance, as presented in Table 3. Tim-tim had superior balance score over the other varieties when it was produced in wet process, but the score was declined when it was processed using honey natural method (dry process). Furthermore, the balance score of Ateng Super and Multi-varieties appeared to be higher when semi wash process was applied. Moreover, honey natural process seemed to produce lesser balance score over the other processing methods across the varieties, except for Ateng Super which produced a higher balance score compared to that of the other two methods. Balance which defines as an equilibrium of flavor, acidity, aftertaste and body is considered as prominent quality standard of specialty coffee. Therefore, wet process, which is assumed could preserve intrinsic quality of each local variety yields higher score than the natural method.

3.6 Body attributes

Body of single origin arabica coffee produced from this research varied from 7.87 to 8.27 (very good to excellent quality) with an average score of 7.99±0.08 (excellent). ANOVA showed that both factors have significantly influenced ($P \leq 0.01$) the body attributes as seen in figure 4 and figure 5. Figure 4 shows that Tim-tim yielded the lowest score for the body while Ateng Super and Multi-varieties were claimed as the varieties with the highest body, which were significantly different (P≤0.01) from Tim-tim. Hulupi et al. [18] also reported that Borbor variety had higher body score than Tim-tim. This was due to the genotype source of each local variety, which were habitually and indigenously selected by farmers, which should be explored in detail in the future. However, for Multi-varieties, the logical reasons could be resolved, as the sample consisted of a mix of the other three varieties.

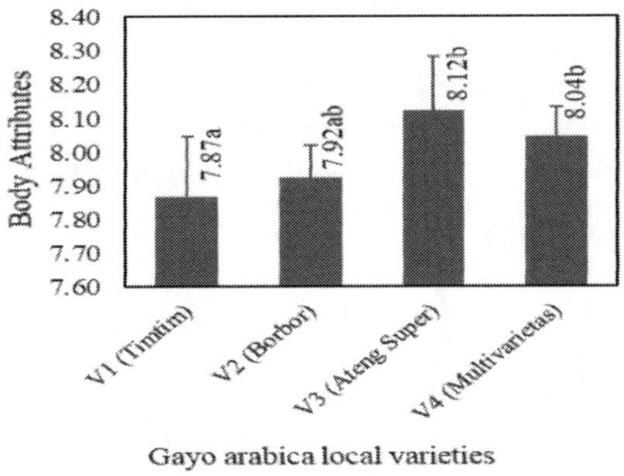

Figure 4. Influence of local varieties (P≤0.01) on body attribute of Gayo Arabica coffee, $DMRT_{0.05}$=0.21; 0.22; 0.23 (numbers which followed by different alphabet showed significant differences).

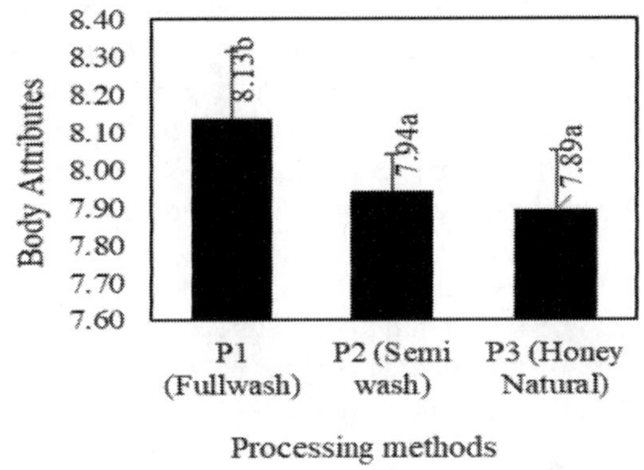

Figure 5. Influence of processing methods (P≤0.01) on body attribute of Gayo Arabica coffee, $DMRT_{0.05}$=0.21; 0.22 (numbers which followed by different alphabet showed significant differences).

3.7 Overall attribute

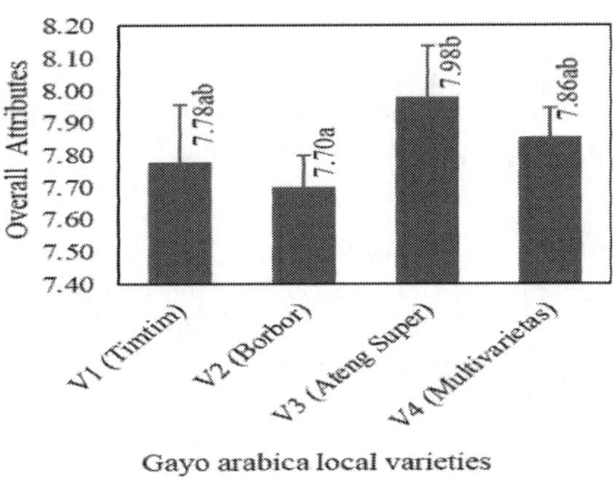

Figure 7. Influence of local varieties (P≤0.01) on overall attributes of Gayo Arabica coffee, , $DMRT_{0.05}$=0.21; 0.22; 0.23 (numbers which followed by different alphabet showed significant differences).

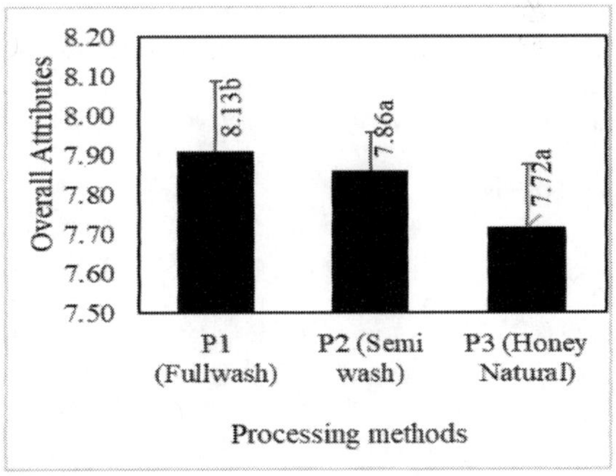

Figure 8. Influence of processing methods (P≤0.01) on overall attributes of Gayo Arabica coffee, $DMRT_{0.05}$=0.21; 0.22 (numbers which followed by different alphabet showed significant differences).

Overall score was in range of 7.60 to 8.00 with an average score of 7.83±0.10 (very good to excellent quality) based on SCAA quality scale. Then ANOVA assumed that both factors have significant influences on overall attributes, as presented in figure 7 and figure 8. Figure 7 shows that Ateng Super variety was assumed superior in terms of overall attributes, as well as other sensory attributes, but it only showed significant differences to that of Borbor variety. Borbor which was considered as variety with lowest overall attribute, showed no significant differences to those of Multi-varieties and Tim-tim.

Moreover, figure 8 shows the effect of coffee processing method on the overall attributes of Gayo arabica coffee single origin. As overall attribute measures the level of homogenization of the sensory attributes over the samples in five cup assessments, full wash processing methods showed the highest score over the other two methods in this research. As proposed by several researchers [13], [17], [14] and [23], wet process techniques, either full or semi wash methods are able to minimize coffee bean defects during green bean processing compared to dry process or honey natural process.

4 Conclusion

It can be concluded that the three local varieties of Gayo arabica coffee yield a specialty-coffee quality with specific characteristics notes, which are varied across the samples. Full wash and semi wash methods appeared to maintain the intrinsic quality of Tim-tim and Borbor and produces a "very good specialty" quality-classification compared to that of honey natural process for both mentioned varieties. In general, Ateng Super variety showed the highest score (85.36±0.14), while full-wash processing methods produced the highest cupping quality (84.79±0.03) compared to others. The Q-grader also remarked that Gayo Arabica coffee has distinguished flavour such as chocolate, dark-chocolate, caramel, nutty and spicy in all experiments. Furthermore, honey natural processing method can improve balance score of Ateng Super variety. However, the method tends to produce coffee with unpleasant fragrance remarks such as sour, slightly aged and over fermented. Further investigations related to chemical composition need to be carried out in order to determine the relationship between processing methods and chemicals associated with cupping performance of each local Gayo Arabica coffee variety.

5 Acknowledgements

All involved researchers in this research wish to gratefully acknowledge Ministry of Research and Higher Education and University of Syiah Kuala for supporting valuable facilities and funding through program "Penelitian Unggulan Unsyiah", in 2017. Moreover, special thanks go to Gayo Cupper Team, in Takengon, for its commendable support through sampling and performing the cupping test.

6 References

[1] ICO - International Coffee Organization (2018): *World Coffee Consumption* [Accessed: 10.08.2018] Http://www.ico.org/price.asp?section=statisticsworld_coffee_consumption

[2] Sheti S 2017 *A Surprising New Trend In Coffee* Forbes [Accessed: 05.08.2018] Https://www.forbes.com /sites/oracle/2018/08/07/with-artificial-intelligence-sometimes-less-is-more/#3811bec86dfe

[3] Wright T and Rahmanullah A 2017 *Indonesia coffee annual report 2017* [Accessed: 05.08.2018] Https://gain.fas.usda.gov/.../Coffee%20Annual_Jakarta_Indonesia_5-15-2017.pdf

[4] ICO - International Coffee Organization (2018 [a]): *Exporting Countries: Total Production.* [Accessed: 10.08.2018] Http:// www.ico.org/price.asp?section=statistics

[5] Baggenstoss J 2008 *Coffee Roasting and Quenching Technology – Formation and Stability of Aroma Compounds* (Switzerland: Eidgenossische Technische Hochshule Zuerich-ETH) Dissertation

[6] Wei F and Tanokura M 2013 *Chemical Changes in the Component of Coffee Beans during Roasting* In: *Coffee in Reader and Diseases Prevention* (London: Elsevier Inc)

[7] Flament I and Bessiere-Thomas Y 2002 *Coffee Flavor Chemistry* (Chichester: John Wiley and Sons Ltd)

[8] Farah A, Monteiro M C, Calado V, Franca A and Trugo L C 2006 Correlation between cup quality and chemical attributes of Brazilian coffee *Food Chemistry* **98** 373-380

[9] Muzaifa M and Hasni D 2016 Exploration study of gayo specialty coffee (*Coffea arabica l.*): chemical compounds, sensory profile and physical appearance *Pakistan Journal of Nutrition* **15** 5 486-491 DOI: 10.3923/pjn.2016.486.491

[10] Abubakar Y, Hasni D, Widayat H P, Muzaifa M, and Mahdi 2017 Quality of gayo arabica coffee as affected by farm altitude and variety *2nd International Conference on Science and Technology* (Mataram: University of Mataram)

[11] AOAC 2000 *Official Methods of Analysis. 16th Edition* (Washington D.C.: Association of Analytical Chemist)

[12] SCAA 2018 *SCAA Protocols: Cupping Specialty Coffee* (Specialty Coffee Association of America -SCAA) [Accessed on 5 Augustus 2018] Http: www.scaa.org.

[13] Bhumiratna N, Adhikari K, Chambers E 2011 *LWT-Food Science and Technology* **44** 2185-2192.

[14] Sunarharum W B, Williams D J, Smith H E 2014 *Food Research International,* **62** 315–325.

[15] Duarte S C, Pena A, Lini C M 2010 *Food Microbiology* **27** 187-198

[16] Leloup V, Gancel C, Liardon R, Rytz A, and Pithon A 2004 *International Scientific Colloquium on Coffee* (Bangalore, India: ASIC) pp 93–100

[17] Farah A 2012 *Coffee Constituent in Emerging Health Effect and Diseases Prevention* ed Chu Yi-Fang (New York: John Willey and Sons Inc) pp 21-58.

[18] Hulupi R, Nugroho, D and Yusianto 2013 *Pelita Perkebunan* **29** 2 69-81

[19] Muzaifa M, Hasni D, Patria A, Febriani, and Abubakar A *International Journal on Advanced Science Engineering Information Technology* , **8** 1 165-171

[20] Bertrand B, Boulanger P, Sussert S, Ribeyre F, Berthiot L, Descroix F, Joet T 2012 *Food Chemistry* **135** 2575-2583.

[21] Mazzafera P and Padilha-Purcino R 2004 *20th International Scientific Colloquium on Coffee* (Bangalore, India: ASIC)

[22] Clarke R J and Macrae R (Ed) 1985 *Coffee Chemistry*, vol 1 (London: Elsevier Applied Science Publishers)

[23] Lee L W, Cheong M W, Curran P, Bin Y, Liu S Q 2015 *Food Chemistry* **185** 182-191.

IOP Conf. Series: Materials Science and Engineering **523** (2019) 012028 doi:10.1088/1757-899X/523/1/012028

Amino acid and sensory profile of *Kopi Luwak* (Civet Coffee)

M Muzaifa[1,2,*], D Hasni[2], D Yunita[2], Febriani[3], A Patria[2], A Abubakar[4]
[1] Doctoral Study Program of Agricultural Science, Syiah Kuala University, Darussalam, Banda Aceh, 23111, Indonesia
[2] Department of Agricultural Product Technology, Faculty of Agriculture, Syiah Kuala University, Darussalam, Banda Aceh, 23111, Indonesia
[3] Department of Chemistry, Faculty of Mathematics and Natural Science, Syiah Kuala University, Darussalam, Banda Aceh, 23111, Indonesia
[4] Department of Animal Husbandry, Faculty of Agriculture, Syiah Kuala University, Darussalam, Banda Aceh, 23111, Indonesia

E-mail: murnamuzaifa@unsyiah.ac.id

Abstract. *Kopi luwak* (civet coffee) is produced exclusively from coffee beans excreted by Indonesian palm civet or luwak (*Paradoxurus hermaphroditus*) and relatively has a high price. The aim of this research was to identify amino acid compositions of *kopi luwak* (green and roasted beans), its changing forms and its correlation to sensory profile. The coffee beans of *kopi luwak* were collected from six locations in Gayo Highland, Aceh-Indonesia. Amino acid compositions were analyzed by LC-MS and sensory profiles were performed using a cupping test based on Specialty Coffee Association of America (SCAA) procedures. The result showed that green and roasted beans had significant differences on their Alanine, Tyrosine, Cysteine and Arginine contents. Valine, Glutamine, Aspartic acid and Glutamic acid are considered as the main amino acids which have higher percentage in both green and roasted bean. Several amino acids are absence in green bean but presence in roasted bean or vice versa, such as alanine and arginine. Glutamic acid content was the highest among others, both in green bean and roasted bean. The sensory profile of civet coffee showed that all samples from various origins/sources were classified as specialty coffee because the obtained total score of cupping was above 80.

1. Introduction

Coffee is one of the most commercialized agricultural product and considered as the most widely consumed beverage in the world. Coffee is drunk by millions of people every day and has attracted a considerable attention among consumer world-wide mainly because of its health benefits and sensory quality, especially its specific flavor [1-3].

The flavor of brewed coffee is determined by pre and postharvest factors [4], [5]. Numerous investigations with respect to the influence of the "from farm to fork practices" such as species, origin, coffee processing methods which turns out coffee cherries to green coffee bean roasting, and brewing have been published [6-8]. The research has acknowledged that the coffee processing method is

*Corresponding author : murnamuzaifa@unsyiah.ac.id

known as one of the most important aspect that can influence coffee flavor. In coffee bean processing, farmers usually applies three common ways, which include wet fermentation/washed methods, semi dry (washed) and dry fermentation (natural) processing methods [9], [10]. However, there is an uncommon process called 'digestive bio-processing' which is a fermentation step. In some countries like Indonesia, this fermentation of green coffee bean occurs in the gastrointestinal tract of *luwak*, Indonesian name for civet (*Paradoxorus hermaphroditus*), and the product is named as civet coffee or *kopi luwak* [5], [11], [12].

Kopi luwak is known as one of the most expensive coffee available today. This coffee is very hard to find, since its produced only 500-700 kg per year [13]. The price for a single cup of *kopi luwak* is $35-$80 and one pound of beans costs $100 to $600 (http://www.most-expensive.coffee). People have been discussing this rare coffee because of its exclusive harvesting process and originality. The coffee is made from coffee cherries that have been eaten by civet and pass through its digestive system. However, civet is unable to digest the beans and excretes them with its faeces and these excreted beans are collected to make *kopi luwak* [13], [14]. *Kopi luwak* quality is mainly contributed by two factors: selection of the best (the ripest) coffee cherries by civet and chemical changes that occur in the digestive tract of civet which produces specific aromatic chemical compounds [12], [15]. In detail, [16] stated out that caged luwak that regularly fed up with papaya and water melon tends to produce *kopi luwak* with papaya-aroma like, whilst *kopi luwak* from the wild one, tend to have more complex aroma since the wild luwak that able to look up their feeding by themselves.

The diversity of chemical compounds are responsible for the flavor and aroma of coffee. Most chemical studies were carried out on coffee related to chlorogenic acids, caffeine, trigonelline, nicotinic acids, sugar, lipid and volatiles [17]–[20]. The intensive studies on amino acids of green and roasted coffee are yet to be carried out especially for *kopi luwak*. Amino acids have important contributions to beverage quality because they serve as precursors to coffee aroma and flavors. Then during the roasting process these amino acids take part in Maillard reactions resulting various components that contribute to flavour and color of coffee brew [8], [21]–[23]. According to [24], green coffee beans contain a variety of free amino acids where these amino acids are largely trasforming during roasting. To our knowledge, there is no study that has investigated the amino acids composition of kopi luwak and its correlation to its sensory profile. Therefore, this study aimed to analyze amino acid compositions of kopi luwak in both green and roasted coffee bean and its correlation to its sensory profile.

2. Materials and Methods

2.1. Samples Preparation

Six samples of *wild kopi luwak* were obtained from six cultivation areas, four from Bener Meriah and two from Central Aceh. The samples are collected in June-July 2017. The detail location for each sample could be seen in Table 1. The coffee species used in this study was *Coffee arabica*.

Collected coffee beans were cleaned and washed from the *luwak* feces with flowing water, and dried to reach moisture content of 12%. Furthermore, the dried coffee beans were manually dehulled to produce green coffee beans. A half of the green beans were roasted at a medium degree based on the Specialty Coffee Association of America (SCAA) cupping protocol standard method [25] one day prior to analysis. Both, the green beans (unroasted beans) and roasted coffee beans, were stored in a vacuum container at ambient temperature prior to analysis. Prior to analysis, both coffee beans (green and roasted) are grounded with coffee grinder where the resulted particle size should able to passing through US standard 20 mesh (approximately 70-75% of the particles).

Table 1. List of coffee samples

Sampels	Origin/source
A	Arul Badak Bener Meriah, Aceh-Indonesia
B	Jejem 1 Central Aceh, Aceh-Indonesia
C	Wih Pongas Bener Meriah, Aceh-Indonesia
D	Blang Panas Bener Meriah, Aceh-Indonesia
E	Jejem 2 Central Aceh, Aceh-Indonesia
F	Kenawat Redelong Bener Meriah, Aceh-Indonesia

2.2. Amino Acid Analysis

The procedure of amino acid analysis refered to [8] with a slightly modification. Coffee sample was first acid hydrolyzed to convert the proteins into their constituent amino acids. About 0.1 g of ground coffee sample was transferred to a 15-mL glass tube with sealer and 10 mL of HCl 6M was then added. The mixture was heated under vacuum at 110°C for 22 h. The solvent was then evaporated until the solution became a thick liquid of extract. The extract then was re-soluted with 3 ml ultrapure water and centrifuged at 13.000 rpm for 10 minutes. The supernatant was ready to be injected to the UPLC-QToF-MS/MS system (Waters, Massachusetts, USA) equipped with Acquity UPLC BEH C18 column to perform amino acid analysis. The mobile phase consisted of a mixture of 0.1% pentadecafluorooctanoic acid (PDFOA) and 0.1% formic acid in H_2O/Acetonitrile (99.5%/0.5%) (A) and 0.1% pentadecafluorooctanoic acid (PDFOA) and 0.1% formic acid in H_2O/Acetonitrile 10%/90% (B). The injector volume was 5 uL and the flow rate was 0.4 mL/min. The amino acid content of the samples was quantified by comparing their retention times and peak area with these of standard curves. The amino acid content was expressed as milligram amino acid per 100 g DW of coffee beans. All solvents used were HPLC grade.

2.3. Sensory Analysis (Cupping Test)

Evaluation of sensory profile of coffee was carried out using the cupping test, a method published by [25], was commonly used as a standard protocols to measure ten coffee sensory attributes which are fragrance, aroma, flavor, aftertaste, body, acidity, uniformity, clean cup, sweetness and overall characteristics of beverages. Briefly, the procedure started with roasting 100 g kopi luwak (green bean) in medium degree (65 Agtron Scale) at one day prior to cupping test take place. The coffee then grounded and the examination was started by preparing five glass for each samples. Each glass contained 8.25 g of ground coffee and then 150 ml of water at a temperature range of 93-95°C were poured into the uncovered cup of coffee. After 4-5 minutes, the assessment starts with length of time approximately 20-30 minutes.

The steps of the procedure of cupping evaluation were started evaluation of fragrance and aroma. The dry fragrance should be examined during the first 15 minutes. Fragrance score is obtained by lifting the lid to sniff the coffee powder. Then aroma is valued by infusing the coffee grounds with water, the appeared crusts on the surface left unbroken for approximately 3-5 minutes. Then the beverage is stirred three times to break out the crust. Sniffing the aroma is done with spooning the liquor with the back of the spoon, which allows the foam to run down the cup. The score for fragrance is then marked by dry grounds evaluation whilst the aroma by the liquid drinks. Then it continued to flavour, aftertaste, acidity, body, and balance evaluation. It took place when the coffee liquor temperature reached to 160°F (71°C) which is commonly occurred roughly 8-10 minutes from the infusion. The process starts by aspirating the liquor into the mouth in a special way in order to the liquor covers the palatable area as large as possible, especially the tongue and upper palate. By doing so, the retronasal vapors are at its maximum intensity at these elevated temperatures, where flavor and aftertaste are marked at this point. Then as the liquor temperature decreases (160°-140°F), acidity, body, and balance are followed. Then sweetness, uniformity, and cleanliness are evaluated in order as the coffee liquor reaches room temperature (below 100°F) and should be over when the temperature of the sample down to 70°F (21°C). For this evaluation, one sample is placed on five different cups. Then each coffee grader should examine on each individual cup, the maximum marks for each cup are 2 points (total 10 points for each sample). Overall refers to the holistic reflection of integrated rating of samples as perceived by the coffee grader. The overall score is given after the coffee grader assessed all attributes and combined these as "cupper's points". As similar as sweetness, uniformity, and cleanliness, overall is measured by awarding 2 points per cup where coffee grader is assessed 5 cups or each coffee samples in order to reach 10 points maximum scores. Defects refer as negative or poor flavors, which detect during cupping test. Defects divided into two categories, which is taint and fault. Taint is score -2 meaning that this score is given to sample, which has off-flavor, but not overwhelming. Taint commonly notice in aromatic aspects. Another one is the fault, as an off-flavor which cause unpalatable. Defects mostly found in taste aspects, since it affects all cupping attributes, the score is -4. When the evaluation is over, all the scores are marked on the standard SCAA cupping form. The final score is based on the flavor experience of the individual assessor as single judges.

The panellist, named as Q-grader in this descriptive sensory measurement is three certified and trained panellist who professionally do the cupping test at regular basic. Since all the protocols refers to [25], the test form based on its standard and scale used in this cupping test ranged between 6 and 9 (good to oustanding quality). The protocols stated out that to be classified as specialty coffee, the 6 point should be absent and the total score must exceed 80.

2.4 Data Analysis

The collected data was tabulated and processed with Microsoft Excell Version 2010. All data were expressed as mean ± standard deviation. For amino acids compositions, statistical significance was assessed using one-way analysis of variance (ANOVA) from two replications for each sample and continued with a Least Significant Different (LSD) Test ($P \leq 0.05$). For correlation between the amino acid contents and cupping test results, the statistical measurement was conducted using 2015 released SmartPLS v.3.2.3. (SmartPLS GmbH).

3. Results and Discussion

3.1. Amino Acid Compounds

From Table 2, it could be seen that from all amino acids presented in coffee bean, only four amino acids were significantly different ($P \leq 0.05$) between green bean and roeasted bean. Accordingly, alanine, cysteine and arginine in green bean were statistically lower than those in roasted bean, whereas tyrosine was significantly lower in roasted bean than green bean.

Table 2 also presents the types of amino acids available in green bean and roasted bean of *kopi luwak*. The amino acids are classified based on their R side chains, which in detail are divided into six groups. Aliphatic amino acids are amino acids which has non polar R group and hydrophobic sides [26]. For *kopi luwak* green bean, glycine and alanine are not detectable, but after roasting process alanine is present in small contents. In this group, Valine has the highest value followed by Proline in green and roasted bean. Isoleucin presents for both green and roasted bean in neglected amount. This group with Proline specifically is responsible for produce pyrolles, pyridines by reacting with Maillard intermediates during roasting [23].

The second explained group is aromatic amino acids, which classified as well as non-polar R-group. The amount for three mentioned amino acids, Phenylalanine, Tyrosine and Tryptophan are closely in range from 0.418 to 0.467 mg/100 g of green bean. However, in this group, the contents of all amino acids are down to half of it and even disappeared after roasting.

Table 2. Amino acids in green and roasted bean of *kopi luwak* (mg/g of bean)

No.	Amino Acids		Green Bean		Roasted Bean	
1	Aliphatic Amino acids with R Non Polar and aliphatic					
	a.	Glycine	0	±0.00	0.000	±0.00
	b.	L-Alanine*	0	±0.00	0.562	±0.2
	c.	L-Proline	0.758	±0.36	1.100	±0.64
	d.	L-Valine	1.998	±0.91	2.022	±0.82
	e.	L-Leucine	0.773	±0.30	0.588	±0.34
	f.	L-Isoleucine	0.230	±0.08	0.168	±0.06
2	Aromatic amino acids					
	a.	L-Phenylalanine	0.465	±0.19	0.280	±0.24
	b.	L-Tyrosine*	0.418	±0.09	0.000	±0.00
	c.	L-Tryptophan	0.467	±0.29	0.237	±0.12
3	Amino acids with ionized R Polar					
	a.	Asparagine	0.083	±0.06	0.265	±0.21
	b.	L-Glutamine	1.088	±0.37	1.177	±0.61
	c.	L-Serine	0.605	±0.21	0.685	±0.16
	d.	Threonine	0.638	±0.27	0.742	±0.26
4	Sulfuric amino acids					
	a.	Methionine	0.085	±0.06	0.285	±0.27
	b.	Cysteine*	0.133	±0.04	0.815	±0.33
5	Amino acids with R negative ion (acid)					
	a.	Aspartic acid	1.493	±0.33	1.858	±0.65
	b.	Glutamic acid	2.102	±0.37	2.460	±0.92
6	Amino acids with R Positive ion					
	a.	L-Arginine*	0.525	±0.26	0.000	±0.00
	b.	L-Lysine	0.162	±0.06	0.190	±0.08
	c.	Histidine	0.283	±0.12	0.153	±0.05

Note: Mean values of amino acid followed by an asterick (*) showed significant differences ($P \leq 0.05$).

The third group is amino acids with ionized R group and also has polar side which makes this group hydrophilic [26]. In this group, Glutamine has the highest contains and considers among the major amino acids in coffee bean. In this group, majority of amino acids values are not change even after the roasting process, except for Asparagine. Aspargine presents in trace amounts in green bean (0.083 mg/g) and being higher after roasted (0.265 mg/g). For sulfuric amino acids group, Methionine and Cysteine are belong to this classification. Sulfuric amino acids considers as aromatic amino groups [26]. [23] stated that this group is bound with protein in stage of green bean. Then this amino acid degrades during roasting and interact with reduncing sugars and Maillard intermediates to perform intensely aromatic volatiles such as furfurylthiol.

Moreover, amino acids group with positive ion such as Arginine, Lysine and Histidine tend to have less value after roasting process. Even Arginine, which considered as amino acids with highest contains in this group, has zero value after roasting. This finding is similar with [27]. Lastly to discuss is amino acids with negative ion and considers as acid, such as aspartic acid and glutamic acid. Both of these are reported to be the most common amino acids found in coffee and has significantly influenced during coffee flavor formation. These compounds shares major parts of amino acid compositions, for both green and roasted coffee bean.

3.2. Sensory Quality of Kopi Luwak

The sensory quality of *kopi luwak* is done by cupping test. The protocol system evaluates ten attributes sequentially which are aroma/fragrance, flavor, aftertaste, acidity, body, balance, sweetness, clean cup, uniformity and overall. For sweetness, clean cup, and uniformity, the scores were all 10 meaning that these three sensory attributes were in their best characteristic levels (Data were not shown in Table 3). The cupping test score of *kopi luwak* could be seen in Table 3. In Table 3 only seven attributes are measured its correlation with amino acids composition. For sweetness, clean cup and uniformity attributes, Gayo coffee has specific characteristics which has maximum score in these attributes. Therefore, these three attributes are ignored since all the six samples have similar scores.

Table 3. Cup quality of *kopi luwak*

Sample	Sensory Attributes Measured by SCAA Cupping Test Standard							
	Fragrance	Flavor	Aftertaste	Acidity	Body	Balance	Overall	Cup Score Total
A	7.80	8.00	7.80	7.50	8.00	7.80	7.80	84.50
B	8.00	7.80	7.80	7.50	8.00	8.00	7.80	84,75
C	8.00	7.80	7.50	7.50	8.00	7.80	7.80	84,00
D	8.00	8.00	7.80	7.50	7.80	7.50	8.00	84,75
E	8.00	8.00	8.00	7.80	8.00	7.80	8.00	85,75
F	7.80	7.50	7.80	7.50	8.00	8.00	7.80	83,75
Average	7.93	7.85	7.78	7.55	7.97	7.82	7.87	84.50

Table 3 shows that avarage value of sensory attributes (fragrance, balance, aftertaste, acidity, body, balance and overall) reaches very good quality scale and classified as specialty coffee. Coffee, which categorized the specialty coffee, should have a total score above 80.00. The highest total score belongs to sample E (85.75), whilst the lowest is sample F (83.75) with avarage 84,50. Compared to previous research [16] who reported that wild palm civet coffee has total score 85.25, sampel E considers as excellent quality, whilst the others samples are a coffee specialty with very good classification. Based on Table 3 as well, aftertaste and acidity of E are in excellent quality scale and also the highest among others. Acidity is categorized as aroma and taste sensations, describe as delicate and crisp, lush and rich, powerfully tart but sweet at same time. On the other hand, aftertaste or alternatively named as mouthfeel is flavor that lingers after the liquor is swallowed [25], [28]. Compared to another sample, F has the lowest value of flavor among other samples. Flavor is main quality standard in cupping test, which holistically combination aroma and taste, where aroma perceived by volatile compounds inhale through the nose [29] with basic taste perceived by tongue receptors [30].

3.3. Correlation the amino acids and the cup quality of green bean of kopi luwak (Civet Coffee)

Correlation of amino acids and cup quality of *kopi luwak* is could be seen in Figure 1. The data presents in Table 2 statistically correlated with the amino acids contents in Table 1 and performed by Smart PLS v 3.2.3.

IOP Conf. Series: Materials Science and Engineering **523** (2019) 012028 doi:10.1088/1757-899X/523/1/012028

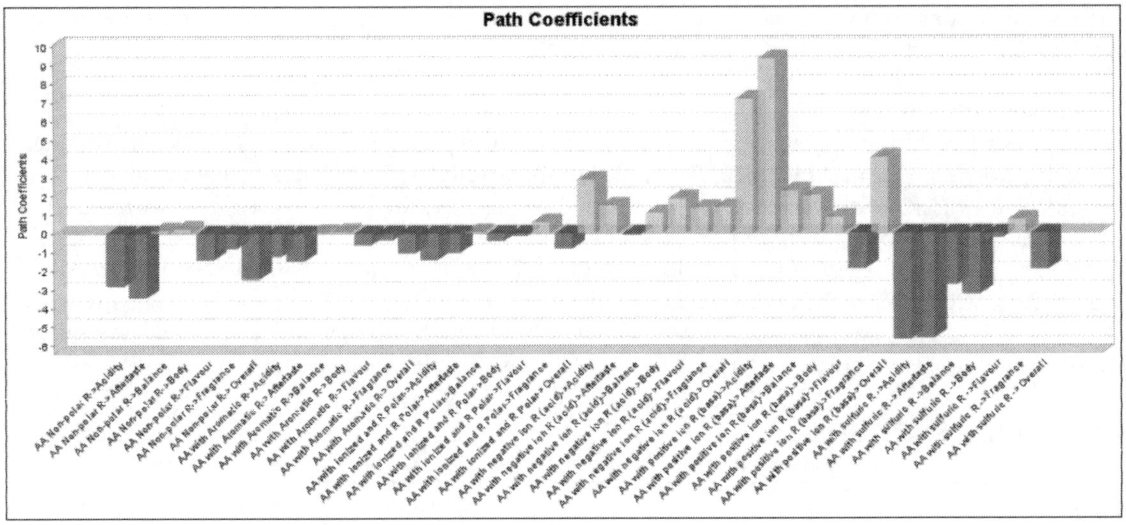

Figure 1. Path Coefficients based on Partial Least Square Result based on contents of amino acids in green bean and the Cupping quality of *kopi luwak.*

In Figure 1, it can be shown the correlation within the amino acids compounds based on its R function with the sensory profile of *kopi luwak*. Even in SCAA cupping protocols, ten attributes are evaluated, here we are only perform the correlation within seven attributes (fragrance, flavor, aftertaste, acidity, body, balance and overall) since the other three presented the similar value over the varied samples. Figure 1 shows that most of amino acid groups performs negative correlation towards the cupping sensory, except amino acids with ionized polar group, amino acids with negative ion R group and positive ion R group which are assumed to have positive correlation based on its bar color and positive value of path coefficients.

First to discuss is correlation within the non-polar group with body and balance attributes. Many reports have been made in terms of the significance of non-polar group which has the hydrophobic sites in their molecules, such as L-Alanine, L-Valine, L-Leusine, and L-Isoleucine towards bitterness [31], [32]. In this cupping test scope, body refers to the combination of bitterness and astringency, whilst balance defines as how well the equilibrium between aftertaste, body and acidity in one cup of coffee. In this study, L-Glysine and L-Alanine are absence in green bean, but L-Valine as the major amino acid presents in this group, shows strong correlation among other amino acid compounds (R^2= 0.984). Therefore it can be stated that L-Valine, continue with L-Leucine and l-Valine performs the body and balance attributes in coffee. [33] reported that the presence of Valine, Leucine, Glycine, Alanin in peptides also influence the bitter taste since these are considered as binding determinants.

For fermentation product like coffee, the metabolic activity of responsible microorganisms during fermentation will classify food quality, generates taste and flavor as well as the food acceptance [32], [34]. Usually, the presence of glutamate in fermented foods such as meat, soy and cheese product will contribute to the umami taste [32]. However in this study, amino acids with negative ion R group which consists of aspartic acid (R^2=0.953) and L-Glutamic acid (R^2= 0.961), shows the strong and positive correlation towards acidity, aftertaste, body, flavor, fragrance, and overall exclude balance. Acidity refers to the feel of crispiness and sweetness which produces bright and dry sensation that elevates the saliva and coffee taste [5]. This might be occurred due to both of aspartic acid and glutamic acids are having R with acidotic ion. In Figure 1 as well presents the correlation between positive on R groups with acidity, aftertaste, balance, body, flavor attributes. In this groups, Histidine claims its major percentage and showed strong correlation (R^2=0.952) than others. Therefore, it can be concluded that might be histamin has major influenced towards sensory profile of *kopi luwak*.

4. Conclusions

Green and roasted bean of *kopi luwak* had significantly differences for the value compounds of L-alanine, L-Tyrosine, Cysteine and L-Arginine. Valine, L-Glutamine, Aspartic acid and Glutamic acid are considered as the main amino acids that have higher percentage in *kopi luwak*. Several amino acids in *kopi luwak* are absence in green bean which later on enable to track its presence in roasted bean or vice versa. Glycine is not available ini both green and roasted bean. Avarage value of each kopi luwak's sensory attributes (fragrance, balance, aftertaste, acidity, body, balance and overall) reaches very good quality scale and classified as specialty coffee. Certain amino acid contributes to develop the sensory quality of kopi luwak. PLS correlation L-Valine which belongs to non polar group positively interferes the body and balance attributes. In group of negative ion R-group, namely aspartic acid and glutamic acid have positive correlation on acidity, body, flavor, fragrance and overall. Lastly mentioned is histamin that belongs to positive aciodotic ion group which also has possitive correlation towards acidity, aftertaste, balance, body, flavor and overall attributes.

5. Acknowledgements

The authors would like to thank The Ministries of Research and Technology and Higher Education of Indonesia for financial support in project "Strategis Nasional Institusi" with contract number 92/UN.112/PP/SP3/2018.

6. References

[1] A. Farah, 2012 *Coffee Constituent in Emerging Health Effect and Diseases Prevention*, C. Fang, Ed. (John Willey and Sons Inc. Blackwell Publishing Ltd.) pp. 21–58.
[2] S. Alves, R. Dias, M. de T. Benassi, and M. dos S. Scholz 2006 *Quim. Nova*, vol. 29, no. 6, pp. 1164–1168
[3] A. Maurya 2016 *Int. J. Pure Appl. Res.*, vol. 1, no. 2, pp. 151–156,
[4] D. Selmar, M. Kleinwachter, and G. Bytof 2015 *Metabolic responses of coffee beans during processing and their impact on coffee flavor," in Cocoa and Coffee Fermentations* (Boca Raton: CRC Press).
[5] W. Sunarharum, D. Williams, and H. Smyth 2014 *Food Res. Int.*, vol. 62, pp. 315–325
[6] L. Batista and S. Chalfoun 2015 *Quality of coffee bean in Cocoa and Coffee Fermentations* (Boca Raton: CRC Press) pp. 114–124.
[7] L. Lee, M. Cheong, P. Curran, B. Yu, and S. Liu 2015 *Food Chem.*, vol. 185, pp. 182–191
[8] W. Dong, L. Tan, J. Zhao, R. Hu, and M. Lu 2015 *Molecules*, vol. 20, no. 9, pp. 16687–16708
[9] A. Teixeira, C. Brando, R. Thomaziello, and R. Teixeira 2005 *The raw bean in Espresso coffee: the science of quality*, A. Illi and R. Viani, Eds. (London: Elsevier Academic Press), p. 91.
[10] G. Duarte, A. Pereira, and A. Farah 2010 *Food Chem.*, vol. 118, no. 3, pp. 851–855, 2010.
[11] M. Marcone, 2004 *Food Res. Int.*, vol. 37, no. 9, pp. 901–912
[12] U. Jumhawan, S. Putri, Yusianto, T. Bamba, and E. Fukusaki 2016 *J. Biosci. Bioeng.*, vol. 122, no. 1, pp. 79–84
[13] I. Smith 2014 *Kopi Luwak Coffee-World most exvensive coffee beans from civet poop or an urban mith?* (iPACT Limited).
[14] M. Muzaifa, A. Patria, Febriani, A. Abubakar, D. Hasni, F. Rahmi, and I. Sulaiman 2016 *Kopi luwak: produksi, mutu, dan permasalahannya*. (Syiah Kuala University Press)
[15] S. P. Putri, U. Jumhawan, and E. Fukusaki 2015 *J. Biosci. Bioeng.*, pp. 33–41
[16] M. Muzaifa and D. Hasni 2016 *Pakistan J. Nutr.*, vol. 15, no. 5, pp. 486–491
[17] F. Mayer, M. Czerny, and W. Grosch 2006 *Eur. Food Res. Technol.*, vol. 211, no. 4, pp. 272–276
[18] S. de Morais, F. de Aquino, P. do Nascimento, and E. Chang 2009 *Quim. Nova*, vol. 32, no. 2, pp. 327–331
[19] A. C. L. Amorim, A. M. C. Hovell, A. C. Pinto, M. N. Eberlin, N. P. Arruda, E. J. Pereira, H. R. Bizzo, R. R. Catharino, Z. B. Morais Filho, and C. M. Rezende 2009 *J. Braz. Chem. Soc.*,

vol. 20, no. 2, pp. 313–321
[20] E. A. Nascimento, S. A. L. Morais, R. Chang, F. J. T. Aquino, B. H. P. Alves, and R. J. C. F. Afonso 2015 *Int. J. Appl.*, vol. 5, no. 1, pp. 26–37
[21] Y. Liu and D. Kitts 2011 *Food Res. Int.*, vol. 44, no. 8, pp. 2418–2424
[22] H. Czech, C. Schepler, S. Klingbeil, S. Ehlert, J. Howell, and R. Zimmermann 2016 *J. Agric. Food Chem.*, vol. 64, no. 25, pp. 5223–5231.
[23] O. Janzen 2013 *Chemistry of coffee* (Hamburg, Germany: Elsevier Ltd.)
[24] M. Murkovic and K. Derler 2006 *J. Biochem. Biophys. Methods*, vol. 69, no. 1–2, pp. 25–32,
[25] SCAA (*Specialty Coffee Association of America) 2009 Coffe facts and statistics* [Online]. Available: www.scaa.org. [Accessed: 21-Oct-2015].
[26] U. Satyanarayan and U. Chakrapani 2013 *Biochemistry with Clinical Concepts and Case Studies*, 4th Revise (New Delhi: Elsevier)
[27] E. Bekedam 2008 *Coffee Brew Melanoidins, Structural and Functional Properties of Brown-coloured Coffee Compounds* (University Wagenigen)
[28] A. Taylor and J. Roozen 1996 *Crit. Rev. Food Sci. Nutr.*, vol. 36, no. 8, pp. 765–784
[29] M. Petraco, 2001 *Beverage preparation: brewing trends for the new millenium, in Coffee: Recent Development*, R. Clarke and O. Vitzthum (Eds. Oxford: Blackwell Science) pp. 140–164.
[30] N. Rawson and X. Li 2004 *The cellular basis of flavour perception: taste and aroma, in Flavour Perception*, 3rd ed., A. Taylor and D. Roberts, (Eds. Oxford: Blackwell Publishing) p. 75.
[31] H. Akitomi, Y. Tahara, M. Yasuura, Y. Kobayashi, H. Ikezaki, and K. Toko 2013 *Sensors Actuators B Chem.*, vol. 179, pp. 276–281
[32] C. Zhao, A. Schieber, and M. Ganzle 2016 *Food Res. Int.*, vol. 89, pp. 39–47
[33] N. Ishibashi, T. Kubo, M. Chino, H. Fukui, I. Shinoda, E. Kikuchi, H. Okai, and S. Fukui, 1998 *Agric. Biol. Chem.*, vol. 52, no. 1, pp. 95–98, 1988.
[34] R. Hutkins 2006 *Microbiology and technology of fermented foods* (Blackwell Publishing)

IOP Conf. Series: Materials Science and Engineering **523** (2019) 012029 doi:10.1088/1757-899X/523/1/012029

Sensory evaluation with descriptive method on Aceh local rice (*Oryza sativa* L.) mutant M_6 using Gamma-ray irradiation

R Hayati[1,*], Efendi[1], F Yunita[1]
[1]Agrotechnology Department, Faculty of Agriculture, Syiah Kuala University, Banda Aceh 23111, Indonesia

E-mail: ritanabila@yahoo.com

Abstract. Aceh is the richest province in Indonesia in the context of diversity with the local rice varieties. Therefore, it requires conservation efforts within this issue. The weakness parameters of local rice are long-lived and low production. One of the technologies that are been used to obtain the characteristics of short-lived local rice and high production is by using gamma-ray irradiation technology. This research aims to know the quality of sensory evaluation with descriptive method on Aceh local rice Mutant M_6 by gamma-ray irradiation. The Aceh local rice Mutant M_6 that has been used in this study were I-CKU-6, UF1 B4-9-2017, O-SGP, O-10D, S-103, S-97, S-01 (conventional and organic cultivation), O-39 e, O-10 d, O-57 e (organic cultivation), S-97, S-84 dan UFi-B (conventional cultivation). The approach that has been used in this research was using Completely Randomized Design (CRD) non factorial with three replicates. The result showed that local rice with irradiated gamma-ray (M_6) had a very significant effect on the flavour, color, texture, shape, and overall acceptance of rice by conventional and organic cultivation. The local rice O-SGP (5.82) is the local rice that has been received by panelists, followed by local rice S-103 (5.75), O-10D (4.95) and S-97 (4.70), while the aroma was favored by panelists is O-SGP (4.92). The S-84 local rice which was grown conventionally was preferred by panelists with a value of 9.85.

1. Introduction

Rice (*Oryza sativa* L.) is an important commodity because it is the staple food crop in Indonesia and also in several other countries [1]. The rice has a high nutrient content of 360 calories carbohydrates, most carbohydrates of rice consist of starch (89-90%), and a small parts are pentose, cellulose, and hemicellulose, and 6.8 g of protein, mineral content such as 6 g of calcium, 0.8 g of iron and 0.2 g of vitamin [2]. The rice development of local Aceh varieties are starting from Sanbei varieties that have a high-yield potential. The weakness of the Sanbei varieties is that it has a relatively longer plant life, causing the flowering period reached 96 days and it will be lead to a long harvest time.

Due to this issue, it is necessary the improvement efforts to assemble high-yielding varieties on early maturity, the high-yield potential and resistant to various abiotic and biotic stress if planted in different locations, one of them is by improving the plants through mutation and induction using of gamma-ray irradiation[5]. The application of gamma-ray irradiation shows a different genetic variability such as a shorter crops, early maturity, and high-yield potential. The mutation technique has also been shown to be highly functional in increasing of rice yields [4,5,6,7].

* Corresponding author : ritanabila@yahoo.com

On the other hand, the other important factors to be noted in the development of new varieties are not only based on a high-yield result, but also the quality of rice and rice flavour are more important to note[8]. The way to assess the rice quality is by using a sensory evaluation. There are many methods were used in sensory evaluation, one of them is descriptive analysis. The advantages of those method are to get a method that can be performed by a trained panelist not only by an expert, but also able to give a clear explanation for all the sensory product, and can be done on products that have been exist in the market, herb, an idea or a new product that have not being in competitor yet[9,10,11]. The purpose of this research is to evaluate the sensory characteristics of the Aceh local rice mutant M_6 with gamma-ray irradiation by using descriptive method.

2. Materials and Methods

2.1. Time and place of research

This research was conducted at the Laboratory of Seed Technology, Agrotechnology Department; and the Laboratory of Food and Agricultural Product Analysis, Department of Agricultural Product Technology, Faculty of Agriculture, Syiah Kuala University in Darussalam Banda Aceh. This research took place from November 2017 to February 2018.

2.2. Tools and materials research

The tools that has been used in this research were analytical scale, folio envelope, plastic, seed blower, mortar-wood, Canon camera of type EOS700. The materials used in this research were the rice which obtained from Industry and Seed Technology Laboratory, Agrotechnology Department Faculty of Agriculture, Syiah Kuala University, it was grown from rice cultivation at the Main Seed Centers, Samahani Aceh Besar. This rice was a Sanbei variety genotype M6 by gamma-ray irradiation with doses 250 Gy for 36"15' in the Center of Isotopes and Radiation Technology produce which is the one of nuclear energy agency facilities in Jakarta on March 10, 2015 and generate M1 genotype, while this research using of 8 genotype M6 strains. The strains are irradiated by gamma-ray which consists of 7 strains namely: I-CKU-6, UF1B4-9-2017, O-SGP, O-10 D, S-10 3, S-01, S-97 and 1 strain of Sanbei first elder that is 6-ISNB-1 as a comparison (the strain was not irradiated by gamma-rays). Each replication in this research was used the rice per strain, taken from the best of 5 clumps and then each clump is taken for 5 best panicles.

2.3. Experimental design and data analysis

This research used a Completely Randomized Design (CRD) non factorial with 3 replications. The treatments studied were genotype consisting of Aceh local rice mutant M6, namely: I-CKU-6, UF1 B4-9-2017, O-SGP, O-10D, S-103, S-97, S-01 (conventional and organic cultivation), O-39 e, O-10 d, O-57 e (organic cultivation), S-97, S-84 and UFi-B (conventional cultivation). The data obtained were analyzed by F test, using the mathematical model as follows:

$$Yij = \mu + \tau i + \beta j + \varepsilon ij \qquad (1)$$

If the F test shows a significant effect between the treatment factors, then it was continued with the test of Honestly Significance Difference (HSD) on level 5% (0.05) both of the treatment level.

$$HSD_{0,05} = t_{0,05}\ (dbg)\ \sqrt{\frac{KTg}{r}} \qquad (2)$$

2.4. Sensory evaluation

The implementation of sensory evaluation was using the human senses as a primary tool for assessing the quality of Aceh's local rice. Sensory evaluation with descriptive method was performed to determine the attributes (flavour, color, texture, shape and overall acceptance) that will be assessed by panelists, and furthermore together with researchers will equate the perception in determining the characteristics based on attributes that have been mutually agreed. This method was used 15 panelists and each panelist has been tested to get the same thoughts in determining the attributes that will generated in this study. The panelists came from students and laboratory staff of the Agrotechnology Department, Faculty of Agriculture, Syiah Kuala University.

3. Results and discussion

Based on the result of F test, the analysis of variance showed that the treatment of genotype Aceh local rice by gamma-ray irradiation (M6) has a very significant effect of the attributes on the flavour, color, texture, shape and overall acceptance of rice. The average sensory evaluation attributes can be found in Table 1 and Table 2.

3.1. The rice (uncooked)

3.1.1. The flavour of rice. Sensory evaluation indicated that panelists prefered to the O-SGP genotype scent with a score of 4.92 followed by other genotypes there are O-10D, S-103, UF1 B4-9-2017, and I-CKU-6 genotypes. According to Juliano[12] the flavour of rice is influenced by its varieties of rice. In addition to being influenced by varieties, the scent or flavour will also change with the length of storage. Unrefined rice by 100% will having a smell bad (stale) after being stored for a long time. Scent changes in storage is faster than color change. It was also added by Haryadi[13], the flavour in the rice can be survive with coating by using maltodextrin.

Table 1. Sensory evaluation attributes on the flavour, color, texture, shape, and overall acceptance of Aceh local rice Mutant M6 by gamma-ray irradiation (conventional and organic cultivation).

Rice (uncooked)	Sensory Evaluation Attributes				
	Flavour	Color	Texture	Shape	Overall Acceptance
I-CKU-6	4.30^{bc}	4.40^{b}	4.27^{c}	4.10^{b}	4.23^{b}
UF1B4-9-2017	4.47^{cd}	3.32^{a}	3.35^{a}	3.48^{a}	3.33^{a}
O-SGP	4.92^{d}	6.13^{d}	5.52^{e}	5.67^{d}	5.75^{e}
O-10D	4.82^{cd}	4.85^{c}	4.53^{c}	5.08^{c}	4.95^{d}
S-103	4.80^{cd}	4.32^{b}	4.52^{c}	4.83^{c}	5.82^{e}
S-97	3.80^{ab}	4.27^{b}	4.92^{d}	4.77^{c}	4.70^{cd}
S-01	3.68^{a}	3.27^{a}	3.30^{a}	4.35^{b}	4.47^{b}
HSD 5%	0.54	0.38	0.31	0.43	0.37

Description: The numbers followed by the same letter in the same column are not significantly different (HSD Test at 5% level)

3.1.2. Rice color. In the color attribute panelists choose the genotype O-SGP, this is caused by the color of the rice is affected by degrees, amylose levels and changes during rice storage, high degrees resulted in more and more epidermis released so that the color of the rice becomes whiter. In addition, the acceptance of the color of a material varies depending on natural factors, geographical, and social aspect of society.

3.1.3. Rice texture. The sensory assessment use of the hedonic test on the preferred texture attribute of rice with the gamma-ray irradiated rice on the M6 genotype yields a value of between 3.30 and 5.52 with somewhat dislike, neutral and rather like. The high texture of the genotype O-SGP is significantly different from the genotype S-01, UF1 B4-9-2017, I-CKU-6, S-103, O-10D, S-97. Texture is the surface appearance of an attribute of a foodstuff produced by a combination of physical properties and chemical properties that are widely accepted by their touching, visualization, and listening. Texture is also defined as the properties of a food that can be observed by the eyes, skin, and the mouth muscles[14].

3.1.4. The shape of rice. In the perspective of rice shape attribute, the lowest score was shown by genotype rice UF1 B4-9-2017, which was significantly different from the genotype I-CKU-6, S-01, S-97, S-103, O-10D, O-SGP. The size and shape of rice are the dominant character derived from the genetic properties of the parent and can be used as the parameter of purity determination of a variety. Long character and shape of rice are influenced by genetic factor, agro ecosystem and soil fertility[15].

3.1.5. Overall Acceptance of Rice. The highest overall attributes acceptance was on S-103 rice which was not significantly different from O-SGP rice with the criteria is rather like. The lowest overall acceptance value of rice is UF1 B4-9- 2017. But it is significantly different from I-CKU-6, S-01, S-97, O-10D rice. This showed that the panelists are selecting rice from gamma-ray irradiation on S-103 and O-SGP rice.

3.2. Cooked rice

Table 2. Sensory evaluation of the shape, texture, flavour, internal and eksternal color, and acceptance a whole rice of Aceh local rice Mutant (M6) by gamma-ray irradiation in conventional and organic cultivation.

Cooked rice	Sensory Attributes					
	Shape	Texture	Flavour	Internal color	Exsternal color	Overall Acceptance
O-39e	11.39 abc	9.27 b	7.76 ab	8.57 a	12.17 ab	11.18 bc
O-10d	8.35 a	7.30 ab	6.76 a	8.46 a	15.19 bc	9.15 ab
O-57e	14.37 cde	6.76 a	6.86 a	8.45 a	10.96 a	10.04 ab
S-97	17.40 e	7.09 a	7.43 ab	10.65 b	13.99 ab	12.23 c
S-84	9.39 ab	8.02 ab	9.85 b	8.80 a	11.20 a	10.09 ab
UF1-B	12.45 bcd	8.58 ab	7.91 ab	8.77 a	12.02 ab	8.77 a
HSD 5%	1.90	1.90	2.67	1.39	3.71	1.94

Description: The number followed by the same letter in the same column are not significantly different (HSD Test at 5% level). O-39e, O-10d, O-57e in organic cultivation, S-97, S-84, UF1-B cultivating conventionally.

3.2.1. The shape of rice. Table 2 shows the highest average of rice shape is shown by S-97 rice strain not significantly different from the M0-Snb and O-57e rice strains, but significantly different from the O-39e, O-10d, S-84 and UF1-B. The lowest rice form was found in the O-10d rice strain which was not significantly different from the O-39e and S-84 rice strains, but was significantly different from the M0-Snb, O-57e, S-97 and UF1-B rice strains.

The size and shape of rice grains (dimensions) are the basis in determining the quality of rice in the international market. Based on the length / width of rice observed ranged from 1.82 to 2.77 mm. For M0-Snb strain, O-57e and S-97 are round, while O-10d, S-84 and UF1-B strain are medium and O-39e are slightly rounded. From the preferred test data of rice form, the best results are shown by S-97 rice strains not significantly different from the M0-Snb and O-57e rice strains. This suggests that panelists prefer a rounded shape of rice rather than oval or slightly rounded shape.

3.2.2. Rice texture. Table 2 shows the highest average of rice texture found in O-39e rice strain which not significantly different to M0-Snb, O-10d, S-84 and UF1-B strains, but significantly different to O-57e and S-97. The lowest rice texture was found on the O-57e rice strain which was not significantly different to M0-Snb, O-10d, S-84, S-97 and UF1-B rice strains but was significantly different to O-39e rice strain. According to Kartika[16] texture is a sensation of pressure that can be observed by mouth (bitten, chewed and swallowed), or touching by fingering.

3.2.3. Rice flavour. Table 2 shows the average level of the highest rice flavour is shown by S-84 rice strain that not significantly different to M0-Snb, O-39e, S-97 and UF1-B, but significantly different to O-10 d and O-57e. The lowest rice flavour was found in the O-10d rice strain which was not significantly different to O-57e, M0-Snb, O-39e, S-97 and UF1-B, but differed to S-84. According to De Mann[17], in the food industry flavour testing is considered important because it can provide results assessment of the product related to the acceptability or absence of a product.

3.2.4. Internal color of rice. Table 2 shows the average internal color of the highest rice is shown by M0-Snb rice strain which are not significantly different to S-97, but significantly different to O-39e, O-10d, O-57e, S-84 and UF1-B. The lowest internal color rice was shown by O-57e rice strain which was not significantly different to O-39e, O-10d, S-84 and UF1-B strains, but significantly different to S-97 and M0-Snb rice strains. According to Winarno[18], the color plays a role in determining the level of consumer acceptance of a product, although the product is of high nutritional value, good taste and good texture but if the color is not attractive it will cause the product is less desirable.

3.2.5. External color of rice. Table 2 shows that the average of the highest eksternal color of rice is shown by M0-Snb rice strain which is not significantly different to the O-10d rice strain, but significantly different to the O-39e, S-97, O-57e, S-84 and UF1-B. The lowest eksternal color rice was found in O-57e rice strain which is not significantly different to the O-39e, S-97, O-57e, S-84 and UF1-B rice strains but significantly different to the O-10d and M0- Snb. There are factors that affecting a white color on rice, the degrees of rice and storage conditions. The longer the rice is pounding the more color is becoming white because many layers of aleuron are lost[19].

3.2.6. Overall accepatance of rice. Table 2 shows the highest average overall attributes acceptance of rice shown by S-97 in rice strain. It's not significantly different to O-39e rice strain, but significantly different to M0-Snb, O-10d, O-57e, S-84 and UF1-B. The lowest of acceptance a whole rice was found on the M0-Snb rice strain which was not significantly different to O-10d, O-57e, S-84 and UF1-B rice strains, but significantly different to O-39e and S-97 strains. The S-97 rice strain had the highest acceptance rate with 12.23. The acceptance testing of overall rice is an assessment of all observed quality factors including shape, texture, flavour, internal and exsternal colors. This test intends to find out the level of panelist acceptance of a product. The best results are shown by S-97 rice strain, this is due to the genotype has the best form, flavour and the best color that suits the tastes and can be accepted by consumers. According to Soekarto[19], the physical and chemical tests also nutritional test can indicate a high quality product, however it would be meaningless if the product can not be consumed because it is uncomfortable or the nature of the sensory is not appetizing or unacceptable to the consumer.

4. Conclusion

The treatment of Aceh local rice genotype by gamma-ray irradiation (M6), which are grown in a conventional and organic cultivation, has a very significant effect on the flavour, color, texture, shape and overall attributes acceptance of rice. On the other hand, the treatment of Aceh local lines mutant rice genotype with conventional and organic cultivation has a very significant effect on a test favorites of shape, texture, flavour, inside and outside color and the acceptance of whole rice. The local rice O-SGP (5.82) has been received by panelists, followed by local rice S-103 (5.75), O-10D (4.95) and S-97 (4.70), while the aroma is favored by panelists is O-SGP (4.92). S-84 local rice which is grown conventionally is preferred by panelists with a value of 9.85.

5. References

[1] Christianto E 2013 *J. JIBEKA* **7**(2): 38-43
[2] Astawan M 2004 *Quality of rice (* Bogor: Research Center for Food Crops)
[3] Soedjono S 2003 *J. Agri.* **22**(2):1-9
[4] El-Degwy I S 2013 *Int. J. of Agri. and Crop Sci.* **5**(23) : 2789-94
[5] Polesi L, Junior MDM, Sarmento SPS, Brazaca SGC 2017 *Rice Sci.* **24**(1): 48-55
[6] Kim B, Woo S, Kim M J, Kwon SW, Lee J, Sung S H 2018 *Food Chemistry* **241**:154-162
[7] Abacar J D, Mio L Z, Xing-cheng Z, Cheng-qiang D, She T, Zheng-hui L, Shao-hua W, Yang-Feng D 2016 *Rice Sci.* **23**(1): 33-41
[8] Yang D S, Lee K S, Kays, S J 2010 *J. Sci. Food Agri.* **15**(2): 595-601
[9] Aminah A 2000 *Principles of sensory assessment.* (Bangi: Publisher Universiti Kebangsaan Malaysia)
[10] Bordonaba J G, Cantin C M, Larrigaudiere C, Lopez L, Lopez R, Echeverria G 2014 *Posharvest Biology and Technolog.* **93**: 49-60
[11] Ummarat N, Arpaia LM, Obenland D 2015 *Posharvest Biology and Technology*. **109**:82-96
[12] Juliano B O 1994 *Criteria and test for rice grain quality. In: Rice Chemistry and Technology.* (American Association of Cereal Chemists, St. Paul, Minnesota)
[13] Haryadi 2008 *Chemical and technology starch* (Yogyakarta: PPS UGM Press)
[14] Lewis M J 1987 *Physical properties of foods and food processing system* (Canada: Cametot Press)
[15] Setyono A and P Wibowo 2008 *Proc. on Rice(Indonesian: Center for Rice Research)*
[16] Kartika B, Hastuti P, Supartono W 1988 *Guidance for the sensory test of food material* (Yogyakarta: UGM Press)
[17] De mann J M 1989 *Principle of food chemistry* (The avi pub co Inc westport Connecticut)
[18] Winarno F G 2004 *GMP in the rice milling industry. In the national workshop efforts to increase the value added of rice processing* (Jakarta:General Company of Bulog In collaboration with Fateta IPB)
[19] Soekarto S T 1990 *Organoleptic assessment for food industry and agricultural products* (Jakarta: Bhratara Karya Aksara)

IOP Conf. Series: Materials Science and Engineering **523** (2019) 012030 doi:10.1088/1757-899X/523/1/012030

Effect of H_2SO_4 concentration on cellulose isolation from palm empty fruit bunches.

N Nilawati[1], R Rahmi[1,*], L S Desiyana[2]
[1]Chemistry Department, Syiah Kuala University, Banda Aceh, Indonesia
[2]Pharmacy Department, Syiah Kuala University, Banda Aceh, Indonesia

E-mail: rahmi@fmipa.unsyiah.ac.id

Abstract. Isolation of cellulose from palm empty fruit bunches had been conducted. Isolation was conducted with H_2O_2 30% and H_2SO_4 with various concentrations (20, 30, 40, and 50%). Hydrolysis and bleaching processes were performed for 90 minutes. The obtained cellulose was characterized by Fourier Transform Infrared (FTIR), X-Ray Diffraction (XRD) and Scanning Electron Microscope (SEM). FTIR analysis confirmed the typical bands of cellulose in the sample. XRD patterns showed increasing H_2SO_4 concentration on hydrolysis process increased crystallinity of cellulose. However, at H_2SO_4 concentration more than 40% the crystallinity of cellulose reduced. It was due to the reduction of crystalline part of cellulose.

1. Introduction

Indonesia is a country that has the largest palm oil plantation in the world. In 2015, Indonesia has an area of 6,735,300 hectares of oil palm plantations spread over 22 provinces with 31.07 million tons of palm oil production per year. Oil palm plantations in Indonesia are expected to continue to rise, as it is a potential producer of vegetable oils. This palm oil processing industry produces a very large amount of solid waste. Several types of solid waste were palm empty fruit bunches, coir, mud, palm shells, etc., Palm empty fruit bunches are the largest solid waste. The average production of palm empty fruit bunches is 22% to 24% of the total weight of fresh fruit bunches processed at the palm oil industry.

Palm empty fruit bunches are composed of several important components that can be utilized and processed into other materials with more economical value. The constituent components include cellulose, lignin, holocellulose, hemicellulose, water and other extractive substances. One of the most important components of palm empty fruit bunches that can be converted to be another valuable product is cellulose. Cellulose is a natural polymer that can be used as raw material for bioplastics [1-4], bioethanol, and food packaging materials. Palm empty fruit bunches contains 38,76% or about 37,50% of cellulose with fiber content reach 72,67%.

* Corresponding Author: rahmi@fmipa.unsyiah.ac.id

In order to obtain pure cellulose from raw material, some methods can be applied such as chemical method, mechanical method and biological method. One of chemical methods is strong acid hydrolysis. This method is the most commonly method for cellulose purification where strong acid easily hidrolyze the amorphous part of the cellulose fiber and hemicellulose.

In our previous work [5], cellulose had been isolated from palm empty fruit bunches using HCl. However, the crystallinity of cellulose obtained was very low. Crystallinity is very important to show the level of purity of cellulose and it will affect the mechanical properties of cellulosic products [6]. Rashid (2017) reported that cellulose isolation process from cotton wool using different concentrations of acid resulted cellulose with different crystallinities [7]. Therefore, in order to study the effect of acid concentrations on cellulose isolated from palm empty fruit bunches, the hydrolysis process was conducted with several concentrations (20, 30, 40 and 50%) of H_2SO_4 in this work. The obtained cellulose was characterized by XRD, FTIR and SEM.

2. Materials and Methods

2.1. Materials

Palm empty fruit bunches were collected from Aceh. Sulfuric acidand hydrogen peroxide were analytical grade.

2.2. Isolation of cellulose

Palm empty fruit bunches were washed with distilled water and dried. Dried palm empty fruit bunches were cut into small pieces. The small pieces of palm empty fruit bunches were hydrolized using H_2SO_4 with various concentrations (20, 30, 40, and 50%). Where 3g of the small pieces of palm empty fruit bunches were added into beaker glass containing 26 mL of H_2SO_4 and stirred for 90 min. The hydrolysis process was stopped by adding frozen distilled water to the mixture. The mixture was filtered and rinsed several times until neutral pH was reached. After hydrolysis process, the hydrolized sample was bleached with H_2O_2 30% for 1h at room temperature. The obtained cellulose was dried in a oven for 6 h at 40°C.

2.3. Characterization

a. Fourier Transform Infrared (FTIR)

FTIR analysis was performed using Cary 630 Fourier Transform Infrared Spectrometer (Agilent Technologies). The sample was place in a set holder and spectral scanning was conducted in a wave number ranged from 4000 to 600 cm^{-1}

b. X-Ray Diffraction(XRD)

X-Ray Diffraction (XRD) analysis was done to determine the crystallinity of cellulose. XRD patterns were recorded using the Shimadzu XRD-700 Series X Ray Diffractometer. It was operated at 40 kV and 30 mA producing $Cu_{K\alpha}$ with λ =0.154 nm using a step size of 0.02°/minute in the range of 2θ =10–70°. The crystallinity of cellulose can be determined using a formula :

$$C = \frac{I_{200}-I_{min}}{I_{200}} \times 100\% \qquad (1)$$

I_{200} was found between scattering angles of $2\theta = 22°$ and 23°. I_{min} was obtained between $2\theta = 18°$ and 19° [6]

c. Scanning Electron Microscopy
SEM analysis was conducted to determine the morphology of cellulose. Cellulose was affixed to the set holder with double adhesive, and coated with gold or palladium metal in a vacuum. Micrographs were obtained with JSM-6510A/JSM-6510LA (Analytical/Analytical low vacuum SEM) at 500× magnification.

3. Results and Discussions

3.1 XRD

In order to study the crystallinity of cellulose obtained from palm empty fruit bunches, the XRD analysis was performed. The results were shown in Figure 1. Figure 1 showed typical XRD patterns of cellulose at 2θ=15° and 22°. Figure 1(a) was XRD pattern of palm empty fruit bunches without hydrolysis process. Figure 1(b), 1(c), 1(d) and 1(e) were XRD patterns of cellulose isolated from palm empty fruit bunches using H_2SO_4 20, 30, 49 and 50% , respectively. Compared with hydrolized cellulose, the XRD pattern of palm empty fruit bunches without hydrolysis process showed broader peaks. It indicated the crystallinity was lower than hydrolized cellulose. Increasing H_2SO_4 concentration increased the crystallinity of cellulose. However, the crystallinity of cellulose decreased at H_2SO_4 concentration 50%. The increasing in crystallinity was due to the reduction of some components contained in the sample such as lignin, pectin and the amorphous part of cellulose during hydrolysis process. Cellulose contains amorphous and crystalline part. However, in high acidic concentration the cellulose will be decomposed and resulting lower crystallinity.

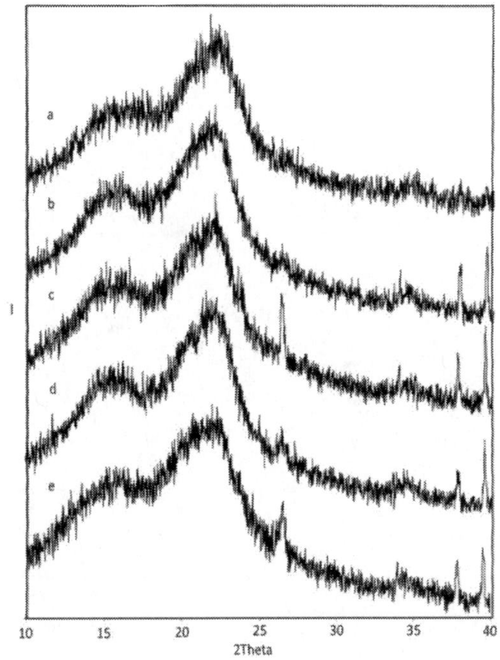

Figure 1. XRD patterns of celluose without hydrolysis (a), hydrolized with H_2SO_4 20% (b), 30% (c), 40% (d) and 50% (e).

Table 1. The yield and crystallinity of cellulose isolated from palm empty fruit bunches (initial weight was 3g)

Cellulose	Yield (g)	Crystallinity (%)
H_2SO_4 20%	1.564	61
H_2SO_4 30%	1.532	63
H_2SO_4 40%	1.512	66
H_2SO_4 50%	1.492	65

Hydrolysis at high concentration of acid caused the chain opening of cellulose that change the arrangement of polymer chains and cellulose was more easily degraded. The reduction of these components resulting reduction of the yield of cellulose obtained as shown in Table 1. The increasing H_2SO_4 concentration decreased the yield of obtained cellulose. Based on Figure 1 and Table 1, the highest crystallinity of cellulose was obtained at H_2SO_4 concentration of 40%.

3.2 FTIR

FTIR analysis was conducted in order to study the functional groups contained in the cellulose. The results were shown in Figure 2. Figure 2 showed FTIR spectra of palm empty fruit bunches (a) and cellulose isolated from palm empty fruit bunches (b).

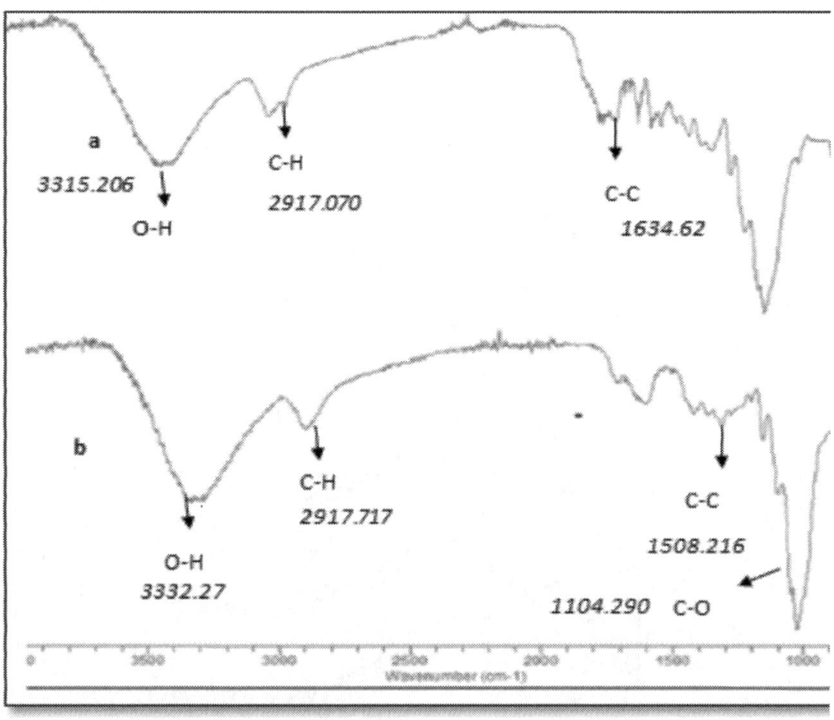

Figure 2. FTIR Spectra of (a) palm empty fruit bunches (b) cellulose isolated from palm empty fruit bunches

Figure 2b showed absorption band at wave number of 3332.273 cm^{-1}attributed to -OH vibration. The absorption band of –OH vibration was also found in FTIR spectrum of palm empty fruit bunches with lower wave number (3315.206 cm^{-1}). It was due to the removal some components of palm empty fruit bunches after hydrolysis process. Absorption bands at wave numbers 1300-1400 cm^{-1} in FTIR spectrum of cellulose can be assigned to -O- which linked with carbon chain in the cellulose. The absorption bands at 2917.717 and 1023.053 cm^{-1} corresponded to $-CH_2$ and C-O vibrations, respectively. These results showed typical FTIR spectrum of cellulose. The results obtained in accordance with research conducted by Isroi [8].

Compared with FTIR spectrum of cellulose isolated from palm empty fruit bunches, the FTIR spectrum of palm empty fruit bunches showed some differences. The differences were due to the reduction of lignocellulosic component after hydrolysis process. The result was in agreement with FTIR spectrum of cellulose isolated from palm empty fruit bunches using alkaline [9] and cellulose isolated from sugarcane bagasse [10].

3.3. SEM

The morphological observations were performed using Scanning Electron Microscope (SEM). SEM images of cellulose isolated from palm empty fruit bunches were shown in Figure 3 (magnification 500x) and 5 (magnification 2000x).

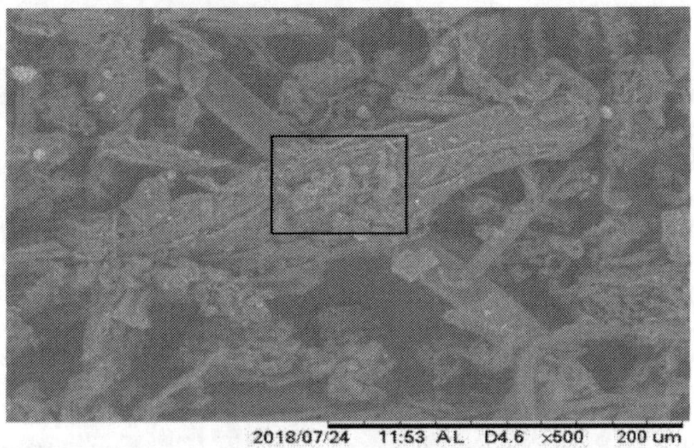

Figure 3. SEM image of cellulose isolated from from palm empty fruit bunches at 500x magnification.

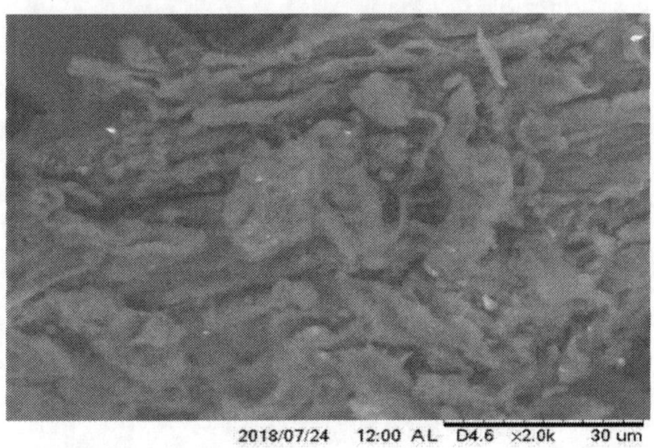

Figure 4. SEM image of cellulose isolated from palm empty fruit bunches at 2000x magnification.

Figure 3 showed the existence of various particle sizes of cellulose. SEM of cellulose particles showed that they are irregular in shape and polydisperse. The different in particle size was found due to hydrolysis process that could reduced the particle size.

SEM images also showed amorphous and crystalline part of cellulose as shown in Figure 3. At magnification 2000x showed amorphous part of cellulose. Figure 3 and 4 indicated the cellulose has been successfully isolated from palm empty fruit bunches. The results are consistent with results reported by Xie [11] for cellulose isolated from bamboo using microwave liquefaction combined with chemical treatment and ultrasonication.

4. Conclusions

Isolation of cellulose from palm empty bunches with variation of H_2SO_4 concentration has been successfully conducted. FTIR analysis confirmed the functional groups contained in cellulose isolated from palm empty fruit bunches. XRD analysis showed crystallinity of cellulose influenced by H_2SO_4 concentration used in hydrolysis process. The highest crystallinity was found at H_2SO_4 concentration of 40%. SEM images confirmed the morphology of cellulose isolated from palm empty fruit bunches with different particle sizes due to hydrolysis process. These results indicate palm empty fruit bunches can be used as a potential source of cellulose.

5. References

[1] Cao X, Chen Y, Chang P R, Muir A D, and Falk G 2008 *Express Polymer Letters* **2**(7) 502-510.

[2] Kampangkaew S, Thongpin C and Santawtee O 2014 *Energy Procedia* **46** 318-325.

[3] Vignehwaran N, Ammayappan L and Huang Q 2011 *Original Journal* **1** 137-142.

[4] Wattanakornsiri A, Pachana, K, Kaewpirom, S, Sawangwong P and Migliaresi C 2011 *Journal of Science and Technology* **33**(4) 461–467.

[5] Rahmi, Lelifajri, Julinawati and Shabrina 2017 *Carbohydrate Polymers* **170** 226-233

[6] Ahvenainen P, Kontro I and Svedström K 2016 *Cellulose* **23**(2) 1073-1086 .

[7] Rashid M, Gafur M A, Syarafat M K, Minami H, Miah M A J and Ahmad H 2017 *Carbohydrate Polymers* **170** 72-79

[8] Isroi I, Cifriadi A, Panji T and Syamsu K 2017 *IOP Conference Series Earth an Environmental Scienc*e **65** 012011

[9] Pradana A, Ardhyananta H and Farid M 2017 *JURNAL TEKNIK ITS* **6**(2) 413-416

[10] Aprilia S, Ambarita C A, Karmila, Armando A and Guswara Y S 2017 *Oriental Journal of Physical Sciences* **2**(2) 103-108

[11] Xie J, Hse C, Hoop CFD, Hu T, Qi J and Shupe T F 2016 *Carbohydrate Polymers* **151** 725-734

IOP Conf. Series: Materials Science and Engineering **523** (2019) 012031 doi:10.1088/1757-899X/523/1/012031

Low density composite board from sugarcane residue and polymer of high density polyethylene

Sofyana[1,*], Hisbullah[1], P N Alam[1], Zuhra[1]
[1]Department of Chemical Engineering, Faculty of Engineering, Syiah Kuala University –Banda Aceh, Indonesia

E-mail: sofyana71@unsyiah.ac.id

Abstract. Bagasse is one of biomass wastes often found in the sugar industry. In large quantities, such wastes are a problem for environmental sustainability because it is not handled properly. Likewise, plastic wastes are an environmental problem with difficulty coping. Both of these wastes can be integrated into a product with economic value by processing it into a composite board. The research aims to produce low-density composite boards and to study the process variables that influence its manufacturing process, namely the compressing temperature and Maleic Anhydride (MAH) as a coupling agent. HDPE plastic waste will be tested as an adhesive for the manufacture of composite boards. The composite board is made by preparing the raw material with the drying of the sugarcane residue and the dissolving of the plastic waste using xylene solvent. Furthermore, the mixing process is carried out with various coupling agents by 2, 4, 6, 8, and 10%, as well as the hot press temperatures by 160, 180, and 200°C. The results showed that the optimum value of the hot press temperature of the composite board was 160°C with 10% coupling agent concentration. The value of the thickness expansion was in the range of 0.0021–0.0153%, the water absorption was 0.067–0.640%, and the average density was 449.80 kg/m^3. Its MOE was 92.188–459.901 Mpa, the MOR was in the range of 1.92–3.07 Mpa, and the DSC analyses were 5.3 and 23.52 KJ/g at 2% and 10% MAH concentrations, respectively. The physical and mechanical tests of the composite board obtained have fulfilled the standard of SNI 03-2105-1996 with the exception for MOR.

1. Introduction

Sugarcane residue or bagasse is one of biomass waste produced from sugar industry. Its production ranges from 35 to 40% of sugarcane processed. In developing countries, it is commonly used as fuel in sugar industries. Bagasse contains 43.8% cellulose, 28.6% hemicelluloses, 23.5% lignine, 1.3% dust, and 2.8% other components. It can be utilized as raw material in production of plastic, paper, xylitol, furfural, animal feed, biogas, ethanol, glucose, and composite board [1-3].

Composite board is board made from lignocelluloses material combined with synthetic plastic adhesive. The commonly used lignocellulose is sawdust while adhesive utilized is urea formaldehyde. As its abundant availability in line with the mushrooming in street sugarcane drink, thorough research on the use of bagasse as a potential raw material for composite board is worthwhile.

* Corresponding author : sofyana71@unsyiah.ac.id

The previous research show on binderless insulation board shows that the use of bagasse can enhance the quality of the product as compared with the board based on coconut fibre. Bagasse based insulation board at density of 350 kg/m^3 performs satisfactorily for construction material as thermal insulation. Due to its environment-friendly process as no chemical adhesive is used, this product may compete its conventional counterpart where urea formaldehyde adhesive, which is dangerous for health and environment, is a common practice [1].

The main factors affecting characteristics and quality of composite board are density, form and humidity of particle, the pressing cycle, hot press duration and temperature, and quality and type of adhesive and many studies have been conducted to study the types of resins and boards. Plastic waste can be used as potential adhesive instead of urea formaldehyde. [4-9]. The utilization of recycle plastic of thermoplastic type for composite board production, such as polyethylene (PE), polypropylene (PP), polyvinylchloride (PVC), and high density polyethylene (HDPE) can at some degree solve environmental problem. The use of recycle plastic can decrease the board production cost [1], [10], [11].

2. Materials and Methods

2.1 Materials.

Materials used in this study were bagasse as a *filler*, white plastic bottle waste of HDPE type as matrix, and *maleic anhidrate* (MAH) as *coupling agent.*

2.2 Methods

Bagasse is prepared to have particle size of 120 – 150 mesh and density of 250 – 450 kg/m^3.The amount of coupling agent applied is , 4, 6, 8 dan 10 %w/w). Hot press temperature is varied at 160, 180 and 200 °C. Pressing time is 30 minutes. The ratio of filler to matrix is 50:50 %w.

2.2.1 Preparation of bagasse as raw material. Bagasse is sun-dried to have 10% water content. It is then dried using oven at temperature of 80 °C to obtain the water content of 6-7% [4]. After that, it is sieved to suit the relevant particle size.

2.2.2 *Preparation Board.* The size-reduced HDPE of 50 g is put in a three-neck flask and 20% xylene of 20 ml is added. The materials are then heated at temperature of 100-200 °C. After the HDPE is completely melted, the prepared bagasse is added and stirred for ±15 minute up to homogeny. The mixture is then unloaded into a container and the solvent is let to vaporize for 24 hours. After that, 30 minutes of hot pressing is implemented. After the temperature gets stable, the composite is ready to be formed as specimen suitable for testing.

2.2.3 *Board Testing.* The ASTM D-638, is used as standard for tensile test. Tensile Strength ASTM D-638, Strength ASTM D-790, and DSC (*Diffrential Scanning Calorymetri*) are used for tensile, break modulus, and thermal tests, respectively.

3. Result and Discussion

3.1. Board appearance

The physical characteristics of composites such as colour, surface condition, compactness and fragility can be implied by visual appearance. The difference appearance between the composites with pressing temperature 160 ^{0}C and 200 ^{0}C caused by the fact that the filler experiences overheated such that it turns deep brown at 200 ^{0}C. Beside that, type of filler used will also affect the visual appearance, physical and mechanical characteristics. Since bagasse contains much cellulose, its colour turns deeper as temperature increases during the pressing process. Also, distinct smell results during the process [1].

IOP Conf. Series: Materials Science and Engineering **523** (2019) 012031 doi:10.1088/1757-899X/523/1/012031

3.2Physical Properties

Thickness expansion, water absorbing capacity, and density are three physical properties adopted in this work. Thickness expansion indicates the dimension stability, while the water absorbing capacity indicates the capacity in absorbing water during soaking process. The concentration of coupling agent and hot press temperature provide real impact to the three properties, as shown in Figures 3 and 4.

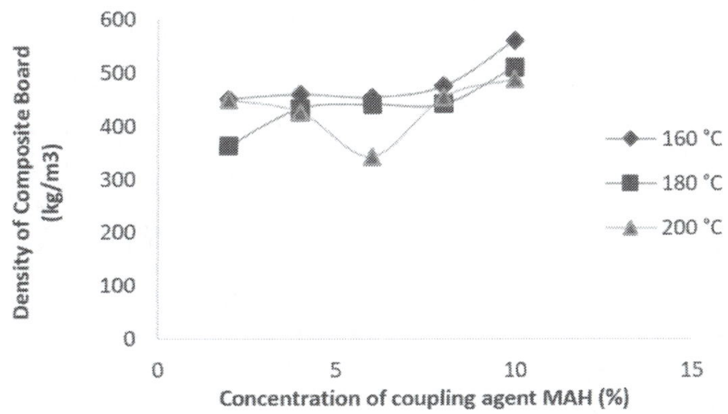

Figure 1. The effects of coupling agent concentration and temperature on the composite density.

Results shown in the figure suggest that the higher the temperature implemented the lower the density obtained. The degradation of composite board quality can occur during heating process. Excessive heating can cause filler and matrix to damage where voids are formed in the board and hence, the density decreases [12].

For the thickness expansion and water absorbing capacity, their low values are obtained at temperature 160 ^{0}C. At low pressing temperature, the composite is denser and of no void. Higher pressing temperature, e.g. 200 ^{0}C, can cause overheating on filler such that the adhesive cannot completely bind and hence, voids occur and water can easily penetrate into the board. The thickness expansion is related to the absorption of water by the composite. High water content can weaken inter-particle bonding and then the composite bulges. Hence, the dimension of the board changes [13].

The concentration of coupling agent affects the board capacity of water absorption in which the higher the concentration of coupling agent used the lower the capacity of water absorption. When more coupling agent, i.e. matrix, is used, voids between bagasse (filler) particles will be filled with it. Consequently, penetration of water can be hampered.

3.3Mechanical Properties

The mechanical properties adopted here in the modulus of elasticity (MOE) and modulus of rupture (MOR). Figure 5 shows the effects of temperature and concentration of coupling agent on MOE and MOR. The previous research suggests that the higher the temperature of pressing process and the longer the pressing time the higher the value of MOE [14]. In this work, as shown in the figure, however, this phenomenon does not appear. Over-curing and over-heating are suspected as the responsible factors. These factors make the filler degraded resulting in void formation, and hence, the decrease in MOE. Besides, the thickness of the composite also affects the level of MOE in relation to temperature [2].

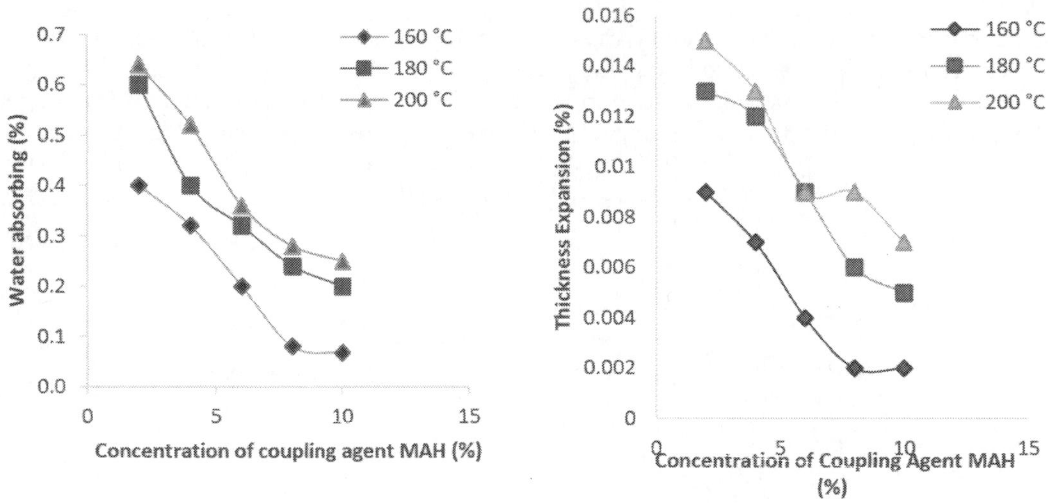

Figure 2. The effect of coupling agent (MAH) concentration on water absorbing capacity and thickness expansion.

The value of MOE, as shown in the figure, increases as the increase in the coupling agent concentration. The adhesive behavior between matrix and filler increases due to estherification processs between anhydride of maleic anhydride and hydroxyl that forms the filler. As a result, the bond between matrix and filler intensifies and hence, the elasticity of the composite increases. The higher the pressing temperature the higher the value of MOE. This phenomenon is also unseen in this work [5]. The same reason as mentioned above for the value of MOE is also applicable here. Temperature near 200 ^{0}C bagasse particles start to degrade leading to void formation [1]. In that condition, HDPE melts and the humidity of bagasse disappears leading to burning of hemicelluloses and cellulose and lignin degradation. The value of MOR obtained in this work ranges between 1,927 – 3,073 MPa, while the minimum allowable value for MOR as per SNI 03-2105-1996 standard is 9,881 Mpa. In this respect, therefore, all of the composite board produced in this work does not fulfill the standard.

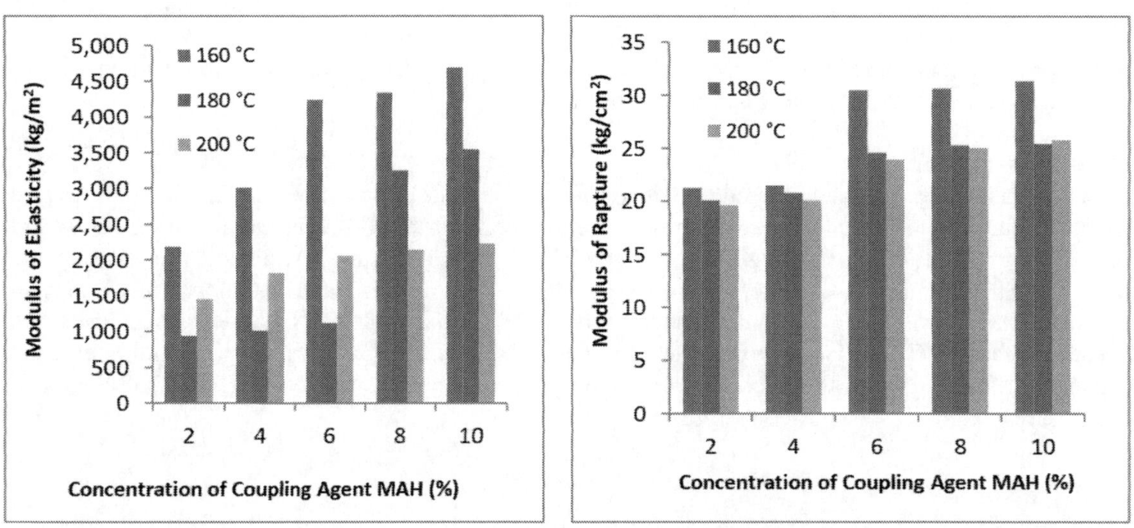

Figure 3. The effects of temperature and concentration of coupling agent on MOE and MOR

IOP Conf. Series: Materials Science and Engineering **523** (2019) 012031 doi:10.1088/1757-899X/523/1/012031

In general, the comparison between the values of physical and mechanical properties tested in this work and the standard of SNI 03-2105-1996 is given in Table 1.

Table 1. The Comparison of Physical and Mechanical Properties

Parameters	SNI 03-2105-1996	This Work
MOE (Mpa)	min 0,15	92,188 – 459,901
MOR (Mpa)	min 9,881	1,927 – 3,073
DSC (kj/g)	-	5,30 - 23,52
Density (kg/m^3)	250 – 450	342,18 – 560,49
Thickness Expansion (%)	Max 12	0,002 – 0,015
Water Absorbing Capacity (%)	-	0,067 – 0,640

Melting point and amount of energy needed to melt the bagasse-waste-based HDPE composite are given in Table 2. The amount of coupling agent affects enthalpy and melting point. Higher amount of it will need more enthalpy since the resulting composite will be more compact. However, due to its higher plastic content, the melting point will be lower.

Table 2. Enthalpy and Melting Point of The Composite

Hot Press Temp. (°C)	Coupling Agent Conc. (%)	Enthalpy (kJ/g)	Melting Point (°C)
160	2	5,300	325,440
	10	23,520	244,570

4. Conclusion

The optimum condition of the composite board based on physical and mechanical properties in this research is obtained at hot press temperature of 160 °C and coupling agent concentration of 10%. The capacity of water absorption obtained ranges from 0.067 to 0.640%, the thickness expansion from 0.002 to 0.015%, and density from 342.175 to 560.488 kg/m^3. In terms of MOE, the composite board resulting from this work fulfills the standard of SNI 03-2105-1996 but MOR does not meet the standard.

5. References

[1] Panyakaew,S., dan Fatios,S *2011 Energy and Buildings* **43** 1732–1739.

[2] Iswanto,A.H., Azhar, I., Supriyanto, dan Susilowati, A 2014 *Agriculture, Forestry and Fisheries* **3(2)** 62-66.

[3] Rosmeika,Sutiarso, L., dan Suratmo,B 2009 *Jurnal Engeneering Pertanian* **8** 2.

[4] Dashti H., Salehpur Sh,Taghiyari HR, Akbarifar F, Heshmati S 2012 *J Nanometer Biostruct* **7(30)** 853-60

[5] Kelly MW 1997 *For. Ser. Gen. Tech. Rep. FPL* **10** 66

[6] Maloney TM 1993 *Modern Particleboard and dry process fibreboard manufacturing*. 2nd Ed. (San Francisco: Miller Freeman) P.689.

[7] Stockel F, Konnerth J, Moser J, Kantner W, Gindl-Almutter W 2012 *Wood Sci. Technol* **46** 611-20

[8] Taghiyari HR, Rangawar H, Farajpour Bibalan O 2011 Bioresource **6(4)** 4067-75]

[9] Valenzuela J, Von Leyser E, Pizzi A, Westemeyer C, Gorrini 2012 Eur *J Wood Prod*.

[10] Doost-hoseini, K., Taghiyari, H.R., dan Elyasi, A 2014 *Composites: Part B* 58 10–15.

[11] Maulana,F 2012 *Hasil Penelitian Industri* **25** 2

[12] Hasan, S., Hartono, R., Sucipto, T., dan Iswanto,H.A 2013 *Jurnal Teknologi Pengelolaan Limbah* **16** 3.
[13] Fauziah, Wahyuni, D., dan Lapanporo, P., B 2014 POSITRON **8** 2
[14] Massijaya,Y.M 1997 *Developmentof Board Made From Waste Newspaper (*Tokyo : Dissertation at Tokyo University*)*

IOP Conf. Series: Materials Science and Engineering **523** (2019) 012032 doi:10.1088/1757-899X/523/1/012032

Characterizing mode choice behaviors of the evacuees during emergency evacuation using a Logistic regression model

S Sugiarto[1,*], Lulusi[1], M Isya[1], Taufik[2], M Haiqal[3]
[1]Department of Civil Engineering, Syiah Kuala University, Banda Aceh, Indonesia
[2]Alumni, Department of Civil Engineering, Syiah Kuala University, Banda Aceh, Indonesia.
[3]Architecture and Planning Department, Syiah Kuala University, Indonesia

E-mail: sugiarto@unsyiah.ac.id

Abstract. The most crucial transportation components in studying individual decision-making behavior are mode choice that used by evacuees. Especially in typical emerging countries which private motorized transport is mostly used in daily activities. Mode choice evacuation behavior is viewed as one of the essential aspects in a transportation disaster planning. This study is aiming to investigate an individual's preferences for selecting motorized and non-motorized modes for their evacuation during imaginary tsunami evacuation. The Stated Preference (SP) experimental data were collected in Kuta Raja sub-district, one of the most exposed sub-district in Banda Aceh. Binary logistic regression was conducted to evaluate the effects of several factors quantitatively and to identify the dominant factor that affected evacuation mode choice. The empirical result depicts that mode choice used by evacuee during an emergency evacuation is suggestively associated to the socio-demographic attributes such as male gender, age, government employee, low-income household, household member and motorcycle owned. It found that the medium-income households substantially rely on motorcycle mode for their evacuation. The more motorcycle owned within the household, the more likelihood of using motorcycle for their evacuation. The male gender and government employee also has a statistically positive correlation to the explanatory power of motorized transport used to evacuate.

1. Introduction

Indian Ocean Tsunami on December 26, 2004 event, the magnitude was 9.0 and epicenter is 255 km from Banda Aceh, the provincial capital of Aceh. The earthquake itself caused damage and Casualties. The subsequent tsunami killed at least 108,100 people and left 127,700 people missing in Indonesia [1]. Indonesia has a lack of disaster information and familiarity caused a postponement in the response of citizens to the tsunami and had led to abundant casualties and missing people. In contrast to developed countries, for instance in Japan, a study conducted by [2], they had a significant effort in the response of citizens, and it caused a substantial effect on reducing casualties by taking advantage of what they learned from disaster prevention education. It was called the "Kamaishi Miracle," and the story spread around the world. Further investigation of the evacuation behaviors by [3], they concluded that the fatality ratio in convinced region was predominantly affected by the force of the tsunami and the residents' characteristics.

*Corresponding author: sugiarto@unsyiah.ac.id

IOP Conf. Series: Materials Science and Engineering **523** (2019) 012032 doi:10.1088/1757-899X/523/1/012032

The most crucial transportation components in studying individual decision-making behavior are mode choice that used by evacuees. Especially in typical emerging countries which private motorized transport is mainly used in daily activities. A study by [4] argued that excessive car evacuation could cause severe traffic congestion, and could lead to large numbers of casualties during tsunami disasters. They investigated the possible factors that lead to unnecessary car evacuation can ensure smoother tsunami evacuations and mitigate casualty damages in future tsunami events. The result advised that the distance of evacuation acted as a principal factor in selecting car evacuation relative age and sex variables. Further explorations on car usage behavior during the evacuation were done by [5, 6]. They revealed that car usage, which exceeded the road capacities of local communities, caused severe traffic congestion and tsunami casualties.

In the context of the emerging world, the problem with a high demand for private-motorized transport in tsunami evacuations created traffic wrongly and intended to the complexity of evacuation behaviors. Seeing the rapid urbanization and problem of private-mode dependency, the investigation of factors that influencing mode choice is crucially need to be deliberate. It is, therefore, this study is aiming to investigate an individual's preferences for selecting motorized and non-motorized modes for their evacuation during imaginary tsunami evacuation. The Stated Preference (SP) experimental data were collected in Kuta Raja sub-district, one of the most exposed sub-district in Banda Aceh. The modeling framework used in this study is a binary logit (BL) regression model. The BL model is a simple and open form discrete choice model that widely used the behavioral arena. In the behavior modeling, the BL has been commonly used to analyze behavioral research relating travel and transport behaviors. For example in Thailand [7] used BL to examine tsunami evacuation behavior related to transportation disaster response.

Further implementation BL model by [8, 9] to explore travel behavior related to individual preferences in voting congestion charging policy in Jakarta. Perhaps a more recent study by [4] conducted BL analysis to assess the effects of numerous factors and to identify the leading factor that exaggerated evacuation mode choice in Tohoku, Japan. Thus, in this study, the BL model is implemented to examine the effects of individual travel behavior factors and household's socio-demographic characteristics and to determine the most significant influencing factors that affected evacuation mode choice. The empirical result from this work might be of specific help in the plan of a more effective policy for the evacuation planning in particular for the Banda Aceh region. Furthermore, a better understanding of local characteristics is essential for successful transportation-mobility managements policy implementation to gain a better life for society [10, 11].

The rest of this report is structured as follows. In the next section gives a brief description of the SP survey, the distribution of SP survey and describe the formulation of the BL model. In section 3 and 4, we then present model estimates and discuss the findings. Lastly, the conclusions of this work are obtainable at the end of this report.

2. Study Area

This study focuses on tsunami evacuation based on a hypothetical scenario. The tsunami harshly struck in the city of Banda Aceh. The Stated Preference (SP) method is used in this study. The SP method is widely used in travel behavior studies [see for example 10, 11, 12]. The questionnaire survey was conducted in March 2018, approximately fourteen years after the 2004 event.

Table 1. Summary of SP questionnaire

Description	Detail
Time of survey	March 2018
Target location	Kuta Raja sub-district of Banda Aceh, Aceh Province
Distribution methods	A paper-pencil based direct interviews and Collected by the enumerator
Number of questionnaires	100

The target respondent was residents who have experienced the 2004 event. A paper-pencil based on the direct interview was distributed to respondents in Kuta Raja sub-district, one of the most exposed sub-district in Banda Aceh as shown in Figure 1.

The respondents were asked general features of evacuation behaviors if they will face a tsunami event in the future such as departure times, evacuation mode, evacuation place (e.g., escape the building, tall building). Furthermore, the question related to individual socio-demographic characteristics such as gender, age, income, marital status, household's membership, the number of a car owned and the number of motorcycles owned. The questionnaires were distributed to 100 and used in this study. The SP survey is summarized in Table 1, which gives the time of the surveys, target locations, distribution methods, number of samples and sampling distribution.

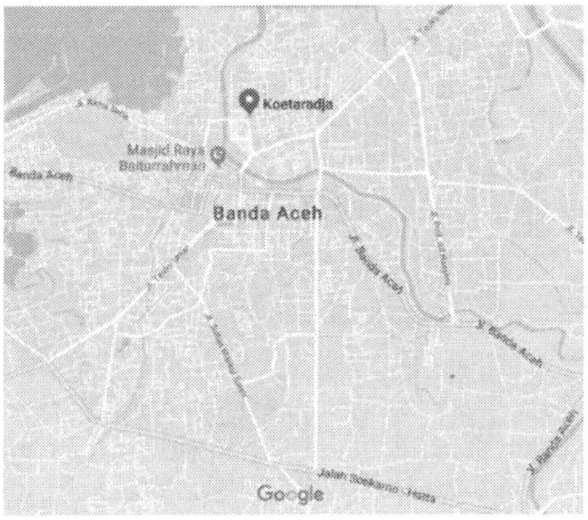

(a) Banda Aceh city (map data © 2018 Google)

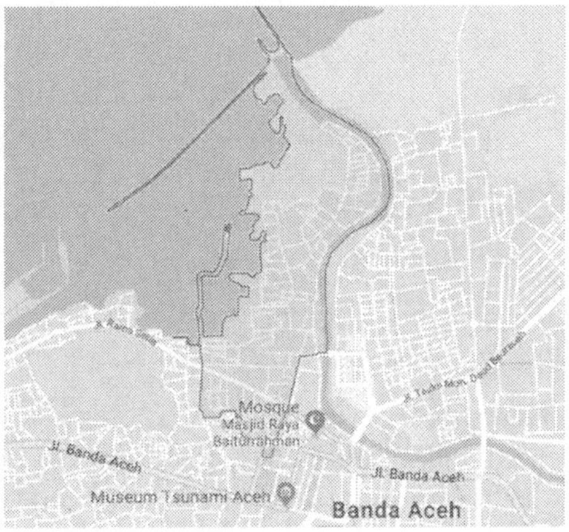

(b) Kuta Raja sub-district

(c) Kuta Raja sub-district and its villages

Figure 1. Study area

Table 2 shows the distribution of respondent's socio-demographics. It can be seen that young people (≤30 years) is accounting for few sample set, accounting for less than 5% of respondents. The gender distribution is slightly skewed toward females (65%). As regards employment status, our sample indicates 68% of employed respondents, and less than 10% of the sample consists of students. As for the income, it is found that up to 49 % has a low to medium monthly income. Additionally, the distribution of samples reveals that close to 50% of respondents are renting their house.

Table 2. Description of respondents' socio-demographics

Item	Category	Share	Item	Category	Share
Gender	Male	35%		IDR 1 or less	1%
	Female	65%	Monthly	IDR 1 - 2	5%
Age	20 years or less	2%	Income	IDR 2.1 - 4	43%
	21-30 years	3%	(in	IDR 4.1 - 6	36%
	31-40 years	19%	million)	More than IDR 6	15%
	41 years or more	76%			
Occupation	Working	68%	Driver's	Renting house	49%
	Student	9%	license	Owned house	51%
	Unemployed	23%			

3. Measurement

The SP questionnaire were employed to capture the effect of socio-demographics respondent to the their mode choice behavior towards emergency evacuation. The dependent variable of the mode choices used by respondents to evacuate if a tsunami will happen in the future. Two choices of the mode are offered to the respondent to select one between two choices (i.e., motorized transport and non-motorized transport).

Table 3 Distribution of the dependence variable (mode choice for evacuation)

Variable	**Distribution of a 4-point Likert score (%)**	
	Motorized	**Non-Motorized**
Mode Choice Set	89	11

Table 3 depicts that the dependent variable looks like to be normally distributed. A Kolmogorov-Smirnov test is conducted to ensure the data normally distributed with 5% significant level of error. The mean intention to select motorized in the sample set was 0.89 (SD = 0.31). It depicts that people in Banca Aceh have a strong intention to use motorized transport for their evacuation.

4. Binary Logistic Model

A modeling approach of discrete choice model in used to explore the behavioral characteristic among variables of socio-demographics and mode choice selected for emergency evacuation. A simple formulation of binary logistic (BL) regression is used to delve more deeply related respondent's intentions to choose their mode during evacuation.

The modeling BL choice behavior is formulated for the following choices: respondent to select one between two choices, that is motorized transport and non-motorized transport for their evacuation. The probability that evacuee i chooses motorized π_i is modelled based on the random utility maximization theory as shown in Equation (1) and (2). More advance and detail of the binary choice and maximum likelihood formulation can be referred to [9]. In this study unknown parameters are estimated using maximum likelihood estimator and implemented in NLogit 5.

$$logit(\pi_i) = \log\left(\frac{\pi_i}{1-\pi_i}\right) = \beta x_i \quad (1)$$

$$\pi_i = Pr(Y_i = 1|\mathbf{x}_i) = \frac{exp(\beta x_i)}{1+exp(\beta x_i)} \quad (2)$$

Table 4 Distribution of the dependence variable (mode choice for evacuation)

Variable	Classification of variables (parameter coding)
Mode Choice (MC)	Binary choices ○ 1 used motorized transport ○ 0 otherwise
Gender (GD)	Dummy of gender ○ 1 male ○ 0 otherwise
Age (AG)	Nominal ○ 1 if 20 years or less ○ 2 if 21-30 years ○ 3 if 31-40 years ○ 4 if 41 years or more
Occupation (OCP)	Dummy variable of government employee ○ 1 if government employee ○ 0 Otherwise
Income (INC)	Dummy variable of monthly household' income ○ 1 if low income household (income < 2 million IDR/month) ○ 0 Otherwise
Household Member (HM)	Nominal ○ 1 if has <3 household members ○ 2 if has 3-5 household members ○ 3 if has 6-10 household members ○ 4 if has more than 10 household members
Motorcycle Owned (MO)	Dummy of having motorized transport within household ○ 1 if has motorcycle ○ 0 otherwise

5. Result and Discussion

The BL regression is conventional method to determine the best of fit model among dependence variable and predictors using calibration data set. The exogenous variables was set as binary choice as Yi = {1: used motorized transport, 0: used non-motorized transport}. According to the equation (2) Y represents response variables, and xi is a vector of explanatory variables such as socio-demographic characteristics and individual travel behaviors. Variable notation, definition and empirical settings for both response and regressor used in this study are summarized in Table 4.

The stepwise calibration method is used to select the best fit of the model. The maximum likelihood estimator (MLE) is implemented to calibrated BL parameters. To ensure our model is a statistically acceptable the 5 % level of error is applied for t-test significantly. However, several variables still be selected within 10% level of error as it is for behavioral analysis rather than for predicted model. The result of the calibrated parameters and its significant level can be seen in Table 4 including the Goodness of Fit (GoF) calibrated model.

Table 5 exhibits the parameter results of estimation of mode choice model considering socio-demographic attributes. Calculated indicates of model performance include log-likelihood at convergence (LL (b)) and McFadden Pseudo R-Squared. The McFadden Pseudo R-Squared indicate shows substantially the best fit to our data set, as shown at the bottom of Table 5.

Table 5 Calibrated parameters and Goodness of Fit of mode choice model

Predictors	Coefficient	T-value	P-Value
GD	4.52	2.58	0.009
AGE	1.78	2.29	0.020
OCP	2.29	1.72	0.085
INC	-11.17	-2.67	0.007
HMEM	0.78	1.72	0.084
MO	3.44	1.84	0.066
Dependent Variable			MC
Number of Sample			100
LL (0)			-34.65
LL(β)			-15.72
$\chi 2$ (5df)			37.86
$\chi 2$ Significance level			0.000
McFadden Pseudo R-Squared			0.546

Examining the calibrated parameter results presented in Table 5, the parameter of the male gender dummy has a positive sign for the selected motorized transport mode. This indicates that the male gender plays an important role in influencing an individual's decision. That is, the male gender the more likely they are to use motorized transport compared to the female gender. It seems that the positive correlation also linked to others individual's socio-demographic attributes such as age, government employee dummy variable and a number of household member.

Interestingly, the variable of dummy low-income household has a statistically negative association to the motorized transport mode. It seems that low-income household tends to used non-motorized transport as they may not have owned motorcycle or car in the household. However, the mobility attributes of motorcycle held have a substantially positive correlation to the motorized transport mode. This tendency is found to be larger medium-income households (monthly income range from 2 – 4 million IDR) has at least one or two motorcycles. It is presumed that such household finds it necessary to use a motorcycle for their evacuation.

6. Conclusion

This paper reports on investigation and characterizing of an individual choice behavior relating to the mode choice used to evacuate during emergency evacuation. Using data from a Stated Preference survey collected in 2018, a binary logistic model is used to capture the impact of observable such as demographics, household characteristic and mobility attributes in examining mode choice. We find that there is strong association between choice of motorized transport and observable regressors.

The specific empirical finding performed by Logistic model reveals that the male gender plays an important role in influencing an individual's decision to choose motorized transport. A positive correlation also linked to the attributes of age, government employee dummy variable and a number of household member. Low-income household has a statistically negative association to the motorized transport mode. It seems that low-income household tends to used non-motorized transport as they may not have owned motorcycle or car in the household. While the mobility attributes of motorcycle held have a substantially positive correlation to the motorized transport mode. It is presumed that such household finds it necessary to use a motorcycle for their evacuation.

Indonesia has lack of disaster information and familiarity caused a postponement in the response of citizens to the tsunami and had led to abundant casualties and missing people. An investigation of the evacuation behaviors plays the most crucial components decision-making behavior. Our result confirmed previous research by [4] that typical emerging countries which private motorized transport is mainly used in daily activities. An excessive motorized transport evacuation could cause severe traffic congestion, and could lead to large numbers of casualties during tsunami disasters. Therefore, the empirical findings of this work may offer general assistance in efforts to the government for the suggested establishment the guidance for evacuation planning, especially deal with complex transport problem. There are widely used statistical approaches that can analyze engineering phenomena such as the probabilistic applications for seismic hazard study done by [13]. Furthermore, a better understanding of local characteristics is vital for successful engineering policy implementation to gain a better life for society [14-17].

7. Acknowledgments

The Syiah Kuala University financially supported this study under Contract No. 289/UN11/SP/PNBP/2018. Furthers thanks go to the staff and students of Transportation System Laboratory, Civil Engineering Department of Syiah Kuala University for their substantial help with the data collection.

8. References

[1] Imamura, F. 2009. *Journal of Earthq. Tsunami* **3**, 59.

[2] Suppasri, A. Hasegawa, N. Makinoshima, F. Imamura, F. Latcharote, P. Day, S. 2016. *Front. Built Environ*. **2**.

[3] Katada, T. Kanai, M. 2016. *J. Disaster Res*. **11**, 845–856.

[4] Makinoshima F, Abe Y, Imamura F, Machida Gaku., Takeshita Y. 2017. *Geoscience*, **7**, 112.

[5] Hara, Y. Kuwahara, M. 2015. *Transportation Research Part A Policy Pract*. **75**, 1–15.

[6] Makinoshima, F.; Imamura, F.; Abe, Y. 2016. *Coastal Engineering Journal*, **58**,1640023.

[7] Charnkol T. Tanaboriboon Y. 2006. *IATSS Research*. **30** (2), 83-96.

[8] Sugiarto, S., Miwa, T., Sato, H., Morikawa, T. 2014. Congestion charging: Influence of public consciousness on acceptability in Jakarta Metropolitan Area. *Proceedings of 21st World Congress on Intelligent Transport Systems*, Detroit, USA.

[9] Sugiarto, S., Miwa, T., Morikawa, T. 2018. *Transportation Letters: The International Journal of Transportation Research*, **10**(1), 52–63.

[10] Sugiarto, S., Miwa, T., Sato, H., Morikawa, T. 2017a. *Journal of Urban Planning and Development*, **143**(2), 1-12.

[11] Sugiarto, S., Miwa, T., Morikawa, T. 2017b. *Transportation Research Part A Policy Pract*, **103**, 36–53.

[12] Toşa, C., Sato H., Morikawa, T. and Miwa, T. 2018. *Journal of Transport Geography*, **68**, 78-93.

[13] Setiawan, B., *Indonesian Journal on Geoscience*, **4**(2), 81-96.

[14] Saleh, S.M., Sugiarto, S., Hilal, A., Ariansyah, D. 2017. A study on the traffic impact of the road corridors due to flyover construction at Surabaya intersection, Banda Aceh of Indonesia. AIP Conf. Proc. **1903**, 060005.

[15] Aggraini, R., Sugiarto, S., Pramanda, H., 2017. Factors affecting trip generation of motorcyclist for the purpose of non-mandatory activities. *AIP Conf. Proc*. **1903**, 060011.

[16] Sugiarto, S., Miwa, T., Morikawa, T. 2018. *Case Studies on Transport Policy*, Article in press.

[17] Saleh, S.M., Sugiarto, S., Anggraini, R.., 2019. *Open Transportation Journal*, 13(1), 17-24.

IOP Conf. Series: Materials Science and Engineering **523** (2019) 012033 doi:10.1088/1757-899X/523/1/012033

Approaching model of Manning's coefficient due to an Effect of density and height of vegetation in open channel

M Rizalihadi[1,*], **Ziana**[1], **N Shaskia**[1]
[1]Civil Engineering Depertment, Syiah Kuala University, Banda Aceh, Indonesia.

Email: maimunrizalihadi@unsyiah.ac.id

Abstract. The presence of vegetation growing in the channel can cause increasing the Manning's coefficient. The magnitude of Manning's coefficient is depended on the characteristic and type of vegetation. The purpose of the research is to study and model the effect of density and height of vegetation on Manning's coefficient in open channel. The study was conducted in open channel with 15.5 m length, 0.5 m width and 1.0 m height. At the centre part of the channel is planted with Elephant grass (*Pennisetum Purpureum*). The effect of different density and height of vegetation were run with fixed discharge to measure the velocity and water profiles in order to obtain Manning's coefficients. The effect of those variables on the developed model was evaluated using statistical analysis of variance and paired t-test with significance level of 0.05. The results showed that Manning's coefficient is increasing as increasing density and height of vegetation. If compared to un-vegetated channel, Manning's coefficient increase in between 4.92-54.07%. Based on statistical analysis, it is found that $F_{calculation}$=367.776 > F_{critic}=19.452, $t_{calculation}$=12.298 and 22.464 > t_{critic}=2.571 and R^2=0,967, showing that the density and height of vegetation significantly influence on the Manning's coefficient with strong correlation variables in the model.

1. Introduction

Mostly irrigation channel or any other water bodies in Indonesia and Aceh specifically, the vegetation can be found growing on the channels. A lot of type and kind of vegetation grow in the channel, for example, water grass (*Acorus Gramineus*), water hyacinth (*Eichhornia Crassipes*), watercress (*Ipomoea Aquatica Forsk*), elephant grass (*Pennisetum Purpureum*) and jeringau or sweet flag (*Acorus Calamus*) [9].

The presence of vegetation in the channel on one side can prevent slope stability of channel from erosion, but on the other side, the existence of plants in the channel can increase the flow resistance making increase loss energy in the channel. A lot of articles has been fully published by [2-5,10] related to flow resistance due to the presence of vegetation. The results of the study indicated that the presence of plants either flexible, stiff or rigid characteristic, and either submerged or partially submerged vegetation in the water, may increase the flow resistance and increase the Manning's coefficient. And even, [3] has completely published values of Manning's coefficient for types of *shrubs and woody* vegetation.

However, the prediction of the vegetation resistance is very complex since there are many different species with their unique characteristics changing during the season, may vary significantly from place to place, and may also change in time [9]. This inhomogeneous character of the vegetation in the field

*Corresponding author: maimunrizalihadi@unsyiah.ac.id

make hardly to take into account in a model of the equation. Moreover, the difference of type dan characteristic of vegetation might result in different flow resistance, although these vegetations are equally flexible, rigid or stiff vegetation [4,9]. Due to the importance of vegetation effect in the channel roughness, therefore, it is necessary to do further research by studying for different characteristic and type of vegetation in the channel to make the more reliable approach of manning's coefficient model in general application.

The objectives of this research is to study the effects of vegetation density and height of Elephant grass (*Pennisetum Purpureum*) on Manning's coefficient, develop the approaching model and test statistically the effect of those two variables on Manning's coefficient.

This research is a continuation of research as mentioned in [6,9]. These studies were laboratory experimental based, and both research used Elephant grass (*Pennisetum Purpureum*) as vegetation in the channel. [9] has published research about the effect of vegetation height on flow resistance, resulting in increasing flow resistance in the channel as increase of vegetation density, which was indicated by increasing of vegetated Manning's coefficient (n_v) in between 0,0083-0,0196. Furthermore, [6] has also published a similar study by examining the effect of vegetation density on flow resistance in the channel, resulting in increasing n_v as increasing vegetation density in between 0,0141-0,0207. These showed that both effect of vegetation height and density separately gave an effect on manning's coefficient. In fact, in nature, sometimes it is found that the vegetation in the channel are grown in different height and density, therefore it is necessary to study the combination of two condition of vegetation height and density. This is the primary object of this research.

2. Manning's coefficient

2.1. Original Manning's equation

One the most commonly used equations governing open channel flow is known as the Manning's Equation. It was introduced by the Irish Engineer Robert Manning in 1889 as an alternative to the Chezy equation. The Manning's equation is an empirical equation that applies to uniform flow in open channels and is a function of the channel velocity, flow area and channel slope [1].

Flow velocity is dependent on the amount of friction between the water and the stream channel. Smoother channels will have less friction and, therefore, faster flow. Channel roughness contributes to turbulence, which dissipates energy and reduces flow velocity. The following equation is the Manning Equation, which is used to calculate average flow velocities in open-channel systems:

$$V = \frac{1}{n} R^{2/3} S^{1/2} \tag{1}$$

In this equation, V is the average flow velocity, R is the hydraulic radius, or the ratio of the cross-sectional area of the stream divided by the perimeter of the channel in contact with water (wetted perimeter), S is the slope of the water surface, and n is the Manning roughness coefficient. From the mathematics, it follows that a greater n-value will result in a smaller value for flow velocity.

2.2. The Manning's coefficient

The Manning's coefficient (n) as in the equations 1 is normally derived from the roughness of materials forming channels. In fact, [1] stated that in nature the flow resistance is influenced by many factors, so the total Manning's coefficient can be obtained using the equation 2 below.

$$n = (n_0 + n_1 + n_2 + n_3 + n_4) m_5 \tag{2}$$

In some cases, the values of n_1, n_2, n_3 can be ignored due to not significant and m_5 is 1 for straight channel, so Equation 2 is dominated by the base roughness of channel and due to vegetation, as in Equation 3.

$$n = n_o + n_v \tag{3}$$

where: n_0=the base roughness; n_1= the variation of channel surface; n_2=shape variation of channel section; n_3=obstacles due to the structure; n_4= due to vegetation; and m_5=correction factor for channel bend.

For specificse, [3] has published specific formula in determination of resistance due to shrubs and woody vegetation, as in Equation 4. This formula as a base to calculated the Manning's coefficient.

$$n = n_0 \sqrt{1 + \left(\frac{C_D \sum A_i}{2gAL}\right)\left(\frac{Kn}{n_0}\right)^2 \left(\frac{A}{P}\right)^{\frac{4}{3}}} \tag{4}$$

where: A_i = individual vegetated area (m^2), A = channel area (m^2), C_D = drag coefficient of vegetation; g= gravity (m/det^2); K_n = conversion Manning's equation (m$^{1/3}$/d); L= length of channel (m); n_0 = base roughness of channel; P = wetted perimeter (m).

3. Research method

3.1. Experimental design

Experiments were conducted in a 15.5 m long, 0.5 m wide and 1.0 m deep glass-walled flume. The slope of flume is set with fixed slope of 0%. Discharge of 5.53 l/s is conducted through one V-notch gate to maintain a constant water level of 45 cm. Bed flume was layered uniformly with fine sand with 15 cm thick. At the center of 3 m long of flume is planted with Elephant grass (*Pennisetum Purpureum*).

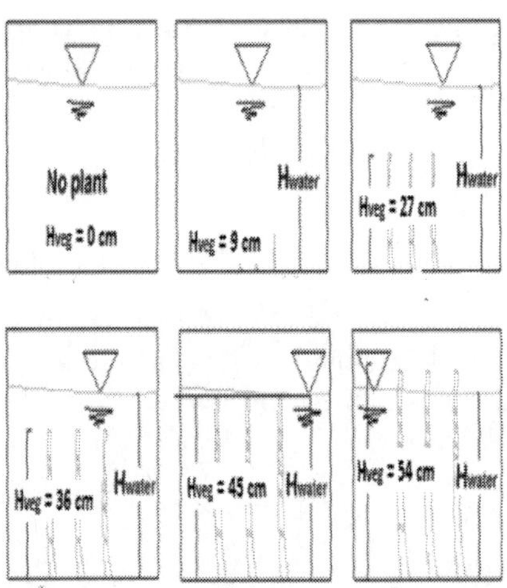

Figure 1. Variations of vegetation height.

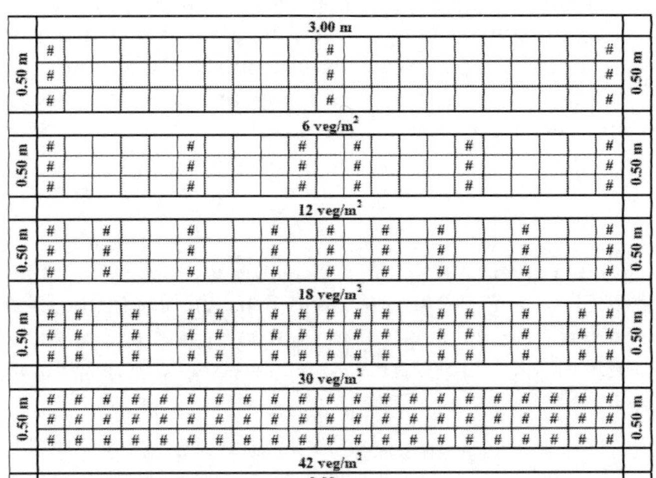

Figure 2. Variations of vegetation density.

The vegetation is set in the channel with a density (VD) of 0, 6, 12, 18, 30 and 42 veg/m^2 and combined with vegetation height (VH) of 0, 9, 27, 36, 45 and 54 cm. For more detail, the setting of vegetation in the channel can be seen in figure 1 and 2.

3.2. Series of Measurements

The series of measurements under fixed discharge are run on the flume with a different variation of vegetation density and height. Totally, there are 36 series of measurements and analysis in this research, completely shown in the Table 1. The velocity of flow was measured using microcurrent meter at upstream, centre and downstream of flume. The velocity was measured at the bottom, 0.2h, 0,6h, 0.8h and surface level to obtain the profile of velocity distribution and the mean velocity. The slope of energy was measured using point gate at every 0,50 m as long 14 m of channel. And the manning's coefficient was analyzed based on these parameters using the method below.

Table 1. The series of experimental design.

Vegetation Density in (veg/m²)	**Vegetation Height in (cm)**					
	0	9	27	36	45	54
0	$(S_f, V_m, n)_{1,1}$	$(S_f, V_m, n)_{1,2}$	$(S_f, V_m, n)_{1,3}$	$(S_f, V_m, n)_{1,4}$	$(S_f, V_m, n)_{1,5}$	$(S_f, V_m, n)_{1,6}$
6	$(S_f, V_m, n)_{2,1}$	$(S_f, V_m, n)_{2,2}$	$(S_f, V_m, n)_{3,2}$	$(S_f, V_m, n)_{2,4}$	$(S_f, V_m, n)_{2,5}$	$(S_f, V_m, n)_{2,6}$
12	$(S_f, V_m, n)_{3,1}$	$(S_f, V_m, n)_{3,2}$	$(S_f, V_m, n)_{3,3}$	$(S_f, V_m, n)_{3,4}$	$(S_f, V_m, n)_{3,5}$	$(S_f, V_m, n)_{3,6}$
18	$(S_f, V_m, n)_{4,1}$	$(S_f, V_m, n)_{4,2}$	$(S_f, V_m, n)_{4,3}$	$(S_f, V_m, n)_{4,4}$	$(S_f, V_m, n)_{4,5}$	$(S_f, V_m, n)_{4,6}$
30	$(S_f, V_m, n)_{5,1}$	$(S_f, V_m, n)_{5,2}$	$(S_f, V_m, n)_{5,3}$	$(S_f, V_m, n)_{5,4}$	$(S_f, V_m, n)_{5,5}$	$(S_f, V_m, n)_{5,6}$
42	$(S_f, V_m, n)_{6,1}$	$(S_f, V_m, n)_{6,2}$	$(S_f, V_m, n)_{6,3}$	$(S_f, V_m, n)_{6,4}$	$(S_f, V_m, n)_{6,5}$	$(S_f, V_m, n)_{6,6}$

3.3. Analysis of measurement data

There are 3 locations that be an object of measurements to analyze the distribution and mean velocity, namely: at the upstream, at the middle and the downstream of the channel. The velocity distribution is measured using a current meter on the flow surface, depth of 0.2h, 0.6h, 0.8h and a channel base. The mean velocity is calculated using the Equation 5 as below.

$$V_m = \frac{V_p + 3.V_{0,2h} + 2.V_{0,6h} + 3.V_{0,8h} + V_d}{10} \tag{5}$$

where: V_m = mean velocity (m/s); V_p = velocity at the surface (m/s); $V_{0.2h}$ = velocity at depth 0.2h (m/s); $V_{0.6h}$= velocity at depth 0.6h (m/s); $V_{0.8h}$ = Velocity at a depth of 0.8h (m/s); V_d = velocity at the base (m/s).

3.3.1. Slope of energy

Loss of energy can be calculated using the Bernoulli equation [1]. This value is obtained from the measurement of water depth and velocity in upstream (H_u) and downstream (H_d). The total energy losses can be written as in the equation below. Then the slope of energy (S_f) can be calculated by the Equation 7, where L is the length of the vegetation channel.

$$H_u + \frac{V^2{}_u}{2.g} = H_d + \frac{V^2{}_d}{2.g} + h_f \tag{6}$$

$$S_f = \frac{h_f}{L} \tag{7}$$

3.3.2. The Manning's coefficient

The coefficient of Manning (n) is obtained by arranging equation 1 to be as equation below, where V_m is mean velocity.

$$\mathrm{n} = \frac{1}{\mathrm{V_m}} \mathrm{R}^{2/3} \mathrm{S}_f{}^{1/2} \tag{8}$$

The value of n in Equation 8 is the total Manning's coefficient produced by channel roughness (n_o) and vegetation roughness (n_v) as in Equation 3. Then the Manning's coefficient of vegetation can be obtained using Equation 9.

$$n_v = n - n_o \tag{9}$$

3.4. Model development

3.4.1. Regression model

In this research, multiple linear regression is applied in developing the model. The dependent variable is Manning's coefficient of vegetation (n_v) and independent variable are vegetation height (V_H) and density (V_D), so the model can be formulated as in equation 10. The coefficient of matrix (β_0, β_1 and β_2) is obtained using least square method and concisely is written as in equation 11.

$$n_V = \beta_0 + \beta_1 V_H + \beta_2 V_D \tag{10}$$

$$\begin{bmatrix} n & \sum_{i=1}^{n} V_{H_i} & \sum_{i=1}^{n} V_{D_i} \\ \sum_{i=1}^{n} V_{H_i} & \sum_{i=1}^{n} V_{H_i}V_{H_i} & \sum_{i=1}^{n} V_{H_i}V_{D_i} \\ \sum_{i=1}^{n} V_{D_i} & \sum_{i=1}^{n} V_{H_i}V_{D_i} & \sum_{i=1}^{n} V_{H_i}V_{D_i} \end{bmatrix} \begin{bmatrix} \beta_o \\ \beta_1 \\ \beta_2 \end{bmatrix} = \begin{bmatrix} \sum_{i=1}^{n} n_{V_i} \\ \sum_{i=1}^{n} n_{V_i}V_{H_i} \\ \sum_{i=1}^{n} n_{V_i}V_{D_i} \end{bmatrix} \tag{11}$$

3.4.2. Statistical test of regression model

Coefficient of determination (R^2). This parameter is to see whether or not there is a statistically significant relationship between the dependent variable and the variable added to the equation. The value of R^2 is in the range of 0-1. The more R^2 tend to 1, the stronger correlation between independent and dependent variable in the model

Analysis of variance (Anova). Anova is applied to test the effect of vegetation height and density on Manning's coefficient in the model regression. These variables are tested using two ways anova method with the confidence level (α) = 5%. If $F_{calc.} > F_{stat.}$ then reject H_o and accept H_a, it can be concluded that the vegetation height and density influence the Manning's coefficient. On the other hand, If $F_{calc.} < F_{stat.}$ then accept H_o and reject H_a, it can be concluded that the vegetation height and density do not influence Manning's coefficient.

T-test. This test is applied to measure the contribution the vegetation height or density in the model. These variables are tested using two-tailed test with the confidence level (α) = 5%. If $t_{calc.} > t_{stat.}$ then reject H_o and accept H_a, it can be concluded that the vegetation height or density influence the Manning's coefficient. On the other hand, If $t_{calc.} < t_{stat.}$ then accept H_o and reject H_a, it can be concluded that the vegetation height or density do not influence Manning's coefficient.

4. Results and discussions

4.1. The profile of velocity distribution and mean velocity

The velocity on various vegetation height and density has been measured and collected at the downstream, centre and upstream in every depth as mentioned in above method. A part of measurement is depicted figure 3 and 4. Figure 3 is the distribution of velocity profile for various vegetation height on vegetation density of 42 veg/m^2. Meanwhile, figure 4 is the distribution of velocity profile for various vegetation density on vegetation height of 45 cm. As seen in both figure that the velocity decreases with increase in height and density of vegetation. The profile of velocity distribution can draw the mean velocity in the channel. The presence of plants in the channel make water flow to be blocked, and arise backwater flow. This condition makes the increasing flow resistance, so at the block layer due to vegetation will decrease the mean velocity.

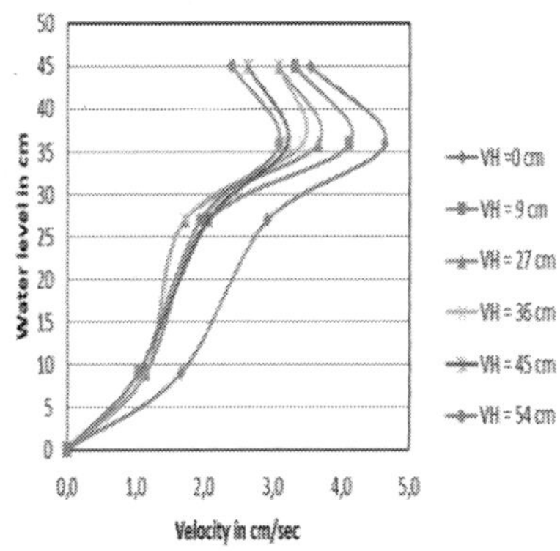

Figure 3. velocity distribution on variation of vegetation height (VH) and density (VD) of 42 veg per sq. m

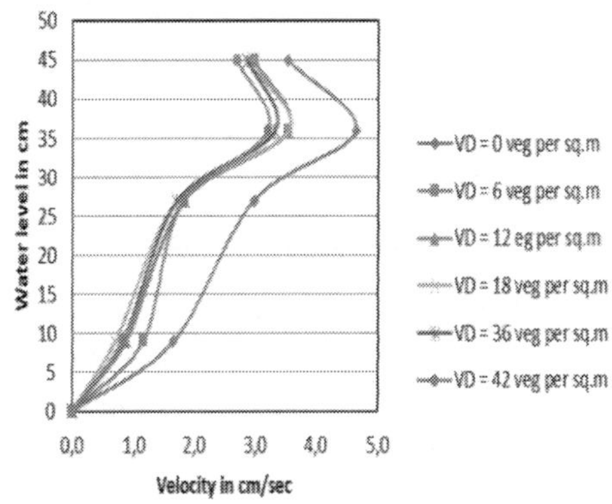

Figure 4. velocity distribution on variation of vegetation density (VD) and height (VH) of 45 cm

The higher density and height of vegetation, the slower mean velocity in the channel, or it can be explained that the mean velocity tend to decrease as increase the density and height of vegetation, as can be seen in table 2. If compared to un-vegetated channel, the velocity is reducing by 11.87-25.42 %. Surely, this condition will affect flow system in the channel.

Table 2. Mean velocity on vegetation density and height.

Vegetation density (VD) in veg/m^2	Vegetation height (VH) in cm					
	0	9	18	36	45	54
0	2,480	2,480	2,480	2,480	2,480	2,480
6	2,480	2,185	2,081	2,047	2,007	1,967
12	2,480	2,152	2,038	2,004	1,964	1,924
18	2,480	2,109	1,995	1,961	1,921	1,881
30	2,480	2,092	1,978	1,944	1,904	1,864
42	2,480	2,078	1,964	1,929	1,889	1,849

4.2. The manning's coefficient due to vegetation density and height

As discussed above that the mean velocity tends to decrease as increase the density and height of vegetation. This clearly is shown in the figure 5. The graph tells us that how the mean velocity keeps decreasing as increasing the density and height of vegetation. Based on the analysis using equation 8, the total manning's coefficient due to vegetation height and density are obtained. The results showed that total manning's coefficient (n) tends to increasing as increasing density and height of vegetation. Maximum total manning's coefficient is 0.040 and minimum is 0.0587. If compared with un-vegetated channel (n_o=0,0381), the manning's coefficient increase 4.92% up to 54.07%.

Table 3. The Manning's coefficients due to vegetation density and height.

Vegetation density (VD) in veg/m²	Vegetation height (VH) in cm					
	0	9	18	36	45	54
0	0,000	0,000	0,000	0,000	0,000	0,000
6	0,000	0,002	0,006	0,008	0,012	0,014
12	0,000	0,004	0,009	0,011	0,015	0,016
18	0,000	0,005	0,010	0,011	0,016	0,018
30	0,000	0,008	0,013	0,014	0,019	0,020
42	0,000	0,008	0,013	0,015	0,020	0,021

In addition, using equation 9 the manning's coefficient due to various vegetation height and density are obtained as tabulated in table 3 and clearly explained as in figure 6. The figure shows that the manning's coefficient due to vegetated channel (n_v) increase as increase vegetation height and density. These values are inversely proportional to the velocity as seen in figure 5. as the greater vegetation height and density, the wider flow area which is block by vegetation making flow resistance increase, so that the velocity becomes decreased while manning's coefficient becomes increased. An additional manning's coefficient due to vegetation in the channel (n_v) should be considered in designing channel is in between (0,002-0,021)

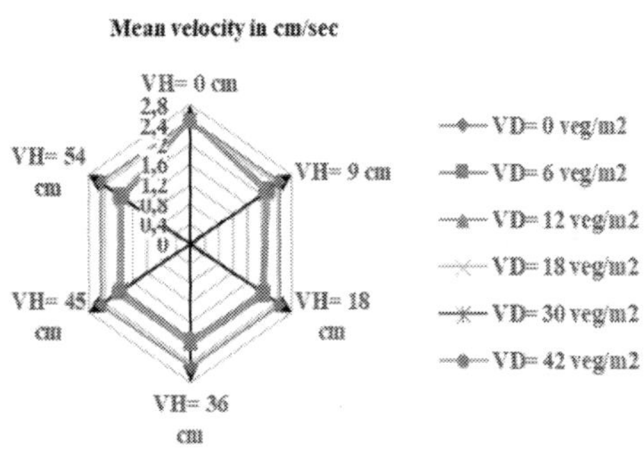

Figure 5. Mean velocity on variation of vegetation height and density

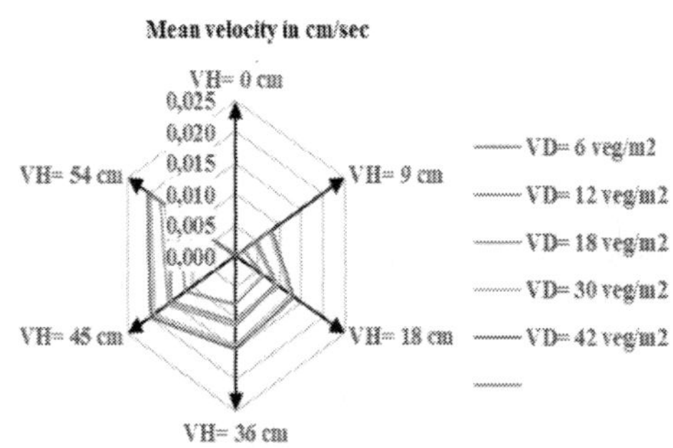

Figure 6. The Manning's coefficient on variation of vegetation height and density.

4.3. The approaching model development

The multiple linier regression is applied to develop the approaching model. The dependent variable is manning's coefficient of vegetation (n_v) and independent variable are vegetation height (V_H) and density (V_D). Based on equation 12, the coefficient of matrix (β_0, β_1 and β_2) is obtained and the using equation 12, the approaching model can be developed as in equation 12.

$$n_V = -0.00107 + 0.01227\ V_H + 0.00019\ V_D \quad (12)$$

Based on statistical analysis, it is found that $F_{calculation}$=367.776 > F_{critic}=19.452, $F_{calc.}$> $F_{stat.}$ then reject H_o and accept H_a, it can be concluded that the vegetation height and density influence the Manning's coefficient. In addition, based on t-test, it is found that $t_{calculation}$=12.298 and 22.464 > t_{critic}=2.571, $t_{calc.}$> $t_{stat.}$ then reject H_o and accept H_a, it can be concluded that both vegetation height and density significantly effect on the manning's coefficient. And lastly, it is obtained that determination coefficient (R^2=0,967), showing that the density and height of vegetation significantly strong correlate to the manning's coefficient of the model.

5. Conclusion

The result of this research can be concluded that the presence of Elephant grass (Pennisetum Purpureum) with various vegetation height and density inside the channel can affect the mean velocity. The greater vegetation density and height result in the lower velocity in the channel. The mean velocity in the channel decrease of 11.87-25.42 % as the effect of vegetation density and height. This effect on increasing the total coefficient of manning (n) ranging from 0.040-0.0587, so that there is an additional vegetated manning's coefficient (n_v) in between (0,002-0,021), compared to the unvegetated channel (n_0 = 0.0381) or an increase 4.92% up to 54.07%. So it can be concluded that vegetation density and height in the open channel can affect the value of Manning coefficient. It is suggested that further research is required to include additional characteristics such as; diameter and leaf width of the vegetation or other types of vegetation so that the flow resistance due to the presence of vegetation can be modeled more general and perfect for the development of hydraulic science in the future.

6. Acknowledgments

The writers would like to thank Dian, Desy, Hari and Hydrotechnic Laboratory technicians of Civil Engineering Department, Syiah Kuala University, Aceh, for carrying out experiments. The writers appreciate the constructive comments and suggestions provided by AIC reviewers and Committee for giving us to attend the Seminar and helping in improving the quality of this article.

7. References

[1] Chow VT 1959 *Open-channel hydraulics* (McGraw-Hill, NY)
[2] Freeman GE, Rahmeyer WH and Copeland RR 2000 *Determination of resistance due to shrubs and woody vegetation* Technical Report, ERDC/CHL TR-00-25.
[3] Hakim J O E and Salama MM1992 J. Irrig. Drain. Eng. **118**
[4] Järvelä J 2002 J. Hydrology **269**
[5] Kim J, Ivanov VY and Katopodes N D 2012 *Water Resources Res.* **48**
[6] Rizalihadi M and Afrianti D 2015 Kehadiran Rumput Gajah (Pennisetum Purpureum) di saluran terhadap tahanan aliran *Prosiding Konferensi Nasional Teknik Sipil 9,* Universitas Hasanuddin, Makassar, in Indonesien
[7] Rizalihadi M and Ihsan M 2016 Pengaruh ketinggian tanaman Pandan Tikar (Acorus Calamus) Terhadap Tahanan Aliran Pada Saluran Terbuka *Prosiding Konferensi Nasional Teknik Sipil 10,* Universitas Atmajaya, Yogyakarta, in Indonesian
[8] Rizalihadi M and Safiana D 2015 *Procedia Engineering* **125**
[9] Rizalihadi M, Ziana, Shaskia N Asharly H 2018 The effect of ratio between rigid vegetation height and water depth on the manning's coefficient in open channel, *IOP Conf. Series* **352**
[10] Tsujimoto T, Kitamura T, Fuji Y and Nakagawa H 1996 *Journal of Hydroscience, Hydraulic Engineering* **14**

IOP Conf. Series: Materials Science and Engineering **523** (2019) 012034 doi:10.1088/1757-899X/523/1/012034

Deformation and crack analysis of tunnel structure subjected to static distributed load using Pseudoshell model

M Hasan[1,*], Husaini[2], N M Abdullah[3]
[1)]Dept. of Civil Engineering, Syiah Kuala University, Banda Aceh, Indonesia,
[2)]Dept. of Mechanical Engineering, Syiah Kuala University, Banda Aceh, Indonesia,
[3)]Engineering Consultant, Banda Aceh, Indonesia,

Email: muttaqin@unsyiah.ac.id

Abstract. The objective of this research is to study the deformation and crack in shotcrete of tunnel structure subjected to distributed static load. Tunnel is modeled as two dimensional finite element model using commersial finite element software, namely ATENA (*Advance Tool for Engineering Non Linear Analysis*) version V5. Non linear material models for concrete and reinforcing bars are adopted. For deformation and crack analysis, pseudoshell model is adopted in which fictitious crack opening is assumed occur first in pseudoshell then propagate to the tunnel structures. Three variations of pseudoshell thickness of 30 mm, 60 mm and 100 mm are used. The development of crack opening and propagation of crack and its displacement due to the increasing of static load is therefore analyzed. The analytical results show that the dummy load and displacement at the crack opening width of 1 mm for the pseudoshell thickness of 30 mm, 60 mm and 100 mm are 48 kN/m and 48,33 mm; 55 kN/m and 42,75 mm; 67 kN/m and 35,5 mm respectively. The relationships between dummy load, displacement and crack opening width are closely linear.

1. Introduction

With the increasing of infrastructures development such as tunnel to support transportation network, it is important to analyze the deformation and crack of structures during their serfice life. Crack will markedly reduce the durability of concrete segment, and finally influence the entire durability of tunnel [1]. Therefore, study on crack form, distribution and origin will be helpfull to strengthen the weak part of segment, and effectively improve the service life of tunnel.

Most research focus on stress and displacement of tunnel structures but not crack problem[2][3]. Some of researcher studied the crack formation of tunnel structures during construction period such as the reasons of cracks in segment in process of manufacture and erection process [4] and the using of hybrid method to research the crack behavior of shotcrete tunnel during 50 days after construction [5]. During service life, tunnel structures are subjected to static load due to soil pressure and traffict load such as movement of trains. There is a lack of research on the deformation and crack formation of tunnel structures during their service life. Therefore, in this research, the deformation and crack formation and propagation of tunnel structures during their service life subjected to static distributed load using pseudoshell model by two dimensional finite element method is studied.

[*]Corresponding author: muttaqin@unsyiah.ac.id

With the understanding of deformation and crack of tunnel strauctures during their serfice life, we can predict how long the serfice life of the structures and when the structures start to have crack. Therefore, the repairing and strengthening of the structures can be performed at the right time to extend the service life of the structures and to prevent the failure of the structures.

2. Literature Review

2.1 Shotcrete tunnel structure and material model

If high flexibility is required during tunnel excavation (e.g. in difficult ground condition and/or in urban areas), the New Australian Tunneling Method (NATM) has proved to be a powerfull mode of construction. In NATM method of tunnelling, shotcrete is used as a mainly material in tunnel. Shotcrete lining is a coating applied to the freshly dug tunnel cross section which has a thin layer and has a flexible shell. In general, shotcrete layer consists of two layers of reinforcement in the form of a grid. As known from reinforced concrete design, the present of reinforcement leads to a distribution of cracks finally resulting in a stabilized crack pattern. Cracking of shotcrete is caused by bending moments in the shell induced by heterogeneous soil and rock conditions. Moreover, shrinkage and thermal gradients (both in space and time) result is tensile loading favoring the development of cracks [5].

Hybrid methods combining the in situ measurements of displacements by means of monitoring system with a constitutive law for shotcrete were developed [6]. They allow the quantification of stress states in shotcrete shells. The employed material model for shotcrete has to account the chemical, thermal and mechanical cross effects [7]. Later, the proposed material model was extended towards considerations of creep and towards modelling of brittle failure [8,9].

The structural model proposed by Lackner and Mang [5] was aiming at real-time applications of the hybrid method on site. It is characterized by several simplifications concerning the deformation and stress state of the shotcrete shell as follows:

1) The change of the longitudinal curvature of a shell section is negligible.
2) A smooth displacement field is assumed. As long as the tunnel shell does not collapse (which would result in a localization of displacement), this is reasonable assumption.
3) Perfect bond is assumed between the shotcrete shell and the surrounding rock.
4) During the deformation, the thickness of the shell is assumed to be constant. Because of the small thickness of the shell, this assumtion is reasonable.
5) Lines which are perpendicular to the interior surface of the tunnel shell right after installation, remain perpendicular to the interior surface during the deformation. Because of the small thickness of the shell, this assumtion is reasonable.
6) Only axial stresses in the circumferential and longitudinal directions are considered. Since shotcrete shell are loaded predominantly by biaxial stress states stemming from circumferential and longitudinal deformations, this assumtion is reasonable.

An axial strain in circumferential direction is determined by approximating the displacement field locally by means of a linear finite beam elements. The axial strain in the longitudinal direction is determined by means of interpollation between the measured displacement components in the longitudinal direction [5].

The thermochemomechanical material model for shotcrete developed by Lackner and Mang [5] is formulated within the framework of thermodynamics of chemically reactive porous media. It is based on a macroscopic description of phenomena on the microlevel of the material by means of state variables. In the material model, two external varibles (strain tensor and absolute temperature) and six internal varibles (degree of hidration, tensor of plastic strains, vector of hardening variables, viscous strains, macroscopic flow strains and viscous slip) are used [7,8].

The ductile behavior of shotcrete subjected to a multi-axial state of compressive stresses is accounted for by hardening Drucker-Prager model. This material model is suitable for the simulation of predominantly biaxial stress state. Microcracking of shotcrete is modeled by means of the maximum tensile stress (Rankine) criterion. According to assumption 6) above, two Rankine criteria are employed to control the tensile stress of shotcrete in the circumferential and longitudinal directions. The use of two hardening forces allows consideration of cracking in the circumferential and longitudinal directions as two independent processes. Hence, in contras to isotropic softening, the strength in one direction remains unchanged when the crack in the other direction is opening.

As stated above, the present of reinforcement leads to a distribution of cracks finally resulting in a stabilized crack pattern. The formation of a stabilized crack pattern is accounted for by the average crack spacing. The interaction between the reinforcement bars and the surrounding shotcrete, the so-called tension-stiffening effect, results in an increase of a stiffness and strength of the composite material. The tension stiffening effect is considered within the material model for shotcrete by increasing the fracture energy by the factor γ, with $\gamma > 1$. Both the crack spacing and the factor γ depend on the material properties of shotcrete and steel, the geometric properties such as the shotcrete cover, the effective tension height for shotcrete and the reinforcement ratio [5].

2.2 Pseudoshell model

In analyzing the structural behaviors and assesing the structural safety of aging tunnel such as railway tunnels, roadway tunnels, and waterway tunnels of hydraulic power facilities, available analytical approaches can be classified primarily into two categories. One is to model the time dependent behavior of the surrounding geological materials with a simplified frame model for tunnel lining [10-12], in which hinges are introduced into the frame to model cracking in the lining concrete. The other is to model the lining with simplified boundary conditions, mostly with spring supports [13]. In the latter approach, the smeared crack method is often used to model the non linear material behavior due to the occurrance of cracks in the lining. As is often observed during tunnel inspections, aging tunnels are often plagued with various kind of cracking problem, where the crack-mouth-opening displacements (CMODs) of the two longitudinal cracks that were located respectively at the spring line and in the arch area rearched several milimeters. In most the cases, due to the lack of data measured on the cross-sectional deformation, these crack patterns and CMODs serve as important indices for evaluating the safety of these underground structures [14].

Shi [14] proposed a pseudoshell model which is based on the beam theory that ensured the uniqueness of the solution on deformation when the beam subjected to the same ratio of load to flexural regidity for analyzing the crack on tunnel lining. By stiffening the lining in the pseudoshel model to increase its flexural rigidity, theoretically it is still possible to obtain the true lining deformation under certain conditions. To achieve this, crack analysis is indispensable. By applying relevant loads to the pseudoshell with the aim of reproducing the observed cracking behavior, craking analysis is carried out based on the extended fictitious crack model (EFCM). When the tip a crack reaches the peseudoshell, a plastic hinge is introduced into the shell, allowing a rigid body rotation to take place at the crack surfaces while the crack analysis is continued using a crack-opening displacements (COD) controlled algorithm. Hence, the cross-sectional deformation of the tunnel can be obtained at any specified CMODs. Next, a quasi loosening zone model is used to calculate the ground pressure. Based on the assumption that the ground deformation must equal to the cross-sectional deformation of the tunnel, the external load can be obtained by ajusting the extent of the loosening zone through iterative computation. To fasilitate crack analysis in pseudoshell model, the EFCM employs two modeling techniques. Before a crack penetrates the lining, crack analysis is carried out by the crack-tip-controlled modeling method. As soon as the tip of a crack reaches the pseudoshell, a COD-controlled algorithm is required for the crack analysis.

3. Numerical Method

3.1 Numerical model

The numerical model used in this study is Sieberg tunnel, Austria. The tunnel has 10.42 m wide and 8 m high as shown in Figure 1. The tunnel constructed by shotcrete having the thickness of 330 mm. The shotcrete is reinforced by steel wiremesh. The diameter of wiremesh is 6 mm and spaced at 100 mm. The thickness of shotcrete cover is 30 mm. The compressive strength, Young modulus and Poisson'ratio of concrete are 39.6 MPa, 40.8 GPa and 0.2, respectively. The tensile strength, Young modulus and Poisson' ratio of wiremesh are 440 MPa, 200 GPa and 0.3, respectively.

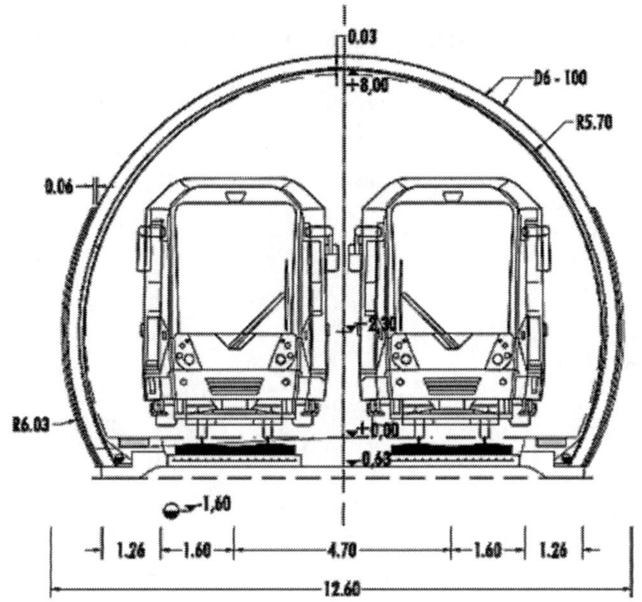

Figure1. Cross section of Sieberg tunnel, Austria

The tunnel is modeled as two dimensional (2D) finite elemen (FE) model using commersial finite element software, namely ATENA (Advance Tool foe Engineering Non Linear Analysis) version V5 [15]. A 5.8 m long distributed load is applied on the left side of the tunnel using pseudoshel model as shown in Figure 2. Three variations of pseudoshell thickness of 30 mm, 60 mm and 100 mm are analyzed. The steel material for pseudoshel is adopted.

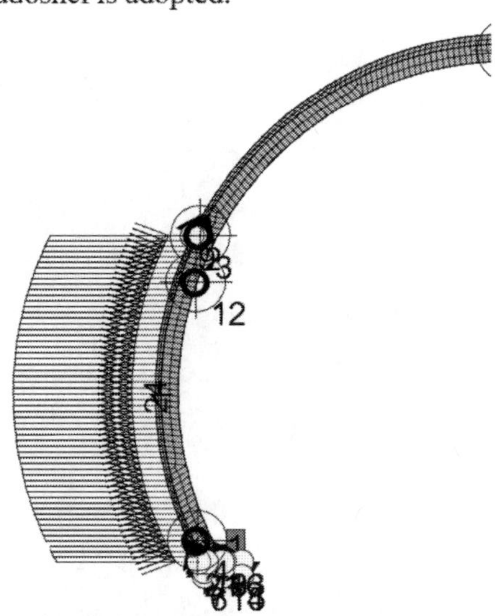

Figure 2. 2D FE model of tunnel with 5.8 m long distributed load on the left side (only half part of FE model is shown in the figure)

3.2 Constitutive models

Concrete is modeled as a plane stress element as shown in Figure 3. Stress-strain models shown in Figures 4 and 5 are adopted for concrete and steel materials, respectively [15,16,17]. For pseudoshell, an elastic isotropic plane stress element is used. A smeared crack with fix crack model (Figure 6) is adopted in the analysis. For cracking modeling, the Rankine fracture criterion is adopted. The criterion is as follows:

$$F_i^f = \sigma_{ii}^t - f_{ii}^t \leq 0 \tag{3.1}$$

where: F_i^f = Rankine fracture criterion, σ_{ii}^t = elastic parameter, and f_{ii}^t = concrete tensile strength.

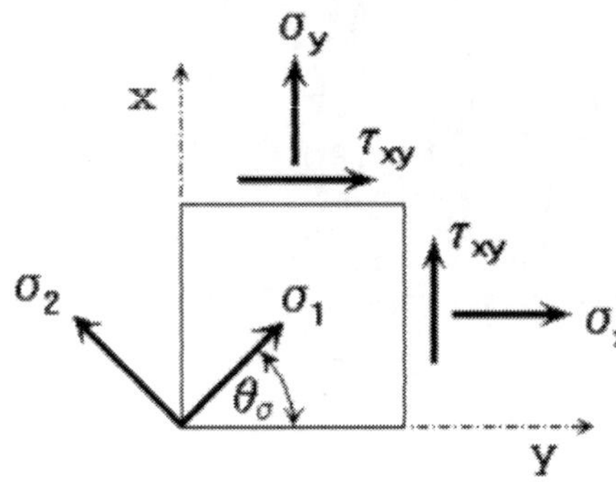

Figure 3. Plane stess element for concrete

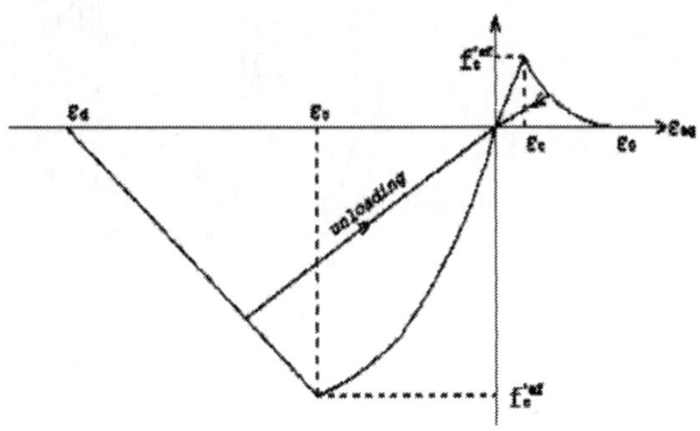

Figure 4. Stress-strain model for concrete

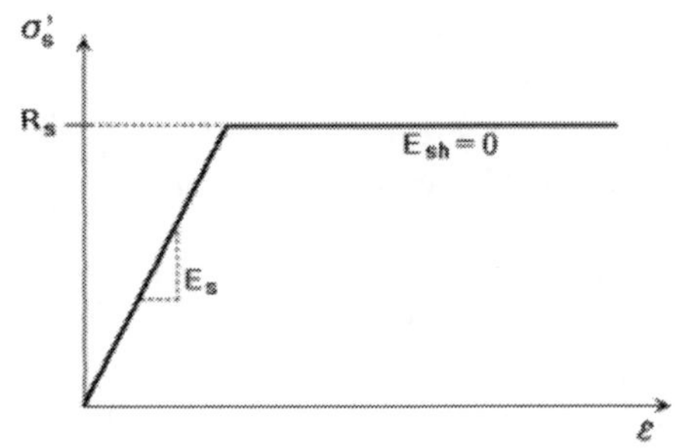

Figure 5. Stress-strain model for steel

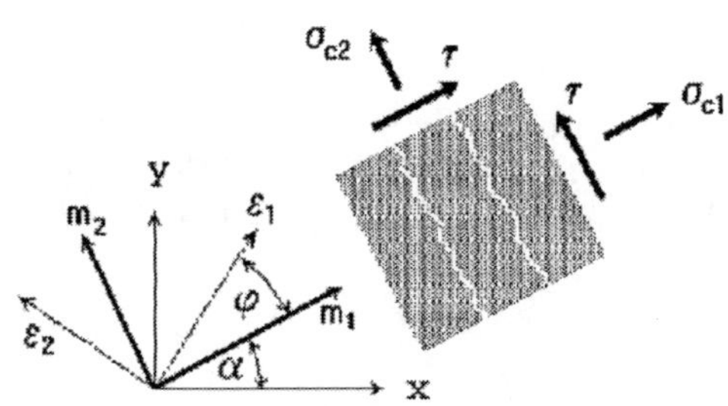

Figure 6. Fix crack model

3.3 Analytical procedure

The analysis is conducted by loading controll using ATENA software [15]. The loading increment for each step is 0.1 kN/m. The modified Newton-Raphson method is used as the confergence algorithm. The fictitious crack opening is assumed occur first in pseudoshell then propagate to the tunnel structures. The maximum crack mouth opening displacement (CMOD) of 1 mm is set. The displacements and crack are monitored at 100 monitoring points.

4. Numerical Results

From 100 monitoring points in this study, the maximum displacements for all variations of pseudoshell thickness occur at the monitoring point 9 as shown in Figure 7. Figure 8 shows the relationship between displacement and CMOD of tunnel structure at monitoring point 9. From this figure, it can be seen that the relationship is almost linear. The thicker the pseudoshell thikcness is the lower the displacement is. The displacements at the crack opening width of 1 mm for all variations of pseudoshell thickness are shown in Table 1.

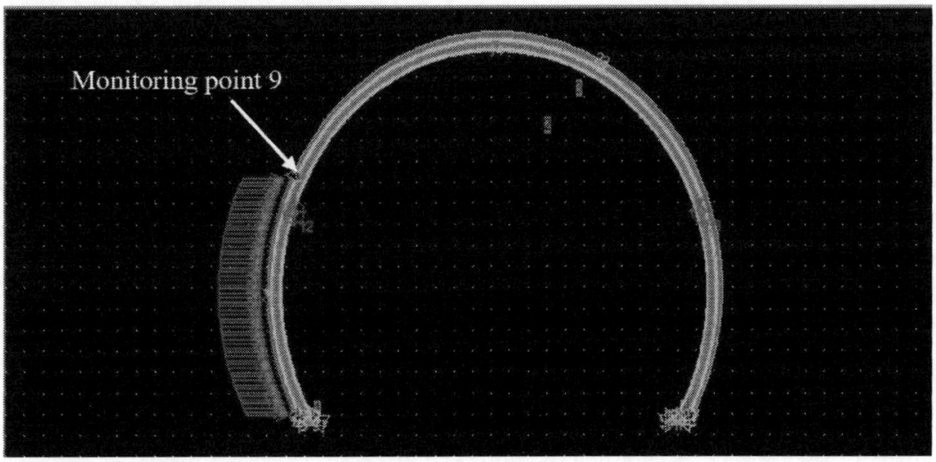

Figure 7. Monitoring point 9

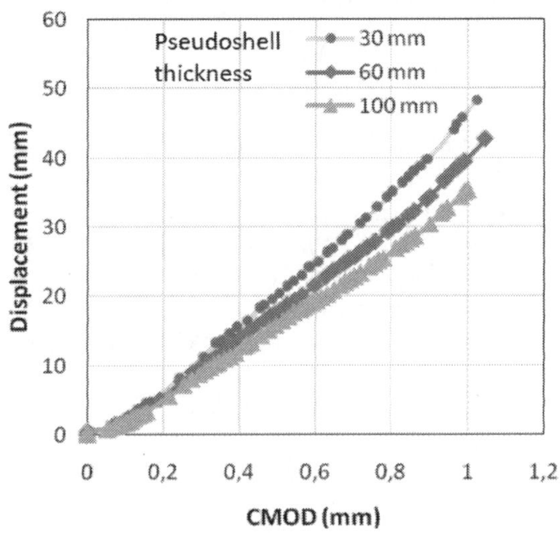

Figure 8. Displacement and CMOD relationship at monitoring point 9

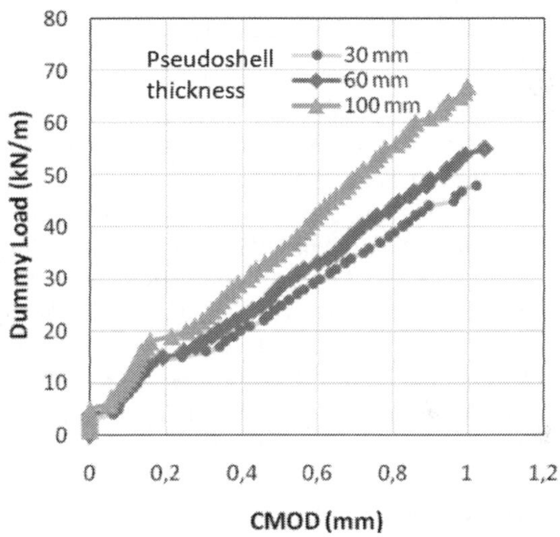

Figure 9. Dummy load and CMOD relationship at monitoring point 9

Figure 9 shows the relationship between dummy load and CMOD of tunnel structure at monitoring point 9. From this figure, it can be seen that the relationship is almost linear. The thicker the pseudoshell thikcness is the higher the load is. The displacement at the crack opening width of 1 mm for all variation of pseudoshell thickness is shown in Table 1.

Table 1. The displacements and dummy loads at the crack opening width of 1 mm

Pseudoshell thickness (mm)	Displacements at crack opening width of 1 mm (mm)	Dummy loads at crack opening width of 1 mm (kN/m)
30	48,33	48
60	42,75	55
100	35,5	67

Figures 10 and 11 are the displacement and dummy load at the crack width of 1 mm as a function of pseudoshell thickness. While the maximum displacement decreases linearly with increasing of pseudoshell thickness, the maximum dumly load increases.

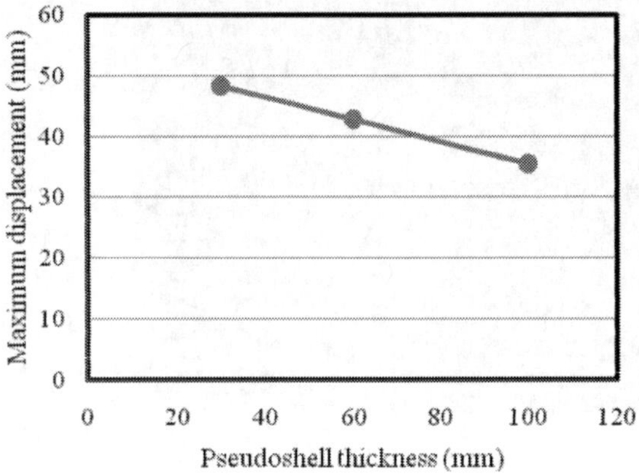

Figure 10. Displacement at the crack width of 1 mm and pseudoshell thickness relationship

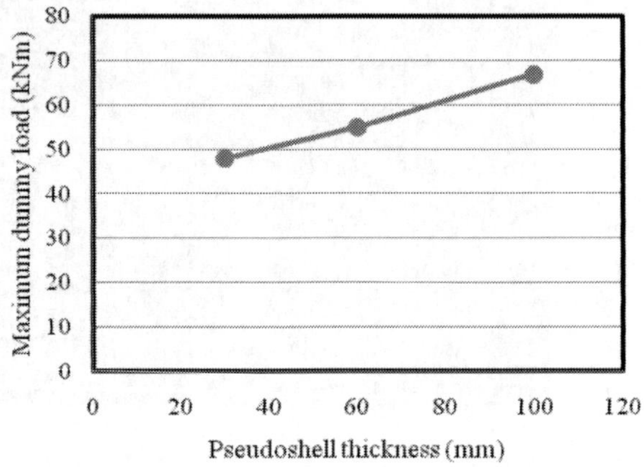

Figure 11. Dummy load at the crack width of 1 mm and pseudoshell thickness relationship

5. Conclusions

Analysis of deformation and crack of tunnel structure subjected to distributed static load was carried out. The following conclusions are drawn from this research.

1. The maximum displacements of the tunnel at the crack opening width of 1 mm for the pseudoshell thickness of 30 mm, 60 mm and 100 mm are 48,33 mm, 42,75 mm and 35,5 mm, respectively. The maximum displacement decreases with increasing of pseudoshell thickness.
2. The dummy loads of the tunnel at the crack opening width of 1 mm for the pseudoshell thickness of 30 mm, 60 mm and 100 mm are 48 kN/m, 55 kN/m and 67 kN/m, respectively. The dummy load increases with increasing of pseudoshell thickness.
3. The relationships between dummy load, displacement and crack mouth opening displacement are closely linear.

6. Acknowledgments

This work was supported by grant provided by Syiah Kuala University Research and Community Service Center (LPPM) under contract number of 48/UN11.2/PP/PNBP/SP3/2018.

7. References

[1] Chen J S and Mo H H 2009 *Tunneling and Underground Space Tech.* **24** 91-102
[2] Bloom, C B, van der Horst M E J and Jovanovic P S 1999 *Tunneling and Underground Space Tech.* **14** 217-224
[3] Lee K M, Hou X Y and Ge X W 2001 *Int. J. Num. and Anal. Meth. Geomech.* **25**: 365-390
[4] Zhu W B and Ju S J 2003 *Mod. Tunn. Tech* **40** 21-25
[5] Lackner R and Mang H A 2003 *Eng. Fract. Mech.* **70** 1047-1068
[6] Hellmich C H, Mang H A and Ulm F J 2001 *Com. and Struct.* **79** 2103-2115.
[7] Hellmich C H, Ulm F J and Mang H A 1999 *ASCE J. Eng. Mech.* **125** 692-701
[8] Sercombe J, Hellmich C H, Ulm F J and Mang H A 2000 *ASCE J. Eng. Mech.* **126** 284-291
[9] Lackner R, Hellmich C H and Mang H A 2002 *Int. J. Num. Meth. Eng.* **53** 2358-2388
[10] Kitagawa T and Inagaki D 1993 *J. Geotech. Eng.* **463** 105-114
[11] Jiang Y, Esaki T and Yokota Y 1993 *Proc. Tunn. Eng.* **3** 17-24

[12] Tasaka Y, Uno H, Ohmori T and Kudoh K 2000 *J. Geotech. Eng.* **652** 73-90
[13] Yin J, Wu Z, Asakusa T and Ota H 2001 *J. Struct. Mech./Earthquake Eng.* **668** 17-27
[14] Shi Z 2009 *Crack Analysis in Structural Concrete – Theory and Application.* (Butterworth-Heinemann, Burlington), USA
[15] Cervenka, V 2013 *ATENA Program Documentation.* (Cervenka Consulting), Prague
[16] Husaini and Zuhaimi 2016 *International Journal of Technology* **7** 456-462
[17] Hasan M, Husin S and Nursaniah C 2016 *Key Engineering Materials* **711** 302-309

IOP Conf. Series: Materials Science and Engineering **523** (2019) 012035 doi:10.1088/1757-899X/523/1/012035

Changes in the unit price of work for reinforced concrete construction based on building sites

Mubarak[1,*], Abdullah[1], M Riza[2], Y Hayati[1]
[1]Dept. of Civil Engineering, University of Syiah Kuala, Jl. Syekh Abdurrauf 10, Banda Aceh, Indonesia
[2]Faculty of Engineering, University of Syiah Kuala, Jl. Syekh Abdurrauf 10, Banda Aceh, Indonesia

E-mail: mubarak@unsyiah.ac.id

Abstract. The unit price of work is the price required to complete each unit of quantity of work that consists of the unit price for the wage of labour, materials, and equipment used. Changes in unit prices are possibly determined by the different sites in which a building is constructed. For budget planners or construction executors, a good understanding of the changes in the unit price of work is required to identify the reasonable amount of budget needed. Given these circumstances, this study was conducted to analyse and identify the pattern of changes in the unit price of work based on the locations of a building in the Province of Aceh. This research focuses on the unit price of reinforced concrete work for the building structure components, as the work that has the most significant proportion of cost in building construction. The pattern of change is described in the form of price indexes by applying the price in Banda Aceh City as a reference for other cities/districts. The analysis results in two patterns of the unit price index, which is below or above of the reference index. This condition follows the pattern of the unit price of materials and wages. The ease of access to a region also determine the changes in the price index.

1. Introduction

The accuracy of each estimate is determined mainly by how much project information has been obtained at each stage of the project. Some previous studies suggest that estimation accuracy may be influenced by the availability of adequate information relating to design documents, historical data on previous construction project prices, accuracy and reliability of cost information [1]. A good understanding of the scope of works and the availability of sufficient project information will provide more accurate estimation results [2] [3].

The change and accuracy of estimation values on a particular type of building can also be influenced by the project location factor [4]. Location factors are needed to adjust the whole element of project costs from a defined scope of construction work based on one geographical location to another geographic location. This factor accommodates differences in productivity and labour costs, equipment, materials, distribution, duties, taxes, procurement processes, technical/design aspects, and project administration [5]. Estimating the price of a building needs to consider the geographically differentiated location conditions by setting a factor adjustment to make the estimation value generated more rationally [6]. Location factor considerations in the estimation process are often associated with the exchange rate

* Corresponding author: mubarak@unsyiah.ac.id

and time factors due to the difference in the cost of building a construction project when implemented at different times and locations [7]. Geographically, construction costs may vary due to distance variations of project sites and suppliers, climatic and weather conditions, and market conditions. Construction built in the city centre can have variations in price when compared to constructions built on the remote area. In general, the further the location of a construction project, the more expensive it will cost to transport materials, labour, equipment to the job site.

Cost estimation at the conceptual stage are prepared to provide preliminary information on budget requirements for a type of construction and at the same time act as a justificatory to declare the project can be continued or not continued. This requirement encourages the process of developing some estimation methods involving location factors to be more accurate, flexible, manageable, and rapidly updated. A number of studies have developed various estimation methods using locations as variables, such as in studies related to location cost adjustment factors [6-9], development of the construction cost index [10, 11] or related to the development of cost models [12, 13].

The process of estimating the price of a building can use a variety of methods according to the stages of the project cycle or adjust to the needs of an estimate. One of the commonly used estimation methods is to use estimates based on the unit price (UP) of work. The unit price of work is the unit of price used to estimate the value of a work based on the amount of material, labour wage, and equipment used. In most construction works in Indonesia; unit prices have been established through SNI standards or other guidelines issued by the relevant technical institutions. For example, Indonesian National Standard SNI 7394: 2008 on the procedures for calculating the unit price of concrete work for the construction of buildings and housing [14] and the Ministry of Public Work Regulation No. 26/PRT/M/2016 on the unit price analysis of work for public services work. [15].

The Province of Aceh consists of several cities and districts with varying geographic and distance conditions from the capital of Banda Aceh. This difference in itself affects the difference in material prices, labour wages, and equipment costs. Furthermore, the unit price of work is also affected due to the difference in prices. This study was conducted to analyse and identify changes that occur in the unit price of construction works by using a price index based on the location aspect of construction. This research is focused on the structural components of reinforced concrete in building construction. The selection of structural components of a building caused by this component has the highest proportion of cost compared to other building components.

2. Methods

2.1. Object and data collection

This study uses the object price of the work unit for the structural components of a building. The component of the structure is a reinforced concrete construction with work components consisting of concrete, reinforcement, and formwork. The study area consists of 23 cities and districts within Aceh Province.

The data used in this study is secondary data in the form of basic price for the material, labour, and equipment for the year 2018 issued by the Governor of Aceh [16]. Requirements for each work related to the object of research using SNI standard [14] and Minister of Public Works Regulation [15].

2.2. Analysis

This research begins by identifying the type of material, labour, and equipment needed for each work. The identification process is done by referring to the requirement of each unit price of work. Data on material, labour and equipment prices that have been collected in advance are grouped for 23 regions and three types of reinforced concrete work.

Analysis of unit price of work (UP) for each type of work (i) and region (j) is done by summing the cost for material (M), labour (L), and overhead/profit (OP) as shown in equation 2.1.

$$UP_{ij} = \sum(M + L + OP) \quad (1)$$

The UP used in this study can be distinguished by the work components of reinforced concrete and the structural components of the building. The UP for the components of work consisted of 4 UPs for concrete work, 2 UPs for reinforcement work, and 5 UPs for formwork. The UP for building structural components consisted of the construction of the foundation, tie beams, columns, beam, and slab. The list of UP standards used is shown in Table 1.

Table 1. List of standard unit price for reinforced concrete constructions

UP for reinforced concrete work components [15]			UP for building structures components [14]
Concrete (per m3)	Reinforcement (per 10 kg)	Formwork (per m2)	
Anl. No. A.4.1.1.7 (f'c = 19.3 MPa)	Anl. No. A.4.1.1.17a (plain bar)	Anl. No. A.4.1.1.20 (for foundation)	Anl. No. 6.28 (for 1 m3 foundation)
Anl. No. A.4.1.1.8 (f'c = 21.7 MPa)	Anl. No. A.4.1.1.17b (deformed bar)	Anl. No. A.4.1.1.21 (for tie beam)	Anl. No. 6.29 (for 1 m3 tie beam)
Anl. No. A.4.1.1.9 (f'c = 24.0 MPa)		Anl. No. A.4.1.1.22 (for column)	Anl. No. 6.30 (for 1 m3 column)
Anl. No. A.4.1.1.10 (f'c = 26.4 MPa)		Anl. No. A.4.1.1.23 (for beam)	Anl. No. 6.31 (for 1 m3 beam)
		Anl. No. A.4.1.1.24 (for slab)	Anl. No. 6.32 (for 1 m3 slab)

The unit price index is analysed by setting a price for Banda Aceh City as the reference price for other regions with a value of 1.000. Changes in the index occurring in a region (UPI_i) resulted from comparing the price difference in the observed region (C_i) and prices in reference city of Banda Aceh (C_{BA}). The price index is calculated by equation 2.2.

$$Unit\ Price\ Index\ (UPI_i) = \frac{C_i}{C_{BA}} \quad (2)$$

3. Results and discussions

3.1. The unit price of work

The reinforced concrete work unit prices are grouped based on the work components and structural components of a building. The price component of each reinforced concrete work consists of material prices, labour wages, and overhead/profit. The standard used the assumption of work is accomplished by using no mechanical equipment, or only using simple working tools. The results of UP analysis are shown in Table 2.

The unit price for concrete works appearances consistent with the highest prices appearing in Simeulue District, a district located on an island separated from the mainland of Sumatra island. The lowest price occurred in Aceh Tamiang District, a district located on the border with North Sumatra Province. The extent of unit price disparity for both areas reached 40%. For reinforcement and formwork, the highest and lowest work unit prices tend to vary across districts. The highest price disparity that occurs is 44.7% for reinforcement work and 39.7% for formwork work.

Table 2. The unit price for the reinforced concrete work components

City/ District	Unit Price (in thousand rupiahs)										
	Concrete (per m^3)				Reinforcement (per 10 kg)		Formwork (per m^2)				
	A.4.1.1.7	A.4.1.1.8	A.4.1.1.9	A.4.1.1.10	A.4.1.1.17a	A.4.1.1.17b	A.4.1.1.20	A.4.1.1.21	A.4.1.1.22	A.4.1.1.23	A.4.1.1.24
Banda Aceh	1,144	1,165	1,201	1,212	162	191	299	325	540	560	622
Sabang	1,171	1,193	1,232	1,244	164	219	315	342	529	548	600
Aceh Besar	1,144	1,165	1,201	1,212	162	212	283	307	492	510	557
Pidie	1,124	1,146	1,183	1,195	169	205	285	309	483	500	**537**
Pidie Jaya	1,108	1,130	1,168	1,179	177	206	292	317	504	523	575
Bireun	1,102	1,124	1,162	1,173	184	221	314	341	565	585	663
Bener Meriah	1,114	1,136	1,172	1,184	188	235	303	329	526	546	608
Aceh Tengah	1,152	1,175	1,214	1,226	**216**	**264**	314	341	554	574	664
Gayo Lues	1,159	1,184	1,228	1,241	194	237	333	362	**593**	616	**703**
Aceh Tenggara	1,157	1,179	1,218	1,230	164	222	341	372	566	587	632
Lhokseumawe	1,084	1,106	1,143	1,154	156	185	**342**	**373**	593	616	681
Aceh Utara	1,112	1,133	1,170	1,181	157	193	341	372	577	600	654
Aceh Timur	1,132	1,154	1,191	1,202	150	**185**	341	372	592	**617**	674
Langsa	1,201	1,222	1,259	1,270	156	199	339	370	590	615	672
Aceh Tamiang	**1,006**	**1,027**	**1,064**	**1,076**	**149**	192	312	339	531	552	597
Aceh Jaya	1,105	1,127	1,164	1,176	164	200	286	310	508	527	590
Aceh Barat	1,038	1,060	1,097	1,109	163	199	323	352	551	572	633
Nagan Raya	1,090	1,111	1,148	1,159	163	206	312	340	542	563	625
Aceh Barat Daya	1,090	1,111	1,148	1,159	165	199	285	309	509	528	592
Aceh Selatan	1,134	1,156	1,194	1,206	166	199	257	277	463	480	545
Subulussalam	1,163	1,186	1,227	1,239	174	222	**248**	**267**	**458**	**473**	568
Aceh Singkil	1,252	1,277	1,319	1,332	184	222	278	300	508	526	618
Simeulue	**1,408**	**1,437**	**1,486**	**1,502**	200	251	264	285	491	508	601
Min	**1,006**	**1,027**	**1,064**	**1,076**	**149**	**185**	**248**	**267**	**458**	**473**	**537**
Max	**1,408**	**1,437**	**1,486**	**1,502**	**216**	**264**	**342**	**373**	**593**	**617**	**703**

The unit price standard of building structural components combines the use of concrete, reinforcement, and formwork in one analysis at a time. The analysis is expressed in a cubic metre (m^3) unit of concrete for each structural component with the specified number of reinforcement uses. The use of reinforcement for structural components is 150 kg for the foundation, 200 kg for the tie beam, 300 kg for the column, 200 kg for the beam, and 150 kg for the slab. The analysis results are shown in Table 3. The highest work unit prices on the foundation, tie beam, and column components were seen in Aceh Tengah District, while for beam and slab components were seen in Gayo Lues District. Both districts are geographically located in the central region of Aceh Province and reside in areas with topographic forms dominated by hills or mountains. The lowest unit price for foundation components appeared in Aceh Tamiang District and for four other structural components appeared in South Aceh District.

Table 3. The unit price for the reinforced concrete structure of the building

City/ District	Unit Price (in thousand rupiahs)				
	6.28 (foundation)	6.29 (tie beam)	6.30 (column)	6.31 (beam)	6.32 (slab)
Banda Aceh	5,200	5,698	7,271	7,930	7,916
Sabang	5,439	5,969	7,531	7,987	7,928
Aceh Besar	5,126	5,596	7,013	7,535	7,470
Pidie	5,267	5,738	7,099	7,580	7,477
Pidie Jaya	5,419	5,900	7,370	7,873	7,823
Bireun	5,636	6,166	7,898	8,495	8,535
Bener Meriah	5,630	6,135	7,708	8,220	8,203
Aceh Tengah	**6,152**	**6,681**	**8,441**	8,913	9,004
Gayo Lues	5,955	6,523	8,385	**8,957**	**9,017**
Aceh Tenggara	5,520	6,099	7,739	8,247	8,153
Lhokseumawe	5,343	5,920	7,690	8,298	8,280
Aceh Utara	5,344	5,922	7,629	8,178	8,117
Aceh Timur	5,262	6,135	7,708	8,220	8,203
Langsa	5,388	5,964	7,706	8,355	8,306
Aceh Tamiang	**5,043**	5,570	7,116	7,651	7,565
Aceh Jaya	5,179	5,661	7,193	7,710	7,696
Aceh Barat	5,295	5,840	7,493	7,993	7,971
Nagan Raya	5,278	5,803	7,423	7,962	7,939
Aceh Barat Daya	5,174	5,651	7,177	7,717	7,709
Aceh Selatan	5,082	**5,511**	**6,922**	**7,384**	**7,388**
Subulussalam	5,212	5,627	7,144	7,506	7,629
Aceh Singkil	5,579	6,046	7,690	8,143	8,247
Simeulue	5,906	6,348	7,942	8,389	8,492
Min	**5,043**	**5,511**	**6,922**	**7,384**	**7,388**
Max	**6,152**	**6,681**	**8,441**	**8,957**	**9,017**

3.2. The unit price pattern for reinforced concrete work components

The work unit price index is obtained based on the amount of the work unit price in each district/city reviewed in the previous section. The index is the ratio of prices in a city/district to prices in the area of reference. The value of 1.0000 is given as a price index in the city of Banda Aceh which is a reference for the price index in other districts/cities. The index values obtained can generally be categorised as 'upper reference' index ($UPI > 1.000$) and 'below reference' index ($UPI < 1.000$). The result of the index analysis of reinforced concrete work unit price based on job component is shown in Table 4, while the average index pattern provided in Figure 1.

For concrete work component, the unit price index in all cities/districts is positioned in the two categories. However, condition with index values is the same as reference areas appear for concrete work in Aceh Besar District. It occurs due to material prices and labour wages for these two regions are relatively the same. The highest index value for concrete work occurred in Simeulue District and the lowest in Aceh Tamiang District. Of all the areas analysed, eight cities/districts are included in the category of the area with the upper reference index, and 13 cities/districts are included in the category of the area with the below reference index. In general, cities/districts with the below reference index category are geographically located on the east/western coast of Aceh province, while cities/districts with upper reference index categories are located in the middle region or separate islands.

Table 4. The unit price index for the reinforced concrete work components

City/ District	Unit Price Indices										
	Concrete (per m^3)				Reinforcement (per 10 kg)		Formwork (per m^2)				
	A.4.1.1.7	A.4.1.1.8	A.4.1.1.9	A.4.1.1.10	A.4.1.1.17a	A.4.1.1.17b	A.4.1.1.20	A.4.1.1.21	A.4.1.1.22	A.4.1.1.23	A.4.1.1.24
Banda Aceh	**1.0000**	**1.0000**	**1.0000**	**1.0000**	**1.0000**	**1.0000**	**1.0000**	**1.0000**	**1.0000**	**1.0000**	**1.0000**
Sabang	1.0231	1.0240	1.0254	1.0258	1.0150	1.1458	1.0536	1.0535	0.9804	0.9774	0.9654
Aceh Besar	1.0000	1.0000	1.0000	1.0000	1.0016	1.1119	0.9480	0.9459	0.9110	0.9106	0.8964
Pidie	0.9827	0.9837	0.9852	0.9856	1.0469	1.0749	0.9545	0.9518	0.8945	0.8925	0.8644
Pidie Jaya	0.9686	0.9700	0.9721	0.9728	1.0937	1.0769	0.9770	0.9746	0.9341	0.9329	0.9252
Bireun	0.9634	0.9648	0.9671	0.9678	1.1361	1.1599	1.0495	1.0498	1.0463	1.0446	1.0666
Bener Meriah	0.9740	0.9747	0.9758	0.9762	1.1647	1.2312	1.0132	1.0121	0.9751	0.9746	0.9784
Aceh Tengah	1.0073	1.0085	1.0105	1.0111	1.3336	1.3850	1.0513	1.0514	1.0275	1.0243	1.0679
Gayo Lues	1.0131	1.0166	1.0221	1.0238	1.2007	1.2428	1.1146	1.1160	1.0992	1.0992	1.1301
Aceh Tenggara	1.0111	1.0123	1.0141	1.0147	1.0154	1.1612	1.1429	1.1462	1.0490	1.0472	1.0161
Lhokseumawe	0.9478	0.9492	0.9514	0.9521	0.9679	0.9704	1.1442	1.1474	1.0991	1.0992	1.0948
Aceh Utara	0.9717	0.9725	0.9737	0.9741	0.9694	1.0093	1.1412	1.1447	1.0691	1.0703	1.0512
Aceh Timur	0.9893	0.9901	0.9913	0.9917	0.9250	0.9717	1.1412	1.1447	1.0977	1.1014	1.0848
Langsa	1.0499	1.0490	1.0477	1.0473	0.9673	1.0452	1.1349	1.1389	1.0932	1.0971	1.0809
Aceh Tamiang	0.8791	0.8818	0.8859	0.8873	0.9216	1.0065	1.0433	1.0441	0.9848	0.9854	0.9604
Aceh Jaya	0.9662	0.9674	0.9692	0.9698	1.0149	1.0478	0.9568	0.9539	0.9420	0.9405	0.9497
Aceh Barat	0.9071	0.9096	0.9134	0.9146	1.0101	1.0438	1.0821	1.0839	1.0211	1.0203	1.0183
Nagan Raya	0.9523	0.9536	0.9556	0.9562	1.0101	1.0814	1.0454	1.0460	1.0053	1.0051	1.0046
Aceh Barat Daya	0.9523	0.9536	0.9556	0.9562	1.0238	1.0438	0.9538	0.9512	0.9441	0.9425	0.9515
Aceh Selatan	0.9911	0.9923	0.9942	0.9948	1.0251	1.0448	0.8598	0.8541	0.8586	0.8565	0.8773
Subulussalam	1.0161	1.0181	1.0212	1.0222	1.0765	1.1608	0.8289	0.8216	0.8494	0.8439	0.9133
Aceh Singkil	1.0946	1.0960	1.0981	1.0988	1.1412	1.1608	0.9293	0.9244	0.9423	0.9385	0.9939
Simeulue	1.2309	1.2335	1.2375	1.2388	1.2403	1.3140	0.8837	0.8761	0.9108	0.9068	0.9666
The lowest index	**0.8791**	**0.8818**	**0.8859**	**0.8873**	**0.9216**	**0.9704**	**0.8289**	**0.8216**	**0.8494**	**0.8439**	**0.8644**
The highest index	**1.2309**	**1.2335**	**1.2375**	**1.2388**	**1.3336**	**1.3850**	**1.1442**	**1.1474**	**1.0992**	**1.1014**	**1.1301**

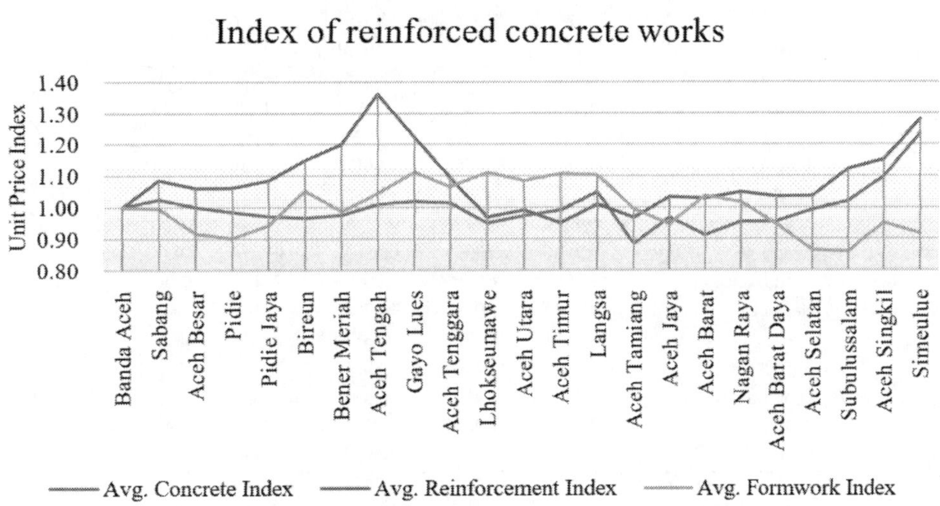

Figure 1. Reinforce concrete work index pattern

IOP Conf. Series: Materials Science and Engineering **523** (2019) 012035 doi:10.1088/1757-899X/523/1/012035

For reinforcement works, all the unit price indices are placed into the two categories. The highest index occurred in the district of Aceh Tengah, while the lowest index occurred in the district of Aceh Tamiang and Aceh Timur. In reinforcement work, generally, almost all cities/regencies are in the upper reference index category (74% for A.4.1.1.17a and 83% for A.4.1.1.17b), while the rest are in the category below reference index. Steel material for reinforcement is generally supplied from Medan City in North Sumatra Province. Cities or districts under below reference category is the closest regions to the supply centre of the material.

Related to formworks, the highest index of 5 types of formwork occurred in several areas, namely in Lhokseumawe City, Aceh Timur District, and Gayo Lues District, while the lowest index appeared in Subulussalam City and Pidie District. There are 9 to 11 cities/districts that are categorised below reference index, while the upper reference index categorises the rest. The characteristics of a location associated with the distance to the supplies sources can directly affect the price changes.

3.3. Unit price pattern for building structural component

The unit price index for building structural components demonstrate a particular pattern of changes in each region. The pattern of changes occurring at the structural component price index can be explained as a combined superposition of three unit prices consisting of the price of concrete, reinforcement, and formwork. The index changes in the cities/districts studied are shown in Table 5, and the pattern of index changes is illustrated in Figure 2.

Table 5. The unit price index for the building structural components

City/ District	Unit Price Indices				
	6.28 (foundation)	6.29 (tie beam)	6.30 (column)	6.31 (beam)	6.32 (slab)
Banda Aceh	**1.0000**	**1.0000**	**1.0000**	**1.0000**	**1.0000**
Sabang	1.0460	1.0475	1.0358	1.0072	1.0015
Aceh Besar	0.9858	0.9821	0.9645	0.9502	0.9437
Pidie	1.0129	1.0071	0.9764	0.9559	0.9445
Pidie Jaya	1.0420	1.0355	1.0137	0.9928	0.9882
Bireun	1.0839	1.0822	1.0864	1.0712	1.0782
Bener Meriah	1.0826	1.0768	1.0602	1.0366	1.0362
Aceh Tengah	1.1832	1.1726	1.1610	1.1239	1.1375
Gayo Lues	1.1451	1.1448	1.1533	1.1295	1.1391
Aceh Tenggara	1.0616	1.0704	1.0644	1.0400	1.0299
Lhokseumawe	1.0275	1.0391	1.0577	1.0464	1.0460
Aceh Utara	1.0277	1.0394	1.0494	1.0313	1.0254
Aceh Timur	1.0120	1.0768	1.0602	1.0366	1.0362
Langsa	1.0362	1.0467	1.0599	1.0536	1.0493
Aceh Tamiang	0.9699	0.9776	0.9788	0.9648	0.9556
Aceh Jaya	0.9961	0.9935	0.9893	0.9723	0.9723
Aceh Barat	1.0183	1.0249	1.0306	1.0080	1.0069
Nagan Raya	1.0149	1.0184	1.0209	1.0040	1.0030
Aceh Barat Daya	0.9950	0.9918	0.9871	0.9732	0.9738
Aceh Selatan	0.9774	0.9672	0.9521	0.9311	0.9333
Subulussalam	1.0024	0.9875	0.9826	0.9466	0.9638
Aceh Singkil	1.0729	1.0611	1.0577	1.0269	1.0419
Simeulue	1.1358	1.1140	1.0924	1.0579	1.0728
The lowest index	**0.9699**	**0.9672**	**0.9521**	**0.9311**	**0.9333**
The highest index	**1.1832**	**1.1726**	**1.1610**	**1.1295**	**1.1391**

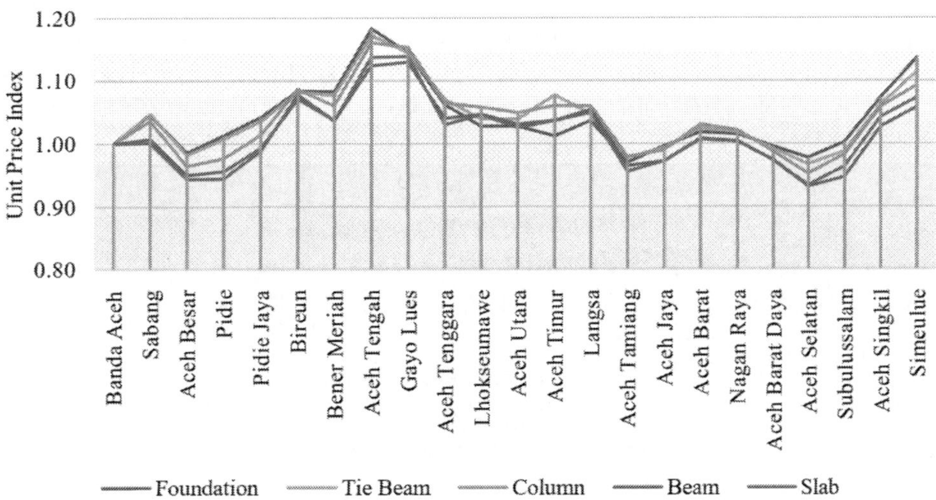

Figure 2. Reinforce concrete structural components index pattern

The price index for the five structural components shows a tendency that majorly of cities/districts reviewed (65% to 78%) are in the upper reference index category. These conditions illustrate the implications of the material and labour supply process for the completion of the works. When considered on the material aspect, materials are generally sourced from local production or materials trade in from other regions. Materials produced from the local area are generally in the form of aggregate materials for concrete work and wood materials for formwork work. Materials for reinforcement and other manufactured materials mostly must be imported from outside the province of Aceh, such as from the Province of North Sumatra or even from factories located in Java Island. The difference in the distribution distance between the material source and the location of the building becomes the factor causing the price difference material in each region. In view of labour aspects, construction labour generally comes from outside the province of Aceh, especially for labour in many government projects. The limitation of skilled labour in each region requires project implementers to recruit workers from outside with the consequence of increasing costs needed for the workforce mobilisation to each project location.

4. Conclusions

Changes that occur in the unit price of work have been studied using the object price unit of reinforced concrete work. This research makes a geographical factor of a region as the base in the study of price change which expressed in index form. Changes in unit price are seen from the side of change based on the components of reinforced concrete work and the structural components of a building.

Changes in the unit price index of work in each district/city in Aceh Province can be classified under the category reference index and the upper reference index. Index classification of the price of a region may vary by work types. Work in one region may be categorised as below reference index, while other works are considered in the category of the upper reference index. The price analysis process is done based on the standard that sets the same value of material and labour requirements, so the differences that arise are only related to the factor of material price change and labour wage in each region.

Implementation of price index information of work unit needs to be renewed, considering the extent of potential changes related to economic aspect in a region. The price index information will remain accurate and useful as a cost estimation guideline if within a specified period it is continuously adjusted to changes in material prices and labour costs. Changes in distribution patterns related to supply and demand will directly affect the rise or fall of its price.

5. Acknowledgement
This research was supported for funding by PLK grant from Research and Community Services Institute, Syiah Kuala University, with the contract No. 73/UN11.2/PP/PNBP/SP3/2018 of Fiscal Year 2018. The authors are also express high appreciation to all parties who have contributed to this research from data collection to completion of the research.

6. References
[1] Ogunlana S and Thorpe A 1991 *Building and Environment* **26** 77
[2] Azman M A, Abdul-Samad Z and Ismail S 2013 *Int. J. Proj. Man.* **31** 994
[3] Hatamleh M T, Hiyassat M, Sweis G J and Sweis R J 2018 *Engineering, Construction and Architectural Management* **25** 113
[4] Humphreys K K 2005 *Project and Cost Engineers' Handbook* (New York: Marcel Dekker)
[5] AACE 2006 *AACE International Recommended Practice No. 28R-03*
[6] Migliaccio G C, Guindani M, D'Incognito M and Zhang L 2013 *J. Constr. Eng. Manage.* **139** 858
[7] Choi J, Yun S and de Oliveira D P 2016 *Can. J. Civ. Eng.* **43** 1075
[8] Migliaccio G, Zandbergen P and Martinez A 2015 *J. Manage. Eng.* 2015 **31** 04014037
[9] Zhang S, Bogus S M, Christopher D L and Migliaccio G C 2017 *J. Constr. Eng. Manage.* **143** 04016087
[10] Xu J and Moon S 2013 *J. Manage. Eng.* **29** 10
[11] Remer S A, Lin S, Yu N and Hsin K 2008 *Int. J. Production Economics* **114** 333
[12] Bala K, Bustani S A and Waziri B S 2014 *Journal of Engineering, Design and Technology* **12** 518
[13] Fachrurrazi, Husin S, Tripoli and Mubarak 2017 *Proc. Engineering* **171** 282
[14] SNI 7394:2008 on Procedures for calculating the unit price of concrete work for the construction of buildings and housing
[15] Ministry of Public Work Regulation No. 26/PRT/M/2016 on the unit price analysis of work for public work services.
[16] Governor of Aceh Regulation No. 028/993/2017 on the pricing standard of construction materials/services for Aceh Government.

IOP Conf. Series: Materials Science and Engineering **523** (2019) 012036 doi:10.1088/1757-899X/523/1/012036

Investigating public perceptions and its implication toward Trans Koetaradja policy considering latent motivation

S Sugiarto[1,*], S M Saleh[1], R Anggraini[1], M Merfazi[2]
[1]Department of Civil Engineering, Syiah Kuala University, Banda Aceh, Indonesia
[2]Public Works and Spatial Planning Department, Banda Aceh, Indonesia

E-mail: sugiarto@unsyiah.ac.id

Abstract. Urban bus system so-called Trans Koetaradja (TK) has been considered and piloted as a way to mitigate motorcycles and cars dependency in Banda Aceh, capital of Aceh Province. It is assumed that the effectiveness of the such proposed policy would be useful if the public acceptance of this policy has a positive response. Therefore, this study intends to investigate the attitudes of the public in responding to such proposed policy. The data were collected within line 3 and line 4 of TK that is City Center - Mata Ie and City Center - Ajun - Lhoknga. The data collection method used was Stated Preference (SP) with a total of 220 samples collected using a stratified random sampling method. The SP questionnaire contains information about socio-economic, travel behavior, and respondents' perceptions representing their attitudes towards TK policy. People are dependent on their vehicles due to public transport challenging to use. The psychological indicators such as like driving a car & motorcycle and TK has limited accessibility have significant manifested to the latent variable of Inhibition of freedom of movement/activity. These sign representing the attitudes of the people in Banda Aceh would be dependence to their own vehicles and directly affected to the rejection to use proposed bus system of the proposed policy. Furthermore, male commuters who have low education level and income households more likely to autos dependent and more feeling inhibition of freedom of movement/activity if they use public transport.

1. Introduction

Public transportation as an element of the urban transportation system plays a critical role. Public transportation must be able to provide convenience for all communities in all their activities and be able to reach every urban area. Adequate public transportation will be able to increase road capacity and reduce traffic density, especially in urban areas so that it is excellent if supported by careful planning. As a result of the increase in public income and the high number of private vehicle ownership in Banda Aceh resulted in severe transportation problems, including congestion. Public dependence on private transportation modes is a significant factor in congestion in Banda Aceh. For this reason, the Aceh Government applies the policy of urban transport system with a mass transit system, namely the operation of the Bus Trans Koetaradja as a policy to reduce congestion impacts. TK is planned to operate at several lines within Banda Aceh City (See Fig. 1) and the TK Line can be seen in Figure 2.

Several studies have proven that bus reform policy such as bus rapid transit (BRT) in Jakarta has emerged as an economical transit alternative with significant potential for Jakarta [1,2]. Furthermore, providing proper public transportation would mitigate traffic congestion, reduce CO_2 emissions, and

*Corresponding author: sugiarto@unsyiah.ac.id

IOP Conf. Series: Materials Science and Engineering **523** (2019) 012036 doi:10.1088/1757-899X/523/1/012036

improve the effectiveness of energy consumption [3,4]. A tactic to advance a well-performing public transportation service could have a special place in the overall strategy of transportation planning [5].

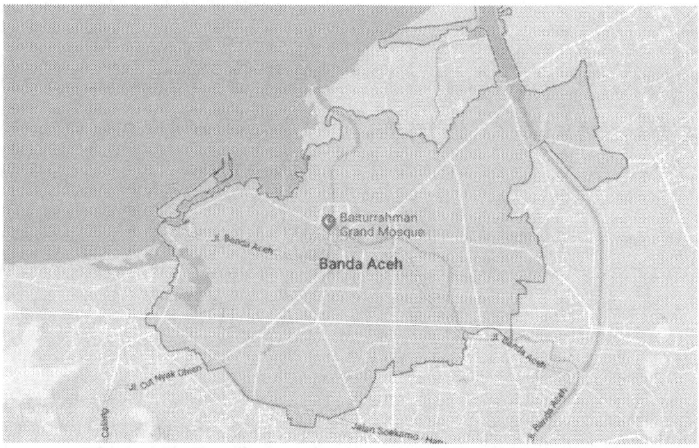

Figure 1. Banda Aceh city (map data © 2018 Google)

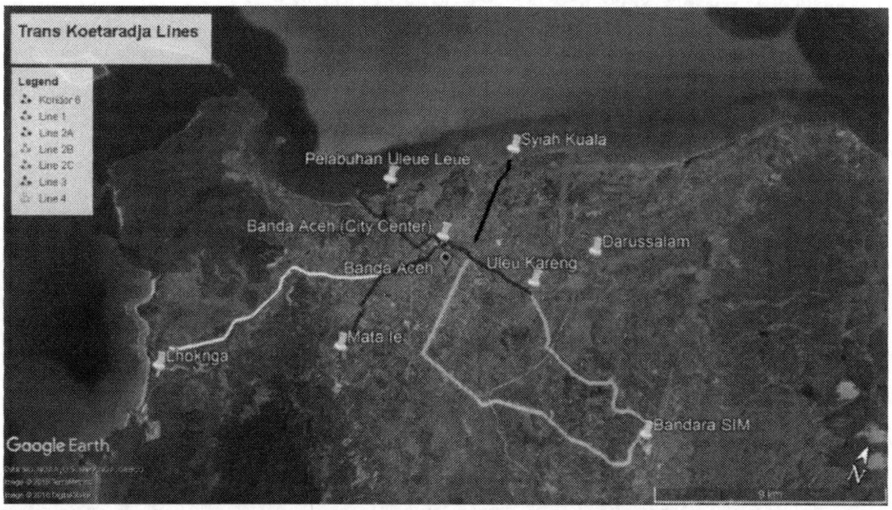

Figure 2. Study area & Trans Koetaradja Lines (overlay on map data © 2018 Google)

Understanding the social aspect related to the travel behavior that influencing the public acceptance in particular transport policies is a very crucial issue in order to ensure the effectiveness of the specific transport policy proposed by the government. For example [6] pointed out that determinants such as problem awareness, policy fairness, and perceived effectiveness are important factors affecting the acceptability of transport pricing proposal in Jakarta, capital of Indonesia. Research in the emerging world conducted by [7, 8, 9, 10] examined respondents' psychological motivations to assess the effects or observable and unobservable attributes of respondents' choice behavior. They found that determinant factors such as recognition, fairness, and consciousness of congestion problems and inhibition of freedom of movement have the most substantial encouragement on causal acceptability to transport measures in Jakarta. Existing literature mentioned above pointed out that travel behavior is significantly linked with the personal judgment that substantially associated with perceived benefit or effectiveness of the particular policy. Most profound understanding to the public need is crucial to any investigations for ensuring the efficiency of bus reform policy. Therefore, this study intends to investigate the attitudes of the public in responding such proposed policy by examining public perceptions of the TK policy especially within line 3 and 4 of TK, namely the corridor of the City Center - Mata Ie and the City Center - Ajun – Lhoknga TK's lines. Through this research, it is also expected that the information on the extent of public perception of TK policy will be obtained as input for relevant agencies in making decisions.

2. Literature Review

2.1 Trans Koetardja Policy
As a step to improve transportation services in Banda Aceh, the Aceh Government through the Department of Transportation, Communications, Information, and Telematics plans for urban bus system transportation, the Bus Trans Koetaradja. The Department of Transportation, Communication, Information and Telematics (2016) stated that the objectives of Trans Koetaradja policy were:

1) Resolve urban transportation problems such as congestion, parking chaos, traffic accident, and pollution;
2) Improve mass mobility services conveniently and comfortably;
3) Support the movement of resources for economic growth;
4) Realizing the Banda Aceh as a Green City.

2.2 Public Attitudes
There have been wide-ranging studies endeavoring to understand the determinant factors that influence public attitudes in the transport policy arena. For instance research by [6] pointed out that determinants such as problem awareness, policy fairness, and perceived effectiveness are important factors affecting the acceptability of transport pricing proposal in Jakarta, capital of Indonesia. More evidence by [8] found that infringement of freedom, policy fairness, problem awareness, and perceived effectiveness are essential determinants that significantly contribute to the acceptance of transport pricing policy. Research in the emerging world conducted by [7, 8, 9, 10] examined respondents' psychological motivations to assess the effects or observable and unobservable attributes of respondents' choice behavior. They found that determinant factors such as recognition, fairness, and consciousness of congestion problems and inhibition of freedom of movement have the most substantial encouragement on causal acceptability to transport measures in Jakarta.

2.3 Structural Equation Modelling (SEM)
Structural Equation Modeling (SEM) is a multivariate statistical analysis technique that combines aspects of factor analysis and path analysis with the aim of confirming measurement models and structural models built on specific theoretical studies [11]. SEM has become one of the most popular methods in multivariate analysis, especially in the social sciences. SEM has typical characteristics which include: 1). There are two types of variables: latent variable and observed variable (manifest); 2). There are two models: the structural model and measurement model, and 3). There are two types of errors: structural error and measurement error.

The SEM is consist of latent variable and observed variable. The latent variable is the critical variable of concern is the latent variable, latent variables are abstract concepts, such as people's behavior, attitude, feeling, and motivation. Latent variables can be observed indirectly and imperfectly through their effects on the observed variables. While the observed variable is variables that can be observed or measured empirically and are often called indicators or manifest variables, observed variables are effects or measures of latent variables. In the survey method using a questionnaire, each statement on the questionnaire represents an observed variable.

2.4 Multiple Indicators Multiple Causes (MIMIC) model
Multiple Indicators Multiple Causes (MIMIC) is one of the branches of SEM modeling. The MIMIC has two part modeling such as exploratory factor analysis (EFA) and confirmatory factor analysis (EFA).

The SEM is a flexible linear-in-parameters multivariate statistical technique that allows handling a large number of endogenous and exogenous variables. The latent constructs (factors) represent theoretical, abstract concepts or phenomena such as attitudes, behavioral patterns, cognition, social experiences, and emotions that cannot be observed or measured directly [12]. There are two types of analysis based on common factors namely exploratory factor analysis (EFA) and CFA. The Multiple Cause Multiple Indicator (MIMIC) model is combined both EFA and CFA [13, 14]. The MIMIC model as initially proposed by [14] systematically consists of a structural equation model and a measurement model, respectively given by:

$$\eta_i = B\,\eta_i + \Gamma z_i + \zeta_i, \text{ and } y_i = \Lambda\,\eta_i + \zeta_i \quad (1)$$

Where y_i is a vector of observable psychological indicators variables, z_i is a vector of explanatory variables that cause ηi, B, Γ and Λ are matrices of unknown parameters to be estimated, and the terms ζ_i and ζ_i are vecta or or of measurement errors. The MIMIC modeling approach has been widely applied in travel behavioral analysis, for instances can be seen in the previous studies done by [7, 8, 9, 10].

3. Data Collection and Empirical Setting

Primary data was obtained from a field survey by distributing questionnaires that were carried out stratified random sampling to the public transport and private mode users within line 3 and 4 of TK lines, that is City Center - Mata Ie and City Center - Ajun – Lhoknga lines. Respondents were commuters or people who carry out daily activities in the area with the proposed activities such as work, trading activities, school or college, and others. Questionnaires used stated preference techniques distributed manually and collected by eight surveyors and distributed within seven sub-districts such as Kuta Raja, Baiturrahman, Jaya Baru, Banda Raya, Darul Imarah, Pekan Bada, and Lhoknga. The secondary data used in this study included technical data on Trans Koetaradja, Map of regional location for the introduction of the area and location of sampling, population data, previous studies, related literature from related agencies.

A MIMIC framework is applied to model and analysis the collected data. The regression parameters were calibrated by applied maximum likelihood estimator (MLE) and performed using Lisrel 9.2. This method aimed to test a model hypothesis and find out the vector of indicator variables (y) connected by a latent variable (y) with a covariate (x) included in the analysis calculation. The MIMIC model tested the consistency of structural theory and measurement through data and had the purpose of estimating parameters (coefficients, variances) and assessing the fit of the model.

The variables set were defined as attributes of a person or object that had a variation between one person to another or one object with another object. There were five variables that were built representing latent variables (unobserved variables) such as suitability of *Trans Koetaradja* (η), Familiar or understanding of *Trans Koetaradja* policies ($\eta 2$), dependence on private modes (car and two wheeled-vehicle) ($\eta 3$), awareness of private mode problems (car and two wheeled-vehicle) ($\eta 4$), and inhibition of freedom of movement caused by *Trans Koetaradja* ($\eta 5$) which represented fourteen observed questions (observed variable) related to psychological perception (indicator). Socio-demographic attributes and travel behavior (x), indirectly affected the indicator (y).

4. Result and Discussion

4.1 Measurement model (confirmatory factor analysis/CFA part)

The results of the measurement model calibration are presented in Table 1 using the MIMIC approach. The latent variables such as policy suitability, familiarity, and understanding of policies, dependence on private cars, awareness of transportation problems and inhibition of freedom of movement are measured by using several indicators. As expected, most of the indicators in this latent constructs group have a significant factor loading (t-value > 1.96; at a significant error of 5%) to the measured latent constructs.

There are five latent variables built and represent fourteen indicators. In Table 1, there are twelve observed variables with loading factors ≥ 0.5 and t-value ≥ 1.96. The twelve measurement models are explained that the calibrated parameters are standardized (standardized loading factor) by adding a value of 1 for the first indicator within a group of a latent construct.

Table 1. Calibrated parameters of the measurement model (CFA part)

Latent Variable		Indicators	Coefficient	t-value
Latent1	Conformity of TK policy	TK is the right policy	1.00	3,64
		TK is a policy that is accepted by the community	1.61	
Latent2	Familiar/understanding about TK policy	Good government and community interaction	1.00	3,74
		TK can reduce environmental problems	0.68	
Latent3	Cars or Motorcycles dependency	Private vehicles are needed in everyday life	1.00	3,53
		Public transportation is needed in everyday life	1.33	
		Public transport is difficult to use in everyday life	3.54	5.13
Latent4	Awareness of problem with Cars or Motorcycles dependency in the society	Heavy traffic is in the city area	1.00	2,87
		Walking and cycling are difficult and dangerous in the city area	0.87	
Latent5	Inhibition of freedom of movement/activity	TK can prevent freedom of driving, especially at bus stops	1.00	2,88
		Like driving a car & motorcycle	1.03	
		TK has limited accessibility	1.38	3.01

Table 1 illustrates the correlation between individual perception as indicators and the latent constructs such as "Conformity of TK policy (latent1)", "Familiar/understanding about TK policy (latent2)", "Cars or Motorcycles Dependency (latent3)", "Awareness of problem with Cars or Motorcycles Dependency in the society (latent4)" and "Inhibition of Freedom of Movement (latent5)". Looking to the latent variables of latent1 & latent2, the indicators of the TK is a policy that is accepted by the community and good government, and community interaction has a significant positive sign to both of latent variables. It seems that the government has given an understanding of policy to the public and people believe that this policy could be appropriate policy to reduce traffic congestion in the city centers of Banda Aceh. The latent variable of "Cars or Motorcycles dependency" has significantly influenced by two indicators such as public transportation is needed in everyday life and public transport is difficult to use in everyday life. People are dependent on their vehicles due to public transport challenging to use. Furthermore, the latent5 revealing that the indicators of like driving a car & motorcycle and TK has limited accessibility have significant manifested to the latent variable of Inhibition of freedom of movement/activity. These sign representing the attitudes of the people in Banda

Aceh would be dependence on their vehicles and directly affected the rejection to use proposed bus system of the proposed policy.

4.2 Structural model (Exploratory Factor Analysis/EFA part)
Table 2 reveals the correlation among observed variable of socio-demographics to the latent constructs. The variable of dummy commuter shows us a positive correlation to the all latent variables except latent4 has signed a negative magnitude. The dummy variable of male, the dummy variable of education and the dummy variable of low-income households have also a negative alerted to the latent variable 4 "Awareness of problem with Cars or Motorcycles dependency in the society". This means that a commuter, a male, has lower education and has a lower income household more likely to less awareness about a problem with Cars or Motorcycles dependency in the society. While those variables have a positive correlation to the latent variable of "Cars or Motorcycles dependency" and "Inhibition of freedom of movement/activity". It seems that they are more likely to autos dependency people and has feeling inhibition of freedom of movement/activity if they use public transport.

Table 2. Calibrated parameters of the structural model (EFA part)

Observed Variables	Variable Setting	Path Loading Coefficients				
		Latent1*	Latent2*	Latent3*	Latent4*	Latent5*
Var1	Dummy commuter: 1 yes, 0 otherwise	0,10	0,16	0,49	-4,47	0,80
Var2	Dummy gender: 1 male, 0 otherwise	-	-	-	-0,30	-
Var3	Dummy age; 1 ≥ 30 years, 0 otherwise	-	-	-0,06	0,83	-0,07
Var4	Dummy education; 1 certificate ≤ College, 0 otherwise	0,06	0,13	0,39	-3,72	0,67
Var5	Dummy job status: 1 working, 0 students	-	-0,01	-	-0,14	-
Var6	Dummy income; 1 if ≤ 3 million, 0 if ≥ 3 million	-	0,08	0,29	-3,30	0,51

* *variables are significant at 5%*

5. Conclusion
Our work discussed in this research contributes to an investigation of the impacts of TK policy proposed by the Aceh Government. We examine the social impact in mainly related to the behavioral travel analysis by focusing on the public attitudes toward evaluating the proposed policy. The result of this research highlight that people in Banda Aceh has revealed that people in Banda Aceh agree to the adoption of TK but the TK itself has limited accessibility concerning the route services. Therefore, they tend to use their vehicles to travel.

Our modeling result using the MIMIC modeling approach reveals that the indicators of the TK are an accepted policy by the public and good communication among public and government have a significant positive sign to the latent constructs of "Conformity of TK policy", "Familiar/understanding about TK Policy". It appears that the government has given an understanding of policy to the public and people believe that this policy could be appropriate policy to reduce traffic congestion in the city centers of Banda Aceh. The latent variable of "Cars or Motorcycles dependency" has significantly influenced by two indicators such as public transportation is needed in everyday life and public transport is difficult to use in everyday life. People are dependent on their vehicles due to public transport challenging to use. Furthermore, the indicators of like driving a car & motorcycle and TK has limited accessibility have significant manifested to the latent variable of Inhibition of freedom of movement/activity. These sign representing the attitudes of the people in Banda Aceh would be dependence to their vehicles and directly affected the rejection to use proposed bus system of the proposed policy.

Regarding the structural equation results, the observed variable of dummy commuter shows us a positive correlation to the all latent variables except to the latent variable "Awareness of Problem with Cars or Motorcycles Dependency in the Society". The dummy variable of male, having primary education, and having low-income households also have a negative alerted to the latent variable of the "Awareness of Problem with Cars or Motorcycles Dependency in the Society". While those variables have a positive correlation to the latent variable of "Cars or Motorcycles dependency" and "Inhibition of freedom of movement/activity". It seems that male commuters who have low education level and income households more likely to autos dependent and more feeling inhibition of freedom of movement/activity if they use public transport. Above empirical findings suggest that the attitudes of people have significant influencing in adopting and changing people travel behavior. Moreover, the empirical result from this study could provide valuable insight related to the people behavior concerning the such proposed policy. These findings suggest that mobility management measures would be appropriate to change the behavior of private-mode users toward public transport use based on public attitudes in responding to urban bus system policy. This effort might be of precise assistance in designing a more efficient policy by encouraging a bus ridership in Banda Aceh which is private vehicle dependency (cars & motorcycles) is severe in the recent years [15, 16].

6. Acknowledgments

The Syiah Kuala University financially supported this study under Contract No. 131/UN11.2/PP/PNBP/SP3/2018. Furthers thanks go to the staff and students of Transportation System Laboratory, Civil Engineering Department of Syiah Kuala University for their substantial help with the data collection. All remaining oversight in this study is own.

7. References

[1] Ernst, P.J. 2005. *Journal of the Transportation Research Board*, **1903**, 20-26.

[2] Susilo, Y.O., T.B. Joewono, W. Santosa, and D. Parikesit. 2007. *IATSS Research* **31**(1), 59–68.

[3] Shapiro, R.J., K.A. Hassett, and F.S. Arnold. 2002. *Conserving energy and preserving the environment: The role of public transportation*. (The American Public Transportation Association).

[4] APTA. 2007. *Public transportation: Benefits for the 21st century*. (The American Public Transportation Association).

[5] Fujii, S., Van, T.H. 2009. Psychological. *Journal of Public Transportation*, **12**(1), 97-110.

[6] Sugiarto, S., Miwa, T., Sato, H., Morikawa, T. 2014. Congestion charging: Influence of public consciousness on acceptability in Jakarta Metropolitan Area. *Proceedings of 21st World Congress on Intelligent Transport Systems,* Detroit, USA.

[7] Sugiarto, S., Miwa, T., Sato, H., Morikawa, T. 2017a. *Journal of Urban Planning and Development*, **143**(2), 1-12.

[8] Sugiarto, S., Miwa, T., Morikawa, T. 2017b. *Transportation Part A* **103**, 36–53.

[9] Sugiarto, S., Miwa, T., Morikawa, T. 2018a. *Transportation Letters: The International Journal of Transportation Research*, **10**(1), 52–63.

[10] Sugiarto, S., Miwa, T., Morikawa, T. 2018b. The tendency of public's attitudes to evaluate urban congestion charging policy in Asian megacity perspective: Case a study in Jakarta, Indonesia. *Case Studies on Transport Policy, in press*.

[11] Jöreskog, K. G. A 1969. *Psychometrika*, **34**, 183–202.

[12] Bowen, N.K., Guo, S., 2012. *Structural Equation Modeling*. Oxford University Press, New York.

[13] Hauser, R.M., Goldberger, A.S., 1971. *Social. Methodol.* **3**, 81–117.

[14] Joreskog, K.G., Goldberger, A.S., 1975. *J. Am. Stat. Assoc.* **70**, 631–639.

[15] Saleh, S.M., Sugiarto, S., Hilal, A., Ariansyah, D. 2017. A study on the traffic impact of the road corridors due to flyover construction at Surabaya intersection, Banda Aceh of Indonesia. *AIP Conf. Proc.* **1903**, 060005.

[16] Aggraini, R., Sugiarto, S., Pramanda, H., 2017. Factors affecting trip generation of motorcyclist for the purpose of non-mandatory activities. *AIP Conf. Proc.* **1903**, 060011.

The influence of the using waste tire rubber and natural ziolite as Asphalt and Cement replacements to compressive strength of Semi-Flexible Pavement

Hamzani[1,*], Munirwansyah[2], M Hasan[2], S Sugiarto[2]
[1]Doctor of Engineering Candidate, Syiah Kuala University, Banda Aceh, Indonesia,
[2]Department of Civil Engineering, Syiah Kuala University, Banda Aceh,

Email: hamzani.hasbi@gmail.com

Abstract. The objective of this research is to study the compressive strength of semi-flexible pavement by using waste tire rubber and natural zeolite as asphalt and cement replacements. The mix design of porous asphalt is started with a determination of optimum asphalt proportion with open-graded aggregate as specified in *Australian Asphalt Pavement Association* (*AAPA*) 2004. The asphalt was then replaced by 3 %, 4 % and 5 % of waste tire rubber and the Marshall test were conducted. The cement mortar with 0 %, 5 %, 10 %, 15 % and 20 % natural zeolite was injected to porous asphalt specimens. The compressive strength test was conducted based on ASTM C670-91a at the age of the specimens of 14 days. The test results showed that the higher compressive strength is reached at the 15 % natural zeolite and 5 % waste tire rubber, which is 15,43 MPa.

1. Introduction

Semi-flexible pavement (SFP) is a combination of two types of rigid pavement with flexible pavement that has several advantages [1]. SFP is made with a mixture of open-graded asphalt matrix (20-30% void ratio) and filling it with selected cement grout [2]. Mortality percentage of 80% of total voids with 3% asphalt content is the optimum composition in the semi-flexible mixture [3]. The use of natural zeolite in the asphalt concrete binder coarse (AC-BC) mixture as a filler can improve the stability of the soaking conditions in the ash of the stone [4].

Furthermore, rutting can be reduced if the asphalt stiffness of hot mixture increases at high temperatures. The service life of the pavement will increase if the appropriate asphalt binder modifier is adopted [5]. The additive of waste tire recycling (WTR) to the base of the asphalt binder can improve the asphalt binder properties, such as reducing construction and maintenance costs, improving the resistance to permanent deformation, reducing fatigue damage and potential thermal cracking, reducing pavement thickness and reducing the possibility of cracking reflective [6].

Indonesia is one of the countries with the most significant natural zeolite reserve in the world and spread almost in every region especially Sumatra and Java. Natural zeolite as an added material is in addition to competitive regarding cost when compared with synthetic zeolites that are only produced abroad, also due to natural zeolite deposits in Indonesia of 400 million tons.

Based on the above description, research is required to develop PSF by modifying asphalt with waste tire rubber material and modifying cement with Aceh's natural zeolite as mortar. This research was conducted to find out the compressive strength of PSF mixture due to the substitution of waste tire rubber on asphalt and natural zeolite substitution on cement as mortar material with a series of testing in the laboratory.

*Corresponding author: hamzani.hasbi@gmail.com

The purpose of modified polymer asphalt uses a technology of warm asphalt mixture with natural zeolite additive materials. This study was conducted to determine the resistance of deformation due to natural zeolite additives on a warm asphalt mixture using asphalt modified series of testing in the laboratory. The use of zeolite on concrete/mortar can improve its compressive strength and durability [7].

This study aims to find out and evaluate the effect of substitution of used waste tire rubber on asphalt and substitution of Aceh natural zeolite on mortar as a mixture of PSF. In detail the purpose of this study are as follows:

a. To obtain the composition of PSF mixture substance using waste tire rubber and Aceh natural zeolite.
b. To analyze the compressive strength of PSF mixture using 60/70 penetration asphalt and Aceh natural zeolite as cement replacement.

2. Materials and Method

2.1 Materials

Open-graded split originated from river stone located in North Aceh is used as fine and coarse aggregate. The properties of fine and coarse aggregate are shown in Table 1 and 2, respectively. The aggregate gradation is shown in Figure 1 which is satisfied the requirement of Australian Asphalt Pavement Association (AAPA) 2004 Standard [8]. Asphalt as a binder used in this study is penetration asphalt penetration 60/70 produced by PT. Pertamina. While the cement used in this study is Portland type II cement produced by PT. Semen Andalas Indonesia. The substitution materials are grated waste tire rubber and natural zeolite. The grated waste tire rubber is sieved to pass no.50 sieve before it is used in this study. The natural zeolite originated from Peukan Bada Aceh Besar District of Aceh Province is used. Natural zeolite is grinded to pass no. 200 sieve and chemically activated using chloride acid (HCl). After chemically activated, the natural zeolite is purified by Aqua D'Mineral.

Table 1. Fine aggregate properties

No.	Type of test	Test Method	Result	Specification	Unit
1	Specific gravity	SNI 03-1970-1990		Min. 2,5	gr/cm
	Bulk		2,521		
	SSD		500		
	Apparent		2,589		
2	Absorption	SNI 03-1969-1990	1,890	Max. 3	%

Table 2. Coarse aggregate properties

No.	Type of test	Test Method	Result	Specification	Unit
1	Specific gravity	SNI 03-1970-1990		Min. 2,5	gr/cm
	Bulk		2,340		
	SSD		500		
	Apparent		2,467		
2	Absorption	SNI 03-1969-1990	1,937	Max. 3	%

2.2 Experimental program

The tests in this study are based on Bina Marga Specification 2010 for hot mix [9] using open-graded aggregate as specified in AAPA Specification 2004. Mix proportion of porous asphalt mixture is based on Marshall method to obtain Optimum Asphalt Content (OAC). After OAC was obtained, the specimens are prepared with 3 variations of waste tire rubber as asphalt replacement. The content of waste tire rubber used is 3 %, 4 % and 5 % of OAC.

The specimens then put in room temperature for 24 hours. After 24 hours in room temperature, the specimens have injected the mortar. The proportion of mortar is 1 cement: 2 sands by volume. The water used in mortar is 0,4 of cement weight. A natural zeolite is used for cement replacement. Four variations of zeolite content, which are 0 %, 5 %, 15 % and 25 % of cement weight are used. Before injected in the specimens, the viscous test of mortar was performed. To inject the mortar in the specimens, the specimen is inserted in the tube, then mortar was filled on the tube and vibrated using the table vibrator for 10 seconds. Compression test using a 200 kN capacity strength test apparatus was conducted on the specimens at the age of 14 days at room temperature.

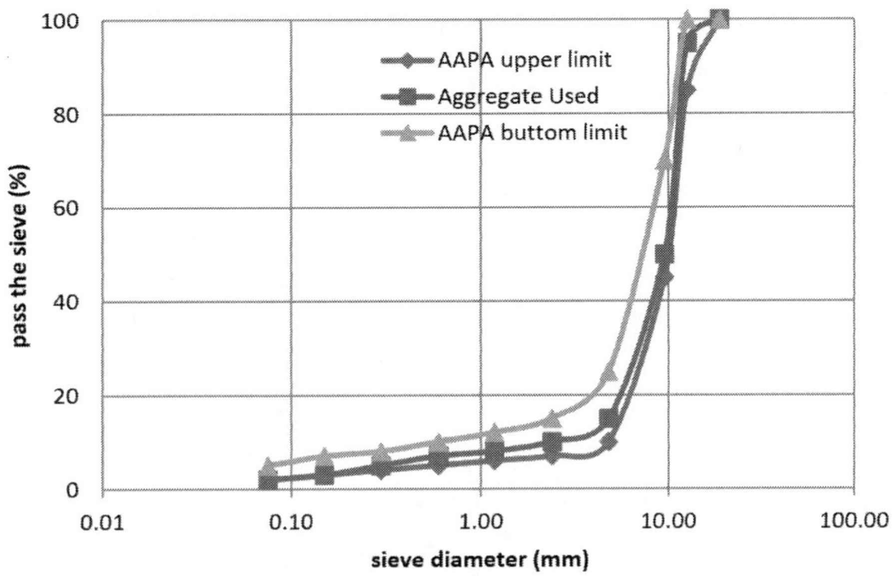

Figure 1. Gradation of aggregate

3. Result and analysis

3.1 Marshall test result for determination of OAC

Based on Marshall test the stability, flow, and Voids in The Mix (VIM) were obtained as shown in Table 3 below:

Table 3. Marshall Test Result for determination of OAC

Parameter	AAPA Specification 2004	Asphalt Content				
		3%	**3.50%**	**4%**	**4.50%**	**5%**
Stability (kg)	Min. 500	503	512	470	362	356
Flow (mm)	2 - 6	3.97	5.6	4	5.78	5.53
VIM (%)	18 - 25	24.95	26.7	22.64	18.43	13.88

Based on Table 3, it is found that optimum asphalt content is 3%. Therefore for the following investigation, the asphalt content used is 3 %.

3.2 Compressive Strength

The compressive strength of semi-flexible pavement for all variation of waste tire rubber and zeolite content is shown in Table 4. Based on the results in Table 3.2 the compression test results of PSF mixture showed that the substitution of used waste tire rubber in asphalt and substitution of Aceh natural zeolite in mortar cement could increase compressive strength at certain limits, after which the value of compressive strength decreased.

Table 4. Test results of compressive strength on any variations of with waste tire rubber and zeolites

No	Zeolite (%)	Waste Tire Rubber (%)	Compressive Strength (MPa)
		3	12.53
1	0	4	11.57
		5	8.1
		3	12.53
2	5	4	14.08
		5	13.5
		3	12.53
3	15	4	14.46
		5	15.43
		3	7.13
4	25%	4	6.36
		5	8.1

3.3 Discussion on Marshall Test Results for the determination of OAC

Stability is an essential factor to measure the quality of the porous asphalt mixture in the traffic load, the higher the stability value indicates the pavement the higher the stiffness value, but at certain limits there is still flexibility that is the value of flow. Void values in the mix (VIM) is a requirement that needs to be met that is about 20% -30%, cavity level is associated with the volume of mortar filled so that it affects PSF mixed compressive strength. Of these three important parameters all meet the requirements of AAPA (2004) so that porous asphalt pavement can be used.

3.4 Discussion on Compressive Strength Results

The PSF mixed strength test results that the mixture with the substitution of waste tire rubber in hot asphalt and natural zeolite substitution on mortar cement has increased the value of compressive strength. The increase of compressive strength value of PSF mixture obtained varies considerably, in zeolite substitution with 0% variation and used tire substitution 3%, 4%, 5% value of compressive strength decreases. While on the variation of 15% zeolite with the variation of tire used 5% mixed PSF increased value of maximum compressive strength and after that happened decrease. The effect of compressive strength value of PSF mixture on variations of waste tire rubber in hot asphalt and zeolite on cement as mortar material can be seen in Figure 2, Figure 3, and Figure 4.

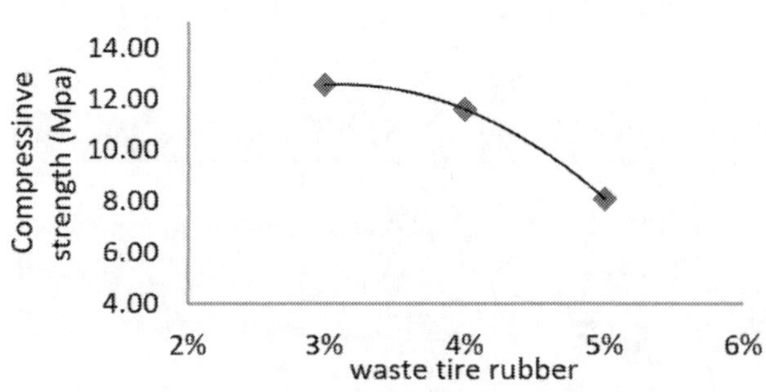

Figure 2. Relation of compressive strength content to waste tire rubber (zeolite 0%)

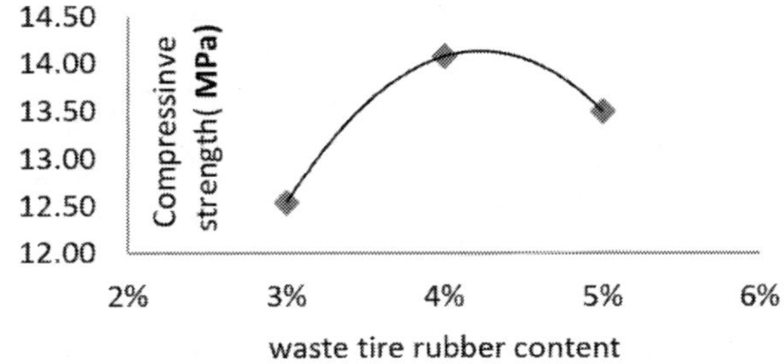

Figure 3. Relation of compressive strength content to used tire rubber (zeolite 5%)

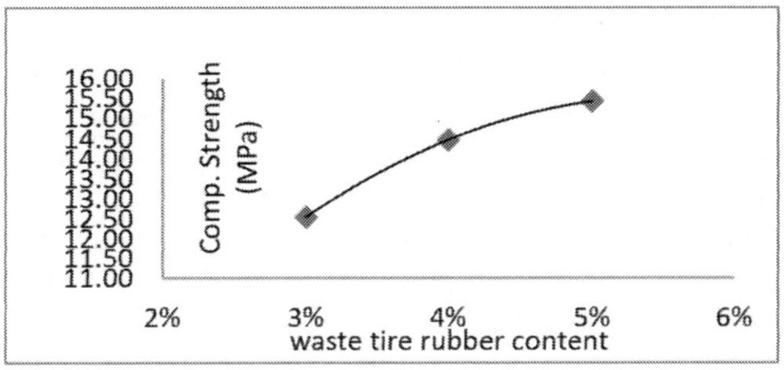

Figure 4. Relation of compressive strength content to used tire rubber (zeolite 25%)

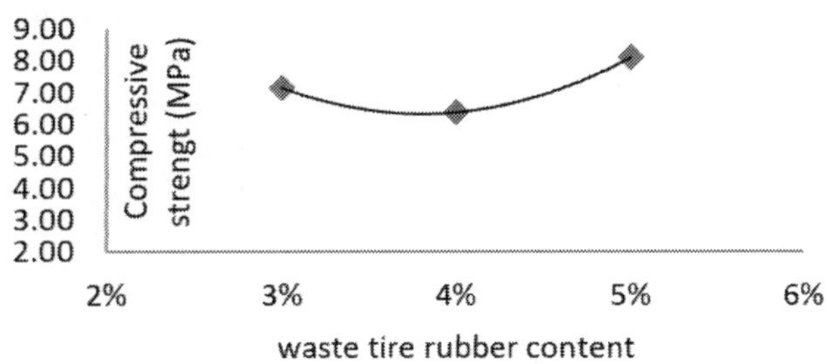

Figure 5. Relation of compressive strength content to used tire rubber (zeolite 25%)

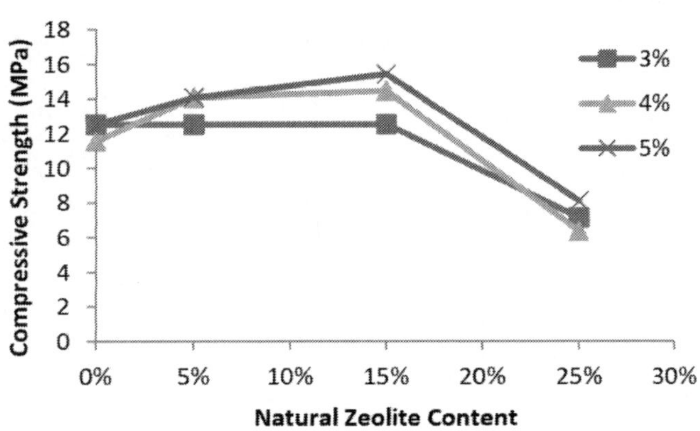

Figure 6. Relationship of compressive strength values to zeolite content

IOP Conf. Series: Materials Science and Engineering **523** (2019) 012037 doi:10.1088/1757-899X/523/1/012037

4. Conclusions

a. The substitution of the used tire powder on the oil asphalt in the porous asphalt mixture can increase the cavity level in the mixture, thereby affecting the compressive strength value of the PSF mixture.
b. Substitution of Aceh's natural zeolite on cement as mortar can increase the strength of the compressive strength value of Semi-Flexible Pavement mix.
c. The maximum compressive strength value of 15.43 MPa was obtained at the level of 5% of used tires and 15% of Aceh's natural zeolite, after a decrease

5. References

[1] Zoorob S E, Hassan K E, Setyawan A 2002 Effect of cementitious grout on the properties of semi-flexible bituminous pavement, *Proceedings of the Fourth European Symposium on Performance of Bituminous and Hydraulic Materials in Pavements,* Nottingham, United Kingdom, 11-12 April, pp 105-112

[2] Nadiah M H, Karim, M R, Mahmud, H B Koting S 2014 *Adv. Mater. Sci. Eng.* **2014** 529305

[3] Setyawan A 2013 Asessing the Compressive Strength Properties of Semi-Flexible Pavements, *Proceeding The 2nd International Conference on Rehabilitation and Maintenance in Civil Engineering Asessing Procedia Engineering* pp. 863-874

[4] Saleh A and Suparma L B 2015 Perancangan Laboratorium Pada Campuran Asphalt Concrete-Binder Course (AC-BC) Dengan Menggunakan Aspal Pen 60 / 70 Dan Zeolit Alam Sebagai Filler, *Annual Civil Engineering Seminar 2015*, Pekanbaru, pp 978-979

[5] Casey, Donnchadh, McNally, Amanda Gilchrist M D 2014 Resources Conservation and Recycling **52** 1167-1174

[6] Ge D, Yan K, You Z, Xu H 2015 *Constr. Build. Mater.* **126** 66–76

[7] Hasan M, Husin S and Nursaniah C 2016 *Key Engineering Materials* **711** 302-309

[8] Anonim, 2004, Open Graded asphalt Design Guide (Australian Asphalt Pavement Association), Australia

[9] Kementerian Pekerjaan Umum Direktorat Jenderal Bina Marga 2010 *General Specification Division 6 Asphalt Pavement* pp 6.1 – 6.5 (in Bahasa Indonesia)

IOP Conf. Series: Materials Science and Engineering **523** (2019) 012038 doi:10.1088/1757-899X/523/1/012038

The effect of Soil-structure interaction on Multi-Storey building resonance and Dynamic Shear modulus for Pidie Jaya Aceh earthquake

M Munirwansyah[1], **R P Munirwan**[1,*], **M Sungkar**[1], **Z Melinda**[1]
[1]Department of Civil Engineering, Syiah Kuala University, Banda Aceh, Indonesia

Corresponding E-mail: r.munirwan@unsyiah.ac.id

Abstract. Earthquake resistance building design must be considered as a design standard in the future for Pidie Jaya - Aceh government as 7th December in 2016 earthquake in Pidie Jaya Aceh which destroyed many major and minor infrastructures. Therefore, soil dynamic parameters must be obtained for design purposes in Pidie Jaya reconstruction and rehabilitation in the future. To avoid the structural failure during the earthquake process, the value of structural vibration frequency (f) should not similar to the value of natural soil vibration frequency (f_n). This article aims to determine the soil dynamic parameter of soil layers for disaster mitigation purposes. Several existing buildings namely Dayah Mudi Samalanga, Baitul Muttaqin Mosque, Cubo Bridge, Regent Office Building, Local Lawyer Office and around Pidie Jaya fault were chosen for soil sampling locations. Dynamic parameter of G_{max} and V_s are the result of this research. Moreover, the f_n value was calculated by Kramer method. Comparison of fn values was performed by simple modeling off values based on SNI 1726-2012. The highest value of the void ratio (e) was obtained in Baitul Muttaqin Mosque soil sample which is 1.84. The highest f_n value is at Pidie Jaya fault, which is 2.01 H_z and the lowest f_n value is at Baitul Muttaqin Mosque, Pidie Jaya, which is 1.301 H_z.

1. Introduction

Soil where construction foundations rests performance a significant part in geotechnical engineering rules. Many natural disasters play some important roles in changing dynamic parameters and bearing capacity of soils like an earthquake. Earthquake effect of the building must be considered deeply in construction design as it affects not only the soil but also above structures which widely known as soil-structure interaction (SSI). [1] explain that at least three-four factors that caused by earthquake damages namely source of the earthquake, path characteristics, local geotechnical characteristics, and quality of structural design. Some other articles also conduct research regarding dynamic soil parameters caused by an earthquake [2] [3] [4] and [5]. Moreover, [2] explained a control how to determine dynamic parameters of soil to applied in design among soil-structure interaction (SSI) framework. [6] researched earthquake hazard analysis for safety design purpose in Aceh due to some earthquake events.

Pidie Jaya of Aceh – Indonesia in Sumatera Island experienced 6.4 Mw earthquake on 7th December 2016 which caused loses and fatalities. [7] noted that the earthquake happens in the depth of 15 km below earth surface and coordinate at 5.29° N, 96.22° E. The shaking followed by several aftershocks for a few days later. Around 48 times of following earthquake impacts recorded in that day and 17 times of aftershock in the following day [4]. Meuredu, Manohara, Samalanga and Bireun are

---

*Corresponding author: r.munirwan@unsyiah.ac.id

the major affect area of the earthquake. Pidie fault was the cause of this Pidie Jaya 6.4 Mw earthquake [8] as can be seen in Figure 1. In fact, Pidie fault is not the most active fault in Aceh. The main faults in Aceh are Tripa Fault, Aceh Fault and Seulimum Fault as can be seen in Figure 1. From 2004 to 2012, at least three large earthquakes recorded which are larger than 8 Mw in Sumatera Island [9].

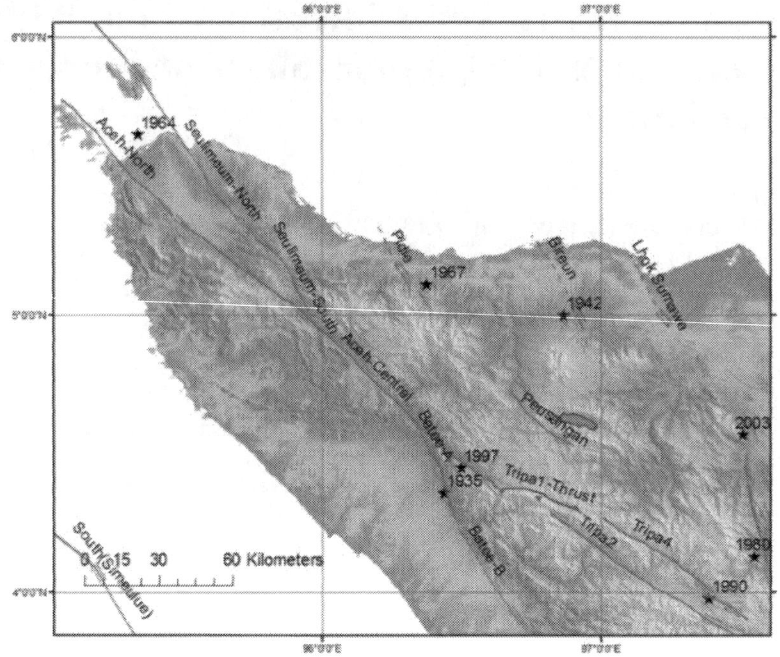

Figure 1. Aceh Faults along Aceh Province (PUSGEN, 2016 [8])

The earthquake wave vibration to the building foundations analysis (SSI) have to concern two factors namely structural vibration frequency (f) and natural vibration frequency (f_n). The two factors also depending on the value of surface peak horizontal acceleration of soil (a_{max}). [1] describe that site conditions and amplification parameters need to be analyzed in order to obtain the surface peak horizontal acceleration (a_{max}). From [10], vibration transfer functions, soil natural frequencies, ratio of damping and shapes of the construction are the limitation in analyze structural vibration frequency (f). There are also several soil-structure vibration frequency analyses from [3] [11] [12] [13] and [14]. The soil-structure model due to seismic damage can be explained by Figure 2. Shear modulus (G), Poisson's Ratio (v) and soil friction angle are soil parameters that need to obtained for SSI analysis. To avoid structural failure during the earthquake process, the value of structural vibration frequency (f) should not equal to the value of natural soil vibration frequency (f_n).

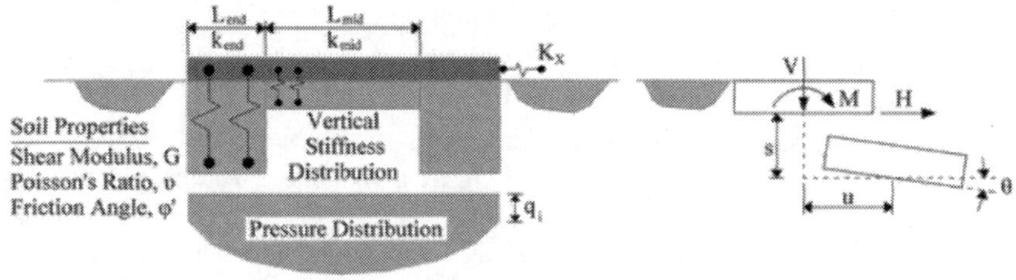

Figure 2. Soil-Structure Model of Seismic Damage (Sotiriadis et al., 2017 [14])

2. Material and Methods

This study was based on experimental works in Syiah Kuala University Soil Mechanics Laboratory Indonesia. Several soil samples were taken around the study area along with their coordinate. Several existing buildings namely Dayah Mudi Samalanga, Baitul Muttaqin Mosque, Cubo Bridge, Regent Office Building, Local Lawyer Office and around Pidie Jaya fault were chosen for soil sampling locations. Some physical and mechanical properties of soil test were then performed namely water content, unit weight of soils, Atterberg limit tests, grain size distribution test and void ratio test. Mechanical tests of soil were also conducted to obtain soil cohesion and friction angle. All of the soil experiments were follow the ASTM standard. The maximum shear modulus (G_{max}) were then calculated in order to obtain the value of soil natural vibration frequency (f_n).

The bearing capacity of soil and wave transfer along the soil layers were influenced by soil-structure interaction (SSI) for dynamic loads. [12] maximum shear modulus (G_{max}) must be obtained first to calculated the value of resonance frequency and vibration amplitude in structural foundations. Soil-structures interaction due to dynamic loads effect the wave propagation and bearing capacity of soil. The empirical equation of G_{max} correspondence with site condition with the influence of fine content (f_c) soil gradation, void ratios (e) and soil plasticity (PI) have been introduced by [12] in equation (1) (2) (3) and (4).

$$G_{max} = 0.0012f_c^3 - 0.1995f_c^2 + 8.4718f_c + 273.86 \text{ (kPa)} \quad (1)$$

$$G_{max} \text{ (vertical)} = 1067,3 \exp^{-1,51e} \text{ (kPa)} \quad (2)$$

$$G_{max} \text{ (horizontal)} = 811,9 \exp^{-1,11e} \text{ (kPa)} \quad (3)$$

$$G_{max} = -0.04PI^2 + 2.7PI + 289.81 \text{ (kPa)} \quad (4)$$

The meaning of natural soil vibration frequency is a occurrence at which a systematically tends to oscillate in the nonappearance of any forces which are drive and damp. Natural soil vibration frequency (f_n) was then calculated from equation (5) (6) and (7) as introduced by [15]. Unit weight of soil (ρ) in kg/m3, ω_0 (H_z) as fundamental frequency, Vs as shear wave velocity (m/s), and H (m) is depth of soil.

$$f_n = \frac{\omega_0}{2\pi} \quad (5)$$

$$\omega_0 = \frac{\pi \times v_s}{2H} \quad (6)$$

$$v_s = \sqrt{\frac{G_{max}}{\rho}} \quad (7)$$

Structural vibration frequency (f) were calculated by equation (8) and (9) as Indonesia standard procedure of SNI 1726-2012 [16]. Ta (s) is a fundamental period, and N is a number of multi-story building which is less than 12 levels of building. The value of f_n should not equal to f to avoid resonance which will lead constructions to fail. Resonance ratio of building (R) to identify construction resonance level then calculated with equation (10) where f_b (H_z) is building frequency and f_t (H_z) is natural soil frequency.

$$f = \frac{1}{T} \quad (8)$$

$$T_a = 0,1 \text{ N} \quad (9)$$

$$R = \frac{fb - ft}{ft} x100\% \quad (10)$$

3. Results and Discussion

Table 1 shows the results of dynamic soil parameter around Pidie Jaya. The highest void ratio value was in Baitul Muttaqin Mosque for 1.84, however the highest G_{max} value was in Pidie Fault for 166,598.51 kg/ms². It may because that the void ratio value in Pidie Fault was the lowest for 1.23. Natural soil vibration frequency (f_n) Of Pidie Fault of 2.013 Hz was the highest along with other dynamic soil parameter like shear wave velocity and fundamental frequency which effect the natural soil vibration frequency.

Table 1. Dynamic Soils Parameter Results

No.	Location	e	Gmax (kg/ms²)	ρ (kg/m³)	Vs (m/s)	ω_o (Hz)	f_n (Hz)
1.	Cubo Bridge	1.27	156,833.91	2.55	7.84	12.31	1.961
2.	Pidie Jaya Fault	1.23	166,598.61	2.57	8.05	12.64	2.013
3.	Local Lawyer Office	1.29	152,168.33	2.59	7.67	12.03	1.916
4.	Regent Office	1.47	115,953.28	2.61	6.67	10.46	1.666
5.	Dayah Mudi	1.54	104,322.53	2.55	6.40	10.04	1.599
6.	Baitul Muttaqin Mosque	1.84	66,319.73	2.45	5.20	8.17	1.301
	Average	1.44	127,033	2.55	6.97	10.94	1.743

Resonance in building able to lead construction to collapse if natural soil vibration frequency (f_n) is equal to structural vibration frequency (f). Therefore, modelling simple multi-storey construction with maximum 12 level of structural were then applied as SNI 1726-2012 with fundamental period approach as can be seen in Table 2.

Table 2. Structural Frequency Model with Fundamental Period Approach

N	Ta (s)	f (Hz)
1	0.100	10.000
2	0.200	5.000
3	0.300	3.333
4	0.400	2.500
5	0.500	2.000
6	0.600	1.667
7	0.700	1.429
8	0.800	1.250
9	0.900	1.111
10	1.000	1.000
11	1.100	0.909
12	1.200	0.833

The meaning of natural soil vibration frequency should not equal to structural vibration frequency is that the difference between these two parameters should not zero. The structures will failure/ collapse if the two parameters are equal. Taller building will have lower structural frequency but the period value will be higher. In contrast, lower building will have high structural frequency and low period value. The comparison is based on closest construction around and the natural soil vibration frequency was from calculation. The resonance ratio will show the resonance probability as shown in Table 3.

Table 3. Comparison Between Natural and Structural Frequency

No.	Location	Building Condition	f_n (H_z)	f (H_z)	N	difference	Resonance Ratio (%)	Resonance Probability
1.	Cubo Bridge	Failure	1.96	2.00	5	0.04	1.97	High
2.	Pidie Jaya Fault	No Failure	2.01	2.00	5	0.01	0.64	High
3.	Local Lawyer Office	No Failure	1.92	2.00	5	0.08	4.19	High
4.	Regent Office	No Failure	1.67	1.67	6	0.00	0.02	High
5.	Dayah Mudi	Failure	1.60	1.67	6	0.07	4.06	High
6.	Baitul Muttaqin Mosque	Failure	1.30	1.25	8	0.05	4.06	High

The comparison of f_n and f for resonance in Table 3 display that all the location has high probability of resonance. However, in reality not all the construction collapse during and after earthquake even the construction with difference close to zero. It may have explained that resonance not the caused of building failure. Furthermore, other technical problems may take account like low bearing capacity, construction strength, liquefaction and good design plan of buildings.

4. Conclusion

To conclude, building failure may also cause by resonance in building if natural soil vibration frequency (f_n) is equal to structural vibration frequency (f). Construction can fail if f_n and f value are similar. In general, higher construction will have lower structural frequency but the period value will be higher. On the other hand, lower construction will have high structural frequency and low period value. Comparison of f_n values was performed by simple modelling of f values based on SNI 1726-2012. The highest value of void ratio (e) was obtained in Baitul Muttaqin Mosque soil sample which is 1.84. The highest f_n value is at Pidie Jaya fault, which is 2.01 H_z and the lowest f_n value is at Baitul Muttaqin Mosque, Pidie Jaya, which is 1.301 H_z.

5. Acknowledgments

This research is partly supported by Indonesia Ministry of Research and Technology and Lembaga Penelitian dan Pengabdian Masyarakat (LPPM) Syiah Kuala University Banda Aceh Indonesia for Professor Research Scheme 2018 No: 287/UN11/SP/PNBP/2018 on 29 January 2018. The authors would also like to thank you for any supports and appreciate also to geotechnics laboratory staffs and assistants for all their assistances.

6. References

[1] Sitharam T G, Vipin K S, James N 2018 *Recent Advances in Soil Dynamics Relevant to Geotechnical Earthquake Engineering* Advances in Indian Earthquake Engineering and Seismology pp 203-228.

[2] Andersen K H 2015 *Cyclic Soil Parameters for Offshore Foundation Design* Frontiers in Offshore Geotechnics III, London 1, 5-84

[3] L'Heureux J S, Long M 2017 *J. Geotechnical and Geo. Engineering* **143** Issue 6.

[4] Munirwansyah, Munirwan R P, and Yunita H 2018 *International Journal on Advanced Science, Engineering and Information Technology* **8(3)** pp 870-875.

[5] Setiawan B, Jaksa M, Griffith M, and Love D 2018. *Soil Dynamics and Earthquake Engineering* **110** pp 244-261

[6] Munirwansyah, Irsyam M, Munirwan R P, Yunita H, and Usrina M Z, 2018 *Geotechnical Approach for Occupational Safety Risk Analysis of Critical Slope in Open Pit Mining as Implication for Earthquake Hazard* IOP Conference Series: Materials Science and Engineering Volume **352** Conference 1

[7] Supendi P, Nugraha A D, and Wijaya T A 2017 *Jurnal Geofisika* Vol. **15** No. 03 pp 17-20
[8] PUSGEN 2016 Laporan Gempa Pidie Jaya Provinsi Nanggroe Aceh Darussalam Indonesia (Pidie Jaya Earthquake Report Nanggroe Aceh Darussalam Province Indonesia) Indonesia
[9] Irwansyah E, Winarko E, Rasjid Z E, and Bekti R D 2013 *Journal of Physics*: Conference Series Vol. **423** 012067
[10] Liu K S, and Tsai Y B 2010 *Earthquake Spectra* **26** Issue 2 pp 371-397
[11] Zania V 2014 *Soil Dynamics and Earthquake Engineering* **59** pp 8-20
[12] Munirwansyah 2002 *Penentuan Modulus Geser Pasir-Berlempung untuk Kondisi Regangan Kecil dengan Uji Kolom Resonansi (Determination of Shear-Clay Modulus Shear For Low Strain Condition With Resonance Column Test)* Ph.D Dissertation, (Pascasarjana ITB), Bandung
[13] Torabi H, and Rayhani M T 2014 *Computers and Geotechnics* **60** pp 9-19
[14] Sotiriadis D, Kostinakis K, and Morfidis K 2017 *Bulletin of Earthquake Engineering* **15** (9) pp 3581-3610
[15] Kramer S L 1996 *Geotechnical Earthquake Engineering* Pearson Education: Singapore
[16] Badan Standardisasi Nasional 2012 *Tata Cara Perencanaan Ketahanan Gempa untuk Struktur Bangunan Gedung dan Non Gedung (Earthquake Resistance Planning Procedures for Structural and Non-Structural Building)*, SNI 1726:2012 Jakarta

IOP Conf. Series: Materials Science and Engineering **523** (2019) 012039 doi:10.1088/1757-899X/523/1/012039

Comparison of earthquake design parameters of low rise building structures based on SNI-1726-2002 and SNI 1726:2012 at 23 districts in Aceh Province, Indonesia

T Saidi[1,*], M Hasan[1], Fakhrurrazi[2,3]
[1]Dept. of Civil Engineering, Syiah Kuala University, Banda Aceh, Indonesia,
[2]Alumni, Dept. of Civil Engineering, Syiah Kuala University, Banda Aceh, Indonesia,
[3]Ministry of Public Work, Republic of Indonesia.

Email: taufiq_saidi@unsyiah.ac.id

Abstract. This paper presents a comparison of earthquake design parameters, i.e. response spectrum, modal mass contribution, base shear and natural vibration period, of low rise building structures based on-SNI-1726-2002 and -SNI 1726:2012 at 23 districts in Aceh Province, Indonesia. A four-story reinforced concrete frame structure is analyzed. The results show that the maximum spectral acceleration based on SNI 1726:2012 is higher than based on SNI-1726-2002 for 15 districts in Aceh. The modal mass participation for those two standards is almost the same. The base-shear force of SNI 1726:2012 is much higher than of SNI-1726-2002. The structural natural period of both SNI-1726-2002 and SNI 1726:2012 is in a good agreement.

1. Introduction

The Indonesian Government has issued several special regulations to anticipate the effects of the earthquake on buildings and minimize the damage of buildings due to an earthquake. One of the regulations is SNI-1726-2002 concerning Earthquake Resistance Design Standards for Building Structures [1]. This standard uses the Indonesian earthquake zoning map which is divided into 6 zones from the lowest zone (zone 1) to the highest zone (zone 6) of seismic intensity.

Since the issuance of SNI-1726-2002, there have been a couple of major earthquake events in Indonesia that have magnitudes greater than the maximum magnitude of previous estimates, such as the Sumatra-Andaman Earthquake in 2004 (Mw = 9.2) and Nias Earthquake in 2005 (Mw = 8.7) ([2], [3], [4]). The earthquakes have caused death of thousands of lives, collapsed and damaged thousands of infrastructures, and trillions of rupiah for rehabilitation and reconstruction. In the last few years a new analytical method has been developed that can accommodate a three-dimensional earthquake source attenuation model (3-D). It can better illustrate the wave propagation attenuation compared to the 2-D model used for the preparation of SNI-1726-2002 earthquake maps. Further intensive research on current attenuation functions and recent studies of active faults in Indonesia further strengthens the need to improve Indonesia's current earthquake map. Therefore in 2012, the government issued the latest regulations/standards namely SNI 1726:2012 concerning Design Procedures for Earthquake Resistance of Building and Non-Building Structures [5].

*Corresponding author: taufiq_saidi@unsyiah.ac.id

IOP Conf. Series: Materials Science and Engineering **523** (2019) 012039 doi:10.1088/1757-899X/523/1/012039

In this standard the earthquake zoning map is made based on the bedrock acceleration of the short period (0.2 seconds) "S_s" and the bedrock acceleration of 1 second period "S_1". Furthermore, in design of building structures, it was agreed that the new Indonesian earthquake map would be prepared based on the most recent seismicity data, the latest research results on seismotectonic conditions in Indonesia, and using the 3-D model with reference to International Building Code (IBC) 2009 ([2], [6]).

One of the areas with high earthquake intensity in terms of earthquake zoning maps in both SNI-1726-2002 and SNI 1726:2012 is Aceh Province. This makes the engineers need a special attention when designing building structures in Aceh Province. The difference in earthquake zoning maps in both standards for each district causes a difference in the maximum spectral response value so that a separate analysis is needed to determine the effect on the structure response behavior in the form of basic shear force (V) and natural vibration period (T). This research was conducted with the aim to get a comparison of the spectra response graph, natural vibration period and basic shear force from the three-story building structures which is calculated based on the two regulations at 23 districts in Aceh Province.

2. Literature Review

2.1 Earthquake map

The earthquake design load in SNI-1726-2002 uses a 10% probability value exceeded in 50 years or had a 475 year return period. Whereas SNI 1726:2012 uses earthquake maps with a probability value of 2% exceeded in 50 years or has a return period of 2475 years. In SNI 1726-2002, earthquake maps are divided into 6 regions. Region 1 represents the lowest seismic value and region 6 represents the highest seismic value while the earthquake area according to SNI 1726:2012 is determined based on the parameter value of the bedrock acceleration of short period (0.2 seconds) S_s and bedrock acceleration period of 1 second S_1 ([1], [5]).

2.2 Soil Classification

Ideally an initial study to define the seismic site classification must be carried out prior to the analysis as shown in [7] and [8]. There are many factors contributed to seismic site classification i.e. basin geometry [9]. Due to some constraints, the used seismic site classification of this study is based on the adopted standards. Based on SNI 03-1726-2002, the soil is classified into hard soil, medium soil, soft soil and special soil [1]. On the contrary SNI 03-1726-2012 classifies the soil into several site classes, as the basis for determining the F_a site coefficient (amplification factor at 2s) and site coefficient F_v (amplification factor at 1s) [5].

2.3 Response Spectrum

According to SNI-1726-2002 design response spectrum C-T is divided into 6 earthquake areas with 3 types of soil namely hard soil, medium soil and soft soil. In the design response spectrum the value of C is the earthquake response factor expressed in gravitational acceleration (g) and T is the vibration natural period expressed in seconds. For T = 0 the value of C becomes the same as spectrum acceleration A_o. Base rock acceleration and peak ground acceleration values are also determined as minimum accelerations that must be taken into account in designing the building structures to ensure minimum robustness of the building structures.

In SNI 03-1726-2012, the response spectrum C-T is calculated based on the following parameters:

(a) Bedrock acceleration. Bedrock acceleration is determined respectively for acceleration response of 0.2 seconds (S_s) and 1 second (S_1) in seismic ground motion maps and expressed in decimal numbers on gravitational acceleration (g).

(b) Site classes. Land site classes are determined whether they are included in SA, SB, SC, SD, SE and SF

(c) Site coefficient. Site coefficients and response parameters of the maximum earthquake acceleration spectra considered risk-targeted (MCER) at the ground surface required a seismic amplification factor for a period of 0.2 seconds and a 1 second period. SNI 1726:2012 Article 6.2 states that the amplification factor includes vibration amplification factors related to the acceleration that represent short period vibrations (F_a) and vibration amplification factors related to the acceleration that represent vibrations of 1 second period (F_v). Spectra response parameters in short periods (S_{MS}) and 1 second period (S_{M1}) adjusted to the influence of site classification must be determined by the following formula:

$$S_{MS} = S_S\, F_a \tag{1}$$

$$S_{M1} = F_v\, S_1 \tag{2}$$

where S_S = MCER earthquake acceleration response parameters are mapped for short periods; S_1 = MCER earthquake acceleration response parameters are mapped for a period of 1 second.

(d) Design spectra acceleration. The design spectra acceleration parameter*s* for short periods (S_{DS}) and 1 second period (S_{D1}) must be determined through the following formula:

$$S_{DS} = 2/3\ S_{MS} \tag{3}$$

$$S_{D1} = 2/3\ SM1 \tag{4}$$

$$T_S = (S_{D1}) / S_{Ds} \tag{5}$$

$$T_0 = 0.2\ (S_{D1}) / S_{Ds} \tag{6}$$

where: S_{DS} = design *spectrum* acceleration parameter in short periods, S_{D1} = design *spectrum* acceleration response parameter for a period of 1 second, T_S = *f*undamental period of structure*s* in short periods (0.2seconds)

(e) Design response spectrum acceleration. The S_a value for a period*s* smaller than T_0 is calculated by the following equation:

$$S_a = S_{DS}\ (0.4 + 0.6\ T / T_0) \tag{7}$$

In periods greater than or equal to T_0 and smaller than or equal to T_S, the S_a value is equal to S_{DS}. For periods larger than T_S, the response spectrum of the design acceleration is taken based on the following equation:

$$S_a = (S_{D1}) / T \tag{8}$$

where: T = fundamental period of structure*s and* S_a = *design response* spectrum acceleration.

All parameters are plotted into the graph and produced a design response spectrum. The acceleration of the bedrock at 0.2 seconds and 1.0 second is first set. The acceleration of the bedrock is then amplified so that it becomes acceleration at ground level or below the building structures for T_0, 0.2 seconds, T_s, and 1.0 second. These values are used as earthquake loads in building structure design [10].

3. Method

3.1 Aceh earthquake map and analytical model

This study uses an earthquake map that is juxtaposed with the administrative map of the Aceh Province so that it can be determined in an outline of the boundaries of each district in both earthquake maps SNI-1726-2002 and SNI 1726:2012. The analytical model used in this study is a four-floor reinforced concrete building structure that has been built in Banda Aceh.

3.2 Stage of research

The stages of research in this study can be described [11], as follows:

1) Collecting Aceh Province Administration Maps and Indonesian Earthquake Maps;
2) Comparing the Aceh Province Administration Map to the Indonesian Earthquake Map so that the boundaries of each district are known in the earthquake map;

IOP Conf. Series: Materials Science and Engineering **523** (2019) 012039 doi:10.1088/1757-899X/523/1/012039

3) Determine the earthquake area of each district to choose the response of SNI-1726-202 spectra and determine the S_s and S_1 values to calculate the response of SNI 1726:2012 spectra;
4) Make the graph of the response spectra for 23 districts based on SNI-1726-2002 using data from the earthquake area that has been determined;
5) Graph the response spectra for 23 districts based on SNI 1726:2012 using the help of Indo spectra software;
6) Model the building structure into SAP2000 software;
7) The input load is evenly distributed, centralized and the response spectra as a dynamic load structure in SAP2000 in accordance with the provisions of SNI-1726-2002 and SNI 1726:2012;
8) Analyze and make a comparison of earthquake design parameters.

4. Results and Discussion

4.1 Response spectra
Response spectra for all 23 district in Aceh Province is shown in Table 1. The higher in response spectra in SNI 1726:2012 is due to adjustments to the large earthquakes those occurred after SNI 03-1726-2002 was set, as well as changes in the value of the earthquake return period from the 475 year return period to a 2475 year return period. In addition, the lower in SNI-1726-2002 response spectra is due to that SNI-1726-2002 still uses a calculation method of decreasing earthquake intensity (attenuation) in 2 dimensions by assuming that the entire region has the same depth of earthquake propagation rock so that the resulting earthquake map is still lacking in detail. In contrast SNI 1726-2012 uses the attenuation method in 3 dimensions so that each point being reviewed will have a different spectral value even though it is in one type of earthquake zone, depending on the extent and depth of the media of earthquake propagation. As results, there are several districts that experienced a decline in SNI 1726:2012 spectra values. In the case of Banda Aceh, the used response spectra in this study is lower than the response proposed by [12].

Figure 1 shows one of the comparison of response spectrum according to SNI 1726-2012 and SNI 03-1726-2002 for Aceh Tengah District.

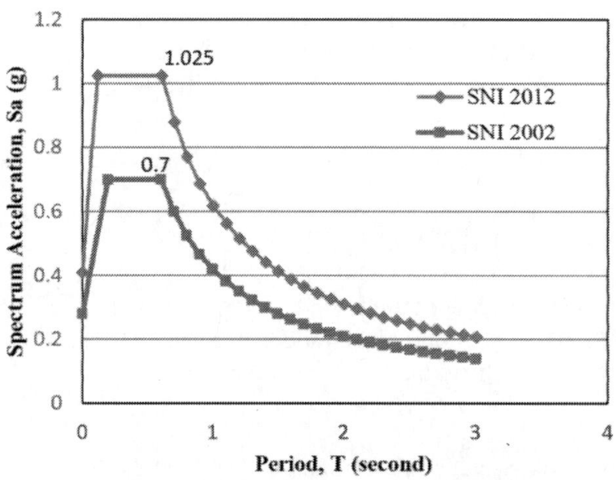

Figure 1. Comparison of the response spectrum of SNI-1726-2002 for Aceh Tengah District

Table 1. Comparison of SNI 1726:2012 and SNI-1726-2012 spectra value (1/2)

District	Maximum spectral acceleration (g) according to SNI 1726:2012 (A)	SNI-1726-2002 (B)	A/B
Banda Aceh	0.906	0.830	1.092
Aceh Besar	0.946	0.830	1.140
Pidie Jaya	0.733	0.700	1.047
Sabang	0.820	0.700	1.171
Pidie	0.897	0.830	1.081
Aceh Jaya	0.830	0.830	1.000
Bireuen	0.681	0.700	0.973
Aceh Barat	0.841	0.830	1.013
Lhokseumawe	0.578	0.550	1.051
Aceh Utara	0.611	0.700	0.873
Bener Meriah	0.683	0.700	0.976
Aceh Tengah	1.025	0.700	1.464
Gayo Lues	0.867	0.830	1.045
Nagan Raya	1.090	0.830	1.313
Aceh Barat Daya	0.848	0.830	1.022
Aceh Tamiang	0.620	0.700	0.886
Aceh Selatan	0.821	0.830	0.989
Langsa	0.585	0.550	1.064
Aceh Tenggara	1.003	0.830	1.208
Subulussalam	0.829	0.830	0.999
Aceh Timur	0.693	0.700	0.990
Aceh Singkil	1.000	0.900	1.111
Simeulue	1.000	0.900	1.111

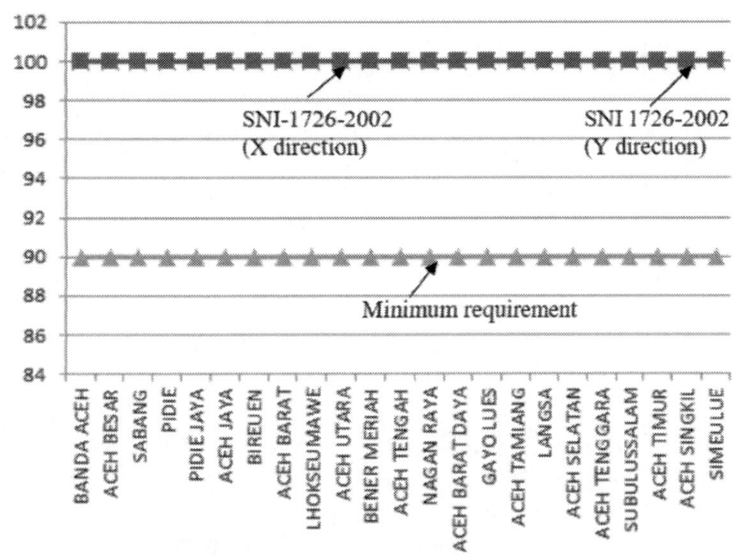

Figure 2. Effective modal mass participation based on SNI-1726-2002

IOP Conf. Series: Materials Science and Engineering **523** (2019) 012039 doi:10.1088/1757-899X/523/1/012039

4.2 Minimum effective modal mass participation

Based on the results obtained, the value of minimum effective modal mass participation in the model studied for the all districts are 99.98% and 99.99% in X and Y directions, respectively as shown in Figure 2 and Figure 3. These values are satisfy the provisions of SNI-1726- 2002 and SNI 1726:2012 which requires mass participation at least of 90%.

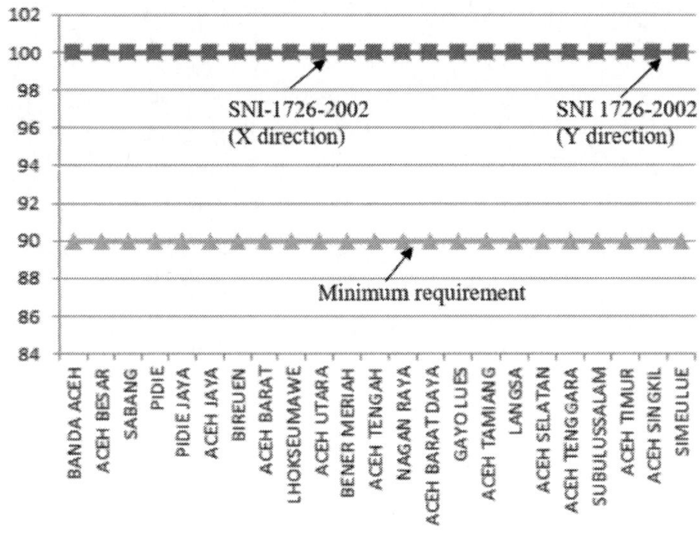

Figure 3. Effective modal mass participation based on SNI-1726-2002

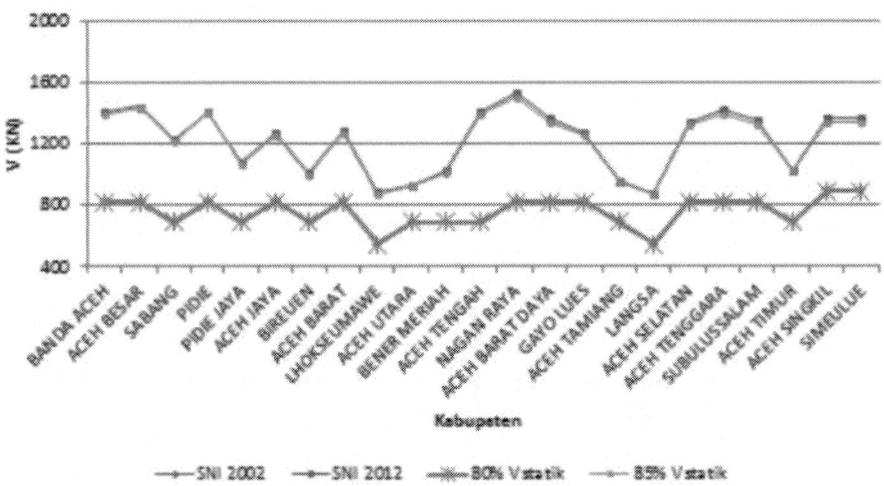

Figure 4. Base shear force in X direction

4.3 Base shear force

Figure 6 and 7show base shear force in the direction of X and Y directions for all districts according to SNI 1726:2012 and SNI-1726-2002. The increase in base shear force of SNI 1726:20012 is due to three reasons: firstly because of an increase in the response spectra value on SNI 1726:2012 so that the load received by the structure becomes larger, secondly because of the increase in the value of building safety factors for residential buildings which is 1.0 on SNI -1726-2002 becomes 1.5 in SNI 1726:2012, where

IOP Conf. Series: Materials Science and Engineering **523** (2019) 012039 doi:10.1088/1757-899X/523/1/012039

the greater the priority factor, the design base shear force of the building will also increase and the thirdly is the increased input combined load.

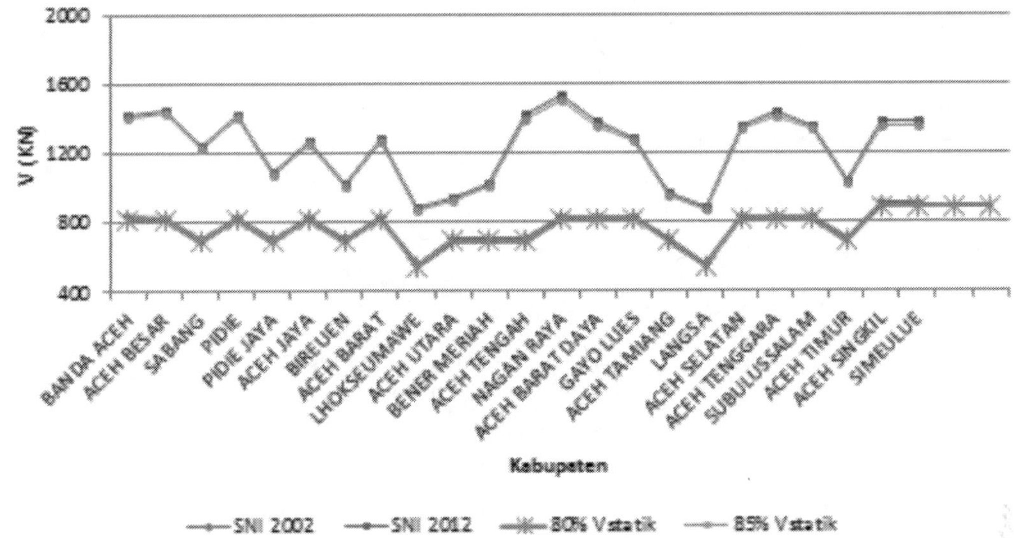

Figure 5. Base shear force in Y direction

4.4 Structure natural vibration period

Figure 6 and 7 show structure natural vibration periods for all districts and their comparison with requirement of SNI-1726-2002 and SNI 1726:2012, respectively. From those two figures, it can be seen that the structure natural vibration periods for all districts in Aceh Province are in accordance with the requirement of those two standards.

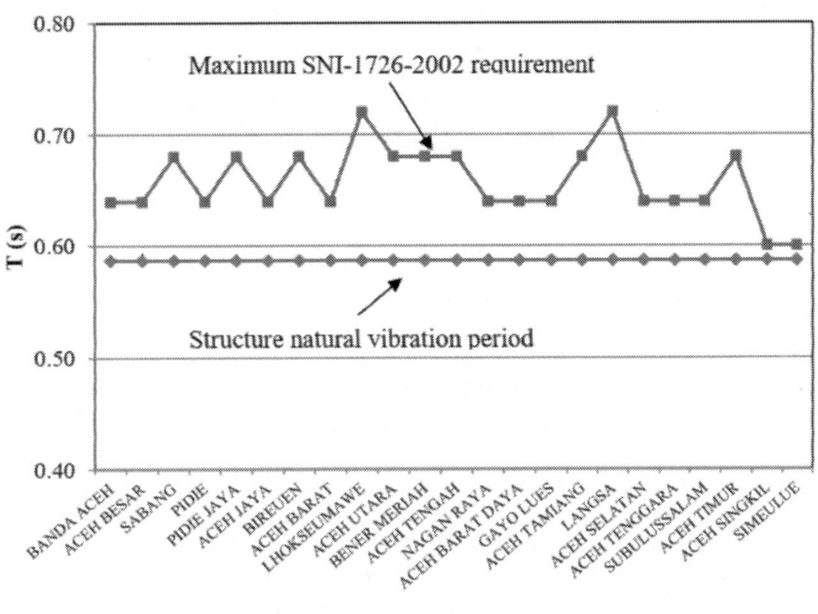

Figure 6. Structure natural periods and their comparison with SNI-1726-2002 requirement

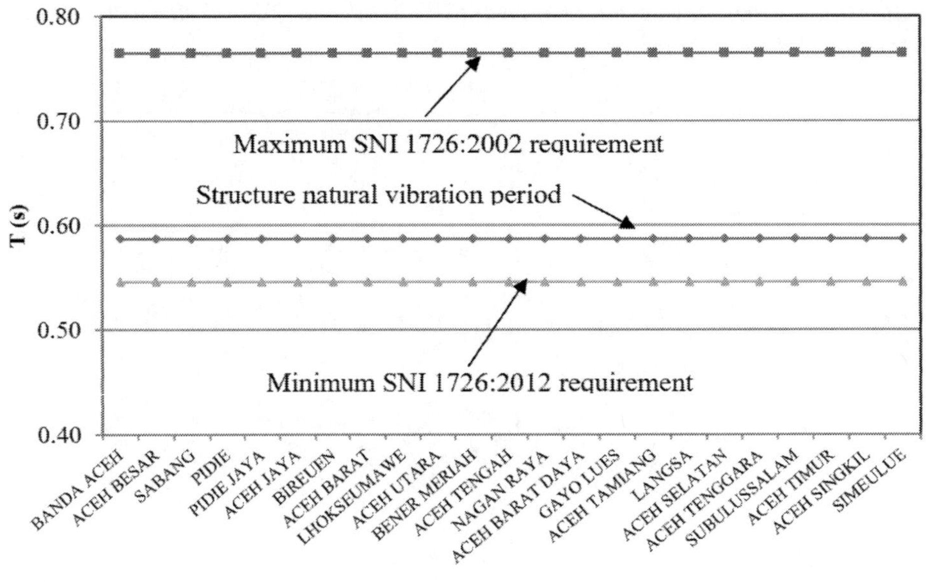

Figure 7. Structure natural periods and their comparison with SNI 1726:2012 requirement

5. Conclusions

Based on the results of this research, the following conclusions can be drawn:

1. Among 23 districts in Aceh Province, 15 of them which are Banda Aceh, Aceh Besar, Pidie Jaya, Sabang, Pidie, Aceh Barat, Lhokseumawe, Aceh Tengah, Gayo Lues, Nagan Raya, Aceh Barat Daya, Langsa, Aceh Tenggara, Aceh Singkil and Simeulue Districts have higher SNI 1726:2012 based maximum spectral accelerations compared to those of SNI-1726-2002. Aceh Jaya District has similar maximum spectral acceleration for both standards. The remaining 7 districts which are Bireun, Aceh Utara, Bener Meriah, Aceh Tamiang, Aceh Selatan, Subulussalam and Aceh Timur have lower SNI 1726:2012 based maximum spectral accelerations than those of SNI-1726-2002.
2. The modal mass participation and structure natural vibration period for all 23 districts satisfy the requirement of both SNI-1726-2002 and SNI 1726:2012 standards.
3. Base shear force for all districts calculated according to SNI 1726:2012 higher than that of SNI-1726-2002

6. References

[1] SNI-1726-2002 *Design Standard for Earthquake Resistance of Building Structures*. (Dept of Settlement and Regional Infrastructures), Bandung (in Bahasa Indonesia)

[2] Irsyam M, Sengara W, Aldiamar F, Widiyantoro S, Triyoso W, Hilman D, Kertapati E, Meilano I, Asrurifak M, Ridwan M, and Suhardjono 2010 *Indonesian Earthquake Map Revision Study Results Summary* 2nd edition (Public Work Ministry), Bandung (in Bahasa Indonesia)

[3] Setiawan B and Saidi T 2012 *Proceeding of 2nd Annual International Conference Syiah Kuala University 2012 and 8th IMT-GT Uninet Biosciences Conference*, 22-24 November 2012, AAC-Dayan Dawood - Banda Aceh, Indonesia

[4] Saidi T, Muttaqin and Sarana D 2015 Studi Komparasi Perilaku Respons Struktur Gedung Beton Bertulang yang Dianalisis Berdasarkan Respons Spektra Event Gempa Aceh Tahun 2010 – 2013 dan Respons Spektra SNI 1726:2012, *Proc KoNTekS 9*, pp 741-748

[5] Sarana D, Saidi T and Muttaqin 2015 *Jurnal Teknik Sipil* **4**(3) 225-236

[6] SNI 1726:2012 *Design Procedures for Earthquake Resistance of Building and Non-Building Structures* (National Standardization Agency)

[7] Setiawan B, Jaksa MB, Griffith MC and Love D 2018 *Boll. di Geofisica Teorica ed Applicata* **59** (1) 27–46
[8] Setiawan B, Jaksa MB, Griffith MC and Love D 2018 *Soil Dyn Earthq Eng.* **110** 244–261
[9] Setiawan B, Jaksa MB, Griffith MC., and Love D 2018 *Eng Geo.* **243** 145 – 159
[10] International Code Council 2009 *International Building Code (IBC)* 1st edition
[11] Budiono B, dan Lucky S 2011, *Studi Komparasi Desain Bangunan Tahan Gempa Dengan Menggunakan SNI 03-1726-2002 dan RSNI 03-1726-201x* (Penerbit ITB), Bandung
[12] Saidi T, Aulia TB, Setiawan B, Abdullah N, and Hasan M 2018 MATEC Web of Conferences **197**, 10001 (2018), 4pp

IOP Conf. Series: Materials Science and Engineering **523** (2019) 012040 doi:10.1088/1757-899X/523/1/012040

The potential utilization of natural materials as a wall covering of the building in heat reduction

K Jeumpa[1,*]
Doctoral Study Program of Technology and Vocational Education, Faculty of Engineering, Padang State University, Indonesia

E-mail: ipajeumpa@gmail.com

Abstract. This paper aims to examine the potential of natural coco fibre to be utilized as a wall covering in building construction for sun radiation protection. Engineering innovation in building construction to control the building's temperature is important. Lowering the building's room temperature can be carried out by modifying the wall construction materials. Thus, the usage of an air conditioner which is not environmentally friendly can be reduced. The wall modification can give more protection and dampen the sun's heat, hence the thermal comfort can be obtained. Natural materials, moreover, can be used in the wall modification to reduce the room temperature. In this study, a comparative analysis was carried out between two materials: (1) the coco fibre coated with cement mixture; (2) the mixture of cement and sand, both were used as materials to plaster the wall. The cement mixture acted to bind the compacted coco fibre sheets. The material was evaluated after the cement mixture had been hardened. The test was carried out by providing heat on one side and measuring the temperature on the other side of the material. The results were significant between the two materials. This indicates the potential of coco fibre as a heat insulation in the wall construction materials that can minimize solar thermal radiation into the room.

1. Introduction

Sunlight is the main source of heat in the building, the heat of sunlight mainly enters the room through roofs and walls. About 83% of the sun's infrared rays going through from the roof and building walls are absorbed and emitted into the room by radiation, conduction and convection. In humid tropical climates, building coating is important to protect the occupants from the sun's heat, also reducing the radiant heat forwarded into the building. Therefore, one of the functions of the building coating is to control or to reduce the heat load from solar radiation into the building (transmission). The entry of solar radiation into the building can be through a vertical cover that is a wall or a horizontal or sloping cover that is roof. Cooling the space in this way is classified as passive cooling system that relies heavily on air movement as a hot carrier medium in the morning to late afternoon to reduce room's temperature, (Cook, 1985; Giovani, 1994) in [1].

The rise of earth's temperature due to global warming has increased the usage of air conditioner as the demand of comfortable room's temperature has increased as well. This increases the energy use thus further worsen the global warming. Bourdeau, 1999 revealed the fact that 50% of the energy absorbed in a building is only for refrigerators, and at least, 30% of a country's total energy is used in housing. This figure is only applicable in developed countries that are more manageable. Hence in developing countries, the figure will be bigger. If this is not handled strategically, it will devastate the sustainability of the national development. Based on these factors, innovations in building construction are desired to conserve energy and environment.

The heat absorbed in the wall propagates into the room and increases the room's temperature. Walls can be made to dampen the sun's heat and protect the indoor space from the heat. Therefore, one way to inhibit heat from the outside is through the wall modification.

* Corresponding author: ipajeumpa@gmail.com

Non-heated rooms can reduce the usage of air conditioner. The reduced usage of air conditioner means to save energy and also save the earth and the environment.

Some researchers had studied on environmentally friendly innovation of building construction materials to combat global warming. For instance [3] had conducted a research on lightweight concrete panels focused on the pearlite as a concrete panel mixture as a heat insulator. Likewise, [4] had conducted research on mortal mix materials focused on the utilization of pumice breksia as the main ingredient of instant mortar as a heat reducer. While Hary Wibowo had conducted a hot conductivity study between styrofoam and rice coir, focusing on the comparison between styrofoam particleboard and rice coir particleboard as heat insulators. Similarly, foreign researchers had studied building materials for heat muffling. For instance, [5] had studied the addition of retro-reflective materials on the wall, which was found to be capable to improve the building's temperature control by reflecting solar radiation back in the opposite direction. Furthermore, [6] conducted a study determining the thermal conductivity of periwinkle shells (seashells) reinforced cement as the building wall construction materials. Meanwhile, [7] conducted a study on assessing the impact of energy performance of hygrothermal behavior of plant-based insulation products in buildings, predicting indoor climatic conditions, and preventing unexpected degradation risks.

Environmental awareness can be realized with the use of materials derived from nature as a form of energy conservation and environmental protection. For example, coconut coir (coco fibre) is minimally utilized in Indonesia, for the most part it is considered as a mere waste. Coco fibres are widely available in Indonesia, as they can be obtained from the coconut harvest each year. Not to mention, 35% of the coconut total weight is consisted of coco fibres.

In this regard, this research is urgently important to evaluate the potential of coco fibre, as a natural material, to reducing heat in the building. The main question of this research is how potential is the coco fibres to act as a heat insulator. This research was conducted with the aim to evaluate the potential of coconut coir, which later can serve as a the building wall coating to reduce heat from the external environment.

2. Materials And Methods

The materials of this research included: (1) coir reinforced cement; and (2) a mixture of cement and sand (commonly used wall plaster material). A compacted coconut fibre sheet was coated with a cement mixture that also acted to bind the coir. The process of the material preparation can be observed in Figure 1 below.

Figure 1. The process of making coir test material

This research is experimental in nature to study the potential of coconut coir as heat insulator. In this research, the two materials were compared by their ability to inhibit the heat propagation. The test was carried out after the mixture had dried and hardened. The test was carried out by providing heat on one side and measuring the temperature on the other side of the materials respectively. The heat applied to the material surface was 60 °C for 2 hours.

3. Results And Discussion

Environmental awareness can be realized with the use of materials derived from nature as a form of energy conservation and environmental protection. Coconut fruit consists of fibres (coir), shell (the hard layer), coconut water, and coconut meat. The fibres can be found on the outer layer of the coconut protecting the shell. Meanwhile the shell and the membrane attached in the inner side, serve to protect the seeds. Inside the coconut, there is liquid containing many enzymes known as coconut water. The solid phase, attached on the inner shell which keeps growing as the coconut getting ripe, called coconut meat. Coconut coir provides protection for the coconut from the external interference including from the heat of sun. Based on this fact, coconut coir is considered to be potential for heat insulation, hence this research to evaluate the application of coconut coir application as a building wall coating is necessary.

This study compared the materials with and without the coir. The test results can be observed in the graph in Figure 2 and Figure 3 below.

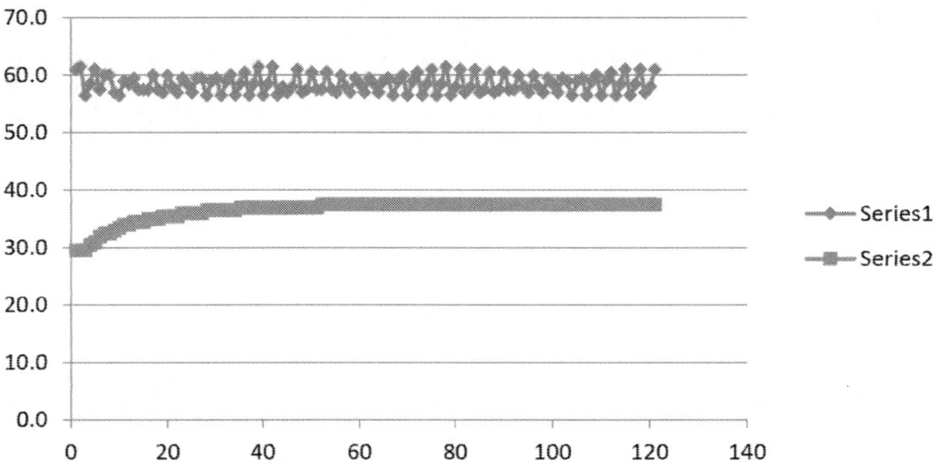

Figure 2. The test result of the material without coir

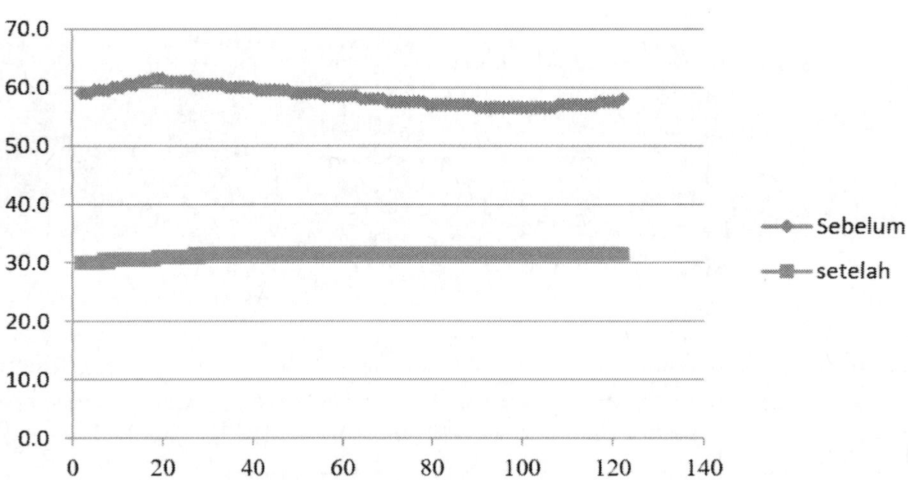

Figure 3. The test results of the material with coir

The graph in Figure 2 shows the result of the test material without coir. The y axis shows the temperature on the material surface expressed in degree Celcius. The x-axis shows the length of time the heat is applied during the test, expressed in minutes. Series 1 exhibits the amount of heat applied on one side of the material expressed, while the series 2 indicates the amount of heat measured on the other side of the material. The graph explains, when the given heat was 60 °C, the temperature measured on the other side was initially around 30 °C. Within the 40th minute, the heat increased rapidly on the other side, indicates a quick heat transmission. A constant heat reached around 39 °C at the 120th minute.

The graph in Figure 3 shows the test results of the coir reinforced material. The y axis shows the temperature on the material surface expressed in degree Celcius. The x-axis shows the length of time the heat is applied during the test, expressed in minutes. Series 1 exhibits the amount of heat applied on one side of the material expressed, while the series 2 indicates the amount of heat measured on the other side of the material. As opposed to the material without coir, the material with coir exhibits no rapid increase of the temperature on the other side at the first 40 minutes (30-31 °C). This indicates a good insulation properties reinforced by the coir towards the material. The constant temperature of 31 °C reached under 40 minutes of the treatment. A significant difference of the heat applied and transmitted can be observed in the test results. However the coir reinforced material is more capable to reduce the heat as opposed to one without coir reinforcement.

This study may still be far from perfection because the test equipment required is still limited. The heat given is still too high (60 °C) because it is rather difficult to set the tool to produce heat near the temperature of the actual environment. Therefore further study is required to measure the effectiveness of coir reinforced material in heat insulation.

4. Conclusion

From the measured temperature compared with the two materials, it can be concluded that, the coir material has the potential to be utilized as a building coating insulation material which serves to reduce heat transmission from the outside environment.

5. Acknowledgments

This research is a preliminary study for my dissertation, which will be used as input in future research. Thank you to Mr. Prof. Selamat Triono and Mr. Rusnardi Rahmat, Ph.D as promoters and co-promoters and Mr. Tulus Ikhsan Ph.D who are willing to review this paper. This research will later be presented to those with an interest in buildings and observers of energy-efficient buildingsThank you countless to those who helped in this research.

6. References

[1] Mintorogo, D Santoso, Widigdo, Wanda K, Yuniwati, Anik 2013. *Efektivitas Styrofoam Sebagai Isolator Panas pada Atap Miring di Surabaya* (Project Report. Petra Christian University).

[2] P S Zalukhu, I Irwan, D M Hutauruk, 2017. *Journal of Civil Engineering, Building and Transportation*, **1**, 1.

[3] S A Yuswita dkk, 2011. *Jurnal Ilmu Pengetahuan dan Teknologi*, **29.**

[4] S Agus dkk, 2013. *Pemanfaatan Pumice Breksia Sebagai Material Utama Mortar Instant Peredam Panas Untuk Mendukung Teknologi Bahan Bangunan Gedung Ramah Lingkungan* (Lembaga Penelitian dan Pengabdian Masyarakat, Universitas Negeri Yogyakarta).

[5] X Meng, T Luo, Z Wang, W Zhang, B Yan, J Ouyang, E Long. 2017. *Energy and Buildings Journal* **127** (488-498)

[6] Omubo-Pepple V.B, Opara F.E, Ogbonda C, 2010. *Journal of Engineering and Applied Sciences* **5** (2): 178-180.

[7] D Samuel, E Arnaud, B Christophe, L Frédéric 2015. *Journal of Building Physics* **39** (5) pp. 421-439.

The effect of rainfall interception loss by Palm-Oil tree towards flood discharge in Seunagan watershed of Nagan Raya district of Aceh Province

A Yulianur BC[1,*], Azmeri[1], Khairuddin[2]
[1]Lecturer, Faculty of Civil Engineering, Syiah Kuala University, Darussalam, Banda Aceh, Indonesia
[2]Officer, Water Resources Department of Aceh Province, Lueng Bata, Banda Aceh, Indonesia

E-mail: fian_7anur@yahoo.com

Abstract: Flood discharge comes from the surface run-off affected by factors such as evaporation, evapotranspiration, infiltration and interception. This research aims to predict the magnitude of interception demonstrated by a palm-oil tree and its effects toward flood discharge. This paper covers the calculations for rainfall interception loss by a palm-oil tree, designed areal rainfall, and designed flood discharge. During the calculations, the watershed is divided into two sub-watersheds. Rainfall data were obtained from Meulaboh, Beutong and Alue-Bilie station. Areal rainfall distribution obeys the pattern of Gumbel distribution. Areal rainfall of 100 years returns period for each sub-watershed are 230.81 mm and 197.59 mm before the inclusion of interception effects, while after interception effects are included the areal rainfalls are 152.94 mm and 189.09 mm. The calculated effective rainfall using the SCS (soil conservation service) method where the curve number (CN) before interception effects are 72.19 and 73.15 while 81.54 and 73.99 after the interception is accounted. Flood discharge is calculated using the method of unit hydrograph synthetic SCS. Calculation results show that designed flood discharge of 100 years return period before and after the effects of interception were 3,042.72 m3/s and 2,521.23 m3/s, respectively. It shows that rainfall interception loss by the palm-oil tree can reduce the flood discharge.

1. Introduction

Rainfall within a particular watershed will become runoff, and runoff is affected by factors such as interception. Hydrologist conducted numbers of researches [5],[6],[9],[13],[14] in modeling the interception of rain by various types of plants. Rao (1986) conducted a rain interception by cashew-plant at Kottamparamba India in 1983 and 1984. The percentage of rain that intercepted by this plant was 31%. Kaimuddin (1994 commenced a prediction modeling study of rain interception by Pines (*Pinus merkussi),* Dammar *(Agathis loranthifolia)* and Needlewood *(Schima wallichii)* in the educational forest of mount *Walat, Sukabumi, Indonesia.* He found that the rain interception demonstrated by Pines, Dammar, and Needlewood were 15.7%, 14.7% and 13.7%, respectively. Other than that, higher rainfall inflicting higher interception and older plants demonstrated higher interception.

*Corresponding author: fian_7anur@yahoo.com

This research aims to identify the rain interception contributed by the palm-oil tree. Simultaneously, accurately predicting the surface runoff of *Seunagan* watershed by perceiving those achieved interception by the palm-oil tree. The research was conducted on 10 years old palm-oil tree, rainfall were measured based on stemflow and throughfall. To determine the magnitude of interception, the amount of rainfall is deducted by the amount of stemflow and the amount of throughfall. The calculated interception value was then being input in designed flood discharge calculation for Seunagan watershed. Synthetic unit hydrograph soil conservation service (SCS) method was utilized to calculate the designed flood discharge where the effective rainfall as the surface runoff were influenced by land cover and soil type. Correspondingly, switching land cover into palm-oil tree contributes towards the magnitude of curve number.

2. Literature Review

2.1 Rain Interception

Rainfall interception is a process of rain falling on the surface of vegetation, detained for an immediate period, then being vaporized into atmosphere or absorbed by pertinent vegetation. Interception process occurs continuously during the rainfall and post-rainfall until the surface of the vegetation canopy dried out. During rainfall, some of the water never reach the ground, therefore it does not contribute in ground humidity, runoff water or groundwater. That water will return to the atmosphere as intercepted water by a canopy, litter, and lower plantation [1]. This is called rainfall interception loss.

Rainfall on the surface of vegetation canopy reaches the ground surface through two mechanisms, namely throughfall and stemflow. Throughfall is the portion of precipitation which immediately falls to the ground surface through the space between canopy/leaves or drips through leaves, stems and branches Stemflow is the rainfall which reaches the ground surface by flowing inside the plant stem.

2.2. The Measurement of Rainfall Interception

Measuring interception in canopy scale is performed by the volume balance approach. The measurement of rainfall interception in trees using volume balance approach is illustrated by Figure 1 as follows:

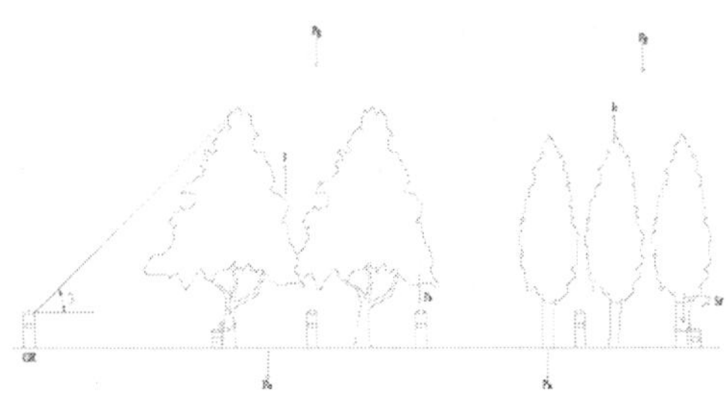

Figure 1. Measurement of rainfall interception in the plantation (Asdak, 2004)

According to the above Figure 1, an equation can be established to determine the magnitude of interception demonstrated by tree plantation during a particular plot of research. The equation is formalized as follow [1]:

$$I_c = P_g - (T_h + S_f) \tag{1}$$

Where I_c is canopy interception, P_g is the total rainfall (gross precipitation) (mm), T_h is throughfall (mm), and S_f is stemflow (mm).

2.3 Measurement of Rainfall

Rainfall measurement is performed to determine the amount of rainfall which not pertained by plants. Daily and weekly rainfall are initial components that are required to be determined using rainfall measurement gauge in an open field without any hindrance by any plants.

2.4 Measurement of Throughfall

The most integrated method in calculating throughfall is by using an apparatus called plastic-sheet gauge or by installing a plastic sheet beneath forest stands which collect water from throughfall as well as stem flow water. The arrangement of plastic is designed in such a way so that the water gathered from throughfall and stemflow are directed into gathering tank [1].

2.5 Measurement of Stemflow

Stemflow is obtained by installing zinc plates forming a circle covering the plants circumference so that the flow from branches and stem could be fully collected into gathering tank or recorded by tipping bucket gauge [1].

2.6 Thiessen Method.

Thiessen Method is one method to calculate the areal rain with the following formula [7],[15],[16].

$$P = \frac{A_1P_1 + A_2P_2 + A_3P_3 + ... + A_nP_n}{A_1 + A_2 + A_3 + ... + A_n} \quad (2)$$

where P is average areal rainfall, P_1, P_2, ...Pn are rain in station 1,2,.....n, while A_1, A_2, ...An are a representative area of station 1,2, 3....,n.

2.7 Estrem Value Distribution

Designed rainfall is calculated according to the type of data distribute matching. Among other distributions are as follows.

2.7.1 Log normal and log Pearson type III distribution. The equation of lognormal and log Pearson type III distribution are[15],[16]:

$$Log\ R_T = Log\ R + K_T S_d \quad (3)$$

Where R_T is designed rainfall for T years return period (mm), R is average rainfall (mm), K_T is frequency factor, S_d is standard deviation (mm). Frequency factor (K_T) for lognormal distribution is determined based on C_v and return period, while frequency factor (K_T) for log Pearson type III distribution is determined based on C_s and return period.

2.7.2 Gumbel Distribution. The estimation of rainfall by Gumbel distribution is as follows [15],[16]:

$$R_T = R + K_T S_d \quad (4)$$

$$K_T = \frac{Y_T - Y_n}{S_n} \quad (5)$$

$$Y_T = -Ln\left(-Ln\frac{T-1}{T}\right) \quad (6)$$

Where R_T is designed rainfall for T years return period (mm), R is average rainfall (mm), K_T is frequency factor, S_d is standard deviation (mm), T is the year of the return period, Yn is reduced mean Sn standard deviation, Y_T is reduced variate.

2.8 Alternating Block Method

Alternating Block Method (ABM) is one method to form a designed rainfall hyetograph [7]. ABM requires hourly rainfall intensity data concentrated at concentration time. The Mononobe formula is one of the formulas for calculating the hourly rainfall intensity, as follows [16].

$$I_t = \left(\frac{R_{24}}{24}\right)\left(\frac{24}{t_c}\right)^{2/3} \quad (7)$$

Where I_t is T years return period rainfall intensity (mm/hour), R_{24} is T years return period maximum daily rainfall (mm), t_c is the concentration-time that can be calculated by the Kirpich Formula, as follows[7].

$$t_c = 0.0195\left(\frac{L}{\sqrt{S}}\right)^{0,77} \tag{8}$$

Where t_c is concentration-time (minute), L is river length (m), and S is the river slope.

2.9 SCS Effective Rainfall

Effective rainfall or excess rainfall are parts of rain which become direct flow in the river. The soil conservation service (SCS) [7] has developed a method to calculate effective rainfall within heavy rain as expressed:

$$P_e = \frac{(P-0,2S)^2}{P+0,8S} \tag{9}$$

$$S = \frac{25400}{CN} - 254 \tag{10}$$

$$I_a = 0.2\,S \tag{11}$$

$$F_a = \frac{S(P-I_a)}{P-I_a+S} \tag{12}$$

Where P_e is the depth of effective rainfall (mm), P is rainfall depth (mm), S is the maximum water retention potential of the soil, which mainly cause by infiltration (mm), *CN Curve Number* is a function of watershed characteristic, I_a is initial abstraction, and Fa is continuing abstraction.

2.10 SCS Synthetic Unit Hydrograph.

The method of SCS synthetic unit hydrograph is a non-dimensional hydrograph where its ordinate describes the comparison between the discharge and the peak discharge while its abscissa describes the ratio of the interval between time and peak discharge time. Designed flood discharge is calculated based on designed rainfall while considering the watershed condition. Data Analysis forming unit hydrograph is performed by using the equation [7].

$$Q_n = \sum_{m=1}^{n\leq M} P_m U_{n-m+1} \tag{13}$$

Where Q_n is direct runoff discharge of the pulse of n, P_m is effective rainfall of the pulse of m; and U is unit hydrograph ordinate.

3. Research Methodology

This research is conducted in *Desa Jati Rejo Kec. Kuala Pesisir Kabupaten Nagan Raya.* The primary data used was the data collected from the research location consisting rainfall data which measured by a rain gauge, stemflow and throughfall, while secondary data are obtained from corresponding institutes/offices.

3.1 Rainfall Interception Research Method

The research is conducted towards ten years old palm-oil tree, the height of 7.50 m, canopy diameter of 11 m, and stem diameter of 85 cm. The research data analysis was conducted to determine the rain interception by a palm-oil tree which measured in the field. The conducted data analysis were rain interception analysis, rainfall data, throghfall data, stemflow data and the relationship between rainfall and stemflow. Eventually, rainfall measurement results, throughfall, and stemflow are utilized to calculate the magnitude of interception based on volume balance approach.

3.2 The correlation between rainfall and interception

The statistical approach will reveal the relationship between rainfall and interception which illustrated by the correlation graph of interception and rainfall where the regression equation will be obtained to show that relationship. The data of rainfall, after subtracted by the magnitude of interception will be the parameter in calculating the flood discharge at the watershed of *Seunagan Kabupaten Nagan Raya.*

3.3 Rainfall Analysis

The analysis of rainfall data in this research was established based on ten years of rainfall history from 2 rain gauge stations, namely Meulaboh and Beutong. From each rain, the station is calculated designed rainfall of 100-year return period based on suitable distribution. Determination of distribution types performed by Chi-Square test and Smirnov-Kolmogorov test. Then this designed rainfall is used to establish the areal designed rainfall by using the Thiessen Method. Furthermore, formed designed rainfall hyetograph by using ABM for rain that has not and has been reduced by the interception of rain by palm trees. After that for both formed effective designed rainfall hyetograph by using SCS method.

3.4 Analysis of flood peak discharge.

Synthetic unit hydrograph SCS method was utilized to calculate the discharge caused by rain that has not been and has been extracted by the value of rainfall interception loss. Based on the land cover map, Seunagan watershed is assumed to have varieties of land cover thus its flood surcharge should be calculated to obtain its peak surcharge value. Furthermore, mapping the terrain which has the potentials to turn into the palm-oil field based on the hillside slope map, land cover map, and soil type map at Seunagan watershed also based on the requirement of compatibility for the palm-oil tree. The value of interception will be included as one of the variables in the calculation of effective rainfall, the change of land cover also influences the potential retention values (S), where curve number (CN) will alter due to the change of land cover. Furthermore, the analysis will provide the peak surcharge of the flood after the effect of interception. Then a comparison is made between designed flood discharge hydrograph before and after rainfall interception by oil palm trees.

4. Result and Discussion

4.1 Designed Rainfall

In this research, due to its vast area, the watershed is divided into two sub-watershed. According to the calculation results, the watershed area is fathomed to have the area of 960.55 km^2. Designed rainfall is calculated to ascertain the magnitude of rainfall for return periods of 100 years. Meulaboh stations calculation for designed rainfall using Normal distribution result as 279.46 mm. While Meulaboh stations calculation for designed rainfall using Gumbel distribution result as 197.59 mm. The calculation of areal rainfall in Seunagan watershed using Thiessen Polygon method, for sub-watershed one extent of influence by Meulaboh station is 183.81 km^{2,} and Beutong station extent of influence is 269.23 km^2, while sub-watershed two only influenced by Beutong station by 507.51 km^2 with the overall area of the watershed of 960.55 km^2. Thus, designed rainfall for sub-watershed 1 is 230.81 mm and for sub-watershed two is 197.59 mm. After interception effect, designed rainfall for sub-watershed 1 is 152.94 mm and for sub-watershed two is 189.09 mm.

4.2 Determination of Curve Number (CN).

The types of land cover on Seunagan watershed is consist of primary dry forest, secondary dry forest, bushes, swamp grove, plantation, rubber plantation, rice fields, settlement, open field and water body. While the highest percentage of land cover is dominated by primary forest and secondary dry land, which are 39.02% and 24.06% respectively.The calculation presented composite CN value, where composite CN value is a merge of varies CN values multiply by the percentage of the type of soil towards varies type of land cover. Composite CN values for sub-watershed 1 is 72.19 and for sub-watershed 2 is 73.15. Any change in the composition of land cover types due to alteration of land function into palm-oil tree field also altered the value of CN, watershed outlet 1 has suitable palm-oil tree area of 290.78 km^2 with CN value of 81.54 while watershed outlet 2 has suitable palm-oil tree of only 43.04 km^2 with CN value of 73.99.

IOP Conf. Series: Materials Science and Engineering **523** (2019) 012041 doi:10.1088/1757-899X/523/1/012041

4.3 Designed Flood Discharge

The designed flood discharge calculation for the 100-year return period was carried out after first determining the effective rainfall hyetograph on conditions that have not been and there has been an interception of rain by the oil palm trees and also includes CN values before and after the oil palm trees exist. Then using the SCS hydrograph unit method obtained a designed flood hydrograph. Based on flood hydrograph before the influence of rain interception by oil palm trees, sub-watershed 1 have peak flood of 100-year return period is 1,484.80 m^3/sec, and for sub-watershed 2 is 1,557.91 m^3/sec. Thus, the magnitude of discharge passing the whole watershed outlet is 3,042.72 m^3/sec. After the effect of rain interception by oil palm trees, sub-watershed 1 has peak flood of 100-year return period is 1,030.82 m^3/sec, and for sub-watershed 2 is 1,490.41 m^3/sec. Therefore, the magnitude of surcharge passes the watershed outlet after the interception is 2,521.23 m^3/sec. Figure 2 and 3 shows a comparison of designed flood discharge hydrograph before and after rainfall interception by oil palm trees. This picture shows that conversion of land to oil palm plantations will reduce flood discharge due to the interception of rain by oil palm trees.

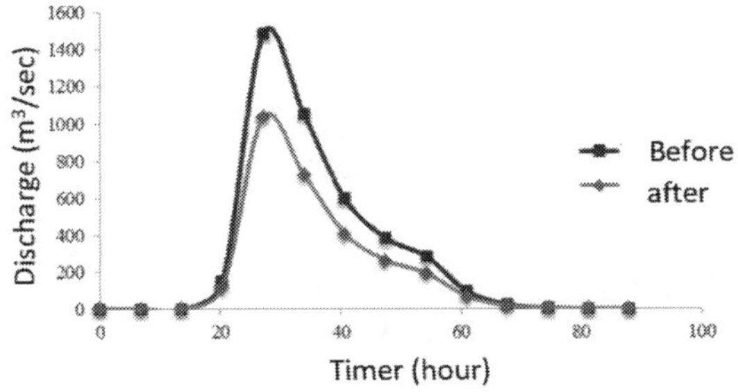

Figure 2. Flood Hydograph of Sub-watershed 1

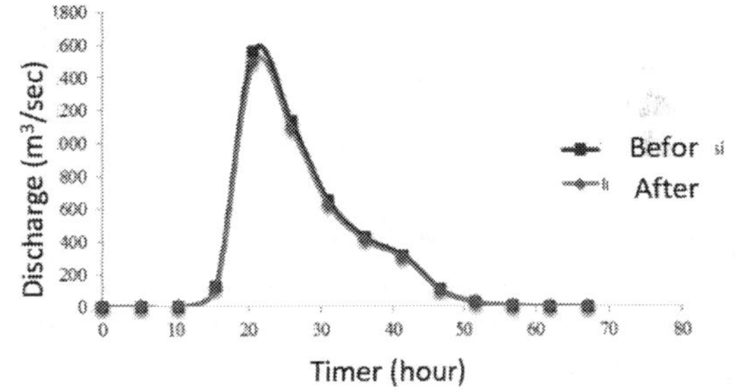

Figure 3. Flood Hydograph of Sub-watershed 2

5. Conclusion

Designed rainfall of 100-year return period for sub-watershed 1 is 230.81 mm and for sub-watershed two is 197.59 mm. After interception effect, designed rainfall for sub-watershed 1 is 152.94 mm and for sub-watershed two is 189.09 mm. Composite CN values for sub-watershed 1 is 72.19 and for sub-watershed two is 73.15. Any change in the composition of land cover types due to alteration of land function into palm-oil tree field also altered the value of CN, watershed outlet 1 has suitable palm-oil tree area of 290.78 km^2 with CN value of 81.54 while watershed outlet 2 has suitable palm-oil tree of only 43.04 km^2 with CN value of 73.99. Before the influence of rain interception by oil palm trees, sub-watershed 1 has peak flood of 100-year return period is 1,484.80 m^3/sec, for sub-watershed 2 is 1,557.91 m^3/sec, and the whole watershed outlet is 3,042.72 m^3/sec. After the effect of rain interception by oil palm trees, sub-watershed 1 has peak flood of 100-year return period is 1,030.82 m^3/sec, for sub-watershed 2 is 1,490.41 m^3/sec, and the watershed outlet after the interception is 2,521.23 m^3/sec. The conversion of land to oil palm plantations will reduce flood discharge due to the interception of rain by oil palm trees.

6. Reference

[1] Asdak, C 2004, *Hidrologi dan Pengelolaan Daerah Aliran Sungai* (Gajah Mada University Press), Yogyakarta.

[2] Balai Pengelolaan Daerah Aliran Sungai 2014, *Peta Tutupan Lahan DAS Krueng Seunagan Kabupaten Nagan Raya Tahun 2012*, (BP DAS Banda Aceh).

[3] Badan Meteorologi dan Klimatologi 2009, *Data Curah Hujan Stasiun BMKG Bandara Cut Nyak Dhien Kabupaten Nagan Raya tahun 2000 sampai 2009*, Aceh.

[4] Badan Investasi dan Promosi Aceh, 2014, *Detail Profil Proyek (Detail Plan of Investment) Komoditi Kelapa Sawit di Nagan Raya.*

[5] Bruijnzeel, LA 1990, *Hydrology of Moist Tropical Forests and Effects of Conversion: a State of Knowledge Review*. Humid Tropics Programme, (IHP-UNESCO, Paris, and Vrije Universiteit, Amsterdam).

[6] Cameron, JL 2007, *Influence of Crown Traits and Leaf Arrangement on Rainfall Interception, Throughfall, and Stemflow in Five Tropical Tree Species* (The University of Winnipeg).

[7] Chow, et. al, 1988, *Applied Hydrology* (McGrow-Hill Company) Singapore

[8] Dinas Perkebunan dan Kehutanan 2014, *Data Luas Perkebunan Kelapa Sawit di Kabupaten Nagan Raya* (Dinas Perkebunan dan Kehutanan Kabupaten Nagan Raya).

[9] Kaimuddin 1994. '*Kajian Model Pendugaan Hujan pada Tegakan Pinus merkusii, agathis loranthifolia dan Schima wallichii di Hutan Pendidikan Gunung Walat Sukabumi'*, Tesis, (Program Pasca Sarjana ITB) Bogor.

[10] Mangoensoekarjo, S dan Semangun, H 2005, *Manajemen Agribisnis Kelapa Sawit* (Gajah Mada University Press), Yogyakarta.

[11] Pahan, I 2013, *Kelapa Sawit Manajemen Agribisnis dari Hulu hingga Hilir* (Penebar Swadaya), Jakarta.

[12] Pangudijatno, G dan Purba, P 1987, 'Kesesuaian Lahan dan Keterkaitannya dengan Pembangunan Perkebunan Kelapa Sawit', *Prosiding Lokal Industri Kelapa Sawit*, Vol. 1, Medan 24 – 25 November 1987

[13] Pelawi, SF 2009, 'Intersepsi pada Berbagai Kelas Umur Tegakan Kelapa Sawit *(Elaeis Guineensis*)', Skripsi, (Fakultas Pertanian Universitas Sumatera Utara), Medan.

[14] Rao, AS 1986, *'Interception Losses of Rainfall from Cashew Trees', Journal of Hydrologi.*

[15] Soemarto, CD *1987. Hidrologi Teknik (*Usaha Nasional) Surabaya.

[16] Triatmodjo, B 2009, *Hidrologi Terapan* (Beta Offset) Yogyakarta.

IOP Conf. Series: Materials Science and Engineering **523** (2019) 012042 doi:10.1088/1757-899X/523/1/012042

Risk impact on cost and time from the factors of contractor's managerial and operational

S Husin[1,*], Abdullah[2], M Riza[2], M Afifuddin[2]
[1]Engineering Doctoral Study Program, Univ. Of Syiah Kuala, 23111 Banda Aceh, Indonesia
[2]Faculty of Engineering, Univ. Of Syiah Kuala, 23111 Banda Aceh, Indonesia

E-mail: saifulhusin@unsyiah.ac.id

Abstract. Aceh Province over the last 20 years has experienced different events that can be divided into 3 (three) phases. 1st Phase is the phase of the conflict (2000-2004), 2nd Phase is the post-earthquake and tsunami rehabilitation and reconstruction phase of Aceh (2005-2009), and 3rd Phase is the post-rehabilitation and reconstruction phase (2010-present). Events that occurred during the last 20 years are certainly likely to provide risks to community activities in the province of Aceh. One of the activities affected by the event is the construction project work. The implementation of construction works affected by events in Aceh Province over the last 20 years is highly vulnerable to risks that impact on achieving project objectives such as cost and time. This study analyzes the impact of the contractor and operational managerial risk factors on the cost and timing of construction. Data were obtained from questionnaires distributed to 15 large qualification companies in Aceh Province. Testing data is done by using the validity test and reliability test. Data that has been valid and reliable then analyzed by using Severity Index (*SI*). The variables of a managerial risk factor with *SI* to the highest cost in 1st Phase are F5 (incompetent Engineer), 2nd Phase is F3 (lack of contractor experience), and 3rd Phase is F5 (incompetent Engineer). From the operational risk factor, the highest cost *SI* in each phase is the G3 variable (electrical disorder). The managerial risk factors with the highest time *SI* in 1st Phase are F6 (lack of top management support), 2nd Phase is F5 (incompetent engineer), and in 3rd Phase is F3 (lack of contractor experience). From the operational risk factor, the highest time *SI* at each phase is G3 (electrical disorder).

1. Introduction

The construction project is a series of complex, non-routine activities, limited by time, budget, resources and carried out according to specifications, with the ultimate goal of cost, time and quality determined [1,2]. To achieve the ultimate goal of the implementation of the work can be influenced by various risk factors, both internal and external. Internal factors are factors that come from problems in the construction project, whereas external factors are factors that originate from problems outside construction projects which, if they occur, may pose a risk to the project [3,4]. The risk is something that arises from the emergence of these factors which can be either a negative or positive impact [5,6]. Positive impacts will certainly be directly accepted by the construction worker because they will

* Corresponding author: saifulhusin@unsyiah.ac.id

contribute to the achievement of the project's objectives, but if the risks that arise cause negative impacts will certainly be avoided because it will disturb targeted results until the failure of the objectives of the project.

In the case of the objective to increase the positive impact and reduce the negative impact of the occurrence of risks to the implementation of construction required construction management [1,2]. Construction management is required to implement the management process effectively and efficiently to achieve the project objectives optimally [1,6,7,8]. From the aspect of risk factors, construction management comes from internal risk factors. The construction management managed by the company's resources is one of the internal factors that greatly affect the success of a company, especially the construction company [9,10,11]. Internal risk factors that directly affect project management include the contractor's managerial and project operations. To analyze the risks from the aspects of construction and operational management will be the different result of analysis because of the happening and felt by the contractor in each implementation area is different.

Construction work in each different work area will experience different risks. This is caused by events or events occurring in each region will have an effect on the occurrence of the risk. Aceh province over the last 20 years has experienced different events that can be divided into 3 (three) phases. 1st Phase is the phase of the conflict (2000-2004), 2nd Phase is the post-earthquake and tsunami rehabilitation and reconstruction phase of Aceh (2005-2009), and 3rd Phase is the post-rehabilitation and reconstruction phase (2010-present). Events that occurred during the last 20 years are certainly likely to provide risks to community activities in the province of Aceh. One of the activities affected by the event is the construction project work. The implementation of construction works affected by events in Aceh Province over the last 20 years is highly vulnerable to risks that impact on achieving project objectives such as cost and time.

Research related to construction project risk in Aceh Province has been conducted related to the assessment of a number of risk factors to the objectives of the construction project in the form of cost, time, and quality. The results are related to project resources [12], external factors [13], managerial and operational factors [14], contracting and design factors [15], and financial factors and methods of construction [16]. To continue the research, this study analyzes the impact of the contractor and operational managerial risk factors on construction cost and time. Data were obtained from questionnaires distributed to 15 large qualification companies in Aceh Province. Testing data is done by using the validity test and reliability test. Data that has been valid and reliable then analyzed by using Severity Index (*SI*). The variables of a managerial risk factor with *SI* to the highest cost in 1st Phase are F5 (incompetent Engineer) (0,510 / Medium), 2nd Phase is F3 (lack of contractor experience), and 3rd Phase is F5 (incompetent Engineer). Of the operational risk factors, *SI* highest cost in each phase is variable G3 (Electrical Disorders). Managerial risk factors with the highest time *SI* in 1st Phase are F6 (Lack of Top Management Support), 2nd Phase is F5 (incompetent Engineer), and in 3rd Phase is F3 (Lack of Contractor Experience). Of the highest operational risk factor, *SI* time in each phase is G3 (Electrical Disorders).

2. Research Methods

The research method discusses the methods of data collection, testing of questionnaire instruments, and risk analysis methods to achieve research results.

2.1 Data collection

The data in this study were obtained from questionnaires distributed to large qualification construction companies in Aceh Province. The questionnaire contained a number of questions aimed at obtaining information on the characteristics of respondents and companies, the possibility of risk events, and the impact of risks on the cost and timing of construction implementation of the contractor and operational managerial risk factors felt by construction executors in Aceh Province. Internal risk factors of contractor managerial consists of 9 variables and operational consists of 11 variables, as shown in Table 1.

Table 1. List of managerial and operational risk variables

Risk Factors	Code	Variable
Managerial Contractor	F1	Lack of project manager experience
	F2	Lack of communication and coordination between parties involved in the project
	F3	Lack of contractor experience
	F4	Losing data / documents
	F5	Incompetent engineer
	F6	Lack of top management support
	F7	Project planning and control is not good
	F8	Unclear authority, duties, and responsibilities (unclear delegation)
	F9	Unregistered project documents
Operational	G1	Lack of supervision of subcontractors and suppliers
	G2	Lack of supervision of the work execution schedule
	G3	Electrical disorder
	G4	Difficulty to establish temporary facilities
	G5	Number of jobs that are not according to plan
	G6	Changes in construction work due to difficulty implemented
	G7	Changes in the performance of suppliers/contractors
	G8	Repairs due to repetitive work
	G9	Bad location conditions and difficult to reach
	G10	Lack of telecommunication network provision
	G11	Late getting permission to do work

The data collection was conducted by collecting the results of the distribution of questionnaires on 15 companies from a population of 20 large qualification companies that had been involved in the implementation of construction projects in Aceh Province over the last 20 years. Data from the company is obtained from the list of major qualification companies that exist in the Construction Services Development Agency 2016. The position of respondents filler the questionnaire is a minimal company employee with intermediate positions such as director, manager, and senior engineer.

2.2 Testing of instruments

To measure the accuracy and determination of respondents in providing answers in the research instrument is done testing the perception of respondents. The respondent's perception test is done to know the validity (accuracy) and reliability (determination) of research instrument before the data in use. The instruments used hereinafter in this study are those that have met the valid and reliable criteria, based on the results of validity and reliability tests that have been done.

2.2.1 Instrument validity test

Validity is the level of validity of the measuring instrument used. The instrument is said to be valid indicating the measuring instrument used to obtain the data is valid or can be used to measure what should be measured [17]. The validity of the questionnaire is measured by calculating the correlation between the data on each statement with the total score using the product moment correlation formula (Equation 1). To specify a valid item or not use the following criteria:

1. If t_{count}> t_{sig}, then the question item is valid.
2. If t_{count}> t_{sig}, then the question item is invalid.

To obtain t_{count} value used the formula as follows:

$$r = \frac{n(\Sigma xy)-(\Sigma x)-(\Sigma y)}{\sqrt{\{n\Sigma x^2-(\Sigma x)^2\}\{n\Sigma y^2-(\Sigma y)^2\}}} \quad (1)$$

$$t_{count} = \frac{r\sqrt{n-2}}{\sqrt{1-r^2}} \quad (2)$$

With r_{xy} = correlation coefficient, Σx = total item score, Σy = total total score and n = number of respondents.

To obtain a t_{sig} value is to determine the desired significance level. In this case, a significant level is taken at 5%. Significant means convincing or meaningful. In this study significant meaning that the proven hypothesis in the sample can be applied to the population. If it is not significant, then the conclusions of the sample cannot apply to the population (no generalization) or apply only to the sample only. A significant level of 5% or 0.05 means taking the wrong risk in making a decision to reject the wrong hypothesis as much as 5% and correct in making a decision at least 95% (trust level). To obtain the t_{sig} value required product moment values as shown in Table 2. From the product, moment values can be concluded that the t_{sig} value taken in this study is t_{sig} > 0,514 because the number of respondents in the study there are 15 respondents (n = 15) with a significant level of 5%. Therefore, the Instrument item is considered valid by comparing it with t_{sig}, therefore if $t_{count} \geq 0.514$ then the instrument is deemed valid.

Table 2. Product moment values

n	Significant Level		*n*	Significant Level		*n*	Significant Level	
	5%	1%		5%	1%		5%	1%
3	0,997	0,999	27	0,381	0,487	55	0,266	0,345
4	0,950	0,990	28	0,374	0,478	60	0,254	0,330
5	0,878	0,959	29	0,367	0,470	65	0,244	0,317
6	0,811	0,917	30	0,361	0,463	70	0,235	0,306
7	0,754	0,874	31	0,355	0,456	75	0,227	0,296
8	0,707	0,834	32	0,349	0,449	80	0,220	0,286
9	0,666	0,798	33	0,344	0,442	85	0,213	0,278
10	0,632	0,765	34	0,339	0,436	90	0,207	0,270
11	0,602	0,735	35	0,334	0,430	95	0,202	0,263
12	0,576	0,708	36	0,329	0,424	10	0,195	0,256
13	0,553	0,684	37	0,325	0,418	12	0,176	0,230
14	0,532	0,661	38	0,320	0,413	15	0,159	0,210
15	0,514	0,641	39	0,316	0,408	17	0,148	0,194
16	0,497	0,623	40	0,312	0,403	20	0,138	0,181
17	0,482	0,606	41	0,308	0,398	30	0,113	0,148
18	0,468	0,590	42	0,304	0,393	40	0,098	0,128
19	0,456	0,575	43	0,301	0,389	50	0,088	0,115
20	0,444	0,561	44	0,297	0,384	60	0,080	0,105
21	0,433	0,549	45	0,294	0,380	700	0,074	0,097
22	0,423	0,537	46	0,291	0,376	800	0,070	0,091
23	0,413	0,526	47	0,288	0,372	900	0,065	0,086

Table 2. Product moment values

n	Significant Level		n	Significant Level		n	Significant Level	
	5%	1%		5%	1%		5%	1%
24	0,404	0,515	48	0,284	0,368	1000	0,062	0,081
25	0,396	0,505	49	0,281	0,364			
26	0,388	0,496	50	0,279	0,361			

Source: Sugiyono (2010)

2.2.2 Instrument reliability test

Reliability test is a reliability or consistency of measuring instrument in measuring what to be measured, meaning whenever the measuring instrument is used will give the same result [17]. So that reliability is a stability or consistency of respondents in answering things related to the forms of questions or statements that are the dimensions of a variable and arranged in a form of a questionnaire. Commonly used reliability analysis is Cronbach Alpha (*C-Alpha*) analysis. The test using the coefficient C-Alpha must be greater or equal to 0.6 is a value that is considered to be able to test whether or not the questionnaire is used. The formulas used are as follows:

$$r = \frac{k}{(k-1)}\left[1 - \frac{\sigma_b{}^2}{\sigma_1{}^2}\right] \quad (3)$$

with r = instrument reliability, k = number of questions, σ_b^2 = number of grain variance and σ_1^2 = varians total. The formula for calculating grain variance and total variance is:

$$\sigma_b{}^2 = \frac{JKi}{n} - \frac{JKs}{n^2} \quad (4)$$

$$\sigma_1{}^2 = \frac{\sum xt^2}{n} - \frac{(\sum xt^2)}{n^2} \quad (5)$$

with $\sum xt$ = the total number of respondents' answers, $\sum xt^2$ = squared the total number of respondents' answers, *Jki* = sum of squares of whole grains and *Jks* = the sum of the squares of the subject.

2.2.3. Methods of data analysis

The data analysis used is to obtain the result of possible risk occurrence (frequency) and impact of risk (severity). Therefore, the Severity Index (*SI*) analysis is needed to analyze the possibility of risk occurrence and Severity Index to analyze the impact of risk.

Severity index shows the index of the impact of risk from the emergence of risk factors. For the calculation of severity index analysis use the formula in the following equation [12].

$$Severity\ Index\ (SI) = \frac{\sum_{I=1}^{5} a_i n_i}{5N} \quad (6)$$

with i = category index of respond, the a_i = weight associated with the value of the i-th response, n_i = frequency of impact of respondent i as a percentage of total respondent for each factor and N = total number of respondents. Measurement of severity is done using a Likert scale with criteria shown in Table 3.

Table 3. *SI* scoring criteria and scales

Qualification	Likert Scale	Assessment Scale
Very Low	1	0,000≤ *SI* ≤0,125
Low	2	0,125< *SI* ≤0,375
Medium	3	0,375< *SI* ≤0,625
Haight	4	0,625< *SI* ≤0,875
Very Hight	5	0,875< *SI* ≤1,000

Source: Majid and McCaffer (1997)

3. Results and Discussion

In this section will be presented the results of data processing and data analysis based on research methods. The discussion is directed to the problem of the impact of risks on construction costs and timing of contractor and operational managerial risk factors.

3.1 Characteristics of respondents and companies

Characteristics of data obtained from the distribution of questionnaires divided into two, namely the characteristics of the questionnaire data from the respondents and the characteristics of the questionnaire data from the company. Results of the company characteristic questionnaire. The company experience in the field of construction has more than 15 years of experience, the number of projects ever handled, almost entirely handled projects of more than 10 projects with the dominant project type of road and bridge projects, the average value of projects worked every year more than Rp. 10,000,000,000 - Rp. 50,000,000,000 with the actual project time being completed every year for 6-12 months.

Respondent data obtained from the results of respondents' answers on questions regarding respondents data. Respondent data are grouped on behalf of respondent, position, gender, the age of the respondent, last education, and years of service. Table 5 shows the results of data processing of respondent characteristics. The position of respondents is dominated by the director, male sex, with an average age above 30 years, recent education domination undergraduate (*SI*) and with average work experience more than 7 years. Based on the results obtained enough reliable respondents to fill out the research questionnaire.

Table 4. Characteristics of respondents

Characteristics of Respondent	Category of Measurement	Amount	(%)
Personel position	Director	5	33,3
	Manager	7	46,67
	Other	3	20,00
Last education	SMA	3	13,33
	Diploma	2	13,33
	Bachelor (S1)	9	60,00
	Post Graduate (S2/S3)	1	6,67
Working Experience in personnel	>2-4 years	1	6,67
	>4-7 years	1	6,67
	>7 years	13	86,67

3.2 Instrument test results

Research results were obtained after data collection and data processing. The data obtained in the form of responses of respondents from the questionnaires that have been disseminated. The questionnaire was addressed to 20 respondents which is a large qualified construction service company in Aceh province, but the respondents obtained as many as 15 companies. This happens because the company address listed on the LPJK is not appropriate and there is no reply from the respondents. This chapter describes the results of research in the form of characteristic data, validity test, reliability test, descriptive statistical analysis, Severity Index analysis (*SI*), and discussion.

3.2.1 Test results validity

Validity test is used to determine whether or not a question item in the questionnaire. This test is performed on each questionnaire statement and the results are compared with rsig = 0,514.

Table 5. The result of validity test

Impact	Code Var.	Value Range t_{count} Per Phase			Information
		1st Phase	2nd Phase	3rd Phase	
Cost	F1-F9	0,530–0,828	0,517–0,768	0,578–0,753	Valid
	G1-G11	0,524–0,884	0,520–0,804	0,520–0,861	Valid
Time	F1-F9	0,518-0,839	0,530-0,797	0,515-0,737	Valid
	G1-G11	0,523-0,889	0,520-0,795	0,520-0,726	Valid

Based on Table 5 can be seen all items of the question under study have value t_{count} greater than rsig. Thus the validity test on the contractor and operational managerial risk factors with the time phase of the review indicates that the questionnaire data collection in this study is entirely valid so that it can be executed to the next analysis process.

3.2.2 Reliability test results

After doing the validity test the researchers conducted a reliability test that aims to determine the level of reliability of research instruments so that the measuring tool remains consistent when measured at different times. This calculation uses the Cronbach Alpha formula. The value of a variable is said to be reliable if the Cronbach Alpha value exceeds the value of 0.6. Reliability test is done jointly to all questions and the results are compared with the value of 0.6.

Table 6. The results of the reliability test of cost impact data

Impact	Risk Factor	Value Range t_{count} Per Phase			Information
		1st Phase	2nd Phase	3rd Phase	
Cost	Managerial Contractor	0,77	0,76	0,82	Reliabel
	Operational	0,82	0,86	0,85	Reliabel
Time	Manajerial Contractor	0,79	0,80	0,92	Reliabel
	Operstional	0,80	0,82	0,87	Reliabel

In Table 6 we can see that the reliability test for each variable indicates that the C-Alpha value for all variables of the contractor and operational managerial risk factors in the analyzed data is greater than 0.6. Thus the reliability test on the contractor and operational managerial risk factors with the time phase of the review indicates that the questionnaire data collection in this study is all reliable so that it can be executed to the next analysis process.

3.3 Result of analysis of Severity Index (SI)

From the result of risk impact analysis using *SI* which has been summarized in Table 7, from contractor managerial risk factors there is no variable with "high" severity scale. The variable with the highest scale for cost and time severity only impacts the risk on a "medium" scale.

The results of the *SI* analysis on the cost of contractor risk managerial factors in 1st Phase and 2nd Phase of all variables have a "medium" impact scale on construction costs. In 3rd Phase, from 9 (nine) contractor risk managerial variables there are only 2 (two) variables with a low cost "cost" scale of the variables F9 and F8, while the other variables have a "medium" impact scale on costs. The results of the *SI* analysis of the time on the contractor's managerial risk factors in 1st Phase and 3rd Phase of the 9 (nine) variables there is only 1 (one) variable with the scale of "low" impact on construction work time is variable F9. In 2nd phase all the variables of the contractor's managerial risk factors show the "medium" impact scale over time.

Table 7. Results of Severity Index analysis (*SI*)

Var. Code	1st Phase		2nd Phase		3rd Phase	
	SI Cost	*SI* Time	*SI* Cost	*SI* Time	*SI* Cost	*SI* Time
F1	0,410/Medium	0,480/Medium	0,450/Medium	0,490/Medium	0,400/Medium	0,470/Medium
F2	0,430/Medium	0,510/Medium	0,400/Medium	0,480/Medium	0,400/Medium	0,450/Medium
F3	0,450/Medium	0,510/Medium	0,490/Medium	0,490/Medium	0,430/Medium	0,530/Medium
F4	0,390/Medium	0,400/Medium	0,410/Medium	0,400/Medium	0,390/Medium	0,440/Medium
F5	0,510/Medium	0,520/Medium	0,430/Medium	0,510/Medium	0,430/Medium	0,480/Medium
F6	0,480/Medium	0,550/Medium	0,450/Medium	0,450/Medium	0,390/Medium	0,470/Medium
F7	0,440/Medium	0,510/Medium	0,400/Medium	0,470/Medium	0,410/Medium	0,470/Medium
F8	0,430/Medium	0,440/Medium	0,400/Medium	0,440/Medium	0,360/Low	0,400/Medium
F9	0,390/Medium	0,370/Low	0,400/Medium	0,430/Medium	0,370/Low	0,370/Low
G1	0,450/Medium	0,450/Medium	0,470/Medium	0,490/Medium	0,450/Medium	0,440/Medium
G2	0,430/Medium	0,470/Medium	0,410/Medium	0,440/Medium	0,430/Medium	0,450/Medium
G3	0,590/Medium	0,560/Medium	0,570/Medium	0,590/Medium	0,520/Medium	0,550/Medium
G4	0,430/Medium	0,470/Medium	0,400/Medium	0,430/Medium	0,440/Medium	0,410/Medium
G5	0,350/Low	0,390/Medium	0,360/Low	0,390/Medium	0,350/Low	0,370/Low
G6	0,400/Medium	0,370/Low	0,430/Medium	0,440/Medium	0,400/Medium	0,430/Medium
G7	0,410/Medium	0,470/Medium	0,440/Medium	0,410/Medium	0,450/Medium	0,480/Medium
G8	0,440/Medium	0,370/Low	0,400/Medium	0,360/Low	0,400/Medium	0,410/Medium
G9	0,510/Medium	0,510/Medium	0,470/Medium	0,470/Medium	0,440/Medium	0,410/Medium
G10	0,370/Low	0,400/Medium	0,410/Medium	0,410/Medium	0,410/Medium	0,400/Medium
G11	0,480/Medium	0,450/Medium	0,370/Low	0,400/Medium	0,360/Low	0,370/Medium

The results of the *SI* analysis on the cost of operational risk factors in 1st Phase of 11 (eleven) variables are only 2 (two) variables with a "low" impact scale on costs, ie, variables G5 and G10. In 2nd Phase and 3rd Phase, there are 2 (two) variables with a "low" scale ie variables G5 and G11 while all other variables have a "medium" impact scale on construction costs. The results of the *SI* analysis of the time on the operational risk factor in 1st Phase there are only 2 (two) variables with the "low" impact scale on the construction work time ie the variables G6 and G8. In 2nd phase there is 1 (one) variable with a "low scale" that is G8. In 3rd phase there is only 1 (one) variable with a "low scale" is the variable G5.

From the calculation of *SI* cost and time that have been summarized in Table 7 for each variable, it shows that in 1st Phase, 2nd Phase, and 3rd Phase the risk value to cost and time tends to decrease from 1st Phase to 3rd Phase. This is because from phase to phase the conditions in Aceh Province are stable enough so that the influence of these risk factors on the cost and time of construction is reduced.

4. Conclusions

The results of this study show the assessment of the impact of risks on costs and time derived from contractor and operational managerial risk factors. Project risk is one of them influenced by the assessment of the impact of risk. From the risk impact assessment using Severity Index (*SI*) analysis can be known how big the impact caused by the variable of a risk factor to the project target. In this study, the targets of the projects studied are cost and construction time.

Based on the assessment of the impact of risks on construction costs and timing, from contractor and operational managerial risk factors there is no variable with a "high" impact scale on cost and time. The highest severity value is in the "medium" impact scale range of costs and time. Nevertheless, the variables on these risk factors still have a different effect on the cost and timing of the construction in the three phases of the review.

5. Acknowledgment

The authors would like to thank the relevant parties who have assisted this research and to the data giver of respondents from the large qualification construction company in Aceh Province who has been involved.

6. References

[1] PMI 2013 A Guide to the Project Management Body of Knowledge (PMBOK Guide) (5th Ed). *Project Management Institute*

[2] Sádaba, S.M., Ezcurdia, A.P., Lazcano, A.M., and Villanueva, P. 2014 *International Journal of Project Management,* **32** (2014) 327-340

[3] Pheng, L.S., Junying, L., and He, S. 2009 *KSCE Journal of Civil Engineering,* **13**(2):85-95

[4] Frame, J.D., 2003 Managing Risk in Organizations: A Guide for Managers, *Jossey-Bass,* NY.

[5] Clayton, M., 2011 Risk Happens, *Marshall Cavendish, United Kingdom*

[6] Diéguez, A. I. I., Cazorla, A. S., and Luque, R. A., 2014 *Procedia-Social and Behavioral Sciences* **119** 407-416

[7] Dey, P. K., 2012 *Production Planning & Control,* **23**(12), 903-921.

[8] Wu, D., & Olson, D. L., 2008 *International Journal of Production Economics,* **114**(2),646-655.

[9] Das T.K. and Teng B.S., 2000 *Journal of Management.* **26** 31-61

[10] Wernerfelt, B. 1984 *Strategic Management Journal,* **5** 171-180

[11] Rumelt, R.P. 1984 Towards a Strategic Theory of The Firm. In R. B. Lamb (Ed.), *Competitive Strategic Management:* 556-570. Englewood Cliffs, NJ: Prentice-Hall.

[12] S.Husin, Abdullah, M. Riza, M. Afifuddin and P. Zalbania 2016 Risk Level of Labor Factors, Materials and Equipment on Construction Projects in Aceh Province, *National Conference of Civil Engineering 10, Atma Jaya University of Yogyakarta,* 26-27 October 2016.

[13] Mubarak, S. Husin, and Syarafina, 2016 External Risk on Construction Project Implementation in Aceh Province, *National Conference of Civil Engineering 10, Atma Jaya University of Yogyakarta,* 26-27 October 2016.

[14] Tripoli, A. T. Bulba, Fachrurrazi, and C.A.W. Mastura, 2016 Identify the Risk of Implementation at Construction Project in Aceh Province, *National Conference of Civil Engineering 10, Atma Jaya University of Yogyakarta,* 26-27 October 2016.

[15] Nurisra, Mahmuddin, N. Malahayati, and I. Sari, 2016 Project Construction Risks Stemming from Contract and Planning Factors in Aceh Province, *National Conference of Civil Engineering 10, Atma Jaya University of Yogyakarta,* 26-27 October 2016.

[16] Fachrurrazi, F. Maulina, M. Jamil, and H. Amalia, 2016 Cost Risk Comparison to Construction Contractors in Three Periods in Aceh Province, *National Conference of Civil Engineering 10, Atma Jaya University of Yogyakarta,* 26-27 October 2016.

[17] Sugiyono, 2010 Research of Quantitative and Qualitative and RND, *CV. Alfabeta,* Bandung.

[18] Abd. Majid. M.Z and Mc Caffer. R. 1997 *Journal of Management in Engineering.* **13,** 91 (1997)

IOP Conf. Series: Materials Science and Engineering **523** (2019) 012043 doi:10.1088/1757-899X/523/1/012043

Single microtremor method for estimating site fundamental frequency at a site in the historical city of Byblos - Lebanon

N Tumur-Uyal[1], **B Setiawan**[2,*], **M S Oucherif**[3]
[1] Department of Geology & Geophysics, National University of Mongolia, Mongolia
[2] Program Study of Geological Engineering, Faculty of Engineering, Syiah Kuala University, Indonesia
[3] Geoexplo, Cheraga, Algiers, Algeria

E-mail: bambang.setiawan@unsyiah.ac.id

Abstract. A single ambient noise measurement has been found to be useful in investigating the near-surface geology in particularly for estimating site fundamental frequency. This single microtremor method is very attractive and has been successfully applied in many projects as the method is non-destructive, low cost, feasible for urban environments, fast, and deep penetration. Furthermore, this single microtremor method has been successfully applied in many projects. This paper presents the application of the single microtremor method to estimate the site fundamental frequency at a site in the historical city of Byblos, Lebanon. Pre-processing of industrial origin detection was carried out. Several durations of the measurements were investigated to obtain the most stable results. The site fundamental frequency was validated using one-dimensional site response analysis. The results are presented in this paper.

1. Introduction

Since the simplicity, convenient and affordable method, single microtremor application has been widely used in earthquake engineering-related research. Single point microtremor method is generally used for the following purposes in geotechnical earthquake engineering such as preliminary seismic site classification, local site effects, estimation of bedrock depth, and sub-surface profiling ([1]; [2]; [3]; [4]). The successful application of the single microtremor method (H/V – horizontal/vertical spectral ratio) can be found in Ref. [3]; [5]; [6]; [7]; [8]; [9]; and [10]. This paper presents the application of a single microtremor method for estimating site fundamental frequency at a site in the historical city of Byblos – Lebanon.

Empirically, there is a strong correlation between the damage caused by earthquakes and the near-surface geology setting of the affected area, which is called the site effect ([11]; [12]; [13]; [14]). Generally, seismic wave velocity is higher at rocks than at soils, therefore, to carry the equivalent amount of energy the seismic wave velocity decreases and the amplitude of the seismic wave expands at soils. This causes a couple of effects: firstly, the ground displacement at soil sites is larger than at rock sites, and secondly, the duration of the ground trembling is longer at soil sites than at rock sites. In the case of the seismic-wave frequency matches the fundamental frequency of the ground layer, resonance will occur ([15]; [16]). These site effects have played a principal role in defining the

*Corresponding author: bambang.setiawan@unsyiah.ac.id

damage to structures, i.e. Mexico ([15]; [17]), Kobe [16], and Izmit [18]. Thus, recognizing the ground response to the seismic waves is important to understand how the seismic waves will affect the structures founded on it.

2. The historical city of Byblos and its seismic sources

The city of Byblos is located on the Lebanese coast approximately 40 km north of the Lebanon capital of Beirut. The city area is about 5 km^2. The city of Byblos' population is approximately 40,000 people. The 8000-years-old site of Byblos city is considered to be one of the oldest continuously inhabited cities in the world. This old site of Byblos city is a UNESCO World Heritage Site (see [19]). The seismic sources for the city of Byblos have been investigated by [20], as shown in Figures 1 and 2.

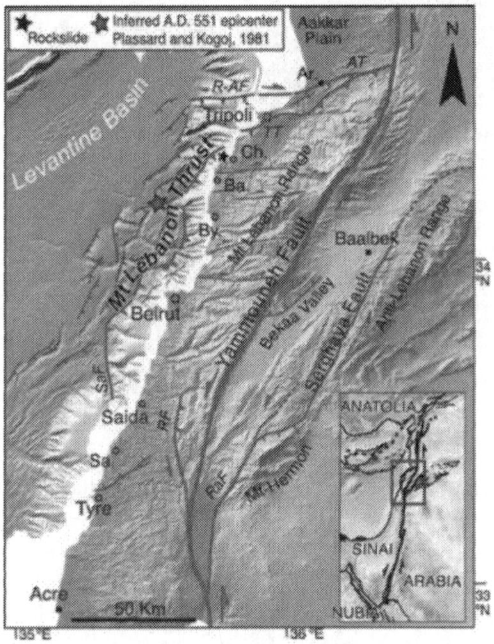

Figure 1. Active faults around the Byblos Abbreviations: Ar—Arqa; Ba—Batroun; By—Byblos; Ch—Chekka; Sa—Sarafand. AT—Aakkar thrust; TT—Tripoli thrust; R-AF— Rankine-Aabdeh fault; RF—Roum fault; RaF—Rachaya fault; SaF—Saida fault [20].

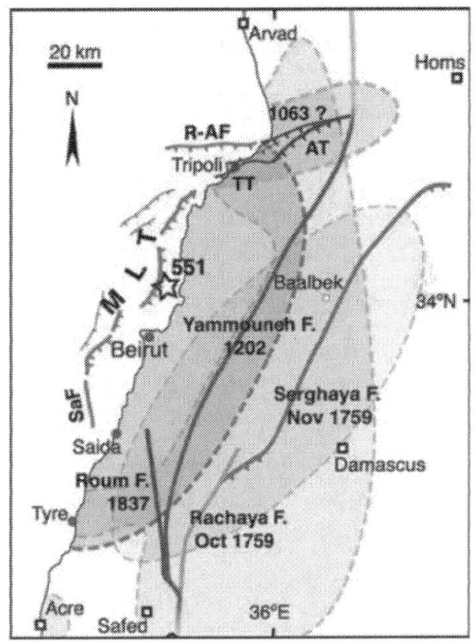

Figure 2. Most likely sources of AD. 551 seismic event (open star—inferred epicenter). Blue color corresponds to AD. 551 seismic event with MMI of VIII (from Sieberg, 1932 in [20]).

As shown in Figures 1 and 2, there are at least two main seismic sources for the city of Byblos which are Mt. Lebanon Thrust (MLT) and Yammouneh Fault. In AD 551 one of a seismic event has struck along the MLT [20]. This event has been estimated for more than VIII Mercalli Magnitude scale at the city of Byblos [20].

Figure 3 shows the city of Byblos-Lebanon. The red mark in Figure 3 is the point of the measured location of this study which is along the Sea Side Road of the city of Byblos. The distance of the point is about 600 m from the coast of Lebanon.

Figure 3. An ambient noise measurement site at Byblos-Lebanon (red mark)

3. Methods

This section describes the equipment, data acquisition, and data analysis used in this study, which is briefly explained, as follow:

3.1. Equipment

The equipment used for the microtremor measurement is Lennartz LE-3D/5sec (0.2 – 50 Hz), as shown in Figure 4. This seismometer with a lower cut off frequency of 0.2 Hz is perfectly suited for a seismic monitor, H/V measurements and seismic microzonation. Essential features and characteristics of this seismometer are a robust instrument, very low power consumption, quick and easy installation (plug and measure), and uniform sensitivity and transfer function. Detailed main features of the seismometer are summarized in Table 1. The used digitizer in this study is CityShark II (Figure 5).

Figure 4. Lennartz LE-3D/5 sec.

Figure 5. CityShark II N° 76.

Table 1. The most important characteristics of the used seismometer.

Parameter	Description	Remarks
Eigen-period	5 seconds	Lennartz LE-3D/5 sec
Upper-frequency limit	50 Hz	
Transduction factor	800 V/(m/s) (differential)	
Output signal	3 analog voltages, max. +- 7V	
Power supply	+10 ... +16V DC, typically 7 mA @12 V	
Dimensions	195 mm diameter, 165 mm height (excluding handle)	
Weight	6.5 kg	
Temperature range	-15 ... +60 °C	
RMS noise @ 1 Hz	< 1 nm/s	
Dynamic range	> 140 dB	

3.2. Data acquisition
The original set of microtremor measurements were carried out at 18:40 pm on 22 September 2016. The weather was generally calm with no strong winds or rain. Measurement location was close as possible to the downtown of the city of Byblos. Care was taken to avoid manholes, foundations or other underground structures. Data was recorded for 15 minutes. The field data sheet was used during the data acquisition in this study.

3.3. Data analysis using H/V spectral ratio method
H/V spectral ratio method is an experimental technique to assess the dynamic characteristics of the ground surface due to any dynamic loads, i.e. seismic loading. This H/V method is based on the analysis of the spectral ratio between the Fourier amplitude spectrum of the horizontal (H) and vertical (V) components of the recorded ambient vibration. Detailed data analysis using H/V spectral ratio method can be found in [21].

4. Results
This result section presents the field data conditions, the HVSR curve, and calculated site fundamental frequency. A further short discussion is elaborated in this section, also.

4.1. Field data
The field measurement was carried out at 18:40 pm on 22 September 2016. The measurement location is Byblos City center (Latitude of 3778866N and Longitude of 744291E) as shown in Figure 3, above. The factual facts recorded in the field data sheet during the ambient noise measurement are summarized in Table 2. About 15 minutes noise data were collected at the dense urbanization area of the city of Byblos.

4.2. H/V spectral ratio curve and site fundamental frequency
This section presents the results of the H/V spectral ratio technique and examines the site fundamental frequency. The H/V curve of this study is shown in Figure 6. The site fundamental frequency obtained from the H/V analysis suggests a frequency of 1.5 Hz. Another interesting observation associated with the H/V curve is the presence of the secondary peak of 3.0 Hz.

IOP Conf. Series: Materials Science and Engineering **523** (2019) 012043 doi:10.1088/1757-899X/523/1/012043

Table 2. A summary of the field data sheet.

Parameter	Description	Remarks
Latitude	3778866N (34 deg 7 min 19.65 sec)	UTM Code 36S
Longitude	744291E (35 deg 38 min 55.608 sec)	
Station type	CityShark II	
Sensor type	Lennartz 5 sec	
Gain	256	
Sample frequency	200 Hz	
Weather conditions	Wind (None); Rain (None); Approx. temperature (28 degrees)	
Ground type	Earth (gravel); Dry soil	
Artificial ground/sensor coupling	None	
Urbanization	Dense	
Continuous noise sources	None	

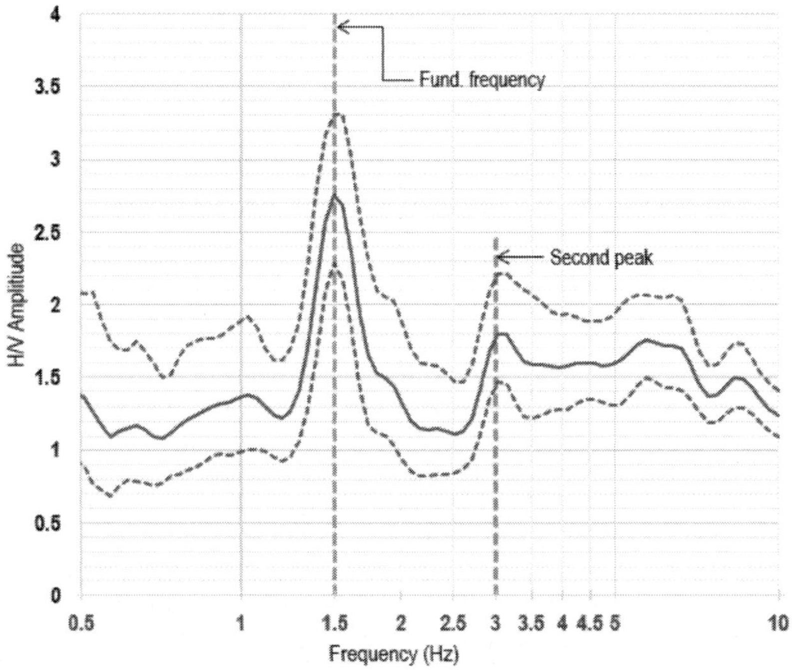

Figure 6. H/V spectral ratio curve.

4.3. Discussion

In practice, the relationship between the H/V peak(s) and the resonance frequency of the shear wave has been summarised by [22] from [23] and [24], as follows: 1) The H/V ratio of a body wave-field always exposes a peak around the fundamental shear wave frequency, for high impedance contrast sites; 2) The H/V ratio shows peaks at the shear wave harmonics, in the case of horizontally stratified media; 3) The amplitude of the first H/V peak is likely to be fairly well correlated with the shear wave amplification, in the case of horizontally stratified media with high impedance contrast. These findings are very useful for interpreting the experimental H/V analysis in this study. The result of the H/V curve of this study suggests 1) the measured site is composed by relatively horizontally stratified layer, 2) the measured site has a moderate to low impedance contrasts (< 4-5), and 3) the amplification of the shear wave of the measured site is about 2.5. However, gathering the additional representative sub-surface information to assist further interpretation of this H/V curve is suggested.

The fundamental period of the site (≈0.65s) is obtained by taking the reciprocal of the fundamental frequency of 1.55 Hz. This result suggests a significant seismic amplification for 16 to 20 meters height buildings in the city of Byblos using the general rule for building natural periods as suggested by [25] and [26].

The H/V curve of this study shows two peaks of 1.55 Hz and 3.15 Hz. These two-peaks of H/V curve suggest the presence of two different impedance contrasts at two different depths at the measured site. The first peak can be correlated to a deep structure and the second one is for a shallow structure ([21]; [27]; and [28]).

Additional measurement using a different sensor of Geospace GS11-3D was carried out to validate the repeatability of this result. Figure 7 presents the comparison of the H/V curves obtained from both Lennartz LE-3D/ 5 sec (red lines) and Geospace GS11-3D (grey lines). The comparison suggests a similar curve pattern.

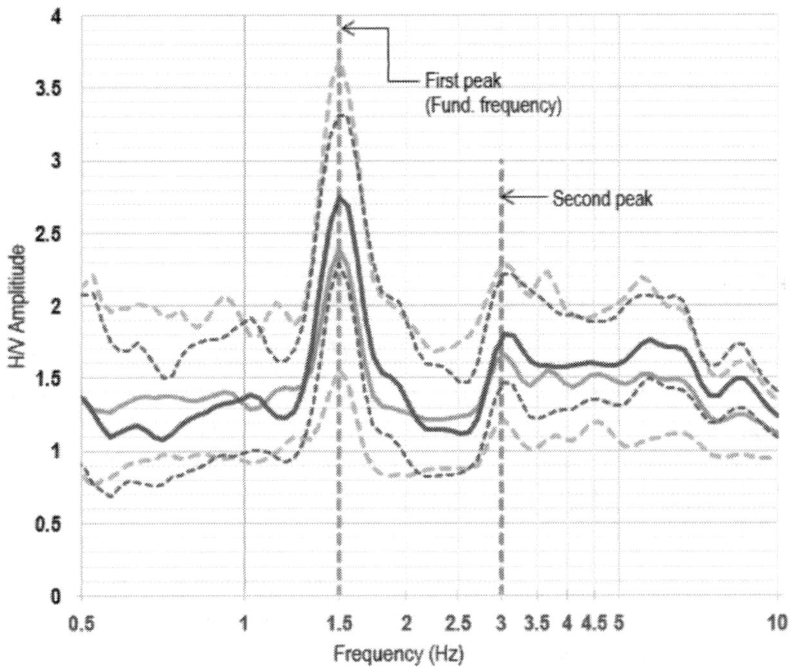

Figure 7. Comparison of H/V spectral ratio curves.

5. Conclusion

From the literature on various research has demonstrated the applicability of single point microtremor measurement and analysis for site effect study. This single microtremor analysis has several advantages, as follow: 1) effortless in-situ application (no extra source needed, simple equipment setting) and quick interpretation of measurements; 2) cost-effective (short duration measurements, large regions may be surveyed in small amount of time) and very fast application; and 3) provides good estimates of predominant soil period and spectral shapes. This single microtremor measurement and analysis have been carried out for the city of Byblos which is vulnerable to seismic hazard. The result suggests a site fundamental frequency of 1.55 Hz at the measured site. Therefore, it can be concluded that single point microtremor measurements provide an easily applicable, cost-effective and fast approach to obtain information on local site conditions if sufficient consideration is given to the limitations.

6. References

[1] Mucciarelli,M and Gallipoli MR 2006 First European Conference on Earthquake Engineering and Seismology. In *Proceedings of the 13th ECEE & 30th General Assembly of the ESC*, Geneva, Switzerland, 3–8 September 2006, p. 211.
[2] Parolai S, Facke A, Richwalski SM and Stempniewski L 2005 *Nat. Hazards* **34** 217–223.
[3] Setiawan B, Jaksa MB, Griffith MC and Love D 2018 *Soil Dyn Earthq Eng.* **110** 244–261.
[4] Setiawan B, Jaksa MB, Griffith MC., and Love D 2018 *Eng Geo.* **243** 145 – 159.
[5] Nakamura Y 1989 *Q Report Railw Tech Res Inst.* **30** (1) 25–30.
[6] Picozzi M, Strollo A, Paraloi S, Durukal E, Ozel O, Karabulut S, Zschau J, and Erdik M. 2009 *Soil Dyn Earthq Eng.* 469–82.
[7] Mokhberi M, Davoodi M and Haghshenas E 2007 *Soil Dyn Earthq Eng.* 300–8.
[8] Fah D, Wathelet M, Kristekova M, Havenith H, Endrun B, Stamm G, Poggi V, Burjanek J and Cornou C. 2003 *Report for Network of Research Infrastructures for European Seismology*, Sixth Framework Program, EC Project Number: **026130**, 54 pp.
[9] Harutoonian P, Leo C, Doanh T, Castellaro S, Zou J, Liyanapathirana D, Wong H, and Tokeshi K. 2012 *Soil Dyn. Earthq Eng.* 23–31.
[10] Di Stefano P, Luzio D, Renda P, Martorana R, Capizzi P, D'Alessandro A, Messina N, Napoli G, Todaro S, and Zarcone G. 2014 *Nat Hazards Earth Syst Sci.* **2** 2597–637.
[11] Graves RW, Pitarka A and Somerville PG 1998 *Bull. Seismol. Soc. Am.* **88** 1224-1242.
[12] Bard PY and Bouchon M 1980 *Bull. Seismol. Soc. Am.* **70** 1263-1286.
[13] Bard PY and Bouchon M 1980 *Bull. Seismol. Soc. Am.* **70** 1921-1941.
[14] Setiawan B, Jaksa MB, Griffith MC and Love D 2018 *Boll. di Geofisica Teorica ed Applicata* **59** (1) 27–46.
[15] Booth ED, Pappin JW, Mills JH, Degg MR and Steedman RS 1986 *The Mexican earthquake of 19th September 1985*. A field report by EEFIT (Institution of Structural Engineers).
[16] Brebbia CA, Beskos DE and Kausel E 1996 *The Kobe earthquake: geodynamical aspects*. Computational Mechanics Publications (Southampton) ISBN:185312 4303, p. 165.
[17] Finn WDL and Wightman A 2003 *Ground motion amplification factors for the proposed 2005 edition of the National Building Code of Canada.* (The National Research Council Canada (NRC) Research Press Web).
[18] Ozcep F, Karabulut S, Ozel O, Ozcep T, Imre N and Zarif H 2014 *J. Earth Syst. Sci.* **123** (1) 73 - 89.
[19] Makhoul N, Navarro C, Lee J and Abi-Youness A 2016 *Int. J. Disast. Risk. Reduction* **18** 12 - 22.
[20] Elias A, Tapponier P, Singh SC, King GCP, Briais A, Daeron M, Carton H, Sursock A, Jacques E, Jomaa R and Klinger Y 2007 *Geology* **35** (8) 755 – 758.
[21] SESAME 2004 Guidelines for the implementation of the H/V spectral ratio technique on ambient vibrations measurements processing and interpretation. SESAME European Research Project, *WP12 –Deliverable D23.12* (European Commission – Research General Directorate), Project No. EVG1-CT-200-00026.
[22] Caserta A 2011 *Seismic site effects (Data analysis and modelling)*. Doctoral Thesis. Department of Geophysics, Faculty of Mathematics and Physics (Charles University in Prague), Czech Republic, 74 pp.
[23] Bonnefoy-Claudet S, Cornou C, Bard PY, and Cotton F 2004 *Nature of noise wavefield*. Final Report, WP08, Deliverable D13.08, Site Effects Assessment Using Ambient Excitations

(SESAME) (European Comission – Research General Directorate) Project No. EVG1-CT-2000-00026 SESAME, 50 pp.
[24] Bonnefoy-Claudet S, Cotton F, and Bard PY 2006 *Earth-Science Reviews* **79** 205 - 227.
[25] IS 1893:2002 (Indian standard code of practice for earthquake resistant design).
[26] American Society of Civil Engineers (ASCE) 2005 *Minimum design loads for buildings and other structures: SEI/ASCE 7-05*. Virginia:(American Society of Civil Engineers)
[27] Toshinawa T, Taber JJ and Berril J 1997 *Bull. Seismol. Soc Am.* **87** 356-369.
[28] Gueguen P, Chatelain JL, Guilier B, Yepes H and Egred J. 1998 *Soil Dyn Earthq Eng*. **17**:329–334.

IOP Conf. Series: Materials Science and Engineering **523** (2019) 012044 doi:10.1088/1757-899X/523/1/012044

Childhood activities adults utilizing for the Intervention Reminiscence-Based Screening

E B Tallutondok[1,*]
Lecturer, Faculty of Nursisng, Universitas Pelita Harapan, Indonesia

E-mail: eva.tallutondok@uph.edu

Abstract. Ageing people is a vulnerable population, so it is a global public health concern among nations. Indonesia has tendency to be an ageing structure population on the future. At one public home care in Jakarta, there was an extremely data about decreasing number of healthy older adults in three consecutive years since 2016. Behind in this problem, the reminiscence screening about childhood activities has done in one public home care. It was utilizing by the basic principle of Intervention Reminiscence – Based Screening. the qualitative study with a mix-method design was used in this study via verbal, then data transformed for four stages namely decontextualization, recontextualization, categorization, and compilation. Data have written text and transform for content analysed by surface structure to identify meaning units follow by coding, categorize, and themes. Participants 22 with inclusion criteria such as healthy, over 60-year-old, pass on sensory and cognitive screening. There were found three themes of childhood activities, namely: physical activity, mental activation, and interactive communication relationship. Although the participants were very excited during survey, there was found one participant who has none idea. Therefore, it is a good idea to develop a media for optimizing cognitive function of elderly in Indonesia.

1. Introduction

Ageing people is a vulnerable population. It is a global public health concern due to some degenerative problems among nations in the world. Although the bio-phyco-social and spiritual aspects are healthy condition, the population ageing needs promote and preventive program to maintain the quality of life. This condition is very important to support Indonesia government in order preparing tendency to be an ageing structure population on the future by 2035 – 2050 (1). In the previous study at one nursing home in Jakarta about three years by 2016, 2017, 2018 (Figure 1), there has been decreasing the total number of healthy older adults; namely: from 83 by 2016, fallen to 43 by 2017, and decline to 22 by 2018 (2). Figure 1 shown the graphic about decreasing total number of healthy older adults who have been declining of cognitive function within three years at nursing home in Jakarta by 2016, 2017, 2018. Decreasing number from 83 to 34 to 22, meaning that every year, there was decreasing the number of cognitive problems up to 50% among elderly at nursing home in Jakarta. Therefore, the intervention reminiscence-based screening (IRBS) has emphasizing to trigger brain to remind childhood activities. In the IRBS, there are three screening such as sensory functions for listening and respond verbal ability, cognitive function, reminiscence function especially childhood activities. However, this paper describes only about one step of screening called reminiscence screening about the ability of childhood activities. According to

*Corresponding author: eva.tallutondok@uph.edu

literature review known that declining cognitive function could be happen during ageing process, but the early dementia can be prevented by a reminiscence intervention. Reminiscence is about a real-life memory to serve three tasks such as directive, self, social that predict adaptive versus maladaptive (3) in which has eight factors involved boredom reduction, death preparation, identity/problem solving, conversation, intimacy maintenance, bitterness revival, teach inform (4). It called a non-pharmacological (5) because effective treatment for healthy older adults (6), (7), illness (8) well-being (9), self-esteem (6), and quality of life (10). There advantages of reminiscence can daily use (11) trough multimedia such as photos, movies, sound, music, indirectly smell, teste, touch, and movement (3).

2. Methods

The qualitative study with a mix-method design was used in this study via verbal, then data transformed for four stages namely decontextualization, recontextualization, categorization, and compilation. Data have written text and transform for content analyzed by surface structure to identify meaning units follow by coding, categorize, and themes (12). There were 22 participants were involved in which have inclusion criteria such as healthy, over 60-year-old, pass on sensory and cognitive screening. However, only participants who have pass on sensory and cognitive screening allowed in this reminiscence screening. It was due to obtain a strong design as a product as a media to trigger memory to optimize the cognitive function (13). Each participant was instructed to mention and share about the childhood experience that can still be remembered today. A semi-structured interview was used in this study for exploring the thought or memory about childhood activities. Finally, this study was approved by the ethic committee with the protocol number is 04.1705173 from the Moschtar Riady Institute for Nanotechnology (MRIN).

3. Results and Discussion

One study described that people under 60-year-old have not been declining cognitive function yet, but declining will be started by 61-year-old with influence factors such as education and physical condition (14). Therefore declining function of molecules, cells, blood vessels, and gross morphology due to ageing process affects the function of brain (15), and these can impact for memory (16). Based on cohort cognitive study found that men were slower than women by age 60 and above in experiencing cognitive decline (17), while elderly aged less than 60-year-old only slight cognitive decline (18). However, the brain infarct history was a precipitation factor or a trigger factor for accelerating the process of dementia at the age of over 60 years (2). Based on the IRBS study, there were found zero to seven childhood activities via reminiscence screening, although all participants (n=22) have mean age 68.1 year-old and mostly have a basic level of education background. However all participants have pass on cognitive screening via MMSE with mean score 27.95 and able to listen and respond verbal without hearing aids. Those screening to avoid bias data, because the poor verbal communication associated with hearing loss (19), (20) and declining cognitive function (2), One study state that a low education level as one of predisposing factor be affected performance on movement and may contribute to impairment of cognition (21). While other studies of reminiscence described that gender (23) and age (24) may lead as predisposing factors to influencing the vulnerable people to lack of well-being (22). To prevent early dementia and maintain healthy ageing, the IRBS has found three themes of trigger reminiscence function, namely physical activity, mental activation, and interactive communication relationship (Table 1).

Although there was one study describes that the more ageing and the more decline the cognitive as well as memory function (17), it does not meant no hope for promotion program as a non-pharmacological intervention for elderly as a vulnerable people. Therefore, the three of themes was developed as a simple media in which able to trigger the past memories for slow down evidence of early dementia among health older adults in Indonesia (Figure 1).

Table 1. The contents analysis of reminiscence screening about the childhood activities (n=22)

Meaning Unit	Code	Sub-categories	Category	Theme
Gymnastic, chess, soccer, Ping-Pong, badminton, volley ball, weightlifting, swimming	Exercise and sports	activity require strengthen of muscle	Hard-skills	Physical Activity
Knits	Skills	activity require skills	Soft-skills	
Angklung, listening to music	Hearing and verbal	pay attention to	social and collective memory	Mental Activation
Watching television	Vision	pay attention to	social and collective memory	
Congklak, ular tangga, kasti, rondes, dolls, cooks, riding bicycle, puzzle, marbles, kites, carambola, panjat pinang	Traditional games	motor activities of leisure and recreation in the peer group	social interactive with person	Interactive communication: personal and interpersonal relationship
Gardening	Gardening	motor activities of leisure and recreation in self	social interactive with environment	
Spiritual act	Spiritual act	motor activities of leisure and recreation in God	social interactive with God	

3.1. Physical activity (PA) theme

PA theme is about an activity in which required the strengthen of muscle or skills, both hard and soft skills. In this condition, sensory and cognitive functions have coordination to receive information and perceive a perception. In the physical activity theme, there was came from nine kinds of childhood activities such as gymnastic, chess, soccer, ping-pong, badminton, volley ball, weightlifting, swimming, and knits, there is given code with exercise, sport, and skills. For code as an exercise and sports, so sub-categories has given name by the activity require strengthen of muscle, while code of skills is given sub-categorize as an activity require skills. Both of sub-categorize are categorize as hard and soft skills, so that the given theme is a physical activity. In IRBS study, PA may lead to prevent early dementia among healthy older adults instead of sleeping or being sit alone at nursing home. PA was a mild exercise about standing, clapping hands, and movements about 30 minutes. PA will fuel baseline cortisol daily in which can help control blood sugar levels, regulate metabolism, reduce inflammation, and assist with memory formulation, then it has a controlling effect on salt and water balance and help control blood pressure (20). This kind of physical activity could be appropriated as intervention reminiscence among healthy older adults either at the nursing home or in the community-based. PA has strongly relationship with cognitive impairment due to imbalance gait performance (25). All older adults are needed to encourage for

IOP Conf. Series: Materials Science and Engineering **523** (2019) 012044 doi:10.1088/1757-899X/523/1/012044

physical activity, because it necessary to invest more training, supervision, and applying skills effectively (26). Despite of people survived in functional abilities until 93-year-old, the older adults over 61 should improve daily activities (27). This kind of physical activity could be appropriated as intervention reminiscence among healthy older adults either at the nursing home or in the community-based. The most importance principle that God has created human brain organs perfectly including five dimensions of cognitive specifically namely; attention, memory capacity, executive cognitive function, language, and visual spatial abilities, therefore it must keep healthy by continuing screening. The reason that healthy older adults no change in functional abilities, but the memory capacity is declining in a mild cognitive impairment gradually. Therefore, the participants always encourage to physical exercise for improving cognitive functioning in the elderly either at nursing home or in the community.

3.2. Mental activation (MA) theme

Different with meaning unit of angklung, listening to music and watching television, the code are vision and hearing. The sub-categorize is pay attention to, so that the categorize is social and collective memory. In the MA theme, the participants should pay attention trough the sensory, especially listening to the music and watching television. Mental activation is very important for healthy ageing, it is because of episodic memory decline (28). Meaning that the memory system involves the mindful converting and recovery of contextually-specific information such as perception conceptual among older adult people. More over the authors explain that the ageing group able to change about capacity of memory especially in the contextual information, so that the older adult difficult to integrate pieces of complex memory. According to Sejdic (29) that listening to music and watching television are thought to be a sedentary activity, so that these might be effects on human gait. However, in the IRBS model was understood that angklung, listen to music, and watching television are not only effect for human gait, that influences in brain activation as well. The reason that angklung can pay attention to maintain the cognitive activation, because playing angklung requires attention to coordinate motor and auditory, then it requires every individual to enhance emotional intelligence and thinking power to interact harmoniously with others. This statement was supported (30) (8) that healthy elderly over 60-year-old ought to synchronize between eye, ear, and hand during playing angklung. Therefore, the brain activates need to give more, so that the elderly must screen sensory frequently. According to Fring (31) that watching television as memory task-brain activation, because news or movies can stimuli naturally. The mental activation affected cerebral flow velocity and (32) and it is able to assess brain response by asking 'what are you watching about?' Besides of that, television known can improve a child's ability to speak verbally and physical, emotional, emotional, and psychological development (33). Therefore, each patient should be screened for measuring cognitive function before doing mental activation (34), it is due to mental activation carried out the impulse or an object signal as optimal as a stimulation involves in brain. To stimulate mental activation, known that is not only by watching television, but listening to music effectively stimulus as well the auditory cortex and other dimensions brain to attention and memory the message. From this concern of mental activation theme, knowing that mental activation affected by sensory function such ear as a medium to stimulate the brain. Again in the model of IRBS, all participants must pass on screening not only normal state but also ability to hear without hearing aids within 50 centimeters until one meter. Therefore, the mental activation was used in the model IRBS such as watching video and postcard pictures. In to stimulate the mental activity, the participants can chose the pictures and selected video by own, because of making fun and leisure only. Most participants aged 60-69 years who needs to be optimized by performing mental activation through remembering and revealing kinds of childhood activities either pictures or video. The intervention was very importance to the participants, because the participant were not only happy but also can remind the bitterness on the pass history and ventilated already by crying. The participant was very satisfied in this session and felt free on the rest of live.

3.3. Interactive communication relationship (ICR) theme

Others meaning unit of *congklak, ular tangga, kasti, rondes*, dolls, cooks, riding bicycle, puzzle, marbles, kites, *carambola, panjat pinang* was code in traditional games, while gardening and spiritual act are not changing meaning in code. However, three of codes have a similar sub-categories as a motor activities of leisure and recreation in the peer group, self, and God, the category is social interactive with person, environment, and God. Finally, it was given theme as interactive communication both interpersonal and intrapersonal relationship. The meaning units collected by asking question in stage of decontextualization. For meaning units were broad and vary, so recontextualization has done by adjusted of each content in the coding list. Therefore, the contents remain only *congklak, ular tangga*, puzzle, marbles, carambola only in the coding list, the contents be a motor activity of leisure and recreation. That was different with meaning unit of gardening and spiritual act, the code list has the same things with the meaning unit. However, the sub-categories and category same with the *congklak, ular tangga*, puzzle, marbles, carambola coding. The contents *of congklak, ular tangga*, puzzle, marbles, carambola are visual games and cognitive process. Those traditional game-based communication effective learning for cognitive process dimension and social emotional, so it is important for encourage interaction with others. Moreover, the code list of contents can improve capacity of verbal communication among aged 60-85-year-old. Therefore, the sub-categories were 'motor activities of leisure and recreation' and the category is 'social interactive with' either other people, plants, and God. Finally, the traditional games, gardening, and spiritual act have concluded in the interactive communication. To prevent the declining cognitive functions progressively, aged 60-85-year-old might motivate to communicate and reminisce, because people will lead to withdraw from others due to difficulty to express feeling (35).

Theme Physical Activity

Theme Mental Activation

Theme Interactive communication: personal and interpersonal relationship

Figure 3. Three themes of childhood activities that derives from the reminiscence screening

4. Conclusion

There were found three themes of reminiscence screening such as PA, MA, and ICR. Those theme have developed in the media of IRBS called a trigger memory for preventing early dementia and maintaining cognitive function on keeping health among older adults in the community and the nursing home. The participants were very happy and enthusiastically to follow the activities in which these activities all recall past experiences and become screening-based reminiscence interventions. Therefore, it is a good idea to develop a media for optimizing cognitive and memory function of elderly in Indonesia.

5. Acknowledgment

This research was full supported by Kemenristek Dikti via DRPM Ditjen Penguatan Risbang Republic Indonesia. I am thankful to:

1. Universitas Pelita Harapan via Lembaga Penelitiandan Pengabdian Masyarakat
2. Mochtar Riady Institute for Nanotecnology as Ethics Committee
3. Kantor Administrasi Wali Kota Jakarta Barat
4. Panti Sosial Tresna Wredha Jakarta Barat

6. References

[1] Kementrian Kesehatan RI. 2016. *InfoDATIN Pusat Data dan Indormasi Kementrian Kesehatan RI.* ISSN 2442-7659
[2] Tallutondok, EB., Samaria, D., Januarita, AJ., Warouw, EM., Randa, EO., Anik, U. 2018. *Advanced Science Letters,* **24**(5):3520-3523
[3] Tallutondok, EB., Lanawati, S. 2018. *Advanced Science Letters*, **24**(5):3524-3526
[4] Bluck, S., Alea, N., Habermas, T., Rubin, DC. 2005. *Social Cognition*. **23**(1):91-117.
[5] Webster, J.D., Bohlmeijer, E.T., Westerhof, G.J. 2010. *Research on Aging*. **32**(4) 527–564.
[6] Jo, H.K and Song, E. 2015. *Educational Gerontology*, **41**: 1–13, 2015.
[7] Hallford, D.J., Mellor, D., Cummins, R.A. 2013. *Memory.* **21** (4), 444_457
[8] Meléndez-Moral, J. C., Charco-Ruiz, L., Mayordomo-Rodríguez, T., & Sales-Galán, A. 2013. *Psicothema,* **25**(3), 319-323.
[9] Mahendran, R., Rawtaer, I., Fam, J., Wong, J., Kumar, AP., Gandhi, M., Jing, KX., Feng, L., Kua, EK. 2017. *Trials*,**18**:324,1-11.
[10] Preschl, B., Maercker, A., Wagner, B., Forstmeier, S., Banos, R.M., Alcaniz, M., Castilla, D., Botella, C. 2012. *Aging & Mental Health*. **16**(8), 964–974.
[11] Hallford, D.J., Mellor, D., Cummins, R.A. (2013). *Memory* **21**(4), 444-457
[12] O'Shea, E., Devane, D., Cooney, A., Casey, D., Jordan., Hunter, A., Murphy, E., Newell, J., Connolly, S., Murphy, K. 2014. *Int J Geriatr Psychiatry* **29**: 1062–1070
[13] Elfrink, TR., Zuidema, SU., Kunz, M., Westerhof, GJ. 2017. *BMC Geriatrics* **17**:95
[14] Bluck, S., Alea, N., Ali., S. 2014. *Applied Cognitive Psychology* **28**: 290–300
[15] Bengtsson, M. 2016. *NursingPlus Open*. **2**:8-14.
[16] Schoonemboom, J., Johnson, RB. 2017. *Köln Z* Soziol.
[17] Salthouse, T.A. 2009. *Neurobiol Aging*. **30**(4): 507–514
[18] Peters, R. 2005. *Postgrad Med J* **82**:84–88.
[19] Pudas, S. 2013. Brain charateristics of memory decline and stability in aging: contributions from longitudinal observations (Dissertation). Retrieved from Stockholm University.
[20] Singh-Manoux, A., Kivimaki, M., Glymour, M.M., Elbaz, A., Berr. C., Ebmeier, K.P., Ferrie, J.E., Dugravot, A. 2012. *BMJ* **344**: d7622
[21] Daffner, KR. 2010. *J Alzheimers Dis*. **19**(4): 1101–1122.
[22] Tallutondok, EB., Lanawati, S. 2017. *Faletehan Health Journal*, **4** (5)
[23] Lin, FR. 2011. *Journal of Gerontology:* **66A** *(10):1131–1136*
[24] Bento-Torres, NVO., Bento-Tores, J., Tomas, AM., Costa, AM., Correa, PGR., Costa, CNM., Jardim, NYV., Picanco-Diniz. 2017. *Brazilian Journal of Medical and Biological Research* **50**(4): e5892,
[25] Wadensten B., Hagglund, D. 2006. *Int J Older People Nurs.*1(3):159-67.
[26] Gaggioli, A., Morganti, L., Bonfiglio, S., Scaratti, C., Cipresso, P., Serino, S., Riva, G. 2014. *Educational Gerontology*, **40**: 486–498

[27] Bennett, R., Zaidi, A. 2016. *International Journal on Ageing in Development Countries*, **1**(1):5-19.

[28] Karim, H., Dolatshahee, B., Khodabakhshi, A., Rezaei, M., Kamrani, AA. 2010. *Aging & Mental Health*. **14**(7):881–887.

[29] Lavretsky, H and Newhouse, PA. 2012. *Am J Geriatr Psychiatry*. **20**(9): 729–733.

[30] Beauchet, O., Allali, G., Montero-Odasso, M., Sejdic´, E., Fantino, B., Annweiler, C. 2014. 9 (6). e99318. PLOS.

[31] Bohlmeijer, E.T., Westerhof, G.J., Emmerik-de Jong, M. 2008. *Aging & Mental Health* **12**, No. 5, September 2008, 639–646. Routledge Taylor & Frances Group

[32] Christensen, K., Thinggaard, M., Oksuzyan, A., Steenstrup, T., Andersen-Ranberg, K., Jeune, B., McGue, M., Vaupel, JW. 2013. *Lancet*, **382**:1507-13.

[33] Walsh, M and Crumbie, A. 2007. Watson's: Clinical nursing and related sciences. Ed. 7th. Bailliere Tindall (Elsevier) London

[34] Murman, DL. 2015. *Semin Hear*. **36**(3): 111-121.

[35] Kim, S-Y and Giovanello, KS. 2011. *Psychol Aging*. **26**(3): 678–688.

IOP Conf. Series: Materials Science and Engineering **523** (2019) 012045 doi:10.1088/1757-899X/523/1/012045

The design of road conditions mapping system by utilizing openstreetmap spatial data

Y W Syaifudin[1,*], D Puspitasari[1], Y Ariyanto[1], R Ariyanto[1]
[1] Department of Information Technology, State Polytechnic of Malang, Malang, East Java, 65141, Indonesia

E-mail: qulis@polinema.ac.id

Abstract. Land transportation still becomes a vital role in human and material mobility to date. In developing countries, like Indonesia, the development of land transportation infrastructure is growing very fast, but the level of road damage is also still quite high. Therefore, road users need information on road conditions that will be passed. This research proposes the use of openstreetmap data to develop spatial data of road conditions dynamically and inexpensively, because it is open access and free editable. The focus is to design spatial data of road conditions and to present a visualization of road condition maps based on Openstreetmap. The raw data will be exported to the Spatial Database Management Systems (SDBMS) and then redesigned the data so that the system can add the status of road condition on each road segments. The result is a system design that can store spatial data and present a visualization of the road condition map. With this system, anyone can update the condition data of a road and can access the road condition map through the web and mobile applications.

1. Introduction

Currently, land transportation facilities have a very vital role in social and economic aspects, namely as a distribution channel between one region with another on an island. Therefore, Information on road conditions is needed by the community as road users and the government as the owner and maintainer of the road, considering the percentage of damaged roads in developing countries is still quite high. Information on road conditions is needed when road users will use the road, so they can make a good preparation and plan on the journey, such as choosing an effective path and preparing the vehicle according to the terrain to be traversed. In addition, this information is also very useful to reduce the risks in travel such as accidents, vehicle damage, and congestion.

To map road conditions, spatial data of all existing roads is required. With reference to this data, we can mark the condition of each road segment. The spatial data types (POINT, LINE REGION, etc.) provide us the fundamental abstraction for modeling the structure of geometric entities in space as well as their relationships [1]. Currently, there are quite some spatial data providers, especially road map data, ranging from data collected by local governments to data compiled by global providers, such as Google Maps, Openstreetmap, and ESRI. They can be used in navigation, in research, in the development of location-based applications [2]. In this study, we propose the use of spatial data from Openstreetmap with the following considerations:

* Corresponding author: qulis@polinema.ac.id.

a. Openstreetmap is a collaborative project to create a free editable map of the world. It is also free to be accessible to anyone and anywhere as long as we are connected to the internet.
b. Openstreetmap spatial data can also be freely downloaded by anyone who needs and can be processed for specific needs.
c. Currently, the spatial data of roads network that was covered by Openstreetmap is very complete, up to date, and accurate, especially for Indonesia.

The core product of Openstreetmap is a spatial database, which contains geographic data and information from all over the world [3]. With the availability of spatial data, we can utilize it without the need to do data collection of roads independently. Next, we can add information related to road conditions based on the road data. Because the data can be accessed from anywhere and by anyone, then the input of information related to the condition of the road can be done by the public, so the road condition data will continue to grow quickly, continuously, and dynamically.

2. Methodology

A spatial database is related to objects in space, including points, lines and polygons typical databases grasp various numeric and character types of data, additional functionality needs to be added for databases to process spatial data types [4]. Spatial data represents multi-dimensional data with points, surfaces, and lines, as a list of numbers using a particular coordinate system [5]. To build a system capable of processing spatial data, we need a Spatial Database Management Systems (SDBMS). There are now quite some database software systems that provide the ability to process spatial data, such as Oracle, Postgresql, Mysql, and ESRI [4]. In this study, we chose Postgresql because it is open source, supports multiple spatial functions, and is quite mature when compared to another open-source DBMS. Static road map data from openstreetmap will be exported to SDBMS for processing and further manipulated by system requirements. Furthermore, the data can be presented via web and mobile applications. The architectural model of the mapping system to be constructed is presented in Figure 1.

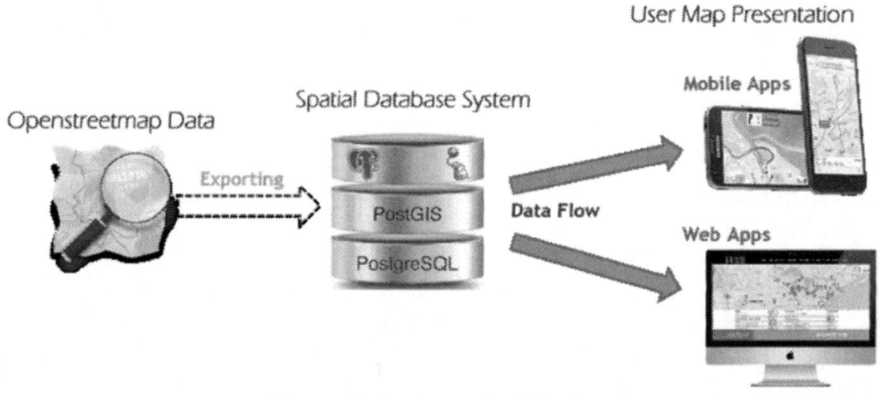

Figure 1. Systems Architecture

2.1. Exporting from Data Source

OpenStreetMap is a web-based project for creating free and open worldwide maps. Openstreet maps are built entirely by volunteers [3] by conducting surveys using GPS, digitizing satellite images, collecting geographic data, and releasing them for public use. It is freely available to all users online around the world [6]. The base map and supporting data can be downloaded for free and open, for further use, processing, and redistribution. This research will use Openstreetmap road spatial data for Indonesia, especially Java region, as a data source. Openstreetmap data can be easily converted into other widely used spatial data formats, such as KLM (Google Earth) and SHP (ArcGIS) [7].

Furthermore, Openstreetmap source data is exported to PostgreSQL which already has PostGIS extension so that it can perform operations for spatial data. Postgresql with this PostGIS extension can next be classified as a Spatial Database Management Systems (SDBMS), a DBMS that can process spatial data. Then the data already stored on this SDBMS can be further processed for the needs of road condition mapping.

2.2. Spatial Data Processing and Presentation

Spatial data from OSM are stored in SDBMS with multiline-string geometry data type. The multiline-string data type is a geometry data type in the form of a collection containing the only line-string. While line-string is a collection of dots that form a set of straight lines that are connected. This data type usually was used to map items such as roads, rivers, trails, and circuit paths. This road geometry data needs to be processed before it can be marked the type of damage. The length of a line varies from tens of meters to thousands of meters, thus requiring the splitting of road data into segments with a maximum length of 5 meters. Each of these road segments will each be marked according to their condition. The steps to present roads condition data is showed in Figure 2.

Figure 2. Spatial Data Processing Steps

Therefore, marking the damage of a road segment can be performed on at least 1 data road segment with a maximum length of 5 meters. Each road segment can be marked by the reporter and stored in a separate status table. The data sets of road segments that have been marked according to their respective conditions are then displayed in a map form on the web browser and mobile device. To display spatial data in the form of maps in web browsers and mobile devices, we can use the Leaflet [8], Node.js [9], and Github libraries [10].

3. The Design

This section describes the type and level of road damage and is extended to database design that refers to the spatial data requirements of road conditions. Implementation of database design requires a segmentation process to break the road into small sections which can then be marked in accordance with the conditions.

3.1. Roads Condition

Technically, road damage indicates a condition in which the structural and functional roads are not able to provide optimal service to the traffic that crosses the road. Generally, road damage is largely due to user behavior, mismanagement and implementation, and inadequate road maintenance.
According to the ministry of public works, through the directorate general of Bina Marga [11], the type of road damage is divided into the following four major categories:

1. **Cracking** is a symptom of damage to the surface of the pavement so that it will cause water on the surface of the pavement to enter the underlying layer and this is one of the factors that will make it widespread.
2. **Distortion** is a change in the shape of the road that occurs due to the weakness of the base soil, less compaction on the base layer, resulting in additional compaction due to traffic load.
3. **Disintegration** is a type of road surface damage that leads to chemical and mechanical damage to the pavement layer. This damage can be in the form of hollow road conditions, the release of grain material of the road, and peeling of the surface layer.
4. **Bleeding/flushing** is a type of physical change of the road as it softens during high temperatures.

The level of road damage will be divided into 3 standard levels, i.e. light, medium, and heavy. This refers to the severity of a type of road damage.

3.2. Database Design

The database design will contain the entity segment of the road as a whole and the segment data entity that is reported in its condition. Street segment entities will be populated with segmented process result data of all road segments taken from Openstreetmap. Each segment will represent a section of the path that can be assigned a status according to its condition. Each segment will contain information such as road id, length, geometry data, region, and road metadata. The relational diagram is shown in Figure 3.

Table : pnm_roads	
field name	data type
osm_id	varchar(10)
code	smallint
name	varchar (100)
oneway	varchar (1)
maxspeed	smallint
geom	geometry

Table : pnm_road_segments	
field name	data type
osm_id	varchar(10)
sid	integer
name	varchar (100)
lgeom	geometry
centroid	geometry
length	double
region	integer

Table : pnm_dmg_segments	
field name	data type
rid	bigint
osm_id	varchar(10)
sid	integer
lgeom	geometry
centroid	geometry
length	double

Table : pnm_dmg_reports	
field name	data type
rid	bigint
user_rpt	varchar(50)
ts_rpt	timestamp
type_id	smallint
level	smallint
information	varchar(100)
photo	varbinary
status	varchar(1)

Figure 3. Table relations

The pnm_roads table represents the data of all paths taken directly from Openstreetmaps. The spatial data path is stored in multiline-string format. Next all the path data is broken down into segments stored in the pnm_road_segments table, where the spatial data of the path segment is stored in line-string format. Next data segments reported by the system status will be stored in the pnm_damaged_roads table. Each user's road damage report will be stored in the pnm_dmg_reports table, including report_id, type of damage, level of damage, related information, image, and status (valid or otherwise). Each road damage report will include one or more road segments and will be stored in the pnm_dmg_segments table. Each stored segment will contain information from line-string geometry, centroid point geometry, and segment length.

3.3. Roads Segmentation

A data path on Openstreetmap is represented in the form of a line string, where the length of each segment varies from tens of meters to thousands of meters. To record the condition of the road, it takes a shorter segment so that the data collection of road conditions will be more detailed. Each path generated from Openstreetmap will be broken down into many segments where the maximum length of a road segment is 5 meters. To segment a path with a maximum length of 5 meters can use the following query:

```
SELECT a.osm_id,(a.ra).path[1] AS sid,a.name,st_segmentize((a.ra).geom,
   0.00004533::double precision) AS lgeom, st_length((a.ra).geom::geography) AS
   length FROM ( SELECT osm_id::bigint AS osm_id,name,st_dump(geom) AS ra
   FROM pnm_roads) a;
```

3.4. Marking of Road Conditions

A road segment may have a damaged status if a report from a user associated with the segment has been validated by the system administrator. Usually, a report will include many segments on a road line. Any incoming reports from users will not be immediately considered valid, so it takes the validation process from the administrator so that the report can be used by other users. The color of the damaged road segment is determined by the type and level of damage. The type of damage represents the type of color, and the level of damage represents the thickness of the color. In this case, there are 4 types of damage so there will be 4 colors that represent each type of damage, namely:

1. Cracking is presented in blue color, with 3 level of damage.
2. Distortion is presented in red color, with 3 levels of damage.
3. Disintegration is presented in green color, with 3 levels of damage.
4. Bleeding/flushing is presented in purple color, with 3 levels of damage.

4. Results of Spatial Presentation

Based on the data design described in the previous section, it is necessary to implement the database design on a PosgreSQL SDBMS with PostGIS extension. In this section will be presented about how the visualization of the segments that have been given status according to the type and level of damage.

4.1. Spatial Presentation

Spatial data already stored in SDBMS can then be accessed and presented via web apps and mobile devices. By using the web and mobile apps, the whole society can easily access this system, either to report or to view damaged road data and position visually. Therefore, with the entire community involved in reporting road damage, the system will be faster and more dynamic in terms of data development when compared to conducting independent road condition surveys.

4.1.1. Web Page Presentation

In developing web applications, we use the Node.js, an event-driven I/O server-side JavaScript environment that allows for JavaScript to operate on the server [12]. The leaflet is the leading open-source JavaScript library for mobile-friendly interactive maps [13]. An OpenStreetmap-based map will be displayed on a web page, as a basis for displaying damaged road data. Further damaged road segments will be added at the top layer, according to the geometry of the segments, including the coordinates, lengths, and locations of the region. The display on the web application is shown in Figure 4.

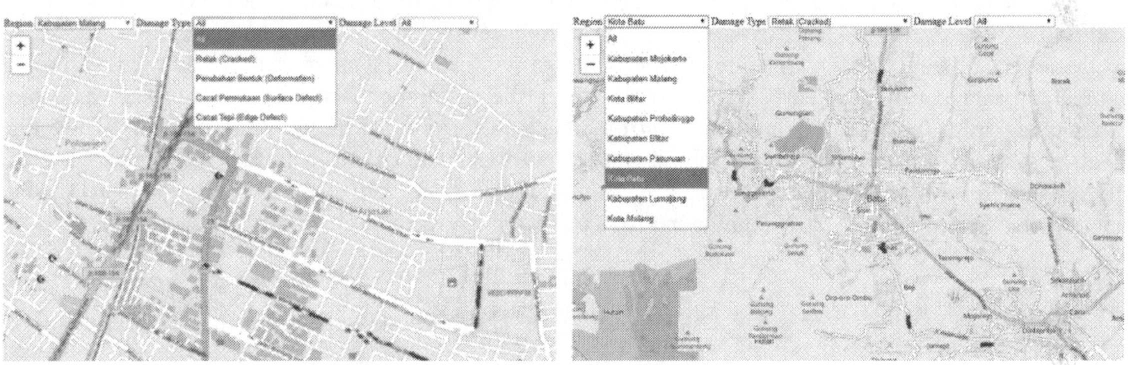

Figure 4. Presentation in Web Application

4.1.2. Mobile Device Presentation

In line with the platform that has been used in web application development, we use the react native libraries from Github to develop Android-based applications. Similar to web applications, applications on mobile devices also display Openstreetmap maps as map bases and then display damaged road segments according to those in the database. This web application will access a web service as an interface with the PostgreSQL database. The view to see the road conditions in the mobile application is shown in Figure 5.

4.2. Discussion

For now, there are still not many applications that utilize and process spatial data on Openstreetmap, whereas Openstreetmap has provided spatial data that is free to access and exploit such as roads, buildings, territorial boundaries, rivers, lakes, and so on. The majority of applications for data collection of road conditions are still dependent on self-survey methods to get spatial data of damaged roads, resulting in slow development and data dynamics. In addition, the information generated can only be accessed by the relevant parties who perform the survey so it can not be accessed by the public.

Prior to 2014, research and development of GIS software the majority used ArcView GIS software where data had to be input through independent surveys, which also conducted data collection and maintenance of roads [14] [15]. Then after that, developed software based on Google Maps but still limited the coverage of its users [16]. Application and data design to map road damage will be used to develop a system that will receive road damage reports from the community and present them in visual form on an Openstreetmap-based map.

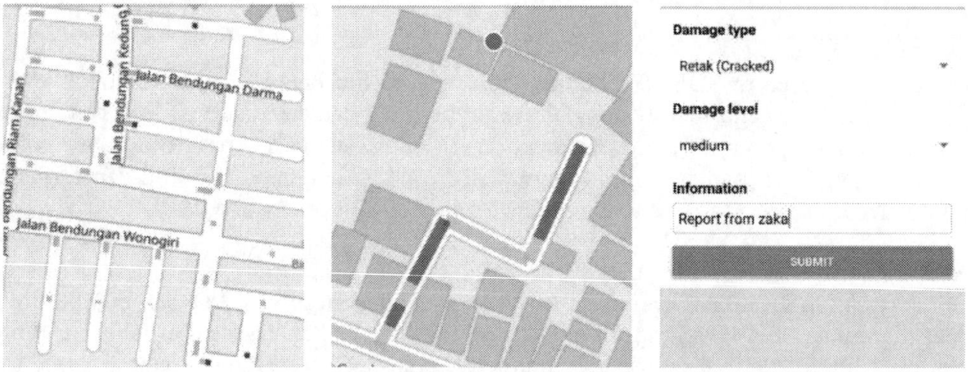

Figure 5 Presentation in Mobile Application

5. Conclusion

The results of this spatial data design have been able to mark a road position in accordance with the conditions of the damage. Each road segment tag will store the road segment id, the type of damage, and the level of damage. Each type and level of damage will produce a specific color on a marked road segment so that it can be clearly visible to the user the difference between the damaged and the undamaged part of the road. Next, the results of this data and application design will be used to develop effective systems for reporting and mapping road conditions. The spatial data that has been collected will be used to develop an alternative path search system to avoid damaged roads. In addition, spatial data will also be used to perform mapping analysis of road improvements and cluster road damage areas.

6. References

[1] Rajan S N, Sinha A K, and Singh J B 2012 *International Journal of Information and Communication Technology Research*. **2** 465
[2] Rifat M R, Moutushy S, Ahmed S I, and Ferdous H S 2011 *Proc. IEEE Student Conference on Research and Development* (Cyberjaya: IEEE Malaysia Section)
[3] Mooney P and Minghini M 2017 *Mapping and the Citizen Sensor* (London: Ubiquity Press) p 37
[4] Gawali T and Wagh R B 2013 *Proc. 4th International Conference on Computing, Communications and Networking Technologies* (Tiruchengode)
[5] Thirunavukkarasu K and Wadhwa M 2016 *International Journal of Computer Science Trends and Technology* **4** 133
[6] Zhang Y, Li X, Wang A, Bao T, and Tian S 2015 *Journal of Urban Management* **4** 135
[7] Zheng J, Zhang Z, Ciepłuch B, Winstanley A C, Mooney P, and Jacob R 2013 *Proc. 21st International Conference on Geoinformatics* (Kaifeng, China)
[8] Agafonkin V 2017 *Leaflet - a JavaScript library for interactive maps* https://leafletjs.com/. [Accessed 12 June 2018].
[9] Nodejs Foundation 2018 *About Node.js®* https://nodejs.org/en/about/. [Accessed 15 June 2018].
[10] Github 2018 *The world's leading software development platform · GitHub* https://github.com. [Accessed 2018 June 18].
[11] Direktorat Jenderal Bina Marga 1983 *Manual Pemeliharaan Jalan No : 03/MN/B/1983* Jakarta: Departemen Perhubungan Republik Indonesia.
[12] Huntley E and Foster M 2018 *Leaflet with PostGIS, NodeJS, and Express* DUSPviz Team
[13] Wang L, Pi R, Zhou X, and Zhou H 2015 *Proc. 4th International Conference on Computer, Mechatronics, Control and Electronic Engineering* (Huangzhou, China).

[14] Zulkifli M B, Yunus M, and Hassan H B 2010 *Journal of Geographic Information System* **2** 215
[15] Piarsa I N, Hadi E S, and Wirdiani N K A 2015 *Int. Journal of Computer Science Issues* **12** 95
[16] Lukito H 2012 *Proc. Seminar Nasional Informatika* (Yogyakarta)

IOP Conf. Series: Materials Science and Engineering **523** (2019) 012046 doi:10.1088/1757-899X/523/1/012046

Online judge MySQL for learning process of database practice course

D Puspitasari[1,*], P P Arhandi[1], P Y Saputra[1], Y W Syaifudin[1], H A Himawan[1], P A Sholihah[1]
[1] Department of Information Technology, State Polytechnic of Malang

E-mail: dwi.puspitasari@polinema.ac.id

Abstract. Database course is the main courses in the department of informatics and computer science. This course aims to provide knowledge to students to build and manage data on Database Management Systems (DBMS) like MySQL. Therefore, the implementation of database processing practices in one of the DBMS such as MySQL is very important as basic skills in the field of informatics. With the growing number of informatics students, lectures cannot correct the results of student exercises quickly. For that, we need a system that can help lecturers to make corrections of query exercise on MySQL automatically, which is called Online Judge MySQL. This application was developed using NodeJS to execute queries that users input and ReactJS to build its interfaces. Testing is done by using this application for the online exam in several classes simultaneously. The results show that this application can correct the test results quickly and lightly.

1. Introduction

Database course is the main courses in the department of informatics and computer science. This course aims to provide knowledge to students to build and manage data on Database Management Systems (DBMS) like MySQL. Therefore, the implementation of database processing practices in one of the DBMS such as MySQL is very important as basic skills in the field of informatics. With the growing number of informatics students, lectures cannot correct the results of student exercises quickly. For that, we need a system that can help lecturers to make corrections of query exercise on MySQL automatically, which is called Online Judge MySQL.

The developed system should be able to assess whether the output generated from the student's query equals the output generated by the correct query. Because the same output can be obtained using different queries. A lot of research has been done to develop an online judge system. Among them is a system that has been created by Miguel A. Revilla [6] entitled "Competitive Learning in Informatics: UVA Online Judge." The system was built based on the needs of the online judge system for various programming languages, as well as referring to the concept of competitive learning that aims to provide lessons about programming through the competition. Some things that have been achieved from this system include discussion forums, a collection of problems, a system for program submission and leaderboard. But this system has a lack of feature; the system did not support query language on MySQL.

There is another system entitled "Rancang Bangun Sistem Basis Data Online Judge untuk Proses Pembelajaran Mata Kuliah Sistem Manajemen Basis Data di Departemen Teknik Informatika ITS." The system can overcome several things in the correction problem, by creating a grader system that can facilitate students to do the task. This is achieved by comparing the student's query with the query that acts as the answer key. The system, then, is using it as a determination of scoreboard scoring through the system. And then displayed it as an evaluation. But in the implementation, there are still some shortcomings such as query restrictions are allowed to run on the system, scoring system through the scoreboard and processing, and the speed of grader execution [8].

*Corresponding author: dwi.puspitasari@polinema.ac.id.

To solve the problem of correction time, asynchronous javascript technology can be used. Asynchronous javascript is JavaScript that is executed as server-side. One of the advantages is the existence of non-blocking techniques. Using the technique, the execution process of an operation does not have to wait for the previous operation to be executed, resulting in a faster and more efficient web application [5]. The asynchronous javascript technology used in this research is NodeJS.

In this paper, the first section will describe concepts that support the research, such as an online judge, NodeJS and ReactJS, JSON, REST. In the second part of this paper will describe the research methodology. In the third section will describe the results and discussion of the research while the fourth section will contain conclusions.

2. Online Judge MySQL

Online Judge, in general, is a server that contains descriptions of problems in different context, as well as data that is set to assess whether a particular solution solves the problems residing on the server [6]. There are several online judge categories currently implemented: Problem Solving Online Judge, HTML & CSS Online Judge, and SQL Online Judge. In this research will be developed SQL Online judge. Just like an online judge for HTML and CSS, online judge for SQL is currently an online course, because there is no rating system in the application. Examples of online course applications for SQL are Codecademy and SqlCourse [6].

Online Judge is an online system that is often used to perform testing of the program in a programming contest. This system can compile, execute, and test on the source code submitted by the participants of the programming contest. The online judge can be used in various languages such as C++, Java, including SQL. In this MySQL database online judge, after the user enters the answer in SQL query form, the system does not compare query text but compares the result of answer query with the result of the key query. The online judge architecture runs entirely on the basis of text-based query execution, in accordance with the rules of the MySQL.

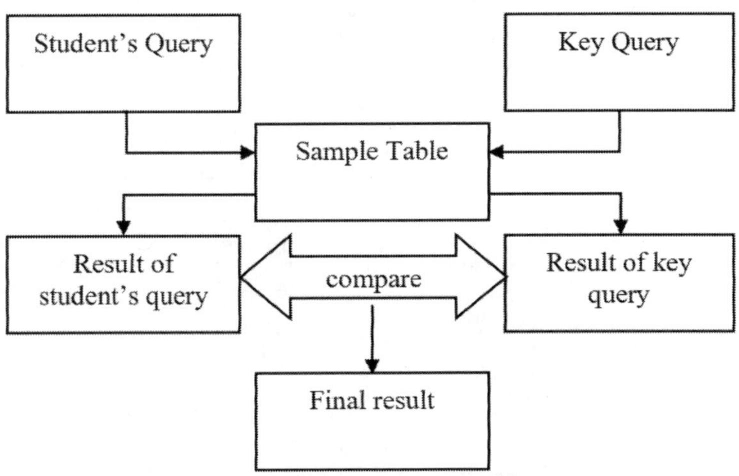

Figure 1. Judging Process Architecture

According to figure 1, the student's query is a query generated by input from the student. The key query is a query generated by input from the lecturer. The sample table is the table, which uploaded by the lecturer as a grading reference. The result of student's query is an object as the result of student's query execution. And the result of the key query is an object as the result of key query execution. The system is then comparing both of query results, and generate the final result. There are three possibilities of the final result Correct Answer, Wrong Answer and Query Error (if the result is an error due to incorrect syntax).

An object used as a comparison because the grading process cannot be done using text because it will produce inaccurate results. The inaccuracy of the results is caused by each query entered by the student may have different syntax and writing, but has the same logic as the expected execution result. By comparing objects, the results obtained are not based on text queries but rather query execution results, which can be more accurate.

The Online Judge MySQL in this research developed using asynchronous javascript technology using the Node.js platform for the backend and ReactJS for the frontend. Node.js is an event-based JavaScript and asynchronous I/O. Unlike most javascript languages that run on the browser. Node.js serves as a server application executed as server-side. One of the advantages of Node.js is the non-blocking technique in which Node.js will execute an operation without having to wait for the previous operation to be executed, resulting in a faster and more efficient web application. Node.js is capable of handling thousands of connections together with minimum resource usage for each process. In addition, another advantage of Node.js is an open source project, so anyone can see the code structure and can also contribute to its development [5].

ReactJS is an open source JavaScript library managed by world-renowned IT companies such as Facebook and Instagram along with developer communities from around the world. This framework is used extensively to develop an interface in a web application (Artemij 2015). When viewed from the architecture of user interface implementation, ReactJS can be said as a component of V (View) in MVC (Model View Controller). ReactJS can use CSS styling features and other CSS frameworks well. For element writing, we can use HTML script inside JavaScript (JSX). ReactJS excellence is when there are data changes, ReactJS will conceptually reload certain parts of the change [1]. This causes the application to be fast and lightly executed.

The system architecture used in this research is REST. REST is a web standards architecture that uses the HTTP protocol in data communications. In the REST architecture, servers that follow the REST architecture provide access to data sources and clients that retrieve data. Each data source is identified using a URI link. REST uses various formats to present data, such as text, JSON and XML. REST uses a stateless HTTP protocol. HTTP commands that can be used are GET, POST, PUT or DELETE functions. The results sent from the server are usually in the form of a simple XML or JSON format without any packet data protocol, so the received information is more readable and parsed on the client side [11].

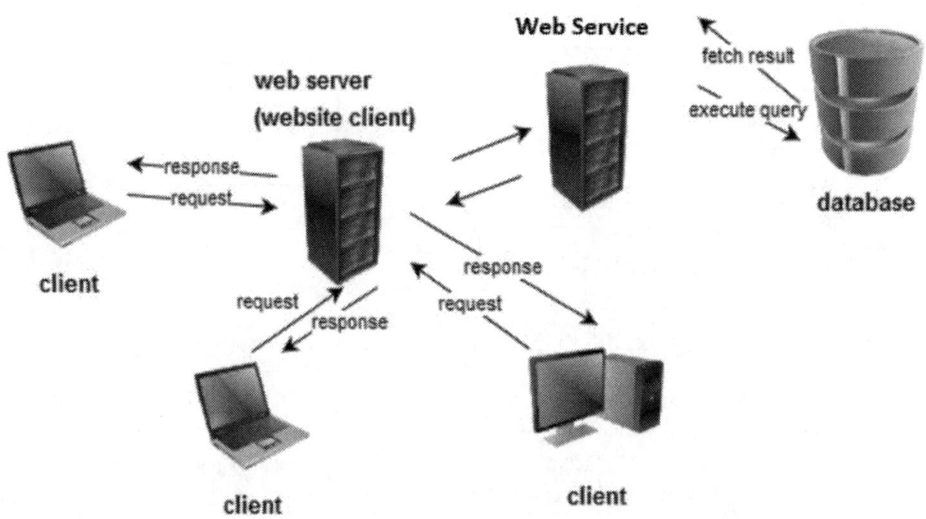

Figure 2. System Architecture

Javascript Object Notation (JSON) is a popular general-purpose data encoding format. Application of JSON has been widely used in database and web service. JSON document structure can optionally

IOP Conf. Series: Materials Science and Engineering **523** (2019) 012046 doi:10.1088/1757-899X/523/1/012046

be limited by a scheme consisting of two things namely the map (mapping of value based on the classification structure type) and lists (grouping value based on the classification of the type). JSON is a lightweight, easy-to-read, and human-written data exchange format, and is easily translated and created by computers. JSON is a text format that does not depend on any programming language because it uses common language styles like C, C ++, C #, Java, JavaScript, Perl, and Python. Because of these traits, making JSON ideal as a language of data exchange.

In this part, will be explained about the system architecture, as shown in figure 2. Client access the website via a browser on laptop or PC connected to the internet. The client computer sends the request, and the webserver sends a response. Through the webserver, data entered by the client is stored into the database. The system is built using Node.js as the backend. Node.js plays a role in the routing process, using the express framework help. Node.js also plays a role in the creation and management of databases and queries, on the MySQL server. While the frontend used ReactJS as a library. ReactJS also play a role to send data received from view to routing express on node.js while the online judge architecture is fully run on the basis of text-based query execution code on the MySQL database.

This research develops Online Judge MySQL which aims to simplify the assessment of database practice, especially on SQL Select command material. In this application, there are features such as managing users, managing the lab and exam schedules, manage classes and class participants, manage a collection of questions and problems, manage assessment. In every practicum activity, there is a scoreboard to see the recapitulation of values from students. In this application, there are three users, namely lecturers, students, and super admin. The data used in this application is managed in MySQL database management system. The structure of the database in this application can be seen in figure 3.

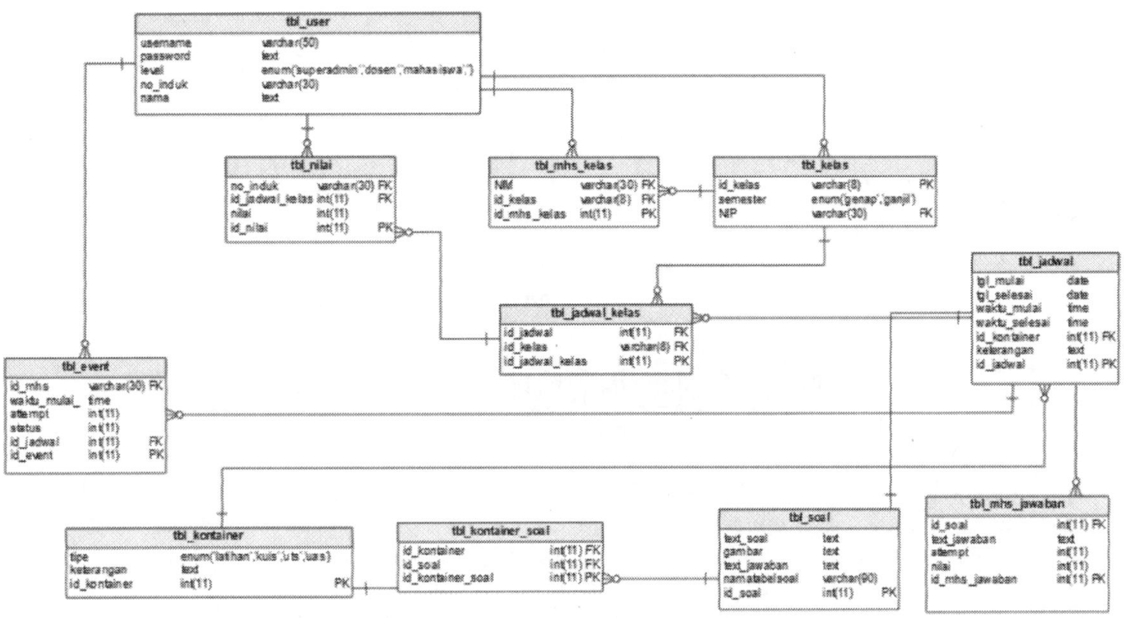

Figure 3. Database Design of Online Judge MySQL

3. Implementation

In order to build this application, we use several support tools consisting of software and hardware. The software that we use in building this application can be seen in table 1, while the hardware that we use to build the application can be seen in table 2.

IOP Conf. Series: Materials Science and Engineering **523** (2019) 012046 doi:10.1088/1757-899X/523/1/012046

Table 1. Software Requirement

Name	Function
Visual Studio Code	As a tool for writing program code
Node,js	As a tool for web-based application development
Apache	As a web server
MySQL	As an application for data storage
Web Browser (Mozilla / Chrome)	As a browser to run web-based applications

Table 2. Hardware Requirement

Hardware	Specification
Processor	Processor Intel Core i 3 2.00GHz
Memory	RAM 4 GB
Harddisk	500 GB

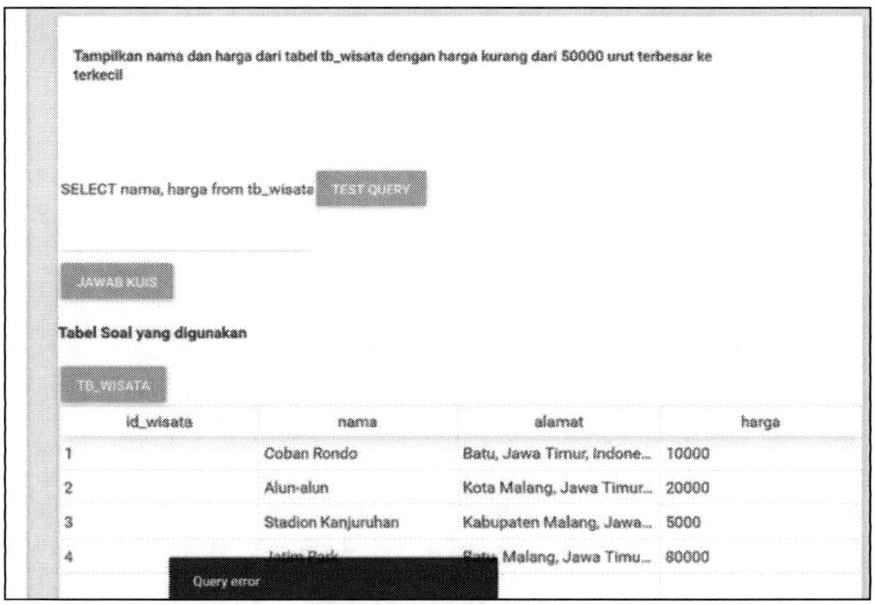

Figure 4. Query Page

4. Experiment results and discussion

To test Online Judge MySQL done two kinds of testing, User Acceptance Test (UAT) and system testing. UAT Testing aims to determine the level of need for online applications MySQL judge and the level of compatibility with the needs of the database practicum especially DML query SELECT material in the Department of Information Technology, Politeknik Negeri Malang. The number of respondents as many as 10 students and database lecturers of as many as 3 people. While for testing the system is focused on the level of success graders in handling the judgment process of the query that allows the occurrence of errors on the system. Grader testing is done by testing two grader versions.

After testing is done, it can be seen that online judge. MySQL application meets the requirement of database material about DML, especially query SELECT. It also can save lecturers time to assess student's work about query SELECT. Then for the grader test, it can be seen that the grader has been able to perform the judgment process appropriately.

5. Conclusion

During the process of design, implementation, and testing can be concluded that online judge application can handle the assessment procedure. It checks whether the student's query equals the key query from the lecturer. Online judge application already met the functional requirements. Grader already meets the needs of the judgment process. In accordance with the User Acceptance Test, the lecturer requires online judge for database practicum material about DML, especially query SELECT.

6. References

[1] Artemij F 2015 *ReactJS Essentials* (PACKT Publishing Copyright)

[2] Roy T Fielding 2000 *Architectural Styles and the Design of Network-based Software Architectures* (University Of California Irvine).

[3] Gilberto C Ricardo, María T S Redriguez, P. Alberto Leyva-Figuerdo, Laura L Mendoza-Tauler 2015 *Methodology for Characterization of Cognitive Activities when Solving Programming Problems of an Algorithmic Nature Olympiads in Informatics Vol. 9.*

[4] Martin K and Alastair R Beresford 2017 *A Conflict-Free Replicated JSON Datatype* (University of Cambridge Computer Laboratory Cambridge).

[5] Muhammad L 2016 *Keamanan Perangkat Lunak pada Bahasa Pemrograman Node.Js untuk Aplikasi Berbasis Web Jurnal Keamanan Perangkat Lunak* Mahasiswa (Program Magister Teknik Elektro Institut Teknologi Bandung).

[6] Miguel A Revilla, Shahriar M, Rujia L 2008, *Competitive Learning in Informatics: The UVa Online Judge Experience Olympiads in Informatics vol. 2.*

[7] Mulyanto, Aunur R et al. 2008 *Rekayasa Perangkat Lunak Jilid 1* (Jakarta: Direktorat Pembinaan Sekolah Menengah Kejuruan).

[8] Demsy I Mustasyar 2017 *Rancang Bangun Aplikasi Sistem Basis Data Online Judge (SBDOJ) untuk Proses Pembelajaran Mata Kuliah Sistem Basis Data di Departemen Teknik Informatika ITS* Skripsi Mahasiswa Teknik Informatika (Institut Teknologi Sepuluh Nopember).

[9] Fauzan N 2013 *Implementasi Web Service pada Aplikasi Kamus Bahasa Indonesia* Jurnal Mahasiswa (Program Studi Teknik Informatika Universitas Muhammadiyah Surakarta).

[10] Tommy P Nugroho 2016 *Aplikasi Pencarian Jalur Terpendek untuk Menemukan Lokasi ATM(Anjungan Tunai Mandiri) Terdekat di Kota Malang Berbasis Android* Skripsi Mahasiswa (Program Studi Teknik Informatika Politeknik Negeri Malang).

[11] Muhamad A Rahman et al. 2013 *Jurnal Teknik Informatika* **2**

[12] Mendel R 2017 *Node.js CS142 Lecture Notes* (Stanford University).

IOP Conf. Series: Materials Science and Engineering **523** (2019) 012047 doi:10.1088/1757-899X/523/1/012047

Grey wolf optimization for track maximum power of photovoltaic system In multiple peak power characteristics

I D Sara[1,*], R Faulianur[1]

[1] Department of Electrical Engineering & Computer, Engineering Faculty of Syiah Kuala University, Jl. Tgk. Syech Abdulrauf No.7, Darussalam Banda Aceh 23111 Indonesia

E-mail: ira.sara@unsyiah.ac.id

Abstract. A Photovoltaic (PV) module has a nonlinear current and voltage characteristic. The maximum power is obtained when the current and the voltage reach its maximum point. This maximum power fluctuates with the changes in irradiance and temperature conditions. To tract the maximum power produced by a PV device under a shading condition, a maximum power point tracking (MPPT) method is required. The grey wolf optimisation (GWO) algorithm has been used to obtain the maximum power point of a PV array. However, how it performs to track the maximum power point of a shaded PV module is still not understood. Therefore this study focuses on the investigation of the GWO algorithm to track the maximum power of a PV module under a partial shading condition. The shaded PV module is simulated based on a one-diode model using the Matlab/Simulink. Different shading conditions are generated to test the performance of the GWO algorithm in tracking the maximum power of a tested PV module. Its performance is also compared to that of the perturb and observe (PNO) algorithm. The findings show that the accuracy of the GWO algorithm is higher than the PNO algorithm. In average, the GWO algorithm is able to track the maximum power point of a shaded PV module with an accuracy of 97%. Comparing to the GWO method, the PNO method is not preferable to track the maximum power point of a shaded PV module

1. Introduction

A photovoltaic (PV) module is a device used to convert sunlight into electrical energy. There are some advantages of using PV modules as electricity generation i.e.: environmentally friendly, easy to maintain and to install and silence. A part from its advantages, it also has a disadvantage, i.e.: its output power fluctuates with changes in the irradiance and temperature conditions. Another condition such as partial shading also contributes to the fluctuation in the power output of a PV module. The relationship between the current and voltage (I-V) characteristic of a solar module is nonlinear. The maximum power is obtained when the current and the voltage output of the PV module reach its maximum point. A maximum power point tracking (MPPT) method is required in order to operate the PV module at its maximum power point.

There have been several methods developed to track the maximum power of a PV module such as a Perturb and Observe (PNO) method, an Incremental Conductance (IC) method and a Hill Climbing

* Corresponding author: ira.sara@unsyiah.ac.id

IOP Conf. Series: Materials Science and Engineering **523** (2019) 012047 doi:10.1088/1757-899X/523/1/012047

(HC) method. All of these methods only work very well to tract the maximum power point of a PV device under normal environmental conditions. However, when a PV device produces a multiple-peak power at one irradiance condition, the optimisation methods fail to find the highest peak of the output power. The output power with multiple peaks occurs when the PV module equipped with bypass diodes experiences a partial shading condition [1], [2], and [3]. Other optimisation method such as a Grey Wolf Optimisation (GWO) [5] and an improved Particle Swarm Optimisation (IPSO) have been studied in [4,6] to exploit their capability in finding the maximum power of a PV device. The capability of this method to find the maximum power point of a PV array under a partial shading condition has been investigated by [6]. The study found that the GWO method performs much better than the PNO and the IPSO methods in obtaining the maximum power point of a PV array under a partial shading condition. However, how the ability of the GWO algorithm to track the maximum power point of a shaded PV module is still not understood. Therefore, this study aims to investigate the ability of the GWO algorithm to track the maximum power point of a PV module with multiple peaks under a partial shading condition. The GWO method will be tested and analysed to obtain the global maximum power point of a PV module under the partial shading condition.

2. Modelling of a Photovoltaic module

A PV module consists of solar cells connected in a series and parallel configuration. The configured cells form a module and then the several modules are connected in series-parallel to form an array. The connection of the PV modules in an array is suited to the requirement of the load condition. A solar cell is generally modelled using a one-diode model as shown in figure 1.

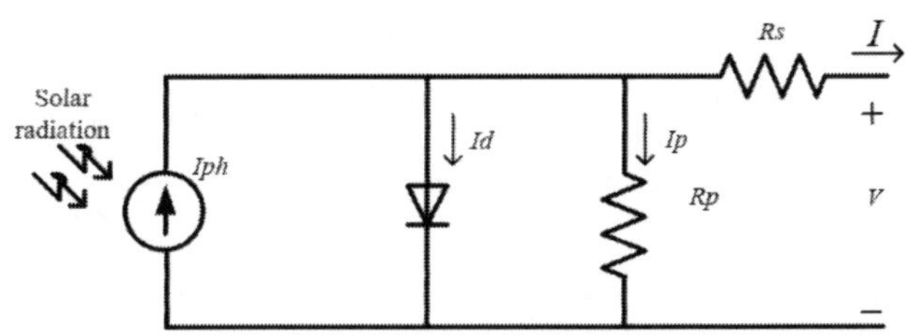

Figure 1. An equivalent circuit of a solar cell based on a one- diode model [7].

As shown in the figure 1, there are several parameters used to represent a solar cell such as a photon current I_{ph}, a diode current I_d, a shunt current I_p which flows through a parallel resistance R_p, and a series resistance R_s. The relationship between the current and the voltage of the solar cell based on these parameters is expressed as follows:

$$I = I_{ph} - I_d - \frac{V + R_s * I}{R_p} \quad (1)$$

$$I = I_{ph} - I_o * \left[\exp\left(\frac{V + R_s * I}{\frac{n * k * T}{q}} \right) - 1 \right] - \frac{V + R_s * I}{R_p} \quad (2)$$

Where I_{ph} is the photocurrent; I_o is the reverse saturation current of diode; q is the electron charge (1,6 x 10^{-19} Coulomb); k is the Boltzmann's constant (1,38x10^{-23} (J/K); T is the module temperature (°K); n is the diode ideality factor.

A PV module can be modelled based on the series and parallel configuration of its cells. The equivalent circuit of a PV module is illustrated in figure 2.

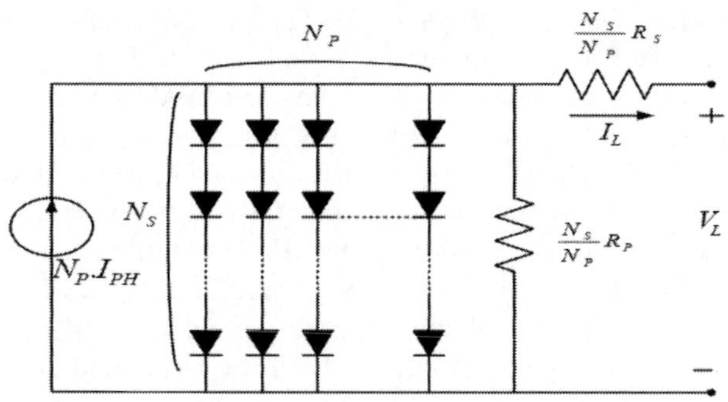

Figure 2. The equivalent circuit of aphotovoltaic module [8].

Where N_s is the number of cells in series; N_p is the number of cells in parallel in a module.

The relationship of the current and the voltage of a PV module is given in (3):

$$I = N_P * I_{ph} - N_P * I_o * \left[exp\left(\frac{V/N_S + I * R_S/N_P}{\frac{n * k * T}{q}}\right) - 1\right] - \frac{V * Np/N_S + I * R_S}{R_p} \quad (3)$$

where,

$$I_{ph} = [I_{sc} + K_i(T - 298)] * \frac{G}{1000} \quad (4)$$

$$I_O = I_{rs}\left[\frac{T}{T_r}\right]^3 exp\left[\frac{q * E_{g0}}{nk}\left(\frac{1}{T} - \frac{1}{T_r}\right)\right] \quad (5)$$

$$I_{rs} = I_{sc}/[\exp(qV_{OC}/N_S knT) - 1] \quad (6)$$

Where E_{g0} is the bandgap energy of a semiconductor material for Si at 300K is 1.12 eV; I_{sc} is the short-circuit current (A); I_{rs} is the reverse saturation current of diode (A); G is the solar irradiance (W/m^2), K_i is the temperature coefficient for the short-circuit current (0.0013 A/°C); T is the module's temperature (°K); T_r is the reference temperature (°K), k is the Boltzmann's constant (1,38x10^{-23} J/K); q is the electron charge (1,6x10^{-19} Coulomb); n is the diode quality constant (from 1 to 2); R_s is the series resistance (Ohm); R_p is the parallel resistance (Ohm), V is the output voltage (V), V_{oc} is the open circuit voltage (V) and I is the output current (A).

3. The Grey Wolf Optimization for The Maximum Power Point Tracking (MPPT)

A Grey Wolf Optimization (GWO) algorithm works based on the hunting behaviour of wolves in nature. The hunting process consists of several steps, i.e.: tracking, chasing and approaching the prey, pursuing, encircling and harassing the prey until it stops moving and finally attacking the prey. Different wolves lead this process. There is a leadership hierarchy of the grey wolves in hunting the prey as illustrated in figure 3 [5]. The first level of the hierarchy is the leader of the wolves and named as the alpha wolf (α). It is considered as the dominant wolf in the group. It has responsibility for hunting and attacking the preys. All of the wolves in the group have to follow his order. The second level is the subordinate wolf or the helper of the alpha wolf. It is called as the beta wolves (β). The responsibility of this helper wolf is to supports the alpha wolf in hunting and making decisions. The beta wolf can replace the alpha's position if the alpha passes away. The third level is the Delta wolves (δ). They are the worker wolves. Their role in the group is to warn other wolf members in case of any dangers, to watch the boundaries of their territory, to support the alpha and the beta in hunting and to safeguard the wounded wolves. The lowest level is the omega wolves (ω). They are considered as the followers and have to follow the orders of all the dominant wolves.

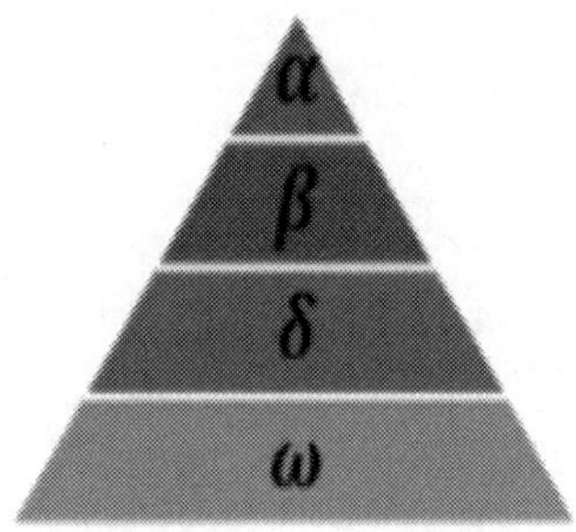

Figure 3. The social hierarchy of grey wolves in hunting their preys [5].

In the grey wolf optimizer, the alpha wolf leads the hunting of the prey. The beta and the delta might involve in the hunting operation occasionally. If a prey is found during the hunting, the three levels of wolves i.e.: α, β, and δ surround the prey. This hunting operation is mathematically modelled as shown in (7) and (8).

$$D = |C \cdot X_p(t) - A \cdot X(t)| \tag{7}$$

$$X(t+1) = X_p(t) - A \cdot D \tag{8}$$

$$A = 2a \cdot r_1 \cdot a \tag{9}$$

$$C = 2 \cdot r_2 \tag{10}$$

Where t is the current iteration, A and C are the coefficient vectors, X_p is the position vector of the prey, X is the position vector of a grey wolf, r_1 and r_2 are random vectors $\in[0,1]$, a is a component which linearly varies from 2 to 0 over the course of iteration. The vector A is a random value in interval [-2a,2a].

The GWO algorithm starts by initializing the number of grey wolf population (α, β, δ and ω), the component of a, the coefficient vectors of A and C. Firstly, it is assumed that the apha, beta and delta wolves know the potential location of the prey (X_p) very well. The search for the prey is applied according to the position of the wolves (X). The wolves diverge each other to search for the prey and converge to attack the prey based on the value of the coefficient vector A. The hunting process depends on the value of the coefficient vector A. If the value of $|A|<1$, the wolves converge to attack the prey and finish the hunt. When $|A|>1$, the wolves diverge from the prey and find the fitter prey. The GWO optimises the value of the alpha, the beta and the delta. The fittest solution is considered as the alpha, then the second fittest solution is the beta and the third fittest solution is the delta. These three best solutions are saved, and their positions are updated according to the position of the best search agents as follows:

$$D_\alpha = |C_1 \cdot X_\alpha - X| \tag{11}$$

$$D_\beta = |C_2 \cdot X_\beta - X| \tag{12}$$

$$D_\delta = |C_3 \cdot X_\delta - X| \tag{13}$$

$$X_1 = X_\alpha - A_1 \cdot (D_\alpha) \tag{14}$$

$$X_2 = X_\beta - A_2 \cdot (D_\beta) \tag{15}$$

$$X_3 = X_\delta - A_3 \cdot (D_\delta) \tag{16}$$

$$X(t+1) = \frac{X_1 + X_2 + X_3}{3} \tag{17}$$

Applying the GWO algorithm for searching the maximum power point of a PV module under one irradiance and temperature condition requires a DC-DC converter. The type of the DC-DC converter used is the Boost converter. The electrical circuit of the combination of the GWO algorithm and the Boost converter in searching for the maximum power point of a PV module is shown in figure 4. The Boost converter is used to increase the magnitude of the input voltage to a certain level. It consists of an inductor, a capacitor, a diode and a Mosfet as a switch. The output voltage of the Boost converter is greater than its input voltage and it is very sensitive to the changes in duty cycle (k) of the Mosfet. The relationship between the duty cycle of the Mosfet and the output voltage is given in (18).

$$V_{out} = \frac{V_{in}}{1-k} \tag{18}$$

Where V_{out} is the output voltage, V_{in} is the input voltage and k is the duty cycle.

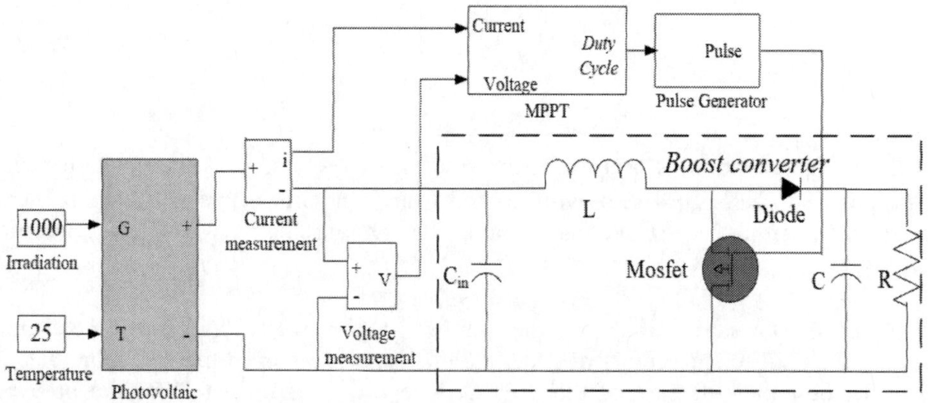

Figure 4. The electrical circuit of a Boost converter combined with a PV module and the GWO algorithm.

The Mosfet used in this study has a switching frequency of 20 kHz, the input capacitor C_{in} has a magnitude of 100μF, and both the inductor L and the output capacitor C have a value of 3.3 mH and 1651μF respectively. The load R connected at the output of the Boost converter has a value of 7 Ω.
The GWO algorithm functions as a controller in the circuit. It is used to find the best duty cycle for the Mosfet. The controller requires two input parameters for searching the best duty cycle of the Mosfet, i.e.: the current and voltage of a tested PV module. These two parameters are used to produce the output power of the tested PV module by multiplying them together. The output power is written as a function of duty cycle of the Mosfet as shown in (19).

$$Power = f(duty\ cycle) \tag{19}$$

The prey of the GWO in this MPPT method is the output power of the PV module. The position of the grey wolves is represented by the duty cycle. The fittest solution of the GWO is the duty cycle of the Mosfet for reaching the maximum voltage. The initial value of the parameter a is set at 2 and its value decreases linearly to 0 during the iteration process. The total number of iterations used is 50.

4. Methods

In this study, a 100 Wp photovoltaic module is selected for the test. The electrical parameters of the tested PV module under standard test condition (STC) are shown in table 1.

Table 1. The Electrical Parameters of a 100 Wp Photovoltaic Module [8]

Parameter	Value
Number of series cell (N_s)	36
Maximum Power (P_{max})	100 Wp
Voltage at maximum power (V_{mp})	18 Volt
Current at maximum power (I_{mp})	5,55 Ampere
Open circuit voltage (V_{oc})	21,6 Volt
Short circuit current (I_{sc})	6,11 Ampere
Standard Test Conditions (STC):	
Solar irradiance	1000 W/m^2
Temperature	25^{0}C

The tested PV module consists of 36 cells connected in series. The number of peak powers produced by a shaded PV module is varied according to the number of bypass diodes connecting the cells in the PV module. A two-peak power of a shaded PV module can be produced by dividing the cells into two groups using two bypass diodes. In total, there are 6 bypass diodes used to obtain different peak power conditions. Different combinations of bypass-diodes connecting the number of cells in a string will form different peak power characteristics of a shaded PV module. The global peak is considered as the highest power gained under a partial shading condition. The tested PV module is modelled using a one-diode model as shown in (3) and simulated using the Newthon-Raphson algorithm. The simulation is done using Matlab/simulink software as shown in figures 5 and 6. The connection of the tested PV module to the circuit, which consists of the combination of the GWO algorithm and the Boost converter, is also simulated using Matlab/simulink software.

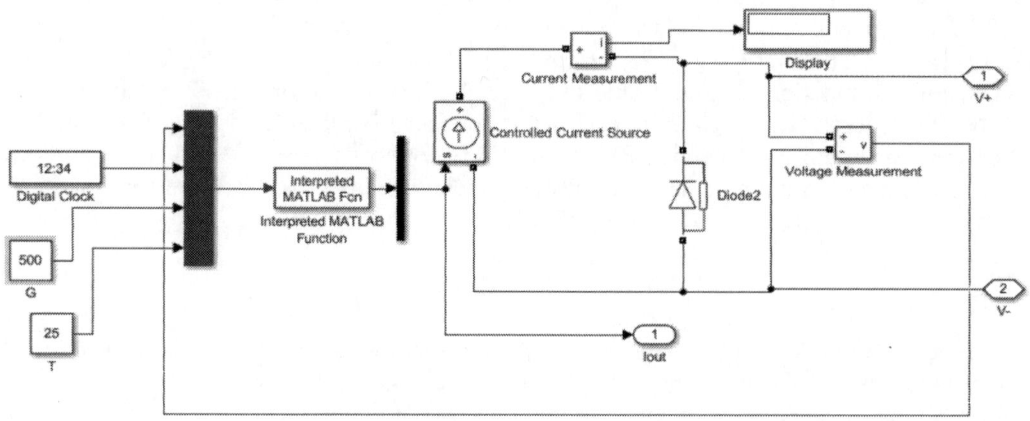

Figure 5. Matlab /Simulink model of photovoltaic consist of 6 series cells.

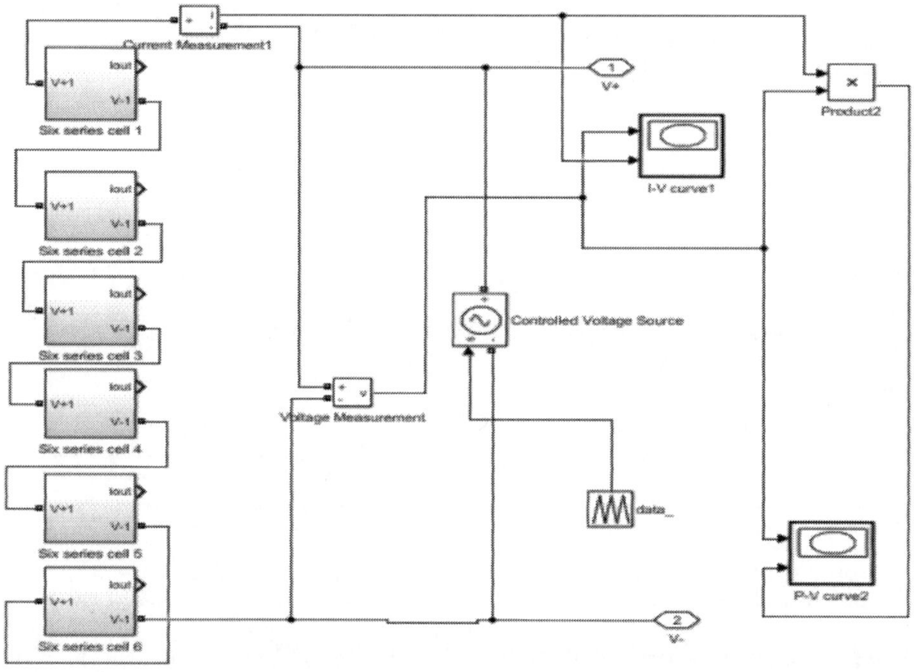

Figure 6. One photovoltaic 100 WP with 6 cells connected to one bypass diode.

This simulation was performed with several scenarios to investigate the ability of the GWO method to obtain the maximum power with multiple-peak power characteristics. Based on reference [7] the multiple peak power is developed in two circumstances:

- Multiple peaks due to the effect of the location and the number of cells undergoing shading.
- Multiple peaks due to the effect of partial shading and temperature.

5. Result of the simulation

The simulation result of how the GWO algorithm obtains the maximum power point of a tested PV module under a partial shading condition is shown in figure 7.

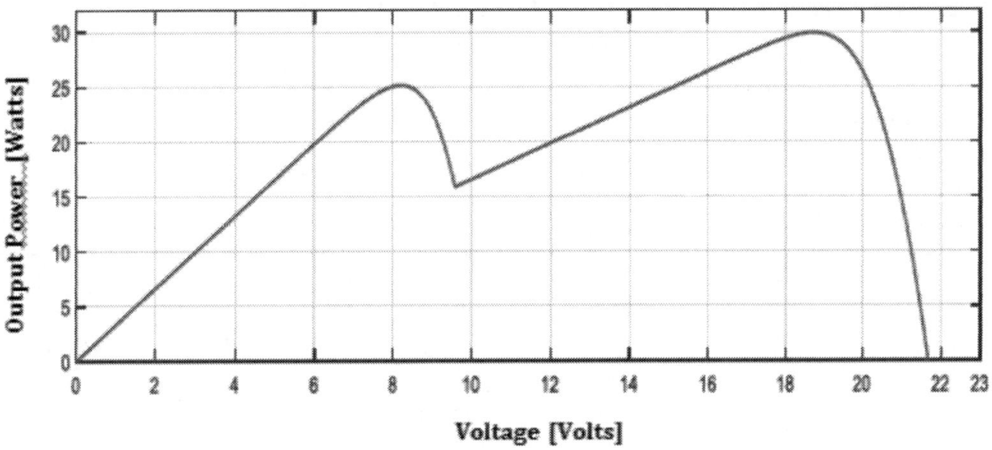

a.

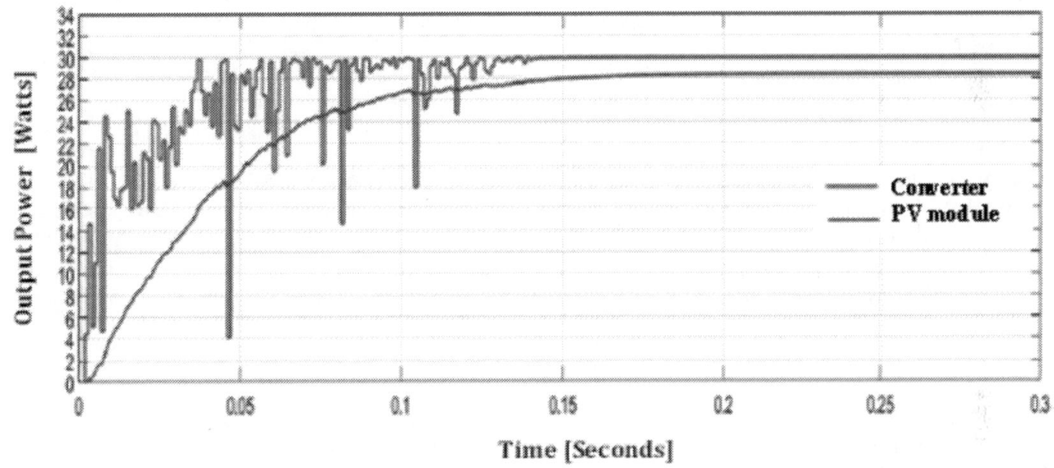

b.

Figure 7. The performance of the GWO algorithm in tracking the maximum power point of a shaded PV module with a two-peak power output at irradiance level of 300 W/m^2.

As shown in figure 7a, the PV module under test produces a two-peak power output under a partial shading condition. The highest peak power is reached at 30 Watts. The GWO is able to reach the same maximum power of a shaded PV module within 0.15 seconds at 30 Watts.

The performance of the GWO method in finding the maximum power point of a PV module at different shading conditions is shown in table 2. Its performance is also compared to the PNO algorithm. In any shading conditions, the GWO outperforms the PNO algorithm in reaching the global peak of the shaded PV module. In average, its accuracy is more than 97% for any shaded condition. The PNO algorithm is not reliable in finding the maximum power of a shaded PV module at any shading conditions with an accuracy of 59% in average. Therefore, it is not preferable for tracking the maximum power point of a shaded PV module.

IOP Conf. Series: Materials Science and Engineering **523** (2019) 012047 doi:10.1088/1757-899X/523/1/012047

Table 2. The comparison of GWO and PNO algorithms in obtaining the maximum power point of a shaded PV module.

Number of peak	Maximum Power (Watts)				Accuracy (%)	
	P-V curve					
	Local peak	Global peak (Maximum Power)	GWO	PNO	GWO	PNO
2	56,25	58,22	56,24	38,50	96,59	66,12
2	56,25	58,22	56,25	38,50	96,59	66,12
2	56,25	58,22	56,25	38,50	96,59	66,12
2	36,92	54,09	53,45	24,40	98,81	45,11
2	36,92	54,09	53,24	24,40	98,42	45,11
2	36,92	54,09	53,71	24,40	99,38	45,11
4	11,52	36,90	35,93	24,40	97,37	66,12
4	11,52	36,90	35,93	24,40	97,37	66,12
4	11,52	36,90	35,93	24,40	97,37	66,12

6. Conclusion

In conclusion, the GWO algorithm outperforms the PNO algorithm in tracking the maximum power point of a PV module under any partial shading conditions. It is also faster in reaching the maximum point power of a shaded PV module within 0.15 seconds. In average, the GWO algorithm is able to track the maximum power point of a shaded PV module with an accuracy of 97%. Comparing to the GWO method, the PNO method is not preferable to track the maximum power point of a shaded PV module.

7. References

[1] Patel H and Agarwal V 2008 *IEEE Trans Energy* Convers **23** pp 302–310
[2] Sera D, Mathe L, Kerekes T, Spataru SV and Teodorescu R 2013 *IEEE J Photovoltaics* **3** pp 1070-1078
[3] Tiong Meng Chung TMC, Daniyal H, Sulaiman MH, Bakar MS 2016 Comparative study of P&O and modified incremental conductance algorithm in solar maximum power point tracking Institution of Engineering and Technology In *4th IET Clean Energy and Technology Conference (CEAT 2016)* pp 43 (6 .)-43 (6 .)
[4] Ishaque K, Salam Z, Shamsudin A, Amjad M 2012 *Appl Energy* **99** pp. 414–422
[5] Mirjalili S, Mirjalili SM, and Lewis A 2014 *Adv Eng Softw* **69** pp 46–61
[6] Mohanty S, Subudhi B, and Ray PK 2016 *IEEE Trans Sustain Energy* **7** pp 181-188
[7] Nguyen XH 2015 *Environ Syst Res* **7** pp 1-10.

IOP Conf. Series: Materials Science and Engineering **523** (2019) 012048 doi:10.1088/1757-899X/523/1/012048

Living together: The phenomenon of house occupancy in Indonesia with particular reference to Balikpapan East Kalimantan

W I Hakim[1,*], **T Y Harjoko**[1]

[1] Department of Architecture, Faculty of Engineering, Universitas Indonesia, Kampus Baru UI Depok, West Java, 16424 Indonesia

Email: wendyhakim@rocketmail.com

Abstract. This study explores housing problem concerning the most widely perceived idea of housing needs that heavily influenced by the technical and macro-scale level as well as the units' tenure and occupancy which is mainly perceived as owning and renting. On the other hand, the housing cannot be seen as such due to the existence of different context of the problem and the various aspirations of occupants. This research highlights the other kind of occupancy phenomenon among Indonesian who can get access to and live in a housing unit without owning or renting –by living and staying with their relatives. In particular reference to Balikpapan East Kalimantan, this study concerns the modus of this specific kind of house occupancy and aims to reveal the housing processes as well as aspirations of the house occupants who live and stay together.

1. Introductions and Background of Study

1.1. Human Settlement and House

The landscapes and its hierarchy change from time to time from natural landscapes into increasingly dense populated settlements [1]. Human settlements can be categorized into groups that reflect their permanent level [2]; it can be in the form of urban, rural and remote settlements [3], and it may refer to occupation in new areas as well as a place of residence that has been inhabited by humans, both large and small scales [4]. A house can be considered as a human settlement in its smallest scale, an architectural product that is constructed and inhabited as the built environment. Furthermore, a house is one of the basic human needs –as a shelter. Although it universally can be found everywhere in the world, there are many differences and uniqueness in its physical and non-physical aspects. Housing is a problem that depends on social and economic realities. Therefore, it needs to be understood in its complexity and dynamics in a changing time context, not just examining the material aspects, but also the cultural background and aspirations of households who will inhabit the house [5].

1. 2. Housing as Macro Problem

Houses providing was developed in countries based on capitalist economics, which saw it as a process of demand-supply between the housing industries and consumers. Generally, in countries that have economies based on market capitalist such as the United States, there is a general ability of a market to allocate goods and resources through the law of supply and demand [6]. Technical, aggregate and macro-scale supply-demand become the dominant

*Corresponding author: wendyhakim@rocketmail.com

IOP Conf. Series: Materials Science and Engineering **523** (2019) 012048 doi:10.1088/1757-899X/523/1/012048

paradigm in viewing housing needs globally. The most well-known housing needs modeling was proposed by Holman in 1975 that essentially sees a net change in the balance between housing stock and the number of households [7]. Holman model was then criticized due to its lack of linkage to economic aspects [7]. Some further studies modified this model by including economic aspects, for instance, it examines the supply-demand of units with a focus on prices, households income, and the number of available units [8] [9] [10]. In accordance, the fulfillment of housing units operates in the matter of tenure status. House tenure refers to an arrangement under which household occupies entirely or some part of a housing unit; the types of it include ownership, rental, etc. [11]. Owned-rental polarization is the principal approach of seeing house tenure and becomes the focus of housing studies in various countries, such as a study conducted in 2013 about some countries in Central and Eastern Europe [12], study in 2016 in New Zealand regarding occupancy patterns from 1986 to 2013 [13], and a study conducted in context of the Poland in 2004 [14]. Further attention to house ownership was carried out by a study concerning United States context in 2007 with regard to income varieties of the owners [15]. Another study that was also conducted in the United States proposed an argument indicated that types of tenure are also conditioned by psychological aspects of the occupants [16].

1.3. Problem and Issue

It was reported by the Indonesian statistical office (Badan Pusat Statistik) in the result of 2010 population census under the topic of family and household structure, that there are some categories of household members' relation with the head of the household which refer to the non-family relationship, stated as "*famili lain*". It is mentioned that not only the presence of children in-laws (1.99%), parent and parents-in-law (2.01%), and grandchildren (4.05%), but also other relatives (3.52%) in households [17]. The widely perceived house tenure status that is discussed above has not yet included kind of occupancy among Indonesian who can get access to housing unit by living and staying together with their relatives and without owning or renting. In developing countries, such as Indonesia, the society is present and constituted by the dual conditions. People, values, and institutions can be either modern or traditional; both can possibly be present co-exist in a matter called dual society [18]. In the context of this dual society, the problem of housing has not been clearly mapped.

1.4. Research Question and Purpose

This study concerns the modus of specific kind of house occupancy namely living together in a housing unit with particular reference to Balikpapan, East Kalimantan, Indonesia. The research question is: why does this kind of house occupancy occur? The purpose of this study is to reveal the housing processes and aspirations of the house occupants who live and stay together.

2. Literature Review: Housing Problem in the Context of Developing Countries and Dual Society

One of a prominent figure in the effort to uncover housing problems in the context of developing countries and dual society is John F. Turner. Through his book titled Housing by People, Turner analyzed housing problem in several Central-South American countries [19]. There were tendencies of failures in housing projects that ware provided by authorities because of a "mismatch" [19]. It happened due to the house is seen as "what it is" by the providers, an assessment that positions a house as an object and merely a building that is used as a residence which can be simply produced with certain physical quantities and standards in a mass and identical manner [19]. In this view, the physical shape of the house itself is the focus, while the house occupants are excluded. Turner presented an argument concerning house as "what it does", that a house plays a role in various aspects of lives of its inhabitants as well as brings satisfaction to its inhabitants -which is not always and even not related to physical problems at all [19]. Turner's study indicates that housing problems are not uniform for everyone, because each person has their own aspects of life. Turner's argument also contains an explanation that in a context of developing countries, the people or the occupants are the factual "developer" and a "provider" of suitable housing for themselves [19]. His arguments also suggest that technical, aggregate, and macro-scale approach to comprehend housing problem does not actually fit with the needs of the occupants. Therefore, there is a necessity to apprehend deeper issue in housing problem. It requires microanalysis of the housing process to reveal various and specific aspirations of house occupants.

3. Research Method

To accommodate microanalysis of the housing process and understand the modus of the occupants, qualitative method and particularly Grounded Research are chosen as the method to explore and reveal the aspirations of the house occupants. Grounded Research is utilized to develop a theory –instead of testing –and helps the researcher to find alternative meaning and understanding of a phenomenon under study [20]. By applying this method, this study begins without any preconceived theory and hypothesis. The informants are 11 households that practice living and staying together with their relatives. They are interviewed and observed to gather the data, focusing on aspirations, experiences, and also the everyday routines in the households during the practice of living and staying together. The informants are members of the households, namely the original members of the households as the hosts and the additional member(s) of the household as the person(s) who come(s) along to live and stay.

4. Findings and Discussion

4.1. Preference of Living and Staying Together Among Kinfolks

Existing studies suggest that the practice of living together in the same house can be seen as post-marital residency both in the form of patrilocal and matrilocal; as discussed in the context of Minangkabau families who do not migrate are commonly live in matrilocal residence [21], Javanese married couples that are expected to set up neolocal household but many of them actually live in matrilocal residence [22], and there are also some patrilocal arrangements exist in Indonesia, namely among Balinese, Endenese, Bataks, and Nias [23]. The recent study published in 2018 mentions that young married couples, in fact, live with one or the other parents [24]. It is also common that elderly parents in Indonesia live with their adult children [25] [26] [27]. In addition, it is common for Indonesian young people to stay at their parental home until they get married [28]. In some Asian countries, most children are expected to live with their parents until marriage, in contrast to children in Western countries who are expected to leave their parental house even before marriage [29].

In the cases that are examined in this study, the original members and the additional member of the household who lives and stay together are bonded as kinfolks. They recognize their relationship with each other as "*saudara*" (relatives). For them, this term is not only referring to people who have blood relations. The data gatherings show that they are affiliated by consanguinity (related as uncle/aunt and nephew and related as cousins), affinity relationship (related as in-laws), and also non-kinship (close family friends). Distinct to the previous studies as mentioned above which demonstrate that living and staying together involve members who have a parent-child relationship, the finding of this study reveals that this practice can also occur among extent members of kinship and furthermore non-kin. As the result of the interviews, they comprehend their practice of living and staying together as a common practice, some of them regard this practice as a tradition that they inherited from their elders, and many of them see it as the preference. This finding is coherence with studies which propose that flexibility of a household's composition and boundaries in South-East Asia including Indonesia is an essential aspect of residency, which a nuclear household may consist other relatives who find shelter and sustenance in it [21] and nuclear family has extensively portrayed the European family, meanwhile in Asia, extended family has been the custom [30].

The previous study which discusses the practice of living and staying together between parent and child stated that whether children obey the tradition, are altruistic or not, they have reasons to live together with their parents and these social and family structures are based on filial piety [31]. On the other hand, there is also filial support that given by the parents to children as discusses that it is common for a young family to live with parents/in-laws until they have enough deposit for housing loan [32]. Comparable to those existing studies, the results of this study also indicate that there are some values and virtues that underlie the practice of living and staying together. Both the original members of the household as the host and the additional member who comes along to stay recognize some obligation and eagerness of assistantship to one another as kinfolks. They mentioned the phrase "*saling menolong sesama saudara*" (help one another as relatives) to explain their reasons for living and staying together. This phrase can indicate the matter of reciprocity. Reciprocity refers to a situation when people share feelings and attitudes mutually that involves not only intersubjectivity and interpersonal interaction but also between groups [33]. As the underlying value of the practice of living and staying together, reciprocity between the host and the one who comes along is especially corresponding with the context of migration as will be described in the next section.

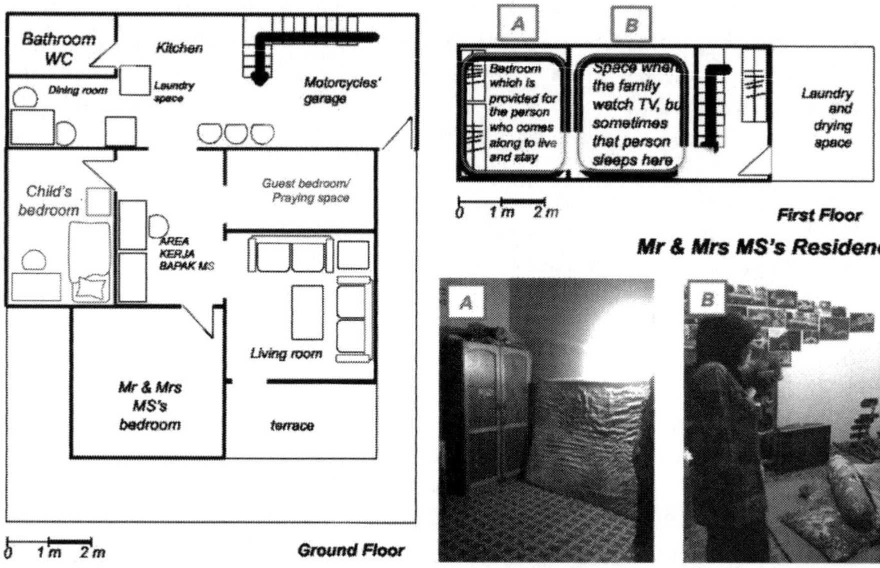

Figure 1. The floor plan and a particular highlight of spaces to sleep that is provided for the person who comes along to live and stay in one of the informants' residence

4.2. Particular Reference of Housing Process in The Context of Chain Migration

Chronologically, the cases of living and staying together that are investigated in this study initiated by some discussion about migration. The follow up of those discussions can be an invitation from the host, request from the person who will come along, or wish from the third party which is affiliated with both of two others (e.g., the parent of the candidate). Both the hosts and the persons who come along are migrants that live and stay in Balikpapan East Kalimantan. The hosts are migrants who arrived initially (migrate to Balikpapan between the mid-1980s and early 2000s) and then inviting some of their relatives to migrate as well as to live and stay in their houses (the persons who come along approximately months to 6 years recent). As mentioned above that there is a willingness to help each other among them, the hosts as previous migrants intend to assist their relatives' youngsters to find employment in Balikpapan as the city of destination. They provide general information about job opportunities in Balikpapan, give the inspiration to migrate, and offer to provide places to stay free of charge. Correspondingly, these offers are matched with the needs of the relatives' youngsters who later will come along, as they are in the middle of employment seeking, have no income, and have not able to afford accommodation yet. In addition, the hosts do not only offer a place to stay but also provide basic needs such as meals.

Statistical data shows that Balikpapan is the biggest migrants recipient area in East Kalimantan concerning inter–provinces in-migration flows [34]. East Kalimantan itself is at top three of highest in-migration recipient in regard to outer island provinces (the two others are Riau and Kepulauan Riau) other than some provinces in Java which are a major destination of internal migration in Indonesia (DKI Jakarta, East Java, and Banten) [35]. The interviews show there is shared and similar opinion among the informants that Balikpapan is an ideal city of destination due to assumed widely open employment opportunity and there are some jobs in mining and oil industries that they associate with a big salary. In the viewpoint of the recent migrants who later come along, they are interested in migrating partly because they are inspired by the accomplishment and sometimes the well-off of the relatives who invite them to migrate to Balikpapan. This finding indicates that migration in this study may refer to a pattern of migration that was called "chain migration". It was explained that in migration there is a movement of migrants who have studied opportunities, provided transportation, given initial accommodation and work; which has been arranged or available by the social relations with previous migrants [36]. Thus, in this case, migrating and living together in the destination city of migration are two things that cannot be separated. The existence and role of relatives who have come and settled in Balikpapan are significant.

5. Conclusion

Beyond the global standard of housing, in the micro level, the occupants have their own idea and preferences concerning how to live as well as how to occupy housing units. To address the research question regarding why living and staying together happen as well as to undertake the purpose of this study that is to reveal the housing process and aspirations of the people who practice this kind of house occupancy, this study have found 2 major findings namely (1) there is a preference for living and staying together among kinfolks that furthermore is based on the value of reciprocity and (2) in particular reference to Balikpapan, living and staying with the relatives refer to housing process in the context of chain migration which the role of previous migrants is enabling the recent migrants to migrate as well as to get access to housing in the city of destination.

6. Acknowledgment

This part of research is funded by DRPM (Direktorat Riset dan Pengabdian Kepada Masyarakat) Universitas Indonesia through Hibah Publikasi Internasional Terindeks untuk Tugas Akhir Mahasiswa UI (PITTA) 2018 No.2550/UN2.R3.1/HKP.05.00/2018

7. References

[1] Wilkie R W 1981 *Proc. Conf. of Latin American Geographers* (Buffalo) **8** (Austin: University of Texas Press) pp 66-89
[2] Roberts B K 1996 *Landscapes of Settlement: Prehistory to the Present* (New York: Routledge) p 120
[3] Newman P, Birrell R, Holmes D, Mathers C, Newton P, Oakley G, O'Connor A, Walker B, Spessa A and Tait D 1996 *Human Settlements in Australia: State of the Environment* (Melbourne: CSIRO Publishing) pp 7-17
[4] Pitzl G R and Pitzl J 2004 *Encyclopedia of Human Geography* (Westport: Greenwood Publishing Group) p 226
[5] Malta F 2006 Sebastiao *J. Focus* **3** 18-22
[6] Rosser Jr J B and Rosser M V 2004 *Comparative Economics in a Transforming World Economy* (Cambridge: MIT Press Books) p 25
[7] Bramley G, Pawson H, White M, Watkins D and Pleace N 2010 *Estimating Housing Need* (London: DCLG Publication) pp 34-35
[8] Bull G 1998 The economics of housing *Housing: The Essential Foundations* ed P Balchin and M Rhoden (London: Routledge) pp 45-69
[9] Glindro E T, Subhanij T, Szeto J and Zhu H 2011 *J. of Central Banking* **7** 163-204
[10] Nordvik V 2006 *J. of Housing Economics* **15** 279-92
[11] OECD 2008 *OECD Glossary of Statistical Terms* (Paris: OECD Publication) p 251

IOP Conf. Series: Materials Science and Engineering **523** (2019) 012048 doi:10.1088/1757-899X/523/1/012048

[12] Augustyniak H, Łaszek J, Olszewski K and Waszczuk J 2013 *To Rent or to Buy-Analysis of Housing Tenure Choice Determined by Housing Policy* (Warsawa: Narodowy Bank Polski Education & Publishing Department)

[13] Statistic New Zealand 2016 *Changes In Home-Ownership Patterns 1986–2013: Focus on Māori and Pacific People* (Wellington: Statistics New Zealand)

[14] Ruoppila S 2004 *European J. of Spatial Development* **9** 1-24

[15] Carter S 2011 *J. of Housing Economics* **20** 159-170

[16] Drew RB 2014 *Believing in Homeownership: Behavioral Drivers of Housing Tenure Decisions* (Cambridge: Joint Center for Housing Studies)

[17] BPS 2010 *Penduduk Menurut Kelompok Umur dan Hubungan dengan Kepala Rumah Tangga* [internet] (Jakarta: BPS) (cited 2017 March 11) Available from: http://sp2010.bps.go.id/index.php/site/tabel?tid=265&wid=0

[18] Harrison D 2005 *The Sociology of Modernization and Development* (New York: Routledge) p 29

[19] Turner J F 1976 *Housing by People: Towards Autonomy In Building Environments* (New York: Pantheon Books) p 38, pp 53-68

[20] Strauss A and Corbin J 1990 Basics of Qualitative Research: Techniques and Procedures for Developing Grounded Theory (Thousand Oaks: Sage Publication)

[21] Dube L 1997 *Women and Kinship: Comparative Perspectives on Gender in South and South-East Asia* (Tokyo: United Nations University Press) pp 98-104

[22] Koentjaraningrat 1984 *Kebudayaan Jawa* (Jakarta: Balai Pustaka) p 137, pp 261-62

[23] Guilmoto C Z 2015 *Population Studies J. of Demography* **69** 299-315

[24] Smith-Hefner N J 2018 courtship and marriage in Indonesia's new muslim middle class *Handbook of Contemporary Indonesia* ed RW Hefner (New York: Routledge) pp 335-45

[25] Johar M and Maruyama S 2011 *J. Health Economics* **20** 87-104

[26] Schroder-Butterfill E 2003 *'Pillars of the Family': Support Provided by the Elderly in Indonesia* (Oxford: Oxford Institute of Ageing University of Oxford)

[27] Cameron L A and Cobb-Clark D 2008 *J. of Population Economics* **21** 1007-33

[28] Utomo AJ, Reimondos A, Utomo ID, McDonald P F and Hull T H 2016 *J. South East Asia Research* **24** 492-509

[29] Yi Z, Coale A, Choe M K, Zhiwu L and Li L 1994 *Population Studies J. of Demography* **48** 65-80

[30] Edlund L and Rahman A 2005 *Household Structure and Child Outcomes: Nuclear vs. Extended Families: Evidence from Bangladesh* (New York: Columbia University)

[31] Beard V A and Kunharibowo Y 2001 *International J. of Population Geography* **7** 17-33

[32] Nilan P and Parker L 2013 *Adolescents in Contemporary Indonesia* (New York: Routledge) p 18, 104

[33] Lefka A 2006 *Encyclopedia of Anthropology* vol 1 ed J Birx (Thousand Oaks: Sage Publication) pp. 1993

[34] BPS 2012 *Statistik Migrasi Kalimantan Timur Hasil Sensus Penduduk 2010* (Jakarta: BPS)

[35] BPS 2012 *Statistik Migrasi Indonesia Hasil Sensus Penduduk 2010* (Jakarta: BPS)

[36] MacDonald J S and MacDonald L D 1964 *The Milbank Memorial Fund Quarterly* **42** 82-97

Spatial extension as a housing strategy in *Kampung Kota*: A case study from Kampung Kingkit, central Jakarta

P Kurniasari[1], **R T Gabe**[1,*], **J Adianto**[1]
[1] Department of Architecture, Faculty of Engineering, Universitas Indonesia, Kampus Baru UI Depok, West Java, 16424

E-mail: rossa.simatupang@eng.ui.ac.id

Abstract. The massive flood of urbanization implicates the shortage of affordable housing for low-income people in the big cities, especially in Jakarta. To live and survive in Jakarta, the *kampung kota* becomes the housing solution rather than problems, due to its ability in providing low-cost housing. While it solves the affordability and the proximity to job location issues, this type of settlement is responsible for several wicked problems to its dwellers, such as space scarcity. Additionally, it creates inevitable health and social problems. In order to overcome these ever-growing problems, the dwellers develop an adaptable social system, which takes form as a spatial extension in the alley for domestic and social activities. This paper takes Gang Kingkit (Central Jakarta) as one of the high-density *kampung kota* settlements to examine the type of spatial extension and how dwellers produce it. Through in-depth interview and direct observation, we discover that there is a social system which allows several types of spatial extension to emerge simultaneously or reciprocally in public space. The social systems are embedded and manifested in the distinctive spatial system, which effectively minimizes the problems of living in *kampung kota*.

1. Introduction

In a living time, low-income people have their own preference in choosing a dwelling. However, in many cases, they are unable to meet all of their preferences due to the limitation of choices. In the end, it is not uncommon for low-income people to live in slums to meet their required preferences. This 'slum area' is often called *kampung kota*. *Kampung kota* is an area within an unstructured and disorganized urban area [1]. As a *kampung kota* means that this area is bound to the condition of *kota* (urban, city, or town) and vice versa [1]. This implies that the existence of *kota* plays a major role in shaping a *kampung kota* since it's a magnetic source of labor. This triggers the massive flood of urbanization. Consequently, *kampung kota* becomes a limited land area and thus not comparable with the *kampung kota* society. Limited and narrow land is one of the main reasons that motivated *kampung kota* society in implementing co-residence as a living strategy.

For *kampung kota* society, sharing a house (co-residence) may be a shortcut in getting a place to live. In addition to being more efficient, co-residence can be the first step in owning a house and being a mediator to help complement each other among the inhabitants [2]. Co-residence is defined as "multigenerational living arrangement where members of a family, reside together in the same household" [3]. However, in this paper, we will discuss co-residence as a situation where a house is inhabited by two or more adults, whether or not blood-related. The way to meet the need for dwelling

*Corresponding author: rossa.simatupang@eng.ui.ac.id

with a co-residence, in various cases, will cause new problems which the inhabitants must manage all the household concern in a limited land. This can trigger a transformation due to the mismatch between the dwelling conditions with the inhabitant needs that must be met [4], one of which is to extend the housing activities out. The extended housing activities is usually a non-private activity, such as cooking and washing clothes, where interaction is started and indirectly increase the level of interaction between neighbors.

The discussed issues in this paper include dwelling extensions that occur due to spatial strategies attempted by *kampung kota* society in response to their condition. The extensions can be done by extending the dwelling space into public spaces, such as terraces and alleys, or in the form of an extension to other residential units. The expected results from this study are to find out why and how *kampung kota* society do the sharing process, including co-residence, as well as why and how the extension is done related to the relationship between the inhabitants and the neighborhood.

2. Extension as a Spatial Strategy in *Kampung Kota*

Dwelling transformation is not unfamiliar to the *kampung kota* society. They tend to do this as there is a lack of options in choosing a dwelling due to their limitations, that they are forced to adapt their preferences. Many factors influence the dwelling transformation. According to Reckless (1973) about control theory, "human behavior is not caused by outside stimuli, but by one's internal needs" [4]. The role of human desire to meet their needs is very strong in influencing the occurrence of the dwelling transformation [4].

However, human needs at every stage of life, from childhood, adolescence, adulthood, to old age, are constantly changing [5]. This ever-changing need will stimulate housing gaps. Housing gap is a condition where there is a mismatch between the existing condition and the needed condition of a dwelling [5]. Housing gaps will develop into housing stress because the action to meet the housing needs cannot be done suddenly [4, 5]. But decision making on housing adjustment is also affected by the status of ownership and how much power the inhabitants have to adjust — the greater the inhabitant's control over a dwelling, the more likely for the inhabitants to adjust, and vice versa. Therefore, the choice arises between tolerating the condition and moving out [4]. There are four types of spontaneous transformation, i.e. alteration, addition, extension, and modification of a house, both in physical form and in a functional way [4]. In the case of *kampung kota* society, the most commonly seen transformation is dwelling extension. Inhabitants in *kampung kota* tend to extend their housing activities outside due to the urge of limited space and the assumption that the alley is a space that can be used for their activities [6]. The orientation and interaction of the building also affect the way to which the extension is done. In some cases, the orientation pattern and the interaction of buildings that form a communal space will limit the extension on the alley and vice versa [6].

3. Methodology

This paper is using qualitative analysis methods. Observations and interviews will be taken in three houses based on the diversity of inhabitant composition, occupancy status, and the uniqueness of the interaction with the neighborhood. Extension mapping will be made to determine which house best represent the extension on the site with its uniqueness. The scope of the location to be used in the observation is the *kampung kota* in Jakarta with permanent and/or semi-permanent building and is occupied by two adults or more per unit. The site selected in the case study is Gang Kingkit, Central Jakarta (Figure 1). The site selection is due to its dense surrounding conditions (16,368.76 people/km^2 in 2016) and not too broad, which is about 0.78 km^2, so the expected interaction with its neighborhood is stronger. The data of the inhabitants and the relationship between them and their neighborhood is collected to know the connection between the extension and the inhabitant's condition. Observation of how the inhabitants use their space to perform their activities then used to reinforce the data.

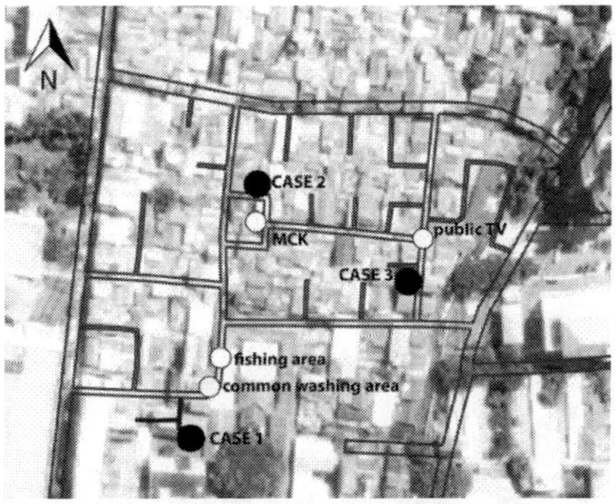

Figure 1. The location of each case in Gang Kingkit, Central Jakarta

4. An Overview of Inhabitants Composition and Dwelling Condition

The site used as a case study is in Gang Kingkit, Kebon Kelapa, Gambir, Central Jakarta. Gang Kingkit is located close to Juanda Station and surrounded by offices, so the site is quite hidden at least for people who are new in this area. Site selection is based on the number of extensions seen in the alleys. Dwelling extension can be seen in every corner of Gang Kingkit, ranging from cooking and washing to the sleeping area. In Gang Kingkit, cooking activities outside is a normal thing as most inhabitants of Gang Kingkit are food traders. The inhabitants who performed most of their activities outside, usually make their canopies that typically made of stretched tarpaulins or asbestos roof. The road width of Gang Kingkit is also not too small (3 meters) so the inhabitants can easily make the alleys in front of their house as their personal space, especially if there is no façade in front of their house.

There are three houses that serve as case studies in this paper, and each has different inhabitant composition and different dwelling conditions. In case 1, there are four inhabitants (three generations) who live in a house with a habitable area of 14 m^2. Mrs. JS, one of the inhabitants of case 1, said that the extension is made because of the intervention by the surrounding buildings. The extended housing activity is cooking on the terrace and washing near publicly available water sources. While in the 2nd case there are eleven inhabitants with a habitable area of 71 m^2. Mrs. ST, Mrs. MR, and Mr. SRP are the 2nd case inhabitants who sell food on the terrace. Due to accessibility need from the kitchen to the trading area, the kitchen located in the back of the house is moved to the terrace, where cooking and washing are carried out as well. In addition, the inhabitants of the 2nd case use MCK (public toilet) also that is located opposite their house.

The 3rd house case is an extreme one as this 29 m^2 house must be inhabited by ten people. Conditions in case 3 naturally force the inhabitants to do most of their household activities outside rather than inside as the house is only enough for them to rest. Activities such as cooking, washing, and drying are performed in front of the house. Moreover, as the house does not have a terrace, the inhabitants of the case 3 use the alley to do their household activities; they even hang utensils such as pans and refrigerators are also placed on the alley side. The neighborhood in case 3 strongly supports such activities as they are close to the communal space and increasingly triggered by the presence of public TV, seating, and canopy. Some of the neighbors even use the facility to sleep along with case 3 inhabitants.

5. Discussion

5.1. Sharing is Greatly Influenced by The Dwelling Condition

Dwelling condition in question is the physical condition of the house and its inhabitants. In case 1 and 3 the main issue is limited land so the probability of sharing will also be higher. This will affect the relationships between inhabitants. How the share takes place is related to what they want to use or do together, depending on their personal needs, wants, and preferences [2]. Based on observations, the relationships between inhabitants in case 1 and 3 are closer than that of the case 2 for which level of sharing is limited to amongst the inhabitants in the same room (eleven inhabitants divided into four rooms). In addition to being affected by land constraints, sharing is also depending on the urgency faced by each inhabitant to meet their needs. In case 2, the urgency experienced is not so great that the sharing level is also limited. Although they are still sharing the space for selling food, they have separate cooking spaces. However, due to the case 3's extensive level of urgency, almost all of the household utensils are used collectively. This resulted in naturally higher interactions between the inhabitants, and their relationship becomes more intimate.

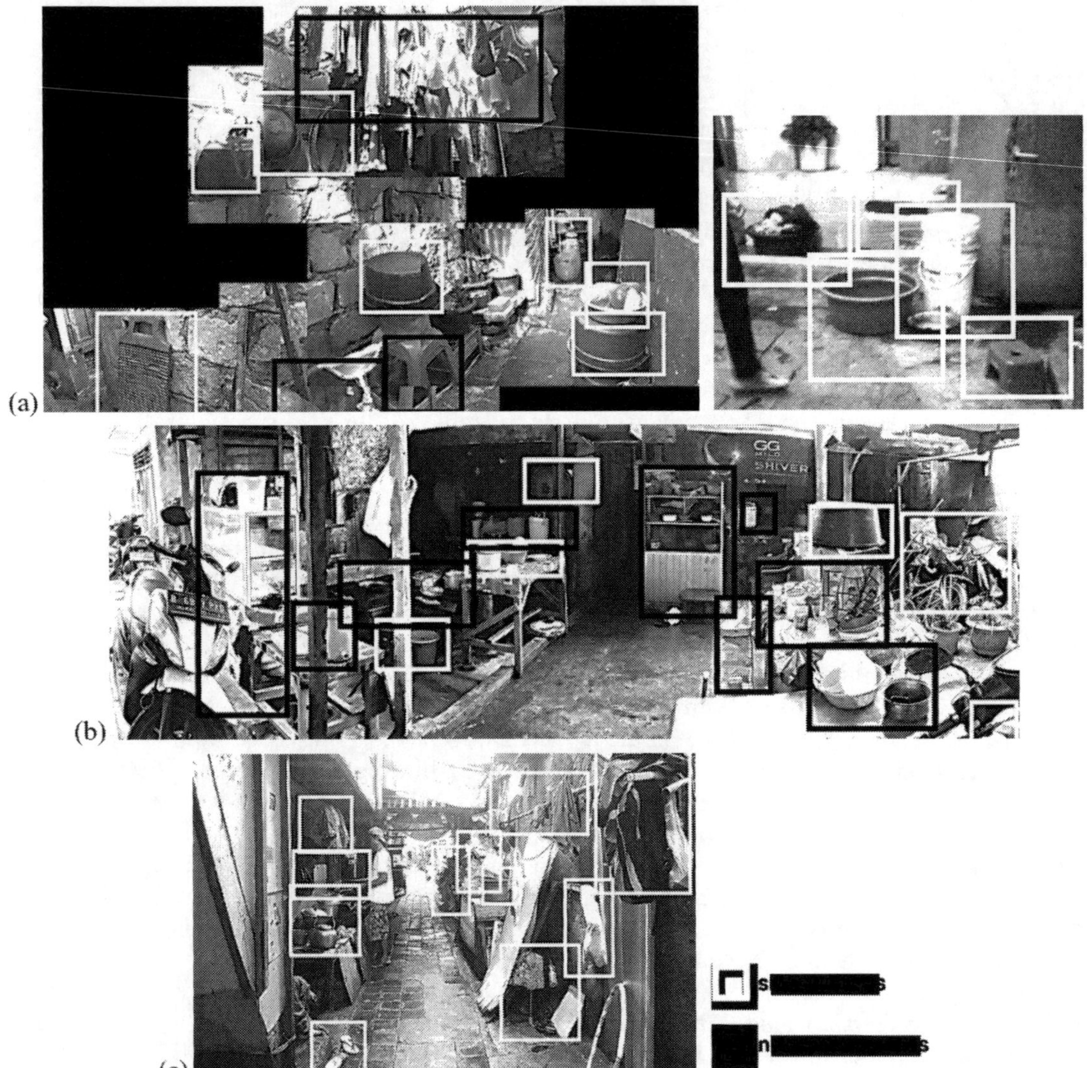

Figure 2. Extension in (a) case 1; (b) case 2; (c) case 3

5.2. There are Always an Unwritten Rules in Sharing

When we share goods or space with others, there are certain limitations, even though each person has a different tolerance to the limitations. According to Orsi and Doskow (2009), there should be agreement on what items will be used jointly and what is not [2]. This is to prevent violations within specified limits. This limitation can be done consciously or unconsciously. The unconscious limitation is usually formed by habit. It is also influenced by the objects capability used by inhabitants. As in the case of 1, 2, and 3, they all use wardrobe with many drawers, so that it provides an opportunity for each inhabitant to have personal space in storing clothes or other personal items. Although these wardrobes are shared, there are limitations by the differentiation of storage space. The limitations in sharing are also influenced by the relationships between inhabitants, especially in the case of space sharing. For example, in case 1 and 3, the sleeping positions of the inhabitants who share one room is determined by the relationship closeness and gender. This is seen in case 3 where a room of 12 m^2 is used as a sleeping room for seven people. The division of sleeping positions in case 3 is generally divided into children (4th generation) and parents (2nd and 3rd generation). A similar case happens in case 2 where the division of the clothesline is based on the closeness of each family members.

5.3. Extension as The Effect of Sharing and Part of Sharing
Despite the background of different conditions, the extension is made due to the mismatch between dwelling conditions and the needs of its inhabitants [4]. The mismatch is triggered by choice to co-reside which is a way to share a dwelling. So in some cases, it can be concluded that extensions are the result of sharing. However, in further observation, it can be seen that the extension is also a way for inhabitants to share on a broader level within the neighborhood. For example, in case 1, 2, and 3, where the occupants use the public water source, this means the inhabitants share the space with neighbors who also use it. In case 3, the sharing level is far-reaching as the inhabitants share not only with the neighbors but also pedestrians and motorists who happen to pass by the alley. The extent of inhabitant's sharing level depends on how far the extension is made in the public space. As part of sharing, extensions also have unwritten rules as they also involve reciprocal relationships with others in the process. For example, in case 1, even though there is no such permanent rule, residents who share the water source have to use these resources at a certain time by their habits. It will form a 'rough schedule' in the use of the resources. In contrast to case 1 where sharing space can only be used by one person at a time, the MCK used by inhabitants of case 2 and 3 has many chambers, so it is not bound by time.

5.4. The Neighborhood Affects How Extension is Done
In the case study, it can be seen that the extension is affected by its physical and social environment. The physical environment can be in the form of resources and the building's layout. As in case 1, washing activities outside are triggered by the building's layout in front of their house that blocks the light so that the washing activity cannot be done in the terrace and also influenced by the near public water resources. This also happens in case 2 and 3, namely the existence of water resources and space in MCK. In case 3, the extension is greatly influenced by the building's layout. The absence of a facade opposite the inhabitant's house gives an opportunity to make such an extension and very possible to form a communal space because it's also supported by facilities. The building layout and growing interaction due to the presence of communal space strengthen the sense of belonging of the entire alley so that the extension can be made more freely. This is in line with what had been stated in [6] that in the road space where buildings are not facing each other, a sense of belonging can be formed so that the inhabitants can claim the road space as their private space.

As for the aftermath of doing an extension, the neighborhood is also affected reciprocally. It can be a positive or negative effect on health and social issues. Doing extensions as a part of the alley in case 3 can improve the social interaction of the neighborhood, but it can also trigger safety issues. According to Mr. UD and Mrs. EL who inhabited the house in case 3, they already lost a bunch of household appliances. This is because there is no space left in the house to store all of the appliances so it must be left outside where it is not protected. Another issue triggered by doing extensions is hygiene and physical health. The lack of awareness and facilities can cause troubles such as a clogged gutter and scattered trash. In another example, constantly sleeping crammed in a room with poor ventilation and lighting can cause physical health like sore limbs and inhibit growth in children.

5.5. Tolerate the Dwelling Condition is The Only Option after Extension is Done

Considering inhabitants' numerous and varied needs in a co-residence, there will be some mismatch that occurs even after the extension is made. In response to those mismatches, the inhabitants must choose between tolerating the dwelling condition and moving out [4]. In this case study, inhabitants choose to tolerate the situation in their way — the diverse responses made by inhabitants in tolerating the dwelling conditions resulted from a different level of tolerance of each member. In case 1, such tolerance is noticed that they spread the bedding only when they want to rest. This is done so that at other times space can be used for other activities such as folding clothes, eating, and receiving guests. While in case 2, when the inhabitants receive guests, they choose to do it on the terrace, so the space inside the house is just for a private matter. In case 3, the inhabitants' tolerance is very high against the incompatibility of their dwelling, even the extension of the private space such as the rest area is also made in response to extreme conditions of the dwelling.

6. Conclusion

The *kampung kota* society has their respective preference in living. However, due to their limited choices, most of them had to live on minimal land, and slum areas often called *kampung kota*. In response to such limited conditions, most *kampung kota* society puts sharing as their strategy to keep up with their daily needs. The form of sharing performed by *kampung kota* society can be in the form of co-residence and sharing housing supports. Although it can greatly support the *kampung kota* society in meeting their needs, co-residence can also lead to a mismatch between dwelling conditions and the needs of its inhabitants. So extension in a functional way is made in response to this situation. By extending the function of the space, the activities can be done anywhere. How far the extension is made depends on the neighborhood, both physical and social environment.

7. Acknowledgment

Expression of gratitude is given to Directorate of Research and Community Engagement Universitas Indonesia (DRPM UI) for administering the 'Hibah Publikasi Internasional Terindeks untuk Tugas Akhir Mahasiswa' (PITTA) 2018 scheme (2461/UN2.R3.1/HKP.05.00/2018), and Faculty of Engineering Universitas Indonesia (FTUI), Departement of Architecture for their support and commitment in this research. Sincere gratitude is also addressed to respondents in Gang Kingkit who participate in this research, especially the inhabitants of each case.

8. References

[1] Sihombing A 2004 *The Transformation of Kampung Kota: Symbiosis between Kampung and Kota, A Case Study from Jakarta* (Depok: Universitas Indonesia) Retrieved from: https://www.housingauthority.gov.hk/hdw/ihc/pdf/phhkt.pdf

[2] Orsi A J & Doskow E 2009 The Sharing Solution: *How to Save Money, Simplify Your Life & Build Community* 1st ed (United States: NOLO) Retrieved from: http://library.uniteddiversity.coop/Money_and_Economics/The_Sharing_Solution.pdf

[3] Keri J S 2008 Co-residence in: Loue S J, Sajatovic M *Encyclopedia of Aging and Public Health* (Boston, MA: Springer) 237 Retrieved from: https://doi.org/10.1007/978-0-387-33754-8

[4] Khan T H 2014 *Living With Transformation* (New York: SpringerBriefs in Geography) 21-34 Retrieved from: https://doi.org/10.1007/978-3-319-00720-5

[5] Seek N H 1983 *Adjusting Housing Consumption: Improve or Move* (England, UK: SAGE) **20** 455-469 Retrieved from: https://doi.org/10.1080/00420988320080811

[6] Putera Y A 2014 *Ambiguitas Ruang Kampung Pluis dalam Perspektif Privat–Publik.* (Bandung: Universitas Katolik Parahyangan) Retrieved from: http://journal.unpar.ac.id/index.php/unpargraduate/article/view/845

IOP Conf. Series: Materials Science and Engineering **523** (2019) 012050 doi:10.1088/1757-899X/523/1/012050

Defensible space in urban housing in Indonesia

M M Muhyi[1,*], **R T Gabe**[1], **J Adianto**[1]
[1] Department of Architecture, Faculty of Engineering, Universitas Indonesia, Kampus Baru UI Depok, West Java, 16424

E-mail: mahdaniar.muhyi@gmail.com

Abstract. Security in the housing is considered as one of the most important things in the fulfillment of occupant's needs. However, there are still many cases of criminality that still occur in urban housing. Defensible space is a concept of residential environment design that is able to prevent crime. Through literature study methods, field observations, and interviews, this paper discusses how defensible space is created in urban housing in Indonesia as well as its effect on the quality of security created between two different types of organic dwellings in Bogor City. The explanation starts from the identification of physical characteristics that create territoriality, natural surveillance, image, and milieu. Then, the social and physical condition are associated with their effects on people's behaviors. The result of our discovery is that physical and social conditions create the whole defensible space condition. In the middle-class urban housing, the environment condition still opens the opportunity for the criminals to commit the crime because the defensible space is not created as a whole. Contrarily, this is created in the lower-class urban housing as a whole. Therefore, the environment condition is free from any crime.

1. Introduction

Housing is considered important to fulfill the human needs to live. According to [1], housing is not just a group of houses, but a system of space where people live, so its presence is very influential on the quality of life of the inhabitants who occupy it. However, there are still many urban dwellings that actually cannot be considered ideal because they have not fully met the needs of the quality of life of their inhabitants. This can be seen from the criminality cases that still occur in urban housing. In fact, good security becomes one of the factors that counts a housing the ideal one. The inhabitants who live there would feel comfortable and their life quality is fulfilled. In addition, in [2], security is one of the important things that should be provided to everyone.

Maslow (1943) in [3] created a diagram of the human hierarchy of needs. Based on the diagram, security becomes a secondary need that is also considered as important in human life as it occupies the second level of human needs. The neighborhood environment as the outer space, which is the extension of the private space of each inhabitant of each house, should become a space that has privacy for a group of occupants in the same neighborhood [4]. However, such conditions sometimes cannot be realized thoroughly due to the lack of residential design and the role of the occupants [4]. According to [5], crime is one form of violation of social norms. This violation is determined within the limits of values upheld in a society. Crime is more likely to occur in locations where there are weak guards, there are many easy targets and is close to where the perpetrators live [6]. Jane Jacobs (1969) in [7] discussed the links

*Corresponding author: mahdaniar.muhyi@gmail.com

between architecture and crime and raised important points about the "eyes of the street". "Eyes of the street" is defined as the supervision made by the person who occupies an area [7].

Defensible space is a concept that allows the prevention of criminal acts [8]. Newman states its definition as *'a model for residential environments which inhibits crime by creating the physical expression of a social fabric that defends itself'* [4]. Environmental design can be used as an approach for crime prevention and gives proven benefits [9]. However, defensible space can only work well if the social conditions of society are in an optimal state because even if an environment has been architecturally designed, it is still unsafe if the social climate is unwell [10]. Therefore, it can be concluded that the behavior formed from defensible space is influenced by physical and social conditions.

Newman reveals some factors that affect crime in a housing, such as a territoriality, natural surveillance, as well as image and milieu. Territoriality can be defined as a claim of space belonging to a person or group of people in which control and defense have resulted from the occurrence of boundary offenses [4]. It is a social interaction that is actualized in an environment and their efforts to maintain its safety of others interferences [11]. An act may be regarded as an offense if one interferes with the territoriality of another. Meanwhile, natural surveillance can be interpreted as the ability to observe the public space in a residential environment so that everyone who occupies it has the feeling of being under the state of supervision [4]. By creating natural surveillance, the crime rate can be reduced even if the occupant is in or away from home and does not require the use of security devices or the role of security apparatus [4]. Lastly, image and milieu are impressions generated from the physical appearance of the building or the residential environment [4]. A crime can be initiated if the stigma invites destruction [4].

Due to the times and different conditions within the society and territory, this theory cannot be implemented in all conditions of occupancy. Not that defensible space can only be well created in a residential environment for the middle-class, as the lower-class also has the same goals and desires to create a safe and comfortable living environment [4]. This statement from Newman was similar to Lee Rainwater's opinion in the discussion of the Pruitt-Igoe case. According to Rainwater (1973), *'...although lower class people may not adhere in action too many middle-class values about neatness, cleanliness, order, and proper decorum, it is apparent that they are often aware of their deviance, wishing that their world could be a nicer place, physically and socially...'* [4].

The defensible space concept can be unconsciously applied in both types of residential environments in their own way. Ultimately, a particular environmental design will create the behavior of the people who occupy it, and in turn, it will affect the quality of security in the environment. Newman (1972) also thinks that the concept of defensible space becomes not only a middle-class society thinking but also an important thing for lower-class society [4]. This theory was issued in 1972 with the object of observation in the form of residential located in the US which is partly flats and is designed by planners. Meanwhile, current physical and social environment in Indonesia is mostly organic or unplanned, so that there can be some differences in the defensible space formed.

This writing aims to examine and aware of the importance of defensible space concept in the neighborhood, the defensible space formed in the urban housing in Indonesia, the physical character formed, and its effect on human behavior who occupy it.

2. Methods

The case study was conducted in Bogor City, taking samples of two types of housing with different economic levels of society (figure 1a). The first case study was Babakan Residence with an upper to middle-income occupants, while the housing for the second case study was Kampung Pulo Geulis populated with lower-income occupants. Dependent variables include both residential types within the same city, the coverage area of physical analysis is not much different (about 16,000 m^2), and does not have a specific gate to enter residential areas (common people are free to pass). Time variable (morning, noon, and night) is not considered as one of the main variables as the analysis is limited to the general physical characters that apply at all times. The type of criminals that become security parameters includes taking the property of another person without the owner's permission (theft, robbery, pickpocketing) and destruction of property without the owner's consent (vandalism).

The strategy is to use a qualitative approach in the form of interview and field observation. The Heads of hamlets (Rukun Warga) were interviewed and one resident of each housing was taken as a sample, with an aim to obtain information on occupant backgrounds, social activities among residents, crime occurrences, as well as perceptions about crime. The sampled residents were chosen by considering how well they know their residential surroundings. Field observation is aimed to see the condition and physical situation of the settlement directly to understand the factors of defensible space that formed in both types of occupancy. The observations of both types of settlements were conducted during weekends and weekdays. The analysis method is referred to Newman's defensible space forming factors, namely territoriality, natural surveillance, image, and milieu.

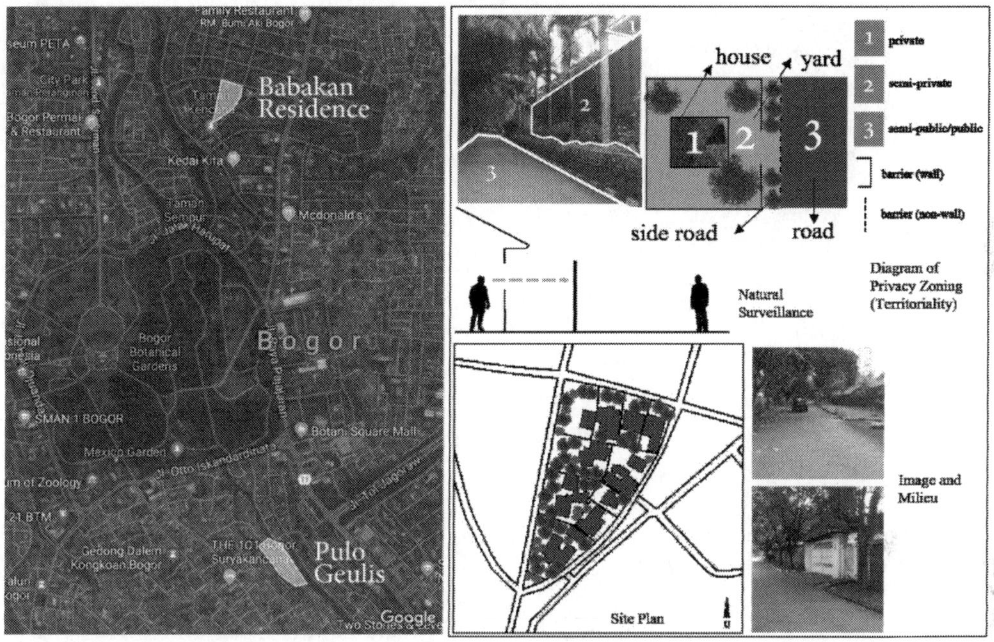

Figure 1. (a) Bogor Map and the location of the areas, (b) Defensible space in Babakan Residence
Source: Google Maps, Muhyi, April 2018

3. Results

The observed Babakan RW 03 Residence has a total area of 280,000 m^2. The physical environmental observation is limited to a block in RT 01 with an area surface of 16,000 m^2 and consists of 10 houses. With the spatial density of the area, the vast majority of vacant land is planted with trees, creating a deserted and quiet space. Inside this residence, there are several public function buildings. The residents of the analyzed housing comprised of 249 families with a total of 675 residents. The residents of RW 03 consist of various professions and ethnic backgrounds. They perform communal activities, namely cleaning the residential area together, Qur'an recitation, and hold a helmet (Rukun Warga) meeting once every 3 months. Some of the crimes that have happened the previous year include mugging, theft, and property destruction (broken glass windows and precious items theft). However, criminality is not entirely gone even though they provide security guards by the next year.

Meanwhile, Kampung Pulo Geulis Baranangsiang RW 04 has a total area of about 33,800 m^2 and is shaped like an island because it is surrounded by the Ciliwung River. The physical observation of the environment is limited to a half of the southern area of 16,000 m^2 and comprises about 176 houses. With the density of the area, most of the land has been built and a few green open areas create a dense and crowded space. Pulo Geulis is inhabited by 681 families with a total of 2,640 residents. RW 04 residents consisted of various professions background but mostly are of Sundanese (Bogor) ethnicity who have long lived there. Communal activities of the residents are held three times a week. Based on the exposure of helmet head, during the last 15 years, there has never been any more crime in Kampung Pulo Geulis. Almost all citizens have been socialized and educated to be alert of crime at all time.

3.1. Defensible Space Principles Analysis in Both Housings

There are some similarities and differences of defensible space aspects found in both housing (figure 1b and figure 2).

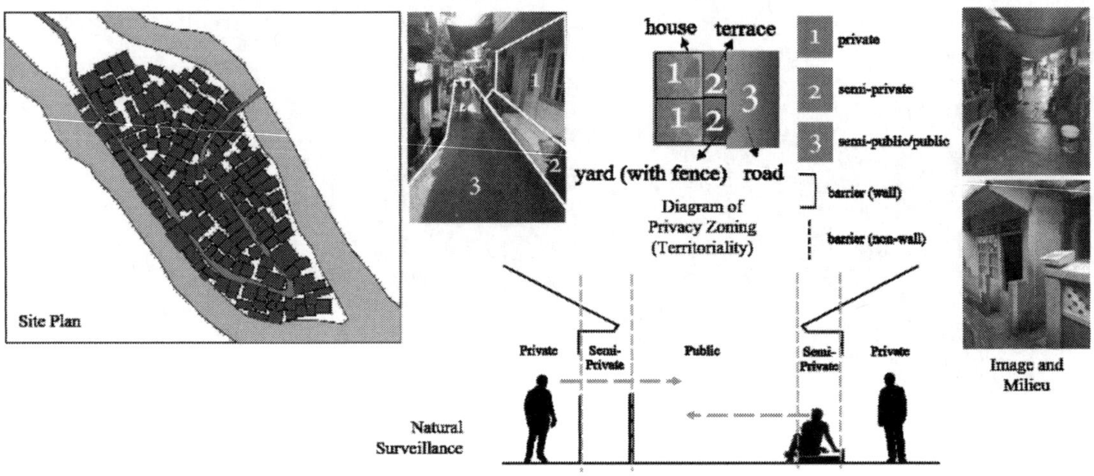

Figure 2. Defensible space in Kampung Pulo Geulis
Source: Muhyi, April 2018

3.1.1. Territoriality. The similarity between both housings is their organic approach, which makes use of the space in accordance with the needs and desires of its inhabitants instead of compositional approach, which gives priority to aesthetical design [4]. Meanwhile, the division of zones between private, semi-private, and public in each occupancy has little difference although both are sufficiently divided. At Kampung Pulo Geulis, the zoning does not limit visual access from the public to private. This is influenced by the social conditions of the *kampung* community who mostly have small houses with preference to outdoor activities and socialize with neighbors. As a result, the streets become communal space between the extension of the activity space of the inhabitants of the house as well as public access roads. Based on Newman statement (1972) in [4], the privacy zoning in site design must be divided clearly to make people know the prohibited zone which should not be entered.

Furthermore, the design of the streets in the residence does not have much interruption than in the *kampung*. Babakan Residence has no road width or material changes. The streets are just closed by portals that are partially left open and passable to anyone. While in Pulo Geulis, there are considerable changes in the design of the road. The main factor influencing the design of the roads in *kampung* is the geographical condition in the middle of the river, so it is only accessible by small bridges. This is not the case for Babakan Residence that is located in the middle of the city and slightly mixed with the city's public facilities. According to the theory, the interrupted circulation or entrance makes the residential area safer by making the criminal hesitates when entering an area that is different from the outside [4]. The accessibility of the housing also is correlated with the security of its environment [12]. Therefore, fewer people are entering Pulo Geulis compared to Babakan Residence.

For the boundaries aspect, there is a significant difference. At Babakan Residence, the boundaries created is clearly only limited the area of private homes. With high and closed fences and CCTV in some houses, it shows that the houses are very restricted from outsiders. However, the overall boundaries of this residential area itself are not so clearly created as there is no boundary separating the area within the residence with the outer area surrounding the residence. In contrast, boundaries in Kampung Pulo Geulis restrict the area of the *kampung* to the outside area with the presence of the river as a very clear delineation, while the boundaries on each house are not so clear. Clear boundaries are more able to exclude strangers [4] hence Babakan Residence fails to keep strangers outside, while Pulo Geulis makes strangers think twice before entering it.

3.1.2. Natural surveillance. Natural Surveillance is formed differently because it is influenced by physical factors related to territoriality and the social condition of the occupants to make the residential area is monitored by the people in it [4]. At Kampung Pulo Geulis, natural surveillance is created through open doors and windows and directly facing the road ahead, semi-private and public fusion zones, and unlimited visual access. In addition, the social condition of the *kampung* community tends to enjoy outside activities and intensely interacts with their neighbors. As a result, it creates supervision from residents who are inside or outside the houses.

Meanwhile, natural surveillance in Babakan Residence is inadequate. High and closed fences cause blockage to house's windows so that residents cannot monitor outside space from inside their houses. In addition, the social conditions of middle-class society tend to barely have activities outside their fences and rarely interact with their neighbors. This resulted in relying monitoring only on CCTV and security guards. Occupants play a passive role in supervising the residence area directly.

3.1.3. Image and Milieu. The image and milieu formed in both dwellings are different. Babakan Residence has an image and milieu that is unsupervised. Although the area is security guarded, the atmosphere of a quiet and poorly supervised space somehow offers the opportunity for the perpetrator to commit a crime in the street area. Meanwhile, the image and milieu of Kampung Pulo Geulis are supervised. Physical and social conditions in this *kampung* give the outsiders a feeling of being watched by the occupants. Therefore, the image and milieu created in both housings affect the possibility of a criminal activity, as Newman (1972) stated in [4].

4. Discussion

4.1. The Reconstruction of Defensible Space Principles

Several things are conflicted with the theory of defensible space. For instance, the theory mentioned that the provision of social facilities becomes one of the efforts to provide communal space for residents and increase the sense of community [4]. The fact is that the inhabitants of Kampung Pulo Geulis eventually perform these activities in the front area of their houses so that almost the entire streets become communal space where social activities are conducted. On the other hand, in Babakan Residence, the inhabitants do not need much activity outside as most of the residents have spacious, comfortable houses, and there are adequate facilities (e.g. children's playroom, relaxation room, etc.). Therefore, it can be concluded that social facilities are not required to be owned by any housing because social activities can be done in the public spaces created by the inhabitants themselves within their house.

Secondly, in the theory, the less number of occupants occupying a residential area the better [4]. Conversely, the increasing number of people who occupy a residential area makes it difficult for residents to separate their neighbors from the strangers that they should be aware of. This is in contrast to what actually happens, that Babakan Residence inhabited by fewer people is more vulnerable to crime than the *kampung* occupied by more people. This is because, in a *kampung*, the occupants interact more intensely with their neighbors compared to Babakan Residence where its inhabitants rarely know each other. Thirdly, Newman (1972) stated in [4] that natural surveillance is better formed in an environment that has a clear and orderly space layout rather than irregular and random space layout. In fact, Babakan Residence with orderly spatial arrangements (grid-shaped streets, well-built houses) is poorly supervised because the windows are blocked by fences that limit the visual access of occupants to oversee external circumstances and the occupants' tendency to stay inside their home. Meanwhile, *kampung* has irregularly 'branched' streets and the houses built randomly so there are many sharp turns and hidden

IOP Conf. Series: Materials Science and Engineering **523** (2019) 012050 doi:10.1088/1757-899X/523/1/012050

corners. However, the whole area is still well supervised due to the intense outdoor activities and open doors and windows.

4.2. Defensible Space Formed in Both Housing

Natural surveillance may cause a contradiction between territoriality physical condition and the goal in the theory of defensible space. This is because natural surveillance is not only resulted from territoriality design but also influenced by the social condition of society itself. Contrarily, some physical properties of housing that should make the environment safe has failed due to social conditions of residents who rarely supervise the outside space. This suggests that physical characteristics do not have to be entirely in line with the characteristics that made up the principle of defensible space theory since defensible space can be formed as a whole.

In the lower-class housing, defensible space is formed as a whole. This causes highly-reduced crime rate in this housing and the area is well supervised. Contrarily, the defensible space in middle-class housing is not formed as a whole. Physical conditions and social conditions cause insufficient natural surveillance and therefore unsupervised.

5. Conclusions

Defensible space is the physical condition of a sheltered environment. In a housing, security becomes one of the important requirement for occupants, so it needs an environmental condition that gives a sense of security for its inhabitants. Based on case studies, some physical characteristics of defensible space in two housing in Bogor City is in line with the principle, and yet there are some which are not supporting the theory of defensible space. This is influenced by the social conditions of the occupants who occupy the housing itself. In Kampung Pulo Geulis, while some of the territoriality aspects are not intact, the natural surveillance remains well-formed as it is influenced by social conditions, that in turn sufficiently supervises the image and milieu. In contrast, although some of its physical characteristics are in accordance with good territorial principles in Babakan Residence, natural surveillance is still poorly established due to social factors of its inhabitants. This resulted in unsupervised image and milieu.

Defensible space is formed from the unity of physical characteristics in a certain environment which then forms the human behavior of its inhabitants. On one hand, the behavior formed in Babakan Residence causes difficulty to the occupant to monitor the circumstances around and still open the chance for the perpetrator to commit the crime as the defensible space is not created as a whole. On the other hand, the behavior formed in Kampung Pulo Geulis drives the inhabitants to supervise the circumstances around at all time and anyone who enters the neighborhood is visible to the inhabitants, so the crime is barely committed in this *kampung* area.

6. References

[1] Payne, Geoffrey K 1977 *Urban Housing in the Third World* (London: Leonard Hill) 46

[2] *UU-01-2011 Perumahan dan Kawasan Pemukiman*

[3] Maslow, A. H 1943 *A Theory of Human Motivation* 377. Retrieved from: https://psychclassics.yorku.ca/Maslow/motivation.htm#r13

[4] Newman, Oscar 1972 *Defensible Space* (New York: Macmillan) 3-108

[5] Sudjono 1974 *Kriminologi: Ruang Lingkup dan Cara Penelitian* (Bandung: TARSITO)

[6] Betchel, R. and Arza Churchman 2002 *Handbook of Environmental Psychology* (New York: John Wiley and Sons) 420-421

[7] Jacobs, Jane 1961 *The Death and Life of Great American Cities* (New York: Random House) 85

[8] Newman, Oscar 1996 *Creating Defensible Space* (Washington, DC: U.S. Department of Housing and Urban Development Office of Policy Development and Research) 9

[9] Irving, Charlotte 2015 *Crime Prevention Through Environmental Design: Application to Christchurch* (Lincoln: Lincoln Planning Reviewt) 7 46 Retrieved from: https://journals.lincoln.ac.nz/index.php/LPR/issue/view/78

[10] Elffers, H. & Danielle Reynald 2009 *The Future of Newman's Defensible Theory* (London: European Society of Criminology and SAGE Publications) 31 Retrieved from: https://www.researchgate.net/publication/249752224_The_Future_of_Newman%27s_Defensible_Space_Theory

[11] Faqih, M. Santosa, H. R. & Zubaidi, Fuad 2013 *Territoriality in the Traditional Context, Psychology and Behavioural Sciences* (New York: Science Publishing Group) 2 89 Retrieved from: http://www.sciencepublishinggroup.com/j/pbs

[12] Hidayati, Z. & Noviana, Mafazah 2013 *Korelasi Akses Perumahan Dan Kriminalitas di Perumahan Kota Samarinda* 1 55 (Samarinda: Jurnal Kreatif Desain Produk Industri dan Arsitektur) Retrieved from: http://e-journal.polnes.ac.id/index.php/kreatif/article/view/71

IOP Conf. Series: Materials Science and Engineering **523** (2019) 012051 doi:10.1088/1757-899X/523/1/012051

Interiorization of public space in a high-density settlement: a case study in *Kampung* Cikini-Ampiun

C Devina[1], J Adianto[1,*], R T Gabe[1]
[1] Department of Architecture, Faculty of Engineering, Universitas Indonesia, Kampus Baru UI Depok, West Java, 16424

E-mail: joko.adianto@gmail.com

Abstract. High-density settlements in an urban area are always confronted with the lack of space issue. This situation leads to the use of public space – specifically the alley as a circulation space – for exclusive uses of the inhabitants, such as cooking, cleaning, and selling. This study will discuss the strategy of making a public space into individual interior space as an analogy of collecting process by Walter Benjamin, from choosing, bring inside, and be presenting. A qualitative method was conducted to trace the process, and the result of interiorization in public space in which case study was taken is located in *Kampung* Cikini-Ampiun, Central Jakarta. The study revealed that the process and the result of individual interiorization in public space are no longer separated from each other and yet they are connected by the inhabitant' social relations.

1. Introduction

Jakarta is the capital city of Indonesia where the core of politic, economic, and cultural movements occurs. This centralized system makes Jakarta one of the highest population density. In 2017, Jakarta's population density reached 15,518 persons/ km^2 [1]. Due to high population density, the land price keeps skyrocketing and left high-density settlements as one of the side effects used by the low-class populations.

Cultural habits, such as *arisan* (social gathering in the form of lottery), and tropical climate as an effect of geographical conditions, certainly influence the way of doing activities. These conditions created unique strategies to confront the limited space issue in high-density settlements. It brings out many phenomena of such space informal usages. There are many household activities that were normally happened inside a house, where space is privately owned, yet happened outside on roadsides.

Regarding this situation, interiority is continuously being challenged in a new, different way. The interesting part highlighted in this paper is the interior as an extension of domestic private space of household that conducted in public owned space due to the limited space issue, that differs the individual strategy compared to the common ones.

Interior is usually understood as a situation being inside. In its relation to the exterior, their definition sometimes undergoes degradation and defined as a rigid relation separated by walls. In fact, the interior and exterior are not in a static-absolute relation, but rather a complementary-dynamic relation [2]. In order to regulate the dynamic relations, a concept of interiority is required.

Based on McCarthy interiority is defined as, "...abstract quality that enables the recognition and definition of an interior" [3]. It also controls the boundary as a point of transition between interior and exterior. Grosz as cited by Atmodiwirjo, Yatmo and Ujung argued that boundary between inside and

*Corresponding author: joko.adianto@gmail.com

IOP Conf. Series: Materials Science and Engineering **523** (2019) 012051 doi:10.1088/1757-899X/523/1/012051

outside does not stand in a fix condition or as a limitation, but an elastic line that makes inside and outside became entities that can be traversed [4]. Interiority makes the understanding of interior became an entity that can be controlled and potentially controlling [3]. In order to control the space, it should be limited and restrained. Those actions are part of interiorization.

Interiorization as an act by a human is more appealing and visible for us to be analyzed rather than the concept of interiority itself. Based on the explanations above, this paper discussed the process of interiorization that occurs in *Kampung* Cikini-Ampiun.

2. Methodology

In this paper, qualitative methods are used to find the process of interiorization. Firstly, actors and activities are set as variables of this research. The data are collected through visual observations and confirmed by conducting interviews with inhabitants. The observation focused on the private objects as traces for domestic activities. Csikszentmihalyi and Rochberg-Halton stated,"...the household as a place where people have (at least some) control over their things, with the home accommodating a complex material ecology of the self." [5] The observation is located in Kampung Cikini as a high-density settlement in an urban area within a lively social interaction between the neighborhoods, which makes it interesting to be furthermore discussed.

The respondents as the actors are selected from three households living side by side in one certain area where many private objects are placed on the roadside, namely RT 12. The selection is based on site mapping and identification of domestic objects in its relation to activities that happened outside the house. The interviews include some basic information about the respondent background, the consideration behind the spatial arrangement, and so on. It took three months to gather all the information before analyzed by the theory below.

Interiorization is an act of making interior by a human. Walter Benjamin explained interiorization as a collecting process – including selecting, bring inside to a system of relations, and presenting. By that, the interior can be defined as a selected exterior. The process of collecting could be interpreted as an organization, operations, and dynamics of interior making as mentioned by Charles Rice refers to Benjamin [6]. The step of selecting is reflected on the chosen spatial elements based on the inhabitants' preferences while bringing inside to a system of relation is seen on the spatial arrangement, how the objects are related one to another. And the last, presenting is a result of the arrangement that can describe the inhabitants based on its value. Here, objects have an important role as a tool for interiorization. So, to see the process of interiorization, spatial arrangement is needed to be seen. As Winton stated, "...in which careful observation of its contents may reveal the traces of everyday life" [7]. Walter Benjamin's collecting also consider interior as "...an art of genre where the fictional framework for the individual's life is constituted in the private home" [6], which makes it suitable for this paper.

In this paper, the public space is seen as an exterior for the inhabitants. Public space itself was defined as a space that outside the control of a person or a group and became a mediator between private space [8]. Here, the alley is standing as a public space that can be accessed by everyone.

3. Gang Ampiun St. in Kampung Cikini

Kampung Cikini is a high-density settlement located in Pegangsaan, Menteng, Central Jakarta. Pegangsaan as an urban village is the highest population density area in Menteng subdistrict with 27,934 inhabitants living in 0.98 km^2 area [9]. The area known as Kampung Cikini itself is the settlement located in RW 01.

Kampung Cikini is barely seen from the main road as it is surrounded by public formal building such as office buildings, universities, and hospital. This condition makes Kampung Cikini a strategic area. Gang Ampiun – an alley in Kampung Cikini across the neighborhood – became the main alternative access for pedestrians and bikers as it connects Cikini commuter line station to RS Cipto Mangunkusumo and other office buildings.

As a high-density settlement, Kampung Cikini – especially in Gang Ampiun – is an animated area with a closed relation between neighborhoods. The majority of the inhabitants came from low-income households, working as online driver, laborer, employee, and housewife. They usually live in houses less than 20 m^2 with the other four to six people. This condition leaves them no other choice than using public space for their exclusive use, by putting some private object for their activity. These domestic activities later developed into selling activities after the inhabitants realized the opportunity offered by Gang Ampiun as a strategic place.

The observation area for this paper took place in RT 12. In this area, there are four individual activities located both on the left and right side of the alley including cooking, selling, bird feeding, and dishing.

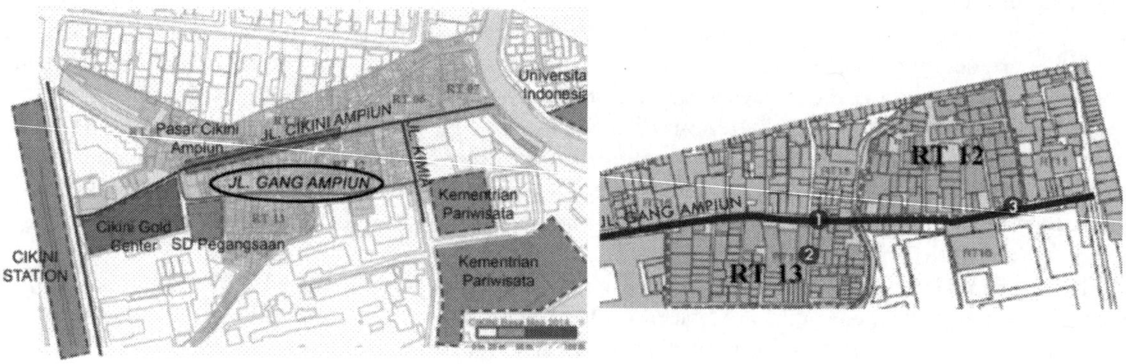

Figure 1. Site location
Source: [9] edited

4. Result

The process of interiorizations that will be discussed in this paper was done by Mrs. Zahra, Mrs. Ratna, Mrs. Aminah, and their families. Mrs. Ratna and Mrs. Aminah live together in one house with their children and husbands, while Mrs. Zahra lives in a house in front of them. The following is an explanation of interiorization process in details.

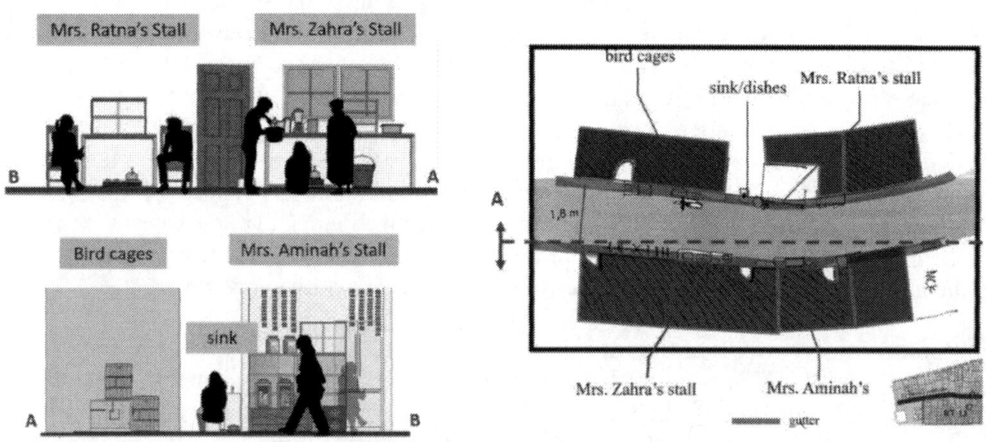

Figure 2. Site layout (right) site section (left)

The first thing that will be discussed in this chapter is the cooking and selling space owned by Mrs. Zahra. It is located in front of her house. Mrs. Zahra lives with her husband, children, sister, brother in law, and nephew in 6 x 4 m^2 area of the house. Facing this limited space of living area and considering children's safety, she decided to move the cooking area in front of her house. Here is where the process of interiorization began.

The process of collecting by Mrs. Zahra: selecting (include separating), bring inside (into the system of relations – layout), presenting. Selecting a step can be seen in the way how Mrs. Zahra chooses the area based on her familiarity, which is the front side of her house. In the process of selecting, some things are added and removed as an idealization of space [6]. In this case, Mrs. Zahra tried to manipulate the environment by putting some rocks on the gutter so that she can put up the interior elements over it. For some occasions, the rock in front of the house door is lifted to give access to the gutter. As an example, right after Mrs. Zahra finished her cooking, she put out the rock to dispose of the residual water into the gutter. This action confirms elimination on selecting step.

The process of selecting has also appeared in tools preferences. Here, Mrs. Zahra chooses two tables as a main interior element, rather than a stall due to the nature of her products. The products she sells are noodles and blended instant drinks. Both noodles and blended instant drinks need some spaces to be made rather than displaying them. Other than economic reason, she chose tables for its compactness. The inhabitant in facing the lack of space issue considered a negative space under the table as a potential cooking area, which leads to the next step of collecting, bringing inside.

Bringing inside to the system of relations could be seen in this case on the way of arrangement, how an object is connected. For example, after been brought inside the tables are transformed into a structure of interconnected action for cooking and selling. Mrs. Zahra arranged the tables side by side. The first table accommodated the preparation and making process, such as blending, while the second table functioned as storage and sometimes displaying some snacks she sells. This layout program is integral to the activities and the preference of the inhabitant which shows her identity and habits as a presenting step.

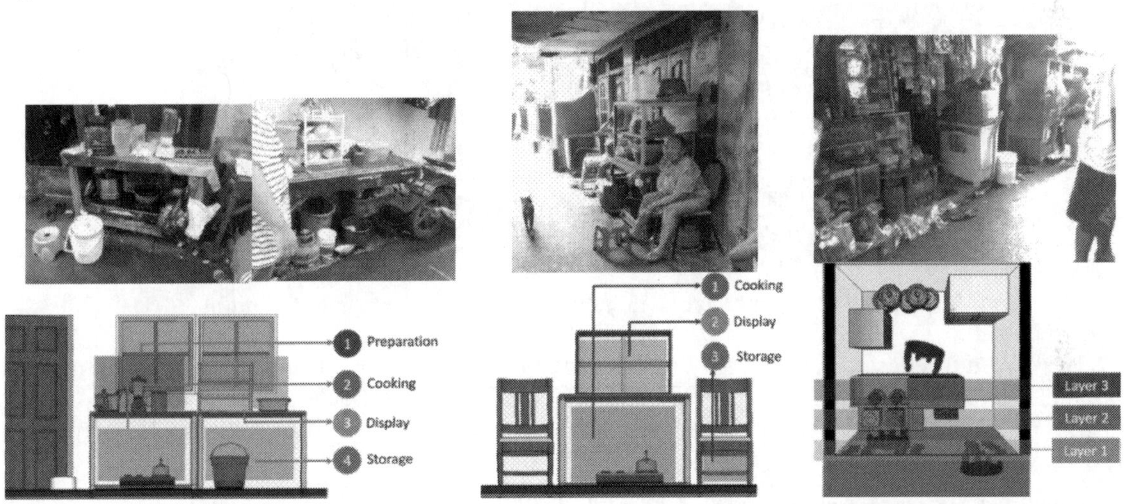

Figure 3. Mrs. Zahra's table arrangement (left) Mrs. Ratna's table (center) Mrs. Aminah's stall (right)

Alongside Mrs. Zahra's, there are a cooking and selling space similar owned by Mrs. Ratna. In selecting step, Mrs. Ratna chooses a table and a display cabinet as her tools. The way she treated the table is similar to Mrs. Zahra. Similarly, the left space under the table is used for cooking space. The display cabinet is chosen to adjust the heavy meal she sells such as rice and side dishes. The meals are made in large proportions simultaneously, then displayed in the cabinets. Unlike the previous inhabitants, Mrs. Ratna creates a vertical layout as she tends to display the foods rather than the making process.

The strategy of interiorization applied to Mrs. Aminah's stall is a system of layering created by the arrangement of the product. The first layer structured by small sachet products. In order to deal with the limited space issue, those products were hung under the ceiling and positioned in the very front of the stall creating a kind of curtain. Unconsciously, the curtain became a tool for giving signals of a visitor by the sound of product frictions.

IOP Conf. Series: Materials Science and Engineering **523** (2019) 012051 doi:10.1088/1757-899X/523/1/012051

The second layer is made by a bigger product that was stored in a jar, like crackers. This layer is more permanent than the first layer, but still open to the intrusions. While the outer can be easily deformed by visitors, the second layer isn't. In order to have the product, the customer should pull and cut the sachets 'curtain'. On the second layer, the customer only allowed to pick the product from the jar, while the jars itself weren't deformed. The last layer – constructed by a display cabinet – is the thickest layer. This layer can only be accessed by the owner, Mrs. Aminah, to take the product inside before it's given to the customer.

Cleaning space on the roadside acted as a temporal space because the duration is very short – five to ten minutes – in every occasion. The existence of the space highly depends on the current situation of the circulation. Here the selecting is seen in time preference. Mrs. Ratna and her family prefer to use the space around 11.00 A.M, 2.00 PM, and/or 7.30 P.M, avoiding crowd circulation of people from going to work and going home. By that, the selected hours became a part of the interior, and the rest is exterior. Another temporary interior is found in bird feeding that occurs in the morning. The movement of the bird created an interior as it crossed the street.

It turned out that Mrs. Ratna, Mrs. Zahra, and Mrs. Aminah considering each other while choosing their product. Each of them sells a different kind of product, so they didn't have to compete with each other. The connection of the interior could be seen in the arrangement by Mrs. Aminah. She put her chair closer to Mrs. Ratna interior space so that they could have some conversations. By that, the interior has indirectly a relation one to another.

Figure 4. The inhabitants gather in front of their house in the night

Every day, the observation spot seems to be a point of gathering between households. In the afternoon – around 2.00 P.M – and the evening – around 7.30 P.M, the other householder of each house went outside one by one sitting in front of their house to have some interactions. The kids are also playing on the alley with consideration to the circulation.

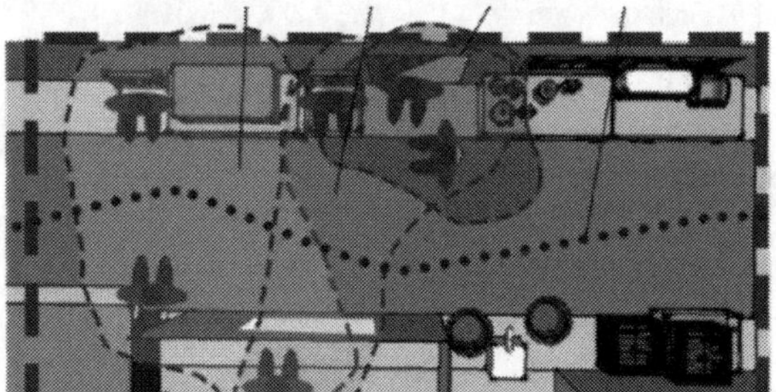

Figure 5. Diagram of the interconnected interior created by interactions

5. Discussion

From the result, we can confirm that interiorization is flexible spatially or temporary as interiority is. McCarthy argued that interiority has a temporary characteristic, because of its elements such as flexible boundaries, activity, and dynamic intimacy [3]. McCarthy gave an example of the differences in term of interior qualities between day and night. As the resulting interior can also become temporary and makes interiorization also depends on time.

The dependency on time makes the individual interior became a trigger for social interactions in a neighborhood, which makes it connected one to another. Though the interconnected interior seems to be inclusive, the alley still allows people to go through it as a public space.

Based on the result we can also see the interiorizations in its position to some issue, such as the lacking space. The form of interiorization they have been doing for years is placing some privately-owned object in front of their house. Their spatial view alongside the issue makes them utilize some odd negative space like under the table became usable. Other than confronting the lack of space issue, interiorization is also considering the ability to control the space as interior related to privacy. This kind of interiorization could be seen in Mrs. Aminah's stall, where the layers became a refuge for her privacy.

The interiorizations by the inhabitants in this case definitely are illegal. It could lead to another issue such as privatization or illegal territorialization. But, here rather than view in a negative way, we take a different approach to see the interiorization as an object to be learned.

6. Conclusions

Interior is not always space inside and limited by the wall. It surely can be formed outside even in public space. In order to make an interior, inhabitant should do the interiorization. Here, collecting as an analogy of interiorization – from selecting, bring inside, to presenting – can also be seen in a high-density area as a strategy to deal with the limited space issue. The inhabitants saw the potential of negative space under the table and change it into a space for cooking and storing. Another occurrence is that the process of interiorization makes an interior in the sense of control, to achieve an ideal space physically and mentally for inhabitations. The senses of control, one of them are aimed to achieve privacy, for example by making a layering system. To make an ideal condition, collecting process should involve elimination in selecting step also. Interiorization also gives a sense of personalization, that can be found in bringing inside and presenting step, to make uncontrollable public space becoming controllable. In this paper, individual interior in RT 12 *Kampung* Cikini is happened to be interconnected to each other by social interactions and makes sense of inclusive living room but still can be accessed by everyone.

7. Reference

[1] Katadata.co.id 2018 (cited 26 May 2018). *Berapa Kepadatan Penduduk di DKI Jakarta*. Available from: https://databoks.katadata.co.id/datapublish/2018/01/25/berapa-kepadatan-penduduk-di-dki-jakarta

[2] Meagher C 2017 *Constructing an Interior Public: uchi and soto in the Japanese Sharehouse*. **14** 114.

[3] McCarthy C 2005 *Towards a Definition of Interiority*. **8** 112–125.

[4] Atmodiwirjo P, Yatmo A and Ujung A 2015 *Outside Interior: Traversed boundaries in Jakarta urban neighborhood*. 79-81

[5] Burrell K 2014. *Spilling Over from the Street. Home Cultures*. **11** 151.

[6] Attiwill S 2013 "interiorizt" in G. Brooker & L. Weinthal *The Handbook of Interior Architecture and Design*. (London: Bloomsbury) p 110-111

[7] Winton A 2013 Inhabited Space: Critical Theories and the Domestic Interior" in G. Brooker & L. Weinthal. *The Handbook of Interior Architecture and Design*. (London: Bloomsbury) p 46

[8] Madanipour A 2005 *Public and private spaces of the city*. (London: Routledge) p56

[9] Adianto J, Okabe A, Ellisa E and Shima N 2016 *The Tenure Security and Its Implication to Self -Help Housing Improvements*. **3** 54

IOP Conf. Series: Materials Science and Engineering **523** (2019) 012052 doi:10.1088/1757-899X/523/1/012052

Co-residence as housing strategy for Betawi families: Case study of Betawi family houses in Cengkareng, West Jakarta

D H Arima[1], J Adianto[1,*], R T Gabe[1]
[1] Department of Architecture, Faculty of Engineering, Universitas Indonesia, Kampus Baru UI Depok, West Java, 16424

E-mail: joko.adianto@gmail.com

Abstract. The number of Betawi families, known as the native tribe of Jakarta, in the capital city is decreasing due to massive urbanization. The fierce economic competition with the resilient migrants has purged them to the suburban Jakarta. In order to live in Jakarta and close to their current job locations, transforming the inherited multi-generational family house to co-residence becomes the favorable option. This paper aims to examine the co-residence as a housing strategy of Betawi families to live in Jakarta. There are certain adjustments in determining the strategy of living together, especially with a lot of relatives. Qualitative research method with in-depth interviews and longitudinal through direct observations were delivered to investigate and discover the meticulous the process of transformations throughout the years. These Betawi families have been living the residences since the 1980s, and both have numerous core family from various generation. This type of living, co-residence, need some strategies such as expanding new spaces, shared spaces and merger-separation of the clean-dirty water and electric channel. Based on the selected two houses of Betawi families in Cengkareng, West Jakarta, this paper poses the argument that co-residence is not only the housing strategy for Betawi families to live in Jakarta but also generating extra incomes to the family as it provides rent rooms for migrants.

1. Introduction

The majority of the academic literature on coresidence living has focused on the implications of this living arrangement for financial within the family. Based on the previous study, coresidence is still the major form of living arrangement for the elderly population in contemporary China, and it is higher in rural areas than in urban areas [1]. Older parents' and children's economic resources are major determinants of choice to co-reside [1]. The better their economic conditions, the less likely they will choose coresidence, or conversely the worse the economic conditions, the more likely they will choose coresidence [1]. Some of these factors represent the needs of the parent and include teen parenthood, a need for financial support or child care, or a general need for parenting or other assistance on the part of the mother [2]. Cultural factors may also play a key role in the formation of three-generational families, as some cultures with a familial orientation may enter into such arrangements due to cultural preferences, not out of need [2].

Betawi people have a habit of providing some land to their children when they get married where they can live [3]. Nonetheless, it is causing the fragmentation of the house and raising population density on Betawi settlement area. Homeowner status may reflect the ability of the parents to give their children a home, which results in prolonged coresidence [4]. Some family members will settle in a parent's

*Corresponding author: joko.adianto@gmail.com.

residence; this caused a type of co-residence living although some others move out of the house. This Betawi family house can accommodate some of the needs of the residents who occupy it. Various changes and adjustments are done on the house as a housing strategy for the sake of family wellness.

It was further assumed that each life cycle stage would result in a different set of housing needs and trigger off a change in housing requirements and the likelihood of a move [5]. For example, following the first move from the family, the needs of house space are likely minimal be adequate. The following marriage and the birth of a child space needs will increase, and the family would move to a larger house. Finally, once the children have left home, the family would likely to move in again in order to reduce housing consumption and the costs of maintenance [5].

Often, the adjustment of needs in the form of parents' residential transformation also occurred in a small scale, such as the addition of sleeping space, changes in space layout and addition of other new space. In this case, there are certainly some adjustments in determining the strategy of co-residence, especially with quite a lot number of relatives in it. The adjustment may not only be a physical transformation of the occupation but also in the form of career mobility or family adaptation [6].

The purpose of this writing is to find out how the stage by stage process of physical adjustment, related to the background of any house changes that occur. Finding out strategies of co-residence of the house inhabited by Betawi families is another purpose.

2. Research Method

The variable that used in this writing is residence in two Betawi families. Residences of both families are basically a parent's residence, which then transformed into a residence for future generations. The transformation of residences in both families has been done several times either on a large and small scale since the 1980s. Data collected during the observation process are plan drawing in each residential transformation, family lineage in the year in which the residence is transformed, the cause of residence changes, the dimension and type of room addition in residence to each transformation, space ownership and responsibility each family member at the residence, mechanical systems in the last transformation.

The method of writing this research is qualitative from the observation at a glance over the past few years, followed by two months in-depth observation mainly with the oldest generation of two Betawi families in Cengkareng, West Jakarta. This direct observation of the location is performed by taking pictures, sketching together with the old generation and interviewing some occupants, including reviewing the transformation stages and background of each change. Based on the interview and the in-depth interviews, a time series plan residence drawing since the first establishment is made that showed changes in each residential transformation.

The way to analyze data is with linking all data, mainly forms of residential transformation with differences in occupants at the residence, to deepen the reasons residence transformation and the way they lived their life at the residence that transformed. This paper also comes with literature studies such as textbooks, journals, and trusted internet sources to discuss the topic more deeply and to gain a better understanding of the topic as well.

3. Result

Commonly, the previous studies discuss coresidence as a strategy for an economic issue. It was told that economic resources are major determinants of choice to co-reside. The better their economic conditions, the less likely they will choose coresidence, or conversely the worse the economic conditions, the more likely they will choose coresidence [1]. Children would be the main beneficiary in this economic case [1]. Moreover, for parents, coresidence is an important way to improve their quality of life. Elderly parents who cannot take care of themselves due to advanced age or losing a spouse, coresidence with their children help them [1].

But this time, coresidence is about the culture of Betawi families. Betawi people have a habit of dividing part of their residential land for each of their children. This condition has occurred continuously since the previous four generations. The oldest generation, in this case, got a residential area of 237 m^2 from the previous generation. Then they built the residence on a small portion of the land. When children grow up and get married, they will share their land little by little to their children. This then triggers the formation of the Betawi community in a particular dwelling and forms a coresidence living.

Both cases studies are located at Bojong Raya St. 001/04, Rawa Buaya, Cengkareng, West Jakarta. The distance between them only around 300 meters. Basically, this area has some of the Betawi families who have lived since a long time ago that made up about 40% of the population in the *Rukun Tetangga 001* (abbreviated as RT 001 or neighborhood). The number of residents in RT 001 is as many as 560 people. Both selected residences were selected as case studies due to the significant transformation that has taken place. The transformation of houses in both families has been done several times either on a large and small scale. The residence of both families is basically a parent's residence which then switched to be the residence for successive generations.

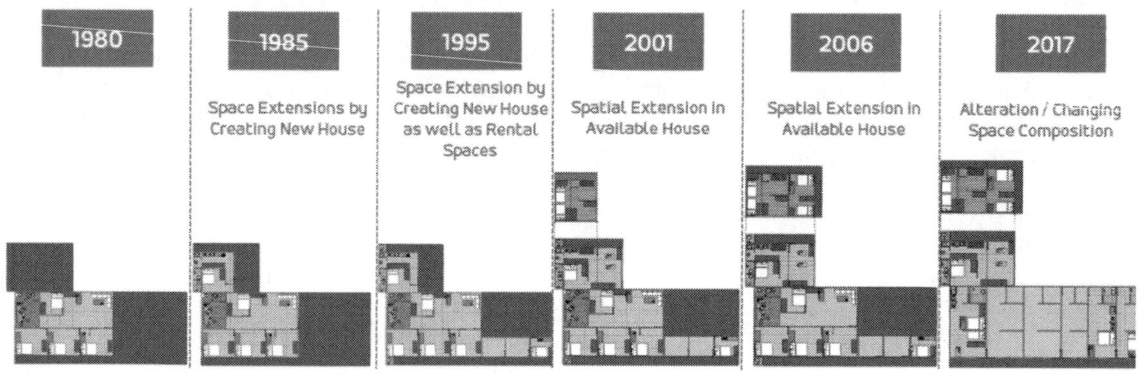

Figure 1. House Transformation in grandpa Saman and grandma Idda family
Source: Personal Graphic, April 2018

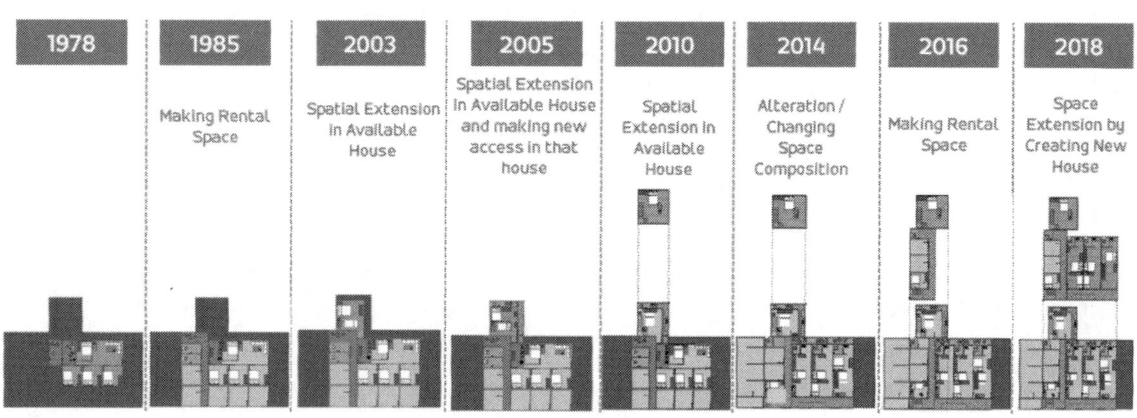

Figure 2. House Transformation in grandpa Mahdi and grandma Rauhil family
Source: Personal Graphic, April 2018

The first family house belongs to Grandpa Saman and Grandma Idda, which was built in 1980 within an area of only 65.5 m^2. The number of residents at that time was 8 people. During the last transformation of 2017, there were 19 people of the family members who occupied a building area of 237 m^2. The second family house owned by Grandpa Mahdi and Grandma Rauhil was built in 1978 with an initial building area of 96 m^2. The last transformation performed in 2018, the building area has an area of 237 m^2 occupied by a total of 20 family members.

IOP Conf. Series: Materials Science and Engineering **523** (2019) 012052 doi:10.1088/1757-899X/523/1/012052

Within face coresidence with four different generations, in this case, it was needed any residence adjustment strategies.

3.1 Strategies of Co-residence in House Transformation of Betawi Families
The following are some strategies of co-residence in the house transformation of Betawi families. Each strategy will describe an example of a representative period from Grandpa Saman or Grandpa Mahdi families. The selection is based on the most significant transformation that takes place in the house.

3.1.1. Expanding new Spaces in order to Get More Spaces. In both case studies, there is an ongoing transformation of space that still exists in the same residence with access that depends on it. This extending new spaces with conjoint accessibility occurred because the family needs more spaces for new residents, also the growth of children. Families will take the initiative to create additional space in certain areas. Extensions generally tend to be horizontal. When Betawi families no longer have free space to extend the space, then vertically extension will be applied. This is indicated by maximizing the available land for residence. This space extension can be done in various habitable areas or service areas. The person in charge of this dwelling is the head of the old occupant's family.

This conjoint accessibility occurred as the service rooms like kitchen and bathroom remain in the main residence. So this conjoint accessibility makes all their activities occurred inside the house. Access that is utilized in this case is the primary residence access that has been made since a long time ago. The location of the access is on the front side of the house. So to go to the new space created in the outdoor area, occupants still have to go through the main access, along with other census families. A few years later, when they have own kitchen and bathroom in their new spaces, they would make new access that connects to outdoor directly, so their access no longer dependent on the main residence.

In both families, the transformations are taken place in the form of space extensions to create a separate residence. This is to improve the quality of space privacy in the family. This generally occurs when the children of the Betawi family enter the marriage phase which then given some portion of the land to become their place to live. By creating new access to this new space, it is expected that the occupants have separate access. It is important to avoid sharing it with other family members in residence. Space extension can be made vertically or horizontally. A door as new access is formed when the extension is done horizontally. Whereas, if the extension is applied vertically to make some space on the second floor, a stair will be created as an intermediary to the new access.

There are two types of stairs created in this case, the shared stairs (located outside) and private stairs (inside the residence). The stairs made outside is connected to the desire to cease the involvement between upper and lower residences while the stairs inside maintain the connection between the upper and lower residences. Separate accessibility between upper and lower residences is due to the existence of two married couple. While merging upper-lower residence access is because the residence is only occupied by one census family. The addition of access directly affects the separation of family accessibility within the residence. The person in charge of this dwelling is the head of each household.

3.1.2 Shared Spaces Together While Co-Residence. Living together in residence with multifamily, sharing space is something that happens. Space consists of two types, namely habitable space and service space [7]. The habitation space is the living room, dining room, and bedroom while service space is including kitchen, bathroom, balcony, terrace, storage room, commercial space and other open spaces [7].

Betawi residents will share the space in residence in the form of a non-private room, both habitable spaces and service spaces. The only private space in the Betawi residence is the bedroom. This is an uninterrupted space even by other family members, except the census family. The habitable spaces used together are the living room. And all service spaces include a kitchen, terrace, and bathroom used together for multifamily. Amount of each of these spaces is only one in both families. Therefore, these Betawi family's rooms tend to be spacious to accommodate many residents in one room.

IOP Conf. Series: Materials Science and Engineering **523** (2019) 012052 doi:10.1088/1757-899X/523/1/012052

Shared space has an influence on household activity in the form of shared responsibility. Some residents have responsibility for the maintenance of existing shared space. The more occupants of the residence, the lighter and fewer the housework assigned to each occupant. Shared spaces also have an impact on shared objects. However, they do not share everything contained in residence, only some furniture such as guest chairs, TV and refrigerator are used together. The objects situated in the private space (bedroom), absolutely belong to each space owner.

3.1.3 Merger-Separation of the Clean-Dirty Water Channel and Electric Channel for Efficient and Effective Live. Transformation is needed in order to fulfill the needs of each occupant. In both case studies, the peak of transformation in the form of residence alteration is made to create private residences for each second generations. Residence alteration is in the form of changes in the composition of the habitation. In both case study, alteration leads to changes in the family's water and electricity systems. The Betawi residence now has them more in number resulting from the dividing, and automatically has more water and electricity channel systems. In reality, Betawi residential water and electricity channels are often combined with other inherited residences.

Merged water and the electric channel are aimed so that the family leads an effective and efficient life. The integration of clean water channels in this case study is based on the residence position. Residences that are parallel to each other simplify the formation of clean water channels in the same stream. The residences whose channel are merged with each other are transformed at the same time. This is due to the water channel system that has been planned to be merged from the beginning. Some water channels of rented dwelling are merged as well. These rented dwelling's water channel system is merged with the owner's residence. For clean water provider, it's the rented house owner who is held responsible for the rented room's clean water consumption. The tenant will have to pay to the owner who will then pay such rental spaces monthly clean water consumption to the clean water provider.

The dirty water channel is generally merged. One septic tank can be used for one, two, or six residences. This shows the effectiveness of septic tank usage. The location of the septic tank itself is adjacent to the existing toilet in the residence so that the flow of dirty water would not be too long.

The electrical system of each rental residence is stand-alone with the separate electric channel except in Mr. Zein's family that is still merged with the other five residences. The residential postpaid electricity payment system shall be paid by the head of the family. Prepaid system is applied to the rental residence and coordinated by the tenants.

The existence of merging on the utility system shows the form of multifamily in these Betawi families, even though each census family has separate residence. In these combined conditions, the costs paid collectively per month by the head of the household coordinated by the youngest one. This show evidence of cost savings resulted from living co-residence.

4. Discussion

This co-residence starts when the oldest generation, in this case, gets inherited land in the 1980s. At this time, the price of land was still very low, and there were not many migrants who inhabited the city of Jakarta, especially in the Cengkareng region. As indigenous people, it is not surprising that they also get the vast and cheap land from their parents. The allocation of this inheritance took place on a chain later in the day when their children entered the marriage phase. The existing land has been fairly designed from the beginning to be distributed evenly to all children.

Parents' income, ownership, and size of the home are indicators of parents' opportunities to offer their children accommodation of living [4]. In this case, the size of the home is one indicator that causes co-residence in the Betawi family. This then became a culture in the Betawi family to do co-residence as a housing strategy. Although culture became the first factor that led to the formation of co-residence living, it was also caused by the low economic ability of the Betawi family. The existence of inherited land that is obtained without paying can reduce the cost of building residences quite rapidly. Even in the journey of residence, various other economic benefits such as cheaper household financing are borne together. From time to time, the Betawi families' occupancy change and tend to have an increase in family members. Increased number of family members has triggered the transformation of dwelling in both families. The addition of family members occurred through marriage, childbearing, and the return of the old occupant to the parent's home [8]. The family life cycle can affect housing transformation. Not only increasing the need for more spaces but changing the housing composition based on the physical and psychological needs of residents.

In the case study, not all children in the Betawi family eventually occupied the land that had been given by their parents. Some children actually stay at another residence. The only factor that causes this is due to the following spouse who also gets heir land with better conditions. As for another inherited land owned, it is used as a rental space construction that produces passive income every month.

With the existence of co-residence, several positive impacts were found. Household financing that is diverse like water, electricity can be borne jointly so that the cost of spending is more efficient. In addition, household responsibilities such as sweeping, mopping and so on, can also be shared so that the workload is lighter. However, this, on the other hand, is relied on by some parties not to take part in caring about household responsibilities. As for the other disadvantages of co-residence living is the lack of privacy in each family due to mixing space. This is the most dominant trigger for physical adjustment, along with changes in composition and needs of family members.

5. Conclusion

Based on the discussion, it can be concluded that as a dwelling's residents face the development of the stages in life and changing needs, these Betawi families make some physical adjustment. Furthermore, the Betawi families living together in a single residence and later would be split into several residences. Basically, co-residence is beneficial to the occupants as they share responsibilities, and utility costs can be divided, so they actually save more. Spatially, co-residence requires special strategies of the building so they can continue occupying the residence. Some adjustments made are extending the spaces with conjoint or separate access, sharing some spaces through co-residence, and merged-separated clean-dirty water and electric channel.

6. Acknowledgments

We would like to express gratitude to Directorate of Research and Community Engagement Universitas Indonesia (DRPM UI) for administering the *Hibah Publikasi Internasional Terindeks untuk Tugas Akhir Mahasiswa* (PITTA) 2018 scheme under grant number 2461/UN2.R3.1/HKP.05.00/2018, and Faculty of Engineering Universitas Indonesia (FTUI), Departement of Architecture for their support and commitment in this research. Sincere gratitude is addressed to respondents who sincerely participate in this research.

7. References

[1] Chen J and Chen Q 2016 *The Journal of Chinese Sociology: Socioeconomic statuses across generations and coresidence in China* **3** 24

[2] Dunifon RE, Guest KMZ and Kopko K 2016 *Grandparent Coresidence and Family Well-Being: Implications for Research and Policy* 110-126

[3] Anisa, Ilham J and Purnama T 2010 *Perubahan Pola Pemukiman Masyarakat Betawi di Condet* **6(1)** 65-72

[4] Isengard B and Szydlik M 2012 *Living Apart (or) Together? Coresidence of Elderly Parents and Their Adult Children in Europe* **34(4)** 449–474

[5] Murdie RA, Chambon AS, Hulchanski JD, and Teixeira C 1999 *Differential Incorporation and Housing Trajectories of Recent Immigrant Households: Toward a Conceptual Framework*

[6] Skobba K R 2008 *The influence of section 8 vouchers on the housing careers of working poor families*

[7] Nguluma HM 2003 *Housing Themselves*

[8] Clark WAV and Onaka J L 1983 *Lifecycle and housing adjustment as explanations of residential mobility* **20(1)** 47-57

Creating atmosphere in hotel interior space with material roles: *Bata Pejaten*

J Lathifa[1,*], N R Kusuma[1], E Arvanda[1]
[1] Department of Architecture, Faculty of Engineering, Universitas Indonesia, Depok, Jawa Barat 16424, Indonesia

E-mail: jessica.lathifa@gmail.com

Abstract. As a major tourist destination in Indonesia, Bali possesses hospitality industries that are competing to present the best experience for guests. One of the hotels, *Katamama*, offers a specific spatial experience through the use of *Bata Pejaten*, handmade bricks traditionally made from mixtures of clay and *paras* sand—a fine stone powder of volcanic material sourced locally in Bali. Meticulous selection of *Bata Pejaten* applied on the 5-star quality hotel became its own uniqueness as these bricks are generally seen as a simple inexpensive material and mainly used for temples construction. As an interior space, it should be recognized not only related to the usability to meet human needs, but also the emotional effect of giving experience. Those spatial experiences are the atmosphere of space that exists between human and their built environment. It is understood through the senses of human, creating an impression and significant meaning to the person. In this case, the material becomes one of the important aspects of spatial atmosphere production. For *Bata Pejaten,* it is their craftsmanship, composition, and organic transformation. The paper aims to explore how *Bata Pejaten* and its properties and qualities generate atmosphere of space. This paper presents a case study analysis on interior spaces characterized by its specific material and authentic substances from volcanic eruptions as the core investigation. This paper demonstrates a design approach from the material perspective. In particular, this study introduces the concept of using materials attentively in creating spatial atmosphere characterized by modest material yet rich in its values.

1. Background

Hospitality industries are dealing to provide best experiences for guests. *Katamama Hotel*, located at the tropical landscape, Bali, Indonesia, is presented as a case study that shows the use of *Bata Pejaten*, a handmade brick traditionally made from mixtures of clay and *paras* sand—a fine stone powder of volcanic material sourced locally in Bali. As a boutique 5-star hotel, *Katamama* used almost 1.500.000 pieces bricks as the major material and this became its own uniqueness as these bricks are generally seen as a simple inexpensive material. *Bata Pejaten* is presented truthfully as the original material and applied continuously from the exterior to the interior space of the hotel. As natural materials, it has shown its compositions through the authentic substances originated from volcanic eruptions and then occasioned the organic transformation of material. Those bricks affect intangible qualities of design not only by upgrading its hotel design value but also creating distinctive experience and wellbeing aspect of the guests.

*Corresponding author: jessica.lathifa@gmail.com.

IOP Conf. Series: Materials Science and Engineering **523** (2019) 012053 doi:10.1088/1757-899X/523/1/012053

The material becomes one of the important aspects to produce spatial qualities i.e. atmosphere of space. The atmosphere of space is a quality that exists between human and their built environment understood through the senses, creating an impression and significant meaning to the person [1,2]. The impression is created as the psychological effects through perceptivity of space. It reminds that the interior space should be recognized to be related for two functionalities: as its usability to meet human needs and also its emotional effect of giving experience. Human body becomes a fundamental instrument that captured those experience influenced by architectural qualities—atmosphere—of a temporary environment [2, 3].

The atmosphere of space occurred between objects existed in a place with human as subjects in imaginative and abstract conditions [4, 5]. Although these sensations are intangible qualities, they are produced by the existence of tangible objects in the terrestrial environment. Therefore, the aesthetic of atmosphere in its approach obtained through two viewpoints, *perceptions aesthetic* and *production aesthetic* [4]. Production aesthetic of the atmosphere refers to the approach from the viewpoint of the object. In practice, the atmosphere can be formed through the elements or objects to become a wholeness of the spatial product; material as an object. This also related to [3]statement that the atmosphere is the encounter of the emotional impression and the quality of the space reproduced by the integration of objective material, the proportion of space, the aging material, the connection of material, as well as connections with places and buildings, rhythm, light, etc. It can be understood that the material, its integration and aging phenomenon, the encounter between the material and its existence in the environment, also with the natural elements—in this case, related to light—will affect the formation of the atmosphere. Therefore, creating atmosphere can be designed through the material based on a thorough understanding the material presence as the object of interior space by its properties and qualities.

The physical properties of material—the colours, textures, shapes, and densities—are seen on the material's surface. Those physical properties may change due to its presence in the terrestrial environment, so it is necessary to understand what is contained and manifested in the material resulted from its interaction with the environment. Its properties are visually measurable and experienced with other senses based on reality while its presence also impressed its character and provide sensual effects as senses and physical pleasure. Those are known as the qualities of materials, which give a psychological effect on the subject [3]. According to the physical properties of a material, the deep understanding of physical properties occurs objectively through three aspects, the *surfaces,* and *substances* of material, within a *medium* relating to the surrounding environment and terrestrial phenomena [6].

The *surface* of a material is the place where encounter happened between its material substances with the immaterial medium in the world [6]. Related to the *medium*, according to [7], material properties—surfaces and substances—then become uncertain attributes of a material; *processual* and *relational.* Therefore, the understanding and assessment of material properties occurred through 3 aspects, including surfaces, substances, as well as the medium which surrounds it in the environment.

From these findings, researchers have attempted to identify atmosphere objectively from the material not only as a material but furthermore, it is the composition of the material itself and/or with other elements; its settings. Those settings to form architecture cannot be separated from the environment of its existence. Then, as a built environment, material connection with the surrounding environment and context needs to be understood comprehensively in the design process.

The existence of built environment cannot be separated by it surrounding. The atmosphere of space must be part of its environment because otherwise, it will ignore the will of the atmosphere [2, 8, 9]. The atmosphere can also be formed and perceived in the object and its relation to the context in which the object is found. In the end, in designing the atmosphere of space, it is not only about organizing the architectural elements but also considering the relationship within the context of the building. This paper presents the identification of *bata pejaten* as the object, its form and space as a spatial element, and its relation to the terrestrial environment to enhance spatial atmosphere in hospitality interior space design context. This paper argues that these experienced qualities can be provided through the use of natural material precisely and allowing connection of material, built environment space, and the surrounding itself, to the natural world.

The method described in the following chapter combines the identification of *bata pejaten* as a natural material, its material arranged as spatial elements, building features, and the built environment as the main evaluation.

2. Methodology

A case study is presented that shows the used of *bata pejaten* as the major material of a 5-star hotel building. First, the objective of the material properties is reviewed through physical condition and collected data by images and comprising photographs. Second, the used of the material and its configuration as formed spatial elements. On the final steps, the observation of the interior space design qualities focused on the applied material on the hotel typology complements previous data as supporting aspect to be considered in built environment space.

2.1. Material Properties

Bata Pejaten as a handmade brick lasts with its manual production prioritizes expertise and its craftsmanship by hand. Its material substances—mixtures of clay with *paras* sand and coated with palm oil— due to the origin of the sand from the volcanic eruptions shown a unique response to nature. Its original colour is terracotta, brownish-red colour, dominated by clay. The results by the young environmental researcher showed that volcanic ash contains major elements of aluminium, silica, potassium, and iron. As the main substance, iron enables the process of oxidation and corrosion of the bricks. Therefore, it affected changes of colours of the material—brownish-red colour gradually turned to black.

Its craftsmanship of material production also affects the character of the material. *Bata Pejaten* possessed 4 different surfaces, ranged from the finest surfaces to rough surfaces (figure 1). As home industry product, not all produced bricks are perfectly neat in their block shapes and its surfaces.

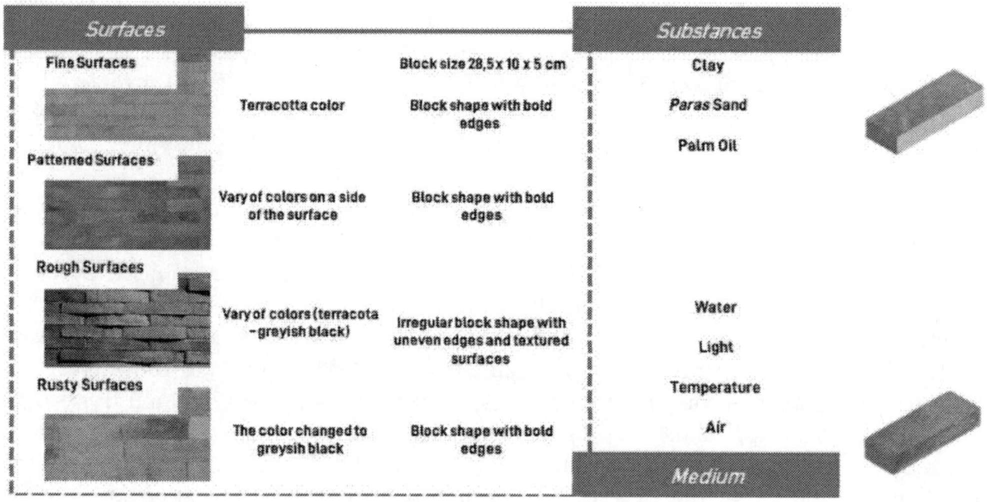

Figure 1. Bata Pejaten 4 different surfaces.

Other than colour change due to corrosion phenomena, inconstant shapes and surfaces, *bata pejaten* also showed varieties of colour due to combustion process on production phase. Those phases resulted in patterned surfaces on one side of the material surfaces. This is another indication that the combustion phase can also affect the colour density of the clay content caused by the heat intensity, direct exposure to the heat source, and the length of burning time.

2.2. Material Qualities
Located in the tropics with high intensity of rainfall, bata pejaten showed its interaction with medium and leads change of physical properties and also affects material qualities. The terracotta colour of brick that created warm atmosphere then affected the perceived quality of space. The presence of bricks in a humid area can 'relieve' the warmth of the brick through the change colour to greyish black, giving a cool colour effect in turn. Changes in the appearance of brick also responded to the variable time of sunlight exposure, which revealed the dynamic qualities that are constantly changing over time.

2.3. Material Configurations
There were 2 techniques found to construct *bata pejaten* in the hotel, (1) ideal brickwork construction; (2) latticework construction. These materials are acclaimed as the final surfaces or widely known as exposed walls. As spatial elements, *bata pejaten* is applied for building features such as wall covering, flooring, and also column of the buildings. For the first techniques, those arrangements become distinctive, as the brick remains very tight without being separated by other materials.

For the second technique, it turned the wall 'breathable', created a connection directly to nature and became a semi-outdoor space. Thus, every area of the hotel with different physical properties and material qualities as well as material configurations created an atmosphere with diverse and dynamic spatial quality.

3. Discussion and Analysis of The Results
In the case study of Katamama Hotel, *bata pejaten* is used from the facade to the interior space, from the lobby through the guest room walls. Chosen areas to be discussed are including the lobby, the restaurant, and the corridor as the access to guest rooms (Table 1). Those selected areas were determined as they are the first areas experienced by guests, which play a major role in creating impressions to guests.

From the very first area, the lobby, variations of brick colours—brownish-red to greyish-black—applied irregularly to the lobby area. The colour composition was not completed after arranged by variations of colours, but it is possible to continue transforming the brick's ability against the corrosion process. Strong tropical climate and hot temperature due to building coastal locations are dimed by the presence of bricks with blackish colour and the natural process of blackening of bricks. Extreme weather along with major sun exposure throughout the day led a new characteristic related to its interaction with nature. It is suggested that the use of natural material could express environmental processes and temporality of nature, such as aging and the passage of time [8].

The second area continued to the restaurant, finest physical properties of brick's surface found in the walls and columns. The use of patterned surfaces combined with strong terracotta colour created modern ambiance and unique atmosphere.

The last area's examination is done throughout the corridor, before reached the lift to access the guest rooms, textured bricks and its irregular shapes applied to this area's walls. Hence, assembly of those properties as the building features affected its spatial qualities. Those contoured uneven walls created a particular sense of depth and intimate atmosphere.

The corridor of guest room's wall showed a different configuration of bricks. The brick arrangement with applied latticework constructions soothes the experience of the intimate atmosphere before. The contrast atmosphere created by the pattern of sun rays which coming through the latticed wall of bricks. That semi-outdoor corridor enhanced human innate existence of its environment: Bali with the tropical climate, strong breezes, and warm temperature of the coast.

The latticework constructions of bricks also affected how the light came into the interior space. The sun rays penetrate through the wall vary in patterns overtimes. This phenomenon also brings one's mind to the memories and gives similar effects of the light rays in the morning through the forest. The guest gets a meaningful experience of space as reminiscent of the temporal nature.

Table 1. The used of *Bata Pejaten*

Area	Material			The Atmosphere of Space	Theory
	Physical Properties	**Qualities**	**Configuration**		
Lobby	 The ideal shape of brick with smooth and matte surfaces, the gradation of colors at random on a face.	 the color contrast created on the brick which exposed directly to sun rays all day compared to to the shaded brick, so that the visual perception of the brick turn out to be calm and relaxing.	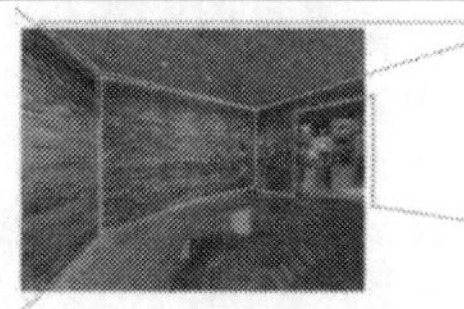 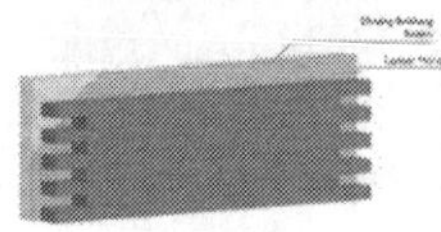 The brickwork configuration created a tight arrangement applied thoroughly as wall and flooring (with the herringbone pattern), the material composition of brick on the wall and wood on the ceiling.	 The consistency of brick application (walls and floors), the material composition (with wood), and manipulating the scale (from high and massive to adapt human scale) created an intimate and relaxing atmosphere for the guest.	The material arrangement of human size, scale, and proportion; Material and the context..
Res-taurant	 The shape of brick with smooth and matte surfaces, and arranged regularly with patterned bricks (diagonal lines with color gradations).	The dominance of the original brick color (red terracotta) with smooth and matte surface creates a warm effect both visually and haptically sensed.	The brickwork configuration in a regularity, combined with patterned bricks, tightly applied on walls and column.	 Arrangement of bricks with regularity, material composition (with wood on the ceiling, terrazzo on the floor) creates a more modern, warm, and attractive space.	Material arrangem ent of colors and textures; Material compositi on.

Corridor (Lift)	 The uneven shape of bricks with rough texture, variety of colors dominated with brownish-red or grayish-black bricks.	The imprecise texture and shape of the brick creates depth effect and stimulates a strong haptic sensation.	The brickwork configuration—without mortar—due to the characteristics of the imperfect brick (the block shape is not precise), the application of those brick create contours on a wall.	 The arrangement of bricks that create contoured walls, the close and narrow distance between walls, and the intensity of light creates a mysterious and intense atmosphere.	The material arrangement of colors and textures; The material arrangement of human size, scale, and proportion.
Corridor (Guest-room)	 The dominance of precise shape of the brick with smooth and matte surfaces, the gradation of colors at random on a face.	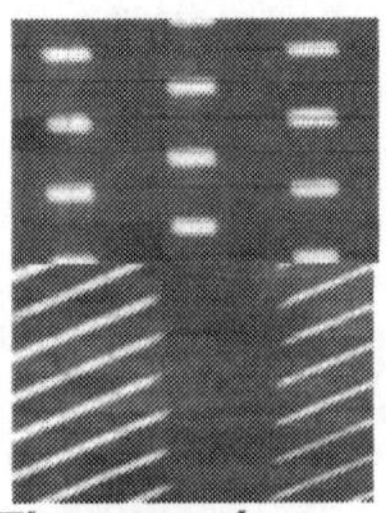 The arranged bricks in the outer areas make a direct contact with the environment thus enhancing the warm temperature' effect of the brick.	 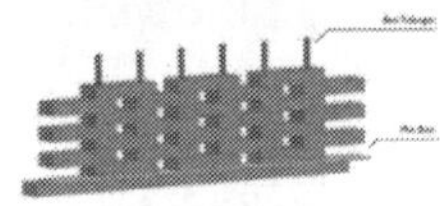 The latticework of bricks applied for the outer wall element constructs the semi-outdoor corridor space, the material composition of bricks on the walls and columns with *jogja* tiles (terracotta colors) on the flooring.	 The contrast ambiance of the space compared to the previous space (corridor in front of the lift) presents a natural and calm atmosphere of the space, blends well with the tropical climate atmosphere of the environment, creating a unified tropical experience.	Material settings to the time, the *medium*; and the context.

Based on case study analysis, *bata pejaten* is capable to create spatial qualities through its physical properties by its surfaces, changes due to the interaction between its substances with the medium, material qualities, and material configurations. In more details:

- *Bata Pejaten* exposed different physical properties and qualities, showed the range of decolouration, textures, and surfaces as an active living material responds to its natural phenomenon.

- As a natural material, *bata pejaten* is able to determine the atmosphere with strong linkages to the surrounding environment
- *Bata pejaten* as a material is able to perform its responses as interaction to nature by its fading colour due to exposure of sunlight, warm to cool effect regarding colour warmth due to direct contact with air and water as the medium, and its changing colour due to corrosion process.
- *Bata pejaten* as applied material for spatial elements
- As a built environment typed such as hotel, the use of natural material created a distinctive experience and provided a good environment for the wellbeing of guests.
- *Bata pejaten* as natural material showed its connection to nature, qualified to improve comfort, tranquillity, and fulfilled hotel function.

4. Conclusion

Through the use of *bata pejaten*—handmade and traditional local brick—fairly and consistently as the main material, physical properties, qualities, and also its configurations able to created authentic space atmosphere in hotel interior space. The impression of the hotel became unique and luxurious in a different perspective because of its craftsmanship in the use of the material as well as in the design process.

In this case study analysis on the Katamama Hotel building, the configuration of material in architecture as a built environment presented the material configuration into spatial elements and its relation with the terrestrial environment. This suggests that those rules created a strong atmosphere by the use of natural materials as well as reactions towards and close relationships with nature.

Its applications gave benefits as lead comfort, tranquillity, and created a distinct sense of place. Those successful material based designs applied to the hotel typology lead the hotel as one of the top 5-star boutique hotels with numerous world awards and guarantees. Thus, the suggestion that I can give is the approach through material perspective requires our understanding to the material content, its interaction with the medium in the environment, and the effect on the surfaces and appearances of the material. Accordingly, having a deep understanding on the material, considering its configuration as a spatial element and referring to the terrestrial environment, the atmosphere of space as an aesthetic production can be created through the aesthetic of the material.

5. References

[1] Caan, S. 2011. *Rethinking Design and Interiors: Human Beings in The Built Environment*. London (Laurence King Publishing).

[2] Zumthor, Peter. 2006. *Atmosphere.* Berlin: (Birkhduser Verlag AG).

[3] Böhme, Gernot. 2014 Urban Atmospheres: Charting New Directions for Architecture and Urban Planning, in Borch, Christian., *Architectural Atmospheres: On the Experience and Politics of Architecture* (p. 42-59). (Basel: Birkhäuser).

[4] Böhme, Gernot. 2017. *The Aesthetics of Atmospheres*. (London: Routledge).

[5] Pallasmaa, Juhani. 2014. Space, Place, and Atmosphere: Peripheral Perception in Existential Experience, in Borch, Christian., *Architectural Atmospheres: On the Experience and Politics of Architecture* (p. 18-41). (Basel: Birkhäuser)

[6] Gibson, J. J. 2015. *The Ecological Approach to Visual Perception* (Classic ed.). (Boston: Houghton Mifflin).

[7] Ingold, T. 2007. *Materials Against Materiality* (Cambridge University Press) UK.

[8] Suriano, M. A. 2016. *On an Architecture of Atmosphere* (Ryerson University).

[9] Böhme, Gernot. 2013. Encountering *Atmospheres: A Reflection on the Concept of Atmosphere in the Work of Juhani Pallasmaa and Peter Zumthor*. OASE Journal for Architecture, 97-99.

IOP Conf. Series: Materials Science and Engineering **523** (2019) 012054 doi:10.1088/1757-899X/523/1/012054

Housing preferences and strategies of Javanese migrants in Jakarta

C Theresia[1], **R T Gabe**[1,*], **J Adianto**[1]
[1]Department of Architecture, Faculty of Engineering, Universitas Indonesia, Kampus Baru UI Depok, West Java, 16424

E-mail: rossa.simatupang@eng.ui.ac.id

Abstract. As the center of development growth, Jakarta has been flooded with massive migration from all over the country, especially Java. With the limited skill to compete in the formal economy sector and soaring house price, living in limited size house becomes the plausible solution for the Javanese migrants. This reality bite serves a severe mismatch between the culture-based housing preferences and available rooms in the occupied house. In order to minimize the mismatch, the migrants deliver various meticulous strategies by utilizing the available resources. This paper examines the housing preferences of Javanese migrants in Jakarta and their strategy to minimize the mismatch between the housing preferences and the occupied houses. Qualitative method and means-end chain analysis are conducted to reveal the housing preferences based on Javanese culture, the delivered strategies, and their connections. The respondent selection encompasses low-cost apartment and high-density slum settlements to represent the common housing type of low-skilled migrants in Jakarta. The result demonstrates that the anchored spirit of togetherness in housing preferences takes form in their unique strategies beyond physical boundaries and materiality.

1. Introduction

Indonesia has been mentioned as one of the world's most diverse nation – for it has around 1.000 communities with more than 400 languages and 17.000 islands [1]. Every community has their characteristics which differ one from another. This is applied to Javanese ethnic as well that is spread all over Java island.

Further discussion then comes into the huge migration of Javanese people to Jakarta – the capital city of Indonesia – which had happened for years and causing them to be the most ethnic population in the city. The amount of Javanese migrants even surpasses the Betawinese ethnic, the original ethnic of Jakarta. The previous study usually discusses migrants who become the minority in their destination places. It shows that migrants tend to live next to their ethnic enclaves [2].

Examining the migration itself can be seen that the main reason is the economic factor. These migrants try to get a better way of living and find moving to the capital city is worth a try. With the limited financial condition, their choices of dwelling are constrained to certain places, such as high-

* Corresponding author: rossa.simatupang@eng.ui.ac.id

IOP Conf. Series: Materials Science and Engineering **523** (2019) 012054 doi:10.1088/1757-899X/523/1/012054

density slum area and low-cost apartment. These two choices fit their financial resources, but not their preferences. Consequently, they make some unique strategies to adapt to the discrepancy.

The discussed issues in this paper will reveal Javanese low-skilled migrants' housing preferences – like the majority in Jakarta – in terms of location, closure to family and friends, housing attribute features. The paper will also explain how culture influences one's housing activities. Further discussion will describe the mismatch occurred and their unique strategies in dealing with the mismatch those are a different one to another, depending on the available resources they have.

2. Materials

2.1Javanese Migrants' Housing Preferences and Choices

Almost everyone has experienced moving from one place to another. This process is called migration, where a family or individual move permanently or semi-permanently [3]. Many factors can affect the process, such as factors in the origin and destination place, obstacles in between, and personal factors [3].

Unfortunately, the migrants often come with limited skill and low competence of education, and most of them are only high school graduates or even lower. With this confined asset, only a few migrants can earn a better living than before in their origin places. Meanwhile, the rest even has to struggle by doing any possible job just to survive. No wonder, living in a comfortable preferred house remains a reverie for some. They also have to deal with the plausible solution, to live in a slum area or a low-cost apartment as mentioned before. The house they choose to live in then termed as a housing choice – which may be distinct from their true preference [4]. The distinction is caused by regulation, supply factor, budget, etc. [4]. Another interesting fact is the evocative data that they have become the most ethnic population in the city, even surpasses Betawinese, the original ethnic group of Jakarta. A study has shown that a newcomer tends to live close to the same ethnic group [2] – a question arises naturally whether it remains important for an ethnic group that has dominated and spread all over the destination city. This will be revealed by a direct interview with the migrants.

2.2Javanese Culture Influences Housing Activities

In examining housing activities, culture is perceived to take an important part in influencing an individual or family housing preference [4]. Further, the culture has to be dismantled first into concrete variables such as lifestyle, family, identity, and value, to find its correlation with housing activity itself [5]. To study further about Javanese culture, there are three forms in every culture namely notion and norms, social system, and physical culture [6]. Notion and norms are inherited from generations. In Javanese culture, this involves an obvious level in the social structure. They have been taught to respect the elderly in certain attitudes, such as specific call while communicating with them. Not only by age, but the structure is also applied by how they label different kind of people, from the aristocracy to the proletariat. The notion is also about having many kids because they believe it will bring warmth to the family [7]. Javanese people are also raised to show care to other people – sometimes knowing in the future they may need it in return.

The second form is the social system, that can be observed in their everyday life. In Javanese tradition, they have a close relationship with neighbors, and they get used to help one another or in their own phrase *gotong-royong* (mutual cooperation). They also have a closer relationship with their direct families such as parents, children, and siblings. Furthermore, *sungkan* (hesitate to be burdensome to others) feeling often occurs among the extended family. Javanese people also have specific ceremonial in every important moment of life, such as *midodareni* when they are getting married, *mitoni* in their 7th month of pregnancy, etc. They also believe these have to be held in their own houses, as respect and gratitude to their parents for raising them. Javanese people also value a gathering with their loved ones is priceless, which prompts them to put aside anything else. There is even a proverb that says in English "even in starving, gathering together is the most important thing." The third form is physical, such as carving and pattern can be directly applied in the dwellings. Together, these three cultural forms influence individual or family's lifestyle, value, norms, perspectives applied in their dwellings.

3. Methodology

In this paper, a qualitative method is used with an interview, direct observation, and means-end chain method. The purpose of using an interview is to reveal their actual preferences of houses despite the limitations. Meanwhile, the purpose of means-end chain method is to reveal the direct influence of tradition toward housing activities, by looking up to the certain chain that ends up with "tradition" on top of the chain. The referred value consists of ten different values, which includes tradition value [8].

Means-end chain method is used to trace spatial attributes – consequences – values. The process starts with identifying the spatial attribute in the dwelling, continuously asking why they are featured to reveal the underlying consequences, to show at last the values upheld by the related person. This method usually starts with spatial attributes, but in a certain condition, it has to be started from the consequences. The certain condition mentioned is a small sized housing that may not contain spatial attributes needed. Starting from consequences or typical activities that the respondents do can unveil the spatial attributes that may be available outside of their houses.

This research is applied to four low-skilled Javanese migrants in Pesanggrahan, specifically, Gateway Apartment where many migrants gathered from all over the country: Manado, Padang, Ambon, Java, and many more. Near this area, there are several industrial places, for example, Cipadu, where many labors are needed. It's not a surprising fact that many low-skilled migrants come and live directly to this area where finding jobs are more likely easy.

Another reason why many migrants live in the Gateway Apartment is the affordable monthly fare, so it becomes a plausible choice for them who mostly comes from the lower middle class. The first respondent rents the apartment unit and another one owns the unit by herself. The third one rents a room in a high-density slum settlement near the apartment, and the last one owns a house in a similar condition. This variation is expected to represent all types of dwelling condition of low-skilled migrants. Despite all the differences of dwelling, all of the respondents earn a living in Gateway Apartment as a food stall owner and a daily-paid worker. The criteria met by these four respondents are as follows: they born and raised in their ethnic area in Java, moved to Jakarta in their productive age and were prompted to find a job to earn a better living.

4. Results

The results are shown in the pictures below. From the means-end chain method can be seen the chains which end up in "tradition" value. The results show a variety of spatial attributes that can be used to accommodate a specific Javanese tradition they uphold; whether it is inside the dwelling or not and whether it belongs to them or their relatives'.

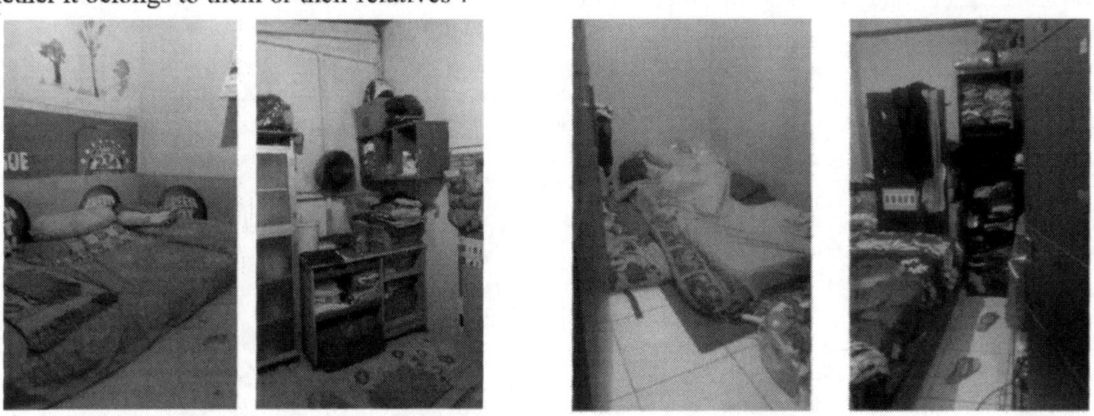

Figure 2. Mrs. RO's (left) and Mrs. YT's (right)

Right pictures in figure 2 illustrate 5 m x 5.6 m apartment unit condition of Mrs. YT. She owns the apartment and has lived here for about ten years with her daughter and 4 grandchildren. Mrs. YT also has a stall on the ground floor that sells fresh veggies. Their activities are pretty much the same, standing by for the whole day in their stall and they only use their apartment unit for sleeping, going to the bathroom, and praying. The rest activities such as eating and gathering with family and guests are performed on the ground floor – in front of their food stalls. They prefer to do all things downstairs due to the bigger space, and more fresh air is available.

Talking about her own preference, Mrs. YT answers simply to live close with her daughters and grandchildren. For specific housing attribute feature, she says nothing but a spacious living room. She explains her strategy to receive her guests on a special day is by opening her door unit and rolling out a mat to the aisle.

Figure 3. Mrs. WT's (left) Mrs. YN's (right)

Next is Mrs. WT's. Her means-end chain schema is quite different, for there is no chain that ends up connected to tradition value. She explains that she is not into tradition kind of things anymore since she decided only to do what is written in the Holy Bible. But, according to the conclusion taken from those four schemas, the value systems are looking similar from one to another that consist of benevolence and security.

Living in a 5 m x 5,6 m sized apartment, those left pictures in figure 3 explain the condition inside the apartment of Mrs. WT. She lives here with her husband, while her children live in her own house in Bintaro. She rents the apartment here to cut time and transportation cost because she has a food stall on the ground floor of the apartment and works there daily. She also tells the reason to rent this apartment is her preference to live close to her workplace. Mrs. WT also shares her preference for a house with a spacious living room so she can easily arrange a family gathering and any kind of celebration. However, living in a small sized apartment is no big deal for her, because she has her own house in Bintaro already – which can be an alternative place for holding any celebration.

The last one is Mrs. YN who lives in her own house of 7m x 11m. She has lived here with two daughters for the last ten years. As captured on the right side of figure 3, her house already has physical boundaries for a guest room, a living room, a dining room, and two bedrooms. She tells that her preference has already fulfilled for living next to her immediate family members: her uncle and brother. Mrs. YN is the owner of a food stall in Gateway Apartment, and she comes home almost late in the midnight. At such a time like this, she always asks her sister in law to come by and accompany her daughters until she arrives. Other than the location consideration, the spatial attributes of her house have accommodated every activity they do in terms of size, furniture, and arrangement needed.

IOP Conf. Series: Materials Science and Engineering **523** (2019) 012054 doi:10.1088/1757-899X/523/1/012054

5. Discussion

5.1 Javanese Migrants' Housing Preference and A Spatial Attribute Features

From the interview, can be concluded that their actual preference is to live close to their immediate family members such as parents, children, and siblings. An extended family member will trigger a hesitant feeling – related to their notion as mentioned before. The reason is they can easily ask or give any help to immediate family members due to their natural bond. Specifically asked about the closure to other Javanese, these four respondents agree that it remains no more important, since they believe other ethnics will act good and helpful – for the same root they have as Indonesian. Next preference, the lower priority they have is to live close to their workplace, so they can manage to get home earlier and have more family time. This shows how they put everything else aside so they can manage time together with their beloved ones.

The migrants do not merely consider the location, but they also have a preference for spatial attributes as well as concluded from the interview and the means-end chain method. For example, according to the interview performed to all respondents, they do have a preference for a spacious room that may gather many people inside. Another concern is how to indulge their guests and make them feel comfortable in their houses. The spatial attributes those are used to accommodate this consideration are different among the migrants due to its subjective nature, and it will depend on the owner's perspective. For example, Mrs. YT deems the best way to indulge their guests is by letting them enjoy fresh air and beautiful view on the ground floor of the apartment, while Mrs. YN finds setting her house to relax is the best one. She does it by putting some green vegetation, a spacious windy living room, and even two standing fans, with particular consideration to the well-being of her guests.

5.2 Culture Affects Migrants' Housing Activities by Direct and Indirect Ways

It is proved from the interview and schemas that culture does influence one's housing activities. Even though there is no physical attribute that is revealing Javanese' culture, but in schemas, there are many chains found that led to the tradition value. These come from spatial attribute such as spacious room for family and friends who come and gather to celebrate certain ceremony of big moments in life in their own residence, in order to strictly follow their Javanese tradition. However, one schema showed a difference– Mrs. WT who said that she does not hold the Javanese tradition anymore and only follow her religion. In her schema, no chain is found to be connected to the tradition value. Nevertheless, the composition of these four migrants' values looked similar.

The exact similar values of these four respondents are benevolence and security value. It makes sense how these two values can be upheld as they may be derived from the notion and norms that have been taught for decades. For example, the behavior to care for other people that encourages them to put on mind about their family and friends' wellness, influences their way to serve their family members, as reflected in their consideration on organizing their dwellings. This is also proved by the means-end chain, that reveals many actual consequences of the spatial attributes which are based on the prominence of their beloved guests. A specific example reflected in Mrs. YN's. The dining table was bought especially for serving her guests well, while she and her family used the ordinary kitchen table for daily routines.

All the above also prove the theory that says we have to dismantle culture first into concrete variables before correlating it with housing activities [9]. It can be concluded from the means-end chain method that the values they uphold and the perspective they have, have been influenced by culture, which next can influence housing preference in terms of location, size, spatial attribute features.

5.3 Different Strategies on Mismatch Occurs Depend on Available Resources Migrants Have

This part can be seen by direct observation in the houses and from the interview. These four migrants have different strategies in dealing with the mismatch that occurs between housing preference and choice they get. In the most limited condition as in Mrs. RO's, she limits her relation to friends and neighbors, so she doesn't have to host any friends in her dwelling. On the other side, her strategy to arrange the

Javanese tradition she keeps on upholding is by coming back home to her origin place. She explains how the lack of relatives in Jakarta, budget, and small sized of dwelling has forced her to go back to her hometown. Meanwhile, in her rented room it can be seen clearly how she compresses all primary activities in the available area included sleeping, eating, taking a rest, etc.

Up next, Mrs. YT and Mrs. WT's strategies are to deal with their small sized apartment units. For the mismatch of spatial attributes that Mrs. WT prefers, it is solved by her own house in Bintaro, as it has a more spacious room inside and perfect for holding any family gathering. She also takes advantage of the spacious area in front of the stall to welcome her close guests. Mrs. YT has the same strategy for welcoming her close family and friends. However, for holding the traditional ceremony and gathering, she makes use of the function hall of the apartment. Their dwellings are merely used for sleeping, taking a bath, and praying. The rest activities are done on the ground floor, they even sometimes do the prayer in the stall area.

Mrs. YN's dwelling – is the most spacious area of all respondents, therefore, not only separation between public and private area, but she can also make separate room to welcome formal guests and closed ones. Talking about the mismatch, the location of her house actually fits her preference which is close to her brother. The spatial attribute features inside also fit her preference of a spacious living room to accommodate many people in a family gathering. Not only that, there seemed to be more specific spatial attributes, like side access, a dining room, many more. These are for the comfort and well-being of her family and friends who visit.

6. Conclusion

From this research, it can be concluded that Javanese migrants' housing preference in terms of location is to be close to their direct family members that include parents, children, and siblings. As in spatial attributes, they prefer a spacious room that is wide enough to welcome their family and close relatives. Specific preference to live close to Javanese neighbors is not important anymore. About the strategies for dealing with the mismatch, they do it differently. In the most limited condition, migrant will compress activities to the primary needs and do it in the available space of the dwelling. The allocation of space will also be minimized by merging private and public area. A dwelling that is wide enough allows a physical boundary to separate between the private and the public area. In the best condition, they have specific attribute features as the owner prefers. About the strategy to overcome the limitation especially in terms of size, they use outside spatial attributes available such as extension right in front of the dwelling, a function room, and the restaurant area if they live in an apartment, or it can be far away in their hometown.

7. Acknowledgments

We would like to express gratitude to Directorate of Research and Community Engagement Universitas Indonesia (DRPM UI) for administering the *Hibah Publikasi Internasional Terindeks untuk Tugas Akhir Mahasiswa* (PITTA) 2018 scheme under grant number 2461/UN2.R3.1/HKP.05.00/2018, and Faculty of Engineering Universitas Indonesia (FTUI), Departement of Architecture for their support and commitment in this research. Sincere gratitude is also addressed to four respondents who kindly participate in this research.

8. References

[1] Goebel, Zane. 2013. The Idea of Ethnicity in Indonesia, *Understanding Society*. Paper 71.
[2] Andersen, Hans Skifter. 2015. Explanations for Special Neighbourhood Preferences among Ethnic Minorities, Housing, Theory and Society, **32**:2, 196-217.
[3] Lee, Everett S. 1966. A Theory of Migration, *Demography*, **3** (1) 47-57
[4] Jansen, Sylvia J.T.; Coolen, Henny C.C.H.; Goetgeluk, Roland W. 2011. *The Measurement and Analysis of Housing Preferences and Choices* (Springer) New York

IOP Conf. Series: Materials Science and Engineering **523** (2019) 012054 doi:10.1088/1757-899X/523/1/012054

[5] Rapoport, Amos. 2000. Theory, Culture and Housing, *Housing, Theory and Society*, **17** (4) 145-165.
[6] Koentjaraningrat. 1986. *Pengantar Ilmu Antropologi* (Aksara Baru)
[7] Koentjaraningrat. 1984. *Kebudayaan Jawa.* (Balai Pustaka).
[8] Schwartz, S. H. 2009. Basic Human Values: An Overview, theory, methods and Applications. *Retrieved July, 28,* 2014.
[9] Beamish, Julia O; Goss, Rosemary Carucci; Emmel, JoAnn. 2001. Lifestyle Influences on Housing Preferences, **28**(1&2).

IOP Conf. Series: Materials Science and Engineering **523** (2019) 012055 doi:10.1088/1757-899X/523/1/012055

Wall finishing materials and heritage science in the adaptive reuse of Jakarta heritage buildings

M Rahmadina[1,*], NR Kusuma[1], E Arvanda[1]
[1] Universitas Indonesia, Depok, Indonesia

E-mail: meutia.rahmadina@gmail.com

Abstract. Given its tropical setting, moisture problems are inherent issues found in walls of heritage buildings in Indonesia that require replacement of wall finishing materials with specific properties. This is where heritage science becomes an important approach in the creation of plasters and paints that best resemble its original and/or provide better performance through innovations. Heritage science possesses great potential in improving evidence-led conservation work in Indonesia by incorporating properties of traditional materials with present technology for cultural sustainability that is gaining urgency in the midst of burgeoning population growth. However, despite its growing application for historically significant heritage buildings, the need for preliminary scientific testing is not commonly acknowledged and applied in the adaptive reuse of small-scale heritage buildings in Indonesia. Therefore, this paper studies two UNESCO-funded pilot projects in Jakarta and tracks the role of heritage science throughout their conversion, and aims to evaluate the constraints, by conducting literature studies and eight interviews with various stakeholders involved. It is found that heritage science is yet a key player in Indonesian conservation works, however, it helps uncover material properties of wall finishing materials that best tackle moisture problems in Jakarta heritage buildings.

1. Introduction

Yung and Chan [1] stated that the adaptive reuse of heritage buildings is part of sustainable urban regeneration that involves the revitalization of otherwise decayed and obsolete buildings for better living environments with environmental, social, and economical benefits. Hence many heritage buildings are found to be converted to commercial buildings by small-scale developments to sustain its life in the long run.

2. Moisture Problems in Heritage Buildings

However, Smith [2] explained that moisture is one of the main sources of problems found in heritage buildings, especially in the tropics such as Jakarta. According to Bertoldi [3] despite being historical artifact themselves, wall finishing materials—such as plasters and paints—need to be replaced to protect the damaged building from further deterioration. According to Ayurdhawan et al. [4], replacement is important as moisture problems can cause damages on the walls as indicated by the presence of loose/flaking paints/plasters, efflorescence, the disintegration of bricks and renders, cracks, rising damp damages, and plant, fungal, and algae invasions.

*Corresponding author: meutia.rahmadina@gmail.com.

IOP Conf. Series: Materials Science and Engineering **523** (2019) 012055 doi:10.1088/1757-899X/523/1/012055

3. Heritage Science

This science is where heritage science serves to bridge the needs of both transforming heritage buildings into commercial spaces whilst protecting them from further moisture damage without losing its value as a historical object. Heritage science was first introduced in 2006 in the UK and is a discipline that combines scientific research and the conservation of heritage buildings. Strlic [5] stated that Heritage Science provides evidence for sustainable conservation and improves the wellbeing of communities surrounding the heritage sites. Brokerhof [6] elaborated that in conservation works, science plays an important role in uncovering traditional knowledge of the past using present technological advances to create new materials innovation for a better future. Heritage science fulfills its role as scientific input can enhance the process of replacing existing damaged materials with material properties bearing the closest resemblance to its original [7]. Based on the author's observation, although frequently applied in practice for monumental heritage buildings, heritage science is a relatively untapped discipline in the commercial adaptive reuse of small-scale heritage buildings in Indonesia [9]. The urgency to scientifically evaluate the composition of existing wall finishing materials and the environmental context of the building to determine compatible replacements are often overlooked due to lack of awareness and financial constraints. On that account, this paper aims to analyse the application of heritage science specific in two small-scale heritage buildings in Indonesia, identify challenges faced by the projects, and track the role of heritage science through a specific case study on the conversion of small-scale heritage buildings in Jakarta into commercial spaces, namely Kedai Seni Djakarte and Historia Café.

4. Research Methodology

The study in this paper began with a literature review to identify the sources of moisture problems in heritage walls, subsequent compatible materials needed to replace deteriorated ones. Based on these theories, a case study was conducted to examine the role of heritage science, using two small-scale adaptive reuse projects in Jakarta. Lastly, a total of eight interviews were conducted to two private building owners, two heritage architects, one heritage expert from a local NGO, one contractor, and two suppliers of breathable plasters and paints involved in both projects. The purpose of the interviews was to gain insights on the application of heritage science in conservation works of both case studies.

Two UNESCO-funded pilot projects are chosen as they best represent the preliminary step taken by the Indonesian government to forge collaboration between private heritage building owners, heritage experts, and NGOs in adaptive reuse projects and made the online documentation accessible to the public. Case Study 1, Kedai Seni Djakarte (two-storeyed building occupying 320sqm) and Case Study 2, Historia café (two-storeyed building occupying 400 sqm) were built in 1913 as part of a larger office complex Batavia Zee en Brand Assurantie Mij by architect P.A. Moojen and 1914 for a trading company Matschappij voor Uitvoer en Comisiehandel (MUCH) by architect Eduard Cuypers company respectively. Both buildings are classified as Heritage Building Grade A—buildings with high cultural significance—as regulated by the Indonesian Provincial Regulation No.9 of the year 1999 regarding the Conservation and Reuse of Built Heritage.

5. Summary of Outcomes

5.1. Causes of Moisture Problems in Heritage Buildings Walls

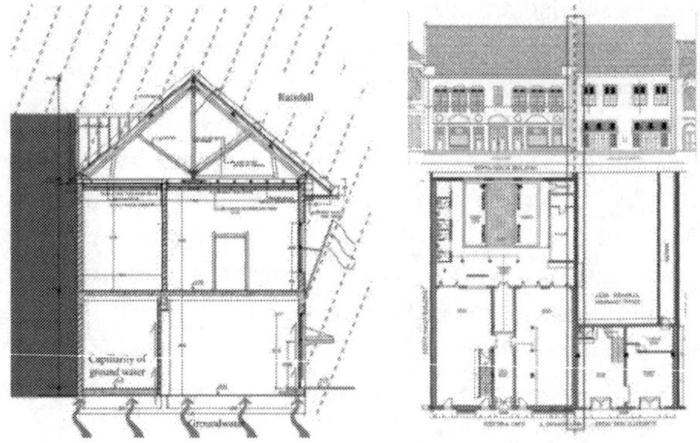

Figure 1. Sources of Moisture in Kedai Seni Djakarte and Historia Café

In the case studies, it was found that water penetration occurred through various sources; through capillarity of ground water, exposure to and leakage of rainfall, and the capillarity of a flooded gap between the two adjacent buildings. In both case studies, leakage was caused by long-term obsolescence which caused the roof to be in a severely deteriorated condition. The existence of a gap with 10cm in length between both buildings was discovered further exacerbate this condition as the top of this gap was not covered with any flashing which periodically caused prolonged flooding and increased water penetration to both buildings. Twenty years of abandonment resulted in both masonry walls containing high humidity.

5.2. Material Properties of Walls Finishing Materials: Plasters and Paints

5.2.1. Breathable Plasters. Reichel, A., Hochberg, A., & Köpke, C. [8] stated that classic plasters of the past are made of quicklime as it possesses good breathability as indicated by its porosity, elasticity, and velocity of capillarity. As members of a local NGO named Pusat Dokumentasi Arsitektur (Center for Architectural Documentation), Abieta, et al. [9] released a conservation booklet guide stating that replacement of plasters must be based on scientific research on its composition to enable the replacement plasters to contain similar properties.

Figure 2. Experimentation and Mixture of New Plaster on Site

Figure 2 shows the experimentation done on site for the composition of plasters used in both buildings. Grounded red brick, lime powder, Portland cement, and sand were bought from a local supplier and experimented manually which resulted in the discovery of its ratio that equals to 2:1:1:2. Interviews conducted with the architects, contractor, and materials suppliers involved revealed that the ratio was a result of a combination and modification of two sources: existing data sheet of plaster used in another heritage building (see table 1), and the empirical knowledge passed down from one construction workers to the other. The purpose was to imitate the plaster bearing the closest resemblance to the original while ensuring its breathability as indicated by the high value of liquid water permeability. High permeability allows moisture inside humid walls to evaporate from the surfaces [2].

Table 1. Data Sheet for Plasters Sampled from Bank Indonesia (2009) with 2: 1 ratio for Sand: Lime Powder

	Material Properties	Results
Physical Aspects	Density	1.5 g/m3
	Porosity	33.2%
Composition	Sand	65%
	Lime Powder	32%
Capillarity		0.256 gr/minute

5.2.2. Breathable Paints.

Three types of paint were used in both case studies, two of which are classified as breathable paints. It was found that acrylic paint was used for the first floor of both case studies in 2012 as private building owners were oblivious towards the inherent moisture problems of the walls. However, following the pilot project in 2015-2016 Silicate Paint and Silicone Resin Emulsion Paint (SREP) were used for the adaptive reuse of both floors on Case Study 1 and only on the second floor on Case Study 2. These two paints were the most frequently purchased in the local market for conservation projects across Indonesia. The former is silicate mineral paint produced by the German company, KEIM, which is regarded as high-quality breathable paints albeit expensive—while the former is silicone resin emulsion paint (SREP) produced only within the local market under the name PT. Archielite Prima Konstruksi.

Table 2. Data Sheet for KEIM Mineral Silicate Paint

Material Properties	Results
pH	11
Liquid Water Permeability	0.12 kg/m^{2}h0,5

On the other hand, the Silicone Resin Emulsion Paint is only distributed within the local market. It is a local home-produced paint which had been used for conservation purposes of several heritage buildings across Indonesia. Due to its local manufacture, the price of this paint is much lower than that of its international counterpart. Hence SREP becomes the second popular choice for authorities working in the field of conservation in Indonesia. Table 3 below is the results of a laboratory test to scientifically examine its material properties.

IOP Conf. Series: Materials Science and Engineering **523** (2019) 012055 doi:10.1088/1757-899X/523/1/012055

Table 3. Data Sheet for Silicone Resin Emulsion Paint (SREP)

Material Properties	Results
pH	9.70
Porosity	3.8%
Capillarity on Bligon	0.016 gram/minute
Capillarity on Brick Wall	0.216 gram/minute

From the tables above, it can be concluded that both types of paint share similar material properties which make them breathable, which are the high pH value and water vapor transmission rate. High pH value indicates the ability of plasters to prevent fungal and bacterial growth that burgeons in an acidic environment. Water vapor transmission rate is the velocity in which water sourced from moisture can vaporize from within the historic walls. All these properties demonstrate the breathability of the paint to withstand further moisture damage. From the data above it can be concluded that both silicate and silicone resin emulsion paint are breathable hence compatible for very humid walls.

5.2.3 Physical Inspection of Current Walls Condition. After having analyzed the substances of plasters and paints used, a physical inspection was conducted by the authors to determine the compatibility of plasters and paints used (See figure 3). In Case Study 2, the first floor (areas painted with acrylic paint) experienced significant damage than the second floor (painted with SREP). The walls with murals painted in acrylic paint experienced the most significant damage as indicated by the presence of severe flaking of paints, efflorescence, disintegration of the bricks (figure 3). This painting demonstrates the non-breathability of the paint which exacerbates the damage. Meanwhile in Case Study 1, although both first floor (painted with silicate mineral paint) and second floor (painted with SREP) experienced overall lesser damage than Case Study 2, cracks and efflorescence can be found in several areas in the first floor due to the existing severe case of rising damp.

Figure 3. Damages observed in Case Study 1 (left) and Case Study 2 (right)

5.2.4 Challenges of Heritage Science Implementation in Jakarta Commercial Adaptive Reuse

5.2.4.1 Lack of public awareness regarding moisture problems and the importance of involving Heritage Science. By collaborating with NGOs and heritage consultants, private building owners were able to acquire vital insights on adequate wall treatment they otherwise would be unaware of. In both case studies, private building owners were initially unaware of the high humidity of the building walls which can be found by the false selection of paints initially. However, after the collaboration with heritage experts during the pilot project, the right decisions are made in terms of breathable wall finishing materials used.

5.2.4.2 Limited financial, technological, and human resources for holistic preliminary scientific research. Kennedy [7] elaborated that Non-destructive testing includes the means of portable laboratory equipment such as near-infrared spectroscopy (NIR), X-ray fluorescence (XRF), hyperspectral imaging, and et cetera. Micro destructive testing, the gathering of very small samples in specific areas, was conducted in both case studies since the advanced equipment for non-destructive testing on site are not yet available in Indonesia as they are very expensive and require qualified human resources to operate them. In both case studies, data regarding the extended context of the site was not analyzed in detail. Measurements were only limited to humidity levels of the masonry brick walls indicated by a small device by the paint supplier that only indicates a high level of humidity only when it is beyond 4.

5.2.4.3 The omission of technical data sheets of wall finishing materials used. Both projects are pilot projects that are well documented and made accessible online at lestarikanbangunantua.org in the form of pdf file. The public can gain an understanding of the conservation process comprehensively through the conservation plan. However, it is found that it did not include extended information on the history of interventions applied to the building, as information is only limited to literature-based studies and highly summarized description of repairs and other changes made to the building.

5.2.4.4 Non-scientific-based knowledge transfer regarding the composition of new plaster. The plaster ratio found to be breathable on two projects based on modification through direct experimentation on site and past scientific examination. Hence composition of plasters is often loosely based on construction workers' empirical knowledge, which should have been scientifically strengthened and modified through the involvement of heritage science.

6. Concluding Observations

It is found that the material properties of wall finishing materials that best combat inherent moisture problems in walls of heritage buildings are those that indicate their high breathability. Breathable plasters and paints are indicated by (1) high pH value and (2) high water vapor transmission rate and capillarity. The plaster ratio was formulated through manual experimentation which was found to be 2:1:1:2 for red brick powder : lime powder: Portland cement : sand. Silicate Mineral Paint and Silicone Resin Emulsion Paint are both proven to be breathable which can be observed by the minimum damage experienced by the surfaces painted. From this study, it is concluded that heritage science is still a very rudimentary discipline in Indonesia when implemented for small-scale adaptive reuse projects for two commercial spaces. Challenges of implementing Heritage Science in the case studies are observed to be (1) lack of public awareness regarding moisture problems and the importance of involving heritage science; (2) limited financial, technological and human resources for holistic preliminary scientific research on existing materials; (3) omission of technical data sheets of wall finishing materials used; (4) non scientific-based knowledge transfer regarding composition of new plaster. However since Heritage Science is already used in conservation projects across Indonesia, it can be potentially developed and integrated into small-scale adaptive reuse projects.

7. Acknowledgments

The work described in this paper was fully supported through a research grant by Universitas Indonesia. The author warmly thanks those who have participated in the interviews.

8. References

[1] Yung E H and Chan EH 2012 *Habitat International* **36** 352-361.

[2] Smith BM 1984 Moisture Problems in Historic Masonry Walls: Diagnosis and Treatment. Washington, USA: (United States Government Printing).

[3] Bertoldi M 2013 *Colours and Materials of Historical Centers: Sustainable Conservation and Management*. Retrieved from http://hdl.handle.net/10589/74128. (Accessed 25 June 2018).

[4] Aryudhawan et. al (n.d.) *Caring for your Heritage Building: Heritage Building Owner's Information*. Available from: https://drive.google.com/file/d/0B-cIhi8pBoccWDNpaVITUGhoYXM/view (Accessed June 20, 2018).

[5] Strlic M 2015 A Brief Theory of Heritage Science (London: National Heritage Science Forum). Retrieved from https://nationalheritagescienceforum.wordpress.com/2015/04/16/a-brief-theory-of-heritage-science-byprofessor-matija-strlic-ucl-institute-for-sustainableheritage (accessed 20 June 2018).

[6] Brokerhof AW 2015 How Can Science Connect with and Contribute to Conservation? Recommendations and Reflections, *Studies in Conservation*, **7**-13.

[7] Kennedy CJ 2015 The Role of Heritage Science in Conservation Philosophy and Practice, The Historic Environment: *Policy & Practice*, **6**:3, 214-228.

[8] Reichel A, Hochberg A and Köpke C 2004 Plaster, Render, Paint and Coatings. Munich, (Germany: Birkhäuser).

[9] Abieta A, et. al. 2011. Pengantar Panduan Konservasi Bangunan Bersejarah Masa Kolonial. Jakarta, Indonesia: Pusat Dokumentasi Arsitektur (in Bahasa).

IOP Conf. Series: Materials Science and Engineering **523** (2019) 012056 doi:10.1088/1757-899X/523/1/012056

Shifting from place to non-place: A case study on the central market of Banda Aceh

F I Qismullah[1,*]
[1] Lecturer, Architecture Department, Faculty of Science and Technology, State Islamic University Ar-Raniry, Jalan Ibnu Sina, No. 2, Darussalam, Syiah Kuala, Kopelma Darussalam, Syiah Kuala, Kota Banda Aceh, Indonesia

E-mail: fitri.qismullah@ar-raniry.ac.id

Abstract. The transformation of Banda Aceh, Indonesia, through the phases of political war and tsunami disaster leads to the alteration of the central urban areas. The impact of the temporary agglomeration is on the functions of public spaces. There has been a paradigmatic shift of culture transformation in this city from a place of local tradition to a multicultural environment opened to change. The complexity of this transformation was especially influential to its central market. By using the two-sided theories and framework of Jacobs and Alexander, and Auge and Koolhas, this study analyses the historical evolution and perspectives of the central market for urban space. Data collection was from field observations and interviews with the vendors and customers in the market. The results imply that without disregarding the importance of heritage and history, stakeholders that have the power to change the central market must be realistic and adjust the condition according to the current people's needs as users of this place. Nevertheless, despite some of the vendors' rejections towards the development of modern markets in the city, this realism of modern markets has been able to solve the problem of other cities by the concept of one-stop shopping for time efficiency and goods availability.

1. Introduction

Much of the recent interest in urban design has focused on the creation and management of the place. This concept of place has been the subject of debate, offered in two different positions of thought by a significant number of theories. It originates from the 1950-60s when the rise of modernist concepts takes place in planning and destroying earlier urban structures. The view of most theorists from this era holds on to the conservative notion of traditional place making for an urban living, accentuating to the reconstruction of the pre-modern textures, which is thought as idealistic and utopian. Although encouragingly attractive and promising, not all of these notions appear to be feasible and acceptable in the current trend of cities, as the needs of the society changes accordingly. In opposition, from the 1980s onwards, new concepts of openness in places have been embraced, adopted from the current needs and conditions of cities. These concepts corroborate the chaotic modern cities with creations of the fresh novel and liberated ideas. Urbanisation as the driving force of this second position is motivated by the increase of urban population in the world.

*Corresponding author: fitri.qismullah@ar-raniry.ac.id.

It has been clear for some time that both the professions of planning and design are concerned with the response of urban development to the new set of this global demographic. This new change has influenced the social, economic, cultural and political forces of cities and is considered as the cause for 'non-places'. Meanwhile, these conditions have precipitated some important transformation of places in cities: the creation of generic cities and loss of identities. This transformation is being reproduced in space through building designs and the urban structure and is mostly seen in the image of Asian countries. Current cities in the world are in a struggle in finding their identity through the inevitable influence of modernization. As such, the discussion on the place is integral to an understanding of the future transformation of cities.

The transformation of Banda Aceh, Indonesia, through the phases of political war (from 1973 to 2005) and tsunami disaster (December 26, 2004) leads to the alteration of the central urban areas. The impact of the temporary agglomeration is on the functions of public spaces. There has been a paradigmatic shift of culture transformation in this city from a place of local tradition to a multicultural environment opened to change. The complexity of this transformation was especially influential to its Central Market or well known by the city as *Pasar Aceh*. This market has gone through a transformation in which intrigues the researcher to conduct a study on how its restoration after the tsunami grasp the concept of place and non-place. The research question formulated is, "What are the perspectives of vendors and customers as users of the public space in relation to the concept of place and non-place on the Central Market in Banda Aceh?" The findings of this study demonstrate a variety of perspectives that can be adopted for future development and design of this market which is beneficial to its vendors and customers as the primary users of this public space.

2. Literature Review

This study discusses and focuses on different views to understanding the concept of place and non-place. The main analysis of place focuses on Jane Jacob's *Death, and Life of Great American Cities* and Christopher Alexander's *A City is Not a Tree*, while further analysis of non-place is discussed through Marc Auge's *Non-Places: Introduction to an Anthropology of Supermodernity* and Rem Koolhaas's *The Generic City*. Each offers varied understandings of the issues as they approach the discourse of 'place' from different perspectives. The oldest of these four texts is Jacob's work to the newest writings of Koolhaas. Therefore, it is important to consider that each author is writing in a different period with different backgrounds influencing the thoughts and ideas. What each and every author testifies is that there has been a historical evolution of the design for urban space and demonstrates a variety of perspectives that can be adopted.

2.1. Jane Jacobs

As a journalist with a close interest in city life and urban structure, Jacobs did not come from a professional world of architecture or planning but stands from a neighborhood activist's view. Two important works of Jacobs include *Death and life of Great American Cities* in 1961 followed by *The Economy of Cities* in 1969. From close observations of cities, her basic notion draws around the factors for the destruction of liveable cities. These ideas were best presented in her most influential piece, *Death and Life of Great American Cities*, which was the first leading attempt to attack modernist planning. The book presents an intriguing perspective on planning and urban design, scathing to argue that the modern tradition brought from the ideas of Le Corbusier had obliterated the former social cohesiveness of crowded and diversified neighborhoods with the development of mega blocks and skyscrapers. The anecdote of urban life permeates throughout Jacob's second chapter *'The Uses of Sidewalks: Safety'* and chapter eleven, *'The Need for Concentration,'* which helps connect with her central theme of vitality and diversity.

In the second paragraph of her second chapter, Jacobs constructs the image of a city. For Jacobs, the image of the city is seen by the full ingredients of the street and their sidewalk as the vital organs of a city. This image was drawn based on her personal experience of living in Boston. As she writes, "Think of a city and what comes to mind? It's streets. If the city's streets look interesting, the city looks interesting; if they look dull, the city looks dull" [1]. She agrees to the idea that city streets should be full of people, although it might consist of strangers. These strangers that are believed as menacing by many people are reversibly thought to be acting as polices, delivering voluntary controls for passive surveillance in public space. However, her argument is crucial in pointing out that a successful city lies in people's feeling of safety and security among these strangers [1]. Jacobs's rhetoric 'eyes upon the street' from the enjoyment of the natural proprietors of the street to watch other people may insure safety on the streets.

Jacobs's chapter *'The Need for Concentration'* appears to be most significant for encouraging density and diversity of cities. Her observations suggest that we need all kinds of diversity to construct a decent city, mixing in a mutualism manner for the development of society and civilization in the city [1]. She believes the essence of urban life originates from the energy of diversity of people and range choices of activities at different times. Her claim for mix uses, however, contradicts the ideas for zoning that most planning had been based. She encourages the mixed zoning combined with the need for concentration of people by her statement "Everyone is aware that tremendous numbers of people concentrate in city downtowns and that, if they did not, there would be no downtown to amount to anything – certainly not one with much downtown diversity" [1]. Jacobs argues that low semi-suburb densities are not a good long-term plan. She quotes "As the city continues to grow, the character that makes these semi-suburbs reasonably attractive and functional is lost…" [1].

Although agreeing to the need for concentration, Jacobs counter-argues for the need to house people in high-rise buildings, which can kill diversity by obstruction [1]. She believes that densities of twenty dwellings to the acre and above are unacceptable as they will always stay as strangers [1]. It was regarded as problems of the cities as echoed in her statement "…between the point where semi-suburban character and function are lost, and the point at which lively diversity and public life can arise, lies a range of big-city densities that I shall call 'in-between' densities. They are fit neither for suburban life nor for city life. They are fit, generally, for nothing but trouble" [1]. She continues that this was especially due to the standardization of the buildings created, such as in most functional designs of Le Corbusier [1]. Her statement "Variations from the past are a foundation to which new variations of the present (and eventually the future) are added" [1], shows her preference for traditional city solutions, that the combination of numerous lively streets, parks, residential and non-residential uses can create a vibrant city to the lifeless and unremarkable high densities and ground coverage of the modern era [1].

Herein, Jacobs is building her critique of modernization whereby the inert image of the modern city is not considered as a liveable place. However, with the increasing population of the world, it is impossible to control and guarantee such an idea of voluntary policing in attempting to spread buildings to low densities for more active liveable streets. The conditions of today's cities are inevitably changing according to human needs. Jacob's argument for the resident's participation as a 'street ballet' is a fantasy in today's everyday life and might only be applicable to her hometown, Hudson Street, Greenwich Village. Her promotion of Greenwich Village as the ideal environment for urban living has drawn the attention of many designers to further investigate the true meaning of a 'liveable' city.

In addition, how applicable are these theories of place from Jacobs that can be applied to such chaotic countries in the East? Culture also plays an important role in the development of cities, as it influences people's behavior in public space. The notion that safety from the number of stores and people in public space is hard to measure nowadays, as individual living does not guarantee social interactions. Nowadays, streets can no more be the measure of a liveable city. What we see and imagine of cities, either through the presentation on post cards, magazines and most of our imagination, cities consist of skyscrapers captured from the view of a bird's eye. The taller and denser the buildings are, the likelier they will be regarded as a busy and vibrant city living.

IOP Conf. Series: Materials Science and Engineering **523** (2019) 012056 doi:10.1088/1757-899X/523/1/012056

2.2. Christopher Alexander

Drawing from Jane Jacobs, Christopher Alexander has written an elegant essay entitled *A City is Not a Tree*. Alexander has a similar conservatism view with Jacobs's fascination for traditional mixed and chaotic urban neighborhood as a 'coherent whole'. He claims "All traditional towns have these features in their growth. But the modern practice of urban development does not have these features. It does not deal with growing wholes at all" [2]. However, he considered Jacobs's solution as unreal in imagining the modern city as some kind of mixture between Greenwich Village and an Italian hill town [3]. Differently to other authors for the concept of 'place,' Alexander discovers the concept of the city in the technique of mathematical relations. Patterns of historical towns are structured in the form of rationale relations of the city's element. Good urban structure of the city is counted based on abstract concepts, as abstract as his essay title in which the tree does not have green leaves but is a set of theories of the city structure [3]. As stated by Alexander: "A set is a collection of elements which for some reason we think of as belonging together", which then the set of elements are called as a system [3].

Alexander terms unplanned cities, which grows spontaneously as natural cities organized in a semi-lattice structure; and planned cities by the professionals as artificial cities organized as a tree structure. He considers the attempt of modernity to create the artificial cities as greatly influential but unsuccessful in terms of creating a 'place' [3]. This is because most artificial cities have not been able to find the basic principle of how cities in the past have worked [3]. Fixed and workable parts of the city system are believed to be unable to be controlled by city planners and designers. The system consists of elements labeled through sets of numbers, creating sub-sets when overlapped or disjoint [3]. Alexander clearly explains in a mathematical voice: "A collection of sets forms a semi-lattice if and only if, when two overlapping sets belong to the collection, then the set of elements common to both also belongs to the collection" [3]. In this axiom, the sub-sets overlap and belong to each other like units. Differently, the tree structure is formed when "A collection of sets form a tree if and only if, for any two sets that belong to the collection, either one is wholly contained in the other, or else they are wholly disjoint" [3]. This condition is more restrictive as each sub-set has its own path separating further at the end.

It is clear that the semi-lattice is a more complex structure than a tree, which leads to separations. This can be seen in the modern planning of zoning in western cities, forming the tree structure. Alexander proposes the mix and overlapping of the semi-lattice as the acceptable form as it tends to grow instead of thought out as in tree structures [3]. Alexander views Le Corbusier and Louis Khan's ideas for identity as tree-like concepts, for example, from their separation of pedestrians with vehicles [3]. Designers are thought to be limited in their capacity to imagine and create the complexity of a semi-lattice structure. However, in the modern society today, the push for development has been stronger than ever in the past. Justification to stop planning the city and accept its natural growth will only create more chaos affecting the economy and social patterns of the city. Planning and building the city might be considered a tree-like approach, but it is the little bits of everyday life in these planned cities that we can then wait to be formed and flourished naturally as the small overlapping sub-sets of the city.

These semi-lattice systems are most significantly proposed through Alexander's book *A Pattern Language* (1977), which was the continuation of the first volume, *The Timeless Way of Building* [4]. His philosophical 253 patterns are presented in the book as guidelines to the solution of problems most cities face today in its design. These patterns are perceived to be the archetypal core of all possible pattern languages, which according to Alexander are the basic ingredients that are "deeply rooted in the nature of things, that it seems likely that they will be part of human nature, and human action, as much in five hundred years, as they are today" [4]. The notion that modern planning is strongly mistaken and 'inhuman' was expressed in his statement that "the languages which people have today are so brutal, and so fragmented, that most people no longer have any language to speak at all – and what they do have is not based on human, or natural considerations" [4]. However, these guidelines contradict to his idea for unplanned cities, as the patterns are used as the basic ideas for planning and

IOP Conf. Series: Materials Science and Engineering **523** (2019) 012056 doi:10.1088/1757-899X/523/1/012056

design. It also raises the question of whether future needs will be the same as the past needs from five hundred years before had been significantly different and some irrelevant and inapplicable for today's design. It is impossible to recreate the overlaps of the historical towns, as it will no longer be suitable for today's concept of 'place.'

2.3. Marc Auge
Marc Auge, clearly described through the title, *Non-places: An Introduction to an Anthropology of Supermodernity* (1992), forms an analysis of current trends leading to the feeling of 'non-place.' Auge's theory of place through this piece investigates the characteristic and elements of non-place. Auge's support to the realism of non-places today is not only relevant to western countries but is the phenomenon, which has already been mushrooming in Asian cities. In opposition to the phenomenology of place, Auge argues that this non-place was created by the driving force of modernization through the former process of pre-modern and modernity. Auge claims that the rise of modernity did not eradicate the essence of ancient places, but instead it is a milieu that represents the survival of an element of the past through the continuing change of the cities [5]. This gives the expression that the essence of modernity indicates movement and alteration. Nevertheless, Auge differentiates the movement concept of modernity and supermodernity following after. Modernity is considered as the continuous temporality of traditional ancient places that was finely expressed through words of the modern poetry, evoking the memory of ancient places [5]. Supermodernity instead is the newest emergence of trends that supersedes the above modernity by covering those historical elements of the ancient places, positioning it as only 'places of memory' [5]. This newness of non-historical places with no concern with identity is what Auge terms a 'non-place', which was created by this condition of supermodernity [5].

This 'non-place' can be characterized in our everyday life nowadays by the so-called 'inhuman conditions' of transit points and temporary living, a dense network of transportation, routines of shopping with silent transactions, all produced with no involvement of human interactions [5]. The main point is that supermodernity is recognized by fast in movement, always changing its structure, as Auge says "It never exists in pure form; places reconstitute themselves in it' [5]. The place will always have a part, which is memorized while non-place will always be continuingly changing. Auge sees these two concepts of place as always mixing with each other and layered in each period [5]. The only measure for today's cities is the existence of transit, interchange, crossroads, passenger, traveler, the housing estate, monument and communication [5], all of which represents the world of strangers that can survive without the need for any deep interactions. The main characteristics of supermodernity are an overabundance of events, space and the world of individuality [5].

Thus, he writes, "What is seen by the spectator of modernity is the interweaving of old and new. Supermodernity, though, makes the old (history) into a specific spectacle, as it does with all exoticism and all local particularity" [5]. Auge's view for supermodernity seems to be realistic, as "non-places are the real measure of our time" [5]. This can be seen from a strong advertisement for branding, which has created familiarity in every city in the world [5], especially in Southeast Asia. Asian countries call for these branding for the sake of their improvement towards being regarded as a developed country. As a result, almost all cities in Asia have the same similarity of tall skyscrapers, and modern glass box buildings with entertainment and branding all over the cities. One can only find the originality of these places in the very narrow street of the cities.

2.4. Rem Koolhas
Unlike Jane Jacobs and Christopher Alexander's understanding of place, Koolhaas has a more profound understanding of the built form in contemporary cities. Amongst his many writings: *Delirious New York* (1978), *Singapore Songlines* and many others represent his realistic way of thinking and accepting the modernistic change of today's most Asian cities. Perhaps, most significantly, his profound interest in Asian cities has made him discover the standardization and generalization of modernity, leading him to focus on the non-place of generic cities. The essay *'The*

Generic City' in his book *S, M, L, XL* (1995), can be considered as the most significant in explaining the phenomena of non-places, which he sees as "a global liberation movement: 'down with character!'" [6]. Similar to Auge, Koolhaas's critique tracks the city dilemmas of abandonment of history, context, identity, and character [6], and it is this issue, which is the central theme of his writing. He writes, "The stronger the identity, the more it imprisons, the more it resists expansion, interpretation, renewal, contradiction" [6]. Koolhaas attempts to explain that this identity may constrain future development.

Much like Auge, Koolhaas argues that the city is part of spatial layers rendering the center inward towards the core of the earth [6]. This city without identity reflects present needs and ability to accommodate anyone, 'it is superficial' in a sense that it can produce any kind of identity and easily change it to something new [6]. This reaffirms his realistic view of today's cities, most particular among the pattern of development in Asian countries [6], where development is fast changing, incorporating different layers of the past even if they are not 'acknowledged'. Thus, Koolhaas heralds the birth of generic cities, which is characterized by 'few and far between emotions' [6] only accommodates necessary movement, of cars in particular, with the sense of speed and transitions of strangers in airports [6]. It is best defined as a 'freestyle' consisting out of three elements of roads, buildings, and nature flexibly changing occasionally at any city [6]. Planning is considered irrelevant, as it makes no difference to place anything, anywhere [6]. It is completely free with no need for context or reasoning, proving that the main point is that the system is successfully operating.

Two other essays from this same book, *S, M, L, XL,* are *'Bigness or the Problem of Large'* (1994) and *'What Ever Happened to Urbanism'* (1994), which also represents his forward thinking and realistic ideas for urban design. Koolhaas supports for modernity from the driving force of new technologies and higher demographic numbers, which has erased the old context of places delivering a new form of 'bigness' with no certain identity as the way leading towards modernization. He critiques the western theories as unsuccessful in dealing with the issue of urbanization and population growth and sees that we are now stuck in the crater of modernity [7]. This means that he is imposing that there is no reverse and therefore the only solution is to be involved. The force to accept modernization has to be uncritical and opened to change. This is particularly found in Asian countries, as they tend to accept change easily.

In comparison with the previous texts, Koolhaas's perspective provided a greater understanding of the real urban structure and architectural form. The emphasis on the generic city, which is the key to spatial form, is the highlight of his discussion. However, what sets Koolhaas's argument apart from others is his assessment of the features of the Generic City, which is more beautiful [6]. Thus, this is a key element to the creation of future cities without the need for history and identity. However, for Koolhaas, the city has not been completely removed from the past. The generic city once had memories of the city that still exist but has been generalized [6]. He writes of history as being paradoxical because it reinforces change and improvement, "Cultures flourish, decay, revive, disappear, are sacked, invaded, humiliated, raped, triumph, are reborn, have golden ages, fall suddenly silent – all on the same site" [6]. This is an important point because it would be incorrect to view the cities as completely removed from the fabrics of past generations.

3. Methods

This study is a qualitative work that intends to investigate the use of the concept of place and non-place on the Central Market in Banda Aceh, Indonesia. The instruments used to collect data were field observations and interviews. Field observations included visual observation and visual recording; meanwhile questions for the interviews were conducted in the form of semi-structured by asking questions to vendors and customers in the market to obtain information on their perspectives towards the sudden change of the Central Market from a traditional culture of shopping experience to a modern building with air conditioner and escalators.

A number of 5 vendors and 5 customers were interviewed, and they were selected randomly based on their availability and consent to participate in this research. In this paper, they are coded as V1 and so forth for the vendors, and C1 and so forth for the customers. The main questions asked inquired about their thoughts towards the change, what has been or has not been yet facilitated, and how they would like the future market to be. After data were collected, the analysis was done by examining the vendors' and customers' opinions and linking them to the two-sided theories and framework of Jacobs and Alexander, and Auge and Koolhas. The results are reported narratively in the next section.

4. Results and Discussion

4.1. Result

A case study is brought upon in this paper, as a justification of the modern trends penetrating to Asian cities, including in developing countries. Banda Aceh, one of an eastern cities in Indonesia, has been transformed through the phases of political war and tsunami disaster, which had led to the alteration of the central urban areas. The occurrence of the phases begun with the endless political conflict since 1873 when Dutch Colonial declared war followed by unfair redistribution of wealth after Indonesia proclaimed independence in 1945 [8], triggering the Aceh Free Movement rebellions since 1976 [9]. This event has turned the city into a place of fear with a local character. This local character rejected western influence due to past traumas and has influenced slow development in the city. Nevertheless, it has succeeded in preserving its own sense of 'place' and created some sort of its own identity which can only be felt through the memories of customers and past generations of vendors.

In 2004, a 9.2 Richter scale earthquake was followed by a devastating tsunami, further welcoming the overwhelming of media, NGO (Non-Governmental Organisation) and agency attentions that urged a peace agreement in 2005 with foreigners entering Aceh delivering aid [10]. The impact of the temporary agglomeration was on the functions of public space as it tends to create a period of multicultural effect of the globalization. Therefore, there has been a paradigmatic shift of culture transformation in this city from a place of local tradition to a multicultural environment opened to change.

The complexity of this transformation was especially influential for the Central Market of the city, *Pasar Aceh*. *Pasar Aceh* was formerly an unorganized traditional market consisting of wooden shops with zinc roofs and narrow alleys in between directly exposed to the sun with no permanent building. It plays a major role in fulfilling the basic and daily needs of the Acehnese (see Figure 1). One of the customers explained the previous market situation as follows:

C4: "This place (i.e., *Pasar Aceh* or the Central Market) used to be very hot because the ceilings were low and the rooftops were made of zinc. The walls were from wooden boards. If it rained, the floors got muddy and wet. The pathway (between shops) only fitted 2 people to walk (side by side)".

Figure 1. Existing traditional market behind the new *Pasar Aceh* building (source: photographic survey 2010).

IOP Conf. Series: Materials Science and Engineering **523** (2019) 012056 doi:10.1088/1757-899X/523/1/012056

Nevertheless, the 2004 natural disaster washed away the former characteristics of the place, in which opportunities then arose for peace and reconstruction. During this phase, the market developed into a new form of an enclosed air-conditioned building with ATM facilities and security guards for protection at the front (see Figure 2). The new built form is in a modern style, funded by foreign aids. The international donors had imposed modern influences upon the Acehnese and the reconstruction period had provided a boost to the local economy. This has increased the significance of *Pasar Aceh* as it is considered as the business center for the local economy. It is operated till late at night (see Figure 3), which was risky and not allowed during the conflict time. One of the vendors and customers informed that:

V1: "*Pasar Aceh* is more comfortable and convenient now. There are air conditioners to cool us from the hot air. Then there are escalators; it is easier to go from one floor to another. The bathrooms are also better; they are now separated for men and women. Before, it was like one bathroom for everyone".

C5: "Now we don't have to be afraid to go to the market at night anymore. Because the building has many floors and a lot of people are still active in the market until night time".

Figure 2. The new *Pasar Aceh* with air conditioner and escalators with security guards at the front (source: photographic survey 2018).

Hence, the new *Pasar Aceh* is ultimately functioned and crowded both during day and night. The perception for convenient shopping in closed buildings such as malls had established a new culture to the former traditional shopping behavior. The new closed building has also become something of a refuge from the former traditional image of the city.

Nevertheless, despite its conveniences, the customers also find some shortcomings, such as the lack of food court for quite a large central market in the city. Customers seek for places to eat to overcome thirst and hunger after shopping, but they find exhaustion instead since the shops and places to eat are disorganized inside and outside of the main building of *Pasar Aceh*. C1 and C2 explained:

C1: "But the problem is there is no food court that is convenient enough for us to buy and enjoy our food during or after shopping. There are a few stalls on the top floor, but who would want to go there if the food available is not much and not enough place to sit?"

C2: "So I finish shopping with my children on the second floor, and we are tired…if we want to eat some *cendol* (i.e., iced sweet dessert of rice flour jelly, coconut milk, and palm sugar syrup), then we would have to go down to the ground floor, go out of the building, and go to the traditional *cendol* stall next to it. If we want to eat something heavier like noodles, we would have to go to another eating place that is across the street or farther down the road. It is very tiring".

Figure 3. Activities continue until night (source: photographic survey 2018).

Furthermore, despite the encouragement from the customers for a convenient modern shopping experience, rejections also came from vendors inside and outside the new modern building (see Figures 2 and 3). Traditional vendors, who sell in shops outside of the main building but near to it (see Figure 1), rejected the supermarket in the new building of *Pasar Aceh* as it decreases the economy beneficial of many small retailers in the advantage of one owner of the supermarket. Similar in rejection, but different in opinions came from new vendors selling their goods on the third floor of the building. They considered themselves not having the same chance of marketing with other vendors on the first floor where customers would be streamed to purchase before having the chance to wander to the next floor. This is as stated by V2 (traditional vendor) and V3 (new vendor in the building) below:

V2: "The big business is devastating us down here. I have been selling clothes all my life to feed my family from this shop. Now the supermarket on the top floor of *Pasar Aceh* also sells clothes…and many customers prefer to go there because they can shop for food and house supplies at the same time in one place".

V3: "I managed to get this small shop on the third floor. But I envy those on the ground floor. Customers would see their shops and go there first. If they find what they need, they would not want to go up here anymore".

4.2. Discussion

As places are a fundamental base of people's memory, there is always a possibility for the challenge to the changes of existing meanings of a certain place towards a new 'becoming.' The debatable concepts of place will continue to be debatable as people's perception and acceptance vary according to the knowledge and culture they experience. Two different concepts of place and non-place have been presented through the analysis of four texts in the literature review of this paper. There are numerous outcomes from Jane Jacobs's and Christopher Alexander's analysis for the concept of place, which originated from the historical era of the 1960s. They both have a good understanding of how the city works which is integral to truly understand the system of a good urban structure. Both analyses have been a critique of modernist planning and orientate their conception to the traditional city system, which has worked well in the past. However, although undeniably well written and intriguing, these conceptions do not provide in-depth thinking from modern architectural and urban design thinking. In some circumstances, the concepts are inappropriate to be fully implemented as some conditions of current cities have changed inevitably. These local solutions might not even be applicable to cities globally around the world as a culture also plays a significant role in the vitality of cities. Dewi and Rauzi even added culture, heritage and social aspects are among the essentials in the reconstruction process of cities after disasters [11].

Although stressing the need for concentration and density, Jacobs and Alexander admire the drama and ballet played by narrow and historical cities. Unlike Auge and Koolhaas, they oppose the developments of skyscrapers and the characteristic of speed and movement from the modern cities, as this may not encourage social living and interactions. Therefore, while Jacobs and Alexander dislikes standardization and generalization of modernity, Auge and Koolhaas is supporting and proposing for supermodernity of individual living and the generic cities, as these are the 'real measure of our time', and being involved seems to be the only solution, instead of reversing and being left behind to moving and developing countries. In relation to the Central Market in Banda Aceh, customers felt comfortable shopping in an air-conditioned building as the temperature in this city is hot and humid for outdoor shopping experiences. Window shopping during the evening is now also convenient as the increase of activities around the market has provided safety for women and children through passive surveillance. Hence rejections from some of the vendors that still prefer local traditions thus with the urge for modern movements are undeniable. Many traditions have changed significantly, and most of them are in the interest and current needs of customers nowadays. As such, although resisting for modern cities which are considered as 'inhuman,' we must be logical and realistic in facing current issues of the inevitable globalization and modernization bringing in the speed of change. New logical and more realistic theories have emerged, which is considered less human but has been widely accepted by most countries in the world, especially in Southeast Asia. This can be seen through the fast development of the Central Market in Banda Aceh, turning it into chaos, but surprisingly, operating successfully.

Urbanization and globalization in Asian countries have brought influence to the development of the city. Over the entire characteristic of today's cities, it is undeniable that it has successfully worked in Asian countries and continues to be developed. There must be a certain reason that this concept is widely accepted and is highly competing for high-rise and other modern development. This can be seen from the case example of the Central Market, as modern movements are closely related to the transition to advanced capitalism. This, I argue, is an attempt to catch up with the speeding movement of modernization. Although it is still uncertain as to what level of modernity it enters because it might range differently from cities to cities. For it has to be clear, that the past pre-modern condition has once been considered as modern to the ancient times. Thus, there is no distinct definition to modernity, but it is the improvement and development that can be measured. There should be no more rhetoric for 'living inside the box' or 'glass house' as the technology of buildings might make it impossible for us to live inside a ball or an egged shaped building. Perhaps in the future people would be able to live and move freely in the floating air. Then where does that leave the term modern, too? Would that lead supermodernity be a historical term too? Perhaps we would need to find another term that suite best with that certain condition in the future. Thus, it would then be too far to reverse cities to the ancient traditional concept of place, as it is now too far for use to reverse to the city living of the ancient cities, there also needs to be a reborn for new creative concepts. Therefore, the main argument put forward is that without disregarding the importance of heritage and history, we must be realistic and adjust our condition according to the current human needs as they are the users of these spaces.

Nevertheless, despite some of the vendors' rejections towards the development of the modern market in the Central Market, Banda Aceh, this realism of modern markets have been able to solve the problem of other cities by the concept of one-stop shopping for time efficiency and goods availability, such as urban sprawl through urbanism. It is to be stressed that the features of modernism are prevailing and pragmatic, highly accepted in Asian countries. It is not a world of dream or fantasy. It is the reality that we must be based upon.

5. Conclusion

The results showed that the process of modern movement in the Central Market in Banda Aceh was encouraged by the increasing welfare condition of the Acehnese during the reconstruction phase, pushing people for further improvement. It can be concluded that the main drive of this change is the economic capacity brought from the entrance of globalization and introduction of modernization to the city. Despite a few flaws, the Central Market in Banda Aceh indicated a discrete desire, as the modern built form is considered successful and acceptable from the crowd of its use. This influence of seduction for the perception of convenience shopping in malls was certainly a form of manipulation of interest and desires from foreigners during the reconstruction phase, bringing changes for the Acehnese. The local sense of 'place' has been shifted to 'non-place,' as the modern change represents the characteristics of a generic city that we see growing in all developing countries across Southeast Asia.

Thus, this study was conducted with a number of limitations. First of all, the samples consisted of only 5 customers and 5 vendors who were interviewed. Second, it focused only on one part of Banda Aceh, which was the Central Market. Therefore, it is suggested that future research can include a larger number of participants and investigate other occupied sections of the city on their shifts from place to non-place. Stakeholders should be taken into consideration as samples as well. Furthermore, questionnaires that can be distributed to a larger number of samples (i.e., customers, vendors, stakeholders) can also be employed to retrieve larger sets of data. The results may substantiate and further contribute to the findings of this study.

6. References

[1] Jacobs J 1984 *The Death and Life of Great American Cities* (Hardmonsworth: Penguin Books)
[2] Alexander C, Naies H, Anninou A and King I 2000 *A New Theory of Urban Design* (New York: Oxford University Press)
[3] Alexander C 1965 A city is not a tree *Architectural Forum* **122** 58-61
[4] Alexander C, Ishikawa S and Silverstein M 1977 *A Pattern Language: Towns, Buildings, Construction* (New York: Oxford University Press) pp. ix-xvii
[5] Auge M 1995 *Non-places: Introduction to An Anthropology of Supermodernity (trans. by John Howe)* (London: Verso)
[6] Koolhaas R 1995 *The generic city* ed J Sigler (New York: The Monacelli Press) pp. 1248-1256
[7] Koolhaas R 1995 *What ever happened to urbanism* ed J Sigler (New York: The Monacelli Press) pp. 958-971
[8] Sofyan I 1999 *The Dutch Colonial War in Aceh* Banda Aceh: The Documentation and Information Center of Aceh)
[9] Zeccola P 2008 A letter from Aceh: Peace and change in Tanah Rencong *Review of Indonesia and Malaysian Affairs* **42** 161-175
[10] Steinberg F 2007 *Habitat International* **31** 150-166
[11] Dewi C and Rauzi EN 2018 *IOP Conf. Ser.: Mater. Sci. Eng.* **352** 1-7

IOP Conf. Series: Materials Science and Engineering **523** (2019) 012057 doi:10.1088/1757-899X/523/1/012057

Influence of material application to wayfinding issue in underground station design

A A Fatinah[1,*], N R Kusuma[1], E Arvanda[1]
[1] Universitas Indonesia, Depok, Indonesia

E-mail: annisa.aulia41@ui.ac.id

Abstract. Getting lost and disoriented due to the lack of legibility of the space are common problems found in underground stations. Wayfinding inside underground stations is often thought as being solely supported by the presence of signage and directory maps as the tools that help users to understand their orientation and route better. However, the influence of materials on wayfinding in underground stations is often overlooked. Hence this paper presents a comprehensive examination of literature studies and an analysis on Dhoby Ghaut Station in Singapore as case studies. This station serves three interchange MRT lines and complex routes, which renders wayfinding issues even more urgent. The goal of this paper is to examine the potential of contrasting the material application for effective wayfinding inside the underground station. To identify aspects regarding the impact of selection and placement of materials applied (on floors, walls, and ceilings of underground stations), literature and case study are carried out. The results indicate that the materials used in underground station influences wayfinding in varying degrees.

1. Introduction

Underground station is a type of underground space that is built to support the public transportation system where people move throughout the space to access another building that is integrated to the station [3]. Underground station provides flexibility and security in mobilization inside the space in which people are able to navigate themselves independently. However, there is some underground station design that still has issues in wayfinding; difficulty in finding the access in and out of the station and to locate specific destination inside the station. The issues are caused by the knowledge gap that people experience which leads to disorientation. To overcome the problems, people perceive the visual information that is provided by the underground station design based on the characteristics of material surfaces. By perceiving the visual information, people are able identify their surrounding and access the space in ease. Accordingly, material application has potential to increase legibility of space and solve the wayfinding issue inside underground station.

2. Background

Wayfinding is a term to describe the spatial experience which people undergoes when they navigate themselves inside a space. According to Passini (1996, 1997), wayfinding is a cognitive ability to assimilate spatial information, make maps to find one's way, make decisions, and execute the three decisions: cognitive mapping, decision-making, and decision execution based on visual perception.

[*]Corresponding author: annisa.aulia41@ui.ac.id

Through the process, people determine their orientation by choosing a point of reference along the route they are about to pick. Therefore, the necessity of sufficient visual information for wayfinding demands interesting settings that allow for gratifying spatial experiences that are accessible and efficient, despite any complexity that different type of space carries (Arthur and Passini, 1992, p.37, 43).

Looking up to the main function of underground station design as a transitional space in which accommodates people's mobility, wayfinding is a process that involves one's visual knowledge about the station in order to identify and access the space easily as an important aspect of wayfinding system in an underground station. Passengers obtain the visual knowledge by perceiving the surface of materials of spatial features of the station design. The visual perception formed gives birth to material experience, which gives an aesthetic, meaning, and emotional value to the human being who interacts with the material.

When interacting with materials, humans use their visual perception of materials to understand material characteristics such as colors and shapes. Therefore, materials are applied significantly on the surface of spatial features by contrasting the materials colours with a consideration of the material's reflectivity and provided artificial illumination [10]. As a windowless environment, the presence of artificial illumination is important to induce surface reflectance of materials inside an underground station. Furthermore, surface reflectance is affected by the amount of intensity and direction of the illumination itself—whether the illumination would cause glare or shadow on an object and thus affects legibility of a space in general. According to Nancy Kwallek on her research with NASA [17], colors of material are the characteristics that exert the most influence on the human perception of color. Not only certain colors, therefore, or certain juxtaposition of color, but also certain degrees of color of materials may have a positive or negative influence on our productivity, including sense of direction, mental lucidity, and psychological level. However, contrasting in material to highlights the spatial features does not take account solely on colors and shapes, but also on the additional aspects such as size, distance, and scale. These aspects create foreground-background relationship between materials with significance. Therefore, material selection should encounter the importance of creating a legible space that is easy to identify.

3. Methodology

This paper aims to identify the potential of material application to solve wayfinding issue in the underground station design. The study of this paper begins with literature study to identify the aspects of wayfinding in underground station and surface properties of materials in which takes part in enhancing legibility and accessibility of space. A study case has been conducted to test the relevancy of wayfinding, materials, and the performance of the wayfinding design of underground station in Dhoby Ghaut Station, Singapore. A direct observation was then carried out to test the relevancy of theories in Dhoby Ghaut Station, Singapore, addressing the superimposed layout and interchange lines problems that may rise as part of wayfinding issue inside the station.

4. Findings

4.1. Background of Case Study

Dhoby Ghaut Station is the largest Mass Rapid Transfer (MRT) in Singapore with three interchange lines. The building is located underground, span from Orchard Road to Penang Road to connect the station to many integrated public places around. The large massing has caused numerous options of route with different complexity for passengers to take inside the station, such as crossing of paths from one room to another and crossing of access from one floor level to another. Hence, it has caused issues on wayfinding in which affects human movement and activities inside the station.

4.2. Findings

4.2.1. Materials and Wayfinding Correlation in Underground Station

Wayfinding is human cognitive ability to process different kind of spatial information (Passini, 1996, 1997). The information is perceived visually and becomes a reference for user to make decision independently, and thus human is able to determine their orientation while moving inside a building (Montello, 2001). One of many types of visual information is the material use on the spatial elements of underground station. Materials facilitate human movement that is associated with wayfinding in a space because the presence of materials is able to increase the readability of a space, thus it is easy to identify (Symonds, P., n.d). This is beneficial for user as it helps them to mark the territory of their activities inside the underground station as per their needs (Abrams, B., 2010).

4.2.2. The Importance of Material Surface Characteristics

Significance of material application on spatial features depends on the characteristics that appear on the surface of materials, such as colors and pattern. Characteristics arise after the substance of material reacts with the subsequent medium of material. In underground station, the medium of material is light from artificial illumination as compensation to the absence of natural light [17]. It reacts with the material substance through reflection and transmission to give material effects, such as reflectivity [10]. Therefore, the level of reflectivity depends on the type of materials applied, as different types of material have different surface reflectance (Anter, K.F., 2000). It affects the way humans capture and perceive the materials as visual information to guide them to determine their orientation inside the station [17]. Different level of reflectivity affects color contrast and pattern helps to form foreground and background besides the scale of the material itself. Accordingly, it forms focal point and key area of a space and thus emphasizes the presence of both spatial and legibility features as guidance and to create continuous navigation experience.

Figure 1. Foreground-background relationship

4.2.3. Material Application on Spatial Features of Underground Station

Materials are applied in interesting arrangement with consideration on pattern and color coordination to highlight spatial features and define their function, such as nodes and landmarks (Wright, 2012). Both of the features become point of references for humans to consider when they choose the most favorable routes because they have the greatest appeal (Alexander, C., 1977, p. 577). Accordingly, material application creates clear articulation of space and thus helps user to navigate themselves inside the station.

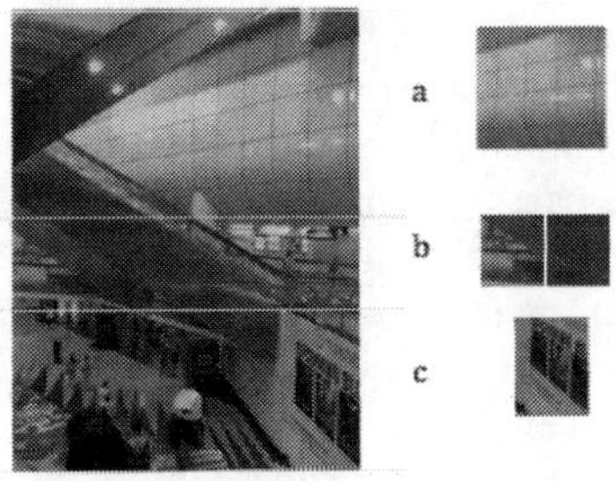

Figure 2. Material scale

Table 1. Mapping of Material Application on Different Location in Dhoby Ghaut Station

Location	Key Plan	Material Application
Concourse Area B2	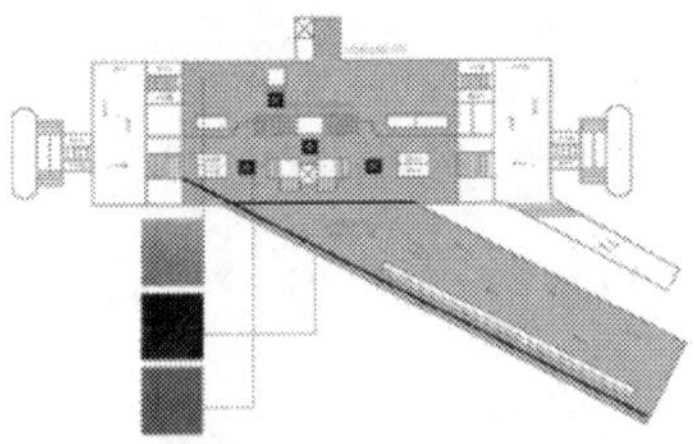	
North South Line (NSL) B3	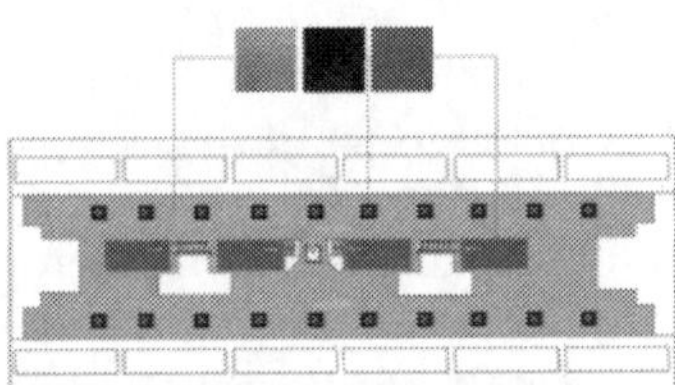	
North East Line (NEL) B5	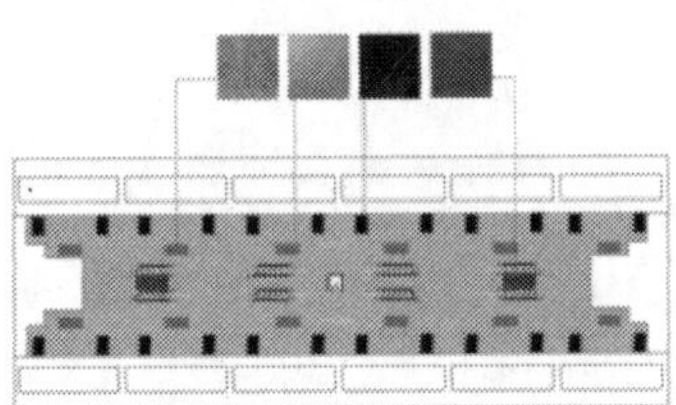	

5. Results

Dhoby Ghaut Station as a train station with the most complex route in Singapore is located 28 meters below the surface of the ground makes the entire room isolated from access to natural light. Therefore, the type of material used is taken into consideration so that the reflection of light from artificial illumination existing in the station space on the surface of the material does not cause glare or shadow which can disturb human sight navigation within the underground station (Wright, 2012). The application of materials is contrast with one and another to form the foreground and background (Lawson, 2005) that is perceived as point of interest and reference when someone is determining his orientation inside the station. Therefore, it can be concluded that contrasting material applications have an effect on solving the wayfinding issue in underground stations.

6. Conclusion

In analyzing the effect of material application on wayfinding in an underground station, an aspect that is studied is the material surface characteristics of materials as visual information perceived by humans as underground station space users. Surface characteristics arising from the reaction between the material substance and the surrounding medium. It is perceived visually by humans regarding wayfinding issue is color. This characteristic arises from the reaction between light as medium and the related material substance. It is then being identified through the process of observation, interpretation, and synthesizing of material of an object that a person sees. The process becomes easy to conduct when the material's presence is significant, whether the contrasting material between one and another in terms of color, scale, or pattern that are applied on spatial elements and legibility features. The significance makes them appear as point of interests of the underground station and thus becomes reference points for human to plan the route he will take. Consequently, it becomes a tool to use as guidance during navigation within the underground station.

7. Acknowledgements

The work described in this paper was fully supported through a research grant called Hibah PITTA by DRPM (Directorate of Research and Community Services) Universitas Indonesia. Hibah PITTA is a grant to support research works done by UI students and academicians for international publication. The author also warmly thanks those who have participated in the interviews.

8. References

[1] A.C. 1977. *A Pattern Language*. New York: (Oxford University Press).
[2] A.D. 2005. *Constructing Architecture*. (Switzerland: Birkhauser).
[3] A.W. 2012. Underground Architecture: Connections Between Ground-Level Public Space and Below-Ground Buildings. (Victoria University of Wellington).
[4] Adams, J. B. 2010. *Wayfinding in Architecture* (University of Florida Scholar Common).
[5] Dhoby Ghaut Information Kit. 2014. *SMRT,2*, 1-15. Retrieved from http//smrt.com.sg/
[6] E. K., O. P., & V. R. 2014. Materials Experience: Fundamentals of Materials and Design. Oxford: Elsevier.
[7] E. S., & J. M. 2012. *Universal Design: Designing Inclusive Environments*. Hoboken, New Jersey: (John Wiley & Sons).
[8] Fleming, R. W. 2017. Material Perception. 365-388..
[9] Fridell Anter, Karin. 2000. What Color is the Red House? Perceived Colour of Painted Facades (Doctoral thesis).
[10] Gibson, J. J. 2015. *The Ecological Approach to Visual Perception*. NY (Psychology Press).
[11] Ingold, T. 2016. Materials against materiality. *Archaeological Dialogues,**14***(1), 1-16.
[12] J. W. 1981. Evaluating Architectural Legibility: Way-Finding in the Built Environment.
[13] K. L. 1960. *Images of The City*. New York: (The M.I.T Press).
[14] L. 2001. *The Language of Space*. Oxford: (Architectural Press).
[15] London Underground Station Idioms. (n.d.).
[16] P. S. 2010. Architecture, Architect, and Wayfinding.
[17] S. H. 2010. Architectural Wayfinding. *Design Resources,01*.
[18] Wastiels, L., & Wouters, I. 2008. Material Considerations in Architectural Design: A Study of the Aspects Identified by Architects for Selecting Materials. *Undisciplined! Design Research Society Conference 2008,*1-13. Retrieved from http://shura.shu.ac.uk/511/

IOP Conf. Series: Materials Science and Engineering **523** (2019) 012058 doi:10.1088/1757-899X/523/1/012058

Active waiting: Potentials of waiting area at airport

M T T Pasaribu[1,*], E Arvanda[1], N R Kusuma[1]
[1] Departement of Architecture, Faculty of Engineering, Universitas Indonesia, Kampus Baru UI Depok, West Java, 16424

E-mail: marythaliatrav@gmail.com

Abstract. For travellers in the airport, waiting is often associated as an unpleasant activity. One of the reasons of the perception is because waiting area is viewed as a space for uncomfortable waiting experience. Bissel explains that the experience of waiting is often conceptualised as a stasis period that bore the passenger to be passive. Regarding to this understanding, Schweizer and Gasparini try to dissect the potential of a waiting space by using active waiting concept as a lens of analysing the space. According to their concept, active waiting is where waiting is seen as an activity that has balance focus between achieving what is expected and how to have comfort when experiencing the process itself. This paper argues, by incorporating this concept to a waiting space, passengers will experience in both passenger system and a new occasional experience. This paper will observe and analyse the relation between spatial elements and the behaviour of passengers at a waiting area within an airport terminal, through the lens of active waiting concept. The experience of waiting can be achieved through three stages of habitation, which is lingering, tarrying, and the state of dwelling. The expected result from this study is to discover the potentials of waiting area at an airport. Findings from this paper may be useful for planning the airport waiting in order to create a more comfortable experience for the body and state of mind of the passengers.

1. Introduction

Waiting in airport terminal is necessary to keep the flow of passengers of different flights in accord to one and another. The duration is also used for the airport operation officers to prepare the aircraft for passengers to board. Aside from the procedural aspects, waiting area is both an extension of a particular place, such as city or country, and the last contact a person has of the particular city. Thus, it is important to assure that passengers can perceive a decent viewpoint of a place trough a safe and comfortable waiting experience. Nevertheless, the experience of waiting is often conceptualised as a stasis period that bore the passenger to be passive. When in reality, it should be alive with so many potentials of activity and experience [1]. Bissel's notion implies the possibility-on what it is to wait, one that does not define waiting as boring neither unpleasant activity, known as the active waiting concept.

This paper will observe and analyse waiting activity, focusing on the spatial elements and the behaviour of passenger through the lens of active waiting concept based of observation made in the 3rd Departure Terminal of Soekarno-Hatta International Airport. The expected result from this study is to find the potentials of active waiting at airport.

* Corresponding author: marythaliatrav@gmail.com

2. Active Waiting

As a transition period between two events, waiting bridges activity that precedes and follows [2]. It indicates a certain plot from the beginning of a trip that ends with an expectation of what to achieve that top the whole journey [3]. This connection is an important part of active waiting. The word 'wait' itself is conceived from the French word 'waitier', that means to watch and from the German word 'wake', that means to guard. The origin of the word brings a sense of anticipation in the act of waiting upon the expectation that it holds [1]. Thus, waiting should be seen as not only an attempt to pass time, but as an attempt to enjoy time while anticipating what is coming out of it [4]. The balance between waiting and the result from the wait is necessary to accommodate an active waiting. This is also how active waiting can be qualitatively measured.

There are several variables that can be used as a tool to observe active waiting in waiting area, which are: interior elements and body movement of passenger. Interior elements are objects in waiting area that act as the medium between the spaces the human body [5]. On the other hand, the human body includes the body parts and its extension in space. In the attempt of waiting, human body show two particular ways of adapting, which is absorption and relaxation. Absorption is the attempt of human body to perceive information through the senses with awareness and anticipated, alerted, fast responding situation. While relaxation is the state of being voluntarily unanticipated to relax and self energize, to prepare it to be for the stage of absorption [4]. The interaction of two observable aspects and the personal variable such as expectation, and the timing set by the aircraft will result to the overall experience of passenger the airport [1].

The observation on the interior elements will focus on its universality, time resistance, and variation, and how this help to accommodate the active waiting and provide a comfortable place to stay and to access the space, with both connection to the important spot as well as some privacy [5]. Meanwhile, the observation on human body will focus on the absorption and relaxation phases, which can be described as the relation between human senses and how it perceives the information around them. From the observation on interior elements and human body, we can see how the spaces are inhabited by the passenger and thus conclude how the active waiting happens in the airport. There are three stages of inhabitation that caused by interaction between human and the surrounding [4], which started with an understanding upon personal expectation and to do list, but without the knowledge on where the activity would specifically take place in the wide area the waiting room provides.

3. Methodology

This paper is based on the observation took place in Terminal 3 Bandara Soekarno-Hatta, in Jakarta. The terminal is a domestic passenger terminal with two kinds of waiting area, a publicly accessed area (non-sterile) and passengers-only area (sterile). The data is based on my observation, on the movement of passenger, whether towards each other and upon the interior element of the waiting area. Terminal 3 is selected due to its recent accomplishment as the most improved airport according to the Skytrax International Airport Awards, which indicated improvement in accommodating passenger in waiting area.

4. Overview of Interior Elements and Body Movement

4.1. The Interior Elements

The waiting room in airport usually is filtered to two groups, the sterile and non-sterile area. The non-sterile is publicly accessed while the sterile are made for less people, which are those who have earned their boarding pass.

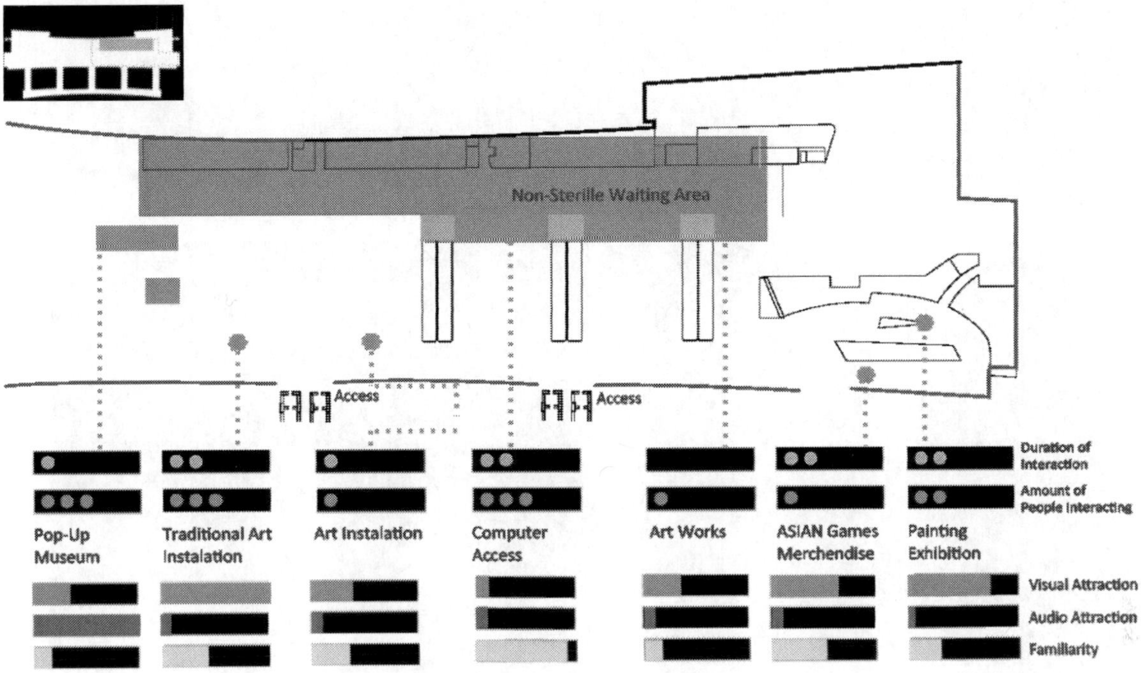

Figure 1. Interior Elements in Non-Sterile Waiting Area

The term interior elements in this paper refer to both fixed features and non-fixed features. But, due to the different context of two airports, the non-sterile and sterile are down to have more non-fixed features, especially the temporary ones. Temporary non-fixed elements have various kind of interaction with the passengers, and in general, the interaction happens in a short amount of time because elements such as installation of exhibition and artworks which are less functional. In the contrary, the sterile waiting area itself is less complex. Instead, the sterile one is filled with more functional non-fixed elements, such as charger booth, television, and furniture to sit on. The seating area has several kinds of shape, colour, material, finishing, size, orientation, height, and grouping. Hence, passengers can choose where to sit according to their preferences.

Both the sterile and non-sterile waiting area use the same type of materials for their fixed interior elements, such as the colour and finishing of their ceiling, flooring, and walls. However, different combination of materials results a different ambience/atmosphere. For example, the way some furniture stands out in the sterile area trigger the sense of anticipation that underlies the active waiting. The little differences, such as height, width, shape, colour, material, finishing, size, orientation, height, distinguish areas, furniture, or groupings for passenger to choose and adjust based on their personal need. The ability of waiting area to accommodate different choices will probably increase the feeling of safe and comfort experienced by passengers.

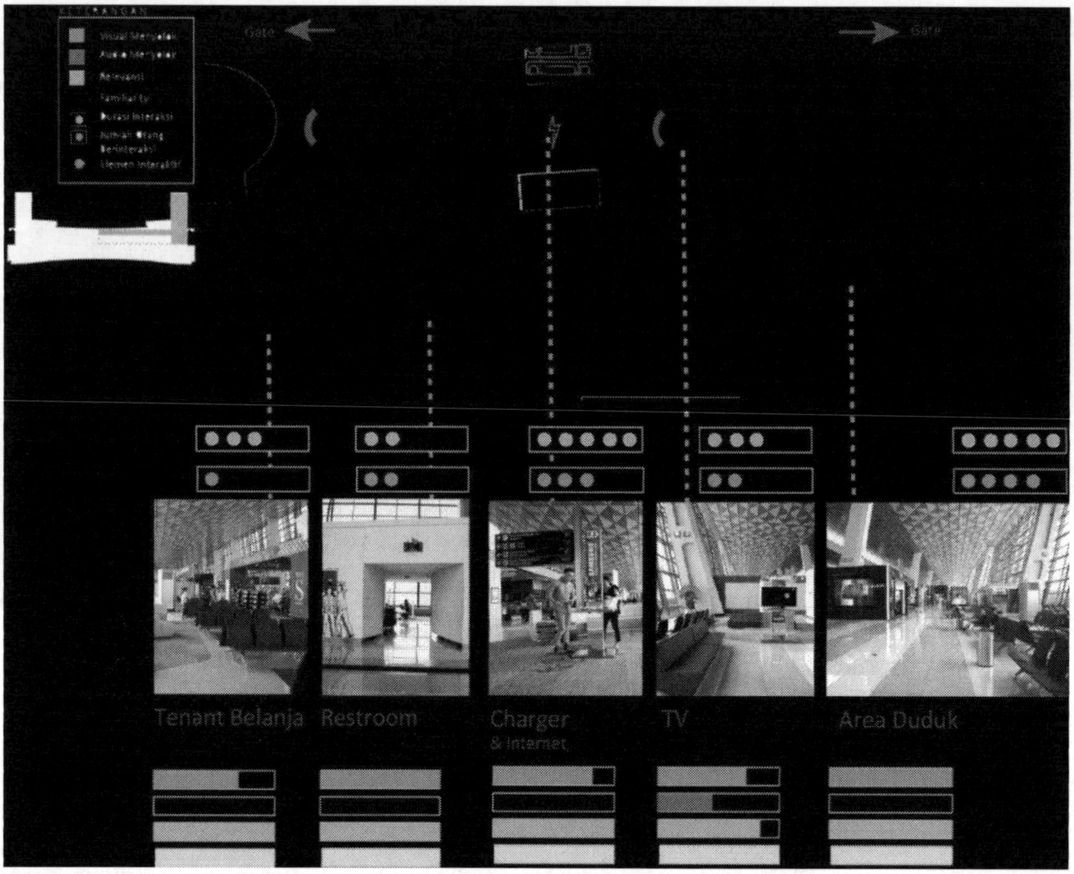

Figure 2. Interior Elements in Sterile Waiting Area

4.2. Movement

When waiting, people tend to be more anxious because they have to get through several procedures for their journey. This state of mind clearly shows in the passengers' body movements and gestures. Specifically, In the non-sterile waiting area, people are more anxious because there are still some procedures to follow before they can enter the aircraft. The passengers are also burdened with their luggage, which are bounded to their being thus make movement more challenging. In the sterile waiting area however, people can be more relax because they have successfully accomplished several important procedures and they only a step ahead to enter the aircraft. One of the differences between the non-sterile and sterile waiting area, is that people have more freedom to move since the baggage is no longer with them.

Their belongings, before they checked their way in, massively influence the movement of these passengers on the non-sterile area. Baggage restrains them to freely contact interior elements around them in non-sterile area. Thus, minimizing the chances of interaction between passenger and the attractions added in the area. How baggage limits the interaction can be seen in figure 2, in which only small amount of people interacting with the attractions with also small amount of durations. This is due to the distance between elements and the seating areas where the baggage is stored, and the locations between each element. The suitcases burdened passenger and keep them in stasis.

Contrarily, in the sterile area, the passengers tend to move less even though they are not burdened with suitcases anymore. This can be the result of a various seating choices that offer comforts to the passengers. Moreover, the state of mind of passengers is more relaxed due to the journey they have passed, and less and leads to the relaxation of the body. The relaxation

Nevertheless, baggage, as a part of a journey still plays a certain role in the movement of passenger in non-sterile area. As stated before, it is important to make sure that the passenger feels safe when they are waiting. The observations at the non-sterile waiting area shown that, most passengers extend their body parts while sitting, as they enter a relaxation phase. Relaxation phase happen when passenger is guiding their luggage, in their most comfortable position. The reason behind this could be an attempt to relax their body parts and reach for the most comfortable position to wait. In addition to that, some of these sittings are followed by extended limb covering certain luggage to show ownership. The urge to keep luggage safe when waiting is an indication of anticipating. To keep passenger actively waiting, it is important to assure that they do not worry too much on guarding their belongings, but aware enough to realize if there is threat. This way, people can rest and guard at the same time.

To be free from the fear of losing their belonging is not the only way to give the safety feeling for passengers. Just like the description of active waiting, which includes awareness of activity in the present as well as the future, providing safety for passengers also means to keep them away from the possibilities to lose their flights. This can be achieved if the waiting area provides a well spread information regarding the flight information, reaching the whole waiting area, including the toilets, corners of the seating area, shops, and dining. Thus, passengers can enjoy doing various activities without fear of missing neither their belongings nor their flights.

5. Overview on Observation

5.1. How Interior Elements Stimulate the Absorption Phase while Waiting

When observing the behavior of passengers, we can see body movement made by passengers. The movements show anticipation and attempts to react quickly in response of new information given to them. These movements include moving into the source of information, for example by getting closer to information board or screen, or closer to the officers. These acts of seeking the source of information are an attempt to keep one self-safe from missing the flight. Nevertheless, it is not possible for a person's body to always be alerted at all times. Thus, there are times when the person feels not necessarily be alerted to gain information regarding the flight and switch the body to do the attempt to not miss the flight by relaxing and reenergizing the body. In between these two phases, it is important for interior elements to be the perfect switch between phases of passenger's body. Theoretically, phase of relaxation is not needed functionally to achieve the goal of waiting, but it is needed to support the absorption phases, in which it is considered needed.

Interior elements in the sterile are mostly objects for seating that adjust the spine upright, and have full access to the auditory and visual attention can keep a passenger informed even when they are not located near the crowds or officers. In my observation, the variation of seating help passenger to perfectly accommodate their self according to the phases they are in. In consequence, this waiting area stimulate and maintain absorption phase of the body and trigger the passenger to successfully board on time voluntarily, without feeling forced neither treated like logistic goods. It means, the comfort of human body and the state of mind are being taken care of as they board, which indicates potentials of active waiting.

5.2. Inhabitation Process in Waiting Area

Time and expectations are very much needed in order to achieve and experience the right amount of absorption and relaxation phase while waiting. The three other aspects form a waiting experience as it responds the existing interior element of the waiting room and affect human body to give another response. This constant cycle of response between time, expectations, experience, body movement and interior elements takes part in the process of inhabiting the chosen waiting area for passengers. In Figure 3 and Figure 4 shown the attempt of experiencing the waiting room. From the two comparable figures, we can learn that the state of lingering, and even the rest of states, in the non-sterile area tends to last longer than in the sterile area. This is due to the how the waiting room is defined by the operational in the sterile waiting area.

IOP Conf. Series: Materials Science and Engineering **523** (2019) 012058 doi:10.1088/1757-899X/523/1/012058

In figure 3, a type of camouflage from a similar toned color surrounding give comfort for the passenger to inhabit the area, even when the object is not common in waiting area. While in figure 4, The place is filled with several elements such as seating and signage that people are familiar of compared to man-made garden on the other side of the passenger's terminal, The process of inhabiting usually occurs when the passenger are looking for the perfect fit in between many slightly different choices. These give a deeper interaction with the space because passenger do try to sense, the information they can perceive in their absorption phase. It is also important to note that these state of lingering, tarrying and dwelling provided by the interior elements in the waiting room are not a process to dwell. The state of dwelling is the highest state of the inhabitation stages where Schweizer believe is a condition where the passenger feel the most comfortable and earn their freedom to actively do what they ought to do during their waiting period, without any limitation or fear [4]. Therefore, the state of dwelling is temporal conditions that can easily disappear once a certain condition made by the airport or the airline take place.

Figure 3. Inhabitation Diagram in Non-Sterile Waiting Area

Figure 4. Inhabitation Diagram in Steril Waiting Area

These give a deeper interaction with the space because passenger do try to sense, the information they can perceive in their absorption phase. It is also important to note that these state of lingering, tarrying and dwelling provided by the interior elements in the waiting room are not a process to dwell. The state of dwelling is the highest state of the inhabitation stages where Schweizer believe is a condition where the passenger feel the most comfortable and earn their freedom to actively do what they ought to do during their waiting period, without any limitation or fear [4]. Therefore, it is safe to say that the state of dwelling top the stages are a temporal condition that can easily disappear once a certain condition made by the airport or the airline take place.

6. Conclusion

The analysis from each point suggests that interior element can provide comfort while also secure passenger by providing personal comfort trough the sense of safety or physical support through the elements. At the same time, interior element is also able to lead both absorption and relaxation phase of human body that help passenger to inhabit the space and experience a meaningful waiting process at its best.

he interior element, in waiting area, holds a very important in provoking potentials of active waiting. It is able to trigger a continuous cycle between body movement of passengers, their expectations, time, and provoke experience of waiting, whereas, the feeling experienced by passenger progress from a state of lingering to the state of dwelling in the top of inhabitation stage. Upon this understanding, we can conclude that either adjustable, or slight variation of interior elements is a step in generating inhabitation as an attempt to accommodate a comfortable waiting experience for the body and the state of mind of passengers.

7. References

[1] Bissel, D. 2007. *Animating Suspension: Waiting for Mobilities*. Routledge , **2** (2), 277-298.
[2] Walker, J.2009. *Human Transit*. Washington, D.C., (USA: Island Press).
[3] Gasparini, G. 1995. *On Waiting*. Time Society .
[4] Schweizer, H. 2008. *On Waiting*. New York, New York, (USA: Routledge).
[5] Herman Miller Company. 2014. *Winning Strategies for Waiting Room*.(Insight).
[6] Transportation Research Board. 2010. *Airport Passenger Terminal Planning and Design*
[7] Center for Air Transportation Systems Research.2009. *Introduction to Airports Design and Operations* (Transportation Research Board)
[8] Kazda, A., Antonin, K., & Robert, C. E. 2007. *Airport Design and Operation* (Vol. 2). Amsterdam, Amsterdam, (Netherlands: Elsevier).
[9] Kesuma, S., Mat, N., Mulyani, S., Risanti, Syahputra, F., & Yuhzan, S. 2012. *American Journal of Economics*.
[10] Perolini, P. 2014. *Interior Environments:The Space of Interiority*. (Zoontechnica)
[11] Nurture. 2008. *Environment Waiting Room*. Michigan, (USA: Steelcase).
[12] Nasional, B. S. 2004. Perancangan Fasilitas Bagi Pengguna Khusus di Bandar Udara. Jakarta, Indonesia.

IOP Conf. Series: Materials Science and Engineering **523** (2019) 012059 doi:10.1088/1757-899X/523/1/012059

Permeable interior: Unfolding threshold space within transit corridor

N R Maheswari[1,*], E Arvanda[1], N R Kusuma[1]
[1] Department of Architecture, Faculty of Engineering, Universitas Indonesia, Depok, Indonesia

E-mail: nikenrahadiani1996@gmail.com

Abstract. At a transit corridor, threshold plays a key role in creating connectivity between the interior of transit facilities and its immediate urban context. *The threshold* is a choreographer of spatial experience and might potentially generate various public activities. As a transitional space, the permeability of threshold space's boundary becomes an important factor for the users' spatial experience and in providing a sense of safety and direction for the pedestrian. Such penetrable properties also help the users predict what is going on in and around the space. This paper suggests that threshold space can be read as an interior space. Therefore, the author uses interior theories as a groundwork for the case study. This paper aims to reveal how permeability in threshold space within transit corridor might have an impact on its atmosphere and the people's experience within. It also argues that permeable threshold can have a significant impact on users' perception of the transitional space. The method used on this paper is a qualitative method, including literature review, case study, and field observation. The findings may be useful in planning threshold space in the future to create a better and safer transit experience.

1. Introduction

At transit spaces, a corridor has an essential role in connecting the interior of a transit facility to its urban environment. Because it gives an important impression of a transit facility and the city that surrounds it, the corridor accommodates the in-between experience. Therefore, it should be designed comprehensively from the beginning. In Indonesia, the regulation in designing transit corridor as pedestrian path has already stipulated in the attachment of Regulation of the Minister of Agrarian Affairs and spatial planning / Head of the national defense agency (Rapermen No. 16 Tahun 2017). However, in it does not elaborate further about specific design requirements and details.

As the connector of the interior and the city, a transit corridor càn be defined as a threshold space. The particular characteristic of a threshold space is its ambiguity because it lies in between the interior and exterior space. Thus, it has the quality of both spaces as the threshold acts as bridge between interiority and exteriority. On the other hand, the threshold can create relations between both. The body of threshold can be created through barrier which creating boundary. These boundaries can create a statement where the threshold is and how much influence the threshold space can give on pedestrians whom in the transit corridor. Therefore, boundary playing an important role when discussing about threshold space.

In the discussion of the threshold space within the transit corridor also, the role of spatial boundaries becomes important in providing users' needs, without disturbing their respective activities and spatial

*Corresponding author: nikenrahadiani1996@gmail.com.

IOP Conf. Series: Materials Science and Engineering **523** (2019) 012059 doi:10.1088/1757-899X/523/1/012059

experience. In that account, this paper aims to look at the potential of transit corridor's boundary its impact on the user's spatial experience and atmosphere. The observation for the case study was conducted at a transit corridor that connects Pasar Senen Railway Station and Senen Main Street, which now is one of the busiest railway stations in Jakarta.

This paper will observe and analyze how boundary helps to create interiority of the transit space and to shape the experience of the people in a transit corridor. Focusing on the boundary as delimiting elements of corridor's interior and exterior. The expected result from this study is to unfold the potential of the boundary (in creating threshold) along the transit corridor.

2. Literature Review

2.1 Interiority and threshold space

According to [2], interiority is explained as an abstract quality that is formed from some dynamic parameters. It is this abstract quality that shapes the awareness of the existence of the interior. McCarthy also argues that interiority is non-absolute and is a transformative concept and depends on various condition according to the context and location. McCarthy stated, "Interiority is not a guarantee of inside location". McCarthy also states that the existence of interiority is always side by side with the exteriority, as 'subject to the trauma of separation'. Exteriority, on the contrary, is an abstract quality that is often associated with outside. Because of it always in line with interiority, the definition of exteriority also depends on how individuals place the perception of interiority in the surrounding space. Meanwhile, the area lies between those two is defined as a threshold. The threshold is not a part of interior neither exterior, however, it sets the limits for both and takes both characteristic. This condition makes threshold can be defined as in-between condition (between exterior and interior), so there can be an ambiguity on its spatial quality. Boettger [1] explained, threshold opens up space and arranges transition, however at the same time threshold also can be read as part of the boundary itself, which seen as a barrier. As threshold acts an element that provides a transition, it also arranges spatial sequence including atmosphere, movement, view, spatial characteristic and spatial quality.

2.2 Boundary in Threshold Space

According to Joedicke in Boettger [1], delimitation defines a the threshold space by giving it spatial volume and body. Delimitation can be seen and elaborated physically through the vertical and horizontal element, or non-physical elements such as odor, temperature, and sound. While Hernes [3] distinguished boundary into 3 (three) categories on how boundary regulates a space, which is "ordering, distinction, and threshold". As of ordering devices, boundary acts as 'tolerance limit', as Harsbarger said [3], for activities/actions and human interaction which creates 'some stability of expectation'. Whilst as a distinction, boundary acts as 'markers of identity' and plays a role in conveying different physical, social and mental characteristics by which space is distinguished from the environment [3]. It also negotiates with time in contributing to make a precise and authentic setting of spatial atmosphere and user's experience. As a threshold, Hernes [3] stated that boundary can become threshold when it has the capability to become crossed by material/resources, human and other material (in this case wind, sun rays, etc.). That refers to the traits of a boundary which is, as March Simon [3] said, 'permeable'.

In defining and analyzing a threshold space, Boettger [1] brings up 6 (six) parameters: delimitation (spatial definition), sequence (spatial sequences), geometry (spatial structure), topography (spatial situation), materiality (spatial design), and furnishing (spatial function). From those six parameters, only delimitation will be discussed further in this paper because it focuses on how the threshold space is defined and perceived in its entirety through its delimiting elements (boundary). Yet actually, studying this parameter offer further understanding of how threshold space regulates activities, sequence, and atmosphere within it.

IOP Conf. Series: Materials Science and Engineering **523** (2019) 012059 doi:10.1088/1757-899X/523/1/012059

3. Methodology

This paper uses a qualitative method to dissect its issue. It begins with a literature review and a case study is selected to analyze the thesis question in context. Direct observation and mapping are used to study the chosen site, which located at Pasar Senen Railway Station, Jakarta, at the west part of the transit corridor. The case study is used to give an understanding of what threshold space is,—its in-betweenness characteristic, and also uncover the role of threshold's boundary as permeable elements in regulating the threshold space.

4. Case Study Analysis

Pasar Senen station is selected because it is one of the major train stations in Jakarta. It is one of the busiest stations, which serves inter-city and inter-provincial routes and it covers various classes from business to economy class, as well as several KRL routes. Therefore the role of the transit corridor that connects visitors to and from the interior of the transit facilities and city space becomes very important in this station.

4.1 Analysis

4.1.1 Physical attributes of the selected path

The northern part of the station is marked by residential and small-scale government offices, as well as privately owned businesses. While in the south, it is marked by a variety of public facilities, densely populated private housing and private business. The western part is marked by shopping centers and public facilities as well as major roads, Senen street. While the eastern part bordering with residential housing and private business. Elongated shape of the station, spanning from west to south. The west side corridor is chosen to be observed.

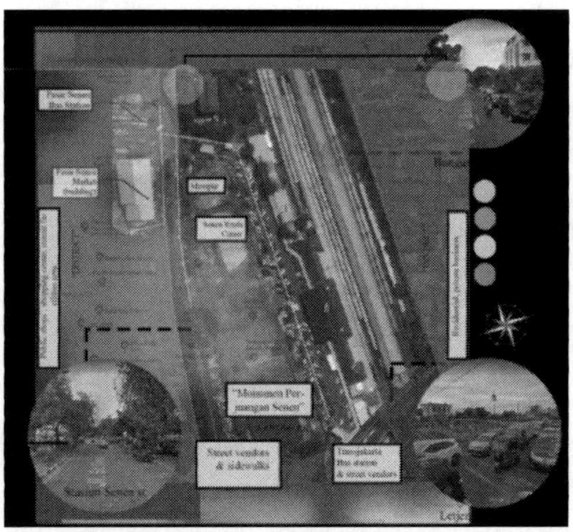

Figure 1. Site of the station and its surrounding.

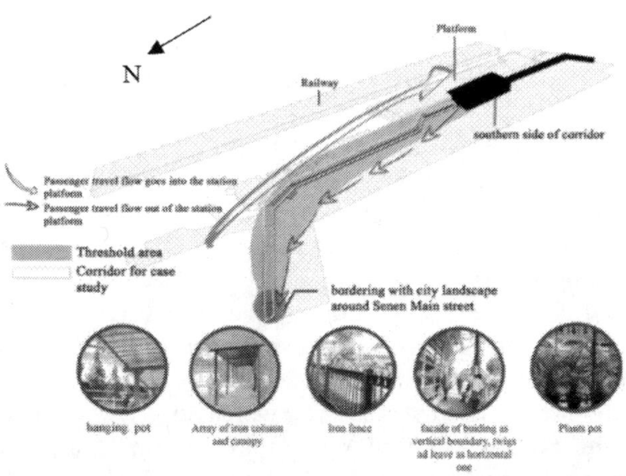

Figure 2. Location of case study and its physical attributes.

4.1.2 The selected points

Due to its elongated form, the transit corridors requires to be observed at several points to obtain more detailed information and break down the pedestrians' experience to the detail. The points are selected base on its physical characteristic or particular interior qualities. Point (1) is the beginning of the path, it is the transition point from the main road, Jalan Raya Senen. Point (2) is where the spatial atmosphere changes because of discontinuation corridor's spatial body. At Point (3) The body of the corridor as a representation of the general transit corridor. And point (4) is the transition point, from the corridor to the interior of a transit facility. The points are being observed by direct observation, mapping,

IOP Conf. Series: Materials Science and Engineering **523** (2019) 012059 doi:10.1088/1757-899X/523/1/012059

photograph, and video to see how pedestrians react on the arrangement of the spatial delimiting elements in the form of transparent boundary.

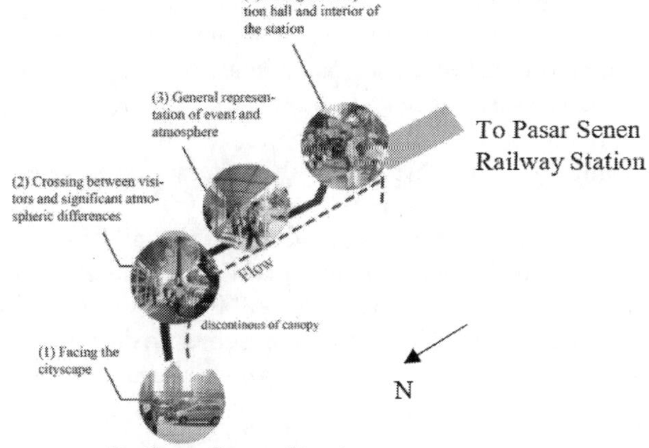

Figure 3. Observation point distribution along the transit corridor.

4.1.3 Analysis of Point 1

This point is where the transition between the exterior (the city) and the interior of the transit corridor begins. At this point, the city slowly diffuses with the interior of the corridor and pedestrians gained focus on its interiority. The interior of the corridor is built from several physical boundaries such as metal roof, the structural columns, and the walking surface. When these physical boundaries allow the corridor's interior to have interchangeable qualities between the exterior factors (sun, wind, noise, etc.) and interior quality, hence it becomes a threshold. At this point, the spatial quality of the city greatly affects the corridor's atmosphere.

Figure 4. Exteriority characteristic elements (of city landscape) on point 1 of transit corridor and the role of the delimiting element.

IOP Conf. Series: Materials Science and Engineering **523** (2019) 012059 doi:10.1088/1757-899X/523/1/012059

At point 1 also, the exterior plays an important part in building the pedestrians' perception. As they walk along the path, the city landscape permeate into the corridor. Pedestrians are also consciously building their mental maps of the landscape while they walk by using the city landscape as a point of reference to determine their position. On the other hand, pedestrian also can determine their position towards the transit facilities by seeing how the layout of buildings arrangement around them. They look for certain architectural features, such as notable architecture as a landmark or the area where public-service motorcycle (ojek) are parking as a node. Unfortunately, in this case study, the surrounding city landscape is quite monotonous and does not really have significant features. No buildings that are really stand out, therefore pedestrians will find a some difficulties in building their mental maps based on the surrounding architecture (figure 5).

Permeable boundary also allows direct interaction between interior and exterior elements. In this case, the direct interaction occurred, for example, through 'ojek' driver who offered their service to the pedestrians. This kind of direct interaction with surroundings will trigger some form of communication (through hand gestures, or voice), and could possibly have some positive impact on pedestrians, presumably they would feel less detached with the surroundings. Although, other kinds of direct interaction such as y.

Figure 5. Different variations of information given by interiority and exteriority characteristic in each point.

Figure 6. Large portion of transparent boundary as threshold allows the nearest city dweller (ojek driver) intervene the interior of corridor and making contact with the corridor's space user.

Permeability in the interior of transit corridors can also affect how users capture spatial sequence and predict what happens in or around the corridor. As a public space, which is always active and crowded, visual access is important in order to calculate time. For example, pedestrians can decide whether they should walk faster or slower, based on their perception of the surroundings. The ease to see what happen

IOP Conf. Series: Materials Science and Engineering **523** (2019) 012059 doi:10.1088/1757-899X/523/1/012059

ahead is also important for the sense of safety, hence, pedestrians could predict how to react if they perceive danger.

4.1.4 Analysis of Point 2

As we get closer to approach the transit facility, the focus shifted from the landscape to the architecture of the facility. Thus, the city becomes the background while the interiority of transit facility slowly permeates. In this point, the inconsistency of the horizontal boundary (the roof) changes the perception of the height limit of space and distorts the continuity of the transit corridor. It is interrupting the rhythm of spatial delimiters. This leads to the influence of the transit corridor's exteriority characteristics expanding and increasing, especially in visual. Thus, the continuity of the transit corridor becomes distracted. This creates a bit of confusion for the user. However, when viewed from the direction of the Senen main street, the discontinuity of canopy as a space delimiter does not feel significant,This is because the loss of some parts of the canopy is replaced by the leaves and branches of shady trees as if a natural 'canopy'. The leaves and twigs are also capable of maintaining color tone on the interior of the transit corridor and the temperature within it, so it is not perceived differently. The only thing that becomes quite prominent was the corridor that becomes noticeably higher.

4.1.5 Analysis of Point 3

At point 3, threshold's boundary plays a significant role in the user's movement. It allows pedestrians to enter and exit the transit corridor at will. Boundary is needed to regulated for movement and emphasize the path. As mentioned above, pedestrians need to feel safe and comfortable when they walk through the transit corridor. Therefore, access should deliberately be controlled to maintain pedestrian's focus and

To control movements, the materials and furnishing elements of the transit corridor plays a significant part in its interior setting. The physical properties of the corridor are not only a form giver but also defining the behavior within (boundary as a distinction). For example, the roof will filter sunlight and protect from rain and people would feel more protected yet in the other hand creating affordance to occupy the space. The path's surface will affect how pedestrians move, the gap between the array of columns will signify freedom of entrance or exit, and furnishing element such as planters will act as an added boundary and will limit pedestrians' access and help filtering the sunlight or dust from infiltrate the corridor.

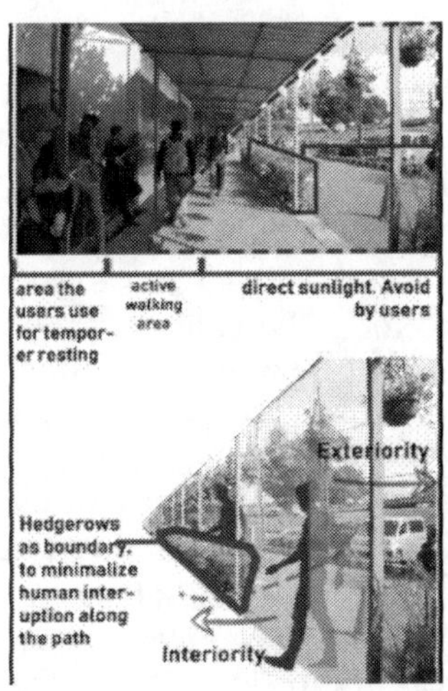

Figure 7. Observation Point (3), gap between array of column as threshold of transit corridor.

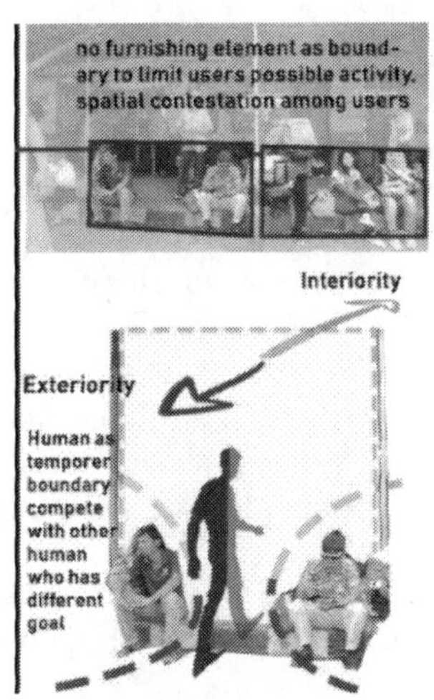

Figure 8. The gap between column occupied by pedestrians as a seat, cause a narrowing space for walk.

During observation, the authors found interesting phenomena. Some physical elements of the corridor actually offer affordance for pedestrians. The different heights between the corridor and the street can afford pedestrians a place to sit, while the columns offer a temporary leaning spot. It shows that people are always looking for comfort in every situation, although their action might potentially interrupt the flow of other pedestrians. Even so, these behavior are interesting to observe because they give clue to what other possibilities we might think about when planning for a transit corridor.

4.1.6 Analysis of Point 4

The interiority of the transit corridor diffuses with the interiority of the transit facility. At this point, pedestrians will get themselves ready to enter their main destination. Ease of visual access will prepare them for what lies ahead, in return, some will respond by moving faster or some will move slower to adapt to the new situation (figure 6). On the other way round, people entering the corridor from inside the transit facility would be looking for a well-defined boundary, although its permeability would help them to see the activities inside, so they can be assured that they are on the right path that will lead them to the city.

4.2 Findings

The analysis from each point suggested that a transit corridor resembles the character of a threshold space. Pasar Senen's transit corridor clearly focusing the pedestrian senses on the interiority of the station. This can be seen from how detailed and progressive the interiority elements of the station are displayed along the corridor rather than its more monotonous exteriority elements. But still, the architectural boundaries such as roof, paved path and rows of columns, is penetrable by exteriority elements. The threshold boundary can be the transition from the outside (the city) to the inside (corridor), the surroundings act not only as a background but also as a point of reference. Certain characteristics from the context form the pedestrians' mental maps. Permeable interior helps pedestrians to understand what kind of place they are in and gives them a better understanding of its immediate context. The permeable interior builds pedestrians' perception of the surroundings that helps them to predict and decide their action within the transit corridor, as experienced in point 1 to 2. Furthermore, physical and non-physical boundaries help the interior to regulate pedestrians' movements along the corridor, as analyzed in point 3. The permeable interior gives better visual access to comprehend what is going on at the transition point between inside and outside, hence people can predict and prepare their next step.

5. Conclusion

Through this study, it can be concluded that arranging boundary or threshold is very important in designing transit corridor. the boundary defines the spatial atmosphere, arranges spatial experience and regulated people's movement by its rhythm, placement, transparency, and many more.

In threshold space, the permeable boundary can help pedestrians to get the sense of what is happening around them, hence, it would help them to orientate, familiarize and navigate their way through the transit corridor. A boundary is not always a physical attribute of space, through the article, it has been shown that boundary can be an intangible aspect such as air, sound or light. In the future, the planning of transit corridor should consider boundary as an important aspect of the corridor, not only as a form giver but also as choreographer of experience.

The permeable boundary allows external factors to easily intrude into the interior of the corridor. It should be noted that for some pedestrians, the intrusion of too many exterior aspects would cause anxiety and threaten the sense of safety. Permeable boundary also gives permission to understand the social condition of urban space where the pedestrian in. In public corridor, the possibility of seeing or making social contact with the inhabitants of the urban space will break the feeling of being disconnected between the pedestrian and surrounding environment.

In planning a public corridor also, we must fully aware of the notion of fear of crime in public space. The corridor design should be able to integrate spatial elements that would overcome anxiety, for example, it should incorporate good lighting, clear pathways, visual connectivity and good surveillance.

For further study, the idea of regulating and choreographing boundary or threshold needs to be raised in public space discourse. Design of public facilities, especially transportation facility, tends to prioritize the basic functions, but have not been too aware of the importance of people's spatial experience in the future. For further research, it would be more interesting if the research is done with a quantitative method. Therefore, parameters regarding pedestrians' experience and their perception could be measured, in the result, the conclusion could have more depth.

6. Acknowledgments

The work described in this paper was fully supported through a research grant by the University of Indonesia. The author thanks the management of Indonesian Railways Company, especially those who involve in Operational Area 1, (PT. KAI DAOP 1) Jakarta, and the Head of Pasar Senen Railway Station who had given the permission and opportunity for the author to do the observation.

7. References

[1] Boettger T 2014 *Threshold Spaces: Transitions in Architecture: Analysis and Design Tools.* (Basel: Birkhauser)

[2] McCarthy C 2005 *Toward a Definition of Interiority. In Space and Culture* **8 No. 2**, 112-125

[3] Tor H 2004 *The Spatial Construction of Organization* (John Benjamins Publishing) pp 83s

IOP Conf. Series: Materials Science and Engineering **523** (2019) 012060 doi:10.1088/1757-899X/523/1/012060

'Lanting' as a way of life: *A legacy of riverine culture and architecture in the present urban life of Sintang City, West Kalimantan*

M S Lubis[1,2,*], T Y Harjoko[1], D Susanto[1]
[1]Department of Architecture, Faculty of Engineering, University of Indonesia, Depok 16424, INDONESIA
[2]Department of Urban and Regional Planning, Faculty of Engineering, Tanjungoura University, Pontianak, INDONESIA

E-mail: miralubis@gmail.com

Abstract. The vast island of Borneo, in the heart of Southeast Asia, is an island famous for its large and powerful rivers. Along with their tributaries and numerous smaller rivers, they represent a key factor in communication and transportation. Sintang is one of the riverfront cities located in the middle stream of Kapuas river in West Kalimantan. In the midst of a rapid urbanization process, 'lanting' or floating houses, the legacy of culture and architecture of past Sintang, are still present and needed by the community. This paper discussed the way of life of lanting community in the midst of vast urbanization of the city. Through the grounded-theory method, this study found that lanting has a significant role as a linkage between the rural and urban community. We elaborated the concept of linkage and depicted the transition process of rural-urban society and the role of river community in the urbanization process of Sintang city.

1. Introduction

The concept of living along the river is essentially part of the oldest human culture. Although most of river settlements in all parts of the world has experienced changes in orientation towards land, in several places along major rivers in Southeast Asia, there are settlements that still reflect the riverine culture as ancestral heritage [1]. In such culture, water can be seen as a steering factor for the development of human settlement along the river, but dynamically water is also affected by the settlement activities. Almost throughout the island of Borneo, the presence of large rivers is a vital means of communication and transportation, and for the growth of culture and economic activity[2]. Kapuas River, the largest and the longest river in Indonesia, plays important role as a unifying element of West Kalimantan region, where traditional settlements, commerce, and economic development have flourished along this river. Sintang is the capital city of Sintang Regency which lies in the middle-stream of Kapuas River and in the junction between Kapuas and Melawi River. Growing from river's edge with strong character of its riverine culture in the form of floating houses or 'lanting' (in local term), Sintang today is becoming more urbanized and land-oriented. Although the life of floating houses is no longer as popular as before, they still continue to survive and adjust to the situation of the increasingly modern Sintang city. So far there are no adequate references and discussions related to this issue, thus, this study will provide overview of the riverine culture and settlements in Kalimantan which have existed for hundreds of years.

2. Methodology

The objectives of this study are: (1) to provide an overview of the definition and typology of lanting and history of the origins of lanting as part of the characteristics of river settlement in Kalimantan; (2) to reveal the way of life of the floating houses inhabitants, which are located between river-oriented and land-oriented settings, and how they evolved along with the process of urbanization and modernization

*Corresponding author: miralubis@gmail.com.

of the city. This research uses a grounded theory approach that emphasizes on the quality of grounded data. Through depth interviews and participant observations, this study reveals how the actors dive into the joys and sorrows of riverine life and how they perceive their existence and their future in the midst of urbanization and modernization process of Sintang city.

3. Literature Review

So far there had been a large number of architectural and multidisciplinary studies related to riverine settlements in general, but only few discussed specifically about floating houses. Based on limited number of floating house studies, I categorized them into two research areas. First category, which is widely practiced by scholars in Europe and North America, is studies on advanced technology innovations of floating building design, that begin blooming since the early 2000s, which were carried out in order to answer the challenges of sea level rise, floods, or land scarcity for residential housings, such as conducted by Koekoek and Gibson [3][4]. Other research in this category which is quite different from Western approach was carried out by Prosun who designed affordable amphibious structures for low-income families in Bangladesh [5]. The second category is studies on traditional floating settlements mostly in Asian countries, which focus more on the historical perspective, vernacular architecture, and socio-cultural society, such as conducted by Travers in Cambodia and Jhearmaneechotechai in Thailand [6][7]. In Indonesia, studies related to floating houses are conducted in several locations, such as in Musi River, Lake Tempe, and in many big rivers in Kalimantan [8][9][10]. The literature reviews have made a number of significant points which respond to the gap identified in this study, so we will elaborate further the existence of floating houses in the midst of urbanization process of the city.

4. **Results**

4.1. Understanding Lanting and River Life

Literally, the word "lanting" comes from the Banjarese language (South Kalimantan) which means: "a raft of arranged bamboos" [11]. People choose to built their houses on the water and live there for a number of reasons, for example, because they do not own property on land, or because their work is related to rivers such as fishery and water transportation. In Indonesia, lanting settlements can be found in several areas that have large rivers such as the Kapuas, Barito, Mahakam and Kahayan rivers in Kalimantan, and in Musi River Palembang, and also found in Lake Tempe, South Sulawesi. In Sintang, there currently remains two main locations of lanting settlements, near Pasar Inpres at Kelurahan Tanjung Puri and near Pasar Sungai Durian. These lanting settlements compose urban functions such as residential, lodging, stalls, restaurants, fuel kiosks, et cetera.

Figure1. Various types of lanting

Lanting has been traditionally built on a wooden raft made of big logs called Tengkawang wood. The process of constructing lanting was not easy. Tengkawang wood which is used as a base platform that can float stably on the water is increasingly scarce because of the deforestation in Borneo island. Only a few of them are left in the remote forests in the river upstream. Meanwhile, to bring these woods from the upstream is also not easy because of quite strict administrative requirements to avoid accusations of illegal logging. So generally the lantings are constructed in the remote upstream close to the source of raw materials, and then brought by tugboats towards Sintang or other destinations.

4.2. Lanting as a legacy of the past

There is a lack of historical documents that describes in detail about the origin and life of lanting in Sintang, while finding a resourceful person from academics or from government institutions who is knowledgeable about the story of lanting was insufficient if not impossible. So we tried to extract this information from the sources we interviewed. There are different versions related to the origin of lanting. According to Edi, A Chinese who owns a coffee shop in Sungai Durian, Chinese people who came to Sintang from Mainland China have an important role in introducing the riverine culture they brought from their homeland. This opinion is in line with the statement of Heidhuis in her book entitled "Golddiggers, farmers, and traders in the Chinese districts of West Kalimantan Indonesia", that generally Chinese who migrated to West Kalimantan were not wealthy merchants, but small traders and underprivileged workers, so when they came to Sintang they lived in their houseboats or built floating houses because they could not afford to purchase land [12]. However, some other sources disagree with this opinion. According to the relative of Sintang Sultanate, lanting had existed long before the first arrival of Chinese traders and miners to Sintang, because historically there were also indigenous people of Malays and Dayaks lived in lanting. Simon Djalil, former head of Sintang Regency confirmed this opinion. He said, there was lack of evidence that in the past riverine culture in the form of floating houses came from Chinese immigrants, although in fact, many Chinese are living in lantings in Pasar Sungai Durian. Even in the past, Pasar Sungai Durian was called 'Pasar Cina' or Chinese Market because there were many Chinese people lived and worked as traders in lanting.

The life of lanting community was very popular in the 1960s, where economic activities and the distribution of goods and services relied heavily on river transportation. Trading activities along the river such as in Sungai Durian or Chinese Market were developing especially after the sultan of Sintang moved the palace from Sungai Durian to Kampung Raja in the northern side of Kapuas River, in order to facilitate the royal authorities in collecting taxes from ships coming from the Melawi River and the Kapuas River [13]. In Tanjung Puri, in the eastern part of Sintang, lanting houses were not as old as those in Sungai Durian because this area used to be relatively empty until the Dutch came in 1820. After entering the territory of the Sintang sultanate, the Dutch also participated in collecting the excise taxes from passing merchant boats so that tax withdrawals were made from all directions, both by the Dutch and by the kingdom. They set up lantings at the river crossing as stations/posts for tax collecting [14].

4.3. Lanting today: river's life in the midst of modern urban life

4.3.1. Docking

In Sintang some lantings serve as jetty for boats. Since the government's dock is inadequate to serve a large number of boats, many lanting owners provide their lantings as docks using rent payment system. The boats coming to Sintang are usually originated from both upstream and downstream of the rivers and tributaries, carrying passengers and goods. The time-length for docking varies, ranging from hours, days, even weeks. Boats carrying passengers from countrysides usually arrive and depart in the same day. Boats coming from distant areas would usually spend nights in Sintang. Usually, they are freight vessels which require a longer time for their loading and unloading. The *kapal bandung* (wooden houseboat) can even lean for weeks, waiting until all the goods they carry are sold out.

4.3.2. Rest Area

Sintang which is located in the middle stream of Kapuas River is a strategic transit hub for both freight and passenger vessels which traveling from upstream to downstream and vice-versa. Some lanting owners run coffee shops, restaurants, fuel kiosks, and public toilets. This floating rest area is needed by the transit passengers since sometimes it is not always easy to go ashore through the quite tall and steep river bank, especially for mothers carrying babies or elderly people. To go ashore, one has to climb small, non-stable wooden footbridges and stairs. In the dry season when the water level is very low, the footbridge should be made longer, while in the rainy season this path becomes very slippery and sometimes coupled with the risk of riverbank's landslide.

4.3.3. Staying Overnight

Some lantings serve cheap accommodation or lodging for commuting villagers by river transportation, who often have to stay for days or even weeks to do various things in Sintang, such as going to the hospital, banks, government offices, etc. I visited several lanting lodges in both Tanjung Puri and Sungai Durian as a participant observer and did some depth-interviews to the lanting owners and some of the visitors. Adam and Husin are the owners of big lodges which located adjacent to each other in a very strategic area close to the market and to the hospital in Tanjung Puri. Their lodges provide affordable accommodation for villagers during their stays in Sintang. The room price is only around 40,000 rupiahs per night, without any strict rules like in formal hotels. The lodging business is run on a sense of solidarity between fellow villagers. The demand for cheap accommodation is very high because people have to allocate more funds for relatively expensive river transportation costs, which could reach 1 million rupiah one way.

4.3.4. Pop-up Market

Pop-up floating market is usually held from early morning until afternoon in Sungai Durian. This vibrant floating market is a meeting place for traders and buyers from Sintang and the surrounding villages. Types of goods traded are diverse, but mainly are fresh products such as vegetables, fruits, meat and fresh fish. Boats coming from Kapuas Hulu in the upper stream usually bring tons of fresh fish to Sintang, while boats coming from big cities such as Pontianak in river's downstream usually bring groceries, clothing, and other household goods. Simon Djalil, former regent of Sintang District, told me that the pop-up market is a representation of Sintang in the past. In 1960s, life in lanting was very popular, and almost all Chinese merchants opened their shops in lanting. It used to be said that what is called 'Pasar Cina' was actually the lanting market. At that time, rubber was a main commodity traded in Pasar Cina since the Dutch era around 1834. Dayak people came from upstream to Pasar Cina selling high quality rubber and other forest products. They then returned back bringing oil and household needs they bought from Sintang.

4.3.5. Temporary dwelling

During observation and interview process I also asked the lanting owners about their feelings and their hopes in the future of living in lanting as their way of life. Surprisingly, almost all of them told me that they hoped that someday they could move to land, so they saw the phase of staying in lanting as only a temporary stay or temporary dwelling. Today living in lanting is more difficult than in the past. This is caused by various factors, ranging from the scarcity of wood for construction materials, the increase of river pollution and the more erratic river behavior (tidal patterns, wind, currents, etc.) as well as psychological factors for being treated as lower class citizens because their houses are 'down there' (water level is always lower than the land). Economic consideration is the driving factor for them to remain living in lanting. Asua, one of lanting owners in Pasar Sungai Durian told me that the number of lanting decrease because more and more people prefer to live on land once they have money. Today there are more lantings for business activities such as lodges, coffee shops, kiosks, etc rather than for residential. People who still remain living in lanting were those who have not been able to purchase houses in land. Another problem that worries the lanting owners was that there is no policy rules that guarantee the legality of lanting as a form of housing in the city. They are always haunted by feelings of anxiety and worry for being evicted by the government since they do not have any legal documents or certificates. It actually happened once or twice in the past, when the governments whose in charge asked the lanting owners to move voluntarily to land or to move their lanting more to the city outskirt because their houses were considered a 'visual disturbance' for the city center.

Another different story came from Suhaimi, a retired civil servant who was one of the owners of Lanting. When he was still working as a civil servant in the 1980s, he built a lanting for his six younger brothers who came from their village hometown to continue their education in Sintang. Staying in lanting could save more money, because they did not have to rent expensive house on land, they did not have to pay for tap water, they can get fish for their food from the river, they could also save the transportation cost because their lanting is located close to major facilities such as schools, markets, banks, hospitals, etc. So they could save a lot of money for monthly expenses. Suhaimi financed his younger siblings until all of them graduated from high school, even one passed the university.

"Alhamdulillah, ada enam orang adik-adik saya yang berangkat dari sini (lanting), dan adek angkat ada dua orang. Mereka selesai semua sekolahnya karena ditolong oleh lanting ini.." -Suhaimi-

From cases like Suhaimi's story and other villagers who use lanting as transit dwelling to enter the urban life, it can be said that lanting is the initial representation of the presence of rural people who immigrate to the city to seek a better life. In a relatively poor physical condition of lanting as dwelling place without legal security, they try to feel at home while keep struggling to continue their education or find a better job in the city, for the sake of better future. Turner (1968) explained such phenomena by arguing that value which lies in the relationship between man and environment is not simply a matter of physical conditions. "A good environment is not necessarily one of the high physical standards; if the inhabitant wants to save money to send his children to school, a poor but cheap house will be better than an expensive one of high material standards that would ruin the family's long-term expectations" [15]. He even wrote: "Even if the dwelling is no more than a shack on a plot of rough land, it frees the owner from rent payments; it also provides him with opportunities for additional income and the chance to invest his small savings, skills, and spare time" [15, p.357]

Lanting as a representation of "being" or *Dasein* in Heidegger's concept, is an attempt of rural people to show his existence in an urban environment no matter how unfamiliar the environment to him. As Heidegger stated (Rose, 2012): "to say that mortals are is to say that in the dwelling they persist through spaces by virtue of their stay among things and locations." As dwelling in an urban context, lanting can also be seen as the practice of marking and claiming, where lanting as *Dasein* exists in an urban setting, but this existence is not seen as considered merely as lanting itself [16]. In this context, lanting is made to mark and claim the site where the urban environment (as the "fourfold") can be disclosed as not simply the environment or unfamiliar environment, but as 'our' environment for them. So in this lanting, rural people as *Dasein* exists *vis-a-vis* the changing and transforming urban life. Through lanting they are struggling to hold onto this relentlessly changing world by claiming it as their own.

5. Conclusions

Lanting or floating houses in Sintang are part of urban settlements, which compose urban functions, such as residential, lodging, stalls, restaurants, kiosks, and so on. There are different versions about the origin of lanting in Sintang. Some stated that lanting was first introduced by Chinese traders, while others believed that lanting had existed long before the arrival of Chinese to Sintang. However, the strategic location of Sintang in the midst of river trade routes has made the lanting community played important role in the development of Sintang in the past.

Today, although the riverine culture is no longer as popular as it was once since the dominance of land transportation, lanting still exist and has evolved into transit functions. From the main activities identified in the daily life of lanting community such as docking, lodging, pop-up markets, and temporary dwelling, it can be said that lanting plays an important role as one of urban elements in Sintang city that represents the urban-rural linkage between Sintang and the surrounding countrysides, in a complex hydro-social system. Lantings and the rivers, which cannot be separated from each other, have become a combination of nodes and lines that open the opportunity of urban-rural interaction which then influences each other. Lanting is also the initial representation of the presence of rural community, especially low-income people, who immigrate to the city to seek a better life, or in other words, lanting is a transitional space from rural life to urban life.

6. Acknowledgment

This work is supported by Hibah PITTA 2018, funded by DRPM Universitas Indonesia No.5000/UN2.R3.1/HKP.05.00/2018

7. References:

[1] Evers H D 2015 *JMBRAS*, **88** Part 1 (2015), pp. 1–xx
[2] Enthoven J J, Yeri P 2013. Sejarah dan geografi Daerah Sungai Kapuas Kalimantan Barat (Pontianak: Institut Dayakologi)
[3] Koekoek M 2010 *Connecting Modular Floating Structures* (TU Delft)
[4] Gibson N 2006 *Architecture in a Northern Flood Plain UWSpace*. http://hdl.handle.net/10012/ 901
[5] Prosun P 2011 *The Lift House: An amphibious strategy for sustainable and affordable housing for the urban poor in flood-prone Bangladesh UWSpace*. http://hdl.handle.net/10012/5787
[6] Travers A 2012 *Floating Communities on Tonle Sap Lake, Cambodia: Spatial patterns and Adaptation Capacity School of Geography, Planning and Environmental Management* (University of Queensland)
[7] Jhearmaneechotechai P 2015 *Journal of Environmental Design and Planning*, **11**, 41-50.
[8] Afdholy A R 2017 *Local Wisdom Scientific Online Journal*, **9** (1): 12-26
[9] Rahman M A 2014 *E-Journal Graduate Unpar Part D – Architecture*, **1** (2)
[10] Iskandar Y, Lahji K 2010 *Local Wisdom* **2** (2) 37 - 45
[11] Mustansyir R 2013 Kearifan dan Kendala Lokal Warga Lanting Sebagai Penghuni Pinggiran Sungai Sambas di Kalimantan Barat Prosiding *The 5th ICSSIS International Conference on Indonesian Studies: "Ethnicity and Globalization"*, 13-15 June 2013
[12] Heidhues M S 2003 *Golddiggers, farmers, and traders in the "Chinese districts" of West Kalimantan Indonesia* (Ithaca: Southeast Asia Program Publications, Southeast Asia Program, Cornell University)
[13] Listiana D 2012 *Tanah Sintang Masa Kolonial: telaah sederhana tentang perubahan status pemerintahan dan wilayah kekuasaa*n (Pontianak: STAIN Pontianak Press)
[14] Sjamsuddin H, Nursam M 2013 *Perlawanan & perubahan di Kalimantan Barat: kerajaan Sintang*, 1822-1942 (Yogyakarta: Penerbit Ombak)
[15] Turner J C 1968 *AIP Journal* November 1968 p.354 - 363
[16] Rose M 2012 Dwelling as marking and claiming Environment and Planning D: Society and Space 2012, **30** 757 – 771

IOP Conf. Series: Materials Science and Engineering **523** (2019) 012061 doi:10.1088/1757-899X/523/1/012061

A review of vertical evacuation on tsunami mitigation case

M Haiqal[1,*], L H Sari[1], Evalina Z[1], P Hasibuan[2]
[1] Architecture and Planning Department, Engineering Faculty, Syiah Kuala University, Indonesia
[2] Civil Department, Engineering Faculty, Syiah Kuala University, Indonesia

E-mail: mhaiqal@unsyiah.ac.id

Abstract. Banda Aceh City was hit by a tsunami at the end of 2004. The experience of this great disaster directly or indirectly has led to many problems that are a must in finding the right solution. It is very important to do in order to minimize the disaster victims of the tsunami in the future. During the rehabilitation and reconstruction period after the 2004 earthquake and tsunami, between 2005 until now, various mitigation efforts have been carried out. One of the tsunami mitigation efforts is to build tsunami evacuation buildings. The city of Banda Aceh is one of the cities in Indonesia that has vertical tsunami evacuation facilities such as escape building. Although it still does not meet the expected availability standards, it can be called the right initial step for Banda Aceh City in the face of the tsunami disaster in the future. However, its existence will be in vain if the community does not use it to its full potential in their daily activities. This study evaluates the effectiveness of the escape building for vertical evacuation based on some supporting literature reviews.

1. Background

The December 26th, 2004 tsunami caused by the 9.1 SR earthquakes on the ocean floor of the Indian Ocean (west of Aceh) hit Aceh. The tsunami caused hundreds of thousands of people dead and missing, destroying thousands of homes, schools, office buildings and everything in front of the tsunami. In numbers, it caused the loss of 60,065 lives and damaged 21,412 houses in the city [1], [2]. After the disaster occurs, various mitigation efforts are carried out, from the emergency response to the rehabilitation and reconstruction period with all capabilities. These efforts involve many parties, both from within and outside the country. The experience is felt directly or indirectly, has spawned many problems that become challenges to find the right formula. This is done to minimize the tsunami disaster victims in the future.

During the rehabilitation and reconstruction period after the 2004 earthquake and tsunami, a period between 2005 and now, various mitigation efforts have been carried out. One of the tsunami mitigation efforts is to build tsunami evacuation buildings. The reason is that of the relatively flat topography of Banda Aceh which is on average only 80 cm in height from sea level [3]. Rescue buildings have been built at several points located in post-2004 tsunami-prone zones in Banda Aceh. Many rescue buildings were built at points prone because the zone has been re-populated either by residents who used to live there or even newcomers ([4], [5]).

*Corresponding author: mhaiqal@unsyiah.ac.id

IOP Conf. Series: Materials Science and Engineering **523** (2019) 012061 doi:10.1088/1757-899X/523/1/012061

2. Literature References

Tsunami comes from Japanese. "Tsu" means harbor, "nami" means wave so that it is generally interpreted as a great tide of the sea in port or other words coastal / waterfront area [6]. According to Ministerial Regulation No EMR. 15 the Year 2011, Tsunami is a wave caused by geological processes under the sea in the form of earthquakes, volcanic eruptions, landslides, and the fall of the meteor towards the sea. Tsunamis occur due to earthquakes at shallow depths, as most of the energy released into the ocean water pools above them. The submarine earthquake strikes a large mass of seawater in a powerful beat. The backwaters are crashing at speeds of up to 800 km / h, as they approach the shore the waves slow down but push upward, slam into the land, and destroy whatever is behind the shore.

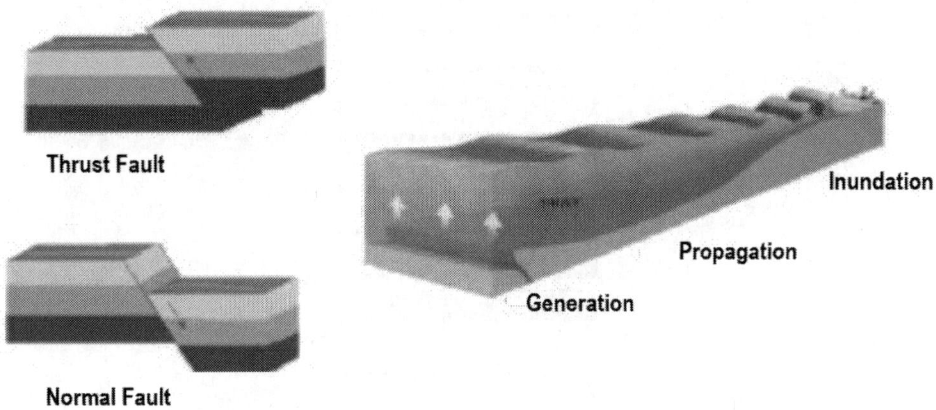

Figure 1. Illustration of tsunami incidence due to fracture.

The earthquake that caused tsunami including:

- Earthquakes centered at sea and shallow (0 - 30 km)
- Earthquakes with a power of at least 6.5 on the Richter scale
- Earthquakes with rising fault patterns or falling faults

Mitigation is a series of efforts to reduce disaster risks, both through physical development and awareness and increased capacity to deal with disaster threats. According to McDonald [7], mitigation includes all activities to prevent emergencies, reduce the likelihood of an emergency, or reduce the deleterious effects of an emergency that cannot be avoided. In the event of a disaster, the most important concern is how to overcome the impact. The more urgent to look for survivors and guide them safely, provide temporary relief and shelter, stabilize the situation, and clear the debris.

In the context of disaster risk reduction, the paradigm shift from disaster management to disaster risk reduction has become one of the government's obligations. The Government is obliged to protect the life and livelihood of the people, including the protection of disasters in the context of realizing common prosperity. To mitigate the potential impacts of a disaster, it is necessary to establish risk identification that is an interaction of threats, vulnerabilities, and capacities. These three components form the basis of the risk assessment generated as an estimate of the impact of the disaster.

The steps that need to be done in the preparation of disaster risk identification in a region include:

a. Identification of Threats
b. Identification of Vulnerability
c. Identification of Capacity
d. Identification of Risks

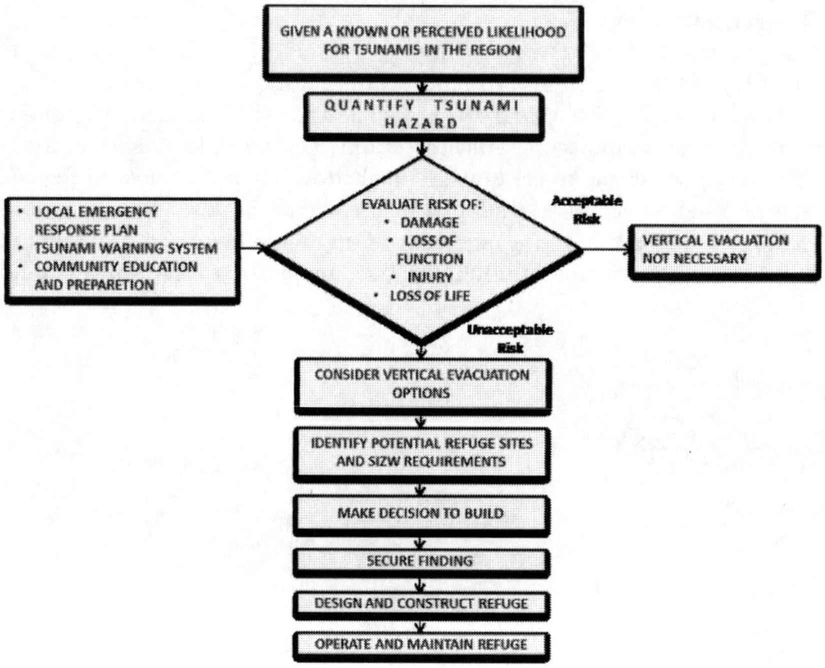

Figure 2. Decision-making and design process for vertical evacuation structures (Source: FEMA 646a [8].

Vertical evacuation structures should be easy to reach and evenly distributed throughout the tsunami inundation zone. Site location will be depending on how much time it takes for people to reach its structure. Vertical evacuation structures not only must able to withstand tsunami waves and debris (built-in trash) but must also withstand previous earthquakes and remain functional. The structure should not look damaged after an earthquake, for example, major cracks in walls or other signs of damage as people will refuse to enter the building [9].

- Types of Vertical Evacuation, according to FEMA 646a [10], are:
  - Natural or artificial highlands
  - Parking Lots Building
  - Community Facilities Building
  - Commercial Building
  - School Building
  - Existing buildings.

- Vertical Evacuation Structure Needs Analysis according to FEMA 646a [11], is as follows:
  - Topographic area;
  - Age and type of construction of existing buildings;
  - Number of residents and visitors in an area;
  - Population diversity due to seasons and other upheavals;
  - Number of vulnerabilities and population size;
  - The readiness of population and visitors, and
  - Emergency management and event response readiness.

- The FEMA P646 guidelines stated that tsunami rescue buildings should have the following design features [12]:
  - A strong system with the spare capacity to withstand extreme power;

IOP Conf. Series: Materials Science and Engineering **523** (2019) 012061 doi:10.1088/1757-899X/523/1/012061

- An open system that allows water to flow through minimal resistance;
- Elastic system, capable of withstanding extreme forces without significant damage;
- A redundant system that can suffer partial damage without experiencing progressive collapse/streak.

3. Methods
The method used in this study is by reviewing the existing literature. In addition to the review, this study will also use observational methods in order compare some of the existing Tsunami Evacuation Building in Banda Aceh with the Tsunami Evacuation Building in Japan. The factors that will be compared and observed is whether the building has been used optimally and its potential in saving human lives when disaster strikes.

4. Discussion
Japan is a country that has experience of the tsunami which often occurred in almost all areas. The latest tsunami occurred on March 11, 2011. The tsunami was caused by the 9.0 magnitude earthquake in the Pacific Ocean Tohoku, Japan. The tsunami killed more than 10,000 people and destroyed towns which also caused an explosion at the Fukushima nuclear power plant. Japan with its technology has many models of mitigation to face the tsunami in the country. One model is a rescue building built with various models and functions.

The Rescue Building at Shirahama Beach Resort in Japan is designed to keep people safe when a tsunami strikes. Concrete Building in Kaifu, Japan, used as a vertical evacuation. Buildings like this are easy to reach, easy to navigate, and also protect the destructive power of the tsunami [13]. In Kise, Japan, rescue building also functioned as a library at the bottom. Buildings like these can be used as community centers and can generate income that can be used for its operations cost.

Figure 3. Tsunami Evacuation Building in Shirahama Beach Resort, Japan

Figure 4.Tsunami Evacuation Building in Kaifu, Japan.

Figure 5. Tsunami Evacuation Building in Kise, Japan

Meanwhile, Banda Aceh is one of the cities where Tsunami occurred back in 2004, after direct observation, not used the Tsunami Evacuation Building to its maximum potential. These tsunami evacuations building also not functioned on a daily basis, this causes the facilities and infrastructure either disfunction or damaged, even if most of these buildings have been designed in accordance to Indonesian national standard and the possibility of site effects. The site effect is controlled by many factors [14]. Incorporating this site effect into building design is crucial for seismic-resistant building ([15], [16]).

Figure 6. Tsunami Evacuation Building in Alue Deah Tengoh, Banda Aceh

Tsunami evacuation buildings in Banda Aceh have fulfilled the criteria of spatial comfort. The spatial comforts are such as good air circulation, sufficient daylight and good outlook [17] of strong structure ([18], [19]). However, from the picture above, it can be seen that the building the building is empty, and some of the rooms look neglected. This building does not function as public facilities for the community. Unlike the Tsunami Evacuation Building in Japan where the ground floor is also functioned as a library. Another evacuation building located in Banda Aceh is Evacuation buildings in Gampong Deah Glumpang. This evacuation building is also aid from Japan. The condition of the building is not functioned in everyday activities. It is unfortunate that this grand building is not used to its full potential.

Figure 7. Tsunami Evacuation Building in Deah Glumpang, Banda Aceh.

This condition can be improved by opening these buildings to the public. In the mitigation process, the opening of the evacuation building to the community can also increase the trust of the surrounding community towards this evacuation building. The desired goal in this method by opening the building to the public is for people to have a certain level of trust and a sense of familiarity towards the building, so that during the disaster, especially when tsunami happen in Banda Aceh, people will immediately evacuate themselves directly to this evacuation building without thinking twice.

Another evacuation building is in Gampong Pie, Meuraxa Subdistrict, Banda Aceh, unlike other evacuation buildings that have special functions as evacuation buildings, it is only functioned as an office and study center. The rescue building belonged to Syiah Kuala University and became one of the university's study centers, the Tsunami and Disaster Mitigation Research Center (TDMRC).

Figure 8. Tsunami and Disaster Mitigation Research Center (TDMRC).

TDMRC can resist earthquake up to 10 Richter scale. Based on the direction of the upcoming wave, the position of this building has taken into account. But unfortunately, the floor below (1st floor) is very massive with partition placement is predicted to inhibit the lunge of tsunami waves in the future. The 2nd floor of this rescue building only has a height of about 5 m from the ground. The 2nd floor can not be used as a safe place for human evacuation if a tsunami hit with level as on December 26, 2014. The 3rd floor of this building has a height of about 10.2 m from the ground surface. Thus this floor can be used as a place of evacuation if disaster tsunami arrived. Likewise, the 4th floor is open which can accommodate refugees who evacuate themselves to the building.

IOP Conf. Series: Materials Science and Engineering **523** (2019) 012061 doi:10.1088/1757-899X/523/1/012061

Figure 9. Museum Tsunami Aceh

The existence of the Aceh Tsunami Museum (MTA) is relatively good in responding to the need for vertical evacuation during the earthquake with the threat of a tsunami. On 11 April 2012 after the M8.5 earthquake occurred, surrounding residents and students and teachers from SMP Negeri 1 Banda Aceh located not far from the Aceh Tsunami Museum to evacuate themselves to the museum. When the panic happens on the main evacuation path on Iskandar Muda road, it is difficult to avoid the occurrence of bootle-neck. The museum became an effective alternative for self-evacuation.

Furthermore, the accessibility of Museum Tsunami Aceh has been equipped with lifts, stairs, and ramps. The museum has 2 units of stairs located in the south and north of the building. The ramp is available starting from the east side of the building which is connected to the central area of the museum by passing over the pool located in the center of this museum, to the 2nd floor at 8 m and continuing to the 3rd floor at the height of 12 m.

5. Conclusion

These buildings should be used optimally; one effort that can be done is to use the building as public facilities. The simplest example is to use the lower part of the building as a library, this example is chosen because the library can attract people, especially children, so it can also increase people's trust towards the building during a tsunami occurs or other disasters. Although Banda Aceh already has 5 evacuation buildings, there are still many things that need to be addressed, for example in addition to the opening of the building to the general public, is the management of building's maintenance and the improvement itself. When associated with public trust, buildings that are well maintained, and look solid from the outside will create a high level of trust in the evacuation process when disaster comes.

6. References

[1] Sari, L.H; Harrid, D.J and Gormley, M. 2013. *Int. Journal for Housing Science*, **37**, (3),161-173, 2013 Published in the United States.

[2] Nurdin, M. 2006. *Journal of Disaster Research*, **1** (1), 116-122.

[3] Tursina, Syamsidik and Shigeru, K, 2017. Projections of Tsunami Inundation Area Coupled with Impacts of Sea Level Rise in Banda Aceh, Indonesia. *AIP Conf. Proc.* **1892**, 100003-1–100003-8.

[4] Harris, D. J., and Sari, L. H., 2015. Assessment of Tsunami mitigation measures. *In Proceedings of 2015 TAU Conference: Mitigating and Adapting Built Environments for Climate Change*

in the Tropics. School of Architecture, Tanri Abeng University, Jakarta, Indonesia, 30-31 March 2015

[5] AtjehLink. 2014. *Ada Bunyi Sirine Tsunami Hari Minggu*, Jangan Panik.

[6] Selena Simmons-Duffin, 2011, History Of Tsunami: The Word And The Wave, https://www.npr.org/2011/03/18/134600508/history-of-tsunami-the-word-and-the-wave

[7] McDonald, Roxanna. 2003. *Introduction to Natural and Man-made Disasters and their Effects on Buildings* (First published, Architectural Press) Oxford, UK.

[8] FEMA. 2009. *Vertical Evacuation from Tsunamis: A Guide for Community Officials* (Federal Emergency Management Agency) Washington DC

[9] Ibid

[10] Ibid

[11] Ibid

[12] FEMA, 2008. *Guidelines for Design of Structures for Vertical Evacuation from Tsunamis*, Federal Emergency Management Agency, Washington DC.

[13] Leonard, G. S.; Evans, N.; Prasetya, G.; Saunders, W.S.A.; Pearse, L.; Monastra, D. and Fraser, S. 2011. *Scoping study for evaluating the tsunami vulnerability of New Zealand buildings for use as evacuation structures* (GNS Science Report) 2011/36 39 p.

[14] Setiawan B, Jaksa MB, Griffith MC., and Love D 2018 *Eng Geo.* **243** 145 – 159.

[15] Setiawan B, Jaksa MB, Griffith MC and Love D 2018 *Boll. di Geofisica Teorica ed Applicata* **59** (1) 27–46.

[16] Setiawan B, Jaksa MB, Griffith MC and Love D 2018 *Soil Dyn Earthq Eng.* **110** 244–261.

[17] Sari, L. H, Yuzni, S. Z, Haiqal, M, and Evalina, Z, 2018. A review of Spatial Comfort in Shophouse in Humid Tropics,, Advancing Knowledge, Research, and Technology for Humanity. *IOP Conference Series: Materials Science and Engineering*, **352**, 1757-8981.

[18] Hasibuan, P, Panjaitan, A, and Haiqal, M, 2018, Compression member response of double steel angles on truss structure with member length variation, Advancing Knowledge, Research, and Technology for Humanity. *IOP Publishing*, **1757**-8981.

[19] Haiqal, M, Kusumawanto, A and Saragih, S, 2016, Effectiveness of Tsunami Evacuation Building as a Tsunami Disaster Mitigation Effort in Banda Aceh, Enhancing the quality of science and engineering through sustainable research and development to face ASEAN Economic Community challenges, *The International Conference of Engineering and Science for Research and Development (ICESReD)*, 2541-4151.

IOP Conf. Series: Materials Science and Engineering **523** (2019) 012062 doi:10.1088/1757-899X/523/1/012062

Effect of sodium sulfate and sodium chloride in two-stage thermal upgrading of low-grade Nickel Lateritic Ore

B Suharno [1,*], C Ramadini[1], R H Shaleh[1], A Shofi[1], F Nurjaman[2]

[1]Metallurgical and Materials Engineering Department, Faculty of Engineering Universitas Indonesia, Kampus Baru UI Depok, 16424 – Indonesia
[2]Indonesian Institute of Science, Lampung, Indonesia

E-mail: suharno@metal.ui.ac.id

Abstract. Selective reduction process and magnetic separation of nickel lateritic ore were conducted through the two-stage thermal mechanism with the addition of sodium sulfate and sodium chloride as the additives agent. The first thermal process was pre-heating at the temperature of 500°C and holding for 90 minutes. Afterward, the reduction continued at the temperature of 1150°C for the same holding time. The mineralogical composition of the reduced samples was determined by XRD. The microstructure of reduced nickel laterite ore was also examined using SEM-EDS. The optimum composition of the additive was the combination of 10% Na2SO4 + 10% NaCl, which produced a concentrate with 5.52% and 85.89% for nickel grade and nickel recovery, respectively.

1. Introduction

Nickel is a precious metal in the infrastructure and technology field, especially in making industries of stainless steel, Ni-based alloy, alloy steels, electroplating, batteries, etc [1]. Nickel was used either in the form of metallic nickel or form of NPI/ferronickel with various of iron content [2]. The extraction process of nickel lateritic requires high intensive energy and usually, nickel is melted to produce low-grade ferronickel with the high amount of slag. Furthermore, this pyrometallurgy process is merely used for low-grade nickel lateritic ore containing nickel grade above 1.5%. Nevertheless, the grade of nickel lateritic ore around the world including Indonesia is about 1.45%. Therefore, this conventional process is ineffective [3]. Recently many researchers focus on the method of upgrading the effectiveness and efficiency of pyrometallurgy process to produce ferronickel by using selective reduction process followed by magnetic separation [4], [5]. Ferronickel is generally used in stainless steel making process so that the result of this process will be able to supply the market needs [6].

In the process of reduction, the addition of the optimum composition and selection of additive could generate the optimum recovery with low reduction temperature. It would be one of the efficient steps in the selective reduction of nickel lateritic ore [7].

In previous research, the sulfur compound had been used as the additive agent in selective reduction process, such as sodium sulfate (Na_2SO_4) [8],[9]. It resulted that the addition of sulfur compound in selective reduction process followed by magnetic separation process could be able to increase the nickel grade and recovery process [4],[10]. Another compound that can be used as the additive is chlorine salt [11],[12], such as sodium chloride (NaCl). Due to the characteristic of chloride

* Corresponding author: suharno@metal.ui.ac.id

which has low melting temperature and volatile properties, chloride compound generally used for extracting metal from its ore through chloridization and volatilization process [13].

The objective of recent work is to extracting nickel oxide of low-grade nickel ore from Desa Morombo, Konawe Utara District, Sulawesi Tenggara Province through selective reduction process followed by magnetic separation with the addition of the combination of sodium sulfate (Na_2SO_4) and sodium chloride (NaCl) as additive and palm kernel shell as the reductant.

2. Experimental

Raw materials in this research are low-grade nickel limonitic ore, palm kernel shell as reductant, sodium sulfate and sodium chloride as additives. The result of XRF analysis for nickel lateritic ore and proximate analysis of the reductant are listed in Table 1 and Table 2, respectively.

Table 1. The Chemical composition of the nickel laterite ore (mass fraction, %)

Element	Ni	Fe	Si	Mg	Al	Ca	Cr	Mn	Co
Mass Fraction (%)	1.4	50.5	16.5	1.81	4.86	0.177	2.68	0.847	0.0662

Table 2. Proximate analysis of palm kernel shell (wt.%)

Reductant	Volatile Matter	Ash	Fixed Carbon	Moisture
Palm Kernel Shell	22.57	21	77	0.43

The ore was dried in the oven at temperature 105°C for 24h to reduced the water content until less than 5 wt.%. Limonitic dried ore and the reductant were crushed and sieved to gain the particle less than 147 μm. Afterwards, the 50gr of ore, 5,5gr of reductant, and additives were mixed and pelletized with various composition of additives content, i.e. 15% Na_2SO_4 + 5% NaCl (code AB), 5% Na_2SO_4 + 15% NaCl (code AC), and 10% Na_2SO_4 + 10% NaCl (code L). The pellets were dried in the oven at the temperature of 100°C for 2h.

Two-stage thermal of selective reduction process was performed in this research. The first stage of reduction temperature was 500°C and the second stage temperature was 1150°C. Each stage was held for 90 minutes. After the selective reduction was done, the pellets were cooling rapidly in water. The pellets then were dried and crushed up to74 μm.

The phase transformation was determined by XRD (*X-ray Diffraction*) to observe the transformation of mineral composition and phase which form after the selective reduction process. Samples were also examined by SEM-EDX (*Scanning Electron Microscopy*) to observe the microstructure of phase and its distribution. Afterward, the wet magnetic separation was conducted by dissolving the reduced ore into the water and separating the concentrate and tailing using 500 gauss magnet. The ratio of reduced ore and water in this magnetic separation process was 1:10.

3. Results and Discussions

Figure 1 shows the iron and nickel grade and recovery of the concentrate in this selective reduction process with various additives.

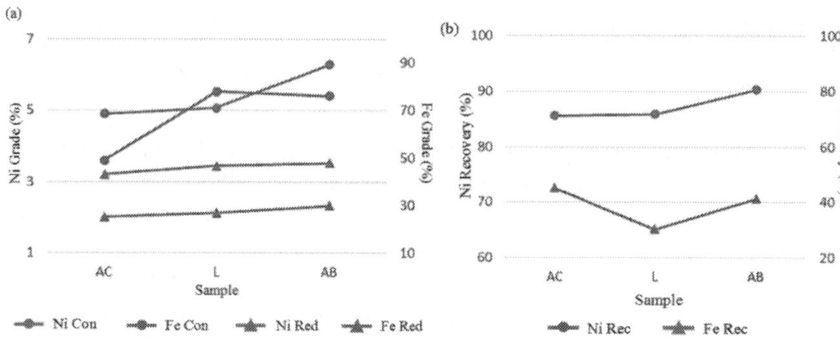

Figure 1. Effect of various additive on (a) grade of nickel and iron, (b) recovery of nickel and iron in concentrate at the reduction temperature of 1150°C

The nickel grade of sample AC is 3.61%. The grade is increased up to 5.53% in sample L and decrease to 5.41% in sample AB. The recovery of iron has the opposite tendency with the nickel grade. When the nickel grade is getting higher, the recovery of iron is decreased. The recovery of iron in sample AC is 45.09% and decreasing to 30.31% in sample L, but it was increased to 41.28% in sample AB.

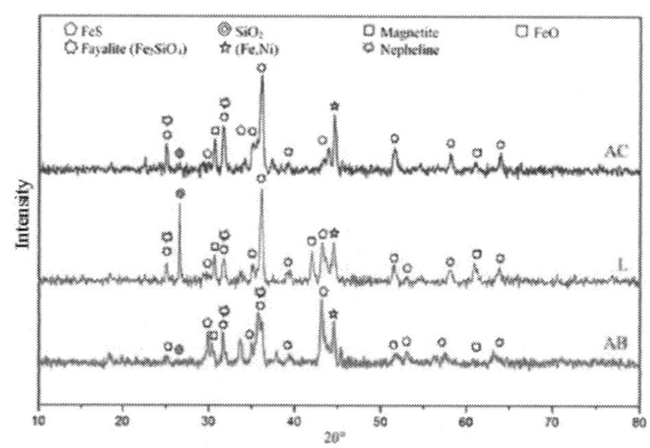

Figure 2. XRD patterns of reduced nickel ore with the variety of additive

Figure 2 shows that the observed phase in each additive composition is kamacite (FeNi), wustite (FeO), magnetite, troilite (FeS), silica, fayalite, and nepheline. The less addition of sodium sulfate in sample AC than sample AB resulted in a small amount of FeS which was found in reduced ore, as shown in Figure 2. The intensity of FeS in sample L is higher than in sample AC due to more addition of sodium sulfate composition in sample AC. The appearance of FeS caused to lower iron recovery due to its non-magnetic characterization. In the magnetic separation process, the magnet will bind magnetic mineral and bring it to concentrate. FeS which has no magnetic characterization will not get into concentrate [8].

The addition of sodium chloride promotes in decreasing iron recovery due to the segregation process. Sodium chloride in pellets reacted with water vapor in it through pyrohydrolysis process to form hydrochloride acid. It reacted with metal oxide to form metal chloride. Nickel chloride and iron chloride absorbed on the surface of the carbon particle, then reduced by hydrogen and carbon monoxide. It produces segregated chloride [12]. The segregation rate of iron chloride is lower than nickel chloride. Therefore, most of the iron chloride is volatilizing before it segregated so that the iron recovery decreased. The excess addition of sodium chloride would result in more intensive of chloridization proses, thus the volatilizing of nickel chloride happened and it will decrease the recovery of nickel.

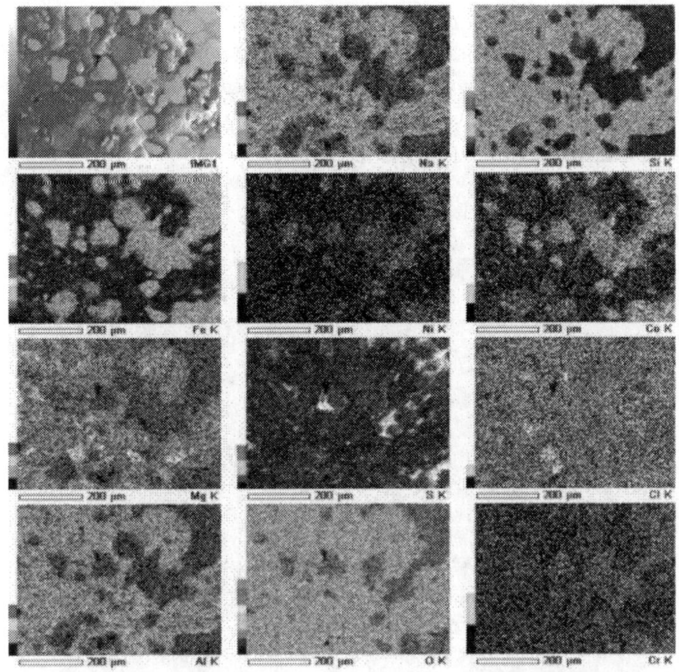

Figure 3. SEM Mapping of Optical Micrograf of Sample L at reduction temperature of 1150°C

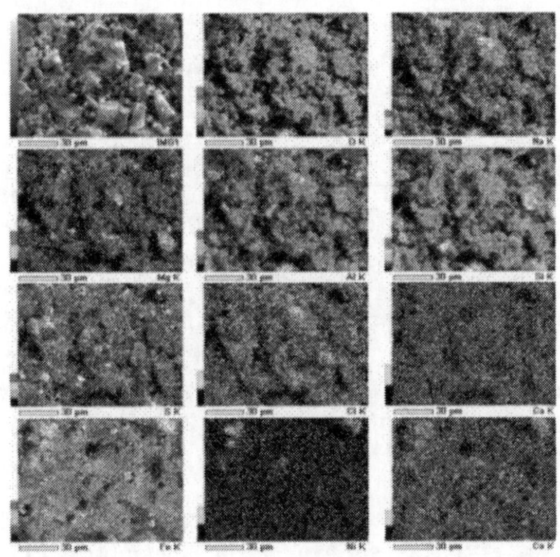

Figure 4. SEM Mapping of Sample L powder at the reduction temperature of 1150°C

The appearance of new FeO peak at the 2θ angle of 42° in sample L indicates that the appearance of FeO is multiplying in the reduced ore. The excess of sodium chloride will bind more FeO to form iron chloride, meanwhile in the excess addition of sodium sulfate will bind more FeO to form FeS. The value of FeO in sample AB and AC are measly and its peak cannot be observed. The increasing of FeS in sample AB is due to the more sulfur in the additive. Silica and wustite are in sample L. It could be caused by obstruction of the reaction which happened to both because of a particular mechanism. This mechanism somehow happened because of the balance composition between sodium sulfate and sodium chloride which added as the additive in sample L. The presence of both phases would be an advantage in the result of magnetic separation process because both of these phases will not get into the concentrate, so the nickel grade in concentrate will increase.

Figure 3 and Figure 4 show that iron can be separated away from the gangue element such as sodium, silicon, magnesium, and aluminum. Iron has its own mapping pattern and concentrated with nickel element to form ferronickel. Iron with sulfur element enfold ferronickel and becoming troilite. Meanwhile, it can be seen a few irons concentrated with silicone and oxygen forming fayalite. Some of the oxygen element concentrated with silicon to form silica and some of them concentrated with iron to form iron oxide. Sodium seems to be concentrated with silicone, magnesium, and aluminum to form a nepheline phase.

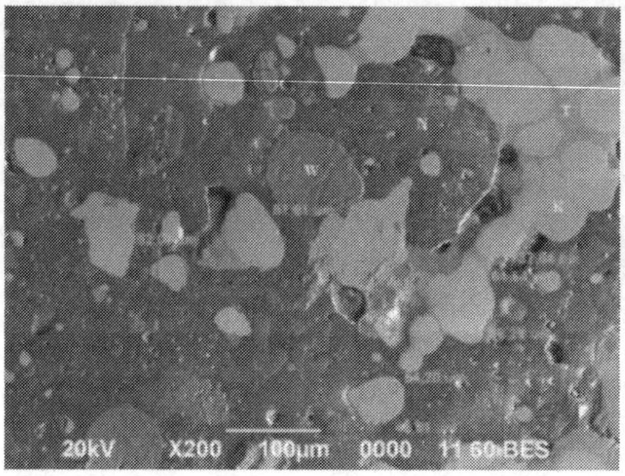

Figure 5. Optical Micrograf and Particle Size of Sample L at Temperature Reduction of 1150°C (K = Kamacite; T = Troilite; W = Wustite; N = Nepheline)

Figure 5 shows the average of ferronickel particle size is 61,75µm. At the temperature of 1100°C, the reaction of NaO-MgO-SiO_2 forms nepheline ($Na_2Mg_2Si_2O_7$). The low melting point of nepheline created liquidus phase at 1100°C, thus promoted the growth of ferronickel particle size [10].

4. Conclusions

The phases which formed in the reduced ore with the addition of sodium sulfate and sodium chloride at the reduction temperature of 1150°C for 90 minutes are kamacite (FeNi), wustite (FeO), troilite (FeS), fayalite (Fe_2SiO_4), and nepheline ($Na_2Mg_2Si_2O_7$). The sulphidation, chloridization, and segregation process will occur because of the addition of sodium sulfate and sodium chloride. It will increase nickel grade and recovery. Furthermore, sulfur in sodium sulfate will generate the grain growth of kamacite particle through the forming of Fe-FeS system. Nepheline phase that occurs in the low temperature can also facilitate the grain growth. The optimum percentage composition of the combination of the additive is 10% Na_2SO_4 + 10% NaCl which generates nickel grade of 5,52% and nickel recovery of 85,89%.

5. Acknowledgment

The authors would say thanks to DRPM UI for their financial support as PITTA grant funds with contract number of 2395/UN2.R3.1/HKP.05.00/2018 that offered to perform this work.

6. References

[1] Mudd GM 2010 *Ore Geol. Rev* **38** (1–2) 9–26.
[2] Kim J, Dodbiba G, Tanno H, Okaya K, Matsuo S, and Fujita T 2010 *Miner. Eng* **23** (4) 282–288.
[3] Norgate T and Jahanshahi S 2011 *Miner. Eng.*, **24** (7) 698–707.
[4] Zhu DQ 2012 *Int. J. Miner. Process.*, **106–109**, 1–7.

[5] Li Q 2010 *Int. Miner. Process. Congr.*, 1549–1556.
[6] Kobayashi K, Todoroki H, and Tsuji H 2011 *Iron steel Int. J.* **51** (1), 35–40.
[7] Elliott R, Pickles CA, Peacey J 2017 *Miner. Eng.* **100** 166–176.
[8] Jiang M, Sun T, Liu Z, Kou J, Liu N, and Zhang S 2013 *Int. J. Miner. Process.* **123** 32–38.
[9] Chen GJ, Shiau JS, Liu SH, and Hwang WS 2016 *Mater. Trans.* **57** (9) 1560–1566.
[10] Rao M, Li G, Zhang X, Luo J, Peng Z, and Jiang T 2016 *Sep. Sci. Technol.* 51 (8) 1408–1420.
[11] Liu WR 2010 *Trans. Nonferrous Met. Soc. China (English Ed.*, **20 (1)**, 82–86.
[12] Zhou S, Wei Y, Li B, Wang H, Ma B, and Wang C 2016 *Metall. Mater. Trans. B Process Metall. Mater. Process. Sci.*, **47** (1), 145–153.
[13] Zhou S, Wei Y, Li B, Wang H, Ma B, and Wang C 2016 *Sci. Rep.*, **6**, 1–12.

IOP Conf. Series: Materials Science and Engineering **523** (2019) 012063 doi:10.1088/1757-899X/523/1/012063

The degree of filler dispersion, rheometric and mechanical properties of carbon black-filled styrene-butadiene rubber composites in the presence of alkanolamide

I Surya [1,2,*], H Ismail[3]
[1]Department of Chemical Engineering, Faculty of Engineering, Universitas Sumatera Utara, Medan, 20155, Indonesia
[2]Sustainable Energy and Biomaterial Center of Excellence, Faculty of Engineering, Universitas Sumatera Utara, Medan, 20155, Indonesia
[3]School of Materials and Mineral Resources Engineering, Engineering Campus, Universiti Sains Malaysia, 14300, Nibong Tebal, Penang, Malaysia

E-mail: indradanas@yahoo.com

Abstract. The degree of filler dispersion, rheometric and mechanical properties of carbon black (CB)-filled styrene-butadiene rubber (SBR) composites in the presence of alkanolamide (ALK) were investigated using a semi-efficient sulphur accelerated vulcanisation system. The ALK was prepared from Refined Bleached Deodorized Palm Stearin and diethanolamine and added into the CB-filled SBR composites at1.0, 3.0, 5.0 and 7.0 parts per hundred rubber (phr). It was found that ALK decreased the scorch time, cure time and minimum torque of the CB-filled SBR composites. The ALK also increased the torque difference, tensile modulus, hardness, resilience and tensile strength; especially up to a 5.0 phr of loading. Due to its plasticising effect, ALK improved the degree of CB dispersion and increased the elongation at break of the CB-filled SBR composites. The crosslink density was increased by improving the degree of CB dispersion and rubber-filler interaction, respectively. The 5.0 phr of ALK was the optimum loading for CB-filled SBR composites which exhibited the highest values in tensile modulus, hardness, resilience and tensile strength.

1. Introduction

Various raw rubbers, rubber chemicals and fillers with their suitable amounts are mixed during the compounding operation [1] in order to achieve rubber composites. The raw rubbers provide rubbery properties to the composites; fillers reinforce the rubbers or reduce the cost of composites products; plasticisers decrease the viscosity, improve deformability and change properties; and accelerators enhance the rate of vulcanisation reaction[2, 3].

By vulcanisation, the weak and plastic raw rubbers are converted into strong elastic rubber vulcanisates. The strength and elasticity can be further enhanced by the addition of a type of reinforcing fillers into the rubber composites. The silica and carbon (CB) are the most famous reinforcing fillers in rubber reinforcement. Their particles are very fine and at a relatively higher filler loading. They tend to interact with each other and form filler agglomeration. The agglomeration will deteriorate the properties of the rubber composites and it is considered as a problem in filler

* Corresponding author: indradanas@yahoo.com

dispersion. In general, to solve the filler dispersion problem, a special additive such as processing aid or dispersant aid is added during compounding[4,5].

Therefore, in this study alkanolamide (ALK) was utilised for solving the filler dispersion problem of CB-filled styrene-butadiene rubber (SBR) composites. The ALK was prepared from refined bleached palm stearin (RBDPS) [6]. The degree of CB dispersion, due to the addition of ALK, was investigated. The rheometric and mechanical properties of CB-filled SBR composites without and with ALK were also investigated.

2. Experimental

2.1. Materials

Styrene-butadiene rubber (SBR), Taipol 1502, was purchased from TSRC Corporation, Taiwan. An N330-grade CB was supplied by the Cabot Corporation. Other compounding ingredients, such as sulphur, zinc oxide, stearic acid, N isopropyl-N'-phenyl-p-phenylenediamine (IPPD), and benzothiazolyl disulfide (MBTS), were supplied by Bayer Co. (M) Sdn. Bhd., Petaling Jaya, Selangor, Malaysia. All materials were utilised as supplied. The ALK was prepared using RBDPS and diethanolamine. The procedure of the ALK preparation was given in our previous report [6 - 9] and itschemical formula is $CH_3(CH_2)_{14}CON(CH_2CH_2OH)_2$.

2.2. Compounding

A semi-efficient sulphur-accelerated curing system was used for compounding. The recipe for the preparation of the CB-filled SBR composites is given in Table 1. The compounding procedure was done in accordance with the American Society for Testing and Material (ASTM) – Designation D 3184 – 80. Compounding was done on a two-roll mill.

Table 1. The composition of the SBR compounds

Ingredients	Content (phr, parts per hundred rubber)
SBR	100.0
ZnO	5.0
Stearic acid	2.0
IPPD	2.0
MBTS	1.5
Sulphur	1.5
CB N330	30.0
ALK	0.0; 1.0; 3.0; 5.0; 7.0

2.3. Rheometric properties

The rheometric properties of the CB-filled SBR composites were determined at 150 °C using a Monsanto Moving Die Rheometer (MDR 2000). The rheometric properties such as scorch time (ts_2), cure time (t_{90}), minimum torque (M_L) and maximum torque (M_H) were measured according to ASTM D2084. The CB-filled SBR composites without and with ALK were subsequently compression moulded using a stainless steel mould at 150 °C with a pressure of 10 MPa using a laboratory hot-press based on the respective curing times.

2.4. Tensile, hardness and resilience properties

Dumbbell-shaped samples were cut from the moulded sheets. Tensile tests were performed at a cross-head speed of 500 mm/min using an Instron 3366 universal tensile machine according to ISO 37. The tensile strength and stress at 100% elongation (M100), 300% elongation (M300), and elongation at break were investigated. The hardness measurements of the samples were performed according to ISO 7691-I using a Shore A type manual Durometer. The resilience was studied using a Wallace Dunlop

Tripsometer according to BS 903 Part A8. The rebound resilience was calculated according to the following equation:

$$\% \, Resilience = [(1 - \cos \theta_2) / (1 - \cos \theta_1)] \times 100 \tag{1}$$

where, θ_1 is the initial angle of displacement (45°), and θ_2 is the maximum rebound angle.

3. Results and Discussion

3.1. The rheometric properties

The rheometric properties of the CB-filled SBR composites without and with ALK are shown in Table 2. The scorch and cure times of CB-filled SBR composites with ALK were lower than that of CB-filled SBR composites without ALK. The higher the ALK loading, the lower was the scorch and cure times. It was due to the amine part of ALK which contributed to the cure enhancement [4, 6, 7, 11].

The torque differences (M_H - M_L)of the CB-filled SBR composites with ALK were higher than that of CB-filled SBR composites without ALK. The ALK increased torque difference up to 5.0 phr and started to decrease beyond the loading.

Table 2. The rheometric properties of CB-filled SBRcomposites without and with ALK

Rheometric properties	ALK loadings				
	0.0	1.0	3.0	5.0	7.0
Scorch time, min.	5.07	4.03	2.89	2.23	2.09
Cure time, min.	12.76	11.15	8.15	6.85	6.13
Maximum torque, dN.m	10.11	11.90	12.47	12.96	11.99
Minimum torque, dN.m	1.29	1.27	1.23	1.21	1.08
Torque difference, dN.m	8.82	10.63	11.24	11.75	10.91

The torque difference relates to crosslinking density of a rubber composite[10, 12-14]. A greater torque difference means a greater crosslink density.

The decrease of the torque difference beyond 5.0 phr of ALK was due to the dilution effect of the excessive amount of ALK which most probably absorbed CB filler and part of curatives consequently decreased the total crosslink density.

3.2. The degree of filler dispersion

The degree of CB dispersion in the SBRcomposites, due to the addition of ALK, was determined quantitatively by Equation 2 [15],

$$L = \eta_r - m_r \tag{2}$$

where: $\eta_r = [M_{Lf} / M_{Lg}]$, and $m_r = [M_{Hf} / M_{Hg}]$; where M_{Lf} and M_{Hf} are the minimum and maximum torques of the filled compounds; and M_{Lg} and M_{Hg} are the minimum and maximum torques of the unfilled/gum rubber compound. A lower value of L, at a particular loading, means a better degree of CB dispersion.

Based on the torques properties data as shown in Table 2, values of L for CB dispersion in the SBR phase is shown in Figure 1. The ALK decreased the L value. The higher the ALK loading, the lower was the value of L, which indicated better CB dispersion. The trends of L values were similar to those of minimum torque (M_L) in Table 2. The minimum torque represents the filler–filler inter agglomeration [16]and the value is used to measure the relative viscosity of a rubber compound [7]. The lower the value, the lower the viscosity of the compound-which not only results in easier processability of filler dispersion but also weakened the filler–filler interaction.

The decreases in minimum torque and L value were due to the function of ALK as an internal plasticiser, as mentioned earlier, which decreased the viscosity of the filled SBR composites. A lower viscosity tent to make easier the filler dispersion and enhanced SBR-CB interaction, respectively. The SBR-CB interaction can be defined as additional physical crosslinks [18-19] and, together with sulphide crosslinks, contributed to total crosslink density [20-21].

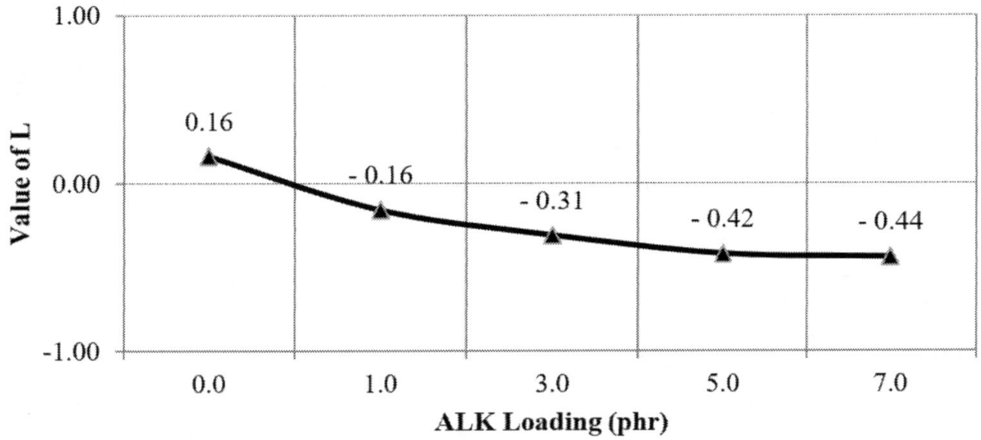

Figure 1. The L values of CB-filled SBR composites without and with ALK

3.3. The mechanical properties

Table 3 represents the M100, M300, TS, EB, hardness and resilience of CB-filled SBR composites without and with ALK. The ALK increased M100, M300, TS, hardness and resilience up to 5.0 phr of loading; beyond the loading started to decrease those properties.

Table 3. The mechanical properties of CB-filled SBR composites without and with ALK

CB-filled SBR	ALK Loadings				
composites	0.0	1.0	3.0	5.0	7.0
M100, MPa	1.31 ± 0.09	1.34± 0.09	1.39± 0.12	1.43± 0.10	1.35 ± 0.09
M300, MPa	3.18± 0.06	3.89± 0.07	4.47± 0.08	4.73± 0.08	3.58± 0.09
EB, %	770.9± 17.5	776.1± 18.9	782.1± 19.8	829.2± 19.2	866.7± 20.2
TS, MPa	18.2 ± 0.5	18.8 ± 0.5	19.1 ± 0.6	19.8 ± 0.6	18.4± 0.8
Hardness, Shore A	55± 0.2	56± 0.2	57± 0.3	58 ± 0.3	56± 0.3
Resilience, %	55.0± 0.4	56.1± 0.4	56.7± 0.6	58.4± 0.5	56.4± 0.6

Tensile modulus (M100 and M300) and hardness of a rubber composite are only dependent on the crosslink density [10]. Resilience is improved to some extent as the crosslink density rises [5]. The enhancements in tensile modulus, hardness and resilience up to the 5.0 of ALK loading were due to a higher crosslink density and the deteriorations in those properties beyond the loading were due to a lower crosslink density. This explanation is in line with the torque difference result in Table 2.

The EB of the CB-filled SBR composites with ALK was longer than that of without ALK. The higher the ALK loading, the longer was the EB. It was simply due to the function of ALK as an internal plasticiser which enhanced the flexibility of CB-filled SBR composites. The ALK provided free volumes which allowed a more flexibility for the rubber chains to move. The higher the ALK loading, the greater were the free volumes and the more flexible were the rubber chains. Presumably, the free volumes were in the layers of excessive ALK[10].

The ALK increased TS of CB-filled SBR composites up to 5.0 phr of loading and started to decrease the TS after the loading. The enhancement in TS up to a 5.0 phr of ALK was due to the plasticizing effect of ALK which softened and plasticised the filled SBR composites achieving a better filler dispersion and greater rubber-filler interaction. The deterioration inTS after the 5.0 phr of ALK was due to the excessive amount of ALK which caused a more pronounced in softening-effect and weaker rubber-filler interaction, respectively.

4. Conclusions

From this study, the following conclusions were drawn:

1) Alkanolamide was as a curative additive which decreased the scorch and cure times of carbon black-filled styrene-butadiene rubber composites. The higher the alkanolamide loading, the lower were the scorch and cure times.
2) Alkanolamide was as an internal plasticiser which decreased the minimum torque but increased the elongation at break of carbon black-filled styrene-butadiene rubber composites. The higher the alkanolamide loading, the lower was the minimum torque but the higher was the elongation at break.
3) The plasticising effect of alkanolamide improved the degree of filler dispersion carbon-black filled styrene-butadiene rubber composites. The degree of filler dispersion further improved with increasing the alkanolamide loading.
4) Alkanolamide increased the mechanical properties of carbon black-filled styrene-butadiene rubber composites. The tensile modulus, hardness, resilience and tensile strength of carbon black-filled styrene-butadiene rubber composites with alkanolamide were higher than those of without alkanolamide.
5) The 5.0 parts per hundred rubber of alkanolamide was the optimum loading for carbon black-filled styrene-butadiene rubber composites.

5. Acknowledgement

The authors would like to thank Universiti Sains Malaysia for providing the research facilities for carrying out the experiment and for making this research work possible.

6. References

[1] Dick J S2001 Rubber Technology: *compounding and testing for performance:* (Hamser)
[2] Surya I, Ginting M and Ismail H 2017 *IOP Conf Ser Mater Sci Eng* **223** 012006
[3] Surya I, Siregar S F and Ismail H 2017 *IOP Conf Ser Mater Sci Eng* **223** 012012
[4] Long H 1985 *Basic Compounding and Processing of Rubber* (Ohio)
[5] Rodgers B 2004 Rubber Compounding: *Chemistry and Applications* (CRC)
[6] Surya I, Ismail H, Azura A R 2013 *Polym Test* **32** 1313-1321
[7] Surya I, Ismail H, Azura A R 2014 *Polym Test* **40** 24-32
[8] Surya I and Ismail H 2016 *Polym Test* **50** 276-282
[9] Surya I and Ismail H 2016 *Plast Rubber and Compos* **45** 287-293
[10] Ismail H and Chia H 1998 *Eur. Polym. J* **34** 1857-1863
[11] Surya I, Siregar S F and Ismail H 2018 *E3S Web of Conferences* **34** 01030
[12] Boonstra B, Cochrane H and Dannenberg E 1975 *Rubber Chem Technol* **48** 558-576
[13] Cochrane H and Lin C 1993 *Rubber Chem Technol* **66** 48-60
[14] The P, Mohd Ishak Z,Hashim A, Karger-Kocsis J Ishiaku U 2004 *Eur. Polym. J* **40** 2513-2521
[15] Lee B 1979 *Rubber Chem Technol* **52 (5)** 1019-1029
[16] Manna A K, De P, Tripathy D, De S and Peiffer D G 1999 *J Appl Polym Sci.* **74** 389-398
[17] Surya I, Ismail H 2018*IOP Conf Ser Mater Sci Eng* **309** 012104
[18] Boonstra B and Taylor G 1965 *Rubber Chem Technol* **38** 943-960
[19] Nunes R, Fonseca J and Pereira M 2000 *Polym Test* **19** 93-103
[20] Kraus G 1965, *Reinforcement of elastomer* (John Wiley & Sons Inc)
[21] Polmanteer K and Lentz C 1975 *Rubber Chem Technol* **48** 795-809

IOP Conf. Series: Materials Science and Engineering **523** (2019) 012064 doi:10.1088/1757-899X/523/1/012064

Experimental study on diesel engine coupled with a catalytic converter run on dual-fuel mode using biogas produced from agricultural waste

H Ambarita [1,*]

[1]Sustainable Energy Research Centre, Faculty of Engineering, University of Sumatera Utara, Medan 20155, Indonesia

Email: himsar@usu.ac.id

Abstract. This work studied experimentally the performance and exhaust gas emission of a diesel engine coupled with a catalytic converter run on dual-fuel mode using refined biogas resulted from agricultural waste. The content of methane on the biogas is 70%. The experiments were carried out at constant load 1500 kW. The engine rotation varies from 1000 rpm to 1500 rpm. The biogas flow rate varies from 0 to 6 L/min. The output power of the engine operated on dual-fuel mode is better than the engine operated with pure diesel. The brake thermal efficiency of the diesel engine with flow rate of 2 L/min and 4 L/min is better than diesel only. The specific fuel consumption of the diesel engine operated on dual-fuel mode is higher than pure diesel mode. The opacity number of the engine operated on dual-fuel mode is better than pure diesel mode. However, the CO number and HC number of the engine operated on dual-fuel mode are higher than pure diesel mode. The diesel replacement ratio is within the range 15.6% % to 74.8%. It is recommended to run the present diesel engine in dual fuel mode with biogas flow rate 2 L/min - 4 L/min.

1. Introduction

Greenhouse gas (GHG) emissions is a big problem for the globe. It will result in global warming and climate change to the globe. To avoid the globe from catastrophe, GHG emission must be mitigated. Many countries have committed to decreasing GHG emission. Indonesia has released the new target to reduce its emission by 29% from level business-as-usual by 2030 using own budget and it can be increased up to 41% using international supports [1]. One of the promising mitigation actions is to level up the renewable energy utilities. There is a strong worldwide commitment to reduce fossil fuel utilization and replacing with renewable energy sources. Several renewable energy sources can be used to replace fossil fuel such as solar energy [2–23], bioenergy [24 – 29], wind energy, and hydropower.

One of the promising mitigation actions for reducing GHG emission by increasing renewable energy utilization is employing biogas to power engines. Biogas is a significant source of biofuel resulted from anaerobic biodegradation of organic materials such as agricultural wastes which are present abundantly in the rural areas. Such areas typically employ diesel engines for agricultural industrial, transportation, and small-scale power plant. Two methodologies can be employed to

* Corresponding author: himsar@usu.ac.id

operate a diesel engine with biogas. The first method is changing the diesel engine to spark engine and run it in pure biogas mode [30, 31] and the second one is dual-fuel mode (diesel-biogas). In the first mode, the CI needs extreme modifications, because the CI engine must be converted into spark ignition engine. The literature shows that CI engine can be operated 100% on biogas with performance comparable with diesel fuel. However, this method will be a problem when the biogas is absence or far from sufficient, the converted CI engine is not easily turned back into pure diesel mode.

In this study, we focus on powering diesel engine with biogas in dual-fuel mode. In this mode, after the compression stage, the charge consisted of biogas and fresh air, a small amount of diesel fuel, called the pilot is put in the chamber. The pilot fuel gets self-ignited and then it becomes the firing source for the initiated biogas fuel. The advantage of a diesel engine operated in dual-fuel mode is it can be operated with a wide variety of gaseous fuels without any significant machine changes [32]. When the biogas is empty or far from sufficient, it is easily put back into pure diesel operation mode. In spite of the fact that a diesel engine run on dual-fuel mode shows its practical use, there are still many problems need to be investigated. The challenges include how to increase the performance and to make the exhaust gases friendly to the environment.

To solve the first challenge, several studies on the improvement of performance diesel engine operated on dual-fuel mode have been reported in the literature. Bedoya et al. [33] studied the effect of mixing system and pilot diesel fuel quality on the performance diesel engine run on dual-fuel mode. Cacua et al. [34] reported the investigation oxygen-enriched air to improve the engine performance. Makareviciene et al. [35] investigated the impacts of methane compositions in a big diesel engine four stokes and four cylinders when operated under dual-fuel (diesel-biogas) mode.

Typically, a catalytic converter is employed to reduce the bad impact of the combustion engine in general. The catalytic converter has been proposed to reduce the emission of a combustion engine since 1975. The catalytic converter converts harmful exhaust gases into a safe compound that can be emitted to the atmosphere and cause less damage to the environment. Rezk et al. [36] reported the effects of the catalytic converter position on the performance and hydrocarbon emissions of a gasoline engine. Michel et al. [37] studied the optimization fuel consumption and pollutant emissions of gasoline-hybrid electric vehicle equipped with catalytic converter. Irawan et al. [38] explored the optimum design of manganese-coated copper catalytic converter on gasoline motor. Those literature are focused the gasoline or spark ignition engines.

The above studies are only focused on a spark engine coupled with a catalytic converter in a gasoline engine. However, only limited studies on a diesel engine coupled with a catalytic converter. Vallinayagam [39] reported a study on emission decrease from diesel engine fueled by pine oil biofuel using selective catalyst reduction and catalytic converter. Lozhkin and Lozhkina [40] proposed a mathematic model of catalysis process in the mode of heat rejection. The model was validated to a catalytic converter with storage device diesel engine of the city bus. Recently, Ambarita [24] reported the study on the effect of catalytic converter on the performance and emissions of a diesel engine run in dual-fuel mode using raw biogas.

In this study, a diesel engine run on duel-fuel mode coupled with a catalytic converter will be investigated. The biogas will be refined. The objective is to explore the effect of catalytic converter on the performance and emission characteristics of the diesel engine when it is run in dual-fuel mode using methane enriched biogas.The results are expected to supply the necessary information to reach GHG emission target and to provide technology in supporting energy for sustainable development which is safe for human health.

2. Method

2.1. Experimental setup

Figure 1 depicts the schematic diagram of the designed experimental setup. It comprises of a small diesel engine, a generator, a series of lamps, biogas container, gas mixer, and measurements sensors. The used diesel engine is typically employed to power a mini tractor in Indonesian agriculture. The specifications of the diesel engine are explained as follows. It is a single-cylinder four strokes engine and designed for diesel fuel. The engine has the highest output power of 4.86 W with a weight of 65 kg. The Bore and Stroke of the engine are 75 mm and 80 mm, respectively. The rated output of the diesel engine is 4.41 kW at a rated speed of 2600 rpm. The compression ratio of the engine is 23:1. A catalytic converter, which is typically used for motorbike purchased in the Indonesian market, is coupled to the CI engine. It is a three-way, Pt-Rh, honeycomb structure catalytic converter.

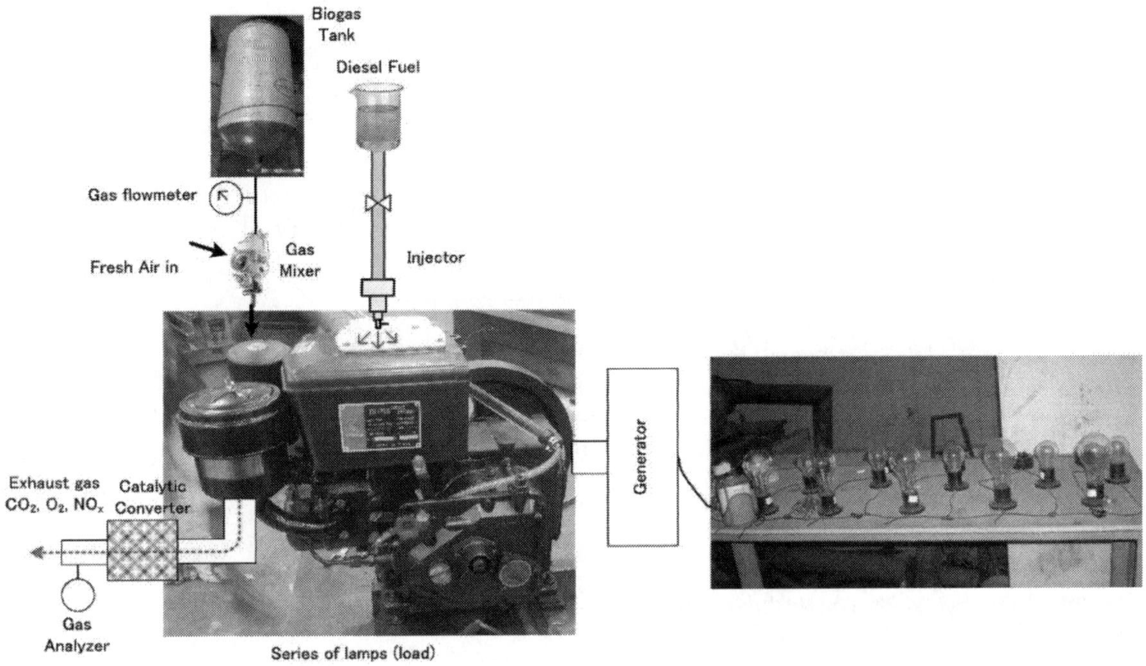

Figure 1. Experimental Apparatus

In this study, the biogas is made of biodegradation of Palm Oil Mill Effluent (POME) by using digester. The biogas digester belongs to PTPN III an Indonesian government-owned Palm Oil Company. The type of the biogas digester is anaerobic mesophilic. The biogas is refined and stored in a particularly designed tank by using a compressor. The methane content of the biogas is 70%. As a note, raw biogas without treatment has methane content of 60%. The methane content of the biogas is estimated using Gas Chromatography. The diesel fuel is obtained from PERTAMINA the fuel retailer in Indonesia belong to the government. Before tested, low heating value (LHV) of the diesel fuel is investigated using a bomb calorie meter. The LHV is 42.64 MJ/kg. The heating value of the biogas will be calculated using the formula suggested by Ludington [41].

The diesel engine is coupled with a single-phase synchronous generator by employing a pulley. The generator specifications are depicted in the followings. The grading frequency, grading voltage, and maximum power are 50/60 Hz, 115-230 V, and three kVA, respectively. The generator is employed to power a series of lights. The features of the exhaust gas such as opacity is characterized using engine smoke meter and HC and CO are characterized by using gas analyzer. The model of the engine smoke meter is HD-410 with main specifications are as follows. The measuring range, absorption coefficient, oil temperature, and operation temperature are 0.00-100%, 0.00 - 21.42 m-1, 0 - 800 rpm, 0 - 100°C, and of -10°C - 40°C, respectively. The model of Gas Analyzer is HG-510 with features as follows. The measurement ranges of CO and HC are 0-9.99% and 0-9999 ppm, respectively. The temperature is collected using thermometer sensor KW 06-278 Krishbow. The accuracy range is±0.5% ± 1°C. Multitaster Meter CD 800Ais used to measure electric charge. Tachometeris employed to measure shaft rotation speed of the CI engine.

2.2. Problem Formulation

The performance of the diesel engine is analysed employing the electrical power, thermal efficiency, and specific fuel consumption. The power resulted by the CI engine P_E (Watt) is estimated by

$$P_E = V \times I \tag{1}$$

where V [Volt] and I [Ampere] are the voltage and the current of the electric power by the generator, respectively. Thermal efficiency is the electric power divided by the energy injected to the engine. For the case of CI engine run on diesel only mode, it is given by

$$\eta = \frac{P_E}{\dot{m}_{diesel} \times H_{diesel}} \tag{2}$$

where H_{diesel} is the heating value of diesel fuel. In the case of engine operated in dual-fuel mode, the thermal efficiency is given by

$$\eta = \frac{P_E}{\dot{m}_{dual} \times H_{diesel} + \dot{m}_{biogas} \times H_{biogas}} \tag{3}$$

where, $\dot{m}_{dual}$ [kg/s] and $\dot{m}_{biogas}$ [kg/s] are the mass rate of the diesel fuel and biogas, respectively. Here, H_{biogas} the heating value of the biogas [kJ/kg].

Specific fuel consumption sfc [g/kWh] is fuel consumed by the engine divided by useful energy. It is a parameter to show how many grams of fuel is burnt to result in 1 kWh of electrical energy. In the case of diesel only mode, it is given by the equation below.

$$sfc = \frac{\dot{m}_{diesel} \times 10^3}{P_E} \tag{4}$$

For diesel engine operated on dual-fuel mode it is defined as:

$$sfc = \frac{(\dot{m}_{diesel} + \dot{m}_{biogas}) \times 10^3}{P_E} \tag{5}$$

To express the ratio of diesel fuel substituted by the biogas, the parameter replacement ratio $[r]$ is used. It is defined as reduced diesel fuel consumed when the CI engine tested on dual-fuel mode and when it is tested on diesel only mode. The parameter is given by the below equation.

$$r = \frac{\dot{m}_{diesel} - \dot{m}_{dual}}{\dot{m}_{diesel}} \times 100\% \tag{6}$$

where $\dot{m}_{diesel}$ [kg/s] is the diesel mass flow rate in diesel mode and $\dot{m}_{dual}$ [kg/s] is the diesel mass flow rate in dual-fuel mode.

3. Results and discussions

In the experiment, the engine speed was operated from 1000 rpm to 1500 rpm and the biogas flow rate was varied from 0 L/min, 2 L/min, 4 L/min and 6 L/min. The load was fixed at 1500 kW. The total number of experiments is 24 tests.

3.1. Performance characteristics

Figure 2 depicts the power resulted by the diesel engine as a function of diesel engine speed. The figure shows that the output power of the diesel engine run operated in dual-fuel mode coupled with a catalytic converter is higher than pure diesel mode. This fact suggests that the catalytic converter does not affect the output power of the diesel engine. In addition, There no significant effect of the biogas flow rate to the output power.

Figure 3 shows the thermal efficiency of the diesel engine run on dual-fuel mode and pure diesel mode. The figure shows that the highest thermal efficiency is shown by the diesel engine when it fueled with biogas with flow rate of 2 L/min. Increasing the biogas flow rate to 4 L/min will decrease the thermal efficiency. The thermal efficiency of the CI engine operated at biogas flow rate 2 L/min and 4 L/min is better than diesel only mode. However, when the biogas flow rate increased up to 6 L/min, the thermal efficiency will be lower than diesel only mode.

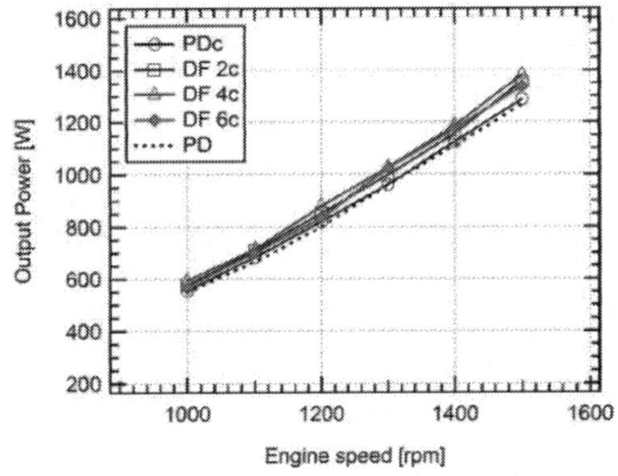

Figure 2. Electric Output power

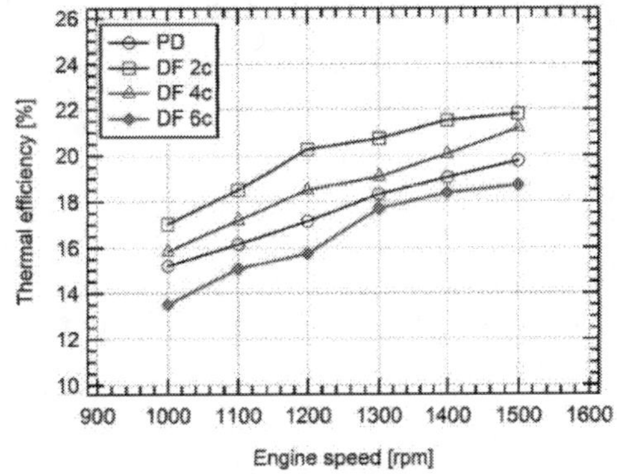

Figure 3 Thermal Efficiency

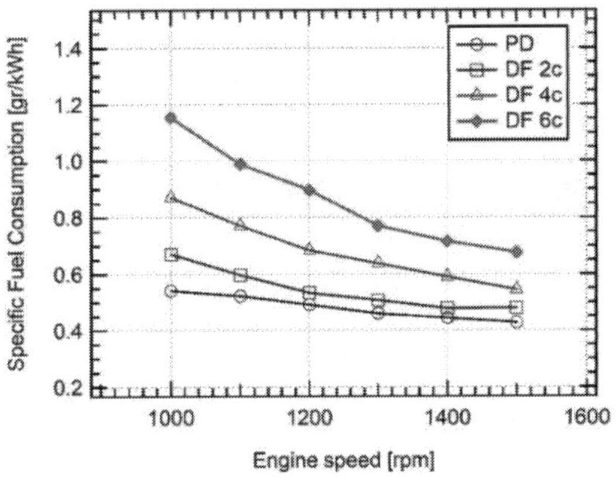

Figure 4. Specific fuel consumption

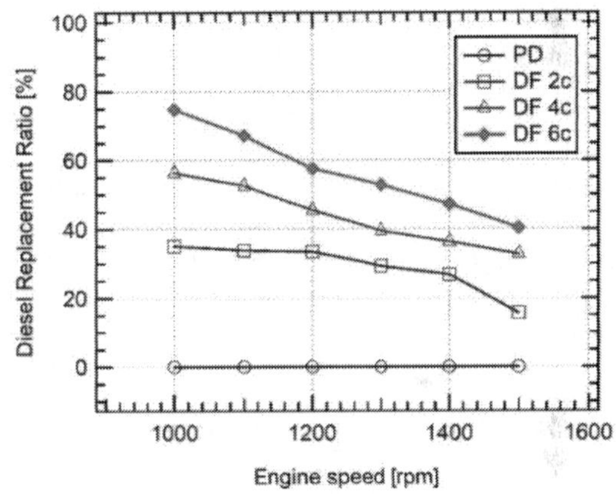

Figure 5. Diesel replacement ratio

The specific fuel consumption of the engine is shown in Figure 4. As a note, a lower specific fuel consumption is better. It was shown that specific fuel consumption decreases with increasing engine speed. This fact reveals that the diesel engine should be operated at high engine speed. Specific fuel consumption decreases with decreasing biogas flow rate.

3.2. Exhaust gas emission

As a note, the objective of installing a catalytic converter to the diesel engine is to reduce negative effect of the exhaust gas. The opacity number of the exhaust gas for all cases are shown in Table 1. The table shows that opacity number of the exhaust gas decrease when the diesel engine run on dual-fuel mode. This is because of some particulates in the combustion chamber decrease by the presence of the biogas.

Table 2 shows the CO emission number of the exhaust gas. It can be seen that the pure diesel mode results in the lowest CO number. The CO emission decrease with increasing biogas flow rate. Table 3 shows the HC number of exhaust gas. It was shown that the lowest HC number resulted by the engine run on pure diesel mode. This is because the diesel fuel is bunt perfectively. By injecting biogas to the combustion chamber, some of the hydrocarbon in the fuel was not burnt perfectively. This makes the HC in the exhaust gas is high. Increasing biogas flow rate will increase HC number.

Table 1. Opacity Number

Engine Speed	Biogas Flow rate			
[rpm]	Pure Diesel	2 L/min	4 L/min	6 L/min
1000	7.7	5.2	6	2.6
1100	9.9	9.3	6.4	5.1
1200	9.4	11.6	9.6	9.6
1300	14.7	12.3	12	13.5
1400	19.1	17.8	13.7	15.9
1500	18.2	31	15.2	18.7

Table 2. CO_2 emission

Engine Speed	Biogas Flow rate			
[rpm]	Pure Diesel	2 L/min	4 L/min	6 L/min
1000	0.02	0.15	0.06	0.03
1100	0.02	0.12	0.04	0.04
1200	0.02	0.11	0.04	0.04
1300	0.02	0.11	0.03	0.04
1400	0.02	0.09	0.04	0.04
1500	0.03	0.02	0.04	0.04

Table 3. HC emission

Engine Speed	Biogas Flow rate			
[rpm]	Pure Diesel	2 L/min	4 L/min	6 L/min
1000	10	248	162	120
1100	6	193	112	115
1200	6	170	94	109
1300	6	116	82	101
1400	6	132	82	98
1500	6	145	82	85

3.3. Diesel replacement ratio

As a note the aim of using biogas in a diesel engine is to replace diesel consumption. The parameter to explore this objective, diesel replacement ratio is used. This parameter should be treated as how much diesel fuel can be reduced when the diesel engine is tested in dual-fuel mode. If the diesel replacement ratio is high, it means a better operational cost. The diesel replacement ratio at different engine rotation speed for all cases are depicted in Figure 5. The figure reveals that diesel replacement ratio starts from 15.6% % to 74.8%. The minimum diesel replacement ratio is depicted by the engine when it is tested in the dual-fuel mode with a biogas flow rate of 2 L/min with a speed of 1500 rpm. The maximum diesel replacement ratio is depicted by the engine when it is tested in the dual-fuel mode with a biogas flow rate of 6 L/min at an engine speed of 1000.

The diesel replacement ratio reduces as the engine speed increases. This is due to higher engine speed, the fuel burning process occurs at a very limited time. This will affect the mass of biogas that can substitute the diesel fuel. For all cases, increasing biogas flow rate will increase diesel replacement ratio. These facts indicate that using biogas in a diesel engine can decrease diesel fuel consumption significantly.

4. Conclusion

In this work a diesel engine coupled with a catalytic converter run in dual-fuel mode has been tested experimentally. The biogas used in the experiment is made of biodegradation of Palm Oil Mill Effluent (POME) by using digester. This POME is known as agricultural waste which has negative impact on the environment. The biogas is refined to increase the methane content. The engine was tested at different engine speed and biogas flow rate. The conclusions of this study are as follows. The output power of the engine run on dual-fuel mode is better than pure diesel mode. Thermal efficiency of the diesel engine with lower biogas flow rate (2 L/min and 4 L/min) is better than pure diesel mode. The specific fuel consumption of the diesel engine run on dual-fuel mode is higher than pure diesel mode. The opacity number of the engine run on dual-fuel mode is better than pure diesel mode. However, the CO number and HC number of the engine run on dual-fuel mode are higher than pure diesel mode. The diesel replacement ratio is within the range 15.6% % to 74.8%. It is recommended to run the present diesel engine in dual fuel mode with biogas flow rate 2 L/min - 4 L/min.

5. References

[1] Ambarita H 2017 IOP Conference Series: Materials Science and Engineering **237 (1)** 012013
[2] Ambarita H, Ronowikarto AD, Siregar RET, Setyawan EY 2018 *Journal of Physics: Conference Series* **978** 012096
[3] Ambarita H, Siregar RET, Ronowikarto AD, Setyawan EY 2018 *Journal of Physics: Conference Series* **978** 012097
[4] Tambunan DRS, Sibagariang YP, Ambarita H, Napitupulu FH and Kawai H 2018 *Journal of Physics: Conference Series* **978** 012099
[5] Ambarita H 2018 *IOP Conference Series: Materials Science and Engineering* **309** 012005
[6] Sitepu T, Sembiring J and Ambarita H 2008 *IOP Conference Series: Materials Science and Engineering* **309** 012007
[7] Ambarita H 2008 *IOP Conference Series: Materials Science and Engineering* **308** 012028
[8] Ambarita H 2017 *ARPN Journal of Engineering and Applied Sciences* **12 (19)** 5357-5365
[9] Ambarita H and Sitepu T 2017 *IOP Conference Series: Materials Science and Engineering* **237** 012014
[10] Sitepu T, Gunawan S, Nasution DM, Ambarita H, Siregar RET and Ronowikarto AD 2017 *IOP Conference Series: Materials Science and Engineering* **180** 012032
[11] Ambarita H and Kawai H 2016 *Case Studies in Thermal Engineering* **7** 36 – 46
[12] Sitorus TB, Napitupulu FH and Ambarita H 2016 *International Journal of Technology* **7 (5)** 910 – 922
[13] Sitorus TB, Ambarita H, Ariani F and Sitepu T 2018 *IOP Conference Series: Materials Science and Engineering* **309** 012089
[14] Sitorus TB, Napitupulu FH and Ambarita H 2017 *Journal of Engineering and Technological Sciences* **49(5)** 657 – 670
[15] Ambarita H 2016 *Case Studies in Thermal Engineering* **8** 346 – 358
[16] Siregar RET, Ronowikarto AD, Setyawan EY and Ambarita H 2018 *IOP Conference Series: Materials Science and Engineering* **300** 012059
[17] Ambarita H, Ronowikarto AD, Siregar RET and Setyawan EY 2018 *IOP Conference Series: Materials Science and Engineering* **300** 012058
[18] Ambarita H, Setyawan EY, Ginting S and Naibaho W 2017 *IOP Conference Series: Materials Science and Engineering* **237** 012011
[19] Setyawan EY, Napitupulu RAM, Siagian P and Ambarita H 2017 *IOP Conference Series: Materials Science and Engineering* **237** 012012
[20] Ambarita H 2017 *IOP Conference Series: Materials Science and Engineering* **180** 012024
[21] Dina SF, Ambarita H, Napitupulu FH and Kawai H 2015 *Case Studies in Thermal Engineering* **5** 32 – 40
[22] Ambarita H 2017 *Journal of Physics: Conference Series* **801** 012093

[23] Siagian P, Setyawan EY, Gultom T, Napitupulu FH and Ambarita H 2017 *IOP Conference Series: Materials Science and Engineering* **237** 012037
[24] Ambarita H 2018 *International Journal of Mechanical and Mechatronics Engineering* **18 (2)** 35-43
[25] Ambarita H 2018 *IOP Conference Series: Materials Science and Engineering* **309** 012006
[26] Arjuna J, Sitorus TB, Ambarita H and Abda S 2018 *IOP Conference Series: Materials Science and Engineering* **309** 012088
[27] Ambarita H 2017 *Case Studies in Thermal Engineering* **10** 179-191
[28] Ambarita H, Widodo T I and Nasution D M 2017 *Journal of Physics: Conference Series* **801** 012095
[29] Ambarita H, Sinulingga EP, Nasution MKM and Kawai H 2017 *IOP Conference Series: Materials Science and Engineering* **180** 012025
[30] Lee TH, Huang SR and Chen CH 2013*Renewable Energy* **50** 342 – 347.
[31] Homdoung N, Tippayawong N, and Dussadee N 2015 *Energy Conversion and Management* **94** 286-292.
[32] Henham A andMakkar MK 1998 *Energy Conversion and Management* **39** 2001 -2009.
[33] Bedoya ID, Arrieta AA and Cadavid FJ 2009 *Bioresource Technology* **100** 6624-6629.
[34] Cacua K, Amell A, and Cadavid F 2012 *Biomass and Bioenergy* **45** 159-167.
[35] Makareviciene V, Sendzikiene E, Pukalskas S, Rimkus A and Vegneris R 2013 *Energy Conversion and Management* **75** 224 – 233.
[36] Rezk KA, Osman M Mand Saeed MN 2004 *Alexandria Engineering Journal* **43 (3)** 313-321.
[37] Michel P, Charlet A, Colin G, Chamaillard Y, Bloch G and Nouillant C 2015 *Control Engineering Practice*. http://dx.doi.org/10.1016/j.conengprac
[38] Irawan RMB, Purwanto Pand Hadiyanto H 2015 *Procedia Environmental Science* **23** 86-92.
[39] Vallinayagam R,Vedharaj S, Yang WM, Saravan CG, Lee PS, Chua KJ Eand Chou SK 2013 *Atmospheric Environment* **80** 190-197.
[40] Lozhkin V andLozhkina O 2017 *Transportation Research Procedia* **20** 412 – 417.
[41] Ludington D, *Calculating the heating value of biogas*, Ditech, Inc. Itchca 2008. NY.

IOP Conf. Series: Materials Science and Engineering **523** (2019) 012065 doi:10.1088/1757-899X/523/1/012065

Effects of welding on the change of microstructure and mechanical properties of low carbon steel

Husaini[1,*], N Ali[1], J K Hamza[1] and S E Sofyan[1]
[1]Department of Mechanical Engineering, Syiah Kuala University,
Jl. Syech Abdurrauf No.7, Darussalam-Banda Aceh, 23111, Indonesia

E-mail: husainiftm@unsyiah.ac.id

Abstract. Steel construction is mostly connected by using a welding technique. The result of the welding process can cause the change of mechanical properties of the welding zone. Therefore, it needs to be reviewed to guarantee the quality of the welding zone. This research aims to assess the effect of welding on the change of microstructure and the mechanic characteristic of welded connection. Low carbon steel was welded by using Shield Metal Arc Welding method (SMAW) with electrode E7016, diameter 2,6 mm. Type of notch used was single V with angle 70° and welding position was 1G. Then, the Charpy impact test, Rockwell hardness and micro-macro structure monitoring on the welding zone including base metal, Heat Affected Zone (HAZ) and weld metal, are conducted. The results show that the highest impact toughness value at metal welding area is 251 joule/mm^2, and lowest impact toughness is at the heat affected area (HAZ) is 119 joule/mm^2. Hardness values of welding zones are found of about 87,6 HR_B for weld metal, 73,9 HR_B for HAZ zone and 67,1 HR_B for the base metal. The microstructure of weld metal zone, HAZ and base metal are included on ferrite and pearlite formed by welding on low carbon steel AISI 1010 observer by using an optic microscope.

1. Introduction

Technological advances and the need to produce robust steel construction have made welding a top choice in the steel connection process. Today, the steel construction involves many elements of welding. This is due to the welding is one of metal connecting techniques that is easy to implement. The welding has been widely used in the construction of ship, bridges, steel frame, pressure vessel, rail, pipeline and so forth [1].

Welding is a process of connecting metal rods using heat energy either with or without filler metal. The type of welding that is often used is electric arc welding or shielded metal arc welding (SMAW) [2]. The electric arc welding (SMAW method) uses the electric arc to melt both the base and filler metal. This welding technique uses a metal electrode covered by a flux. An electric arc is evoked between the base metal and the tip of the electrode. Due to the heat from this electric arc, the base metal and the tip of the welding electrode melt and subsequently solidify simultaneously [3].

The characteristics and conditions around the welding area of the welded metal change during the welding process. These changes affect the mechanical properties and microstructure of the newly welded metal (steel). In addition, these changes induce the internal stress that can decrease the

* Corresponding author: husainiftm@unsyiah.ac.id

IOP Conf. Series: Materials Science and Engineering **523** (2019) 012065 doi:10.1088/1757-899X/523/1/012065

toughness of the welded metal. Therefore, a comprehensive study is needed in order to understand the changes occurred in microstructure and mechanical properties of the welded area.

This study aims to understand the effect of welding on the changes in microstructure and mechanical properties (i.e: toughness and hardness) of low carbon steel. Thus, the impact and hardness tests, as well as the micro and macro structure observations of the welded low carbon steel, are necessarily conducted.

2. Literature review

2.1 Shielded metal arc welding (SMAW)

The covered electrode welding is the welding process by liquefying the base material using heat from electricity flowing through the tip of the flux-covered electrode which melts away during the welding process. In the covered electrode welding, an electric arc-induced between the tip of the electrode and the base metal will generate heat. The heat dilutes both the tip of the electrode and the base metal simultaneously. The electric arc is generated by a welding machine and used the flux covered electrodes. The type of electrode used will greatly determine the welding results [4].

2.2 Covered electrode

An electric arc weld requires an electrode comprising a core metal covered by a chemical mixture (flux). The electrode serves as a trigger as well as a filler material. The electrode coating protects the molten metal from the air environment, producing the protective gas, and stabilizing the arc [1].
The flux material used for the electrode E7016 is low hydrogen potassium, which is also known as the lime type. This flux yields a joint with low hydrogen levels so that the susceptibility of the joint to the crack is very low and its toughness is very satisfactory.
The electrode is classified according to the standard system of American Welding Society (AWS) and American Society Testing Material (ASTM). E7016 electrode allows being applied in all welding positions (vertical, horizontal and overhead) either with the AC and DC current. The electrode helps to yield fine jagged surface, Thus, the existing slag will be easy to clean and the arc can be controlled easily [5].

2.3 Electric current

The amount of electric current required for welding depends on the diameter of the electrode, the thickness of the welded material, the type of electrode used, the connection geometry, the core diameter of the electrode, and the welding positions. The welding zone has a high heat capacity, thus, substantial electric current is required [6].

2.4 Thermal cycle of the welding zone

The welding area can be classified into three zones namely, the weld metal zone, HAZ and the unaffected heat metal zone.
a. The weld metal zone is the part of the metal that melts and solidifies during the welding process. The weld metal comprises of the base metal and the filler material of the electrode.
b. The Heat affected zone (HAZ) is the area of the base metal adjacent to the weld metal zone that undergoes rapid heating and cooling during the welding process, thus, yielding the most critical of the welding joint.
c. The base metal zone is the area of the metal that does not experience both microstructure and properties changes due to the welding temperature. Besides these three main divisions, there is still one area of heat influence, called the fusion line [7].

2.5 The toughness of the welding zone

The toughness of the welded metal depends on its microstructure, similarly for the base metal and the fusion line. During the welding process, the welded metal melts and solidifies, Thus, it contains oxygen and other gases.

The composition of the welded metal depends on the welding process applied. It should be noted that the effects of other elements absorbed during the welding process, especially oxygen [1].

3. The tested material and research method

The material utilized in this study was low carbon steel. The mechanical properties and chemical compositions of the material are shown in Tables 1 and 2, respectively.
Firstly, the notch was made before the welding process. The notch that suits for a 10 mm plate is a single v with an angle of 70° as shown in Figure 1.

Table 1. Chemical compositions of AISI 1010 steel [8].

Matters	C	Mn	P	S	Si
Composition (% weight)	0,08-0,13 max	0,30-0,60 max	0,040 max	0,050 max	-

Table 2. Mechanical properties [9].

Properties	Metric	Unit
Density (ρ)	8,030	kg/m^3
Hardness	60	HR$_B$
Modulus Young (E)	210	Gpa
Shear Strength (τ)	260	MPa
Poisons Ratio (ν)	0,30	-
Yield Strength (σ_y)	305	Mpa

The welding was conducted by SMAW method with the E7010 electrode and 1G welding position for a plat with dimensions of 10 mm x 10 mm. Table 3, summarises the chemical compositions of AWS A5.1 E7016 electrode by weight percentage.

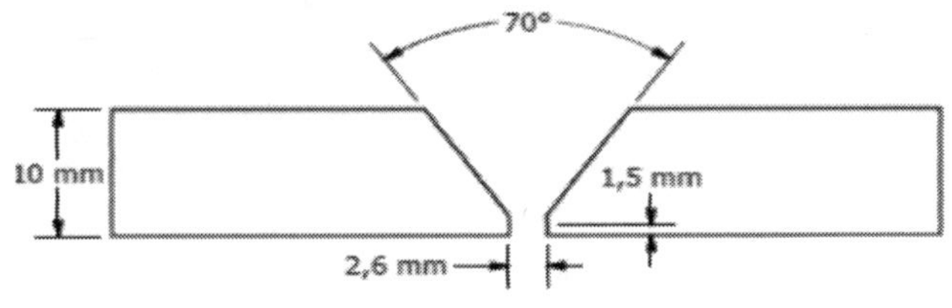

Figure 1. a single v-notch.

Table 3. Chemical compositions of AWS A5.1 E7016 electrode (% weight) [10].

C	Mn	Si	S	P	Ni	Mo	Cr	V
0.12	1.6	0.75	0.035	0.040	0.30	0.30	0.20	0.08

The welded material was cut into a specimen with dimensions as specified by ASTM E-23. Then, the impact strength (toughness) of the specimen was tested. Figure 2 shows the schematic of the tested specimen.

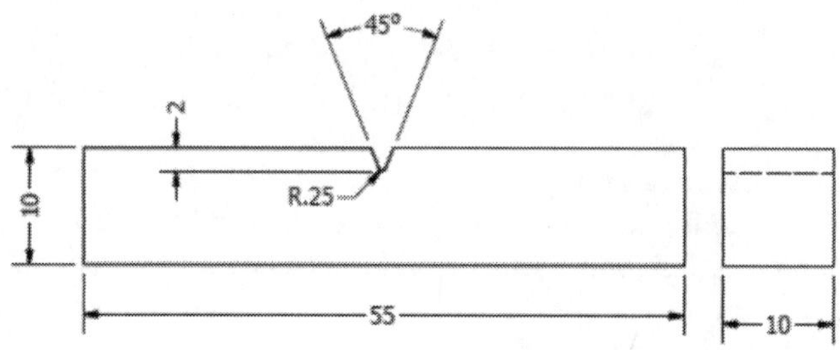

Figure 2. The schematic and dimensions of the specimen

4. Result and discussion

4.1 Chemical compositions

Data obtained from the test of the chemical composition of the specimens was compared with that provided by the AISI 1010 standard for steel. The experimental results reveal that the chemical compositions of the specimen close to that provided by AISI 1010 standard (see Table 4). It is observed that the carbon content of the tested specimen is 0.10%. This value is within the range provided by the AISI 1010 standard namely, 0.08% - 0.013%. While the high-level content of sulfur in the steel can reduce the ductility (steel ductility) and increase the chances of cracking. The sulfur content obtained from the specimen is 0,007% which is a good agreement with that provided by the AISI 1010 standard. It is found that the phosphor content of the specimen is less than the maximum value provided by the AISI 1010 (0,040%). This indicates that the experimental value is in accordance with that suggested by AISI 1010. The Mangan (Mn) content in the low carbon steel can increase the ductility. It is found from the experiment the Mn content of the specimen is 0.27 which is also meet the AISI 1010.

Table 4. Compositions comparison between the specimen and AISI 1010 low carbon steel.

Material	C	Mn	P	S	Si
Low carbon steel Testing result (% weight)	0.10	0.27	0.019	0.007	014
AISI 1010 (% weight)	0.08-0.13 max	0.30-0.60 max	0.040 max	0.050 max	.

4.2 Changes in microstructure

Observation of microstructure using an optical microscope was conducted to find out the changes in the microstructure of the specimen that occurs due to heating during the welding process (SMAW). The result of observation of microstructure in this study has been etched with 2% nital for 1 minute. Figure 3 illustrates the schematic of the welding zone of the observed microstructure. Figure 4 shows the microstructure of the weld metal and HAZ. Figure 5 demonstrates the microstructure of HAZ and Figure 7 shows the microstructure of the base metal.

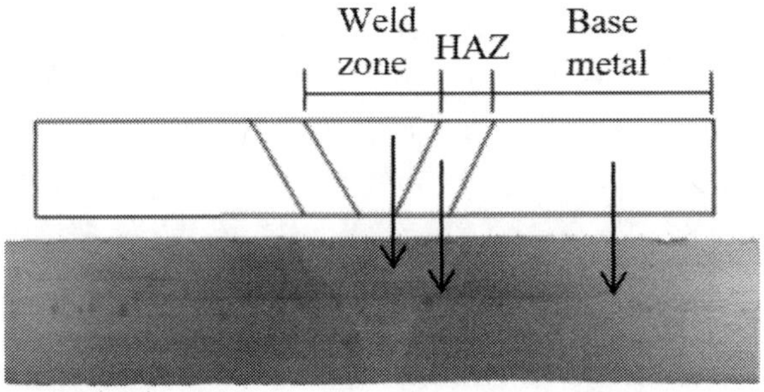

Figure 3. Schematic of welding zone of the observed microstructure.

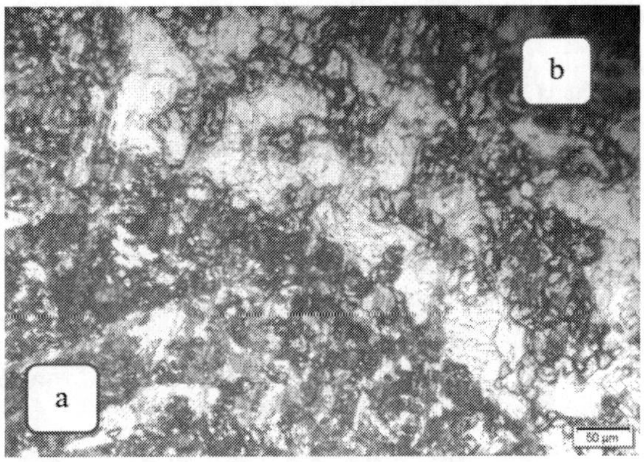

Figure 4. The microstructure of weld metal and HAZ with 20 times magnification a) shows the weld metal zone and b) shows the HAZ.

Figure 4.a illustrates the microstructure of the weld metal zone (formed by the solidified molten metal). As can be seen from the figure, the grain size of the microstructure is finer. Figure 4.b shows the HAZ, while the red line represents the border between the weld metal zone and the HAZ.

Figure 5 shows that the microstructure of the HAZ is coarse ferrite and pearlite. The reason is due to during the welding process, the HAZ adjacent to the fusion line undergoes rapid heating and cooling so that this area is most critical and the peak temperature exceeds the melting point of the steel. Visually, the microstructure of the area close to the melting line is getting coarse. The figure shows that the white ferrite grains (light) are dominant, while the pearlite grains are less (showed by dark colour).

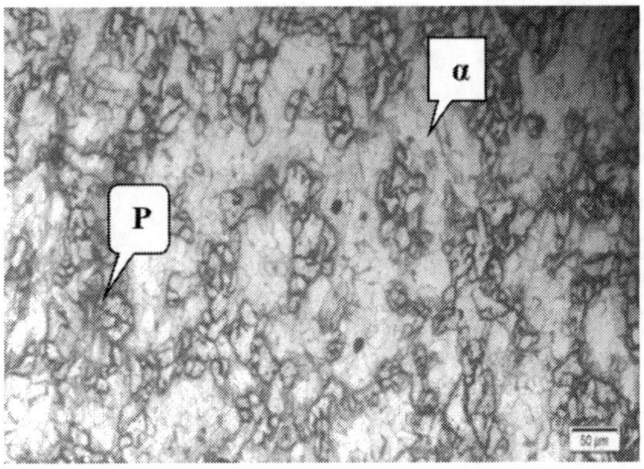

Figure 5. shows the microstructure of the HAZ with 20x magnification, α = ferrite, P = pearlite.

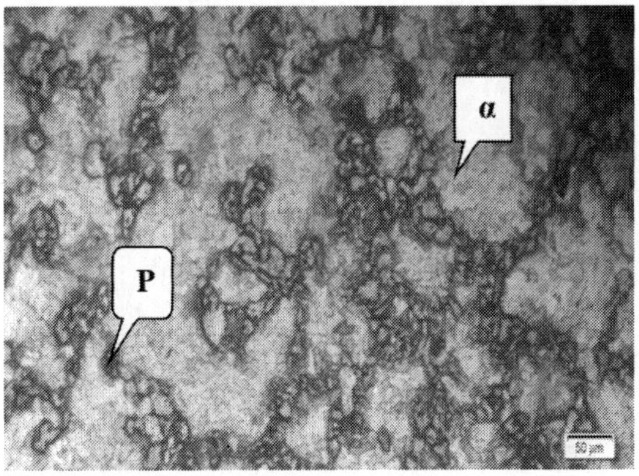

Figure 6. The microstructure of the base metal with 20x magnification, α = ferrite, P = pearlite.

Figure 6 shows that the microstructure of the base metal zone is finer than others. The heat generated during the welding process does not induce the changes in the metal microstructure and properties. As illustrated in the figure, the microstructure is predominantly ferrite which is white in colour. While the pearlite (dark colour) is less. From the microstructure observation, it can be concluded that there is no change in microstructure and grain sizes of the base metal.

IOP Conf. Series: Materials Science and Engineering **523** (2019) 012065 doi:10.1088/1757-899X/523/1/012065

4.3 The hardness of the welding zone

The mechanical properties of the specimen not only depend on its chemical compositions but also its microstructure. The microstructure depends on the process of work, especially in the heat treatment process [11]. The hardness of the specimen (i.e. base metal zone, HAZ, and weld metal zone) from point 1 to 11 (see Figure 7) is summarized in Table 5. It is found the average value maximum of the hardness of the welding zone is 87,0.

Table 5. The hardness of the welding zone.

	Hardness HR_B			
No	Point 1	Point 2	Point 3	Average
1	67,1	66,6	66,7	66,8
2	67,8	67,5	68,7	68
3	70,7	69,1	70	70
4	71,1	71,8	73	72,0
5	79	74,1	81,4	78,2
6	86,2	87,6	87,1	87,0
7	81,8	80,1	80	80,6
8	73,9	73,4	71,6	73,0
9	70,9	70	70,2	70,4
10	67,8	68,4	69,4	68,5
11	66,8	65,8	66	66,2

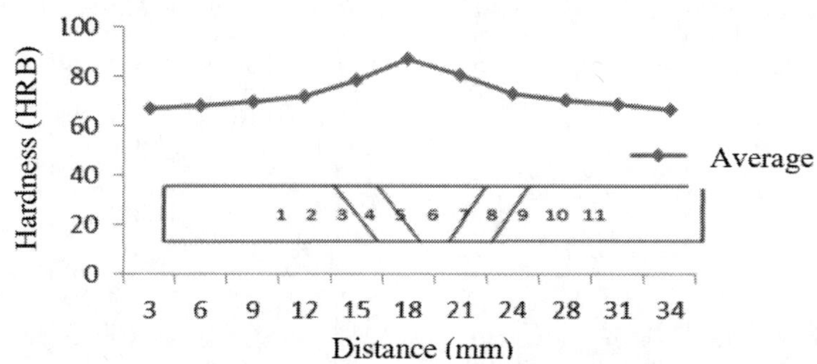

Figure 7. Hardness values of the specimen.

Figure 7 shows the hardness values of the specimen. It is observed that the hardest part of the specimen is at the weld metal zone. This tendency is affected by a higher Mn content in the weld zone than other areas namely 1.6% (see Table 3). As a result, the hardness and toughness of the weld metal zone increase [13]. While the HAZ is slightly harder than the base metal. The reason is due to there is no changes in the microstructure and mechanical properties of the base metal induced by the heat generated during the welding process.

4.4 Toughness of the welding zone

The toughness values of both the weld metal and the base metal are investigated by the impact test. In the test, the energy absorbed (W) by the specimen is measured in Joule and notch impact (K) is in

Joule/mm^2. Table 6 summarises the notch impact values of the weld metal zone, HAZ, and the base metal.

Table 6. The toughness of the specimen of the welded metal.

No	Impact value of welding zone (joule/mm^2)		
	Weld metal	HAZ	Base metal
1	215	119	125
2	251	122	133
3	209	119	140
average	225	120	133

Table 6 demonstrates the result of the impact on each region of the specimen that undergoes the welding process. The result shows that the highest toughness value of the specimen is obtained from the weld metal zone. It is 225 joule/mm^2. While the base metal is 133 joule/mm^2. The lowest value of toughness is gained from HAZ namely, 120 joule/mm^2.

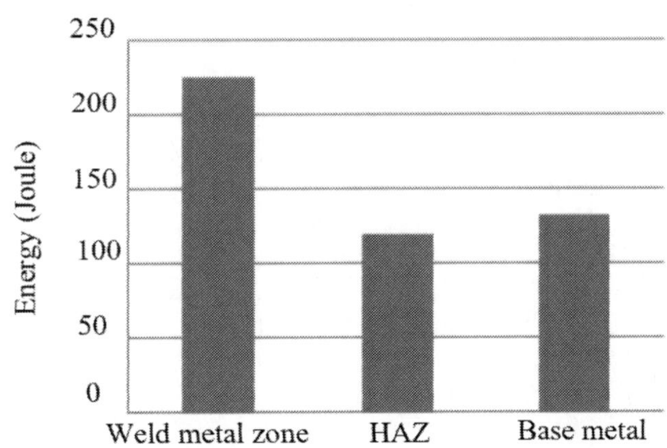

Figure 8. The impact toughness of the welding zone.

4.5 Surface fracture analysis

The observation of macrostructure on fracture surface aims to find out the type of fracture experienced by the welding zone. Therefore, the toughness for the welding zone can be estimated. The higher percentage of fibrous fracture (dimple-shaped) indicates the more ductile the material is. This means the material is tougher. On the contrary, the finer and flatter the fracture surface, the more brittle and fragile the material is. Figures 8-10, show the fracture of the cross-section of the welding zone.

IOP Conf. Series: Materials Science and Engineering **523** (2019) 012065 doi:10.1088/1757-899X/523/1/012065

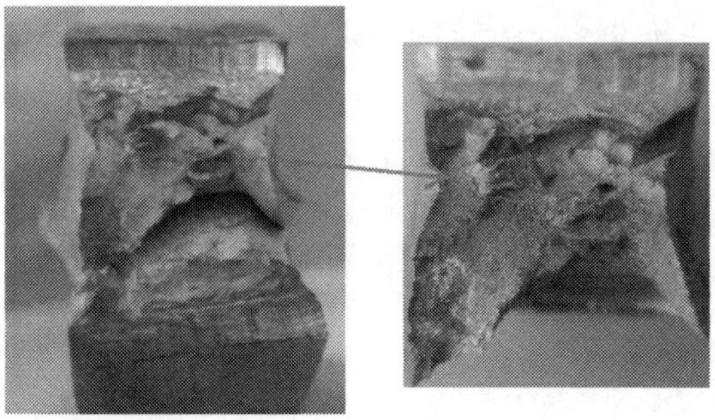

Figure 9. The fracture surface of the weld metal zone.

Figure 10. The fracture surface of the HAZ

Figure 11. The fracture surface of the base metal.

Figures 9-11 illustrate the fracture surface of the welding metal at three different zones. It is observed that the fracture in the welding metal is a ductile fracture. This is characterized by energy absorption, also accompanied by a considerable plastic deformation around the fracture. As a result, the fracture surface appears rough, fibrous, and grey in colour. In addition, the type of fracture occurred not only influenced by the load itself but also affected by the material compositions. Typically, the ductile fracture occurs on the material which has ferrite, pearlite or bainite microstructures which is low carbon steel.

5. Conclusions
From the results and discussions, several findings from this study are concluded as follows:

1. There are the grain size changes on the weld metal zone, HAZ, and base metal. In addition, the highest hardness value is obtained at the weld metal zone and the lowest is at the base metal.
2. The highest toughness value of the specimen is gained from the weld metal zone with a value of 251 joules/mm^2, followed by the base metal, namely 133 joule/mm^2. While the lowest toughness occurs at the HAZ with a value of 119 joules/mm^2.
3. The highest hardness value is gained from the weld metal zone. This is due to Mn matter contained in the welding electrode wire. As a result, it increases both hardness and toughness.

6. Acknowledgment
The authors acknowledge the support provided by the Laboratory of Computational Mechanics, Manufacturing processes Laboratory, Laboratory of Materials.

7. References

[1] Okumura T (Translated to Bahasa by Harsono Wiryosumarto 1981 *Metal Welding Technology* (Jakarta: P.T. Pradnya Paramita)
[2] Sonawan H and Suratman, R 2006 *Introduction to Metal Welding* (Alfa Beta: Bandung)
[3] Soetardjo 1997 *Metal Welding Technology* (Jakarta: Rineka Cipta)
[4] Arifin S 1997 *Electric Welding* (Jakarta: Ghalia Indonesia)
[5] Suharto 1991 *Metal Welding Technology* (Jakarta: Rineka Cipta)
[6] Alip M 1989 Welding Theory and Practice *J. Teknik ITS* **1**
[7] Malau V 2003 *Lecture Module: Metal Welding Technology* (Yogyakarta)
[8] ASTM International 2004 *ASTM A 29/A 29 M-04: Standard Specification for Steel Bars, Carbon and Alloy, Hot-Wrought, General Requirements For, ASTM International* (West Conshohocken: ASTM)
[9] Ridha M and Husaini 2013 *Engineering Material* (Banda Aceh: Syiah Kuala University Press)
[10] Hamid A 2016 Effect of Electric Current SMAW Welding On Low Carbon Steel To The Strength Of The Welded Material *J. Tek. Elektro* **7**
[11] Akhyar H and Husaini 2016 *Int. J. of Metalcasting* **10** 452-6

IOP Conf. Series: Materials Science and Engineering **523** (2019) 012066 doi:10.1088/1757-899X/523/1/012066

Numerical analysis of the stress leading to fatigue failure on a coil spring of the front suspension of a car

T E Putra[1,*], Husaini[1], N Ali[1], I Hasanuddin[1], Hendrayana[1], A Bakhtiar[1]
[1]Department of Mechanical Engineering, Faculty of Engineering, Syiah Kuala University, Darussalam 23111, Banda Aceh, Indonesia

E-mail: edi@unsyiah.ac.id

Abstract. This study discussed the failure of the coil spring of a vehicle using a numerical method. The component was fabricated using ASTM A227. The coil spring was analyzed under three types of conditions, namely, flawless, defect of 0.2 mm, and defect of 3.5 mm. The simulation results show that the maximum shear stress of 745.8 MPa occurs on the spring with a defect of 3.5 mm. This value is close to the allowed shear stress, which is 799 MPa, but exceeded the yield strength of 731 MPa. This results in plastic deformation and contributes to failure due to fatigue.

1. Introduction

Transport vehicles require a good suspension system to dampen vibration, swings and shocks received as they travel along bumpy, hollow, and uneven roads [1]. These conditions are very uncomfortable and may cause accidents. The suspension is also expected to hold the load during some common vehicle maneuvers such as acceleration, braking or deflection while on the road [2]. The coil spring is one of the main components for dampening vibrations and shocks to the load so as to provide comfort and security while the vehicle is in motion [3].

Depending on the condition of their application, coil springs often sustain fatigue failure. This indicates that the tension received below by the coil spring from the maximum stress of the material while sustaining a dynamic load causes fatigue failure [4-8]. The yield strength of the material is also a criterion of failure. Components of automotive suspension must be changed with a traveling distance of 73,500 km, or every five years [9]. The fault of 13.18 % of 24.2 million vehicle tests was recorded [10].

With the development of computing technology, the numerical analysis method has become particularly suitable for use because it will increase the calculation efficiency, the cost-effectiveness as well as save time. Various numerical analysis methods are widely available, but the finite element analysis (FEA) has proven to be reliable in solving problems in the field of continuum mechanics [11].

2. Materials and methods

In this study, a spring made of ASTM A227 material was analyzed. Table 1 indicates the specifications of the material properties that were used as a reference for analyzing the coil spring using FEA. The coil spring was modeled using computer-aided design software. Figure 1 shows the coil spring that was modeled and analyzed by FEA. Table 2 indicates the size and dimensions of the coil spring used for the analysis.

* Corresponding author: edi@unsyiah.ac.id

Table 1. Mechanical properties of ASTM A227 [4,5].

No.	Property	Value
1	Density, ρ	7800 kg/m^3
2	Elastic Modulus, E	200 GPa
3	Shear modulus, G	80 GPa
3	Allowable shear stress, τ_a	799 MPa
4	Yield Strength, σ_y	731 MPa
5	Poisson Ratio, v	0.29

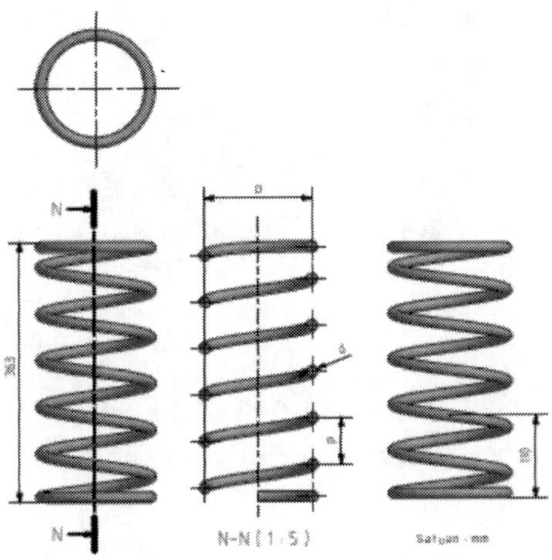

(a) (b)

Figure 1. The modeling of the coil spring. (a) dimensions of the coil spring in 3D; (b) area of defect on the coil spring (unit: mm).

Table 2. Size and dimensions of the coil spring used.

Dimension of coil spring					
Wire diameter, d (mm)	**Outer diameter, D_o (mm)**	**Active coil, N_a**	**Pitch, *p***	**Free Length, L_f (mm)**	**Correction factor, K_w**
14	150	5	65	363	1,26

The analysis of the coil springs was performed when the loading was in a static state. Table 3 indicates that the entire coil spring received the load of the vehicle, five passengers dan goods. All the coil springs were considered to have received an equal load. The loads for the coil spring were classified as minimum and maximum.

Minimum load on one coil spring of the vehicle $= \frac{12258\ N}{4} = 3064,5\ N$

Maximum load on one coil spring of the vehicle $= \frac{15690\ N}{4} = 3922,5\ N$

Table 3. Load received on the coil spring

Load		
Vehicle (N)	**Passengers and goods (N)**	**Totally (N)**
12258	3432	15690

The coil spring was analyzed under different conditions of defects that occurred on the surface of the coil through a FEA. The study was conducted using geometry and similar dimensional values as well as the same material. The coil spring was analyzed under three different conditions, namely, a faultless coil spring, a coil spring with a defect of 0.2 mm, and a coil spring with a defect of 3.5 mm. The shapes of the defect that occurred on the surface of the coil springs are shown in Figure 2.

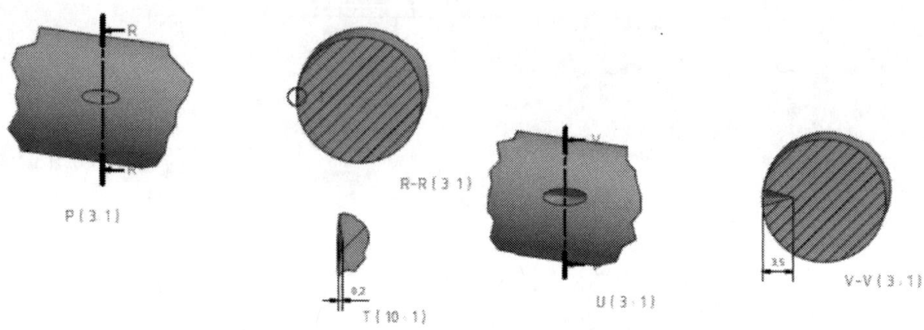

(a) (b)

Figure 2. Shape of the defect formed on the surface of the coil spring: (a) defect up to 0.2 mm; (b) defect up to 3.5 mm.

3. Results And Discussion

Figures 3 and 4 illustrate in detail the occurrence of the von-Mises stress on the faultless coil spring using the FEA, Figures 5 and 6 on the coil spring with a defect 0.2 mm, and Figures 7 and 8 on the coil spring with a defect of 3.5 mm. Based on the results of FEA, the shear stress value of the ASTM A227 coil spring obtained was approaching the allowable shear stress, but its value exceeded the yield strength. Thus, coil springs with such defects undergo plastic deformation and are susceptible to failure.

The magnitude of the defect on the coil spring was affected by the location of the region or occurrence of nodes in terms of the stress distribution and displacement of the magnitude of the defect at each point of the coil. The node or region of maximal stress distribution and displacement magnitude lay at the same point, even though different loads were given, that is, the maximum or minimum load. Likewise, the stress distribution and displacement of magnitude were minimal. The nodes occurred at different points only in the case of minimal shear stress.

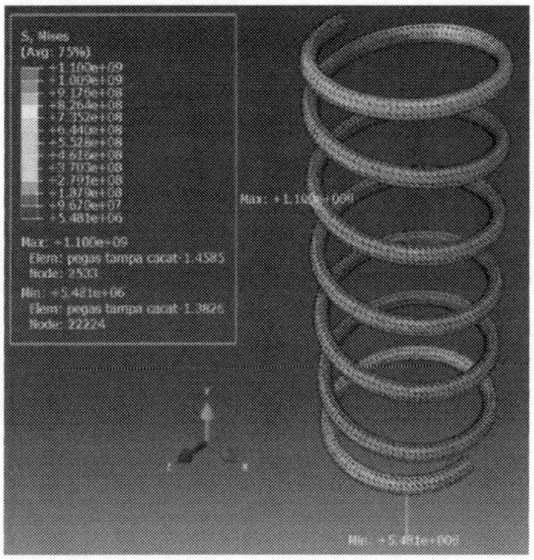

Figure 3. Contour of von-Mises stress on faultless coil spring given the maximum load.

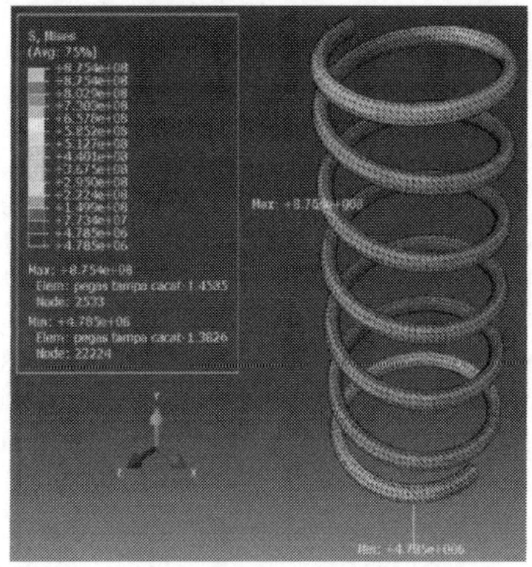

Figure 4. Contour of von-Mises stress on faultless coil spring given the minimum load.

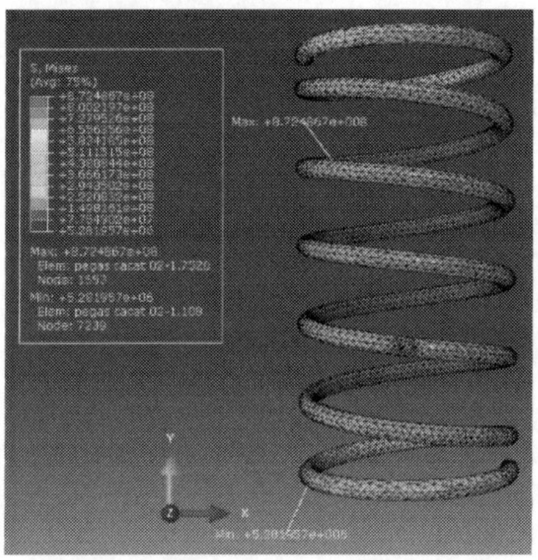

Figure 5. Contour of von-Mises stress on a coil spring with a defect of 0.2 mm given the maximum load.

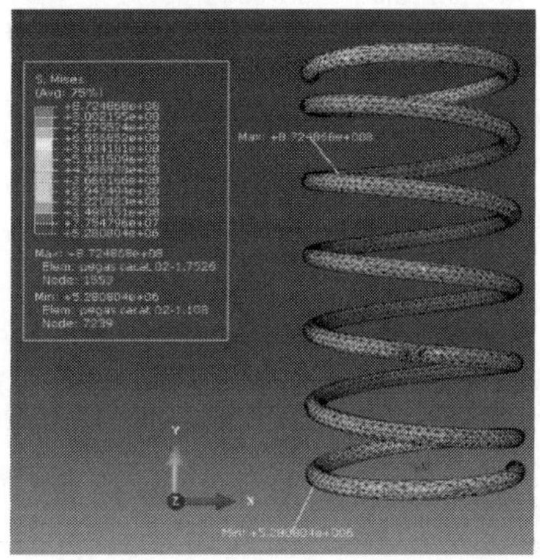

Figure 6. Contour of von-Mises stress on a coil spring with a defect of 0.2 mm given the minimum load.

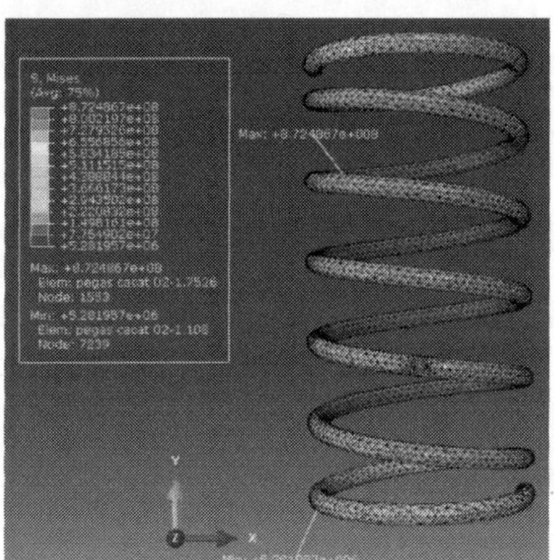

Figure 7. Contour of von-Mises stress on a coil spring with a defect of 3.5 mm given the maximum load.

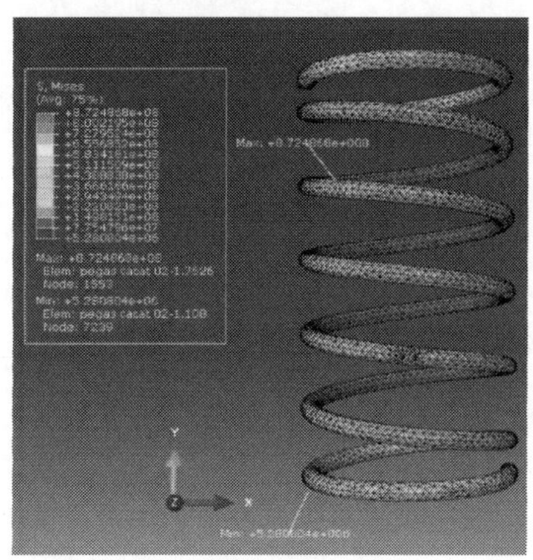

Figure 8. Contour of von-Mises stress on a coil spring with a defect of 3.5 mm given the minimum load.

IOP Conf. Series: Materials Science and Engineering **523** (2019) 012066 doi:10.1088/1757-899X/523/1/012066

4. Conclusion

Some conclusions could be drawn of the stress distribution and magnitude of displacement on the ASTM A227 coil spring as follow:

1. The maximum shear stress that occurred in the coil spring with a defect of up to 3.5 mm with the given maximum load was 745.81 MPa, and with the minimum load was 745.82 Mpa.
2. The minimum shear stresses that occurred in the coil spring with a defect of up to 3.5 mm with the given maximum load amounted to 736.9 MPa, and with the given minimum load was equal to 736.94 MPa. However, it was in the opposite direction.
3. The coil spring with a defect of up to 3.5 mm had undergone plastic deformation and was susceptible to failure, and (d) the condition of the defect in the coil spring affected the location of the occurrence of the node or stress distribution and magnitude displacement.

5. References

[1] Husaini, Putra TE and Ali N 2018 *AIP Conf. Proc.* **1983.**
[2] Prawoto Y, Ikeda M, Manville S K, Nishikawa A, Stinson R D and Solanki K 2008 *Engineering Failure Analysis* **15** 1155.
[3] Daryono 2007 Analisa Umur Pegas Daun pada Suspensi Kendaraan Roda Empat (Universitas Muhammadiyah Malang)
[4] Putra TE, Abdullah S, Schramm D, Nuawi MZ and Bruckmann T 2015 *Mech. Sys. Sig. Proc.* **60-61** 485.
[5] Putra TE, Abdullah S, Schramm D, Nuawi M Z and Bruckmann T 2017 *Mech. Sys. Sig. Proc.* **90** 1.
[6] Putra T E and Husaini 2018 *AIP Conf. Proc.* **1983.**
[7] Husaini, Putra T E and Ali N 2018 *International Journal of Automotive and Mechanical Engineering* **15** 5251
[8] Putra T E and Husaini 2018 *IOP Conf. Series: Materials Science and Engineering* **352**
[9] Roman L, Florea A and Cofaru I I 2014 *Fas. Manag. Technol. Eng.* **1** 289
[10] Hamed M, Tesfa B, Gu F and Ball A D 2014 *Proc.VETOMAC X* **23** 495-505
[11] Pattar S, Sanjar S J & Math V B 2014 *International of Research in Engineering and Technology.*

IOP Conf. Series: Materials Science and Engineering **523** (2019) 012067 doi:10.1088/1757-899X/523/1/012067

Failure analysis of the fracture surface of the crankshaft of a vehicle

T E Putra[1,*], Husaini[1], N Ali[1], H Husin[2], Zulfikar[1]
[1] Department of Mechanical Engineering, Faculty of Engineering, Syiah Kuala University, Darussalam 23111, Banda Aceh, Indonesia
[2] Department of Chemical Engineering, Faculty of Engineering, Syiah Kuala University, Darussalam 23111, Banda Aceh, Indonesia

E-mail: edi@unsyiah.ac.id

Abstract. The purpose of this study was to determine the cause of a fracture in the crankshaft of a 1000 cc gasoline engine. On the broken surface, it could be seen that the cracks started from the direction of the oil hole. A visual examination revealed that there were beach marks on the surface of the fracture, which is usually common in fatigue failure due to a dynamic load. From the test for the chemical composition, it was found that the material was classified as alloy steel. The Rockwell-based material hardness values were different. The highest hardness value of 102.2 HR_B was on the outer shell portion of the crankshaft on the *x* and *y* axes at points 1 and 14. Finally, it was concluded that the reason for the fracture in the crankshaft was fatigue failure, the material hardness that did not match that of the standard, as well as the initial crack contained within the radius of crank pin number 1.

1. Introduction

The crankshaft, which is one of the main components of a piston engine, serves to forward the rotation of the connecting rod to the clutch. In turbo diesel engines, it can be observed that the increase in engine power usually leads to a decrease in the life of the engine components [1]. Cracks generally start from the initial cracks that usually occurs on the surface of a weak material or in stress concentration areas, such as scratches, holes and others resulting from repeated loading. This initial crack develops into a micro crack or propagation that forms a macro crack, thereby culminating in failure [2]. Failure is a condition that is never desired because it causes direct losses. Failures are usually caused by design errors, material defects, and maintenance errors [3].

Failure analysis is a process for determining the factors that cause a component to losing its function. Failure analyses have been performed on crankshafts in several studies, such as by Witek [4]. In the study, failure occurred on crank pin number 2 after traveling a distance of 260,000 km. In 2016, Fonte [5] found that failure occurred on crank pin number 3 after three years and 5000 hours of maintenance. Figure 1 shows the crankshaft parts.

* Corresponding author: edi@unsyiah.ac.id

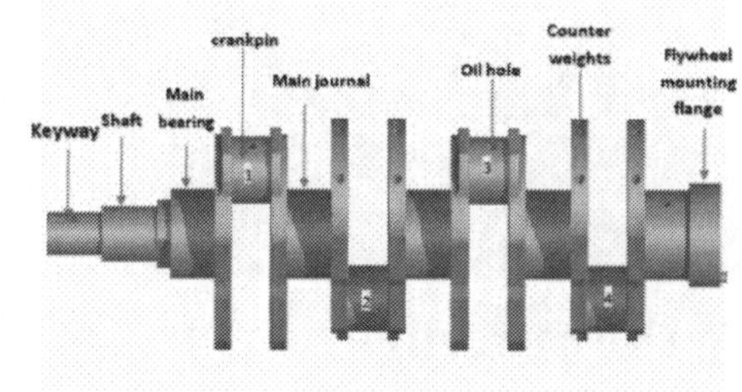

Figure 1. Part of a crankshaft.

Many researchers have concluded that most of the fatigue failures in crankshafts are due to dynamic loading [6-9]. Mistakes, both during disassembly and maintenance, should be avoided to prevent failure of the crankshaft. Scratches, for example, will cause an initial flaw that will lead to failure. Failure usually starts from a critical point experiencing a high stress concentration, such as within the area of the radius, and in errors while mounting the crank pin [6]. Therefore, the purpose of this study was to determine the cause of fracture in a crankshaft.

2. Methodology

In this study, the material used was taken from the crankshaft of a car. First, observations were made to determine the mechanical properties of the material. The visual inspection showed a crack on the crankshaft with an engine rotation speed of 5500 RPM. The crankshaft was broken at crank pin number 1, as shown in figure 2. The visual inspection showed beach marks on the fault surface, as shown in Figure 3, characterizing the cause of the failure.

Figure 2. Failure of the crankshaft.

Figure 3. Failure surface of the crankshaft.

Furthermore, the chemical composition was examined using a PDA-700 spectrometer to determine the elements contained in the material. Then, a hardness test was performed using a Rockwell Type Zwick / Roell ZHR to determine the hardness of the material with reference to the ISO 6568 and ASTM 18 standards with a load of 100 Kgf at the time of testing [10]. The experiments were performed on fourteen different points using a 1/16 indenter ball. Figure 4 shows a sketch of the specimen undergoing the hardness testing. A microstructural test was also performed using the Olympus GX 71 optical microscope with a magnification of 5 to 100 times, to view micro-sized objects and to determine the grain boundaries and the properties of the crankshaft material. Finally, scanning electron microscopy (SEM) was performed to view the broken surface on the crankshaft. The experimental test was necessary to determine whether the initial crack that had occurred on the crankshaft resulting in failure.

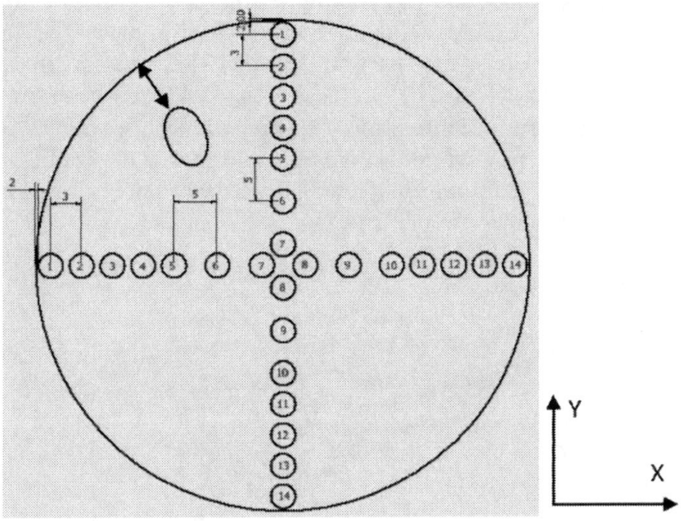

Figure 4. Hardness testing point.

3. Results and discussion

3.1. Chemical composition

From the chemical composition examination results, it could be concluded that the material was alloy steel, as shown in Table 1. According to ASTM A510 [11], the material used was AISI 1060, where the material properties can be seen in Table 2.

Table 1. Chemical composition of the material

Element	Spectroscopy results (%)	Chemical composition (AISI 1060) %
C	0,575	0,55 – 0.660
Mn	0,628	0.60 – 0.90
P	0,008	0.40 Max
S	0,025	0.050 Max

IOP Conf. Series: Materials Science and Engineering **523** (2019) 012067 doi:10.1088/1757-899X/523/1/012067

Table 2. Material properties of AISI 1060 (ASTM A510)

Properties	Metric	Unit
Tensile strength (σ_B)	620	MPa
Posion Ratio (V)	0.30	-
Density (ρ)	7850	Kg/m^3
Yield strength (σ_{ys})	485	MPa
Modulus young (E)	200	GPa
Shear modulus (S)	80	GPa
Hardness, Rockwell B	89	HR_B
Fracture Toughness	36-60	MPa√m

3.2. Hardness test

From the hardness test results, it was found that the highest hardness value of 102.2 HR_B was on the outer shell portion of the crankshaft on the *x* and *y* axes at points 1 and 14. The hardness testing point can be seen in Figure 5. The hardness of the shaft on the outer shell was greater than at the centre to avoid defects and rapid wear during work. The hardness reached 13 mm from the outer surface of the crankshaft. Due to the unsuitable heat treatment process, the crankshaft was harder than the standard AISI 1060 material, which was 94 HR_B, while the average hardness of the crankshaft material that was tested was 100,83 HR_B. Therefore, the crankshaft, which was too hard to overheat up to a depth 13 mm from the outer surface, was prone to failure.

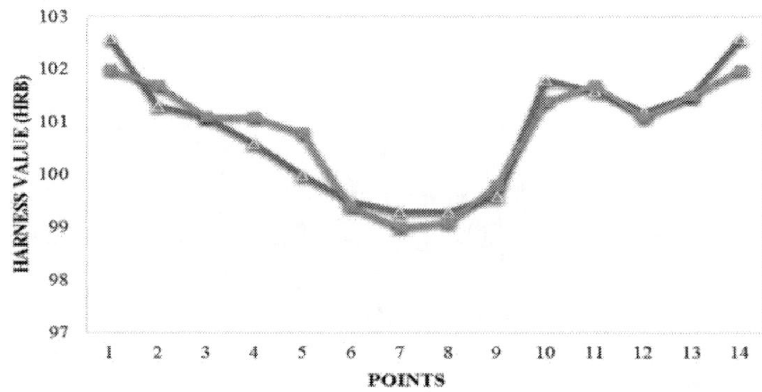

Figure 5. Graph of the hardness values (red is *x*-axis and orange is *y*-axis).

3.3. Observation of SEM

From the SEM test result, a defect around the oil hole was seen at a magnification of 3000 times using an EVO MA10 SEM. Due to maintenance errors and negligence in the installation process, there was a scratch with a sharp object resulting in the defect. The defect can be seen in Figure 6.

After observation of the factography by SEM, the existence of a crack was discovered. The result of a simulation using the finite element analysis revealed that the stress intensity factor was 2.42 MPa √m, which was smaller than the fracture toughness of 50 MPa√m of the material. Therefore, it could be concluded that the crack here did not spread. However, the crankshaft experienced fatigue failure here, as characterized by the beach marks on the crack surface of the crankshaft, as shown in Figure 3 [12]. Another cause was the material hardness that was up to a depth of 13 mm from the outer surface of the crankshaft. As such, the crankshaft was too hard and prone to fracture.

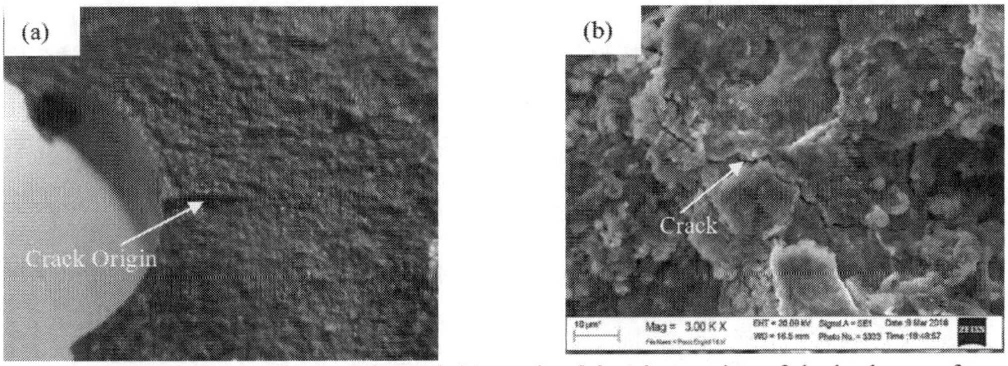

Figure 6. (a) Visual observation results and (b) result of the observation of the broken surface at a magnification of 3000 times.

3.4. Microsctructure

The results of the microstructural test revealed that the matrix was ferrite and pearlite [13] with a mean hardness value of 100.83 HR_B on the *x*-axis and 100.85 HR_B on the *y*-axis. This showed that this material was hypereutectoid steel, which is steel with a carbon content of between 0.02-0.76%. Figure 7 shows the results of the microstructural test with magnifications of 20 and 50 times.

Figure 7. (a) Results of microstructural testing with 50x magnification and (b) results of microstructural testing with 20x magnification.

4. Conclusion

From the results, it was concluded that there were cracks on crank pin number 1 surrounding the oil hole. Over time, the cracks would be enlarged as the crankshaft continued to operate with spin and failure. Based on the simulation results, it could be seen that the maximum stress and strain were on crank pin number 1 due to a defect caused by the occurrence of scratches or the falling of sharp objects when the crankshaft was being installed on the machine.

5. References

[1] Witek L 2016 *Engineering Failure Analysis* **66** 154
[2] Husaini and Zuhaimi 2016 *International Journal of Technology (IJTech)* **3** 456
[3] Husaini, Kishimoto K, Hanji M and Notomi M 2016 ARPN *Journal of Engineering and Applied Sciences* **2** 885
[4] Witek L, Sikora M, Stachowicz F and Trzepiecinski T 2017 *Engineering Failure Analysis* **82** 703

[5] Fonte M, Invante V, Freitas, M and Reis M 2016 *Procedia Structural Integrity* **1** 313
[6] Fonte M, Anes V, Duarte P, Reis L and Freita M 2015 *Engineering Failure Analysis* **56** 109
[7] Putra T E, Abdullah S, Schramm D, Nuawi M Z and Bruckmann T 2015 *Mechanical Systems and Signal Processing* **60-61** 485
[8] Putra T E, Abdullah S, Schramm D, Nuawi M Z and Bruckmann T 2017 *Mechanical Systems and Signal Processing* **90** 1-14
[9] Husaini, Putra T E and Ali N 2018 *International Journal of Automotive and Mechanical Engineering* **15** 5251
[10] ASTM E18–15 2015 *Standart Test Method for Rockwell Hardness of Metallic Materials* (West Conshohocken: ASTM International)
[11] ASTM A519–9 2001 *Standart Specification for Seamless Carbon and Alloy Steel Mechanical Tubing* (West Conshohocken: ASTM International)
[12] Espadafor F J, Villanueva J B and Garcia M T 2009 *Engineering Failure Analysis* **16** 2333
[13] Ridha M, Husaini and Husni 2013 *Material Teknik* (Darussalam: Syiah Kuala University Press)

IOP Conf. Series: Materials Science and Engineering **523** (2019) 012068 doi:10.1088/1757-899X/523/1/012068

Optimal conditions for selective reduction process of Nickel Laterite Ore

F Nurjaman [1,*], **A Sa'adah**[2], **B Suharno**[2]

[1]Research Unit for Mineral Technology, Indonesian Institute of Sciences, Lampung, 35361 Indonesia
[2]Metallurgical and Materials Engineering Department, Faculty of Engineering Universitas Indonesia, Kampus Baru UI Depok, 16424 – Indonesia

E-mail: fajar.nurjaman@lipi.go.id

Abstract. A low-grade nickel laterite ore from Indonesia was selected to investigate the effect of temperature and time of reduction upon the extraction of nickel through the selective reduction-magnetic separation process to produce the ferronickel. The nickel ore was mixed with5 wt.% of anthracite coal and 10 wt.% of sodium sulfate additive. The mixed ore was pelletized into 10-15 mm of diameter. A series of selective reduction processes were carried out into the pellet ores at various temperatures of 950-1150 °C for 60-120 minutes. The optimal selective reduction parameters of nickel laterite ore were observed clearly. The experimental results indicated that the reduction reactions proceed more completely at a higher temperature and longer reduction time. The optimum of nickel grade and recovery in the ferronickel concentrate was obtained at the reduction temperature of 1150 °C for 120 minutes, which was 8.45% and 72.1%, respectively. The increasing of temperatures and times of reduction process resulted in higher of troilite phase formation which suppressed the iron metallization and also increased the ferronickel particle size.

1. Introduction

Nickel is an important metal for stainless steel and alloy steel with extensive industrial applications, especially for modern infrastructure and technology [1]. Nickel is widely used in the form of pure metal or as a ferronickel alloy with varying iron content [2]. Due to the high demand for nickel in the future, there is an increasing interest in lateritic nickel ore as an alternative raw material to replace sulfide nickel ore in nickel production [3].

Nickel laterite ore accounts for up to 70% of total nickel resources in the world. Nevertheless, it was difficult for applying the physical beneficiation process in nickel laterite ore [4]. An effective process known as selective reduction-magnetic separation has been widely studied to produce a high-grade nickel concentrate from Indonesian low-grade nickel laterite ore [5]. Selective reduction of nickel laterite ore was accomplished by the addition of sodium sulfate then followed by the magnetic separation process to separate ferronickel from the gangue minerals [6]. It also has been studied the optimal condition of selective reduction process with variable temperature and reduction time.

The grade of metal in the concentrate increased with the higher reduction temperature due to the increasing degree of metallization [7]. The nickel grade also increased with the longer reduction time because many metallic phases are formed. In addition, the presence of sulfur-based additive increased

* Corresponding author: fajar.nurjaman@lipi.go.id

the kinetic reduction rate and suppressed iron metallization by the formation of troilite (FeS) which increased the selectivity of nickel in magnetic separation process so the nickel grade and recovery were high [8]. Most of the reduction process used bituminous coal as reductant due to its high sulfur content which could improve the formation of troilite [3,9-10]. Nevertheless, the bituminous coal has moderate fix carbon.

In this present work, the reduction process of low-grade nickel ore was carried out by using anthracite containing high sulfur as reductant, which had higher fix carbon than bituminous coal. The effect of the temperature and time of reduction on nickel grade and recovery in ferronickel concentrate was investigated clearly. The size of the ferronickel particle was also observed in this selective reduction process.

2. Experimental

The ore used in this study was from Sulawesi, Indonesia. Table 1 provides the ore composition as determined by X-ray Fluorescence (XRF) analysis. Anthracite coal from Padang, Indonesia was used as the reductant, the composition of the coal used is given in Table 2. The sodium sulfate was used as additives in this selective reduction process.

Table 1. The chemical composition of the laterite ore samples (mass fraction, %)

Element	Ni	Fe	Si	Mg	Al	Ca	Cr	Mn	Co
wt. %	1.4	50.5	16.5	1.81	4.86	0.177	2.68	0.847	0.0662

Table 2. Proximate analysis of anthracite coal (wt.%)

Reductant	Volatile Matter	Ash	Fixed Carbon	Moisture
Anthracite coal	18.25	18.25	60.35	3.14

Both the ore and the reductant were ground to less than 149 μm. The 100 grams of limonitic ore, 5 wt.% of reductant and 10 wt.% of Na_2SO_4 were manually mixed and made into pellets with an approximate diameter of 10-15 mm. The pellets were dried in an oven at 100 °C for 4 h to remove the water, then it placed in a graphite crucible and inserted into a muffle furnace. The reduction temperature was varied in the range of 950-1150 °C and the reduction time in the range of 60-120 minutes.

The reduced pellets were removed from the furnace and quenched into the water, then dried in an oven at 100 °C for 4 h and ground to less than 74 μm in a shaker mill. The reduced pellets were separated in magnetic separation at a magnetic field intensity of 500 gauss magnet to obtain the magnetic product (ferronickel concentrate). The phase transformation of nickel laterite ore in the reduction process was investigated using XRD. The XRF was performed to investigate the nickel and iron grade and recovery in the ferronickel concentrate, while SEM equipped with an EDS was used to observe the phase and its morphological changes after the magnetic separation.

3. Results and Discussion

3.1. The effect of reduction temperature

The grade of metal in the concentrate increased with the higher reduction temperature. This is due to the increasing nickel and iron metallization so metal oxides are reduced to the metal [7]. According to the Fig. 1, the highest nickel grade was obtained at a temperature of 1150 °C where it reached 6.1% with a recovery of 70.3%. The nickel grade increased with the increasing of reduction temperature [8].

Otherwise, the nickel and iron recovery decreased at the high temperature. The nickel recovery was 70.3% during the reduction process at 1150 °C, while the iron recovery was continued to decrease to 17.98%. The decreased of iron recovery was tend to be more significant due to the formation of troilite

(FeS) phases. The formation of non-magnetic troilite phases suppressed the iron metallization which contributes to the increase in the nickel grade in concentrate [2].

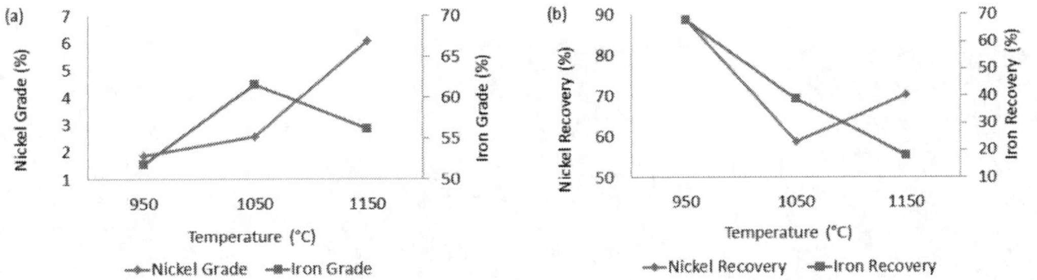

Figure 1. Effect of reduction temperature with the addition of 10 wt.%sodium sulfate on (a) nickel and iron grades and (b) nickel and iron recoveries

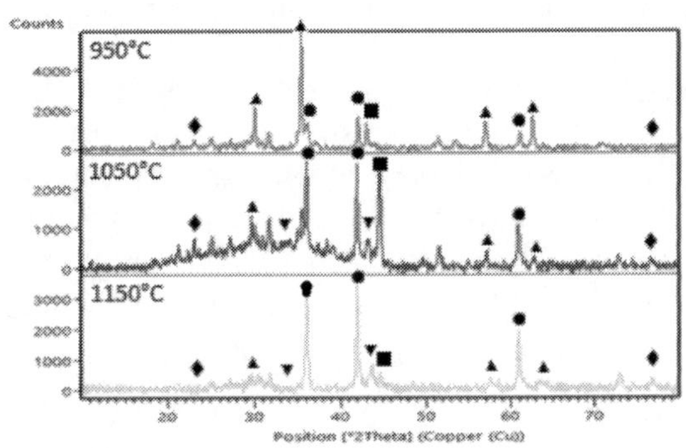

(▼= FeS, ▲= Fe_2SiO_4, ■ = FeNi, ● = FeO, ♦ = SiO_2)

Figure 2. XRD pattern of reduced laterite pellets at various temperature for 60 minutes with the addition 10 wt.% of sodium sulfate.

Table 3. The rietveld refinement calculation of reduced laterite pellets at various temperature for 60 minutes with the addition 10 wt.% ofsodium sulfate.

Phase (wt. %)		FeNi	Fe_2SiO_4	FeO	FeS	SiO_2
Temperature (°C)	950	7.9	44.4	37.3	2	8.4
	1050	6.2	23.2	45.4	16.2	9
	1150	3.6	22	40.6	24.4	9.3

The mineralogical composition of the reduced ore with the addition of Na_2SO4 was wustite (FeO), troilite (FeS), fayalite (Fe_2SiO_4), quartz (SiO_2), and ferronickel (FeNi), is shown in Fig. 2. The peak and composition of wustite increased to a temperature of 1050°C where the reduction of hematite to wustite still occurs, then at a temperature of 1150 °C the peak decreased as some wustite has been reduced to iron and partly reacts with Na_2S to form FeS, as expressed in Eq. (2). The peak of FeS was found at a temperature of 1050 °C and the composition increased until the temperature of 1150 °C. The higher reduction temperature, the more FeS were formed, that suppressed iron metallization so the nickel grade was increased [2,12]. From Table 3, the amount of ferronickel was decreased during the increasing of the reduction temperature due to the transformation of some iron oxide into troilite, thus the small amount of ferronickel that resulted in higher temperature had a more nickel content in concentrate.

IOP Conf. Series: Materials Science and Engineering **523** (2019) 012068 doi:10.1088/1757-899X/523/1/012068

$$\tfrac{1}{4} Na_2SO_4 + CO\,(g) \leftrightarrow \tfrac{1}{4} Na_2S + CO_2\,(g) \qquad \Delta G_T^{\Theta} = -\,36.719{,}98 + 15{,}24\ T\ (J/mol) \qquad (1)$$

$$Na_2S + FeO + 2SiO_2 \leftrightarrow Na_2Si_2O_5 + FeS \qquad \Delta G_T^{\Theta} = -\,81.990{,}96 + 27{,}71\ T\ (J/mol) \qquad (2)$$

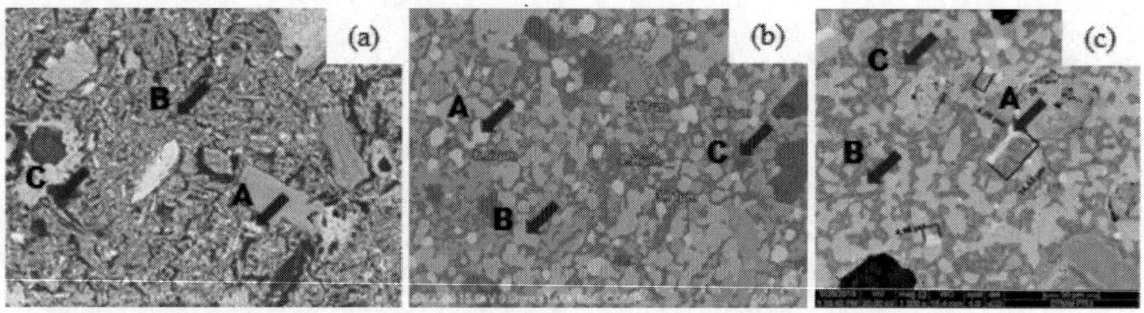

Figure 3. SEM picture of the sample at reduction timeof 60 minutes and the addition 10 wt.% of sodium sulphatewith various reduction temperature of (a) 950, (b) 1050, and (c) 1150°C

The point A, B, and C indicated the ferronickel (FeNi), troilite (FeS), and impurities dominated by Si and O elements, respectively. At a temperature of 950 °C, no measurements of ferronickel grains due to the very small size.The ferronickel particle was growth into 6.83 μm and 14.82 μm at a temperature of 1050 °C and 1150 °C, respectively. During the reduction process, there was a liquid phase caused by the Fe-FeS phase which decreased the eutectic point at 985 °C that caused the agglomeration of ferronickel. At the higher temperature, it accelerated the growth of ferronickel grains [11].

3.2. The effect of reduction time

The metallic phase increasing as the reduction time increased [12]. The more of metal oxide was reduced due to the longer contact time with the gas [13]. At 120 minutes, the nickel grade was 8.45% with the recovery of 72.09%, while the iron recovery was very low. This may occur due to the re-oxidation of the metallic iron when the reduction time was taking too long [12].

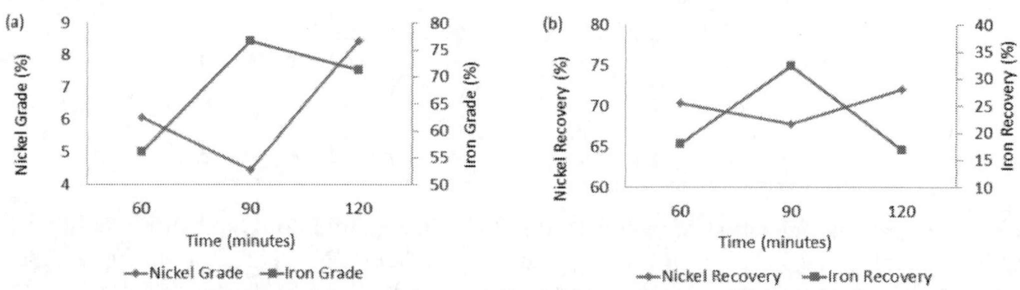

Figure 4. Effect of reduction time with the addition of 10 wt.%sodium sulfate on (a) nickel and iron grades and (b) nickel and iron recoveries

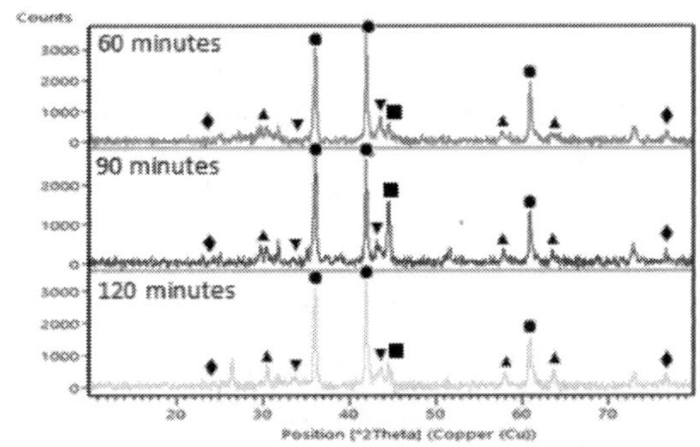

(▼= FeS, ▲= Fe_2SiO_4, ■ = FeNi, ● = FeO, ◆ = SiO_2)

Figure 5.XRD patterns and the rietveld refinement calculation of reduced pellets with various reduction time at the temperature of 1150 ° C and the addition 10 wt.% of sodium sulfate.

Table 4. The rietveld refinement calculation of reduced laterite pellets at 1150 °C in various time with the addition 10 wt.% of sodium sulfate.

Phase (wt. %)		FeNi	Fe_2SiO_4	FeO	FeS	SiO_2
Time (minutes)	60	3.6	22	40.6	24.4	9.3
	90	11.5	30.2	30.1	13.8	14.4
	120	7.1	27	35.6	20.6	9.7

From Fig. 5 and Table 4, the wustite decreased after 60-90 minutes of reduction time due to the reduction of wustite to metallic iron. It is also indicated by the high peak and composition of ferronickelat 90 minutes. However, the wustite increased again after the reduction process at 120 minutes. At a very long reduction time, the reductant would run out and the metallic iron will re-oxidize to iron oxide [14]. It is indicated by the high peak of wustite and followed by the low peak of ferronickel after the reduction process at 120 minutes.

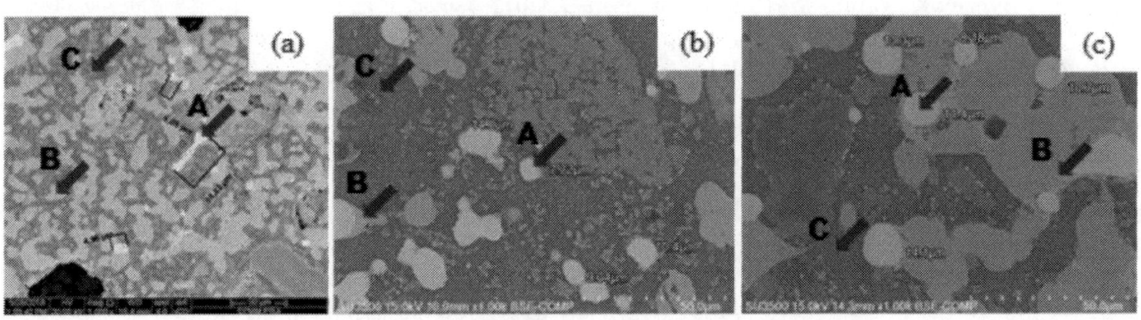

Figure 6. SEM picture of the sample at the temperature of 1150°C and the addition 10 wt.% of $Na_2SO_4$with various reduction time of (a) 60 minutes, (b) 90 minutes, and (c) 120 minutes.

The point A, B, and C indicated the ferronickel (FeNi), troilite (FeS), and impurities dominated by Si and O elements. The size of ferronickel grains of sample 60, 90 and 120 minutes were14.82μm, 15.39μm, and 21.12 μm, respectively. Increasing the reduction time caused the reaction occurred more considerable and led to high metallic phase formation [12].

4. Conclusions

An increase in the reduction temperature led to high nickel grade in ferronickel concentrate. The reduction process at a temperature of 1150 °C resulted in the nickel grade of 6.1% with a recovery of 70.3%. The Na_2SO_4 presence affected the greater formation of troilite which suppressed the iron metallization then resulting in high nickel grade. The reduction of nickel ore until 120 minutes resulting in the highest nickel grade of 8.45%. The long reduction time caused the re-oxidation of metallic iron into iron oxide thus promoting the nickel grade in concentrate. In addition, the high temperature and the long reduction time promoted the growth of ferronickel particle due to the formation of troilite which increased the mass transfer of metals resulting in the agglomeration of ferronickel grains.

5. Acknowledgment

The authors would like thanks to the Ministry of Research, Technology and Higher Education for funding this research through Indonesia's National Innovation System (INSINas) Programme with Grant Number: 06/INS-1/PPK/E4/2018and The Research Unit for Natural Product Technology for utilization of SEM-EDAX equipment in this experiment.

6. References

[1]. Mudd GM 2010 *Ore Geol. Re,* **38** 9-26.
[2]. Li G, Shi T, Rao M, Jiang T, and Zhang Y 2012 *Miner. Eng.* **32** 19-26.
[3]. Elliott R, Pickles C, and Peacey J 2017 *Miner. Eng* **100** 166–176.
[4]. Rao M, Li G, Zhang X, Luo J, Peng Z and Jiang T, 2016 *Sep. Sci. Tech* **51** 1408-1420.
[5]. Li Q, Cui Y, Zhu D, Zhu J, Pan J, Zhang H and Zheng G 2010 *XXV International Mineral Processing Congress*, Brisbane.
[6]. Lu J, Liu S, Shangguan J, Du W, Pan F and Yang S 2013 *Miner. Eng.* **49** 154–164.
[7]. Li B, Wangand H, Wei Y, 2011 *Miner. Eng.* **24** 1556-1562.
[8]. Zhu DQ, Cui Y, Vining K, Hapugoda S, Douglas J, Pan J and Zheng GL 2012 *Int. J. Miner. Process.* **106** 1-7.
[9]. Elliott R, Rodrigues F, Picklesand CA, Peacey J 2015 *Can.Metall. Quart.* 395-405.
[10]. Mayangsari W and Prasetyo AB 2016 *Jurnal Material Metalurgi* **1** 1-68.
[11]. Rao M, Li G, Zhang X, Luo J, Peng Z and Jiang T 2016 *Sep. Sci. Tech.* **51** 1727-1735.
[12]. Li J, Bunney K, Robinson HR and Robinson DJ 2013 *Miner. Eng.* **41** 71-78.
[13]. Jiang M, Sun T, Liu Z, Kou J, Liuand N Zhang S 2013*Int. J. Miner. Process.* **123** 32-38.
[14]. Forster J, Pickles CA and Elliott R 2016 *Miner. Eng.* **88** 18-27.

IOP Conf. Series: Materials Science and Engineering **523** (2019) 012069 doi:10.1088/1757-899X/523/1/012069

The comparison of optimum heuristic and deterministic scheduling rules for job shop scheduling in the manufacture

E Fradinata[1,*], Z M Kesuma[2]
[1]Industrial Engineering Department, Syiah Kuala University, Banda Aceh, Indonesia.
[2]Statistics Study Program, Faculty of Mathematics and Natural Science, Syiah Kuala University, Banda Aceh, Indonesia.

E-mail: edinata69@gmail.com

Abstract. There are some types of scheduling problems in the production line of manufacturing. One of the most multiple usages of the operational machine for doing some jobs is job shop scheduling. The implementation of the job shop problem in this study uses the input of random order of eight job numbers. The raw materials for the job numbers to do the process and the five type machines are considered. The machines work in different functions to do the jobs for some materials. The materials have processed the work in process (WIP) before they are sent to the warehouse. The proposed of this study is to find the optimal condition of job shop scheduling heuristic and deterministic rules to minimize the makespan, tardiness with the heuristic and some deterministic rules in the Lekin software. The heuristic methods are General SBRoutine and the DASH (shifting bottleneck). The deterministic use Critical Ratio (CR) Rule, First come First Serve Rule (FCFS), Longest Processing Time (LPT), Shortest Processing Time (SPT), Weighted Shortest Processing Time (WSPT) and Machine Sequence (MS). The result shows that the minimum makespan is FCFS method with 47 minutes, the tardiness is 27, and for the machine no3. It produces the effective percentage of The contras deviation performance among the jobs of DASH, and compare to the FCFS is as follow, DASH, LPT, MS, SPT, WSPT are increasing 4.26%; 31,91%; 8.51%; 4.26% and 12.77% respectively.

1. Introduction

The scheduling is the important activity to set up planning for scheduling in the manufacturing process due to produce the various products in a limit time [1]. The products must be delivered to the customer in a scheduling order to fulfill the customer requirement [2]. The more accurate planning scheduling will increase the precision of production activity. Most of the industry always tries to implement the short process and minimize the production cost without any change the quality of the product. The minimum process would increase the productivity of time [3].

The job shop scheduling is the specific model of sharing the machine in the use of the task of jobs where it can be handling in multi difference jobs to optimize the exploitation facilities [4]. The job shop scheduling is more applicable used in the manufacturing with the many lot-sizing orders in the short time to fulfill them. The flow shop scheduling protects the bottle next with the sequence scheduling in one machine [5].

* Corresponding author: edinata69@gmail.com

The process in the production area in manufacture the route activity of the material and the worker always consider the time with the schedule set up. The job keeps in the arrival at the workstation to ready to process to obtain the finish goods. The important thing is how to make sure the raw material is not in a big bottleneck in front of the workstation. Precisely, to keep the process with the small tardiness and makespan. The process of the goods in the flow time has the waiting time for the job to do the sequence in the process.

Let say, j is the job then t_j is the processing time, f_j is ready time, d_j is the due date, C_j is the completion time, F_j is the flow time, Lj is lateness, and Tj is tardiness from here we can get the the flow time formula as follow:

$$F_j = C_j - r_j \tag{1}$$

The lateness is:

$$F_j = C_j - r_j \tag{2}$$

The tardiness is the lateness from the job J if it fails to meet its due date, or zero, given as:

$$Tj = \max\{0, Cj - dj\} \tag{3}$$

The variable in the job shop scheduling is as follow:

- Makespan as Cmax
- The Maximum Tardiness as Tmax
- The Total Number of late jobs ΣUj
- The Total Flow Time ΣCj
- The Total Tardiness ΣTj
- The Total Weighted Flow Time ΣWj Cj and Total Weighted Tardiness as ΣWj Tj

In the Lekin software, there are some methods of job shop scheduling [4], such as heuristic (heuristic General SB Routine) and DASH. The heuristic is the analytical technique used to improve performance through the computational process. The deterministic is the mathematical model where the symptoms can be measured with a high degree of certainty. The deterministic rules in the scheduling are such as CR Rule, First come – First Serve Rule, Longest Processing Time, Shortest Processing Time, and Weighted Shortest Processing Time.

2. The Methods

This study runs with the job shop scheduling in the manufacturing to planning the process of the eight jobs and five machines to produce the products from the machines process. The purpose of this study is to explore the comparative methods from different methods, such as the heuristic and deterministic process. There are some rules for the methods of job shop DASH (shifting bottleneck), Critical Ratio (CR) Rule, First come – First Serve Rule (FCFS), heuristic General SB Routine, Longest Processing Time (LPT), Shortest Processing Time (SPT), Weighted Shortest Processing Time (WSPT). There are some workstations to implement the process of this study. There are five machines: grinding machine, lathe machine, drilling machine, grinding machine and cutting machine. The machine has some function to do the jobs. The materials were done with the different functions work in the machine's workstation. The eight jobs were done by the machines sequentially to finish the goods. Some jobs are Connection Pipe (CP/J1), Coppell (CPL/J2), Engle iron (EI/J3), Hollow iron (HI/J4), Multi-Plate Rack (MPR/J5), Ring Pipe (RP/J6), Socket (SC/J7), Standard Swing Arm (SSA/J8). There is job shops scheduling method. At the end of the process of this study is to compare the optimal objectives from the routes and time parameters, and then concludes the result.

3. Result and Discussion

This study used the process of manufacturing with five machines and eight jobs to obtain the job shop scheduling to produce some products. The scheduling is used to plan the process of maximum makespan and maximum tardiness time. The job shop scheduling for the Gantt-chart can be seen in Figure 1. Gantt chart is the representation of the bar chart to illustrate the appropriate planning to the schedule where it shows the starting time activity and the finish time activity from each job.

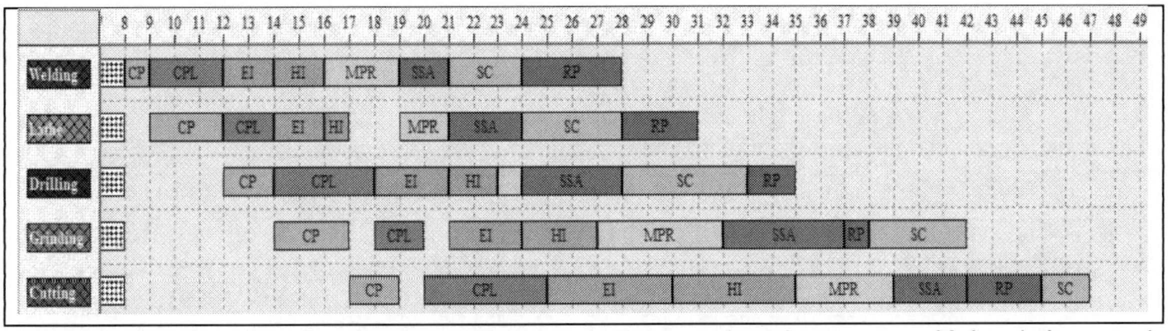

Figure 1.The gantt chart for planning the job shop in the manufacturing process with heuristic general SB routine.

There are some workstations to implement the process of this study. There are five machines: grinding machine, lathe machine, drilling machine, grinding machine and cutting machine. The machine has some function to do the jobs. The materials were done with the different functions work in the machine's workstation. The eight jobs were done by the machines sequentially to finish the goods. Some jobs are Connection Pipe (CP/J1),), Coppell (CPL/J2), Engle iron (EI/J3), Hollow iron (HI/J4), Multi Plate Rack (MPR/J5), Ring Pipe (RP/J6), Socket (SC/J7), Standard Swing Arm (SSA/J8). The matrix of job shop routes are as follow:

Table 1. The matrix of routes for the job shop

Machine	Job								
	CP(J_1)	**CPL(J_2)**	**EI(J_3)**	**HI(J_4)**	**MPR(J_5)**	**RP(J_6)**	**SC(J_7)**	**SSA(J_8)**	**Total**
1Welding M	1	3	2	5	3	4	3	2	23
2Lathe M	3	2	2	1	2	3	4	3	20
3Drilling M	2	4	3	2	1	2	5	4	23
4Grinding M	3	2	3	3	5	1	4	5	26
5Cutting M	2	5	5	2	4	3	2	3	26
Total (minute)	11	16	15	13	15	13	18	17	118

The route of the jobs shows that the sequence of the jobs in a time of process to finished the jobs for each machine. J1 (connection pipe) did the sequence of work with the welding process in 1 hour, then process in the lathe process in 3 minutes, in drilling with 2 minutes, grinding with 3 minutes and at the end do the finishing with cutting in 2 minutes to make it follow the design of the model. The same thing did in the other jobs as well. The total process of the jobs in the workstation machines is 118 minutes.

Table 2. The matrix sequence process for the job shop

No	Machine	Job							
		CP(J_1)	**CPL(J_2)**	**EI(J_3)**	**HI(J_4)**	**MPR(J_5)**	**RP(J_6)**	**SC(J_7)**	**SSA(J_8)**
1	Welding	4	4	4	1	5	5	5	5
2	Lathe	5	5	5	5	4	4	4	2
3	Drilling	3	3	1	3	1	2	2	3
4	Grinding	2	2	2	4	3	3	3	4
5	Cutting	1	1	3	2	2	1	1	1

In the job shop scheduling is very common to manage some jobs for some machines in the dynamic interchange between sharing the machines to produce some product. The sequences depend on each production process[6]. The sharing machines are very partially decided in the group of production to

IOP Conf. Series: Materials Science and Engineering **523** (2019) 012069 doi:10.1088/1757-899X/523/1/012069

manage the capacity of each machine. This is a local decision. The sharing machine can be seen at the table above.

Table 3. The Process time (Pr.tm) and tardiness (T)

Time	Job							
	CP(J_1)	CPL(J_2)	EI(J_3)	HI(J_4)	MPR(J_5)	RP(J_6)	SC(J_7)	SSA(J_8)
Process Time (Pr.tm)	11	16	15	13	15	13	18	17
T (Tardiness)	15	19	23	27	17	23	25	20
wT	30	38	69	108	102	138	150	120

The Process time (Pr.tm) shows that the total of working time for one job to finished the work. In here the J1 show that the 11 hours to finalize the work. The process of tardiness (T) for each job has the variation while the weighted tardiness (wT) has a different value. The minimum makespan and tardiness are on machine lathe machine and job number 1 (J1).

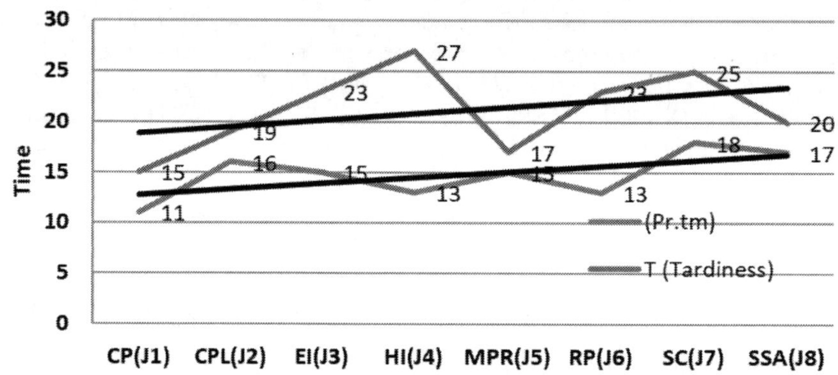

Figure 2. The trend of process time and tardiness for the job shop

The Process time (Pr.tm) and tardiness (T) shows that the process time increase will make the tardiness increase. It means that the sequence time makes the bottleneck at the machine to process. The longer time to process one item makes the queuing to process so the tardiness should be happening on it.

Table 4. The result of job shop scheduling performance of parameters

Methods	Time	C_{max}	T_{max}	ΣU_j	ΣC_j	ΣT_j	$\Sigma w_j C_j$	$\Sigma w_j T_j$
CR	1	47	28	8	280	167	1339	738
DASH	4	49	27	8	281	168	1350	749
FCFS	1	47	27	8	280	167	1344	743
General SB-R	2	47	27	8	282	169	1356	755
LPT	1	62	58	8	381	268	1636	1035
MS	1	51	29	8	304	191	1445	844
SPT	1	49	43	8	278	165	1236	635
WSPT	1	53	47	7	300	188	1190	595

Note: DASH (shifting bottleneck), Critical Ratio (CR) Rule, First come – First Serve Rule (FCFS), heuristic General SB Routine, Longest Processing Time (LPT), Machine Sequence (MS), Shortest Processing Time (SPT), Weighted Shortest Processing Time(WSPT).

There are eight scheduling methods were tried to compare the best result from the planning of the process of making some products. The scheduling methods measured the parameters of Time, makespan (C_{max}), max.tardiness (T_{max}), number of late jobs (ΣU_j), total flow time (ΣC_j), total tardiness (ΣT_j), total weighted flow time ($\Sigma w_j C_j$), total weighted tardiness ($\Sigma w_j T_j$).

The different operations for the job shop model rules are job shop scheduling, such as DASH shifting bottleneck, Critical Ratio Rule, First come – First Serve Rule, heuristic General SB Routine, Longest Processing Time, Shortest Processing Time, Weighted Shortest Processing Time. The First come – First Serve Rule has the best job shop scheduling system on the parameters of C_{max}, T_{max}, ΣC_j, ΣT_j, $\Sigma w_i C_j$, $\Sigma w_i T_j$.

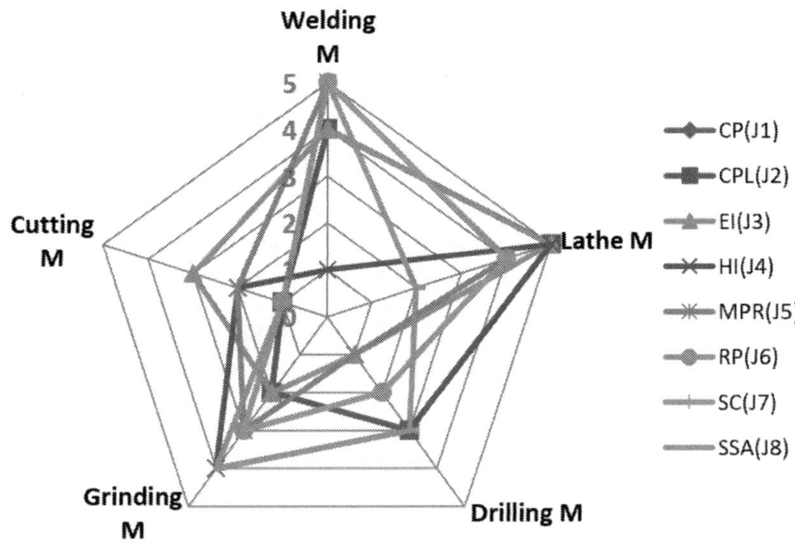

Figure 3. The route of jobs for each machine in a job shop

The sequence jobs on the workstations are illustrated in figure x where one machine can handle more than one job in the sequence time to make the process time to produce the product effectively.

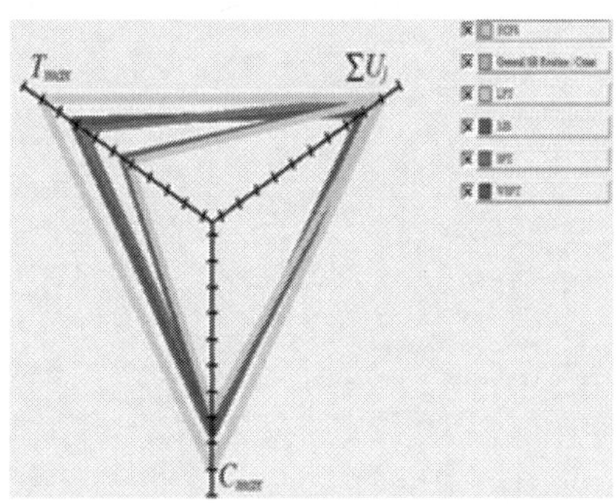

Figure 4. The performance of the relationship between the objectives

The yellow line shows that the performance of FCFS shows the best due to the points in the triangle corner of C_{max}, T_{max}, ΣC_j have nearest to the center value. It illustrates the figure 4. The C_{max} is 47 and T_{max} is 27 where it is close to the general SB routine scheduling.

IOP Conf. Series: Materials Science and Engineering **523** (2019) 012069 doi:10.1088/1757-899X/523/1/012069

4. Conclusion

This paper describes the exploration and analysis of the optimal job shop scheduling for various methods of a certain condition of the job planning. Some methods of the job shops are exploring and compare the performance of the objectives from the process. The heuristic methods are DASH and heuristic General SB Routine and deterministic are CR, FCFS, LPT, SPT, WSPT. The methods are compared the makespan (C_{max}), max.tardiness (T_{max}), number of late jobs (ΣU_j), total flow time (ΣC_j), total tardiness (ΣT_j), total weighted flow time (Σw_jC_j), total weighted tardiness (Σw_jT_j). The result from the objective show that the FCFS shows outperform than the other methods to minimize the C_{max} and T_{max} and FCFS is a very close result from the Heuristic general SB routine (C_{max}) method. The FCFS produce the C_{max} time is 47, T_{max} is 27, ΣU_j is 8, $\Sigma C_j$is 280, ΣT_j is 167, Σw_jC_j is 1344, Σw_jT_j= 743. The total process of the jobs in the workstation machines is 118 minutes with each jobs time are 11, 16, 15, 13, 15, 13, 18, 17 respectively. The contras deviation performance among the jobs of DASH, and compare to the FCFS is as follow, DASH, LPT, MS, SPT, WSPT are increasing 4.26%; 31,91%; 8.51%; 4.26% and 12.77% respectively. In the real situation, the planning of schedules usually has the gap between scheduling planning and the actual. This situation makes the worker in the factory do the adjustment scheduling in the timespan to create the relevance with the target scheduling on the track due to some factors, such as jobs, machines, order, raw material and the condition of worker's scheduling. The scheduling must think what, how, when and who to implement the planning. At last, more accurate planning from the scheduling would increase the productivity in the process of a product in the manufacturer.

5. References

[1] Kassu J and Eshetie B 2015 *Global Journal of Research In Engineering.*
[2] Fradinata E 2012 *ASEAN* (Asian Academic Society International Conference Proceeding Series).
[3] Fradinata E 2013 *ASEAN* (Asian Academic Society International Conference Proceeding Series).
[4] Özgüven C, Özbakır L, and Yavuz Y 2010 *Applied Mathematical Modelling* **34** 1539-1548.
[5] Chen JC, Wirasambada S, Pujawan IN, Wang K, Wu C, Lin C, and Chiu C 2012 *Asia Pacific Industrial Engineering & Management Systems Conference* 1093-1099.
[6] Fradinata E 2013 *ASEAN* (Asian Academic Society International Conference Proceeding Series).

IOP Conf. Series: Materials Science and Engineering **523** (2019) 012070 doi:10.1088/1757-899X/523/1/012070

The relationship between data skewness and accuracy of Artificial Neural Network predictive model

A Larasati [1,*], **A M Hajji**[2], **Anik Dwiastuti**[1]

[1]Department of Industrial Engineering, Universitas Negeri Malang, Indonesia
[2]Department of Civil Engineering, Universitas Negeri Malang, Indonesia

E-mail: aisyah.larasati.ft@um.ac.id

Abstract. The purpose of this study is to investigate the relationship between data skewness in the output variable and the accuracy of artificial neural network predictive model. The artificial neural network predictive model is built using multilayer perceptron and consist of one output variable and six input variable, and the algorithm used is back propagation. Data used in this study is generated by conducting the simulations in 1000 cycles. Three categories of skewness used in the output variables are positive skewness, neutral, and negative skewness. The results show that data skewness does not have a significant effect on the accuracy of the artificial neural network predictive model. These results imply that artificial neural network predictive model has a higher capability to cope with skewed data due to its complexity in the hidden layer.

1. Introduction

Although data mining techniques have been widely adopted in various sectors, data mining models including predictive models built using data mining techniques, are generally still understood as black boxes. Therefore, understanding the arguments behind the predictive model built using data mining techniques is crucial because it is necessary to assess the level of confidence of the model [1]. Understanding model parameters also provide an insight into the conditions of the model that can be used or to convert unreliable models into reliable ones.

One important factor that influences the determination of model parameters is data skewness. Understanding the distribution and variability occurring in the data is generally accomplished by a skewness investigation that is often expressed as a symmetry level of data. Although skewness of data can be visually evaluated through a data plot, still many coefficients need to be considered to ensure data preprocessing performed produce an accurate analysis [2].

The application of machine algorithm, such as the artificial neural network model (ANN), becomes more popular when it links to the problem of skewed (imbalanced) data. Datasets obtained from survey studies are mostly skewed, in which majority of the cases were classified into a certain class and only small number were classified into the other classes (minor). Many existing classification system tend to build a model without considering the proportion of each class. When data is highly skewed, classification model tends to misclassify minority class. Thus, when skewed data is not handled properly, the model tends to have high misclassification rates (poor prediction capability) [3].

* Corresponding author: aisyah.larasati.ft@um.ac.id

IOP Conf. Series: Materials Science and Engineering **523** (2019) 012070 doi:10.1088/1757-899X/523/1/012070

Based on this background, this study aims to explore the relationship between data skewness and the accuracy of the artificial neural network as a predictive model.

2. Literature Review

2.1 Skewness

Skewness is intended to represent how close the data is from the symmetry (or sometimes expressed as how much data distribution to the distribution of normality), where one side of the data distribution is more "stretched" than the other. Although symmetrically distributed data is preferred in statistical processing with the distribution $(-\infty, \infty)$, the concept of asymmetry is also common with distribution $(0, \infty)$. The study of data skewness has been discussed by many statisticians, e.g. Pearson proposed gamma distribution as an alternative to model non-symmetric and non-normal data distribution as well as pareto that is focused on skewed data during data pre-processing [2].

The problem of distribution imbalance in the data classes (skewness) is quite often found in the digital age today. Skewness is often associated with a marginal distribution of unbalanced data. The data collected massively in this era of information technology development is sometimes difficult to understand because of the problem of data skewness and the marginal distribution of unbalanced data [4]. This problem refers to the skewness in the underlying marginal data distribution, which in turn, poses many difficulties in the learning algorithm that was constructed when constructing a model. To solve emerging problems arising from skewness and marginal distribution of data, efforts can be made primarily in two categories: model-oriented solutions (finding appropriate models) and data-oriented techniques (looking for appropriate data processing techniques).

2.2 Artificial Neural Network

Artificial Neural Network (ANN), also known as Neural Network model (NN), is a mathematical or computational model based on neural networks (neurons). Because ANN is designed based on human biological systems, this model is built on the relationship seen in the training data set by adapting synaptic connections that exist between neurons. ANN consists of artificially linked artificial neurons and processes information using a linking approach for calculation. An ANN model consists of input, hidden, and output layer. The hidden layer in ANN contains some nodes that calculate the weights of input based on external or internal information that gets into the network during the learning process. ANN is able to derive the desired information from complicated or incorrect data, to determine patterns and to recognize trends that are actually very complex or almost impossible identified by humans or other statistical techniques. ANN is the best example of adaptive learning that can be designed to perform real time operations with a high level of fault tolerance [5].

In the simplest ANN framework, there is only one hidden layer, a layer containing functions to connect the input layer and output layer. Each layer in ANN can contain more than one variable (node). Each variable (node) is connected by using a one-to-one relationship so that there is a relationship pattern between each node in the input layer and each node in the output layer.

3. Research Methodology

This study is a type of simulation study. Data is generated in 1000 cycle in which each cycle contain 200 data set. The study calculates mean of misclassification rates in every 100 cycles. Resulted data from simulation has negative, neutral, and positive skewness. Generated data in this study is classified to be negatively skewed if it has skewness level < -0.05, and it is classified as positively skewed if it has skewness level > 0.05. Otherwise, data is classified to be neutral.

Each data set is comprised of one dependent and six independent variables. After generating data, this study continued with building Artificial Neural Network Model (ANN) based on that various data skewness. The ANN model is built by splitting each data set into three categories, training (50%), validating (30%), and testing (20%) data set. The accuracy of ANN model is measured by the average value of mean classification rates resulted from each cycle. The misclassification rates mean the proportion of disagreement between predicted target variables compared to the actual target variable. Briefly, the research procedures applied in the study is shown in Figure 1.

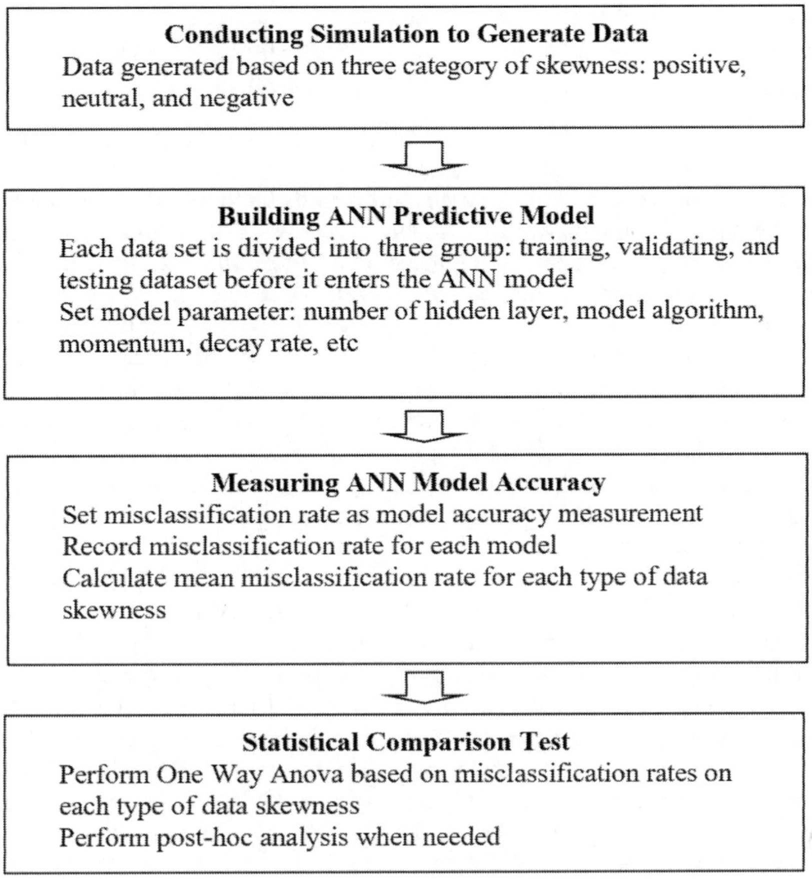

Figure 1. Research Procedures

4. Results and Discussions

Statistic descriptive of the mean misclassification rates results show that negative skewness data has the highest mean and the lowest standard deviation of misclassification rates, while neutral skewness results in the highest standard deviation and the lowest mean of misclassification rates. Detail results of statistics descriptive of mean of misclassification rates are shown in Table 1.

Table 1. Statistic descriptive of misclassification rates

Type of Skewness	Mean	Std. Deviation	Std. Error	95% Confidence Interval for Mean	
				Lower Bound	Upper Bound
negative skewness	.3854	.00728	.00230	.3802	.3906
neutral	.3812	.01704	.00539	.3690	.3934
positive skewness	.3813	.01055	.00334	.3738	.3888
Total	.3826	.01204	.00220	.3781	.3871

IOP Conf. Series: Materials Science and Engineering **523** (2019) 012070 doi:10.1088/1757-899X/523/1/012070

Table 2. ANOVA of Misclassification Rates

	Sum of Squares	df	Mean Square	F	Sig.
Between Groups	.000	2	.000	.379	.688
Within Groups	.004	27	.000		
Total	.004	29			

ANOVA test is conducted to test whether there are significant differences in the misclassification rates due to various skewness level of the dependent variable. The results of ANOVA as shown in Table 2 indicate that p-value = 0.688. Assuming $\alpha = 0.05$, the Anova test results indicate there is no significant difference in the mean misclassification rates caused by various skewness.

Skewness refers to a relative distribution of each category in polynomial data or data distribution toward its mean. Generally, skewed data affect the accuracy of the predictive model including artificial neural network since the performance of an artificial neural network model depend on how well the training data set to teach the model. As explained in [6], skewed data, also known as imbalanced data occurs when certain classes/categories have higher occurrence compared to other classes/category. Since artificial neural network model learns from the pattern of training data set, fewer occurrence data tend to be ignored than the higher one. As a result, the misclassification rates tend to be higher in a skewed data than the uniform distributed data. Thus, the solution to overcome this problem is performing data resampling or less probability sampling for the higher occurrence data until the number of occurrence between each category is equal. However, since in this study we define negative skewed data is data set in which the skewness value < -0.05, neutral data is data with skewness between -0.05 and 0.05, and positively skewed is data with skewness > 0.05, thus there is possibility this study generated simulated data with skewness not really differences although these data are negative, positive or neutral. As a result, the mean misclassification rates are not significantly different. Another possibility is the fact this study generates the same total number of positive, neutral and negative skewed data and applies continued learning, the ANN predictive model has learnt the pattern for positive, neutral and negative skewed data in the same amount thus the resulted mean misclassification rates are not significantly different.

Accuracy based on misclassification rate is not suitable to measure ANN predictive model performance with skewed data [7]. The more suitable accuracy measurement for skewed data is the geometric mean of accuracies per class (*g-mean*), which can be measured as the number of correct classification in a certain category divided by the number of total occurrence in that related category. This study also proposed the use of random over-sampling to cope with skewed data. The accuracy measurement could be not really suitable with the generated data set, thus the mean misclassification rates from positive, neutral, and negative skewed data set are not significantly different.

5. Conclusion

Based on the results, this study concludes that skewness in the dependent variable has no significant effect on the accuracy of the artificial neural network model (ANN). Data that have positive, neutral and negative skewness on the dependent variable have mean misclassification rates that do not differ significantly. One possible reason for this finding is the use of continued learning in the model building step tend to make the pattern of lower occurrence data have learned by the model better. Thus, the ANN predictive model has learned the pattern data at the same level even the data is positively or negatively skewed. Another possible reason for this finding is the mean of misclassification rates is not suitable to measure the predictive model accuracy when the data set is skewed. Thus, the study suggests further investigation to compare the accuracy when the performance measurement is based on mean misclassification rates and the geometric mean of accuracy.

IOP Conf. Series: Materials Science and Engineering **523** (2019) 012070 doi:10.1088/1757-899X/523/1/012070

6. Acknowledgment

The authors acknowledge to the Institute of Research and Community Services (LP2M) *Universitas Negeri Malang* for managing the university research fund (PNBP) 2018 as the financial source and support for this study.

7. References

[1] Ribeiro MT, Singh S, Guestrin C 2016 *Proceedings of the 22nd ACM SIGKDD international conference on knowledge discovery and data mining* ACM 1135–44.
[2] Belzunce F, Mulero J, Ruiz JM, Suárez-Llorens A 2016 *Environ Ecol Stat.* **23** (4) 491–512.
[3] Liu Y, An A, Huang X. 2006. Boosting Prediction Accuracy on Imbalanced Datasets with SVM Ensembles. 107–8.
[4] Abdi L, Hashemi S 2016 *IEEE Trans Knowl Data Eng.* **1** (1).
[5] Gill NS, Mittal P. A 2016 *J Theor Appl Inf Technol.* **87** (1) 1–10.
[6] Barandelaa R, Sanchezb JS, Garcia V 2003 *Pattern Recognit.* **63** (3) 849–51.
[7] Serrano-Guerrero J, Olivas JA, Romero FP, Herrera-Viedma E 2015 *Inf Sci* **311**, 18–38.

IOP Conf. Series: Materials Science and Engineering **523** (2019) 012071 doi:10.1088/1757-899X/523/1/012071

Monte Carlo simulation for predicting the reliability of a boiler in the Nagan Raya steam power plant

I Pamungkas[1], Arhami[1,*], M Dirhamsyah[1]

[1]Mechanical and Industrial Engineering Departement, Engineering Faculty, Syiah Kuala University
Tengku Syech Abdul Rauf Street, No. 7, Darussalam, Banda Aceh, 23111, Indonesia

E-mail: arhami@unsyiah.ac.id

Abstract. The Nagan Raya steam power plant 2x110 Megawatt is one of the power plants owned by PT. PLN (Persero) located in Aceh. The Nagan Raya steam power plant that operates continuously is often failure. An overhaul of maintenance activities that lasted only six months was the cause of frequent disruptions or failures. Performance evaluation in terms of reliability is urgently needed to minimize the failure. This study aims to determine the level of reliability and preventive maintenance time based on Monte Carlo simulation to the nine critical components of the boiler section. Monte Carlo simulations are used to predict precise maintenance time and future reliability values. Monte Carlo simulation results on nine important components of the boiler section, maintenance time is required within 40 days to 86 days of maintenance with a reliability value of 31% to 38%.

1. Introduction

The electric energy industry is currently acting in a very strategic role as electricity consumption increases. Adequate electrical energy industry is needed in community activities and economic growth. PT. PLN (Persero) as the sole manager of the electric energy industry in Indonesia and owned by the Indonesian government in the form of State Owned Enterprises (SOEs) is required to be able to meet the increasing electricity demand every year. Nagan Raya Power Plant 2x110 Megawatt is one of the power plants owned by PT. PLN (Persero) located in Aceh other than ArunLhokseumawe Gas Engine Power Station 184 Megawatt. Total electricity in Aceh reaches 344 MW, while electricity demand in Aceh is currently 325 MW. Means that electricity in Aceh has a surplus. However, power outages are still common if one factory is disrupted or fails, especially the Nagan Raya power plant.

The boiler is one of the parts that has the highest disrupted or failure frequency among other parts. Nagan Raya steam power system is highly dependent on boiler performance. The boiler serves as a water heater using heat resulting from coal combustion to produce high-pressure steam to drive the turbine. Each part of the boiler has many components and if one component is disrupted or failure it will affect the performance of the production system. An overhaul of maintenance activities that lasted only six months was the cause of frequent disruptions or failures. Some studies show that high frequency oh downtime happens due to applied traditional maintenance [1-3]. Performance evaluation in terms of reliability is needed to minimize such disrupted or failure.

The probability of reliability is the probability of a component or system to operate in accordance with the desired function for a given period of time when used under certain operational conditions

* Corresponding author: arhami@unsyiah.ac.id

[4]. Future reliability values are better known to easily prevent damage. Under such circumstances, Monte Carlo simulation is the only time effective way of estimating reliability [5-6]. Monte Carlo simulations are used to predict precise maintenance time and future reliability values. Monte Carlo simulations basically generate uniformly distributed random numbers (0,1) and then convert to non-uniform distributions using inverse transformation methods [7]. The benefits of the Monte Carlo simulation are easy to implement and the results are close to true values. Some reliability studies using Monte Carlo simulations show results close to the actual values[7-9].

The purpose of this study was to predict maintenance time and future reliability values based on Monte Carlo simulation results for each important component of the boiler section.

2. Methods

The object of research observed is boiler section, covering the extent of damage and maintenance activities performed on the machine. Data collection methods used in this study is through direct observation and interviews with the company. The results of this study are expected to be used as a reference for the implementation of scheduled or predictive maintenance. Flow chart of this research can be seen in Figure 1.

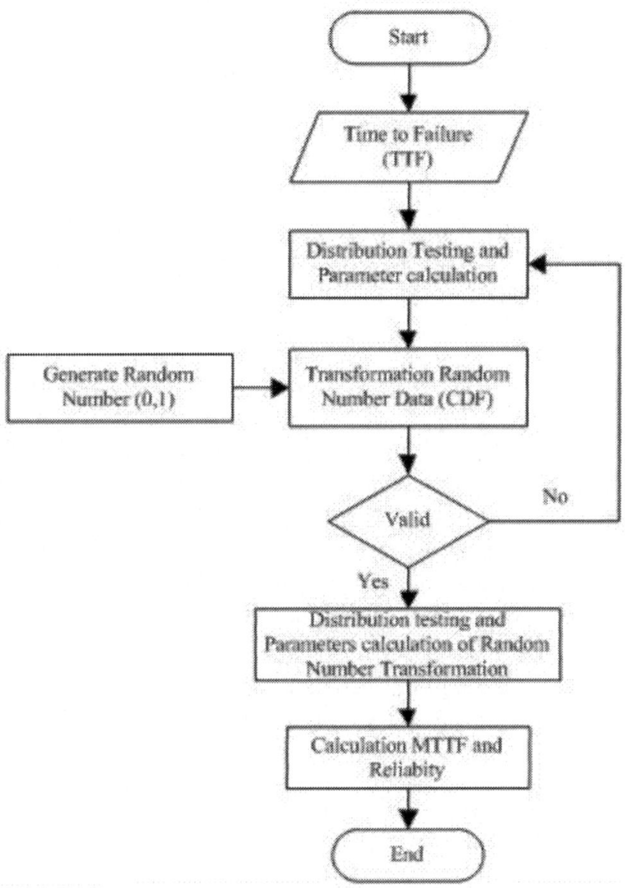

Figure 1. Flow Chart of Monte Carlo Simulation for Reliability

3. Result and Discussion

3.1. Determine the critical Components

Important components of the boiler will be determined based on components that have at least ten failures. The list of component names and the frequency of failure occurring in the period from June 2013 to August 2015 in Figure 2.

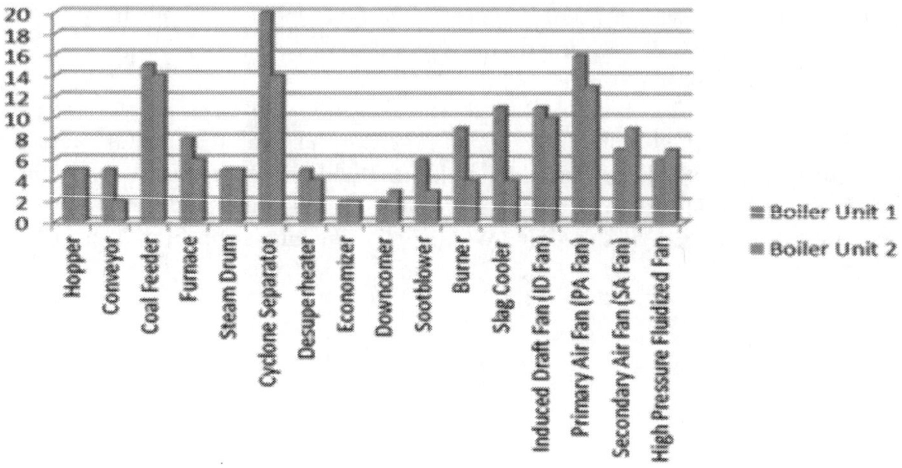

Figure 2. List of Failure Components in Boiler Units 1 and 2

Based on Figure 2, which includes important components in boiler unit 1 is a cyclone separator component, coal feeder, ID fan, PA fan and slag cooler. While in the boiler unit 2 is a cyclone separator component, coal feeder, fan ID and PAfan.

3.2. Distribution Testing and Parameter Calculation for the Actual Time to Failure (TTF)

The distribution test is performed on the actual time for failure (TTF). The distribution types used are normal, lognormal, weibull, and exponential distributions [10]. The selected distribution will be determined using two assessment methods: Anderson Darling (AD) and Pearson correlation coefficient. Anderson Darling's value uses a weighted square distance approach, where the value is determined by the smallest value of Anderson Darling. While the Pearson correlation coefficient value uses the least squares estimation approach, where the value is determined by the greatest correlation coefficient value. The parameters are determined based on the previously selected distribution [11]. Table 1 shows the recapitulation of the best distribution and parameter for the Actual Time to Failure.

Table 1. Recapitulation of the Best Distribution and Parameter for the Actual Time to Failure (TTF)

Boiler	Component	Distributions	Parameter				
			Weibull			Normal	
			θ	β	Γ	μ	Σ
Unit 1	Cyclone Separator	2P-Weibull	45,3291	1,60229	-	-	-
	Coal Feeder	3P-Weibull	41,5813	0,75295	10,7927	-	-
	ID Fan	3P-Weibull	36,1957	0,74912	12,0363	-	-
	PA Fan	3P-Weibull	24,9251	0,89009	19,2769	-	-
	Slag Cooler	3P-Weibull	43,7578	0,58997	9,5246	-	-
Unit 2	Cyclone Separator	Normal	-	-	-	56,3846	36,5654
	Coal Feeder	3P-Weibull	24,7725	0,73746	6,1975	-	-
	ID Fan	3P-Weibull	78,7161	1,36049	7,5417	-	-
	PA Fan	3P-Weibull	46,0149	0,74675	13,1217	-	-

3.3. Generate Random Number and Transformation Random Number Data

Random numbers used are uniformly distributed (0.1) and will be converted to a non-uniform distribution. The inversion transformation method using the cumulative distribution function is the conversion method used [7]. Random numbers are generated using Microsoft Excel programs with function = RAND (). While the inverse transformation process will use several cumulative distribution function equations according to the selected distribution [12]. The transformation process of random numbers in Weibull and Normal distributions will use equations 1 and 2 [12-13].

$$t = \gamma + \theta(-\ln(U))^{\frac{1}{\beta}} \tag{1}$$

$$t = \phi(U).\sigma + \frac{\mu}{\sigma} \tag{2}$$

One hundred random numbers will be generated and then transformed as in Table 2.

Table 2. Results of Random Number Generation and Random Number Transformation

No	Random Number	*Boiler Unit 1*					*Boiler Unit 2*			
		Cyclone Separator	*Coal Feeder*	*ID Fan*	*PA Fan*	*Slag Cooler*	*Cyclone Separator*	*Coal Feeder*	*ID Fan*	*PA Fan*
1	0,6462	27,03	24,63	24,01	29,10	20,27	15,26	14,25	50,35	28,29
2	0,0357	96,09	216,50	192,57	115,66	346,21	64,39	132,93	198,24	243,80
3	0,0536	88,59	183,85	163,79	102,55	279,58	57,36	112,44	180,85	206,92
4	0,3212	49,08	60,04	54,94	48,04	63,82	15,44	35,64	93,98	67,69
5	0,5621	32,13	30,78	29,37	32,69	26,70	7,26	17,92	60,02	35,10
6	0,8474	14,76	14,61	15,32	22,58	11,60	39,03	8,36	28,54	17,26
7	0,4014	42,82	47,63	44,08	41,78	47,02	7,59	28,09	81,16	53,85
8	0,0406	93,73	205,93	183,25	111,45	324,29	62,20	126,29	192,75	231,86
9	0,0197	106,47	266,67	236,85	135,20	454,35	73,80	164,57	222,72	300,59
10	0,3171	49,42	60,76	55,58	48,39	64,85	15,85	36,08	94,69	68,51
...	....	....	....	....	....	....	....	....	....	....
98	0,8099	17,16	16,05	16,57	23,61	12,65	33,63	9,20	32,61	18,84
99	0,1479	67,92	109,09	97,98	70,89	140,73	36,69	65,83	134,27	122,68
100	0,8067	17,36	16,19	16,68	23,71	12,75	33,19	9,28	32,96	18,99

3.4. Validity Test

Validation is considered an important activity as part of the model development process. The validation process is done to ensure that the model developed is accurate enough for the purpose at hand [14]. Validity test in this study is to represent the results of the transformation of random numbers with actual data, whether it is significantly different or not. Validity test in this study uses the Mann-Whitney U method. Table 3 shows the results of the validity test.

Table 3. Recapitulation of the Validity Test

Boiler	**Component**	**Asymp. Sig. (2-tailed)**	α	**Information**
Unit 1	Cyclone Separator	0,612	0,05	H_0 accepted
	Coal Feeder	0,604	0,05	H_0 accepted
	ID Fan	0,708	0,05	H_0 accepted
	PA Fan	0,607	0,05	H_0 accepted
	Slag Cooler	0,724	0,05	H_0 accepted
Unit 2	Cyclone Separator	0,089	0,05	H_0 accepted
	Coal Feeder	0,646	0,05	H_0 accepted
	ID Fan	0,754	0,05	H_0 accepted
	PA Fan	0,679	0,05	H_0 accepted

3.5. Distribution Testing and Parameter Calculation of Random Number Transformation

The test system is the same as distribution testing and parameter calculation for the actual time to failure. The test results will be used to run the simulation. Table 4 shows the recapitulation of the best distribution and parameter for the Random Number Transformations.

Table 4. Recapitulation of the Best Distribution and Parameter for Random Number Transformations

Boiler	Component	Distribution	Parameter Weibull θ	β	γ
Unit 1	Cyclone Separator	3P-Weibull	41,4054	1,10803	0,4299
	Coal Feeder	3P-Weibull	35,4626	0,52600	10,8048
	ID Fan	3P-Weibull	30,8456	0,52340	12,0456
	PA Fan	3P-Weibull	21,7429	0,61891	19,2995
	Slag Cooler	3P-Weibull	35,7479	0,41904	9,5262
Unit 2	Cyclone Separator	3P-Weibull	38,0162	1,23482	4,4035
	Coal Feeder	3P-Weibull	21,0600	0,15549	6,2037
	ID Fan	3P-Weibull	71,2574	0,94089	8,0189
	PA Fan	3P-Weibull	39,1944	0,52179	13,1343

3.6. Calculation of MTTF and Reliability Result of Monte Carlo Simulation

MTTF calculations and the reliability level of the Monte Carlo simulation results are as follows [12].

$$\begin{aligned} MTTF &= \gamma + \theta.\Gamma\left(1+\frac{1}{\beta}\right) \\ &= 0{,}4299 + 41{,}4054 \;\; x \;\; \Gamma\left(1+\frac{1}{1{,}10803}\right) \\ &= 0{,}4299 + 41{,}4054 \;\; x \;\; \Gamma\;(0.9025) \\ &= 40{,}25 = 40 \;\; hari \end{aligned} \tag{3}$$

$$\begin{aligned} R(t) &= e^{-(\frac{t-\gamma}{\theta})^{\beta}} \\ &= \exp\left[-\left(\frac{40{,}25-0{,}4299}{41{,}4054}\right)^{1{,}10803}\right] = 0{,}3445 = 34{,}45\% \end{aligned} \tag{4}$$

The results of calculations on boiler unit 1 cyclone separator components were obtained, 40 days maintenance interval with 34.45% reliability. MTTF calculation results and Monte Carlo Simulation reliability are shown in the Table 5.

Table 5. Recapitulation of MTTF and Reliability Result of Monte Carlo Simulations

Boiler	Component	Monte Carlo Simulation MTTF (Day)	Reliability
Unit 1	Cyclone Separator	40	34,45%
	Coal Feeder	75	38,24%
	ID Fan	68	38,10%
	PA Fan	50	40,73%
	Slag Cooler	83	41,87%
Unit 2	Cyclone Separator	39	31,56%
	Coal Feeder	46	37,68%
	ID Fan	81	38,05%
	PA Fan	86	37,89%

Based on Table 5, the mean time-to-failure (MTTF) of Monte Carlo simulations can be used as predictive maintenance.

4. Conclusion

The result of Monte Carlo simulation to nine critical components of boiler section were obtained. For the component of cyclone separator boiler unit 1 obtained 40 days maintenance interval with 34,45% reliability. Coal feeder boiler unit 1 obtained 75 days maintenance interval with 38,24% reliability. ID fan boiler unit 1 obtained 68 days maintenance interval with 38,10% reliability. PA fan boiler unit 1 obtained 50 days maintenance interval with 40,73% reliability. Slag cooler boiler unit 1 obtained at 83 days maintenance interval with 41,87% reliability. Cyclone separator boiler unit 2 obtained at 39 days maintenance interval with 31,56% reliability. Coal feeder boiler unit 2 obtained at 46 days maintenance interval with 37,68% reliability. ID fan boiler unit 2 obtained at 81 days maintenance interval with 38,05% reliability. While PA fan boiler unit 2 obtained 86 days maintenance interval with 37,89% reliability.

5. Reference

[1] Kumar S, Raj B, Shubham S 2017 *Analysis IRJET4*, 608-613.
[2] Wang Z, Cuntao X, Xianwei L, Yuan-yuan L 2017 *The Computer Journal,* 1-11.
[3] Maleki H, and Yang Y 2017 *Grey Systems: Theory and Application* **7** (1), 111-122.
[4] Patrick O 2001 *Practical Reliability Engineering4th Ed* (USA: John Wiley & Sons, LTD).
[5] Law AM and Kelton DW 1991 *Simulation modeling and analysis* (New York: McGraw Hill).
[6] Sobol IM 1994 A Primer for the Monte Carlo Method *CRC Press* 30-33.
[7] Putri EL, Bahauddin A, Ferdinant PF, 2013 *Jurnal Teknik Industri* 4(1): 352-257.
[8] Kamat, SJ and Riley MW 1975 *IEEE Transactions on Reliability* **24** (1) 73-75.
[9] Hoseinie SH, Khalokakaie R, Ataie M, Ghodrati B, Kumar U 2013 *International Journal of Performability Engineering* **9** (5) 487-494.
[10] Kapur KC and Lamberson LR 1977 *Reliability in Engineering Design* (New York: John Wiley & Sons).
[11] Soesetyo I and Bendatu LY 2014 *Jurnal Tirta* **2** (2) 147-154.
[12] Charles EE 1997 *An Introduction to Reliability and Maintanability Engineering* (Singapore: Thetatayo McGraw-Hill Book Co.).
[13] Priyanta D 2000 *Rekayasa keandalan dan perawatan* (Surabaya, Indonesia: Institut Teknologi Sepuluh November).
[14] Robinson S 2014 *Simulation: The practice of model development and use 2nd Ed*(UK: Palgrave Macmillan).

IOP Conf. Series: Materials Science and Engineering **523** (2019) 012072 doi:10.1088/1757-899X/523/1/012072

Analysis of surface failure of coil spring in passenger vehicle suspension system

T E Putra[1,*], Husaini[1], N Ali[1], M N Machmud[1], A Bakhtiar[1], Hendrayana[1]
[1]Department of Mechanical Engineering, Faculty of Engineering,
Syiah Kuala University, Darussalam 23111, Banda Aceh, Indonesia

E-mail: edi@unsyiah.ac.id

Abstract. This study aims to investigate the failure of a coil spring. The examination of the chemical composition, hardness testing and observation of the fracture surface were performed, and the shear stress occurring on the coil spring was calculated. The results indicated that the inner side of the coil spring was harder than the outer side. Furthermore, the failure was fatigue failure; as shown by the presence of the crack initiation and the benchmarks. The crack initiation occurred because of the stress concentration at one point due to fretting.

1. Introduction

A coil spring is one of the main elastic components in the suspension system of a vehicle. It is expected to not only be able to withstand vibrations but to also withstand a load due to several vehicle maneuvers such as acceleration, braking or turning while on the road. This means that a coil spring must be able to withstand, reduce and absorb the impact from loads, twisting loads and cyclic loads [1-5]. Failures of the coil spring are often due to several factors, namely defects in the raw materials, surface imperfections, improper heat treatment processes and corrosion [1,6-7]. The load acting on the spring can take the form of a tensile force, compressive force or torque (twist force). The springs generally operate with 'high working stresses' and the load varies continuously.

Coil springs often experience fatigue failure in application due to their inability to withstand dynamic loads. Fatigue is a form of failure that occurs in structures undergoing dynamic and fluctuating stresses (for example, bridges, aircraft components, and engines). In this state, failure occurs at a stress level that is lower than the tensile strength or yield at the static load. The term "fatigue" is used because this type of failure usually occurs after repeated stress or strain cycles over a long period of time [8].

Fatigue failure is a brittle fracture mechanism that consists of there stages, namely the initial crack, the crack propagation and the final fracture [9]. According to research conducted by Zhu et. al. [10] with regard to fatigue failure in coil spring, defects due to physical contact and friction caused the coil spring to break. Then, a research conducted by Kosec et. al. [11] found that fatigue failure on a coil spring occurred due to corrosion found in the fault position. Based on the studies mentioned above, it was deduced that several factors could cause failure in the coil spring. Therefore, this study was aimed at examining the cause of failure in a coil spring through an analysis of the broken surface of the coil

* Corresponding author: edi@unsyiah.ac.id

spring. It was expected that the results of the analysis would later explain the cause of the fatigue failure in the spring and recommendations could then be made to those who require them.

2. Materials and methods

In this study, some laboratory tests were conducted. The specimens for the tests are shown in Figure 1. Test on the chemical composition was carried out using a scanning electron microscope and energy dispersion X-ray spectroscopy (SEM, EDS). Furthermore, a hardness test, using the Vickers hardness testing method, was carried out to determine the hardness value of the coil spring. The principle for the testing was based on tracking or indenting the sample with a pyramid-shaped diamond indenter with a slope of about 136°. In this test, the specimen was pressed for 15 seconds with a force of 10 kg, as set on the Vickers test instrument. The test points for the surface toughness of the coil spring can be seen in Figure 2. The specimen consisted of several sides, where *T* was the top, *I* was the inner side, *B* was the lower side and *O* was the outside [7]. SEM images of the specimen were taken of the spring surface that had experienced failure. The analysis of the broken surface was carried out by means of two methods of observation. The first observation was by visual analysis and the second observation was with the aid of the SEM images. Before the testing, the specimen was designed with such a shape as to make it easier for testing.

(c)

Figure 1. Specimens: (a) chemical composition test, (b) hardness test, (c) SEM test.

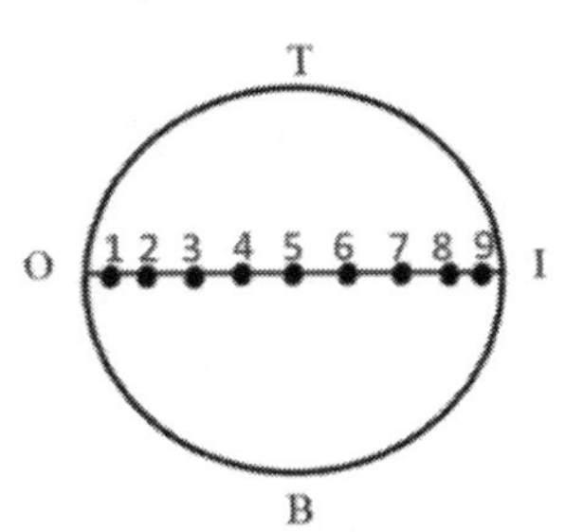

Figure 2. Test points for surface hardness.

The torsion load that occurred in the coil spring was calculated to find the minimum and maximum torsional loads. The maximum twisting load was adjusted to the allowed twisting load of the ASTM A227 in order to ensure that the torsional load acting on the coil spring component was still in a safe condition to prevent fatigue. A Goodman diagram, based on the following equations, was used:

$$\tau_{min} = \frac{8K_w PD}{\pi d^3} \quad (1)$$

$$\tau_{max} = \frac{8K_w PD}{\pi d^3} \quad (2)$$

3. Results and discussion

From the SEM / EDS results, the chemical composition was close to the ASTM A227 standard for the material, as shown in Table 1, especially for elements *C*, *P*, *S* and *Mn*. The elements *Si* and *Cr* did not meet the ASTM A227 standard because the coil springs actually did not have a high hardness as their main characteristic. The main characteristic of the coil springs was their elasticity. The mechanical properties of the ASTM A227 material are shown in Table 2.

Table 1. Comparison of the composition of spring steel specimen with that of ASTM A227

Material	C	Mn	P	S	Si	Cr
Spring steel	0.36	0.6	0	0,02	0.06	0,09
ASTM A227	0.45-0.85	0.30-1.30	0.04	0,05	0.15-0.35	0.03

Table 2. The mechanical properties of ASTM A227

Shear modulus (GPa)	Tensile modulus (GPa)	Density g/cm³	Shear stress (MPa)
11,5x10³	28,6x10³	7700	700

The ASTM A227 standard has a carbon, *C*, 45%-0.85%, while the test result revealed a carbon content of 0.36%, which was within the standard and could be classified as medium carbon steel. The sulfur element, *S*, in a coil spring functions to form inclusions, and high levels of this element can reduce the ductility of the steel and increase the likelihood of cracking. The composition was equal to the ASTM A227 standard, which was 0.02%. The ASTM A227 standard has phosphorus, *P*, content of 0.04%, while the test results revealed a phosphorous content of 0%, which was in accordance with the standard. The function of silicon, *Si*, is to improve hardness. Unsuitable silicon content can cause the coil spring to crack easily. The ASTM A227 standard for silicon, *Si*, content is 0.15%-0.35%, while the test result revealed silicon content of 0.06%. This was very low compared to the ASTM A227 standard, meaning that there was a decrease in the hardness of the coil spring, thereby causing it to be easily deformed. Manganese, *Mn*, functions to increase the ductility of the coil spring. The ASTM A227 standard for the Manganese content is 0.30% -1.30%, while the test result showed a content of 0.60%.

The hardness distribution graph for the inner surface at each point is shown in Figure 3. It can be seen that the distribution tended to be uneven. Point 1 had the lowest hardness value (289 HVN), point 2 had a value of violence that was starting to increase (382 HVN), point 5 had the highest hardness value (455 HVN), and the distribution of violence at the next point tended to be evenly distributed to point 9 (434 HVN). The test results showed that the lower hardness was in the surface area and the higher hardness was further from the surface. This was not in accordance with the distribution of steel in general. The increase in hardness was quite significant on the surface area, and the hardness was lower for areas further from the surface due to heat treatment and hardening of spring surface.

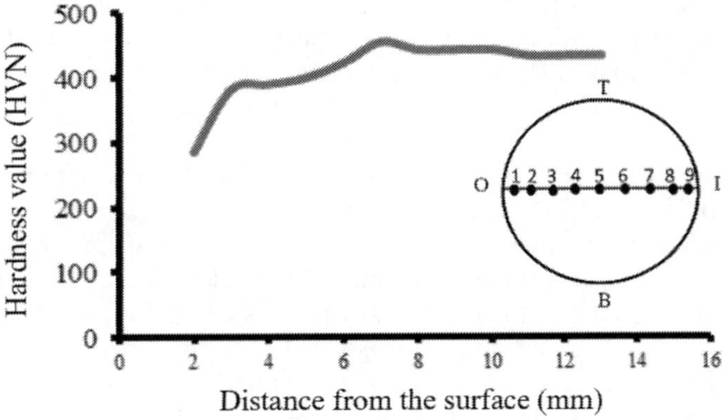

Figure 3. Values of Vikers hardness (HVN)

The testing of the hardness of a cross section of the thread of the spring specimens resulted in an average spring surface hardness value of 345 HVN in the *T* section, 417 HVN in the *I* section, 328 HVN in the *B* section and 429 HVN in the *O* section of the spring, as shown in table 3. These results were in accordance with the hardness value of the ASTM A227 standard, namely 310-543 HVN. The excessively large hardness value caused a decrease in the elasticity of the spring, and when it reached the maximum load, the spring easily experienced cracks.

Table 3. Value of Vikers hardness (HVN) on the outer surface of the coil spring

Section	**Hardness (HVN)**					
	Point 1	**Point 2**	**Point 3**	**Point 4**	**Point 5**	**Average**
T	286	400	421	237	382	345
I	400	446	409	400	432	417
B	231	409	231	381	390	328
O	392	392	434	471	458	429

Based on the results of the visual investigation, the coil spring had a brittle fracture failure, where a broken shape was formed at an angle of $\pm 45°$. It was stated that the fracture that occurred in the coil spring was tired. This fracture started from the surface, spread to the centre and broke as the load exceeded the loading capacity of the material. There was a very clear change in the dimensions or what is known as fretting wear, which occurred due to collisions with other coils; thereby inducing frictional and compressive forces on the coil spring surface. The beginning of the crack occurred at the wear area on the spring surface, which was the outer side of the spring axis, with the direction of the crack propagation being equal to the two surface contact stresses. The presence of wear on the coil spring resulted in crack initiation. Subsequent loading caused the beginning of cracks, which propagated to form a crack area. The crack area was marked by the presence of beach marks. Furthermore, the crack propagation stopped once the remaining spring section was unable to withstand the working load until it broke.

It can be stated that the coil spring, in this study, experienced fatigue fracture, characterized by the presence of beach marks. This fatigue occurred due to changing or repetitive loads. The final fracture

IOP Conf. Series: Materials Science and Engineering **523** (2019) 012072 doi:10.1088/1757-899X/523/1/012072

area that occurred looking wider than the spreading area or the initial crack area, showing the nominal stress that worked on the coil spring had exceeded the limit of the material. The SEM image with an optical enlargement of 30x showed that the crack propagation had spread to the entire surface, thereby forming a chevron pattern, which is a cone pattern, towards the beginning of the crack, as shown in Figure 4 (a). The SEM image with an optical magnification of 40x showed the crack initiation had formed a deep trench, as shown in Figure 4 (b). It is believed that these cracks were corrosion holes caused by stress concentration at one point and changes in the structure of the coil spring. Corrosion was formed due to damage to the protective paint and scars caused by pressure and friction. The next stage continued with the existence of beach marks, and finally, the coil spring broke. Ordinary beach marks were indicated by fine circular lines in the area between the beginning of the crack and the final fracture. Beach marks are more often said to represent the stages of crack propagation. The SEM image with an optical magnification of 500x (Figure 4 (c)) showed the crack initiation area and the characteristics of the fretting wear in the form of sharp holes in the fretting area such that there was a notch that caused the local stress concentration so that this area was very sensitive to the initial formation of cracks. The crack initiation was also accelerated by the shear stress due to the contact action between the coil spring and other specimens.

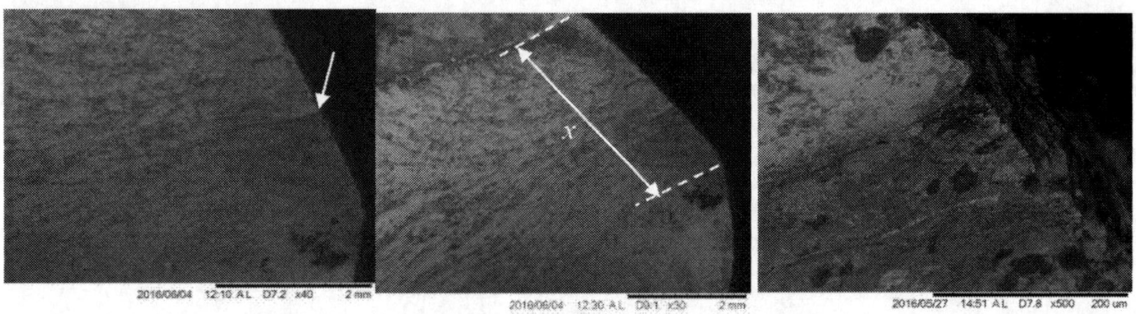

(b) (a) (c)

Figure 4. SEM images: (a) the propagation that formed a chevron pattern (b) fatigue fracture as evidenced by the presence of cracks and the shoreline (c) form of the initial crack of a screw spring and the formation of a notch

To determine the maximum torsional stress, the exact calculation was done as follows. With a thread diameter of 14 mm, an outer diameter of 80 mm, and a number of active coils, the minimum load received by one coil spring was 3064 N, and the maximum load was 3922.5 N. From the results, the minimum twist load received by one coil spring was 257.9 N/mm^2, and the maximum twist load was 329.2 N/mm^2. The ASTM A227 was wrapped under the hot condition so that the allowable torsional stress was 700 N/mm^2. The maximum torsional load was still below the allowable tension value.

The following Goodman diagram shows whether the dynamic load acting on the coil spring component was still in a safe condition against fatigue fractures (Figure 5). The points τ_{min} and τ_{max} were perpendicular to each other on the horizontal axis. A variation of 71.3 N/mm^2 was considered in the torsional stress, while the variation permitted according to the ASTM A227 standard was 700 N/mm^2. This means that the protection against fatigue was $v = \frac{700}{71,3} \approx 9,8$. It can be concluded that this was quite far from the fatigue value of the material, and therefore, the possibility of failure due to torsional stress was quite small. The shear stress received by the coil spring was within the allowable stress, but the spring still failed due to fatigue caused by a creeping crack. Cracking propagation occurred due to the continuous dynamic load and was accelerated by the frictional force and compressive force.

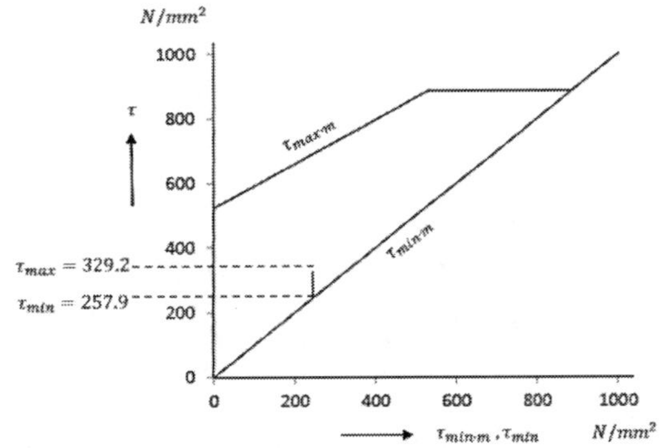

Figure 5. Diagram of Goodman.

4. Conclusion

This study was aimed at investigating the failure of a coil spring. Early cracks occurred in the wear area. After the initial crack was formed, this became the maximum point of stress that forced the crack to spread. With the twist in the coil spring resulting in a tendency to form an initial crack, it can be stated that the coil spring had a fatigue fracture marked by the presence of beach marks.

5. References

[1] Prawoto Y, Ikeda M, Manville SK and Nishikawa A 2008 *Eng. Fail. Anal.* **15** 1155-74.
[2] Putra TE, Abdullah S, Schramm D, Nuawi M Z and Bruckmann T 2017 *Mech. Sys. Sig. Proc.* **90** 1-14.
[3] Putra TE and Husaini 2018 *AIP Conf. Proc.* **1983.**
[4] Husaini, PutraTE and Ali N 2018 *IJAME* **15** 5251.
[5] Putra TE and Husaini 2018 IOP Conf. Series: Materials Sciences and Engineering **352.**
[6] Wahl AM 1884 *Mechanical Springs* (McGraw-Hill Book Company).
[7] Hayes M 2015 *Tech-Spring Report 15 End Coil Failures* (Institute of Spring).
[8] Callister W 2007 *Materials Science and Engineering: an Introduction* (John Wiley & Sons, Inc).
[9] Todinov MT 1999 *Int. J. Mech. Sci.* **41** 357-70.
[10] Zhu Y, Wang Y, Huang Y 2013 *Case studies in Eng. Fail. Anal.* **2** 169-173.
[11] Kosec L, Nagode A, Kosec G, Kovacevic D, Karpe B, Zorc B and Kosec B 2014 *CSEFA* **41** 1-4.

IOP Conf. Series: Materials Science and Engineering **523** (2019) 012073 doi:10.1088/1757-899X/523/1/012073

Failure analysis of a centrifugal pump shaft experiencing plastic deformation using the exact and finite element method

Husaini [1*], **E Saputra** [2], **Husni** [3], **T E Putra** [1], **M Rachman** [1]
[1]Computational Mechanics Laboratory, Department of Mechanical Engineering, Faculty of Engineering, Syiah Kuala University, Darussalam 23111, Banda Aceh, Indonesia
[2]Department of Mechanical Engineering, South Aceh Polytechnic, Tapaktuan 23715, Aceh, Indonesia
[3]Department of Chemical Engineering, Faculty of Engineering, Syiah Kuala University, Darussalam 23111, Banda Aceh, Indonesia

E-mail: husainiftm@unsyiah.ac.id

Abstract. An eight-stage centrifugal pump that was used in the oil and gas industry had plastic deformation on the shaft because of the torque load. Thus, this study was aimed at analyzing the shear stress that occurred on the shaft by an exact and finite element method. The shaft, with a length of 1.5 m and diameter of 0.05 m, was made of SUS 304 material. From the exact and finite element analysis results, the shear stress that occurred on the shaft was higher than the shear stress of the shaft material. Therefore, the pump shaft became plastically deformed and twisted. The results of the finite element analysis showed that the torque, shear stress, and strain were increased when the twist angle on the pump shaft was increased.

1. Introduction

The shaft is one of the components that transmit power by a rotary or axial movement. The load applied on the shaft varies during use, namely, in terms of the tension, torsion, compression, bending, or a combination of these [1,2]. During its operation, the shaft often suffers failure and many researchers have studied the causes of this failure.

A steel shaft has much greater strength and stiffness than one made of a composite material [3]. A torsional load that is not uniformly distributed on the shaft can produce shear stress leading to deformation and may damage some parts of the shaft [4]. The torque load varies with the rotational speed of the shaft [5].

Heat treatment on a welded shaft can protect the shaft from fatigue failure [5-7]. The non-uniform hardness of the shaft surface is one of the causes of shaft failure [8]. Many parameters can causes shaft failure, such as an incorrect fillet radius on the shaft [9], the disproportionate fillet radius on segments of the shaft [10]; a chemical composition that does not meet the standard will affect the mechanical properties [11], improper installation, wrong choice of material, and poor maintenance [12].

* Corresponding author: husainiftm@unsyiah.ac.id

IOP Conf. Series: Materials Science and Engineering **523** (2019) 012073 doi:10.1088/1757-899X/523/1/012073

This study was aimed at analyzing the stress on an eight-stage centrifugal pump shaft that had deformed plastically. This type of centrifugal pump is used in the oil and gas industry to pump diethanolamine (DEA) fluid to decrease the H_2S and CO_2 content of natural gas. This study used the exact and finite element analysis (FEA) methods to analyze the shear stress and strain that occurred on the twisted centrifugal pump shaft [13].

2. Method

2.1. Material Properties

The shaft of the eight-stage centrifugal pump was made from SUS 304/ASTM 240 material [14]. Table 1 shows the mechanical properties of this steel.

Table 1. Mechanical properties of ASTM A240.

No	Description	Value
1	Density, ρ	8000 kg/m^3
2	Elastisity Modulus, E	193 - 200 GPa
3	Shear Modulus, G	83 GPa
4	Maximum Tensile Strength, σ_u	505 MPa
5	Yield Strength, σ_y	215 MPa
6	Poisson Ratio, ν	0.29
7	Shear Strength, τ_a	42 MPa

2.2. Centrifugal Pump Shaft

80% to 90% of the oil and gas industry uses a centrifugal pump [15]. Figure 1 shows the centrifugal pump used in the oil and gas industry that was analyzed in this study. For the FEA, the pump shaft was modelled using computer-aided design software.

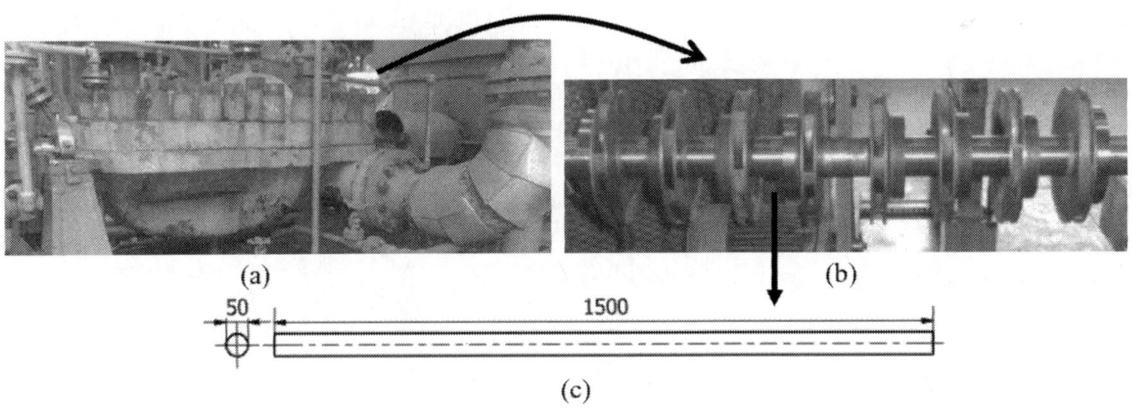

Figure 1. Eight-stage centrifugal pump: (a) lean DEA G3X03 pump used in the oil and gas industry; (b) maintenance of centrifugal pump shaft; (c) pump shaft model.

2.3. Exact and Finite Element Analyses

The purpose of the exact analysis was to measure the torque load and shear stress that occurred on the centrifugal pump shaft. After the exact analysis, a FEA was carried out to compare the results with the data from exact analysis.

The following steps were taken to analyze the centrifugal pump shaft by means of a FEA: (a) torque load of 500 Nm, 1000 Nm, 1500 Nm, 2000 Nm, 2500 Nm, 3000 Nm, 3500 Nm, 4000 Nm, 4015 Nm and 4050 Nm were applied periodically to the shaft to analyze the twist angle that occurred in each torque load condition, and (b) a twist angle of about 8^{0} was applied to the shaft to analyze the torque, shear stress, and strain that occurred.

3. Results And Discussion

3.1. Exact Analysis

The purpose of the exact analysis was to analyze the torque load that occurred on the shaft by using Eq. (1) [16]. In this case, the shaft was twisted by about 8^{0}. The shaft length L was about 1.5 m and the diameter D was about 0.05 m. The shear modulus G of the SUS 304 material was about 83 x 10^9 N/m^2.

$$\theta = \frac{L \cdot T}{I_p \cdot G} \quad (1)$$

$$T = 4542\ Nm$$

Eq. 2 was used to calculate the maximum shear stress that occurred on the shaft [16]. The torque T from Eq. 1 was about 4542 Nm. The radius of the shaft r was about 0.025 m, while I_p was the inertia moment of shaft surface.

$$\tau_{shaft} = \frac{T \cdot r}{I_p} = 186\ MPa \quad (2)$$

Based on the failure theory, Eq. (3) and Eq. (4) were used to analyze the shear stress using the von-Mises and Tresca methods [16]. The yield stress σ_y of the SUS304 material was about 215 MPa.

- Von Mises

$$\tau_{oct} = \frac{\sigma_y}{3}\sqrt{2} = 101{,}352\ MPa \quad (3)$$

- Tresca

$$\tau_{max} = \frac{\sigma_y}{2} = 107{,}5 MPa \quad (4)$$

From the exact analysis, the shear stress that occurred on the shaft was higher than the shear stress of the shaft material (table 1). The shear stress that occurred on the shaft was about 164 MPa and the shear stress of the material was about 42 MPa. Therefore, the shaft was deformed plastically. The shear stress that was calculated using Eq. (1) was greater than the shear stress based on von Mises and Tresca methods.

3.2. Finite Element Analysis

3.2.1. Modelling shaft by Torque Load

In this analysis, a torque load was applied to the shaft periodically to twist the shaft by up to 8^{0}. Figure 2 shows the shaft that was analyzed by FEA.

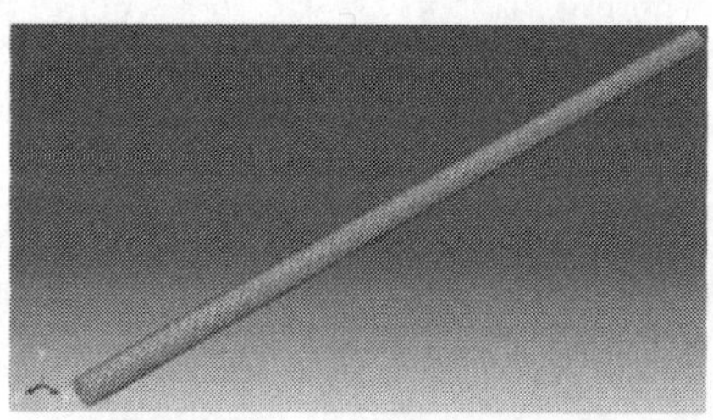

Figure 2. Pump shaft analyzed by FEA.

The twist angle that occurred were dissimilar depending on the torque load that was applied to the shaft. The torque increased with an increase in the applied twist angle. Figure 3 shows the relationship between the torque load and the twist angle. The shaft was twisted for about 8° if the torque load was 3900 Nm.

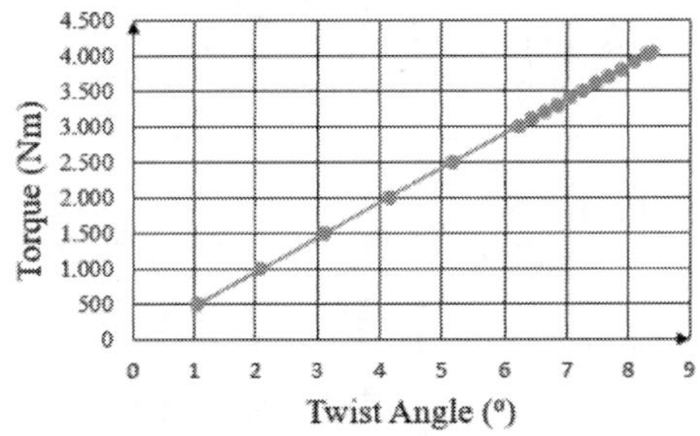

Figure 3. A curve showing the relationship between torque load and the twist angle.

3.2.2. Modelling Shaft by Twisted Angle

In this analysis, a twist angle of about 8° was applied to the shaft. The shear stress that occurred was about 255 MPa and the strain was about 0.034. The shear stress and strain increased with an increase in the twist angle on the pump shaft. As the twist angle increased, there was an increase in the shear stress and strain, as shown in Figure 4.

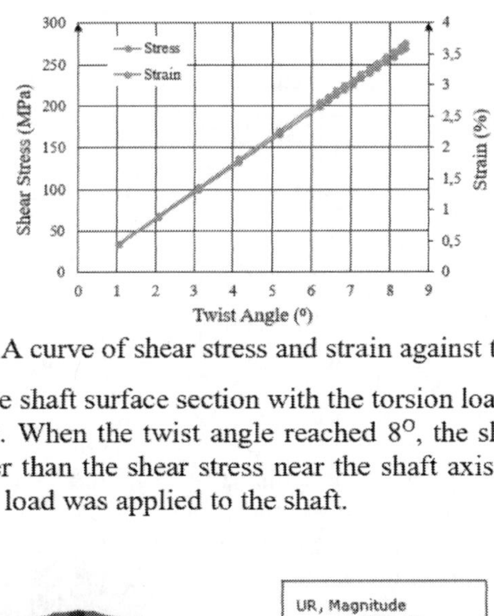

Figure 4. A curve of shear stress and strain against twist angle.

Figure 5 show details of the shaft surface section with the torsion load. Figure 5 (a) shows the shear stress distributed on the shaft. When the twist angle reached 8°, the shear stress that occurred at the outer shaft section was greater than the shear stress near the shaft axis. Figure 5 (b) shows the strain that occurred when the torque load was applied to the shaft.

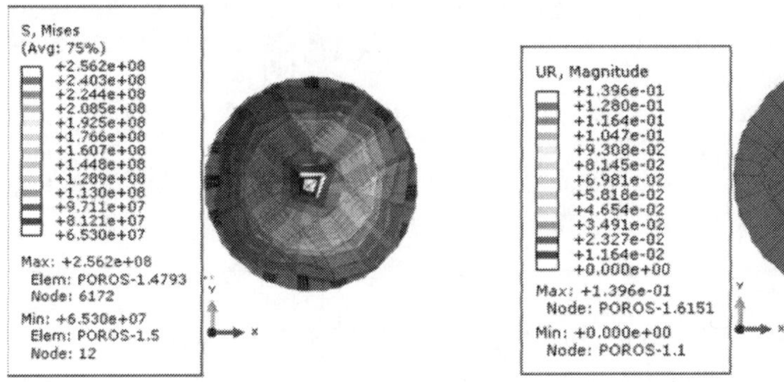

(a) (b)

Figure 5. The condition of the shaft when plastic deformation occurred:
(a) change of stress; (b) change of strain.

Figure 6 shows the shear stress for each node of the shaft obtained by the FEA. The nodes in the red box, which were a part of the shaft that had maximum shear stress, were analyzed. It appeared that each observed node had a different shear stress value. Figure 7 shows the curve of the torque and shear stress to the twist angle that occurred on the shaft. Increasing the twist angle led to an increase in the torque and shear stress that occurred on the shaft. When the twist angle reached about 8°, the shear stress that occurred on the shaft was about 255 MPa and the torque was about 3900 Nm. Both the exact and FEA revealed that the shear stress that occurred on the shaft was higher than the shear stress of the shaft material (table 1). Therefore, the results showed that the shaft had deformed plastically.

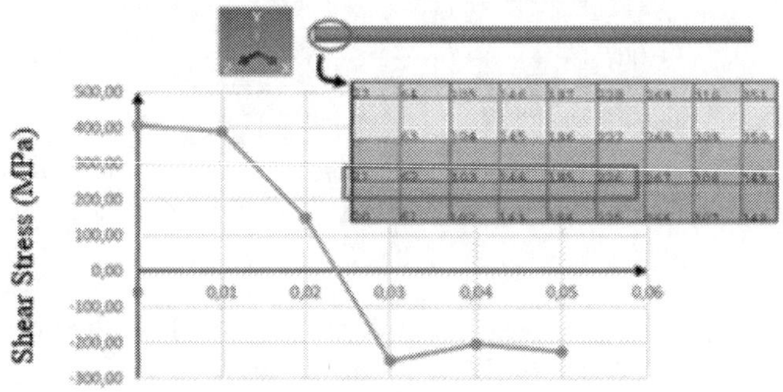

Figure 6. Some observed noded for the shear stress analysis at the end of the pump shaft.

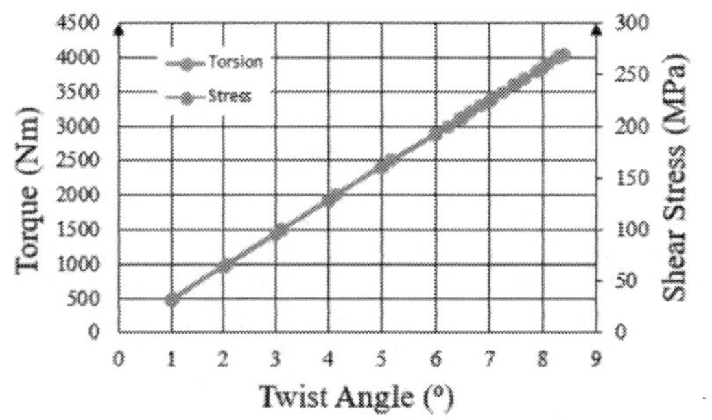

Figure 7. A curve of torque and shear stress to twist angle on the shaft.

4. Conclusions

The following conclusions could be drawn from the results of the exact and FEA on the centrifugal pump shaft that was deformed plastically and twisted about 8° as:

1. The exact and FEA revealed that the failure of the pump shaft that had deformed plastically was caused by the shear stress, which was greater than the shear stress of the SUS 304 pump shaft material.
2. The maximum shear stress that occurred on the twisted pump shaft was about 255 MPa and the maximum strain was about 0.034.
3. The torque, shear stress, and strain that occurred on the shaft increased with an increase in the twist angle on the shaft.

5. References

[1] ASM International 2002 *ASM Handbook: Failure Analysis and Prevention* **11** 2302-2303.
[2] Husaini, Rachman M and Machmud MN 2018 *AIP Conf. Proc.* **1983**.
[3] Kumar A, Jain R and Patil PP 2016 *IOP Conf. Ser.: Mater. Sci. Eng.* **149**
[4] Axinte T and Nutu C 2015 *IOP. Cof. Ser.: Mater. Sci. Eng.* **95**

[5] Fuller RW, Ehrgott Jr JQ, Heard WF, Robert SD, Stinson RD, Solanki K and Horstemeyer MF 2008 *Engineering Failure Analysis* **15** 835.
[6] Moolwan C and Netpu S 2013 *Social and Behavioral Sciences* **88** 154.
[7] Machmud M N, Maulana D and Husaini 2013 *Jurnal Teknik Mesin Unsyiah* **1** 33.
[8] Das G, Sinha AN, Mishra SK and Bhattacharya DK 1999 *Engineering Failure Analysis* **6** 267-276.
[9] Das S, Mukhopadhyay G and Bhattacharyya S 2015 *Case Study In Engineering Failure Analysis* **3** 46-51.
[10] Vogwell J 1998 *Engineering Failure Analysis* **5** 271-277.
[11] Crivelli D, Ghelichi R and Guagliano M 2011 *Procedia Engineering* **10** 3683-3691.
[12] Zambrano O A, Coronado J J and Rodriguez S A 2014 *Case Study In Engineering Failure Analysis* **3** 46-51.
[13] Husaini, Hasanuddin I, Sandi Y B Z and Akhyar H 2018 *AIP Conf. Proc.* **1983.**
[14] ASTM International, ASTM A 240/A 240M – 04 Standard Specification for Chromium-Nickel Stainless Steel Plate, Sheet, and Strip for Pressure Vessels and for General Application (Printed in the United States of America, 2004).
[15] Girdhar P and Moniz O 2004 *Practical Centrifugal Pumps: Design, Operation and Maintenance* (Linacre House, Jordan Hill, Oxford OX2 8DP 30 Corporate Drive, Burlington, MA: Elsevier) pp 10.
[16] ASM International 2002 *ASM Handbook: Mechanical Testing and Evaluation* **8** 278-286

IOP Conf. Series: Materials Science and Engineering **523** (2019) 012074 doi:10.1088/1757-899X/523/1/012074

The effect of cutting speed on dimension accuracy and burr development of high-speed micro drill proses on aluminium

M Dirhamsyah[1], M Tadjuddin[1,*], A Udink[1], Z Yusuf[1], H Y Saiful[1]
[1] Department of Mechanical and Industrial Engineering, Syiah Kuala University, Indonesia

E-mail: m.tadjuddin@unsyiah.ac.id

Abstract. The machining process is a part of the production technology used in almost every manufacturer of industrial products. The world of the industry continues to grow along with technology, emerging machining technology that can produce miniaturization products on of that is a micro-drilling process. This research observed the dimension accuracy and burr development on micro drilling process on Aluminium. The accuracy of the hole dimension and burr development on top and bottom were observed on three different cutting speed. The thickness of aluminum as the workpiece is 0.5mm and drill with a diameter of tungsten carbide drill bit 0.3mm. The results show at the cutting speed of 13,2 m/min, the largest hole diameter was found 0.3117 mm, while the smallest hole diameter is 0.3039 mm measured at a cutting speed of 18.8 m / min. For the top burr height found that the largest number is 0.19 mm and the smallest number at 0.07 mm. While the bottom burr height has the highest value of 0.39 and the lowest at 0.12 mm, it can be concluded that the deviation of hole dimension size and burr height is inversely proportional to cutting speed.

1. Introduction

The machining process is one of the most needed production technologies in almost every product that is produced through the process of casting, rolling, or forming which always requires the final processing by machining. Along with the development of material needs and the quality of machining products, this technology is also constantly developed both in terms of technical (machine capability and cutting tools), the impact of the process on product quality, as well as economic considerations. [1].

One of the fundamental things in the operation of the industry is drilling process which is one of the processes of making holes in a component. When the size of the hole turns small and the smaller the process is called the micro-drill process by using a process that only has a micro-sized hole diameter. This continues to be the focus of the industrial world such as electronics, space, the world of health, chemical reactors, types of filters and much more. The new curing process can be stated as a micro-drill process when it is in size between 1μm - 1 mm in diameter [2].

* Corresponding author: m.tadjuddin@unsyiah.ac.id.

2. Micro-drilling

2.1 Micro-drilling Process

Micro drill is the mention of the word in terms of the micro-drilling process, there are two ways in micro drill operation, namely conventional and non-conventional, the conventional techniques refer to the basic process where there is a tool, and the spindle rotation is installed, this can create micro holes. While the non-conventional techniques are relatively new and are extensively used in the modern world, unlike conventional methods, non-conventional involves various methods such as electrical, chemical, mechanical, thermal and a combination of all, non-conventional techniques that are often used in the modern world namely: Laser, EDM, ECM, SACE, Electron Beam and Ultrasonic Vibration. Micro drills are classified into micro holes measuring 1 μm - 1 mm [2]. In this study, researchers will only use conventional techniques.

2.2 Hole Quality

Failure and damage to the product resulting from the machining process are caused by the following factors: error geometry, thermal error, the hardness of the material, engine vibration, spindle rotation, and contour. All of these factors can exacerbate the quality of the results of the drill [3], as shown in Figure 1.

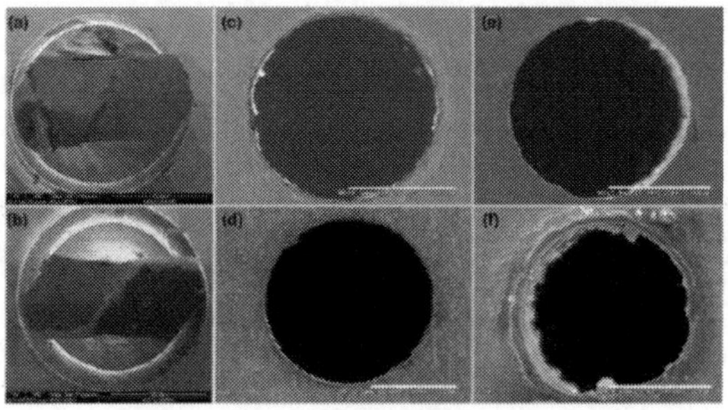

Figure 1. Hole dimension [2].

2.3 Burr Formation

The formation of burrs is a basic phenomenon of the drilling process; this has an impact on the hole surface at the top and bottom, the formation of burrs can create problems in accuracy and quality [2] (see Figure 2). The formation of burrs is caused by several different factors including the properties of the work material, drill geometry and material [4].

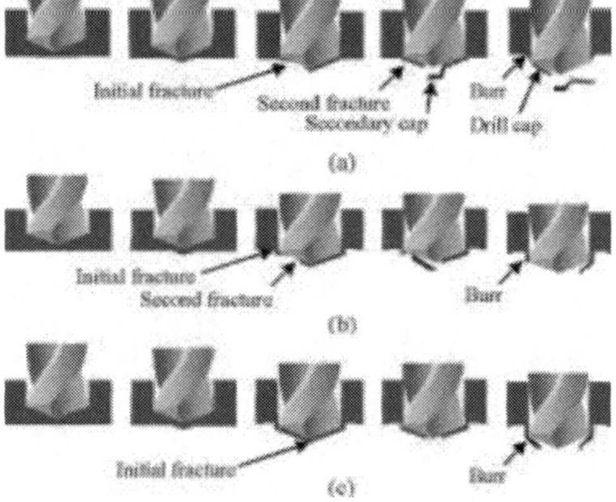

Figure 2. Classification of Burr types [2].

3. Method and Materials

3.1. Setup of machining

The machining process is carried out in this research by adding a high-speed spindle motor with a speed of 24,000 rpm on a CNC milling machine at SMK 2 Banda Aceh. A modified CNC milling machine can be used normally by adding a high-speed spindle. The mechanism can be seen in Figure 3.

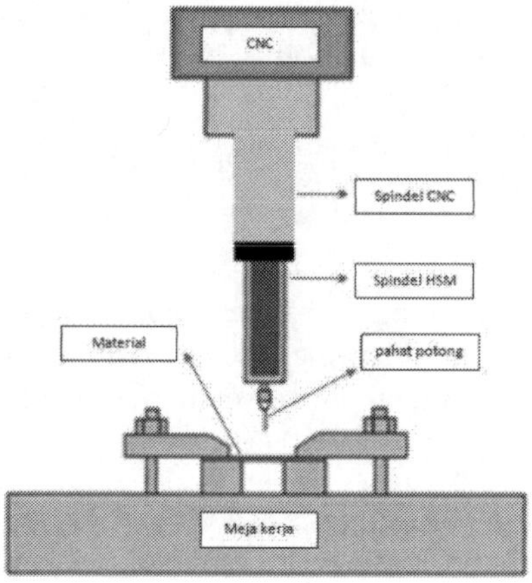

Figure 3. Setup mechanism.

3.2. Parameter of machining

The machining parameters specified in this study are important variables to determine the results of the machining process, the researcher determines the variable of machining which consists of nine different variables can be seen in Table 1.

Table 1. Machining parameters

Tools	Tungsten carbide 0.2 mm drill bits								
Plate material	Brass carbide 0.5 mm thickness								
Spindel Speed	14.000			17.000			20.000		
Plate Number	1	2	3	4	5	6	7	8	9
Feed rate	5	8	10	5	8	10	5	8	10
Repeatation	10	10	10	10	10	10	10	10	10

3.3. Materials

The material that is the object of observation in this study is aluminum material which is obtained in the market in Banda Aceh, with the size cut as specified previously.

3.4. Drilling Tool

In this test the cutting tool used is a tungsten carbide type that has a diameter of 0.3 mm, this tool is generally used for making holes on the circuit board and also for metal materials, this tool is manufactured in Japan by Union Tools company.

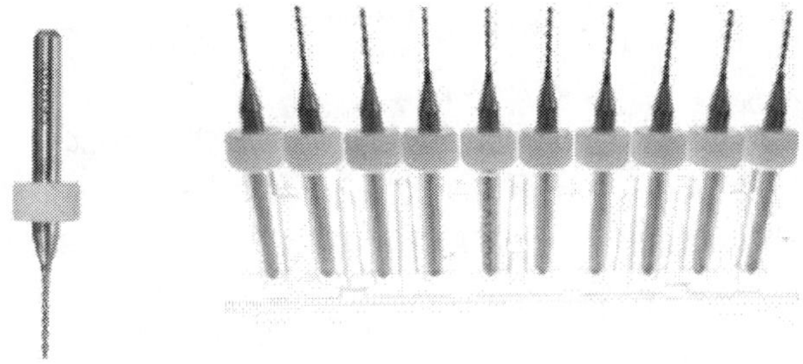

Figure 4. Drilling Tool.

3.5. Hole Measurement

The hole and the quality of the product of the machining process are measured using a stereo microscope. The object magnification of this microscope reaches 50X. In this measurement, the microscope has a limitation; it cannot connect to the PC directly without the help of a camera recorder. So, this microscope is equipped with an additional camera as shown in Figure 6 below. This observation aims to see the accuracy of dimensions and burrs formed in the tests.

4. Result and discussion

4.1. Measurement Data

After stages of the machining process have been done, the aluminum is observed using a stereo microscope to measure the dimensional accuracy as well as the height of the burrs created. The size of each hole is created according to the specified parameters. The measurements can be seen in Table 2.

From the chart and tables above shows that the number and accuracy of the machining process holes are very determined from the spindle speed and feed rate. Where the speed has been determined to start from 14,000 rpm, 17,000 rpm and 20,000 rpm and the feed rate is 5 mm / min, 8 mm / min, 10mm / min. From Figure 5 the machining process recorded as follows, the biggest deviation is 0.03117 mm it was in spindle rotation conditions of 14,000 rpm and feed rate of 8 mm/min while the smallest hole size deviation is at spindle rotation of 20,000 rpm and feed rate of 10 mm / min with a value of 0.03039 mm, this is at a maximum speed and speed with a number.

The results of the holes shown in this study are illustrated in Table 2 and Figure 8, that is the dimensions of the hole graph, this result can show that the dimension value of the hole will increase with increasing feed rate and spindle speed.

Table 2. Results of hole diameter

Spindel speed (rpm)	Feed Rate (mm/min)	Diameter rate of the hole (mm)
14000	5	0,3111
14000	8	0,3117
14000	10	0,3098
17000	5	0,3094
17000	8	0,3076
17000	10	0,306
20000	5	0,3088
20000	8	0,3055
20000	10	0,3039

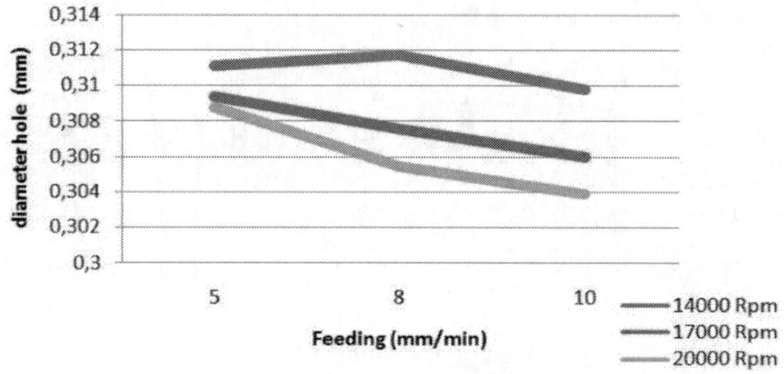

Figure 5. Hole dimension observation.

Figure 6 shows some sample images from the observation. Randomly selected from several options machining parameters.

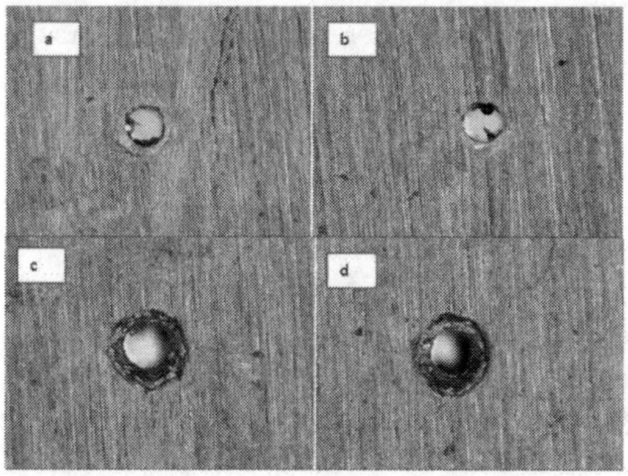

Figure 6. Observation chart of hole dimension.

4.2. Entrance and Exits Burr

In the machining process with the aluminum that was tested, burrs were formed on each hole produced. There were entry burrs and exit burrs. This greatly affects the quality of the result of the machining, as described in Tables 3 and 4.

Table 3. Observation result of entrance Burr

Spindel speed (rpm)	Feed Rate (mm/min)	Burr high entrance (mm)
14000	5	0,19
14000	8	0,18
14000	10	0,14
17000	5	0,16
17000	8	0,16
17000	10	0,07
20000	5	0,18
20000	8	0,15
20000	10	0,07

IOP Conf. Series: Materials Science and Engineering **523** (2019) 012074 doi:10.1088/1757-899X/523/1/012074

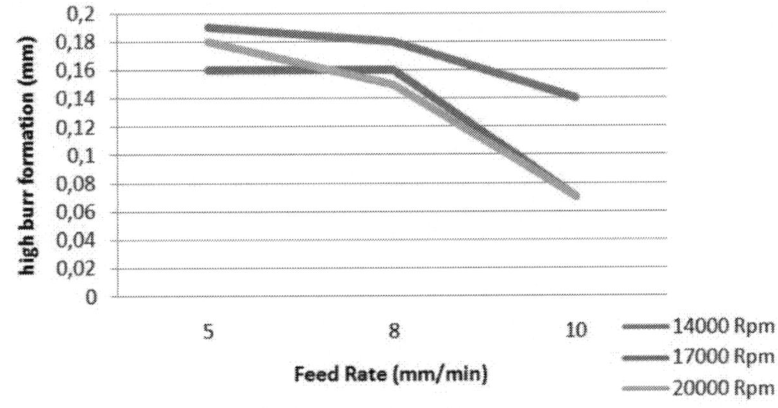

Figure 7. Entrance Burr result.

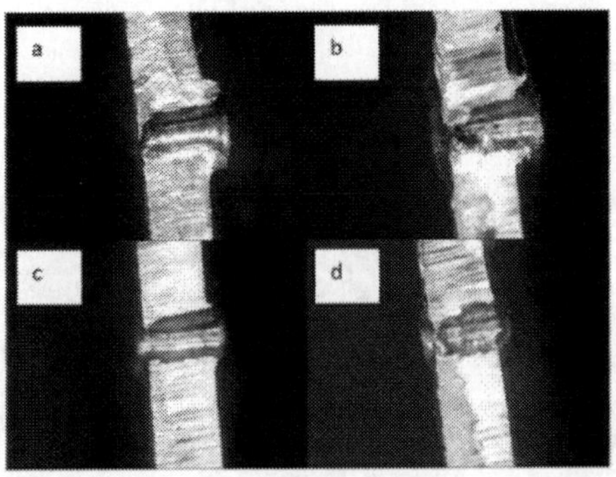

Figure 8. (a) f 5 mm / min 14,000 rpm, (b) f 5mm / min 17,000 rpm, (c) f 10mm / min 17,000 rpm, (d) 10mm / min 20,000 rpm.

From the results of the burr heights measured in table 3 and figure 10 we can see that the highest number is at 0.19 mm, which is spindle speed of 14,000 rpm and feed rate of 5 mm/min. The lowest burr height is at the spindle rotation of 20,000 rpm, and the feed speed of 10 mm/min is shown in Table 3 which is 0.07.

Table 4. Observation result of exits Burr

Spindel speed (rpm)	**Feed Rate (mm/min)**	**Burr high exit (mm)**
14000	5	0,39
14000	8	0,28
14000	10	0,13
17000	5	0,41
17000	8	0,3
17000	10	0,17
20000	5	0,33
20000	8	0,31
20000	10	0,12

Table 4 shows the measurements of the exit burr height. The highest number is at 0.41 mm, which is at a spindle rotation of 14,000 rpm and a feed rate of 8 mm/min. The lowest is at 0.12 mm, with a spindle rotation of 20,000 rpm and a feed rate of 10 mm/min, which is the maximum rotation and rate of the specified parameters.

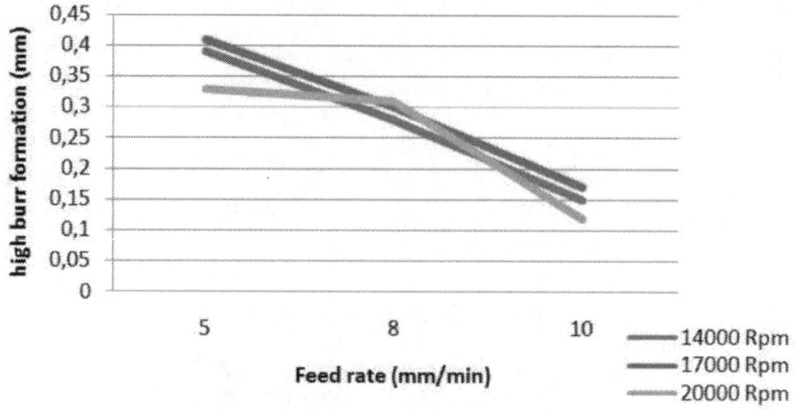

Figure 9. Exits Burr result.

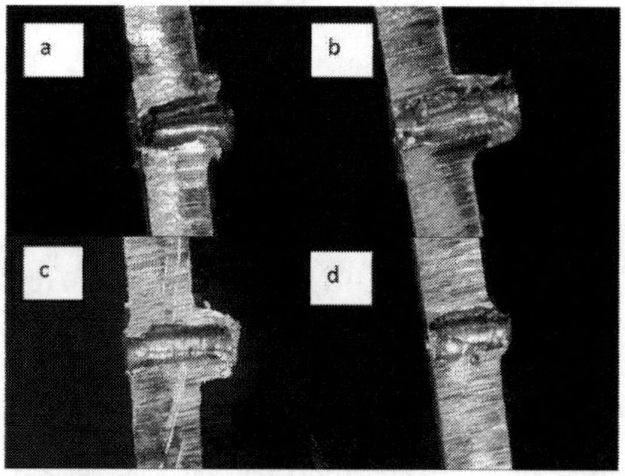

Figure 10. (a) f 5 mm / min 14,000 rpm, (b) f 8mm / min 17,000 rpm, (c) f 10mm / min 17000, (d) 10mm / min 20,000 rpm.

5. Conclusions

The test is carried out on aluminum workpieces measuring at 120 mm wide, 35 mm long, 5 mm thick. There was a total of 9 specimens, with10 holes in each specimen. Based on the test, it was found that different parameters and tool bits affect the accuracy and quality the holes resulted from the drilling process. This is evident in the testing, from the largest hole deviation with a value of 0.3117 mm at spindle rotation of 14,000 rpm and a feed rate of 8 mm/min, while with the smallest hole deviation is at 0.3039,which occurs at a spindle rotation of 20,000 rpm and a feed rate of 10 mm/min. Based on the results of this study, hole accuracy will increase with increasing spindle rotation and feed speed. The same is true with burr formation; the number of burr heights will be lower as the spindle rotation and the feed rate increase.

6. References

[1] Hartono, R. 2016 Jurnal Teknik Mesin **2** (1), Januari 2016 ISSN. Jurnal Teknik Mesin, 2(1)..
[2] Hasan, M., Zhao, J., and Jiang, Z. 2017 Journal of Manufacturing Processes, **29**, 343-375.

[3] Rahman, A. A., Mamat, A., and Wagiman, A. 2009 Modern applied science, 3(5), 221.
[4] Stirn, B., Lee, K., and Dornfeld, D. A. 2001 In Proceedings of the sixteenth annual meeting of the American Society for Precision Engineering (pp. 473-476).

IOP Conf. Series: Materials Science and Engineering **523** (2019) 012075 doi:10.1088/1757-899X/523/1/012075

Mechanical characteristics of marble powder composite materials reinforced Kenaf Fiber against static load

N Fitriadi[1,*], Lindawati[1], I Ardiansyah[1]
[1]Mechanical Engineering Study Program, Politeknik Aceh Selatan
Jl. Merdeka Komplek Reklamasi Pantai, Tapaktuan,Aceh Selatan, Aceh, Indonesia 23711

E-mail: nuzuli@poltas.ac.id

Abstract. Kenaf fibre is one of the strongest natural fibres in the world today. For that purpose, the purpose of this study is to create a new composite of Kenaf fibres combined with marble waste powder. The incorporation of marble powders and Kenaf fibres using a matrix of a polyester resin type has never been done. Thus, this study aims to create and test various composite material composites (mixture of matrix, marble powder and Kenaf fibre) which will produce different mechanical properties and obtain good tensile strength in test specimens with several composition variations. Preparation of test specimens was carried out with 4 (four) compositions of matrix mix and marble powder that is 50:50, 60:40, 70:30 and 80:20 with the weight of Kenaf fibre fixed 2 gram. The largest tensile test value is 63,407 MPa, obtained from 80:20 composition with the largest E value 3.03 GPa. The largest compression test value is 33.49, obtained from the composition of 60:40 with the largest E value of 2.35 GPa. The largest bending test value is 19.80 MPa, obtained from the composition of 80:20 with the largest E value of 36.55 GPa. From the research results of the tensile test, compression test, and bending test which possess good value is between composition ratio 60:40 and 70:30.

1. Introduction

The Kenaf plant (Hibiscus cannabinus L.) is a short-lived herbaceous plant-producing fibers such as hemp, Roselle, and jute. The Kenaf fibers are produced from the bark of the stem. In Indonesia, Kenaf planting spread in Lampung, West Java, East Java, and South Kalimantan with an average productivity of 0.9-1.2 tons/ha of dry fiber [1]. Low productivity leads to high competition at the farm level in obtaining the Kenaf plant [2].

Kenaf fibers have been widely used in various daily life products such as for sacking raw materials, pulp materials, paper, polypropylene composites in the polymer industry and car dashboards. In addition to environmentally friendly, Kenaf fiber is also strong. The mechanical properties of Kenaf pulps derived from Kenaf leather with 6% sodium hydroxide (NaOH) usage have met the NUKP slurry specifications as specified in SNI 6106-1999, ie 4.24 mm long fiber, freeness 760 mL CSF, tensile index 66, 54 Nm / g, crack index 5.72 kPam2 / g and tear index 18.09 mNm2 / g [3].

Various study regarding mechanical properties of Kenaf fiber reinforced-composite have been carried out. Shinji Och (2010), studied the tensile properties of Kenaf fiber extracted from several

* Corresponding author: nuzuli@poltas.ac.id

IOP Conf. Series: Materials Science and Engineering **523** (2019) 012075 doi:10.1088/1757-899X/523/1/012075

Kenaf plants with different characteristics. The study determined how stem diameter and length affect the mechanical characteristics of fibers. It was observed that mechanical characteristics of Kenaf fiber is not ultimately affected by stem diameter and length, however, it is increased asKenaf increase in height. Additionally, M.R.Ishak, et al., (2010) conducted a comparative study of mechanical properties of short Kenaf bast and core fibre reinforced unsaturated polyester composites with varying fibre weight fraction. The result showed that the Kenaf bast fiber reinforced-composites have higher mechanical properties than Kenaf core fiber composites. The composite with optimum fibre content, 20%, provided the highest tensile strength for both bast and core fibre composites. Moh. Yuhazri, *et al.*, (2011) studied the mechanical properties of Kenaf/polyester composites processed through vacuum infusion method. The using of vacuum infusion process in the Kenaf-polyester composite manufactured resulted in a higher strength, low cost alternatives, and environmentally friendly. Furthermore, KotreshSardar, *et al.*,(2014)characterized and investigated the tensile strength of 10%, 20%, 30% and 40 % KFRPC material used as Bio-material for tissue implant. According to the results, it is shown Bio-material with 40% of KFRPC material have good tensile strength and might be suitable for tissue implant application, such as for Tendon replacement. Later, A.E. Ismail and M.A. Che Abdul Aziz (2015) studied the effect of fiber orientations and number of layers on the tensile strength of woven yarn Kenaf fiber reinforced polyester composite. The experiment results showed that fiber orientations significantly affected the tensile strength. Nevertheless, it is not affected by the modulus of elasticity for number of layers. In 2016, X.H. Loh *et al.,* investigated the mechanical properties of Kenaf/polypropylene composite with the variations of fiber weight and length. The result showed a higher fiber length improved the impact properties of the composite. Based on these related studies, it can be seen that Kenaf fibre is feasible to be used with various filler combination for biocomposite material development [10].

On the other hand, South Aceh district is well known for Marble stone production. The Marble stone has been widely produced for various home interior need, home appliances, and as a material for carving art. The activity of processing of raw stone into a valuable product produces marble waste that leads to an environmental problem. Thus, this marble waste should be reused to obtain new products that have economic value. For that purpose, this study is conducted to create a new composite of Kenaf fibres combined with marble waste powder.

2. Method

2.1. Method of Manufacture

Before making specimens, special treatment of Kenaf fibers is first soaked in 1% alkaline solution (NaOH) which serves to remove some contents such as lignin, oil, and protein that can cause decay in the fiber. The composition of the specimen was determined as shown in Table 1. The composition differences focused on the resin and marble powder of 20%, 30%, 40% and 50%, while the use of the Kenaf fiber was with a fixed composition of 2%.

Table 1. The composition of test specimens

No	Specimens type	Polyester Resin %	Marble dust %	Kenaf Fiber %
1	Specimen A	80	20	2
2	Specimen B	70	30	2
3	Specimen C	60	40	2
4	Specimen D	50	50	2

IOP Conf. Series: Materials Science and Engineering **523** (2019) 012075 doi:10.1088/1757-899X/523/1/012075

2.2. Measurement Methods

After the specimens are made with the dimensions as in ASTM standard, some tests are performed as follows:

1. Tension test according to ASTM D 638-03 standard [11].
2. Compression test according to ASTM D 695-02a standard [12].
3. Bending test according to ASTM D790-02 standard [13].

The test equipment used in a servo Pulser test equipment located at Laboratory of Impact Research Center.

3. Results and Discussion

3.1. Tensile Test

Tensile strength testing is one effort to know the material characteristics. Tensile tests were performed on 5 specimens per composition so that the average values as shown in Table 2 can be obtained.

Table 2. The average yield of tensile testing

Composition	Max. Stress (MPa)	Break Point Strain (%GL)	Upper Yield Stress (MPa)	Max. Load (N)	E (MPa)	E (GPa)
50:50	26.47	11.23	26.50	546.16	1588.12	1.59
60:40	33.71	13.42	48.58	513.66	2022.82	2.02
70:30	33.57	14.37	44.20	432.29	2014.32	2.01
80:20	50.53	21.86	42.80	454.62	3032.04	3.03

From the results of the performed test, the maximum tensile stress is 50,53 MPa. This number is to be found from the 80:20 mixture of the polyester resin and marble powder with 2 grams of the Kenaf fiber as shown in Figure 1.

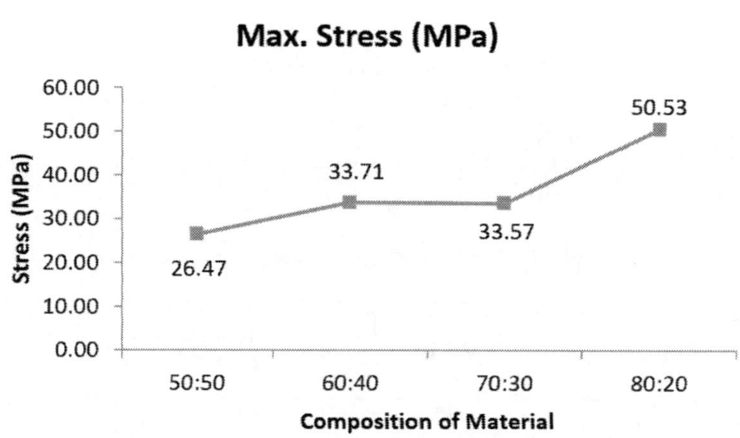

Figure 1. Maximum stress tensile test

From Figure 1, it can be seen that the stress is generally increased by increasing the percentage of resin content. Increasing the percentage of marble powder will reduce the bonding properties of the polyester resin as a matrix to bind the Kenaf fibers as reinforcement. The average of elastic modulus for this specimen is 3.03 GPa.

3.2. Compression Test

Compression strength test is one effort to know the material characteristics. Compression tests were performed on 5 specimens/composition types so that the average values as shown in Table 3 can be obtained.

Table 3. The average yield of compression testing

Composition	**Maximum point Stress**	**Maximum point Load**	**Breakpoint Strain**	**Upper yield Stress**	**Lower yield Stress**	**Elastic modulus**
	MPa	**N**	**%GL**	**MPa**	**MPa**	**Gpa**
50:50	19.03	1141.698	3.624	13.070	12.585	1.61
60:40	23.09	1361.352	4.586	19.518	18.063	2.35
70:30	24.46	1467.78	4.867	21.175	20.069	2.19
80:20	23.88	1432.72	6.102	18.403	17.931	1.96

From the results of this test, it is obtained the best composition on the mixture of polyester resin is a ratio of 60:40 marble powder with Kenaf fiber as much as 2 grams, as shown in Figure 2. Average of elastic modulus for this specimen is 3.03 Gpa. It is known that the more marble powder will increase the compressive strength. However, it is necessary to further study by bending so that the characteristics of this composite material can be known comprehensively.

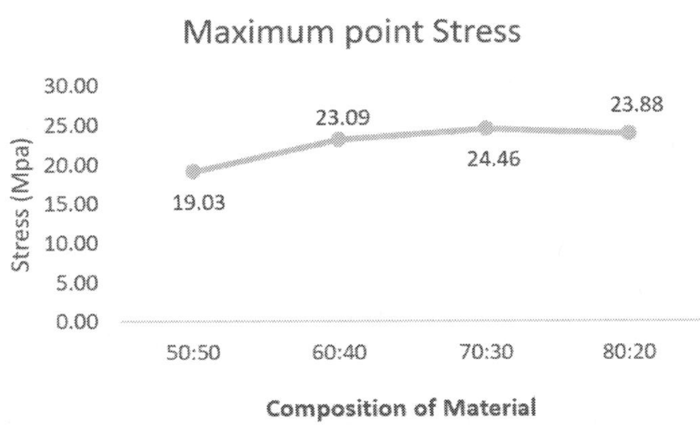

Figure 2. Maximum stress of compression test

3.3. Bending Test

From Table 4 it can be seen that there is a difference in value during bending testing.

Table 4. The average yield of bending testing

Composition	Maximum point Stress	Maximum point Load	Break point Strain	Upper yield Stress	Elastic modulus
	MPa	N	%GL	MPa	Gpa
50:50	9.02	180.34	-4.93	9.65	49.88
60:40	11.15	223.02	4.66	9.20	33.52
70:30	15.91	318.16	12.18	14.80	43.87
80:20	17.05	341.02	8.00	15.47	36.55

The highest value is obtained from a specimen with a ratio of 50;50 with a stress point of 19.80 MPa. While, the lowest value is yielded by a specimen with a ratio of 70:30 with a stress of 13.54. The average stress of this composition is 17.05 MPa, while, the elastic modulus is 36,55GPa. It is known that the more the marble powder will reduce the properties of polyester resin as the matrix. However, the Kenaf fiber as reinforcement enable to increase the bending strength as shown in Figure 3.

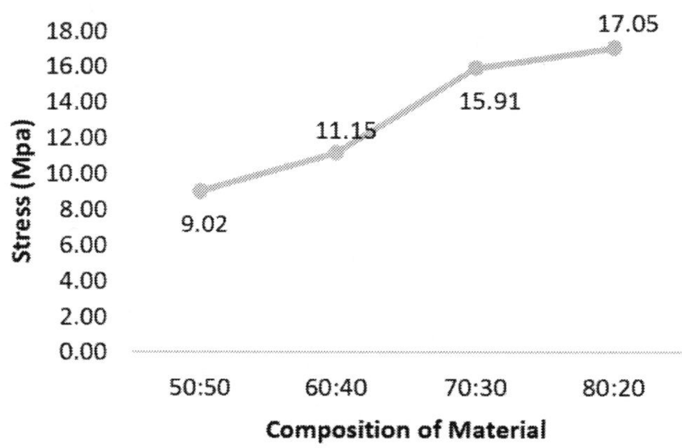

Figure 3. Maximum stress of bending test

From the research of composite tensile test between the mixture of resin and marble powder, it is observed that the highest value is 80:20 with the Kenaf 2 gram with the highest value reaching 58.91 MPa, while the lowest value is 50:50 with the composition Kenaf fiber as much as 2 grams with value 21.35 MPa. The highest elasticity modulus is also possessed by the 80:20 compositions with a 3.03 GPa elasticity modulus value.

The results of the compressive test show that the highest value co-composition is 60:40 with a value of 33.49 MPa with an elastic modulus of 2.35 GPa, whereas the lowest composition is 50:50 with a value of 16.35 MPa and the modulus of elasticity 1.61 GPa.

The result of bending test shows that the composition which has the highest value is 80:20 with a value of 19.80 MPa while the lowest value is 50:50 with a value of 6.125 MPa.

From the result of the tensile, compressive, and bending test, a composition having ideal value will be recommended for the manufacture of the product. From the results of this study the composition that has good and ideal value is the composition of the material is 70:30.

4. Conclusion

The conclusions of 3 tests to obtain the mechanical characteristics of the marble composite material are:

1. The largest tensile test value is 63,407 MPa, obtained from 80:20 composition with the largest E value 3.03 GPa. The largest compression test value is 33.49, obtained from the composition of

IOP Conf. Series: Materials Science and Engineering **523** (2019) 012075 doi:10.1088/1757-899X/523/1/012075

60:40 with the largest E value of 2.35 GPa. The largest bending test value is 19.80 MPa, obtained from the composition of 80:20 with the largest E value of 36.55 GPa.
2. From the research results of the tensile, compression, and bending test, the composition which possesses good value ranges between a ratio of 60:40 and 70:30.

5. References

[1] Natasa aa, Zuhry E, Adiwirman 2016 *JOM Faperta* **3** (2) 1-7.
[2] Sudjindro 2004 Budidaya Kenaf (Hibiscus cannabinus L.) Seri Edisi Khusus Balai Penelitian Tembakau danTanaman Serat. Malang.
[3] Kardiansyah T, Sugesty S 2014 *JurnalSelulosa* **4** (1) 37-46.
[4] Och S 2010 *SRX Materials Science* **2010** 1-6.
[5] Ishak MR, Leman Z, Sapuan SM, Edeerozey AMM and Othman IS 2010 *IOP Conf. Series: Materials Science and Engineering 11 of 9th National Symposium on Polymeric Materials (NSPM 2009).*
[6] Yuhazri M, Phongsakorn Y, Sihombing PTH, Jeefferie AR, Perumal P, Kamarul AM and Rassiah K 2011 *International Journal of Engineering & Technology IJET-IJENS* **11** (1).
[7] Sardar K, Veeresh K and Gowda M 2014 *International Journal of Recent Development in Engineering and Technology (IJRDET)* **2** (6) pp. 104-112.
[8] Ismail AE and Aziz MAA 2015 *Journal of Mechanical Engineering and Sciences (JMES)* **9** 1695-1704.
[9] Loh XH, Mohd Daud MA, and Selamat MZ 2016 *Journal of Mechanical Engineering and Sciences* **10** (2) 2098-2110.
[10] Fitriadi N 2017 *Jurnal Teknovasi* **4** (2) 2017 27-39.
[11] Tension test according to ASTM D 638-03 standard.
[12] Compression test according to ASTM D 695-02a standard.
[13] Bending test according to ASTM D790-02 standard.

IOP Conf. Series: Materials Science and Engineering **523** (2019) 012076 doi:10.1088/1757-899X/523/1/012076

Mechanical behavior of hybrid glass Fiber-Jute reinforced with polymer composite for the wall of the Acehnese boat '*Jalo Kayoh*'

A Akram[1,*], I Hasanuddin[1], N Nazaruddin[2], R Putra[3], MM Noor[4]
[1] Department of Mechanical and Industrial Engineering, Faculty of Engineering, University of Syiah Kuala Banda Aceh, Indonesia.
[2] Department of Informatics, Faculty of Mathematics and Natural Sciences, University of Syiah Kuala, Banda Aceh, Indonesia.
[3] Department of Civil Engineering, Faculty of Engineering, University of Syiah Kuala, Banda Aceh, Indonesia.
[4] Faculty of Mechanical Engineering, University Malaysia Pahang (UMP), Pahang, Malaysia

E-mail: akram@unsyiah.ac.id

Abstract. Hybrid fiber composites are widely used composites in engineering applications, especially for the frame and wall structure of Acehnese traditional boat *Jalo Kayoh*. The purpose of this study is to obtain a hybrid material which is a blend of natural jute fiber reinforced with the synthetic fibers E-glass and polyester as a matrix. The materials potential to become a material for the structure and walls in the traditional Acehnese boat *Jalo Kayoh*. The fabrication of the hybrid composites is carried out manually using an engineered press at a pressure of 25 kg/cm2 for 24 hours. The ratio of jute fiber and E-glass in the composites at each layer are (1:0), (1:1) and (2:1). Tensile testing of the hybrid composites refers to ASTM D 3039 - 00. The results show that the tensile strength of the hybrid composites tends to increase with the increase in the volume of the fiber, with the maximum tensile strength of 4.8 MPa. The increasing number of layers in the jute and E-glass is adding strength to the composites. This will be good to use in the initial material of the structure and walls in the traditional Acehnese boat *Jalo Kayoh*.

1. Introduction

The Acehnese traditional fishing boat, known as Jalo Kayoh [1], is a mean of transportation used by the fishing community to work and perform activities on the water, be it in a river, a lake, the sea, or a fishing pond. Jalo Kayoh is traditionally built using wood as the main material. The scarcity of wood is pushing the production of Jalo Kayoh to switch to low-quality wood materials. This makes the structure and walls of the boat more vulnerable to damage and decay. In this study, the model plan of the Jalo Kayoh has the specification with a length of 4 meters, a width of 0.96 meters and a height of 0.40 meters. Its main function is to help fisherman in their water activities such as short distance transportation, catching fish, crabs, shrimp, and pond rehabilitation.

The durability of the wood used as the main component for the boat can range from 2 to 3 years [2]. It makes wood a less ideal material to use in the long term, so efforts must be made to use synthetic

* Corresponding author: akram@unsyiah.ac.id.

materials to replace wood. As technology advances, researchers have produced breakthroughs in the fields of renewable and environmentally friendly material science (green technology material science). In reducing the use of various petroleum-based products, many developed countries view natural fiber-reinforced composite materials as having excellent potential in replacing petroleum-based products [3]. The use of natural fiber is now categorized as environmentally friendly fiber, easily cultivated, low in price, and has high mechanical strength [4].

A mechanical property of E-glass fibers reinforced with epoxy resin in tensile and bending tests has an increase in strength accompanied by an increase in the weight fraction of fibers [5]. This is followed by an increase in brittleness depending on the increase in the weight percentage, which is accompanied by a decrease in collision energy [6]. Glass fibers can increase mechanical stress concentration when combined with epoxy, polyester, vinyl ester resins thus increasing the threat of brittle fracture [7]. This can be eliminated or reduced through chemical optimization on the composite to avoid damage caused by cracks [8]. Tensile properties can be improved through the use of Kevlar fiber or glass fiber - Hybrid Epoxy for composites, as long as the fiber direction is higher in strength along the fiber direction [9]. It is necessary to develop reinforced natural and artificial fibers hybrid composite material that has the best properties to replace synthetic fibers, so jute fiber and glass woven fibers are used to strengthen matrix material [10].

Various natural fibers have been studied through mixing with polymeric materials, such as kenaf fiber, hemp, flax, hemp, sisal, coir, jute, straw, wood fiber, palm, rice husk, wheat, barley, oats, sugar cane, empty bunches, water pennywort, kapok, banana fiber, pineapple leaf fibers, paper-mulberry, raffia, papyrus, and grass reeds have been used very commonly for research activities [11]. Production with the use of natural fibers, which are cheap and environmentally friendly, can be measured as a source of synthetic fiber substitutes. Application of jute and banana fiber use has been discussed in detail. [12]. Jute fiber is obtained easily and has good mechanical properties, but through the treatment of mechanical properties of alkali medium, it can be increased again [13]. The use of polyester-reinforced straw with a 40% volume fraction produces a high tensile strength of 104 MPa [14].

The purpose of this study was to obtain preliminary data from a combination of glass fiber and jute layer as the basic material for traditional boat making in Aceh Jalo Kayoh. Literature study shows that composite materials are very important to be used for a variety of needs for the manufacture of diverse products, both for car, aerospace, structure, and products used in the marine field. To support the creation of high-performance composites through the use of natural and artificial fibers followed by studying the mechanical properties of composite properties reinforced by natural and artificial fibers (E-glass and jute). Both fibers will be reinforced by polyester resin. The composite will be tested for tensile strength to find out the ideal strength for the initial use of the structure and walls of the traditional Acehnese boat *Jalo Kayoh*.

2. Experimental Details

Jute fiber and glass fiber in this research were prepared for fillers on composites. Fiber-reinforced composites are formed based on different layers. While the test specimens are prepared using the hand layup method, fiber is required weight percentages were added to the resin and mixed properly.

2.1. Materials and fabrication

Jute fiber and E-glass fiber are used as reinforcement. Jute fiber has an area weight of 420 g/m2. Chopped strand mat type E-glass fiber has an area weight of 300 g/m2. Unsaturated polyester resin is used as a matrix. The properties of jute fiber, E-glass, and unsaturated polyester resin can be seen in Table 1. On this side, the composite is made according to the flowchart shown in Figure 1. This figure shows the E-glass fiber and jute fiber are made single and layered. The ratio of the jute layer and E-glass is set to 1:0, 1:1, and 2:1. These specimens are made using the hand lay-up method, where the polymer is mixed with Mepoxe Peroxide hardener in a ratio of 100:1.0; then the ingredients are pressed and left for 24 hours at room temperature. The appearance of the lay-up process for different composites is shown in Figure 2. The volume fraction of fiber (Vf) of jute fiber and E-glass fiber is shown in Table 2.

IOP Conf. Series: Materials Science and Engineering **523** (2019) 012076 doi:10.1088/1757-899X/523/1/012076

Table 1. Mechanical properties of jute fiber, glass fiber, and unsaturated polyester resin

Properties	Jute fiber	Glass fiber	Unsaturated polymer
Density (g/cm3)	1.15	2.4	1.24
Elongation at break (%)	1.4-1.7	0.4	2.1
Tensile strength (MPa)	393-773	2000-3500	42
Young's modulus (GPa)	25.6	67	3.0

Table 2. The fiber volume fraction of four types of composites

Composite	V_f (%)	V_g (%)	V (%)
Jute Fiber	12.5	-	12.5
Jute Fiber / Glass Fiber	16.5	4.2	20.7
Jute Fiber/Glass Fiber/Jute Fiber	19.4	2.5	21.8

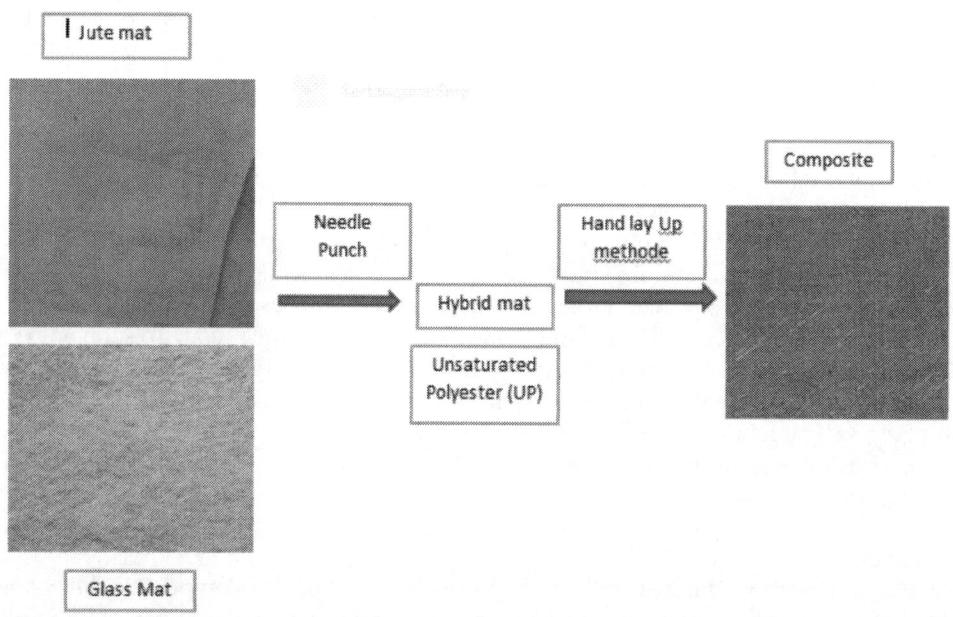

Figure 1. Process flow chart used in the fabrication of different composites using hand lay-up method.

2.2. Measurements

Tensile testing is carried out using tensile tests according to standard D 3039-00. This test is carried out using the Universal Testing Machine (UTM). The final test specimen consists of three replications to be included in each test.

Each layer of jute has a thickness of 0.8 mm and each E-glass fiber has a thickness of 0.33 mm. So according to ASTM standards, lamination thickness varies depending on the number of layers. This combination places the glass fiber layer between the jute fibers in the 3-layer laminate.

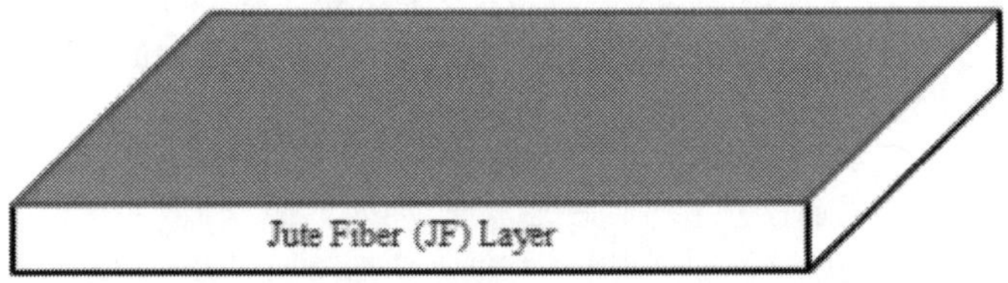

Figure 2. Schematic representation of composite 1.

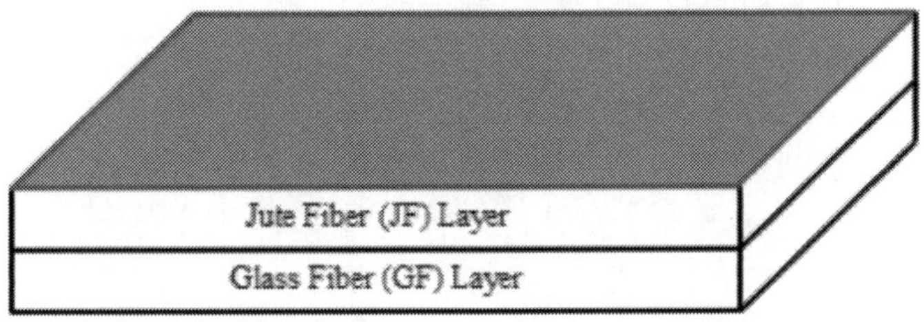

Figure 3. Schematic representation of composite 2.

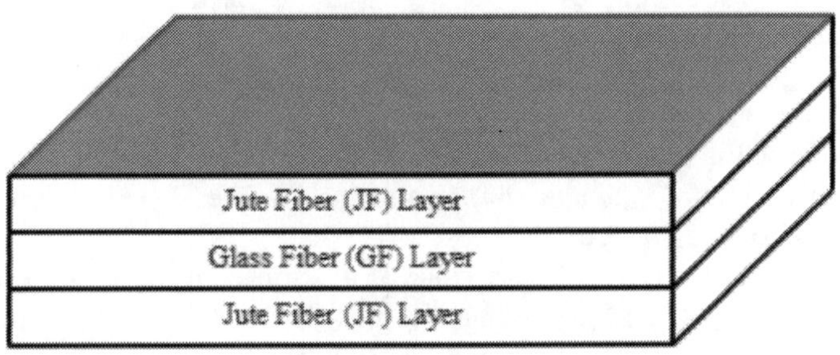

Figure 4. Schematic representation of composite 3.

3. Results and Discussion

Hybrid jute and E-glass fiber composite specimens will be subjected to tensile testing. The results obtained were analyzed, then compared to obtain conclusions from the hybrid composite to show better mechanical properties in the results of all tests. Jute fiber into biodegradable reveals excellent mechanical properties Jute fiber into biodegradable reveals excellent mechanical properties. This will be useful for preliminary data from the planned development of boat wall material replacing wood materials that will be used in general that can be developed towards mass production.

The mechanical properties of several different composite test specimens are shown in Figure. 5. Tensile stress increases linearly with a deflection in the initial stages until the slowing reaches the point of failure. In the case of Jute Fiber (JF), it was observed that the tensile strength reached a minimum value of 0.308 MPa with the lowest flexural deflection. Meanwhile, in the JF/GF-t (t: the jute layer is located on the surface) and JF/GF /JF-t, the two composites show the same maximum tensile strength. It can be observed that JF/GF/JF-t results in a maximum tensile strength lower than the maximum tensile strength of JF. Referring to hybrid composites b (b: jute layer on the top side), similar situations can be observed. In addition, the tensile strength of JF/GF/JF composites is higher than the JF/GF composite value.

Furthermore, from the data shown in the graph in Figure 5, it can be summarized that the average value of tensile stress on the strain that occurs based on the difference in the composite produces a variety of different values. As seen in Figure 6, the tensile strength obtained for JF is around 4,308 MPa, this result is higher than the tensile strength value of JF/GF composite which is 3,771 and lower than JF/GF/JF which is 4,800 MPa.

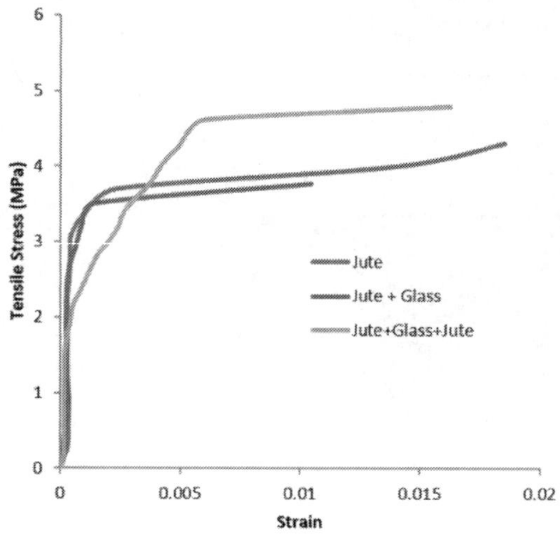

Figure 5. Typical tensile stress curves of different composites.

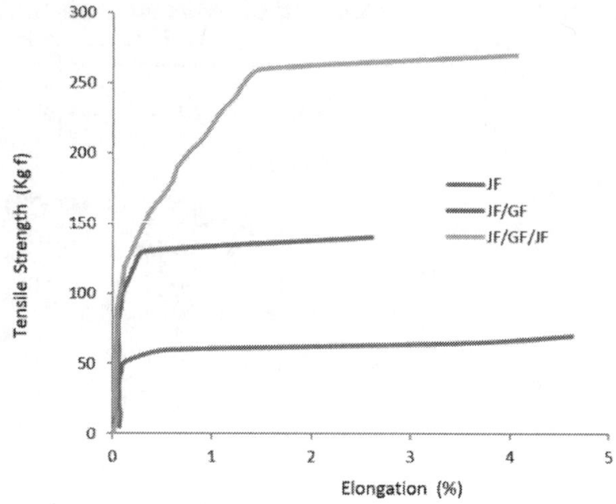

Figure 6. Typical Tensile Strength curves of different composites.

In short, JF showed a significantly higher increase in tensile stress value than the JF/GF sample but also showed a decrease in value compared to the JF/GF/JF sample consisting of two layers of jute fiber and one layer of glass fiber. This indicates that an increase in the fiber jute layer adds to the tensile strength value of the sample being tested.

Meanwhile, the result of the length increase value is a significant increase in the sample consisting of more layers. The JF sample produces a length addition value of 4,625 mm, higher than the JF/GF and JF/GF/JF samples which has a value of 2.61 mm and 4,055 mm.

The results of the observation show that the JF sample which is a sample consisting of single jute fibers has a higher increase in length, but when added with a layer of 300 gram chopped strand mat type glass fiber, the length addition value decreases and this is supported with the JF/GF/JF data samples which produce higher value after adding a jute fiber layer.

4. Conclusion

In this study, the incorporation of jute fibers into a pair of glass fiber reinforced composites results in a hybrid effect on the flexural properties of the composite to be used in the structure and walls of the Acehnese boat Jalo Kayoh. The following points are taken from the experimental results:

1. The tensile stress of different composites shows little variation.
2. It should be noted that JF manifests the highest length increase among the samples tested from different composites, but also shows lower tensile stress than JF/GF/JF composite samples showing more brittle properties.
3. It appears that JF/GF/JF shows higher tensile stress and a better toughness value than the JF and JF/GF samples.
4. The addition of the GF layer to JF results in higher brittle properties; this is indicated by the reduction in length of the sample coated with GF.

5. Acknowledgments
Thank you to Indonesian Ministry of Research, Technology and Higher Education for the funding of research with the No. 081/UN11.2/PP/SP3/2018. Thanks also to the University of Syiah Kuala (UNSYIAH) and Universiti Malaysia Pahang (UMP) for the collaboration and laboratory used.

6. References
[1] Akram, A., Hasanuddin, I., Nazaruddin, N., Syahril, S., Zulfan, Z., and Ahmad, N. 2017 *IOP Materials Science and Engineering*, **352**.
[2] BPS (Badan Statistik Provinsi) Aceh, 2015 *Aceh in Figures* (BPS Provinsi Aceh).
[3] Faruk, O., Andrze, K., Bledzi, H., Fink, P., and Sain, M. 2013, Progress Report on Natural Fiber Reinforced Composite: Macromocular Materials and Engineering
[4] Akova, E. 2013. Development of Natural Fiber Reinforced Polymer Composite: Transfer inovácií 25/201.
[5] Devendra, K., and Rangaswamy, T. 2012 *International Journal of Computational Engineering Research*, **2**, Issue.5.
[6] Singh, S., Kumar, P., and Jain, S.K. 2013 *Advanced Material Letter*, **4** (7), 567-572.
[7] Mobasher, B., Kingsbury, D., Montesinos, J., and Gorur, R.S. 2003 *IEEE Transactions on Power Delivery*, **18** (3).
[8] Lucas, S., Kumosa, M.S., and Armentrout, D.L. 2005 *IEEE Transactions on Power Delivery*, **20** (4).
[9] Srivastava, A. and Choudhary, S. 2013 *International Journal of Mechanical Engineering and Research*. **3**(6), 573-582.
[10] Pujari, S., Ramakrishna, A., and Kumar, M.S. 2014 *International Journal of Current Engineering and Technology*, 2347- 5161.
[11] T.Rajmohan, U.K. Koundinya, A. Premnath, and G. Harish, 2013 *International Conference on Advanced Nanomaterial's & Emerging Engineering Technologies*.
[12] Mahes]aran, T., Velmurugan, M.M. and Mohaideen J, 2013 *An Experimental and Numerical Study of Fracture Toughness of Kevlar- Glass Epoxy Hybrid Composite* 978-1-4673-6150-7113, IEEE.
[13] Gopinath, A. Kumar, S. and Elayaperumal, A. 2014 *Procedia Engineering* **97**, 2052-2063.
[14] Prasad, R., Rao, K. M. M. 2007 **32** (4), 399-403.

IOP Conf. Series: Materials Science and Engineering **523** (2019) 012077 doi:10.1088/1757-899X/523/1/012077

Increase the heating capacity of solar water heaters through two conditions of placing paraffin in copper tubes

D Harun[1], M I Maulana[1], Akhyar[1,*]
Department of Mechanical Engineering, Syiah Kuala University Jln. Syech Abdurrauf No.7 Darussalam – Banda Aceh, 23111, Indonesia

E-mail: akhyar@unsyiah.ac.id

Abstract. Nowadays, Energy is an essential requirement for daily activities. Indonesians use solar energy for various activities such as electricity, water heaters, and dryers. Solar collectors are devices that absorb solar energy and convert it into heat energy. Solar energy heats copper pipes and water passes through heat tubes induces an increase in water temperature. The increase in heating capacity in this experiment was carried out by adding paraffin (as thermal storage material) into a tube in a solar collector. Solar water heaters equipped with solar collectors (called cylindrical trough collectors) have been developed in this experiment. Paraffin placed in the copper tube with two conditions. They are called Collector A and Collector B. The effect of two conditions of placing paraffin in copper tubes on the water-out temperature has been studied in this experiment. The result shows that the highest water temperature found with 56°C in collector A. It is due to the highest solar radiation intensity also record at 1,289 W/m^2 (1:00 PM). While Collector B only produces the water-out temperature of 52°C at 1:00 PM. The conclusion shows that the position of paraffin in tubes with Collector A can increase the water-out temperature in the solar water heating system if compared with the collector.

1. Introduction

Population growth is a primary cause of the increasing gap between current world energy demand and energy supply. Renewable energy such as solar energy is one solution to the energy crisis. Solar energy is potential energy to be managed and developed as a source of further energy reserves, especially for Indonesia as a tropical country whose sun shines throughout the year. Several studies have been observed in the utilization of solar energy for water heaters.

The efficiency of solar collectors with nanofluid-based nanoparticles has been studied from the economics side. The results show that nanoparticles for nanofluid-based solar collectors have economic value if compared to conventional solar collectors. [1]. The potential for solar energy for water heater applications in Inland Norway has been investigated [2]. Solar energy application for water heat systems forced by natural convective flow have been observed using working fluids; it utilized supercritical CO2 [3]. The optimization of integrated solar collectors that integrated with parabolic concentrator (CPC) reflectors to produce hot water has been done by characterizing the geometry of the heater system [4]. The thermal performance of the evacuated-tube solar water heater has been systematically evaluated by placing two collectors side by side, i.e., the water-in-glass and

* Corresponding author: akhyar@unsyiah.ac.id.

IOP Conf. Series: Materials Science and Engineering **523** (2019) 012077 doi:10.1088/1757-899X/523/1/012077

heat-pipes [5]. The highest temperature and sun ray intensity were recorded at 01:00 pm, it observed in an experiment in Aceh-Indonesia after the tsunami. It can affect the results of water heating [6-7].

Thermal storage materials as phase-change materials (PCMs) have been applied to solar heating and cooling systems [8]. The shot peening process can increase the material properties of AISI 304 stainless steel [9]. Solar water heating systems with practical collectors have been optimized through its design [10]. Two water heater designs are built-in storage water heater-plain, and newly designed finned type have been compared for both performances [11]. The use of polymeric materials for glass components and heat exchanger have been reviewed for domestic and industrial applications of solar water heating systems [12-13]. The solar water heating system with a slab has been applied [14]. The relationship between the Nusselt number and friction factor for a solar air heater has been developed [15] and a turbine ventilator application [16]. An Integrated Collector Storage (ICS) to improve the performance of solar water heater has been designed with an emphasis on optics [17]. This experiment aims to investigate the water-out temperature with two conditions of placing paraffin as heat storage material in copper tubes by using solar collectors with cylindrical trough collectors of solar water heater systems.

2. Material and Method

Solar water heaters equipped with cylindrical trough collectors have been successfully built in our laboratory as shown in Figure 1. Copper tubes placed on the top of the solar collector plate with parabolic shaped so that the radiation energy is focused (sun ray in the form of a line) along the copper tube. Solar energy heats copper tubes, and water passes through the heat tube, then the temperature of the water out from those tubes can increase. Furthermore, heat capacity can continue to increase by adding thermal storage material to the copper tube. The material used in this experiment is paraffin for latent heat storage, which applied to solar water heater systems. Paraffin used in this experiment is the standard paraffin available on the Banda Aceh-Indonesia market.

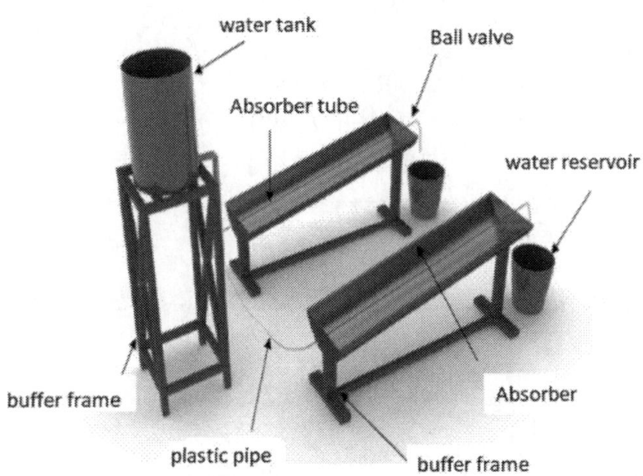

Figure 1. Apparatus of solar water heater developed in this experiment.

The unused drum with 55.5 cm in diameter split in vertically into two parts. Two split drums welded and obtained 170 cm in total length, then its used as a cylindrical collector. Stainless steel plates as reflectors with thickness, length and width are respectively 0.06, 170, and 85 cm, placed on the surface of the drum (parabolic-shaped). Two diameters of copper tubes used as heat absorber are Ø 1.3 cm (1/2 in) and Ø 0.95 cm (3/8 in) with length 190 cm, respectively. Black Styrofoam sheets with 0.9 cm thick are used as a heat insulator. Other instrument used in this experiment consisted of measuring glass, stopwatch, K type thermocouple, thermometer, Solar Power meter, main water tank, valve, PVC pipe, and plastic pipe, storage tanks for water-out. The temperature of water out is measured by two test conditions specified by placing two positions of paraffin in copper tubes. The first is called Collector A where paraffin placed in tube 2. The second is called Collector B where paraffin inserted between tubes 1 and 2 (inside tube 1 and outside tube 2) as shown in Figure 2. Two copper tubes arranged as parallel with diameter outer ½ in and diameter inner 3/8 in, paraffin placed in a different position at both collectors, that are collector A (paraffin in tube 2) and collector B (paraffin in tube 1 outside tube 2).

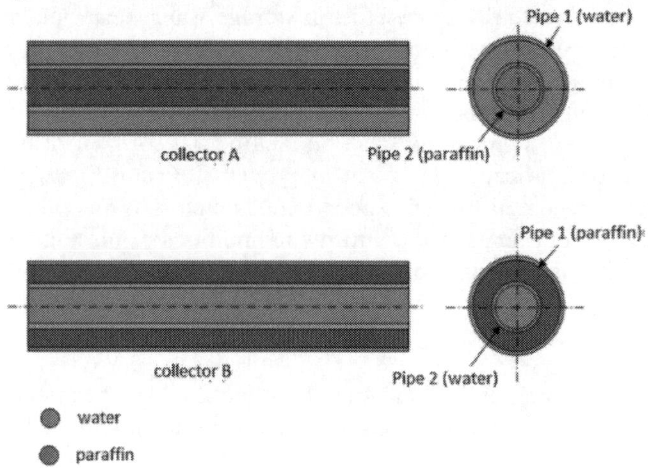

Figure 2. Paraffin placed in different positions.

Temperature recorded at several points in this experiment as shown in Figure 3. There are TA1 and TA2 as thermocouple placed in the surface of copper tubes outside of cylindrical collectors (before water passed solar collector). TB1 and TB2 as thermocouple placed in the surface of copper tubes inside of cylindrical collectors. TC1 and TC2 as thermocouple placed in the surface of copper tubes outside of cylindrical collectors (after water passed solar collector). TP1 and TP2 as thermocouple placed on the surface of the collector plate. TR1 and TR2 as collectors room temperature are recorded. Tf1 as the temperature of water-in, Tf2 as the temperature of water-out, T∞ as ambient temperature.

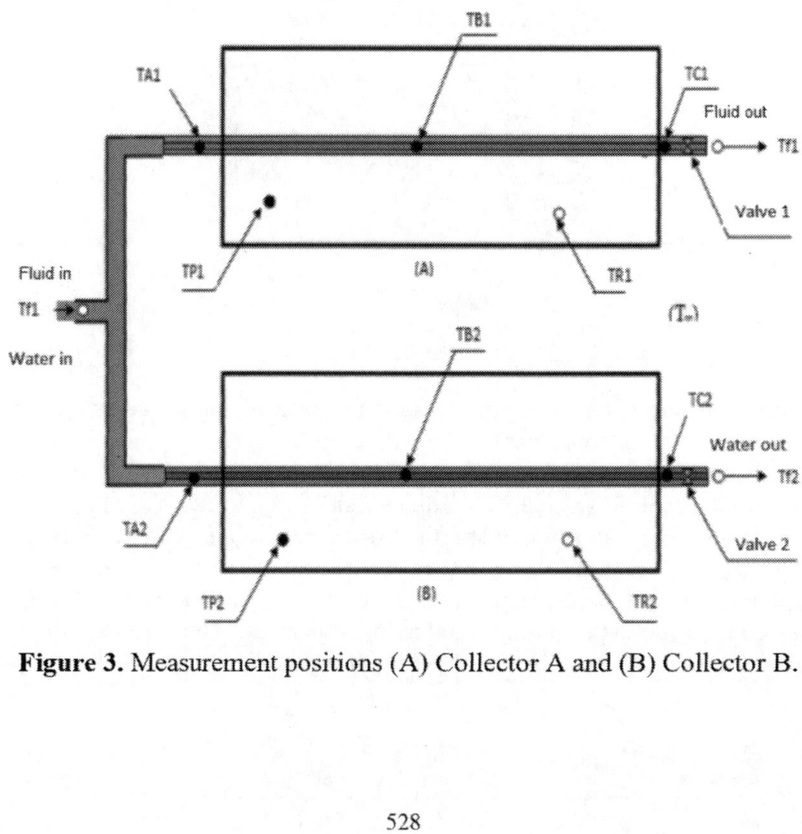

Figure 3. Measurement positions (A) Collector A and (B) Collector B.

IOP Conf. Series: Materials Science and Engineering **523** (2019) 012077 doi:10.1088/1757-899X/523/1/012077

3. Result and Discussion

The effect of the water-out temperature by two conditions of placing paraffin as heat storage material in copper tubes by using solar collectors with cylindrical trough collectors of solar water heater systems has been studied in this experiment. The velocity of the flow rate for both collectors is kept constant at 25 ml/minute.

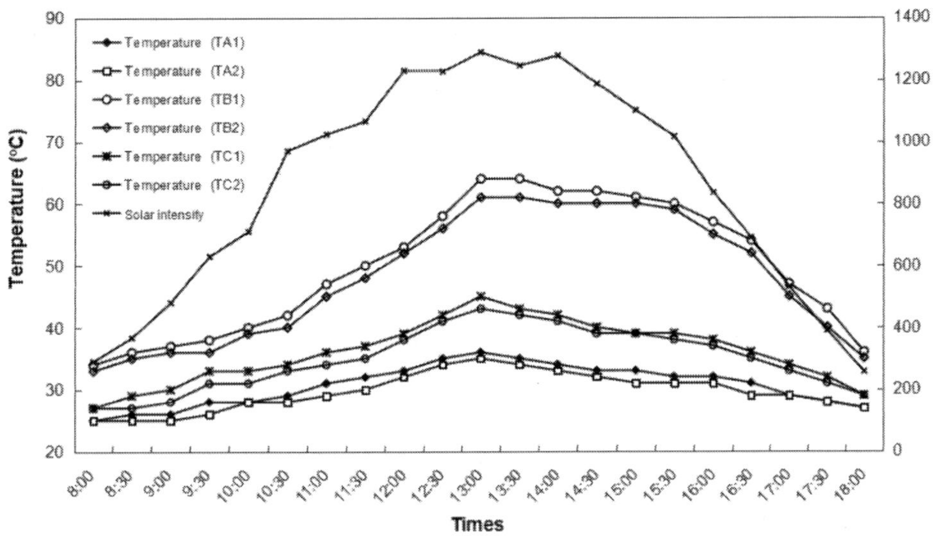

Figure 4. The temperature distribution water-in, water at room collector, and water-out foe Collector A and Collector B against time.

Temperatures recorded on apparatus by variation between collector A (paraffin inside) and collector B (paraffin outside) as shown in Figure 4. The temperature of water-in (before collectors) for both Collector is 25°C; temperature does not affect by heat from solar collectors. The highest temperature of water-out (water after passed solar collector) reaches 64°C. The maximum of solar intensity obtained 1,289 W/m^2 at 13:00 WIB (1:00 p.m.) (Indonesia Western Standard Time). However, the highest temperature of water-out at Collector B found 63°C.

Figure 5 shows the plate temperature, and room temperature for both Collectors (Collectors A and B) have the same temperature is 28°C at 08:00 WIB with a sun intensity 290 W/m^2. It is an initial experimental condition. Starting at 08:30 WIB (8:30 a.m.), plate temperature, and room temperature continued to increase as increased the solar intensity and reached a peak at 13:00 WIB (1:00 p.m.). The plate temperature reaches a maximum that is 77°C while the room temperature is 69°C the difference between the two is 8°C. The water-out temperature in Collector A began to rise started with 32°C, and 31°C for Collector B. The highest temperature reached 56°C at 13.00 WIB for Collector A and 52°C for collector B, the intensity of solar radiation was 1,289 W/m^2. Plate temperature and room temperature at 13:00 - 15:30 WIB are still up and down, while at 16.30 WIB - 18.00 WIB the difference between the plate temperature and the room temperature is not significant, it caused by the solar intensity begun to decline to 257 W/m^2. The temperature increase is not only caused by exposure to sun ray but also due to the heat energy absorb and release by paraffin that placed in the copper tube during the experiment process.

Increasing the temperature of the water out after flowing through the copper tube in the solar collector is caused by several things. First, the parabolic reflector (cylindrical solar concentrator) reflects sun ray focused along the copper tube (focus line) causing an increase in the copper tube temperature. Second, the solar collector chamber which is completely covered with glass at the top causes increasing room temperature, where the heat from room temperature will induce copper tube.

IOP Conf. Series: Materials Science and Engineering **523** (2019) 012077 doi:10.1088/1757-899X/523/1/012077

Third, paraffin in the copper tube act as material to store heat. Copper tubes exposed by sun ray and the heat of room collector absorbed and stored heat into paraffin, subsequent water that passes through copper tubes will accumulate all heat and increase the water-out temperature.

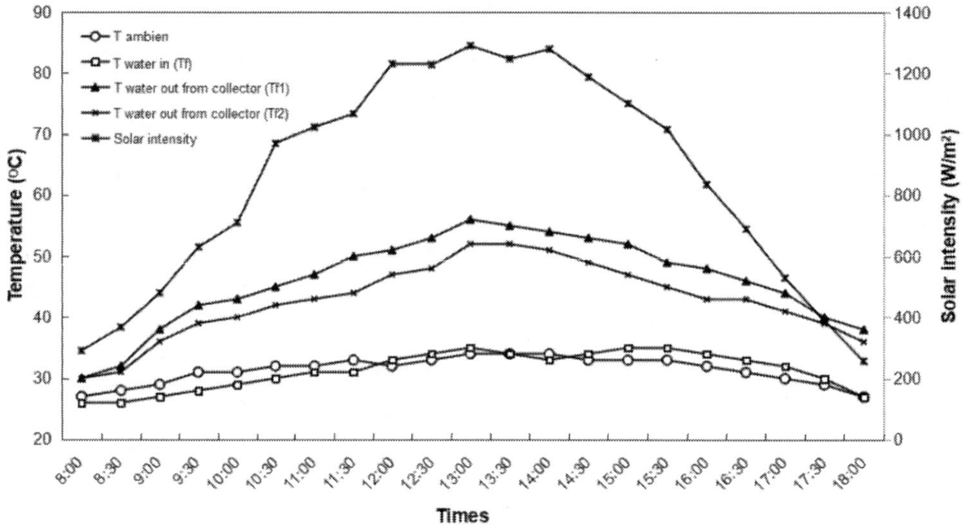

Figure 5. The temperature distribution of water-in and water-out for both conditions (collector A and B).

Comparison of the two positions of paraffin in copper tube affects the water-out temperature are discussed — collector A where paraffin placed at the core of copper tubes appears better compared to Collector B (paraffin covers the water-out channel). The highest water-out temperature obtains in Collector A; this causes by water passing through copper tubes exposed by heat from the sun ray reflector and the heat of the collector space. Heat from water passing will absorb by paraffin which is stored and released in slowly and can affect to the water passing through it. Copper tubes exposed to heat from the sun's reflector and collector room do not directly affect to water passing. The position of paraffin restrains the heat flow that affects the water-out temperature (case of heat transfer).

Parabolic mirror collector that the mirror design allows the collector to focus and collect radiation directly and spread without tracking the sun ray [18]. Absorbent tubes covered with evacuated tubes for reduction of heat losses. There are several types of collectors with high efficiency and can produce high temperatures [19-22]. Classification of solar water heaters is generally divided into two types, and most solar water heaters need a well-insulated storage tank. Solar energy preheats water before entering water heaters with two tank system. However, the backup heater combined with solar storage in one tank system [23,24].

4. Conclusion

The influence of the water-out temperature by two conditions of placing paraffin as heat storage material in copper tubes by using solar collectors called cylindrical trough collectors of solar water heater systems had been intensively investigated. It can be concluded:

- Heat focus line through parabolic reflector as cylindrical solar concentrator is successfully increasing water-out temperature; it developed in this experiment.
- Solar collector chamber with covered by plate glass also induced copper tube and affected to water-out temperature.

- Effect of adding paraffin as the thermal storage material does not significant increasing water-out temperature.
- The position of paraffin has affected to water-out temperature.
- The water out temperature inside the collector for the solar water heater depending on the solar intensity during the experiment.
- The highest temperature of water output obtain at collector A (paraffin in pipe 2) at 56°C, and the solar intensity maximum recorded at 1,289 W/m^2. It is due to paraffin placement in pipe 2 (collector A) can increase the efficiency of the solar water heating system.

5. Acknowledgment

This publication was made possible by a grant from the Research Fund of Ministry of Research, Technology, and Education, Republic of Indonesia and LPPM Syiah Kuala University; the financial support is greatly appreciated. We would like to thank to Mr. Islahul Umam for support during this experiment.

6. References

[1] Otanicar T P and Golden J 2009 *Environ. Sci. Technol.* **43** 6082–6087 .
[2] Hagos D A, Gebremedhin A and Zethraeus B 2014 *Journal of Renewable Energy* **968320** 1-11.
[3] Zhang X, Niu X, Yamaguchi H and Iwamoto Y 2018 *Advances in Mechanical Engineering* **10** (4) 1–15.
[4] Benrejeb R, Helal O and Chaouachi B 2016 *Applied Thermal Engineering* **103** 1427–1440.
[5] Chow T-T, Chan L-S, Fong K-F, Tan R and Z Dong, *Proc. IMechE Part A: J. Power and Energy* **226** (4) 447-461.
[6] Harun D, Akhyar H and Thaib R 2016 *Proceding: International Conference on Engineering and Science for Research and Development (ICESReD)* **25-26**, 149-154.
[7] Farhan A and Akhyar H 2017 *IOP Conf. Series: Earth and Environmental Science* **56** 012002.
[8] Ghoneim A A 1989 *Solar Energy* **42** (3) 209-220.
[9] Iswanto P T, Akhyar H and Utomo F F 2018 *Metalurgija* **57** (4) 295-298
[10] Mills D and Morrison GL 2003 *Solar Energy* **74** (6) 505–511.
[11] Muneer T, Maubleu S and Asif M 2006 *Renewable and Sustainable Energy Reviews* ***10*** (1) 1–23.
[12] Ogueke N V, Anyanwu E E and Ekechukwu O V 2009 *Journal of Renewable and Sustainable Energy* **1** 043106.
[13] Raman R, Mantell S, Davidson J, Wu C and Jorgensen G 2000 *Transactions of the ASME* **122** (2) 92-100.
[14] Raveendrana S K and Shen C Q 2015 *AIP Conference Proceedings* **1674** 020013.
[15] Saini S K and Saini R P 2008 *Solar Energy* **82** (12) 1118–1130.
[16] Harun D, Zulfadhli and Akhyar H 2018 *IOP Conf. Series: Materials Science and Engineering* 352.
[17] Souliotis M and Tripanagnostopoulos Y 2008 *Renewable Energy* **33** (5) 846–858.
[18] Hossain M S, Saidur R, Fayazb H, Rahim N A, Islam M R, Ahamed J U and Rahman M M 2011 *Renewable and Sustainable Energy Reviews* **15** 3801–3812.
[19] Fang G, Liu X and Li H 2010 *Materials Chemistry and Physics* **122** 533–6.
[20] Bourdoukan P, Wurtz E, Joubert P and Spérandio M 2008 *Solar Energy* **82** 1209–19.
[21] Jeffrey K C-F, Su T-L, Jhang P-R, Huang C-Y, Chiu C-H 2011 *Energy* **36** 3554–62.
[22] Chien CC, Kung CK, Chang CC, Lee WS, Jwo CS, and Chen SL 2011 *Energy* **36** 415–23.
[23] Duffie J and Beckham W 2006 Solar engineering of thermal processes (John Wiley & Sons).
[24] Kumar R and Rosen M A 2010 *Applied Thermal Engineering* **30** 1764–8.

IOP Conf. Series: Materials Science and Engineering **523** (2019) 012078 doi:10.1088/1757-899X/523/1/012078

Effect of cutting parameter on tool wear of HSS tool in drilling of Kevlar composite panel

M Iqbal[1,*], S Bahri[1], and A Akram[1]

[1] Department of Mechanical and Industrial Engineering, Faculty of Engineering, Syiah Kuala University, Banda Aceh, Indonesia

E-mail: mohd.iqbal@unsyiah.ac.id

Abstract. The paper reports the tool wear in the drilling of composite panel in various cutting conditions. The drilling process was applied to 4mm thickness Kevlar composite panel by using HSS drill tool (2 flutes, 12 mm diameter, 17 mm length and 118° drill point angle). The process was conducted by using a vertical machining center. The tool wear was measured by using an optical microscope with the max precision 0.01 µm. The drilling process was conducted with a constant feed rate (75 mm/min) and three levels of spindle rotation speeds (1000, 3000 and 5000 rpm). Each cutting condition was applied to two different drill tools. In order to investigate the wear rate, the tool wear was measured in several levels of cutting time (i.e., after drilling of 5, 10, 20 and 30 drill holes). The result of the experiment shows that the cutting speed gives significant effect to the tool wear of HSS drill tool when drilling Kevlar composite panel. Higher cutting speed produced higher tool wear at any cutting time level. The result of the experiment also shows the effect of continuous/ discontinuous cutting. Higher tool wear was found in a direct 20 holes drilling, compare to paused drilling (i.e., start with 10 holes, pause for 20 minutes then continue with other 10 holes).

1. Introduction

Composite is a combination of two or more materials with different phase become a new material with better properties [1]. Composite is reinforced either by synthetic or natural fiber. Natural fiber (jute, knaf, banana, etc.) has some benefit compare to synthetic fiber, i. e. lower production cost and environmentally friendly. However, its lower strength and un-uniform size (compare to synthetic fiber), make natural fiber has limited application as the end product [2]. The most popular synthetic fibers are glass fiber and carbon fiber, because of their strength and economical production cost [3, 4]. Kevlar is synthetic fiber with better properties but relatively more expensive compare to glass and carbon fiber.

Kevlar is a registered trademark for aramid synthetic that related to Numex and Technora. Kevlar (Poly-para phenylene terephthalamide) was first developed by Stephanie Kwolek in 1965 at DuPont Company. Then Kevlar was further developed to be synthetic fiber and other product by DuPont [1]. Kevlar was firstly used as commercial material in 1970 to replace steel fiber in racing tyres. Its tensile strength was five times higher than steel fiber. Since then, due to its mechanical strength, impact absorption, flexibility, and lightweight, Kevlar has been further commercially developed to ship sail,

* Corresponding author: mohd.iqbal@unsyiah.ac.id.

armor and composite panel [5]. Kevlar composite panel has been used as hard, strength, flexible and lightweight component in aerospace and aircraft product [6].

Like any other composite panel, Kevlar required some drilling process to become an end product. Drilling Kevlar composite panel would be a difficult machining process. Its hardness generates a high cutting temperature during the friction between drill tool's cutting surface and hole's surface. High cutting temperature and chemical substance of composite panel lead to an internal heating process that reduces the strength of the cutting tools [7]. Furthermore, its flexibility generates more vibration during the drilling process. Both of the problems increase the wear progress of the drill tool. Since the tool cost and tool replacement time consumes significant production cost and production time, minimizing of the tool wear has been an important issue in machining composite panel. However, the tool wear is not affected only by the nature of the material. The tool wear could be minimized if the correct cutting condition was applied during the machining process [8, 9, 10, 11, 12].

This research aims to investigate the flank wear of HSS drill tool when drilling Kevlar composite panel. A set of experiment was conducted to study the effect of cutting parameter, i. e. spindle speed to the tool wear. The experiment also conducted to investigate the progressive wear rate by drilling several sets of holes and measure the wear separately. The result of the experiment would be used as a reference for machining industry/ workshop in order to minimize the tool cost while maintaining the machining productivity.

2. Methods

The Kevlar composite with 4 mm thickness was prepared as the material of the experiment. The Kevlar composite was cut to rectangle form with 700 mm length and 500 mm width. The material was design to be drilled with 10 rows of holes with 35 mm distance between the center of the rows. Each row consisted of 13 holes with 50 mm distance between the center of the holes. With the particular design, the total of 130 holes could be drilled on the material. The Kevlar composite panel was drilled by using High-Speed Steel (HSS) drill tool. The drill tool was manufactured by Nichi, with 2 flutes, 12 mm diameter, 17 mm tool length and 118° drill point angle. The tools used in the experiment were originally a new tool without any defect or initial wear.

The spindle speed has been selected as the independent variable of the experiment. Three levels of spindle speed have been determined, i. e. 1000, 3000 and 5000 rpm. The feed rate of the drilling process remains constant as 75 mm/min. The drilling process was conducted by using machining center (Agma type A-8 and max spindle speed 8000 rpm) at Laboratorium of Design and Manufacture, Department of Mechanical and Industrial Engineering, Faculty of Engineering, Syiah Kuala University, Banda Aceh, Indonesia. In order to measure the progress of the flank wear, several sets of the holes were set to be drilled continuously, i. e. 5, 10, 20 and 30 holes. After drilling a set of holes, the drilling process took a halt, and the flank wear of the tool was measured and recorded. The tool then used to drill another set of the holes, and similar flank wear was measured and recorded. Two drill tools were used to provide the drilling process for each spindle speed with a different set of holes. The result was compared in order to investigate the effect of the continuous/ discontinuous drilling process on the progress of flank wear.

The flank wear was measured by using an optical microscope (Olympus GX-71 with the max precision 0.01 μm). A special fixture was designed and fabricated to hold the tool during the measurement process. The angle of fixtures could be set to secure the surface of the cutting edge, to make sure that the microscope give a correct reading and the similar position could be repeated precisely. The original tool was put under the microscope before the first drilling process started to identify the original line and surface of the cutting edge. The wear measurement (after drilling) was conducted at a similar original position. The setup of the experiment was shown in Figure 1.

IOP Conf. Series: Materials Science and Engineering **523** (2019) 012078 doi:10.1088/1757-899X/523/1/012078

3. Result and Discussions

The NC programming was developed according to the cutting condition and holes position in the design of the experiment. The Kevlar composite panel after drilling experiment was shown in Figure 2. Each of cutting speed used 2 cutting tools. The first one was used to drill 10 + 10 + 10 holes. Consequently, the experiment produced tool wear data after drilling 10 holes, 20 holes and 30 holes. The second drill tool was used to drill 5 + 15 holes. The data would be the tool wear after drilling 5 holes and 20 holes.

The tool was observed under the optical microscope. The tool wears were captured by the digital camera of the microscope and were further investigated. The images of the tool wear from different spindle speed were shown in Figure 3, 4 and 5. The bright side was the flank wear of the tool. There were some difficulties during the measurement since the original surface has been removed during the drilling process. However, the initial surface has been recognized and recorded before the drilling process. The red line in Figure 3, 4 and 5 was the imaginer edge of the initial cutting surface.

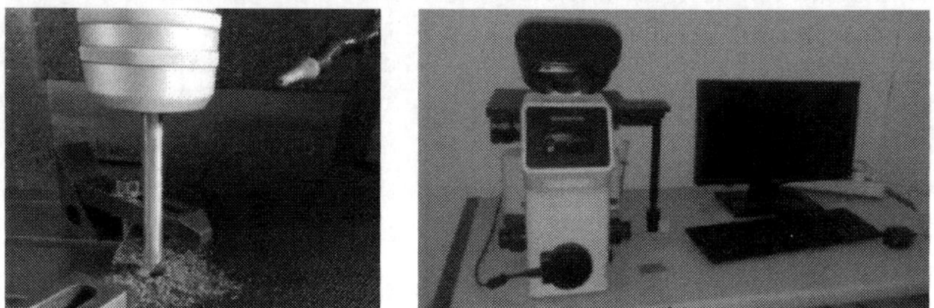

(a) (b)

Figure 1. Setup of the experiment (a) drilling process (b) tool wears measurement process.

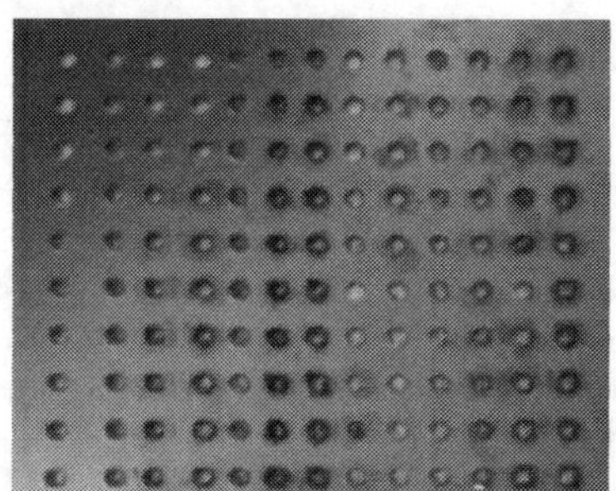

Figure 2. The Kevlar composite panel after the drilling process.

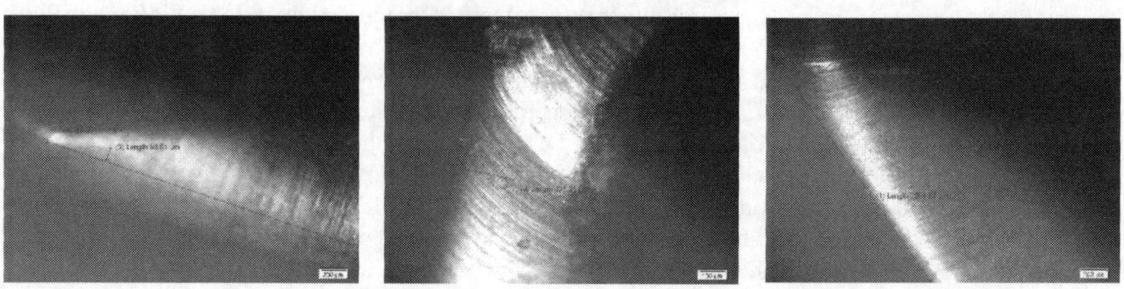

(a) (b) (c)

Figure 3. Figure Tool wears after drilling process with 1000 rpm spindle speed (a) after drilling 10 holes (b) 20 holes (c) 30 holes.

IOP Conf. Series: Materials Science and Engineering **523** (2019) 012078 doi:10.1088/1757-899X/523/1/012078

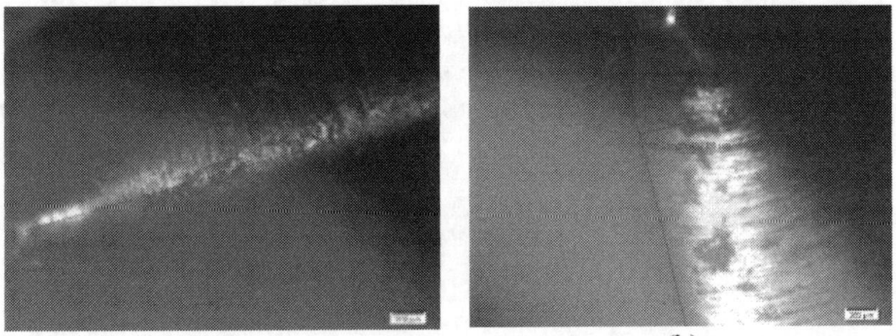

(a) (b)

Figure 4. Tool wear after drilling process with 3000 rpm spindle speed (a) after drilling 10 holes (b) 20 holes.

The result of the experiment was recorded in Table 1. The maximum tool wear of 500 μm has been considered as the limit of the tool life [13]. The tool wear greater than 500 μm was not shown in Table 1 since at that point the tool wear has exceeded the limit and the tool could not be used anymore. Most of the tool exceeded the tool life limit when drilling the 30 holes. Only the drilling with 1000 rpm spindle speed was able to drill 30 holes without exceeding the tool life limit.

Tabel 1. The tool wear data of the HSS tool after drilling experiment

Number of the hole	1000 rpm		3000 rpm		5000 rpm	
	Tool 1	**Tool 2**	**Tool 3**	**Tool 4**	**Tool 5**	**Tool 6**
5		57,11μm		80,98μm		114,93μm
10	98,51μm		186,29μm		209,01μm	
20	238,35μm	423,73μm	432,86μm	469,9μm	474,28μm	575,77μm
30	422,9μm					

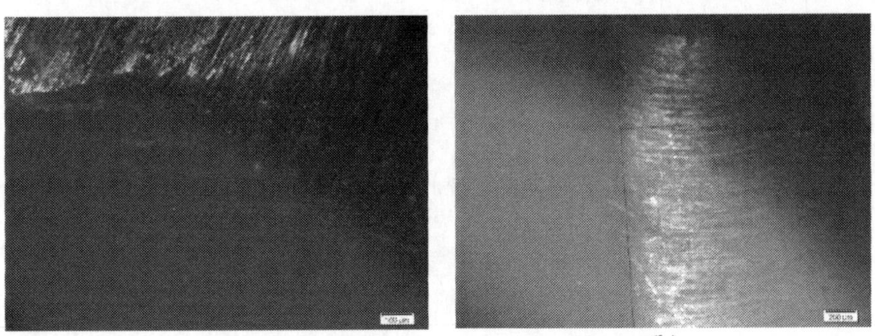

(a) (b)

Figure 5. Tool wear after drilling process with 5000 rpm spindle speed (a) after drilling 10 holes (b) 20 holes.

IOP Conf. Series: Materials Science and Engineering **523** (2019) 012078 doi:10.1088/1757-899X/523/1/012078

The result of the experiment shows that spindle speed gives a significant effect on the tool wear. As shown in Figure 6 the higher spindle speed increases the tool wear at any number of the holes. This was because the higher speed gives more friction between the cutting surface of the drill tool and the surface of the Kevlar composite panel. More friction generates more heat and reduces the strength of the drill tool material. Finally, the wear resistance decreases and the tool wear increases progressively.

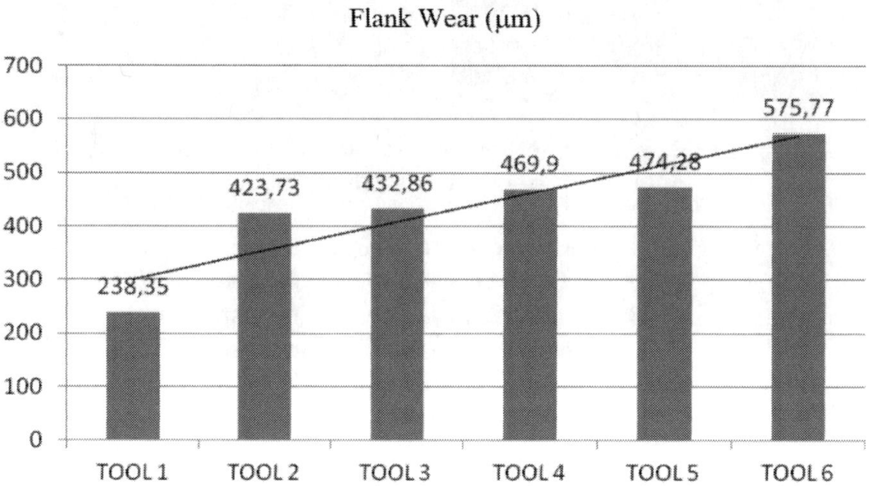

Figure 6. Comparison of tool wears after drilling 20 holes.

The experiment also investigated the effect of continuous cutting to the tool wear. As shown in Figure 6, the tool wear was higher in the second drill tool (tool 2, 4 and 6) compare to the first drill tool (tool 1, 3, and 5) at all spindle speed. The first tools were used to drill 5 holes, pause for 30 minutes, then continue to drill 15 holes. The second tools were used to drill 10 holes, pause for 30 minutes, then continue to drill 10 holes. Even though both of the drill tools were used to drill 20 holes, the result showed that they produced different tool wear. This result showed that continuous drilling produces more tool wear compare to discontinuous one. It was because when the drilling stops (in discontinuous drilling), the tool has released the heat to the environmental for 30 minutes. While in the continuous drilling the temperature increases very fast and there was no opportunity to release the heat to the environment until the 15 holes have been drilled completely.

4. Conclusion

The result of the experiment has been discussed. The spindle gives significant effect to the tool wear. The higher spindle speed increases the tool wear at any number of the holes. The comparison between continuous and discontinuous drilling showed that continuous drilling produces more tool wear compare to discontinuous one.

5. References

[1] Groover M P 2012 *Fundamentals of Modern Manufacturing*, 5th Edition (John Wiley & Sons, Inc).

[2] Elanchezhian C, Ramnath B V, Ramakrishnan G, Rajendrakumar M, Naveenkumar V and Saranavakumar M K 2018 *Material Today: Proceedings*. **5** 1785

[3] Jones F R and Huff N T 2018 *Hand Book of Properties of Textile and Technical Fibers* 2nd Ed **757**

[4] Newcomb B A *2016 Composite Part A: Applied Science and Manufacturing*. **91** P 1 262

IOP Conf. Series: Materials Science and Engineering **523** (2019) 012078 doi:10.1088/1757-899X/523/1/012078

[5] Raja S n, Basu S, Limaye A M, Anderson T j, Hyland C M, Lin L, Alivisatos A P and Ritchie R O 2015 *Material Today Communications*. **2** 33
[6] Dickson A N, Barry J N, McDonnell K A and Dowling D P 2017 *Additive Manufacturing*. **16** 146
[7] Parka K H, Bealb A, Kimb D D W, Kwonc P and Lantripd J 2011 *Wear* **271** 2827
[8] Iqbal M, Akram A and Tadjuddin M 2015. *International Journal of Applied Engineering Research*. **10** N 95 16
[9] Iqbal M, Konneh M, Hanafi M, Abdallah K A and Binting M F 2015 *International Journal of Mechanical And Production Engineering*. V **3** I 4 62
[10] Iqbal M, Konneh M, Said A Y M and Zaini A F M 2014 *Applied Mechanics and Materials Journal*. **446 - 447** 275
[11] Konneh M, Iqbal M and Faiz N M A 2012 *Advanced Materials Research Journal*. **576** 531
[12] Konneh M, Iqbal M, Kasim M A and Isa N M 2012 *Advanced Materials Research Journal*. **576** 535
[13] Yong A Y L, Seah K H W and Rahman M 2007 *Int J Adv Manuf Technol*. **32** 638

IOP Conf. Series: Materials Science and Engineering **523** (2019) 012079 doi:10.1088/1757-899X/523/1/012079

The spare part maintenance of cake breaker conveyor with reliability centered spares method

N Sembiring[1,*], **YP Batubara**[1]

[1] Department of Industrial Engineering, Engineering Faculty, Universitas Sumatera Utara, Medan, Indonesia

E-mail: nurhayatipandia68@usu.ac.id

Abstract. The company that this research conducted is a manufacturing industry that produces CPO. Many machines in the production section need maintenance activities so that existing machines can work without inhibiting the production process. At seed processing station, there is CBC machine that experienced frequent occurrence damage. The machine applies corrective maintenance regardless of reliability machine production components, consequently, often occur sudden machine damage. Calculation of time interval-based maintenance policy scheduled discard task and needs a replacement component (spare part) that should be available are concerned with an optimal using method of Reliability-centered Spares (RCS). Based on the results of the calculations are based on a policy scheduled discard task then obtained a policy to perform maintenance on a screw conveyor components is 91 days, hangers and bearings are 40 days, jig-drilled couplings is 21 days. The result of the method of RCS acquired number of screw conveyor parts needs with this type of repair of repairable spare is 3 units for one year, while jig-drilled couplings with this type of repair repairable, need spare part 27 units for a year, as well as the hangers and bearing with this type of repair of repairable spare part, needs 21 units for one year.

1. Introduction

The current technological developments in Indonesia are running fast and increasingly sophisticated. So that it can be felt in various activities and daily life, especially in the manufacturing industry. In an effort to use production facilities so that the continuity of production can be guaranteed, it is necessary to plan maintenance activities that can support the reliability of a machine [1]. Machine reliability is one of the most important aspects that can affect the smooth process of production, and the products produced [2]. This reliability can help estimate the chance of a machine component to be able to work in accordance with the desired goals in a certain period. Companies engaged in the processing of palm oil which often experiences high problems in its production machinery. This hampers the course of the production process which results in a decrease in production capacity at the seed processing station which has a role to further process fiber produced from the pressing station. Cake Breaker Conveyor machine (CBC) is one machine that has an important role in the separation of palm kernel and shell. When the CBC machine is damaged, it can be ascertained that the next production process will be disrupted. One type of damage that causes the engine to stop is that there are components that cannot be repaired but must be replaced. Maintenance policy is made with Reliability Centered Maintenance (RCM) and spare parts policy using Reliability Centered Spares (RCS) [3]. Therefore, to solve this

* Corresponding author: nurhayatipandia68@usu.ac.id.

problem, it is necessary to apply an engine maintenance system using Reliability Centered Spares (RCS) method so that the procurement of spare parts can be optimized to avoid the cessation of the engine and also to reduce excessive inventory costs.

2. Research method

This type of research is descriptive research, which is a type of research that aims to describe systematically, factually, and accurately the facts and characteristics of a particular object or population [4]. The object of the research observed was the Cake Breaker Conveyor machine (CBC) in the process of producing CPO (Crude Palm Oil) at the company.

3. Result and discussion

3.1. Data collection

3.1.1. Collecting data on the damage of CBC machine components

Data collection based on the determination of critical components is chosen from the frequency of the most damage from each component that is on the CBC machine during the 2014-2017 period. Damage frequency data collected based on historical data in the maintenance section are as follows.

Table 1. Historical data on subsystem damage

CBC Component	Damage Frequency	Percentage (%)	CBC Component	Damage Frequency	Percentage (%)
Discharge Spouts	0	0%	Clamps and Shrouds	1	1%
Conveyor Screw	11	10%	Flange	1	1%
Hangers and Bearing	28	21%	Jig-Drilled Couplings	53	55%
Trough Ends	6	6%	Supporting Feeds and Saddle	6	6%
	Total Damage			95	
	Total Frequency				100 %

3.1.2. Testing the time to repair distribution

Our modeling approach is to consider the time to fix (TTR) as a random variable depending on the many parameters that characterize the subsystem involved. [6] Testing the suitability of the distribution selected using Minitab software which can be seen in the following table.

Table 2. Time to repair (TTR) distribution

Component	Distribution	Anderson Darling Value	P-Value	Chosen Distribution
Conveyor Screw	Normal	0,226	0,76	Normal
	Exponential	2,678	0,003	
	Weibull	0,253	0,250	
Hangers and Bearing	Normal	0,182	0,903	Normal
	Exponential	6,689	0,003	
	Weibull	0,204	0,250	
Jig-Drilled Couplings	Normal	0,732	0,053	Weibull
	Exponential	7,003	0,003	
	Weibull	0,448	0,250	

IOP Conf. Series: Materials Science and Engineering **523** (2019) 012079 doi:10.1088/1757-899X/523/1/012079

3.1.3. Determination of time to repair distribution parameters representing
Determination of parameters based on the TTR distribution that represents can be seen in the following table.

Table 3. Distribution parameters *time to repair* (TTR)

Component	Distribution	Parameter	
Conveyor Screw	Normal	M	5,81818
		Σ	1,67752
		P	0,95302
Hangers and Bearing	Normal	M	5,92857
		Σ	1,78002
		P	0,98264
Jig-Drilled Couplings	Weibull	H	9,36445
		B	2,11346
		Γ	0
		P	0,98847

3.1.4. Calculation of Mean Time to Repair (MTTR)
The evaluation of realistic MTTR for each alternative design is, therefore, a necessity related to the protected design process [7]. Calculations are made based on the selected distribution. MTTR calculation results are as follows.

Table 4. Calculation Results Mean Time to Repair

Component	Distribution	*Mean Time to Repair*
Conveyor Screw	Normal	5,82
Hangers and Bearing	Normal	5,93
Jig-Drilled Couplings	Weibull	8,29

3.1.5. Testing of time to failure (TTF) distribution match
Testing the suitability of the distribution selected using Minitab software with the following results.

Table 5. Distribution of *time to failure* (TTF)

Component	Distribution	Anderson Darling Value	P-Value	Distribution
Conveyor Screw	Normal	0,256	0,641	Normal
	Exponential	2,642	0,003	
	Weibull	0,309	0,250	
Hangers and Bearing	Normal	1,164	0,005	Weibull
	Exponential	1,378	0,04	
	Weibull	0,396	0,250	
Jig-Drilled Couplings	Normal	0,597	0,116	Weibull
	Exponential	5,899	0,003	
	Weibull	0,355	0,250	

IOP Conf. Series: Materials Science and Engineering **523** (2019) 012079 doi:10.1088/1757-899X/523/1/012079

3.1.6. Determination of mean time to failure

Calculations are made based on the selected distribution. MTTF calculation results are as follows.

Table 6. The calculation result of *mean time to failure* (minutes)

Subsystem	Distribution	*Mean Time to Failure*
Conveyor Screw	Normal	2909,3
Hangers and Bearing	Weibull	1280,74
Jig-Drilled Couplings	Weibull	676,78

MTTF of such a system has been obtained using the renewal theory in the condition that the probability distributions about failure and repair times have the respective explicit functions.[8] Maintenance time intervals for scheduled discard task policies are obtained through half of the MTTF (Mean Time to Failure) value with a working day per month which is 24 days and working hours which is 16 hours.[9] Following is the calculation of maintenance time intervals based on scheduled discard task policy for screw conveyor components, hangers and bearings, and jig-drilled couplings, as shown in the following table.

Table 7. Maintenance time interval based on the scheduled discard task

Component	*Mean Time to Failure* (MTTF)	Maintenance Interval (Hours)	Maintenance Interval (Months)	Maintenance Interval (Days)
Conveyor Screw	2909,3	1454,65	3,79	91
Hangers and Bearing	1280,74	640,37	1,67	40
Jig-Drilled Couplings	676,78	338,39	0,88	21

3.1.7. Calculation of non-repairable needs

Non-repairable is a condition when repairing components is difficult to do and is not possible or when repair costs are greater than the cost of purchasing components, including non-repairable components are hangers and bearings. [10-17] Calculations for "λ" t use the following formula.

$$\lambda t = \frac{A \times N \times M \times T}{MTTF} = \lambda t = \frac{4 \times 1 \times 384 \times 12}{1280{,}745} = 14{,}39 \quad (1)$$

so that for hangers and bearing needs are as follows:

Table 8. Number of hangers and bearing component needs

N	fact(n-1)	exp(λt)	$\lambda t^n/n!$	P	P (%)	N	fact(n-1)	exp(λt)	$\lambda t^n/n!$	P	P (%)
0	1	5,62E-07	1	5,62E-07	0%	12	479001600	5,62E-07	164808,9	0,321030	32%
1	1	5,62E-07	14,39162	8,65E-06	0%	13	6227020800	5,62E-07	182451,3	0,423582	42%
2	2	5,62E-07	103,5594	6,68E-05	0%	14	8,7178E+10	5,62E-07	187555	0,529003	53%
3	6	5,62E-07	496,796	0,000346	0%	15	1,3077E+12	5,62E-07	179948,1	0,630148	63%
4	24	5,62E-07	1787,425	0,001350	0%	16	2,0923E+13	5,62E-07	161859,1	0,721126	72%
5	120	5,62E-07	5144,791	0,004242	0%	17	3,5569E+14	5,62E-07	137024,4	0,798144	80%
6	720	5,62E-07	12340,32	0,011178	1%	18	6,4024E+15	5,62E-07	109555,8	0,859723	86%
7	5040	5,62E-07	25371,02	0,025439	3%	19	1,2165E+17	5,62E-07	82983,44	0,906366	91%
8	40320	5,62E-07	45641,28	0,051093	5%	20	2,4329E+18	5,62E-07	59713,32	0,939930	94%
9	362880	5,62E-07	72983,57	0,092115	9%	21	5,1091E+19	5,62E-07	40922,46	0,962932	96%
10	3628800	5,62E-07	105035,2	0,151153	15%	22	1,124E+21	5,62E-07	26770,03	0,977978	98%
11	39916800	5,62E-07	137420,7	0,228395	23%						

IOP Conf. Series: Materials Science and Engineering **523** (2019) 012079 doi:10.1088/1757-899X/523/1/012079

Based on the results of calculations in table 8, the number of spare requirements recommended for companies to be able to meet 95% availability of spare hangers and bearings for a year is 21 units.

3.2. Calculation of repairable needs

Included in the repairable component are jig-drilled couplings and screw conveyors. The formula used is as follows.

a. Needs Calculation of conveyor screw

$$\lambda t = \frac{A \times N \times M \times MTTR}{MTTF} \tag{2}$$

$$\lambda t = \frac{1 \times 1 \times 384 \times 5{,}8}{2909{,}3}$$

$$= 0{,}77$$

For screw conveyor calculations are as follows:

Table 9. Screw conveyor calculations

n-1	exp(-λt)	fact(n-1)	λt(n-1)/(n-1)!	P	P
0	0,46	1	1,00	0,47	47%
1	0,46	1	0,77	0,82	82%
2	0,46	2	0,29	0,96	96%

Based on the results of calculations in table 9, the number of spare requirements recommended for companies to meet 95% availability of screw conveyors for a year is 3.

b. Calculation of need for jig-drilled couplings

$$\lambda t = \frac{A \times N \times M \times MTTR}{MTTF} \tag{3}$$

$$\lambda t = \frac{1 \times 1 \times 384 \times 8{,}29}{676{,}78}$$

$$= 18{,}82$$

The number of spare requirements recommended for companies to meet 95% availability of spare jig-drilled couplings for a year is 27 units.

4. Conclusion

The conclusion of the research that has been done is as follows:

- Based on the results of the determination of critical components, it is known that the cake breaker conveyor machine has eight components, judging from the frequency of the most damage during 2014-2017, there are three critical components namely screw conveyors, hangers and bearings, and jig-drilled couplings.
- Based on the results of determining the maintenance time interval based on scheduled discard task policy, the maintenance of the screw conveyor component is 91 days, hangers and bearings are 40 days, and the jig-drilled couplings is 21 days.
- Based on the results of calculations using reliability centered spares calculation, the number of critical components available in the company is needed for one year, namely for components of screw conveyor components are 3 units, for hangers and bearings is 21 units, and jig-drilled couplings are 27 units.
- Data processing results that have been carried out during the research can be proposed for companies that do not have a fixed machine maintenance schedule in the form of periodic machine component changes so that there will be no sudden engine failure as often happens to the company and avoid the possibility of occurrence delay when replacing engine components that are damaged due to unavailability of replacement components.

IOP Conf. Series: Materials Science and Engineering **523** (2019) 012079 doi:10.1088/1757-899X/523/1/012079

5. References

[1] Kulkarni, M 2013 *International Journal of Performability Engineering*, **9** (3) 321-332.

[2] Rommert, D 2010 *Integrating Reliability Centered* Maintenance and Spare Parts Stock Control.

[3] Alhilman, J 2018 Proceedings of the International Conference on Industrial Engineering and Operations Management Bandung, Indonesia, March 6-8, 2018

[4] Nassaji, H 2015 Language Teaching Research, **19** (2) 129 –132

[5] Brauer, M 2000 Journal of Experimental Social Psychology, **36**, 410–423.

[6] Litvin, Y 2016 *International Journal of Performability Engineering*, **12** (4) 305-320

[7] Gupta, P. 2013 *Journal of Engineering Design*, **24** (7), 499–523

[8] Takemoto, Y 2016 Proc IMechE Part O: *J Risk and Reliability*, **230** (1), 67–74.

[9] Owhor, SC 2015 *American Journal of Engineering Research* (AJER)

[10] Bamrungsetthapong, W 2014 *International Journal of General Systems*.

[11] Ebeling, C. E. 1997. *Introduction to Reliability and Maintainability Engineering*. (London: McGraw-Hill.

[12] Jardine, A.K.S. 2006. *Maintenance, Replacement and Reliability*. (Boca Raton: Taylor & Francis group.

[13] Kumar, S. 2010. *Spare Parts Management – An IT Automation Perspective*. Inggris: Infosys Technologies Limited.

[14] Lechlade, L. 2001. *An Introduction to Reliability Centered-Spares*. Information Science Consultants Ltd.

[15] Louit, D., and Pascual, R. 2006. *Optimization Models for Critical Spare Parts Inventories - A Reliability Approach*. Chile: Pontificia Universidad Católica de Chile

[16] Meilani, D., Kamil, I., and Satria, A. 2008. Analisis Reliability Centered Maintenance (RCM) dan Reliability Centered Spares (RCS) pada Unit Rawmill Pabrik Indarung IV PT. Semen Padang. Universitas Andalas, Teknik Industri.

[17] Moubray, J. 1991. *Reliability Centered Maintenance II*. Oxford: Butterworth Heinemann, Ltd.

IOP Conf. Series: Materials Science and Engineering **523** (2019) 012080 doi:10.1088/1757-899X/523/1/012080

Mechanical properties of glass fiber reinforced polyester resin for use as the wall of the Acehnese boat '*Thep-Thep*'

N Nazaruddin[1,*], A Akram[2], I Hasanuddin[2], M Iqbal[2], R Kurniawan[2], R Putra[3]
[1] Department of Informatics, Faculty of Mathematics and Natural Sciences, University of Syiah Kuala, Banda Aceh, Indonesia.
[2] Department of Mechanical and Industrial Engineering, Faculty of Engineering, University of Syiah Kuala Banda Aceh, Indonesia.
[3] Department of Civil Engineering, Faculty of Engineering, University of Syiah Kuala, Banda Aceh, Indonesia.

E-mail: anzaro@unsyiah.ac.id

Abstract. Fiber composites are very commonly used for engineering applications nowadays. The product has begun to enter the field of boat and ship manufacturing. An Acehnese traditional boat known as *thep-thep,* which has a capacity of 3 GT, is a means of transportation used in Aceh waters. This study focuses on the mechanical properties of the material used for the structure and walls of the *thep-thep*. Specimens are made from materials with E-glass fiber reinforced polyester resin. The hand lay-up method is used in the process of making the specimens, applied with a pressure of 25 kg/cm^2 for 24 hours. This study aimed to obtain higher mechanical properties using composites of layers of chopped strand mat (CSM) fibers and woven roving mat (WRM) fibers with a ratio of 1:0, 1:1, and 2:1, respectively. Tensile testing is done using the ASTM D3039-00 standard. From the tensile test data obtained from several different composite ratios, the highest composite is found at around 49.2 MPa. An increase in the number of fiber layers makes the fiber composites more reliable to be the raw material in the making of the Acehnese boat *thep-thep*.

1. Introduction

The Acehnese traditional boat, known as the thep-thep, is an essential means in Acehnese fishing communities to do activities in the waters, especially in the ocean. Most of the Acehnese fishermen who live in the coastal area still use the Acehnese traditional boat known as Jalo Kayoh [1], but those who depend on their sustenance in the ocean prefer to use the boat thep-thep to catch fish, to travel between islands, and to transport, deliver, and pick up goods from large trawlers that are fishing in the middle of the sea to be delivered to the port. The many uses provided by the boat has made the thep-thep the most widely used boat in almost all Aceh waters. In this study, the thep-thep will be made measuring at 7 meters long, 1.4 meters wide and 1.0 meters high, with a carrying capacity of 3 Gross Tons (GT).

Based on the testing of the mechanical properties of E-glass fiber reinforced epoxy resin in tensile and flexural testing, it was found that the strength will increase with the increasing of fiber weight fraction [3]. This increase will be followed by an increase in the fragility value which corresponds to

* Corresponding author: anzaro@unsyiah.ac.id.

the percentage increase in weight and a decrease in collision energy [4]. Then, the E-glass fiber will increase in mechanical strength when combined with polymer resins such as epoxy, polyester, vinyl ester to increase the threat value to brittle fracture [5]. However, this can be reduced by chemically optimizing some parts of the composite to avoid damage caused by cracks [6].

Fiber composites are widely used in structural construction, which ranges from low strength construction to high strength construction. The advantage of this material is that it is lighter, elastic, corrosion resistant, strong, and easily fabricated. This composite material consists of two parts, the strengthener or filler, and the matrix or binder. The quality of composite materials is strongly influenced by the orientation of fibers and matrices. This study only examined the mechanical properties of E-glass composite composites consisting of a base layer of chopped strand mat (CSM) single fiber composite, woven roving mat (WRM) and a combination of CSM and WRM. The test that is done is tensile testing. The specimens use the D3039-00 standard for the use of fiber orientations 0^0, 90^0, and 0^0.

Based on a literature study, it was found that fiber composites are very effective to be used for a variety of diverse products in various fields, namely automotive, aerospace, and structures and products used in the marine field. To create high-capacity composite products that use artificial fibers, this research is done on the mechanical properties of composite properties reinforced by polyester resin. The composite will be tested for tensile strength to find out the ideal strength for the initial use of the structure and walls of the Acehnese traditional boat *thep-thep*. The purpose of this study is to obtain the basic data of the strength of mechanical properties as the basis of production of the *thep-thep* using synthetic materials.

2. Experimental Details

E-glass fiber in this study was prepared for as the filler reinforcement on the composites. Fiber reinforced composites are formed based on differences between layers. In the preparation of a specimen using the hand lay-up method, it requires fiber which is calculated based on the percentage of weight added to the resin and stirred evenly.

2.1. Materials and fabrication

E-glass fiber is used as a reinforcement of composites. Glass fiber in this study has two types of fiber, the chopped strand mat (CSM) type with a weight of 300 g/m^2 and the woven roving mat (WRM) type with a weight of 300 g/m^2. Meanwhile, the matrix used is an unsaturated polyester resin. The properties of both the E-glass fibers, CSM and WRM, and unsaturated polyester resin can be seen in Table 1. On the table, the composites displayed to follow the form as shown in the flowchart in Figure 1. In the picture, CSM fibers and WRM fibers are made single and layered. The ratio of the E-glass layer is set to 1:0, 1:1, and 2:1. The hand lay-up method is used for making test specimens. The resin is mixed with Mepoxe Peroxide hardener at a ratio of 100: 0.7, then the ingredients are pressed slowly and then left for 24 hours at room temperature. The mechanical properties of the fractions of Glass Fiber Reinforced Polymers (GFRP) can be seen in Table 1.

Table 1. Mechanical properties of jute fiber, glass fiber, and unsaturated polyester resin

Components	Tensile Strength, MPa	Young's Modulus, GPa	Elongation, %	Density, g/cm^2	C/S Area of Single Yarn, mm^2
Glass Fiber	919	113	15	2.54	0.473
Polyester resin	42	30	2.1	1.24	-

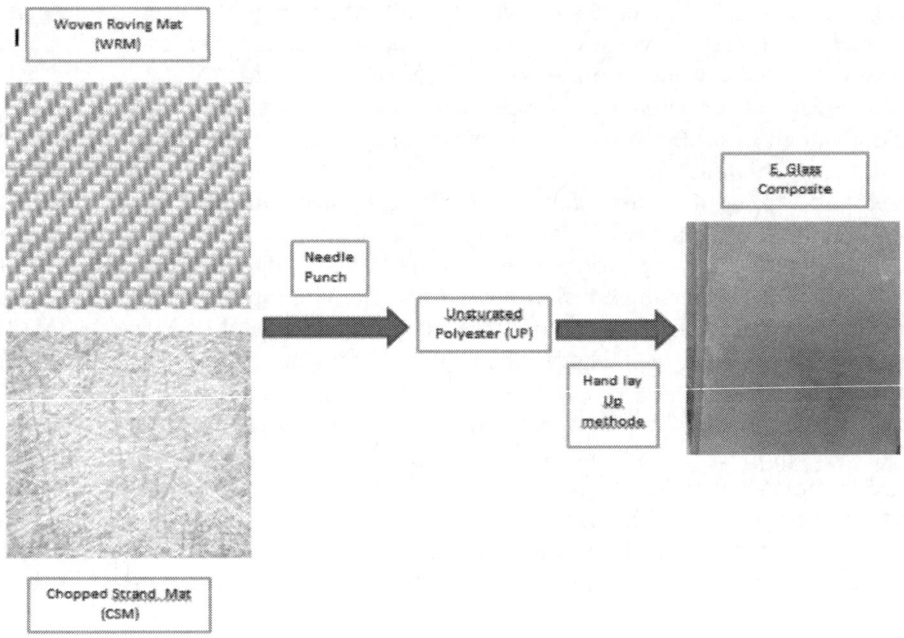

Figure 1. Process flow chart used in the fabrication of different composites using hand lay-up method.

2.2. Measurements

Form specimens from tensile testing are formed by making molds; then the pressure is applied at 25 kg/cm2 for 24 hours. The tensile test is carried out using the D 3039-00 standard. This test is done using the Universal Testing Machine (UTM) machine. The final test specimen consists of three replications to be used in each test.

Figure 2. Schematic representation of composite 1.

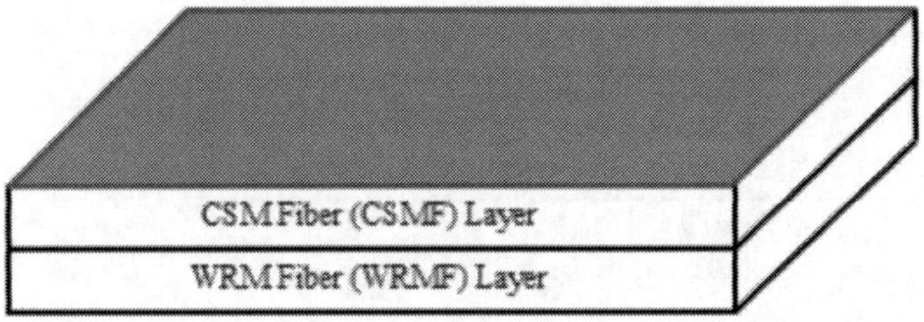

Figure 3. Schematic representation of composite 2.

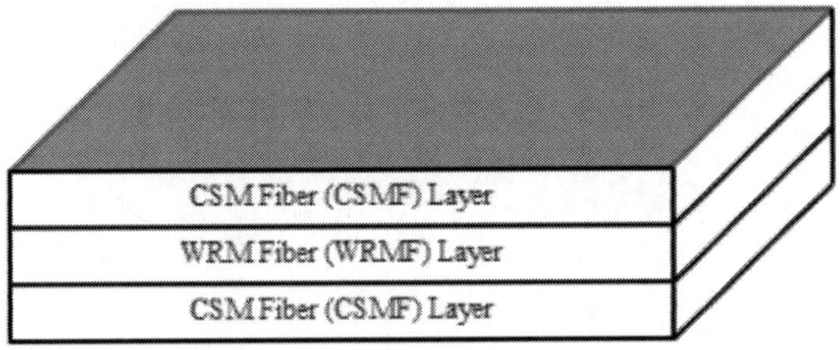

Figure 4. Schematic representation of composite 3.

A pure layer of woven roving mat (WRM) fiber has a thickness of 0.6 mm, while a layer 300 gram/m2 chopped strand mat (CSM) fiber has a thickness of 0.33 mm. So, according to ASTM standards, lamination thickness varies depending on the number of layers. This combination places the WRM layer between the CSM layers in the 3-layer laminate.

3. Results and Discussion

Fiber composites using E-glass fibers is the goal of the experiment in the form of tensile testing. The results of the test are analyzed, then compared with other results to get a conclusion on fiber composites that score better than the rest in all test results. This score will be useful for completing the data planned to develop wood replacement materials for the manufacturing of the structure and walls for the Acehnese traditional boat *thep-thep*.

Some mechanical properties displayed by different composite test specimens are shown in Figure 5. A linear increase in the tensile stress value against the deflection at the initial stage causes the voltage value to gradually slow down to the point of failure. In CSM fibers, it can be observed that the tensile strength can reach 7,385 MPa with lower bending deflection. With the presence of WRM fibers in WRM/CSM-t (t: the layer located on the surface) and CSM/WRM/CSM-t, the two composites show similar maximum tensile strengths. So, it can be concluded that CSM/WRM/CSM-t produces maximum tensile strength higher than the maximum tensile strength of CSM-t. When used as a reference for the fiber composite, a similar situation can be obtained through the results of tensile strength higher than CSM/WRM/CSM-t against WRM/CSM-t composites.

Furthermore, the data shown in the graph in Figure 5 illustrates the average value of tensile stress on the strain that occurs based on the difference in the composite resulting in different variations of values. As shown in Figure 5, the tensile strength obtained for CSM is around 7,385 MPa, this result is lower than the tensile strength value of WRM/CSM-t composite which is 34,462 MPa and even lower than CSM/WRM/CSM-t which is 49,231 MPa.

In short, CSM/WRM/CSM-t showed a significant increase in tensile stress value, higher than the WRM/CSM-t and CSM samples. This indicates that an increase in the fiber layer increases the tensile strength value of the sample being tested.

Figure 6 shows that a given load has a direct effect on the increase in the length of the specimen. The results of the increase in length can be observed significantly in the sample that consists of more layers. The CSM sample produced an increase in the length of 2.99 mm at 120 Kgf of the load. This was lower than the WRM/CSM sample, which experienced an increase in the length of 7.395 at a load of 560 Kgf and a CSM/WRM/CSM sample, which experienced an increase in the length of 8.41 mm at a load of 800 Kgf.

From all the data observed, an analysis was made that the specimen CSM, which was a specimen consisting of 300 gram/m^2 chopped strand mat single fiber, had the lowest length increase value. However, when woven roving mat fiber is added to the layer, the length increase value went up. This is supported by the WRM/CSM specimen data which resulted in higher tensile strength after the addition of another layer.

IOP Conf. Series: Materials Science and Engineering **523** (2019) 012080 doi:10.1088/1757-899X/523/1/012080

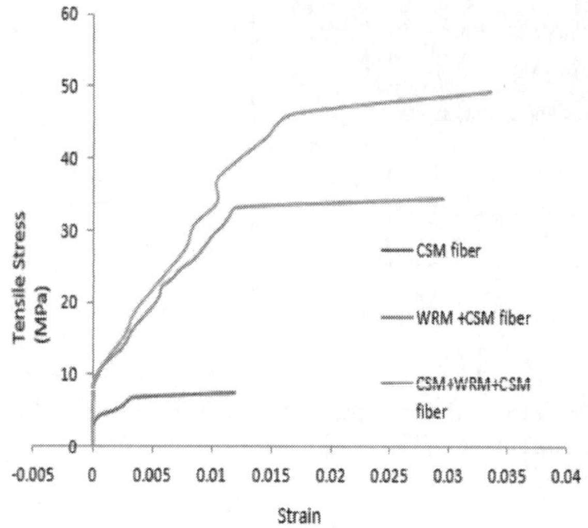

Figure 5. Typical tensile stress curves of different composites.

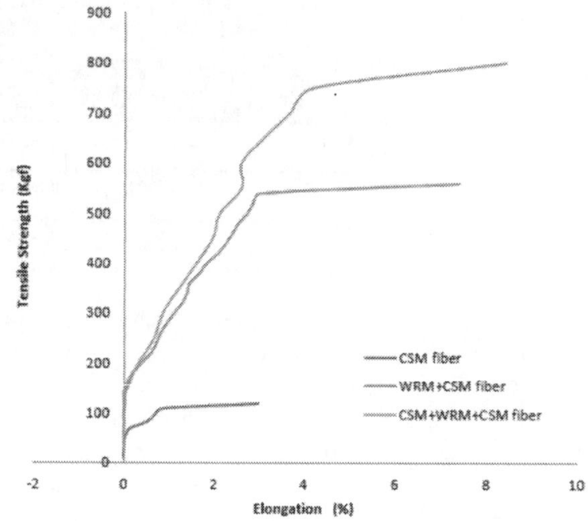

Figure 6. Typical Tensile Strength curves of different composites.

4. Conclusion

In this study, composite reinforced E-glass fibers chopped strand mat type and woven roving mat type produced a composite with a flexural effect that would be used for the manufacturing of frame structures and walls of the Acehnese boat *thep-thep*. There are also points of test results concluding:

1. The tensile strength of synthetic composites shows an increase based on the addition of layers.
2. CSM manifests the lowest length increase among the tested samples from different composites but also shows a tensile strength that is lower than CSM/WRM/CSM synthetic composite specimen.
3. CSM/WRM/CSM specimens show higher tensile stress and have better toughness values than CSM and WRM/CSM specimens.
4. The addition of one CSM layer to the WRM caused the sample to have a higher elasticity. This was followed by an increase in the length increase of the sample layered with CSM.

5. Acknowledgments

Thank you to University of Syiah Kuala (UNSYIAH) for the funding of this research.

6. References

[1] Akram, A, Hasanuddin, I, Nazaruddin, N, Syahril, S, Zulfan, Z and Ahmad, N 2017 *IOP Materials Science and Engineering*, **352**.

[2] BPS (Badan Statistik Provinsi) Aceh, 2015 *Aceh in Figures*, BPS Provinsi Aceh.

[3] Devendra, K and Rangaswamy, T 2012 *International Journal of Computational Engineering Research* **2** (5).

[4] Singh, S, Kumar, P, and Jain, SK 2013 *Adv. Mat. Lett.* **4**(7), 567-572.

[5] Mobasher, B, Kingsbury, D, Montesinos, J and Gorur, RS 2003 *IEEE Transactions on Power Delivery*, **18** (3).

[6] Lucas, S, Kumosa, M.S and rmentrout, DL 2005 *IEEE Transactions on Power Delivery*, **20** (4).

IOP Conf. Series: Materials Science and Engineering **523** (2019) 012081 doi:10.1088/1757-899X/523/1/012081

Analysis of transport workers' postures in the loading process of manual material handling activities by using the photogrammetric method

I Hasanuddin[1,*], R Fahrizal[1], D Asmadi[1]
[1] Department of Mechanical and Industrial Engineering, Faculty of Engineering, University of Syiah Kuala, Darussalam, Banda Aceh 23111, INDONESIA

E-mail: iskandarhasanuddin@unsyiah.ac.id

Abstract. This research was conducted on the loading process of manual material handling of cup mineral water activity at PT. PA, Aceh Besar, Indonesia. With samples of 12 transport workers, the observed loading activity includes the position of lifting, walking, and loading. The loading process is done manually from the warehouse into the truck. This work is done continuously by a transport worker with each transported load reaching 48 kilograms. In a one-time loading process, a transport worker performs as many as 33 trips and the timing for each transport loading process is 20 seconds. With the work done continuously and the loads transported are weighty; therefore there are risks of experiencing musculoskeletal disorders by the transport workers. Thus, questionnaires, interviews, and observations were used as instruments to obtain data. Calculations were further conducted to analyze the postures of transport workers by using the Photogrammetric Method. The results show that the moment force calculation at L5/S1 is greatest in the position of lifting the load at 5.756,81 N; meanwhile, the NIOSH provision is 3,400 N for a normal lift limit. Hence, the solutions proposed to this problem are to reduce the load to 24 kilograms/loading and the utilization of the trolley.

1. Introduction

Human activities in daily life cannot be separated from lifting activities. However, without being realized by humans themselves, there are several ways to lift an object that is not accurate or effective. The way a person performs work activities is called work posture. An incorrect work posture that is carried out daily and continuously will certainly result in an injury that can endanger a person's safety and health. Risks typically to occur in work posture are the Musculoskeletal Disorders (hereafter, MSDs). MSDs are injuries to muscles, nerves, tendons, ligaments, joints, cartilage, and or spinal discs. MSDs are usually the results of any brief or severe incident (such as slip, travel, or fall) that can cause occasional or chronic development [10].

There are several factors why human power is still used in the industrial world. Movement flexibility is a strong reason for the use of human power, especially in manual material handling. Material manual handling activities, which consist of rising, lowering, encouraging, attracting and carrying, are the main causes of several complaints to manual transport workers. Accordingly,

* Corresponding author: iskandarhasanuddin@unsyiah.ac.id.

transport workers most often feel back pain. Factors that affect pain in the back are due to the direction of the load being lifted and the frequency of transfer activity [13].

There are some studies in Indonesia that have focused on MSDs of transport workers. Among them is the material handling manual of rice transport workers [12]. The analysis was done by using the biomechanical methods to identify the risk of spinal cord injury on the rice transport workers. The results revealed that these workers were at risk of causing spinal injury because the value of Lifting Index (LI) is 5.52 when lifting a load of 75 kilograms. This exceeds the legal limit of transport [12].

Another research focused on improving the work posture of cassava transport workers to reduce musculoskeletal complaints. The analysis was done using the OWAS method approach. The results showed that the work postures identified had the potential of going through injuries. The results of the analysis with OWAS was in Category 3, which meant that it needed improvement immediately because the current posture was dangerous. Meanwhile, Category 2 meant that it needed improvement in the future. Improvements were then given and made by the transport workers. They were changes in the workers' postures, which included to tilt the back while working and to stand upright when cutting cassava, draining and cleaning the ends. Thus, after repairs, the results of Category 1 were obtained; this means that there are no longer problems with the musculoskeletal system [18].

Further research was done on ergonomics evaluation of material handling manual activities in the bottles production section. Observations were made to all workers in charge of handling the filled bottles at a company in Bangkalan, Indonesia. The results showed that the evaluation with the revised NIOSH appointment equation in 1991 obtained the lifting index (LI) that is greater than 3. To solve this problem, some modifications were made to the multiple factors of the try method and failed in calculating the Recommended Weight Limit (RWL), but the results also do not provide an LI value below 1. Therefore, the recommendation given is to add a bench and a rolling track at the end of the conveyor belt [20].

The cases described above are among the problems often encountered in material handling manual activities. They often happen due to ignorance and improper work posture that go beyond the recommended lift limits. Most transport workers lack the knowledge on this issue, and some employers even ignore the risks of their workers. When risk and health become issues for workers, therefore further research is essential so that solutions can be provided to these workers. Consequently, the researchers of this study are interested in studying and analyzing the loading and unloading activities done by transport workers at a local distributor company in Banda Aceh. By using the photogrammetric method, it is expected that the analysis is more precise with fewer constraints.

The photogrammetric method is a technique used to analyze an activity visually. This method records an on-going activity and then analyses it based on the equation of the photogrammetric formula to determine the coordinates of the object under study [2]. Based on research using this method, it has been proven accurate in analyzing three-dimensional, and two-dimensional reconstruction needs. The research assessed the effect of camera parameters on 3D object reconstruction by using digital photogrammetric techniques. It revealed that the benefits and results of reconstruction showed a maximum difference of 11 mm or 0.0035 times the size of the object under study. In other words, this digital photogrammetric technique is precise to use for reconstruction needs of both three-dimensional and two-dimensional reconstructions [11].

2. Literature Review

2.1. Laborer

Laborer is those who work in individual businesses and are rewarded on a daily or wholesale basis in accordance with the agreement of both parties (i.e., between employers and employees), both verbally and in writing [9].

IOP Conf. Series: Materials Science and Engineering **523** (2019) 012081 doi:10.1088/1757-899X/523/1/012081

2.2. Ergonomics

Ergonomics is known in Greek, that is from the words *ergos* and *nomos* that have the meaning of "work" and "rules or rules," from these two words in a free sense in accordance with its development, ergonomics is a rule or principle adhered to in the work environment [10].

2.3. Body injury due to work errors

In general, there are three types of body injury due to errors in doing work, namely Cumulative Trauma Disorders (CTS), Repetitive Strain Injuries (RSI), and Musculoskeletal Disorders (MSDs). Explanations on these types of body injury are as the following [10].

a. Cumulative Trauma Disorders (CTS)
   Cumulative Trauma Disorders (CTS) is often called repetitive strain injuries. They are defined as disorders of the muscles, tendons, nerves and blood vessels that are caused or distorted by exertion or repetitive movements [10].
b. Repetitive Strain Injuries (RSI)
   Repetitive Strain Injuries (RSI) is a general term used to refer to several discrete conditions that can be attributed to repetitive tasks, the exertion of power, vibration, and mechanical compression that is sustainable [10].
c. Musculoskeletal Disorders (MSDs)
   Musculoskeletal Disorders (MSDs) are injuries to muscles, nerves, tendons, ligaments, joints, cartilage, or spinal discs. MSDs are usually the result of any momentary or acute event (such as slip, travel, or fall), in addition to reflecting a more gradual or chronic development [10].

2.4. Biomechanics

Biomechanics is the science that applies the laws of mechanics to the structure of life, especially the locomotor system of the body (locomotor = activity where all parts of the body move because of their strength and are generally assisted by gravity). Biomechanics studies the forms and kinds of movements on the basic principles of mechanics and analyses them [7].

Measurement of the torque that occurs on the lower spine (L5-S1) can be done using the following formula:

$$\sum TL5, S1 = 0 \quad (1)$$

$$(F_m \text{ x } E) - \sum(Wxd) = 0 \quad (2)$$

where:
Fm = Force on the spinal muscles (L5-S1)
E = The perpendicular distance of the spinal muscles to the Sacrum center 1 (S1)
W = Human body weight (head, arms, torso, and weight of the weight lifted) (n)
d = The horizontal distance between the L5-S1 center to the center of the body part (head, arms, and torso) and the distance of the L5-S1 to the center of the body mass (cm)

2.5. Segment body weights

Body weight point is the point where the body's weight or body parts work or points that represent a person's body [3].

2.6. Lumbar 5 sacrum 1 (L5/S1)

The lower backbone (Lumbar) is the most critical part of injury. Back pain occurs because the plate (disc) which is cartilage located between the spine (especially the lower part) experiences a force that exceeds that of the plate. As a result, damage to this plate is a permanent pain in the spine. Factors that cause injury include posture or work position that is uncomfortable (incorrect) during work, the weight of the load lifted, and the frequency of transfer activity [6].

2.7. Nordic Body Map (NBM)
Nordic body map is one of the most frequently used questionnaires in the industry to collect complaints on the cause of musculoskeletal disorders. A questionnaire on NBM is equipped with 27 questions related to complaints that are felt in body parts. These include the neck, shoulders, upper back, elbows, lower back, wrists, waist, knees, heels, and feet. This questionnaire is also equipped with pictures of body parts to ease respondents in comprehending the questions asked [8].

2.8. Body posture and work posture
Body posture is a form and structure in the human body that can be observed. In carrying out daily activities, humans have work postures that are formed due to movements produced by themselves. Some movements can even pose risks to the human's health if they are not controlled and regulated [10].

2.9. Manual Material Handling (MMH)
Manual Material Handling (Figure 1) are all the work of lifting loads or objects carried out by workers without other assistive devices to move them from one place to another [20].

Figure 1. Manual Material Handling (MMH).

2.10. Photogrammetric method
Photogrammetric is a technique used to determine the coordinates of a point in a three-dimensional space derived from two or more two-dimensional images. This method is often used in biomechanics [2].

3. Research Method

3.1. Place, time and samples
This research was conducted from March 5 to April 5, 2018. It focused on the loading process of manual material handling of cup mineral water cupping activity at PT PA, Aceh Besar, Indonesia.

With samples of 12 transport workers, the observed loading activity included the position of lifting, walking, and loading. The loading process was done manually from the warehouse into the truck. This work was done continuously by a transport worker with each transported loads reaching 48 kilograms in one loading process. In a one-time loading process, a transport worker performed as many as 33 trips, and the timing for each transport was 20 seconds. With the work done continuously and the loads transported were weighty; therefore there were risks of experiencing musculoskeletal disorders by the transport workers. Complaints by the workers had been given, and they included pain in the shoulders and hips.

Thus, questionnaires, interviews, and observations were used as instruments to obtain data. Calculations were further conducted to analyze the workers' postures during loading by using the Photogrammetric Method.

3.2. Research stages

The research was carried out based on the stages starting from the identification of the problem to the stages of data analysis that derived results as output of this study. The following are the stages of research, presented in Figure 2 the form of a flow chart:

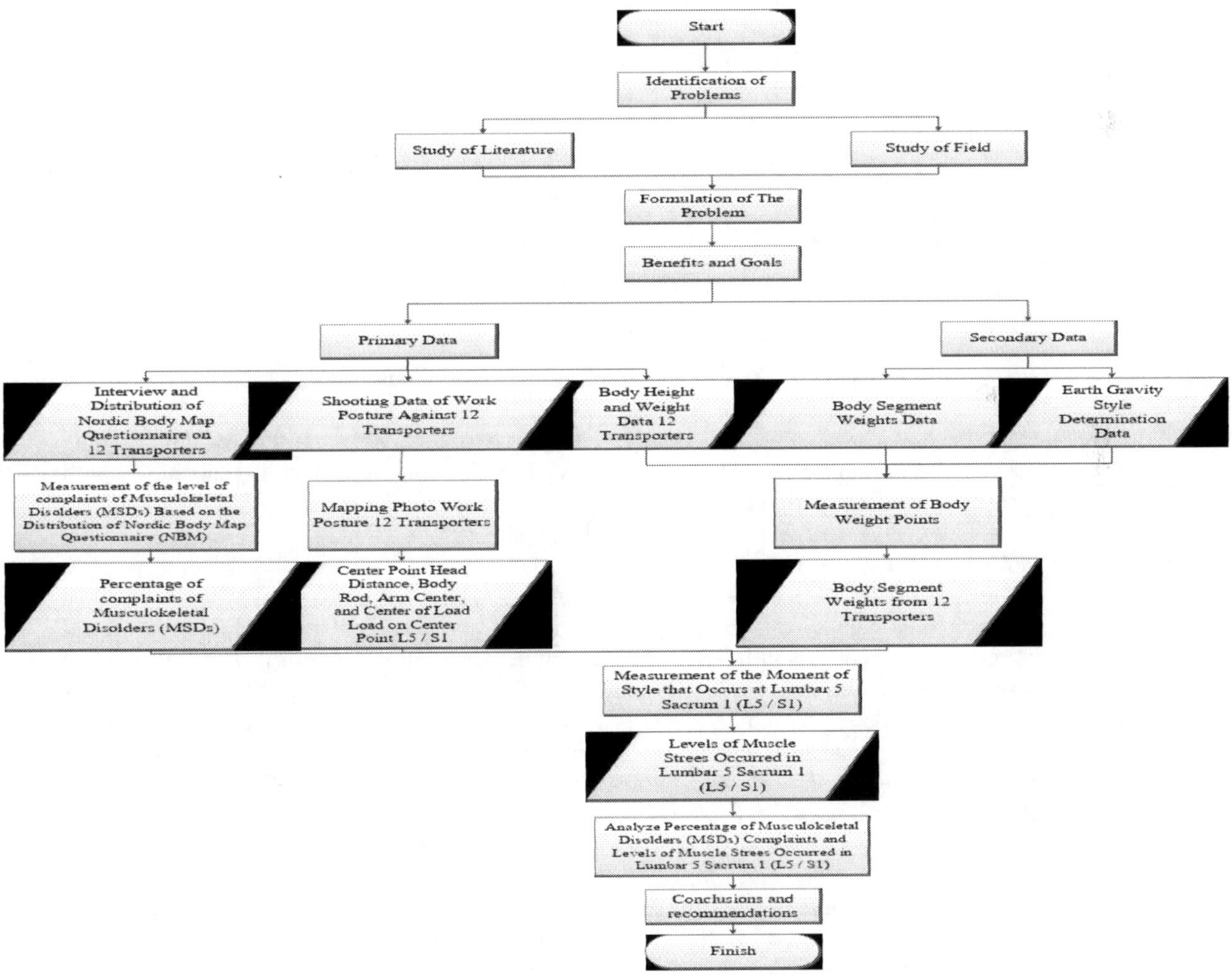

Figure 2. Flowchart of research stages.

4. Results and Discussion

4.1. Nordic Body Map (NBM) Questionnaire

Recapitulation stages were carried out based on the results of distributing the Nordic Body Map questionnaire to 12 transport workers. The goal was to find out the percentage of pain complaints or MSDs. The results showed that the biggest pain complaint and most felt by these workers were at the waist (92%) and back (83%). About 8% of the workers ticked to be "in severe pain" on these body parts. The percentage of complaints of MSDs can be seen in Figure 3.

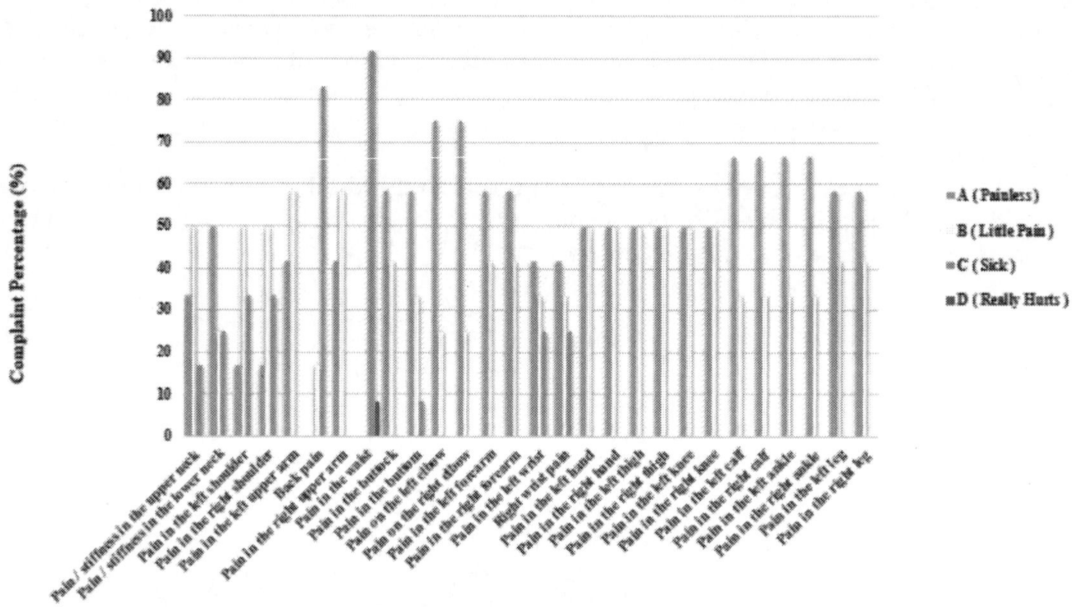

Figure 3. Percentage graph of pain complaints.

In conclusion, the pain complaints felt by the transport workers were caused by the carrying continuously heavy loads of 48 kg/each. This caused severe back pain. Other factors that caused pain are their age factor and working periods.

4.2. Torque on L5-S1

Calculations of torque on L5-S1 included several positions. Namely, the position when they were ready to lift a load, walked with a load, and put down a load. Moment of style obtained in order to find out the level of muscle stress occurs in L5-S1. The magnitude of the torque that occurs in L5/S1 when a worker was ready to lift weights is at an average of 3,956.56 N, with a difference of 556.56 N compared to the provision determined by NIOSH at 3,400 N. Whereas the torque when carrying loads had the average of almost equal to 3.960.66 N. In addition, the moment of force at the position of putting a load had an average of 2766.38 N. The results of the torque calculation for each position can be seen in Table 1.

4.3. Comparison of torque for different lifting weights

Recapitulation of torque comparisons for lifting a load, walking while carrying a load and placing a load with the weight of the loads at 24 kg, 36 kg, 48 kg and 60 kg can be seen in Table 2.

Based on Table 2, it can be concluded that the recommended load limit for material handling manual activities at PT PA is 24 kg/load. This is by taking into account the overall torque of 6766.00 N. Based on this consideration, the load limit of 24 kg/load can reduce the risk of MSDs and can minimize the risks of injury. However, regular monitoring for material handling manual activities to prevent risky things is still necessary.

Table 1. Torque in the all position.

No	Name	Torque in the lifting position		Torque in the carrying position		Torque in the placing position	
		Body Weight (Kg)	Torque (N)	Body Weight (Kg)	Torque (Kg)	Body Weight (Kg)	Torque (Kg)
1	Transporter 1	58	4377,92	58	2985,20	58	2120,76
2	Transporter 2	47	3756,33	47	3657,76	47	2711,34
3	Transporter 3	55	5330,93	55	3266,59	55	1796,81
4	Transporter 4	68	3318,52	68	3439,14	68	3044,89
5	Transporter 5	53	2882,74	53	4003,37	53	2707,63
6	Transporter 6	55	3081,04	55	4008,01	55	3252,45
7	Transporter 7	51	3102,02	51	3987,94	51	3338,30
8	Transporter 8	55	4998,01	55	3706,94	55	2404,22
9	Transporter 9	63	4283,82	63	4565,90	63	2323,93
10	Transporter 10	55	3422,67	55	4585,81	55	2511,56
11	Transporter 11	53	5756,81	53	4488,04	53	4090,79
12	Transporter 12	52	3167,93	52	4833,17	52	2893,87
	Average		3956,56		3960,66		2766,38

Table 2. Comparison of torque for different lifting weights.

Position	Torque (N)			
	24 kg	36 kg	48 kg	60 kg
Lifting	3183,63	3952,73	4378,55	2736,70
Carrying	2035,94	3534,02	2949,87	3066,06
Placing	1546,43	2145,06	2139,33	2635,70
Total Torque	6766,00	9631,81	9467,75	8438,46

4.4. Analysis

4.4.1. Percentage of Nordic Body Map questionnaire (NBM)

The percentage of MSDs complaints level from the questionnaires revealed that the waist was the vulnerable part that can cause serious injury. The cause of the pain was due to the squeezing of the disk (plate) which is a bone of bone that lies between the spine and experiences a force that exceeds that of the plate.

For the pain complaints, most workers felt “sick” at the waist (92%) and back (83%) with “very sick.” These pains caused their risks of MSDs injury that can endanger their health and life. Injuries experienced were not immediately felt by the workers when they conducted material handling manual activities but were felt when they rest at night.

Other factors that affected the level of complaints of MSDs are the age and length of employment of the transport workers. Their age was at vulnerable stages of between 19-57 years old. Their average working periods were 4 months for new employees of transport workers and 4.6 years for old employees of transport workers. Based on the analysis, transport workers who experienced a lot of pain were those aged 36-57 years old with working periods of 2 years. Whereas transport workers between 21-29 years old and working periods between 1.5 - 4.6 years experienced a little more pain. For transport workers aged 19 - 27 years old with working periods of 2-8 months, they claimed to feel no pain but sometimes did felt the symptoms with little pain.

4.4.2. Torque on L5-S1
Based on the calculations of the torque, it was found that the position at risk of causing injury in the L5/S1 section from manual material handling activities was the position of ready to lift a load (bending the body) and walking while carrying a load weighing at 48 kg in one loading process. Whereas for the position of placing a load was still in the safe category. This is because the average moment of force is still acceptable. The position of placing a load did not cause injury to the L5-S1 because when a worker is ready to put on a load, the work posture is close to the center of gravity (L5-S1). Therefore, L5-S1 does not experience excessive pressure when manual material handling activities take place. The magnitude of the torque that occurred in L5/S1 when a worker was ready to lift a load was at an average of 3,956.56 N, where the difference is at 556.56 N to the provision determined by NIOSH at 3,400 N. Meanwhile, the torque when carrying loads had an average equal to 3.960.66 N. In addition, the moment of torque at the position of putting the load had an average of 2766.38 N.

4.5. Proposed improvement
4.5.1. Recommended load limits for lifting
The calculation of the torque on L5/S1 for transport workers by taking into account the load of 24 kg, 36 kg, 48 kg, and 60 kg, it can be concluded that the load limit is still in the safe category. It is still below the standard of normal lift force determined by NIOSH (National of Occupational Safety and Health) (at 3,400 N on L5/S1), which is to lift 24 kg (2 boxes of cardboard) in one loading. This has been clarified earlier in which the moment of force when lifting weights, carrying weights and putting a load are 3,183.63 N for lifting loads, 2,035.94 N when carrying loads and 1,546.43 N for loading positions.

Whereas for transported loads reaching 60 kg, the difference is also not much from the load carried at 24 kg, that is with the torque lifting position at 2736.70 N for walking position, 3066.06 N for carrying position, and 2635.7 N for putting the position of a load. The choice of the recommended lift limit is 24 kg compared to 60 kg because of the total moment of the overall obtained force, the load weight of 24 kg has a total torque of 6766 N.

The load weight of 60 kg has a total torque of 8438.46 N. The magnitude of the load weighing 60 kg can cause injury to certain parts of the body but not to the L5-S1 due to excessive pressure if the transport worker with good and straight standing bodies. For loads that are at 36 kg and 48 kg, the moment value is greater compared to the transported load of 24 kg and has a higher risk of MSDs.

4.5.2. Use of assistive devices in material handling manual activities
The use of trolleys would not be optimal if there are uneven floor surfaces, and this was what was found on site. This situation made it difficult for loading operations when loading goods into the truck. The time needed by three transport workers to lift as many as 12 boxes was 1 minute. When using a trolley, 12 boxes could be transported with a time of 1.5 minutes; this is a difference of 50 seconds with manual lifting. Therefore, the use of trolleys can be considered with the condition that the floor surface of the warehouse must be flat so that the time taken for the use of trolleys can be more efficient or faster than loading items done manually by transport workers.

5. Conclusions
Based on the research results, some conclusions are drawn:

The material handling manual activities that take place showed that there was no standardization for the load transported by workers. Transport workers could transport 48 kg - 60 kg in one loading is thus depended on their physical ability, and not based on the predetermined standards.

The results of data processing on the distribution of Nordic Body Map (NBM) questionnaires to 12 transport workers showed that the highest level of "sick" complaints was at the waist (92%), back (83%), and "very painful" at the waist (8%). Other factors that affected pain complaints were age and years of service. Transport workers who experienced a lot of pain were workers aged 36 - 57 years old with work periods of 2 years.

The calculation of moment force on L5/S1 showed the highest average force moment at 3,956.56 N on weight lifting activity. Whereas for walking activities on carry the burden and put the burden obtained an average of 3,960.66 N and 2,766.38 N. The position of lifting and carrying loads had exceeded the normal lifting load limit of the NIOSH (National Institute for Occupational Safety and Health) standard, which is 3,400 N. The moment of force obtained showed the level of muscle stress in L5/S1.

Therefore, two proposed improvements to be considered are reducing the transport load from 48 kg to 24 kg/loading, and the use of trolley. However, this can only be efficient and optimal if the warehouse fixes their flooring condition to be more flat and smooth.

6. References

[1] Budihardjo I 2002 *Studies of Compressive Forces on L5/S1 during Dynamic Manual Lifting* (Doctoral dissertation) Ames, Iowa State University
[2] Chaffin DB, Gunnar BJ, and Martin BJ 1999 *Occupational Biomechanics 3rd Ed.* (New York: John Wiley and Sons)
[3] Dempster WT and Gaughran GR 1955 *The American Journal of Anatomy* **120** 33–54
[4] Furlanetto TS, Sedrez JA, Candotti, CT and Loss, JF 2016 *World Journal of Orthopaedics* **7**(2) 136-148
[5] Han K, Kwon HJ, Choi TH, Kim JH and Son D *Journal of Cranio-Maxillofacial Surgery* **38**(2) 96–107
[6] Helianty Y, Mona C and Wahyuning CS 2012 *Itenas Rekayasa* **XVI** (1) 57–66
[7] Hidayat RI 2003 *Biomekanika* [Biomechanics] (Bandung: PPS Universitas Pendidikan Indonesia)
[8] Iridiastadi, Herdianto and Yassierli 2016 *Ergonomi Suatu Pengantar Edisi 3* [An Introduction to Ergonomics 3rd Ed.] (Jakarta: PT Remaja Rosdakarya)
[9] Kamus Besar Bahasa Indonesia 2002 *Kamus Besar Bahasa Indonesia Edisi 3* [Indonesian Dictionary 3rd Ed.] (Jakarta: Balai Pustaka)
[10] Kuswana K 2014 *Ergonomi dan K3 Keselamatan dan Kesehatan Kerja* [Ergonomics and Healthy Safety Work] (Bandung: Rosda Karya)
[11] Maherdianta D 2008 *Kaji Pengaruh Parameter Kamera Terhadap Rekontruksi Benda 3D Menggunakan Teknik Digital Photogrammetry Studi Kasus: Rekontruksi Sayap Tengah CN-235* [The Effect of Camera Parameters on 3D Object Reconstruction Using Digital Photogrammetry Technique. A Case Study: CN-235 Middle Wing Reconstruction] (Master thesis) Bandung, Institut Teknologi Bandung
[12] Mas'idah E, Fatmawati W and Ajibta L 2009 Analisa Manual Material Handling (MMH) dengan menggunakan metode biomekanika untuk mengidentifikasi resiko cidera tulang belakang (Musculoskeletal Disorder) (Studi Kasus pada Buruh Pengangkat Beras di Pasar Jebor Demak) [Material Handling Manual (MMH) analysis using biomechanical methods to identify the risk of spinal cord injury (Musculoskeletal Disorder) (A case study on workers lifting rice in the Jebor Demak market)] *Sultan Agung* **XLV**(119) 37-56.
[13] Nurmianto E 1996 *Ergonomi, Konsep Dasar Dan Aplikasinya Edisi 1.* [Ergonomics, Basic Concept and Its Application 1st Ed] (Jakarta: Guna Widya).
[14] Nurmianto E and Ciptomulyono U 2014 *IPTEK Journal of Proceedings Series* **1**(1) 589-594
[15] Barnes, RM 1949 *Motion and Time Study 3rd Ed.* (New York: John Wiley and Sons)
[16] Subana MR and Sudrajat 2000 *Statistik Pendidikan* [Educational Statistics] (Bandung: CV Pustaka Setia)
[17] Sugiyono 2013 *Metode Penelitian* [Research Methods] (Bandung: Alfabeta)
[18] Susihono W and Prasetyo W 2012 *Spektrum Industri* **10**(1) 69-91

[19] Tarwaka, Solichul HB and Sudiajeng L 2004 *Ergonomi (Untuk Keselamatan, Kesehatan Kerja dan Produktivitas) Edisi 1)* [Ergonomics (For Healthy Safety Work and Productivity 1st Ed.] (Surakarta: Uniba Press)

[20] Umami, Khoirul M, Hadi ADR and Agustina F 2014 *Jurnal Rekayasa Sistem Industri* **3**(2) 65-70

[21] Wahyudi MA, Dania WAP, Silalahi LRL 2015 *Agriculture and Agricultural Science Procedia* **3** 195-199

[22] Wignjosoebroto S 2006 *Ergonomi Studi Gerak dan Waktu: Teknik Analisis Untuk Peningkatan Produktivitas Kerja Edisi 4* [Ergonomics of Motion and Time Studies: Analysis Techniques for Increasing Work Productivity 4th Ed] (Surabaya: Guna Widya).

[23] Yanto and Ngaliman B 2017 *Ergonomi (Dasar-Dasar Studi Waktu Dan Gerakan Untuk Analisis dan Perbaikan Sistem Kerja* [Ergonomics (Basics of Time and Movement Studies for Analysis and Improvement of Work Systems] (Yogyakarta: CV Andi Offset)

IOP Conf. Series: Materials Science and Engineering **523** (2019) 012082 doi:10.1088/1757-899X/523/1/012082

Grain-size characteristics of Aceh's coastal deposits

B Setiawan[1,*], S Antonie[2], B G Adhiperdana[3]
[1] Program Study of Geological Engineering, Faculty of Engineering, Syiah Kuala University, 23111 Banda Aceh, Indonesia
[2] Faculty of Sciences, King Abdulaziz University, 21589 Jeddah, Saudi Arabia
[3] Faculty of Engineering Geology, Padjadjaran University, 45363 Jatinangor, Indonesia

E-mail: bambang.setiawan@unsyiah.ac.id

Abstract. This study investigates the grain size characteristics and sediment types of the Aceh's coastal deposits that are primarily based on the sieve analysis. This study is mainly driven by two reasons. Firstly, this study is a preliminary effort to identify coastal sediment in the Aceh coastal areas. Secondly, this information is important to interpret the support mechanism of sediment during transportation and depositional processes. In this study, many sediment deposits from both the east and west coasts of Aceh were collected. Laboratory sieve analysis was carried out to these samples. Then, further analysis of this sieve analysis results is carried out. The analysis reveals the past transport mechanism of this study sand deposits which is river's channel. However, such finding should be investigated further as all the tested sand were taken at the top surface and big main rivers are not indicated at both sand locations.

1. Introduction

Basically, clastic sediment and sedimentary rocks are made up of discrete particles. The texture of sediment refers to the group of properties that describe the individual and bulk characteristics of the particles making up sediment such as grain size, grain shape, grain orientation, porosity, and permeability. These properties collectively make up the texture of sediment or sedimentary rock. Each property can be used to infer something of: the history of sediment, processes that acted during transport, deposition of sediment, and behavior of sediment ([1]; [2]; [3]; [4]; [5]; [6]).

This study investigates the characteristics of clastic sediment of the Aceh's coastal deposits. The investigation is primarily based on the sieve analysis, and the results are presented in this paper. This study is mainly driven by two reasons. Firstly, this study is a preliminary effort to identify coastal sediment in the Aceh coastal area which was produced by ordinary coastal processes such as tidal- and along shore current-related processes or by processes other than tidal- and shore-related processes such as deposits associated with the tsunami event. Understanding this sedimentary deposit is important as Aceh's coastal was affected by the tsunami in 2004 and paleoseismic events may be recognized (see studies by, e.g. [7]; [8]; [9]). Secondly, this information is important to interpret the support mechanism of sediment during transportation and depositional processes [10]. In this study, 10 sediment deposits from both east and west coasts of Banda Aceh were collected. Laboratory sieve analysis was carried out to these samples. Then, an analysis of this sieve analysis is carried out. The analysis reveals the paleohydraulic condition and past transport mechanism of the deposits.

* Corresponding author: bambang.setiawan@unsyiah.ac.id

2. Location

Two separated coastal areas, as shown in Figure 1, were selected to investigate the characteristics of Aceh's coastal deposits. One coastal area is at the western side of Aceh which is Lhoknga area, whereas another one is at the eastern side of Aceh which is Ujung Batee. The Lhoknga area is directly facing the Indian Ocean, whereas the Ujung Batee area is to the Melaka Straits. The Indian Ocean and Melaka Strait have a difference hydraulic current system which may contribute to the deposited sediment.

Morphology of the Lhoknga and the surrounding area is shown in Figure 2 which suggests a wide-open flat to gently sloping ground around the collected samples of this study. The collected samples at the Ujung Batee area are mostly close to the medium to highly sloping ground at the south-west part of the location.

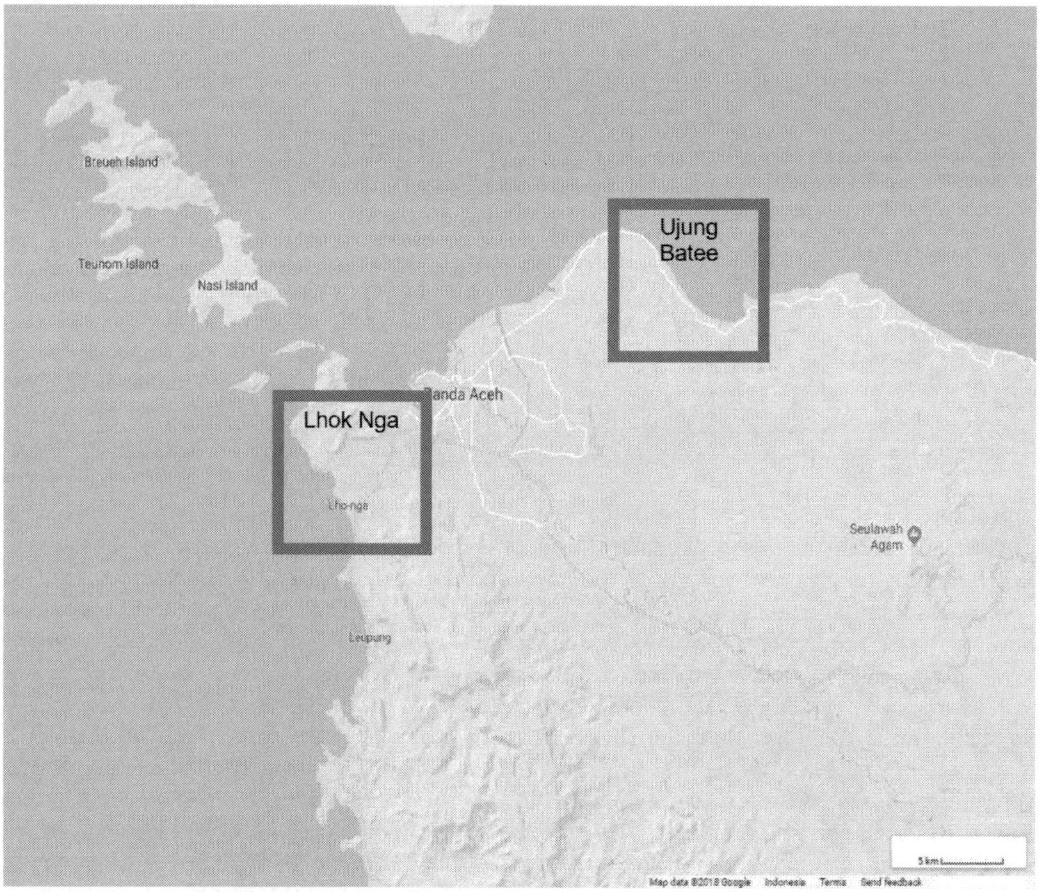

Figure 1. The locality of Lhoknga & Ujung Batee areas

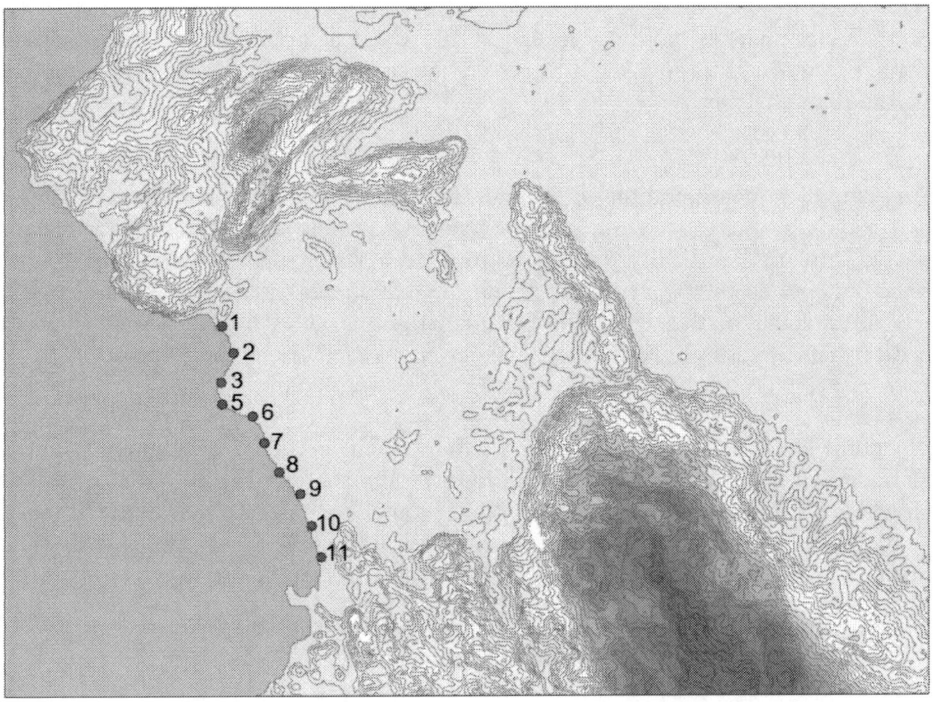

Figure 2. Morphology of the sampling points at Lhoknga area

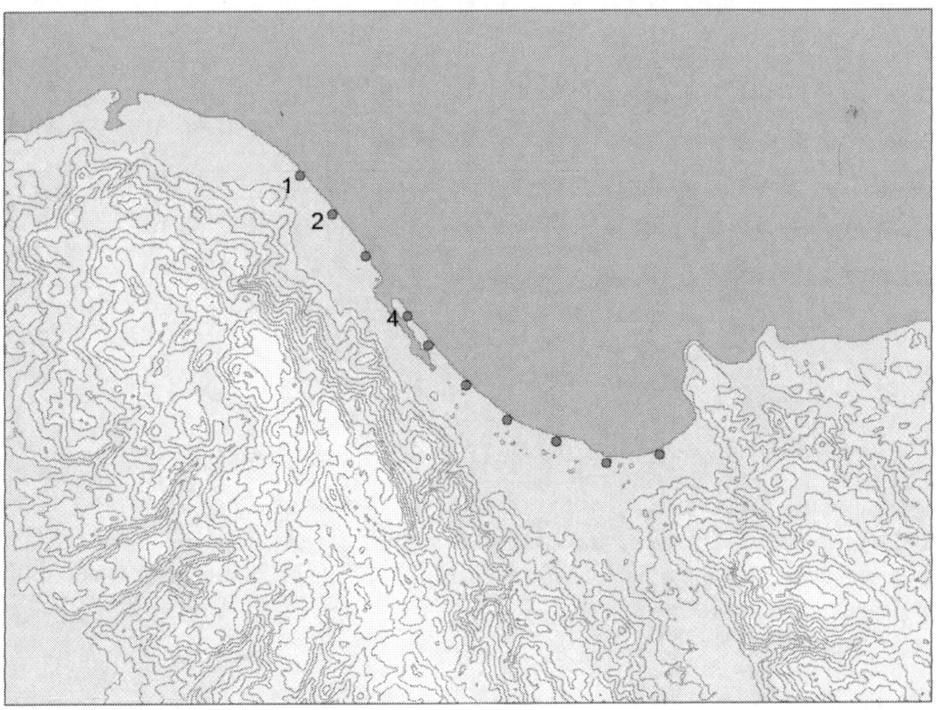

Figure 3. Morphology of the sampling points at Ujung Batee area

3. Methods

Samples for grain size analysis were dry heated at 105°C for at least 24 hours before analysis. Any organics debris (i.e., roots, grass stems, twigs, etc.) in the sample were removed. The samples were not rinsed before the sieve analysis.

3.1. Sieving

Sieve analysis is used to determine the grain size distribution of a sediment sample. This grain size distribution is a bulk property of sediment. After some treatment as mentioned above, a dry loose sample is passed through a vertically stacked set of sieves. The stacked sieves (largest holes on top, smallest on the bottom and follow by a pan) are shaken which can result in any grains that are larger than the sieve holes remain on the screen and the smaller grains pass through it. The distribution of the grain size is determined by weighting the grains collected on each screen are weighed.

3.2. Grade Scales

Grade scales outline limits to a range of grain sizes for a given class (grade) of grain size. The Udden-Wentworth Grade Scale is commonly used by sedimentologists (see Figure 4). This scale uses sets most boundaries to vary by a factor of 2. This scale provides a basis for terminology that describes grain size. e.g., medium sand falls between 0.25 and 0.5 mm. Generally, sedimentologists often use grain size in a unit of Phi (ϕ) which was originally defined as:

$$\phi = -\log_2 d(mm) \tag{1}$$

To make Phi dimensionless, it was later defined as:

$$\phi = -\log_2 \frac{d(mm)}{d_O} \tag{2}$$

where $d_O = 1$ mm. The correlation between ϕ and the millimeter size of soil particle is shown in Figure 4. Phi is the negative of the power to which 2 is raised such that it equals the dimension in millimeters.

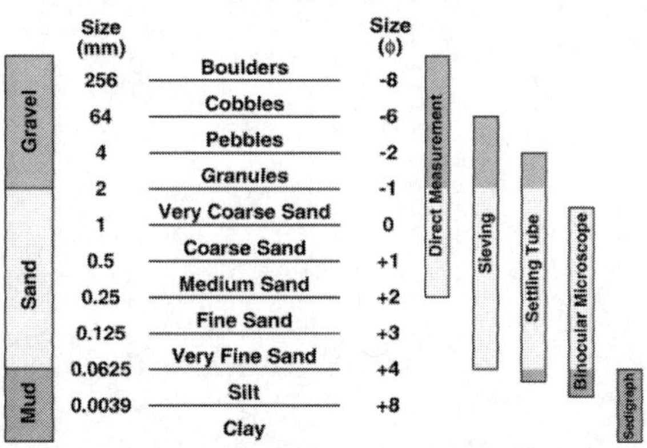

Figure 4. Udden-Wentworth grade scales and measurement techniques

3.3. Describing grain size distributions

To describe the grain size distributions at each tested sample, a distribution curve was developed from the sieve analysis result. This curve is used for further analysis, i.e. various ϕ size, Median of ϕ, Mean of ϕ, standard deviation and skewness (see [11]).

4. Results

Currently, 13 samples (10 Lhok Nga' samples and 3 Ujung Batee's samples) have been sieved. The results of the sieve analysis of this study are summarized in Table 1. The results of each sampling were plotted into a graph from which are all phi values estimated. The summary of all estimated ϕ values is presented in Table 2.

Table 1. Sieve analysis results

No	Sampling Point	Sieve size (mm)								
		2	0.85	0.425	0.25	0.18	0.15	0.106	0.075	Pan
		(gr)	(gr)	(gr)	(gr)	(gr)	(gr)	(gr)	(gr)	(gr)
1	LO 1	0.29	38.78	50.56	69.87	27.2	6.55	2.53	0.1	-
2	LO 2	0.03	8.66	63.76	72.57	32.95	12.3	5.2	0.34	-
3	LO 3	0.93	15.35	25.21	120.4	34.12	0	0	0	-
4	LO 5	4.65	29.78	35.47	33.87	31.35	32.1	27.47	0.77	-
5	LO 6	4.02	34	51.49	51.24	29.35	17.58	9.25	0.25	-
6	LO 7	0.19	24.64	101.43	59.27	9.59	1.66	0.43	0.05	-
7	LO 8	1.83	92.74	70.14	26.41	4.09	1.01	0.9	0.17	-
8	LO 9	21.23	82	42.66	13.63	10.75	7.78	13.14	2.98	0.15
9	LO 10	1.23	14.77	38.97	43.79	38.14	28.92	26.19	2.3	0.3
10	LO 11	0	0.65	1.23	8.86	29.09	42.06	84.85	29.76	2.67
11	UB 1	0.14	3.06	17.08	43.46	44.81	38.53	46.19	3.33	0.18
12	UB 2	1.25	10.99	17.33	46.37	54	34.51	27.84	5.14	0.7
13	UB 4	2.08	28.21	46.28	59.77	32.48	13.2	12.4	1.54	0.06

Table 2. Summary of estimated ϕ from the curves

No	Sampling Point	Estimated ϕ from the curve						
		ϕ_5	ϕ_{16}	ϕ_{25}	ϕ_{50}	ϕ_{75}	ϕ_{84}	ϕ_{95}
1	LO 1	-0.67	-0.25	0.25	1.15	1.8	2.15	2.51
2	LO 2	0	0.4	0.7	1.35	2	2.25	2.75
3	LO 3	-0.35	0.8	1.2	1.5	1.85	2.1	2.25
4	LO 5	-0.75	0	0.5	2	2.75	3.2	3.7
5	LO 6	-0.8	-0.1	0.4	1.25	2.2	2.5	3.2
6	LO 7	-0.35	0.15	0.35	0.75	1.35	1.65	2.1
7	LO 8	-0.85	-0.7	-0.5	0.05	0.8	1.2	1.8
8	LO 9	-1	-0.8	-0.6	0.3	2.3	3.5	8
9	LO 10	-0.25	0.45	0.8	2	2.65	3	3.7
10	LO 11	2	2.35	2.65	3.25	3.8	4	6.2
11	UB 1	0.6	1.25	1.7	2.3	3.05	3.4	3.8
12	UB 2	-0.1	1.05	1.5	2.2	2.8	3.2	3.8
13	UB 4	-0.6	0	0.5	1.4	2.15	2.45	3.5

4.1. Sediment description

Most of the samples at the Lhok Nga are described as Medium SAND, shown in Table 3. Only at LO 7 the sample is very fine SAND. The rest samples at LO 7 and LO8 are described as Coarse to Very Coarse SAND.

IOP Conf. Series: Materials Science and Engineering **523** (2019) 012082 doi:10.1088/1757-899X/523/1/012082

The collected samples at Ujung Batee area are mostly Fine SAND. The Medium SAND is found at UB4. The sediment description is based on the Median or Mean of ϕ ([11]; [12]).

Table 3. Summary of estimated ϕ from the cumulative curve of Phi size

No	Sampling Point	Median of ϕ	Mean of ϕ	SD of ϕ	Skewness	Kurtosis	Soil description
1	LO 1	1.15	1.02	1.08	-0.16	0.84	Medium SAND
2	LO 2	1.35	1.33	0.88	0.00	0.87	Medium SAND
3	LO 3	1.50	1.47	0.72	-0.25	1.64	Medium SAND
4	LO 5	2.00	1.73	1.47	-0.24	0.81	Medium to Fine SAND
5	LO 6	1.25	1.22	1.26	-0.03	0.91	Medium SAND
6	LO 7	0.75	0.85	0.75	0.15	1.00	Coarse SAND
7	LO 8	0.05	0.18	0.88	0.27	0.84	Very coarse to coarse SAND
8	LO 9	0.30	1.00	2.44	0.60	1.27	Coarse SAND
9	LO 10	2.00	1.82	1.24	-0.18	0.88	Medium SAND
10	LO 11	3.25	3.20	1.05	0.16	1.50	Very fine SAND
11	UB 1	2.30	2.32	1.02	-0.02	0.97	Fine SAND
12	UB 2	2.20	2.15	1.13	-0.12	1.23	Fine SAND
13	UB 4	1.40	1.28	1.23	-0.06	1.02	Medium SAND

4.2. Sorting, skewness and interpretation

Grain size analysis determines the strength of currents that transported the sediment. Therefore, the paleohydraulic conditions size can be interpreted. In addition, the variations in sorting and skewness reflect the transport mechanism of the sediment. The result is presented in Figure 5.

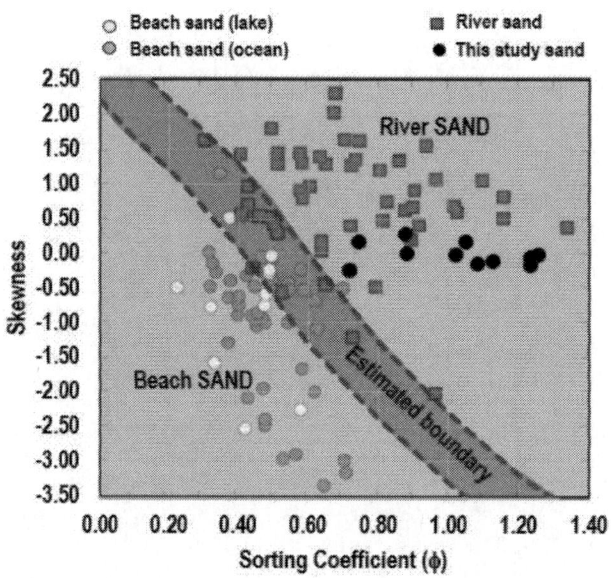

Figure 5. Depositional process interpretation ([11]; [12]).

IOP Conf. Series: Materials Science and Engineering **523** (2019) 012082 doi:10.1088/1757-899X/523/1/012082

5. Conclusions

Grain size characteristics and sediment types of the Aceh's coastal deposits based on the sieve analysis has been investigated. This study is mainly to interpret the support mechanism of sediment during transportation and depositional processes. Thirteen sediment deposits from both east and west coasts of Aceh were collected and tested using sieve analysis. Further particle size statistics analysis reveals the past transport mechanism of this study sand deposits which is river's channel.

6. References

[1] Cantalamessa G and Di Celma C 2005 *Sedimentary Geology* **178** 259–273.
[2] Choowong M, Murakoshi N, Hisada K, Charoentitirat T, Charusiri P, Phantuwongraj S, Wongkok P, Choowong A, Subsayjun R, Chutakositkanon V, Jankaew K and Kanjanapayont P 2008 *Terra Nova* **20** 141–149.
[3] Dawson AG, Shi S, Dawson S, Takahashi T and Shuto N 1996 *Quaternary Science Reviews* **15** 901–912.
[4] Heathershaw AD 1981 *Marine Geology* **42** 75–104.
[5] Mulder T and Alexander J 2001 *Sedimentology* **48** 269–299.
[6] Jiang C, Wu Z, Chen J, Deng B and Long Y 2015 *Procedia Engineering* **116** 771 – 777.
[7] Moore A, Nishimura Y, Gelfenbaum G, Kamataki T and Triyono R 2006 *Earth Planets and Space* **58** 253–258.
[8] Paris R, Lavigne F, Wassmer P and Sartohadi J 2007 *Marine Geology* **238** 93–106.
[9] Paris R, Wassmer P, Sartohadi J, Lavigne F, Barthomeuf B, Desgages É, Grancher D, Baumert Ph, Vautier F, Brunstein D and Gomez Ch 2009 *Geomorphology* **104** 59–72.
[10] Yalin MS 1972 *Mechanics of Sediment Transport*. Pergamon, New York.
[11] Folk RL 1981 *Petrology of sedimentary rocks*. Hemphill Pub Co, 2 edition, 182 pp.
[12] Folk RL and Ward WC 1957 *Journal of Sedimentary Petrology* **27** 3-26.

IOP Conf. Series: Materials Science and Engineering **523** (2019) 012083 doi:10.1088/1757-899X/523/1/012083

Study of temperature loss of Hot Mix Asphalt during transportation

Muhammad[1], A Syuhada[1], S Huzni[1,*], Z Fuadi[1]
Department of Mechanical Engineering, Syiah Kuala University, Banda Aceh, Indonesia

Email: syifaul@unsyiah.ac.id

Abstract-This study was conducted to observe the phenomenon of the temperature loss on hot mix asphalt type Asphalt Concrete Binder Course (AC-BC), on the surface area of the truck wall during transport. The tests were conducted experimentally using thermometers and thermocouples with a 25-ton capacity truck at 153°C loading temperatures. Data retrieval is carried out over time, 20.40,60 to 420 minutes of transport, from the Asphalt Mixing Plant (AMP) plant to the job site with an average truck speed of 30 km/h. Data from the results of the study were analyzed using the descriptive method. The results indicate that the maximum transport distance that meets the technical specifications is at the furthest distance of 165 km, with a minimum temperature of 130,1°C, at a distance of 10 cm from the surface of the truck wall. Loss of average hot mix asphalt temperature to 6 (six) sides of the truck wall surface, floor surface and the top surface, for 420 minutes of transport is (45.2°C-65.7°C). From the phenomenon shows that the hot mix asphalt condition on the surface area of the truck wall is hardened (crusty).

1. Introduction

Asphalt Concrete Binder Course (AC-BC) is often used as an intermediate coating for road construction in Indonesia, including in Aceh province, with an area of 5,677,081 km^2 and 17,372.55 km lengths of national, provincial and district roads [1]. With the huge area and length of the roads, the growth of Asphalt Mixing Plant (AMP) plant is not proportional, the construction requires a huge investment cost [2]. The 140-210 km distance is the furthest distance traveled by road construction practitioners in Aceh to bring hot mix asphalt from Asphalt Mixing Plant (AMP) plant to work site. Therefore, it is very likely that there will be a loss of temperature on hot mix asphalt due to heat transfer during transport, especially around the surface of the truck wall [3]. From this phenomenon, the smoothing temperature required by the technical specifications at a minimum temperature of 130°C is not reached [4].

The long distance and a lot of time required to bring the hot mix asphalt from the Asphalt Mixing Plant (AMP) plant to the work site cause the loss of temperature on the hot mix asphalt, this phenomenon is very contrasting happen around the truck wall, while in the middle area is relatively stable [5]. The temperature loss of hot mix asphalt often occurs due to the influence of wind speed and weather during transport, and daylight is the right time to do pavement work [6]. The adverse conditions of insulation (tarpaulin cover) on the surface of the truck were one of the factors of temperature loss on hot mix asphalt during transport. [7].

From the above phenomenon will the study to observe the rate of temperature loss on hot mix asphalt, type Asphalt Concrete Binder Course (AC-BC) on the surface of the truck wall during transport. It is hoped that this study can provide hot mix asphalt temperature data, on the surface of the truck's wall during transport, so that the is known distance and length of transport times that meet the technical specifications [4].

*Corresponding author: syifaul@unsyiah.ac.id

IOP Conf. Series: Materials Science and Engineering **523** (2019) 012083 doi:10.1088/1757-899X/523/1/012083

2. Materials and Research Method

2.1. Materials and Tools

The material used in this research on hot mix asphalt AC-BC, from Asphalt Mixing Plant (AMP) plant factory production result as per technical specification [4] while the tool used in this study is a thermometer, thermocouple using an intercooler truck with a capacity of 25 ton.

2.2. Research Materials Preparation

Asphalt hot mix AC-BC as the materials used in this research was prepared at Asphalt Mixing Plant (AMP) plant in granular condition at 153°C temperature, as many as 3 (three trucks) with the capacity each of 25 ton/truck, before being transported to the location of the work with mileage of 212 km.

2.3. Data retrieval system

The temperature measurement process in this study was conducted experimentally using thermometer and thermocouple, 15 (fifteen) minutes of completion of hot mix asphalt loading into the truck and after sealing with the tarp cover. Data retrieval was conducted at intervals of 20.40,60 to 420 minutes during transport, from the AMP plant to the work site with an average truck speed of 30 km/h.

2.4. Research Design and Data Analysis

At the making of Asphalt Concrete Binder Course (AC-BC) in this research need to keep some process. The mixture process comprises coarse aggregate, fine aggregate, the filler, must be clean and free of dirt, clay or any other object and dry before it is inserted into the mixer. These materials are reasonably available at the Asphalt Mixing Plant (AMP) plant site according to the aggregate composition for mixing such as 60/70 asphalt binder and additive. The entire process including the mixing process is carried out at high temperatures of 145^{0}C-155^{0}C and loading temperature into the transport truck 135^{0}C-150°C, as per the technical specifications [4]. Data from the results of this study were analyzed using the descriptive method. In this study, there are 15 (fifteen) points of data collection design, of which 6 (six) points are outside the surface of the truck wall, and 9 (nine) points are inside the truck surface as shown in Figure 1.

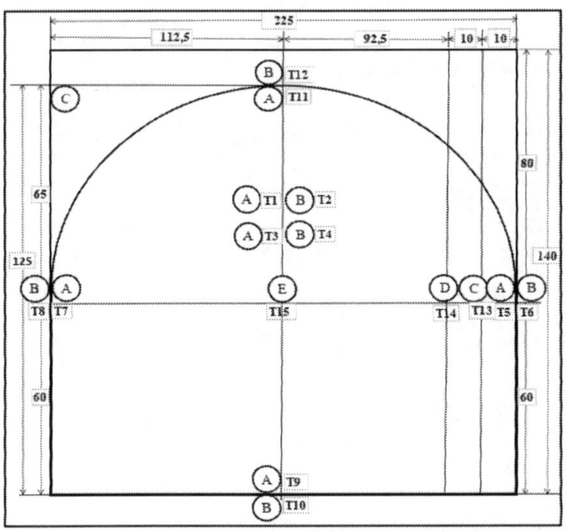

Figure 1. Simulation of thermocouple installation

Caption:

A (T1) = The front wall inside
B (T2) = The outer front wall
A (T3) = Rear wall inside
B (T4) = Outside rear wall
A (T5) = Right inner wall
B (T6) = The outside right-hand wall
A (T7) = Left side wall inside
B (T8) = The outside right-hand wall
A (T9) = Inner floor
B (T10) = Outside floor
A (T11) = Top surface inside
B (T12) = Top outer surface
C (T13) = Distance 10 cm from wall
D (T14) = Distance 20 cm from wall
E (T15) = Distance To the surface

3. Result and Discussion

3.1. Observation of Temperature of Hot Mix Asphalt on Surface of Truck Front Wall

The temperature observation data are obtained based on Figure 1 position of point A (T1) and point B (T2), for 420 minutes of transport, can be seen in Figure 2.

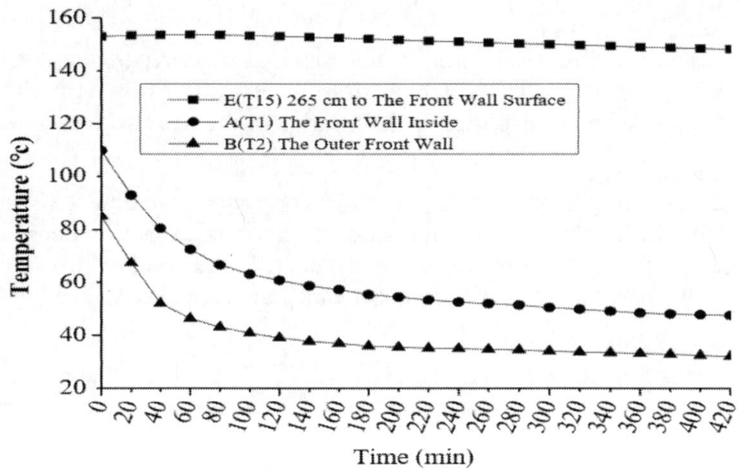

Figure 2. The relationship of the surface temperature of the front truck wall to the time of transport

Figure 2 shows of temperature loss rate of hot mix asphalt on the inner and outer truck front walls for 420 minutes of transport time. Based on Figure 1 of the initial temperature of position E (T15) 153,0^{0}C, to 148,3^{0}C, decrease (by 4,7^{0}C). Position A (T1) 109.9°C, to 47.5°C, decrease (by 62.4°C). Position B (T2) 85.2^{0}C, to 32.0^{0}C, decrease (by 53.2^{0}C).

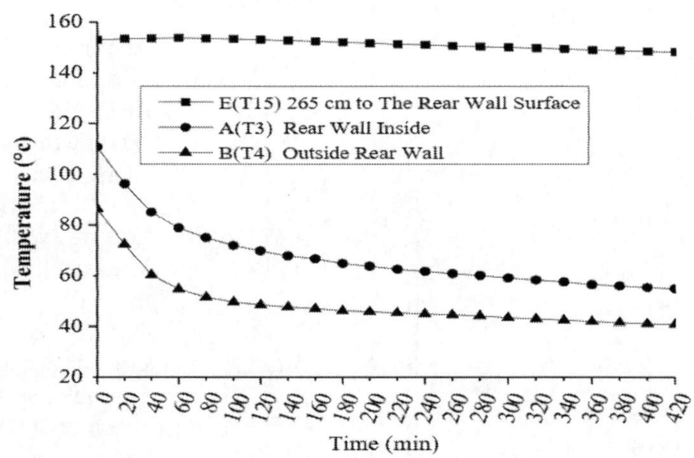

Figure 3. Relation of the surface temperature of the rear truck wall with transport time

The decrease in hot mix asphalt temperature, due to the temperature difference in the middle of the truck with the temperature around the surface of the truck wall inside and outside. At the start of transporting position the temperature E (T15) 153,0^{0}C, position A (T5) 108,5^{0}C, and position B (T6)

85,6^{0}C, the temperature difference (by 44,5^{0}C, and 22,9^{0}C). Arriving at the location of the positioning of temperature E (T15) 148,3^{0}C, position A (T5) 42,8^{0}C, and position B (T6) 34,5^{0}C, the temperature difference (by 105,5^{0}C and 8,3^{0}C). The rate of temperature loss from this hot mix asphalt is decreasing more with the increase of time of transportation.

3.2. Observation of temperature of Hot Mix Asphalt on Rear Surface of Truck Wall

The temperature observation data were obtained from Figure 1 position A(T3) and B(T4), for 420 minutes of transport shown in Figure 3. Figure 3 shows of temperature loss rate of hot mix asphalt on the inner and outer rear wall surface for 420 minutes of transport time. Based on Figure 1 of the initial temperature of position E (T15) 153,0^{0}C, to 148,3^{0}C, decrease (by 4,7^{0}C). Position A (T3) 110.8°C, to 54.9°C, decrease (by 55.9°C). The position B (T4) 86.4^{0}C, becomes 40.9°C, decrease (by 45.5^{0}C).

The decrease in hot mix asphalt temperature, due to the temperature difference in the middle of the truck with the temperature around the surface of the truck wall inside and outside. At the start of transporting position temperature E (T15) 153,0^{0}C, position A (T3) 108,8^{0}C, and position B (T4) 86,6^{0}C, the temperature difference (equal to 44,2^{0}C, and 22,2^{0}C). Arriving at the location of positioning position E (T15) 148,3^{0}C, position A (T3) 54,9^{0}C, and position B (T4) 40,9^{0}C, the temperature difference (by 93,4^{0}C and 14,0^{0}C).Different turbulence factors cause the occurrence of hot mix asphalt temperature difference on the surface of the front truck wall with the back surface. On the front of the truck's trunk, the outer air turbulence is very high, while on the surface of the rear truck, less air flowing turbulence, in other words more laminar flow.

3.3. Observation of temperature of Hot Mix Asphalt on Right Side Wall Surface Truck 10 cm, 20 cm, 112,5 cm

The temperature observation data were obtained based on Figure 1 position A(T5) and B(T6), for 420 minutes of transport can be seen in Figure 4.

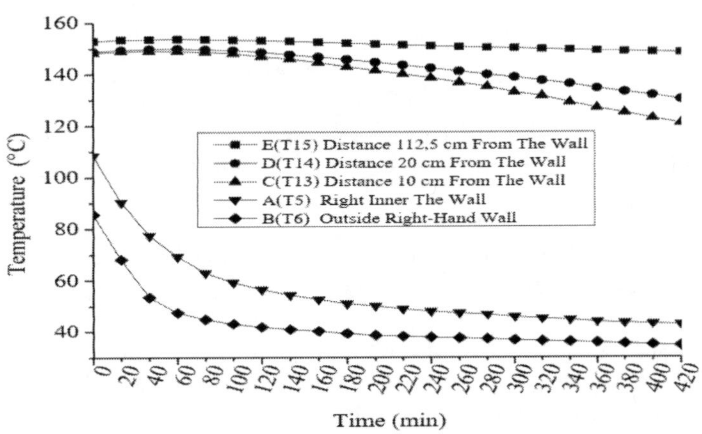

Figure 4. Relation of the surface temperature of the right side wall of the truck, 10 cm, 20, 112,5 cm with a time of transport

Figure 4 shows of temperature loss rate of hot mix asphalt on the inner and outer truck right wall surface on 10 cm, 20, 112,5 cm distance for 420 minutes of transport. Based on Figure 1 of the initial temperature of position C (T13) 148.9°C, to 120.9°C, decrease (by 28.0°C). The position D (T14) at a distance of 20 cm, from an initial temperature of 148.8°C, becomes 130.1°C, decreasing (by 18.7°C). The position of E (T15) at a distance of 112.5 cm, the temperature rate in the first 60 minutes of transport, from a temperature of 153.0°C to 153.7°C, increased (by 0.7°C). It decreases again after the first 60 minutes to 420 minutes of transport from 153.7°C, to 148.3 ° C, decreasing (by 5.4°C).

Position A (T5) 108,5^{0}C, to 42,8^{0}C, decrease (by 65,6^{0}C). The position B (T6) is 85.6^{0}C, to 34.5^{0}C, decrease (by 51.1^{0}C).

The decrease in hot mix asphalt temperature, due to the temperature difference in the middle of the truck with the temperature around the inside and outside surface of the truck wall. At the start of transporting position temperature E (T15) 153,0^{0}C, position A (T5) 108,5^{0}C, and position B (T6) 85,6^{0}C, the temperature difference (equal to 44,5^{0}C, and 67,5^{0}C). Arriving at the location of position E (T15) 148,3^{0}C, position A (T5) 42,8^{0}C, and position B (T6) 34,5^{0}C, the temperature difference (by 105,5^{0}C and 113,8^{0}C). While position temperature C (T13) 148,9^{0}C, position D (T14) 148,8^{0}C, and position A (T5) 108,5^{0}C, the temperature difference (by 40,4^{0}C, and 40,3^{0}C), arriving at the location of positioning position C (T13) 120,9^{0}C, position D (T14) 130,1^{0}C, and position A (T5) 42,8^{0}C, the different temperature of both position (by 78,1^{0}C and 87,3^{0}C). The result of the study at a distance of 20 cm from the surface of the final-temperature truck was 130.1°C still in a safe position, but at a distance of 10 cm after 420 minutes of final-freight time was 120.9°C below the technical specification.

3.4. Observation of Temperature Loss of Hot Mix Asphalt on Surface of Truck Left Wall

The temperature observation data were obtained from Figure 1 position A(T7) and B(T8), for 420 minutes of transport shown in Figure 5.

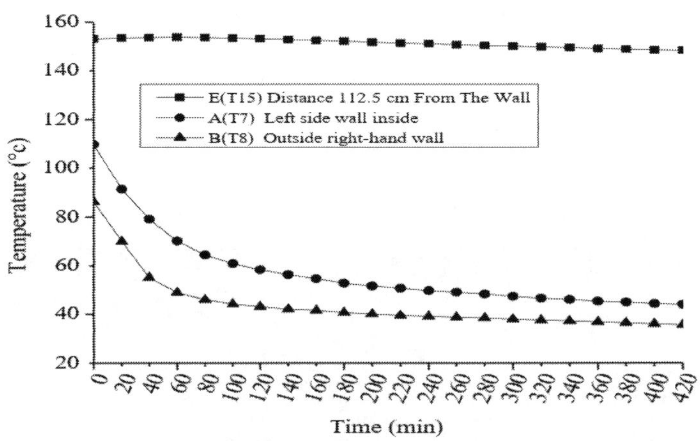

Figure 5. Relation of the surface temperature of the left side wall of the truck with a time of transport

Figure 5 shows of temperature loss rate of hot mix asphalt on the inner left side truck wall surface and outer, for 420 minutes of transport time. Based on Figure 1 of the initial temperature of position E (T15) 153,0^{0}C, to 148,3^{0}C, decrease (by 4,7^{0}C). Position A (T7) 109.6^{0}C to 43.9^{0}C, decrease (by 65.7°C). Position B (T8), 86.2^{0}C to 35.6^{0}C, decrease (by 50.6°C).

The decrease in hot mix asphalt temperature, due to the temperature difference in the middle of the truck with the temperature around the surface of the truck wall inside and outside. At the start of transporting position temperature E (T15) 153,0^{0}C, position A (T7) 109,6^{0}C, and position B (T8) 86,2^{0}C, the temperature difference (by 43,4^{0}C, and 23,4^{0}C). Arriving at the location of position E (T15) 148,3^{0}C, position A (T7) 43,9^{0}C, and position B (T4) 35,6^{0}C, the temperature difference (by 104,4^{0}C and 8,3^{0}C).The final temperature of both the right and left wall surfaces did not occur difference significantly, this is due to the turbulence that occurs on both surfaces of the wall is almost the same. While the temperature loss rate of hot mix asphalt, decreasing along with increasing the time of transportation.

3.5. Observation of Temperature Loss of Hot Mix Asphalt on Floor Surface of truck

The temperature observation data were obtained from Figure 1 position A(T9) and B(T10), for 420 minutes of transport shown in Figure 6. Figure 6 shows of temperature loss rate of hot mix asphalt on the inner floor truck wall surface and outer, for 420 minutes of transport time. Based on Figure 1 of the initial temperature of position E (T15) 153,0°C, to 148,3°C, decrease (by 4,7°C). Position A (T9) 111.9°C, to 53.5°C, decrease (by 58.4°C). Position B (T10), 94.7°C to 45.0°C, decrease (by 49.7°C).

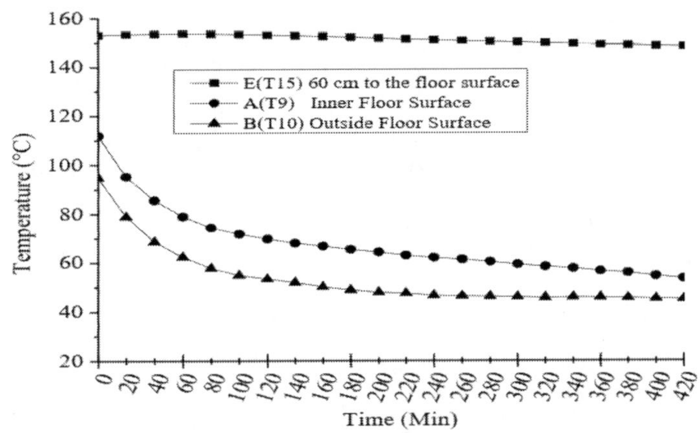

Figure 6. Relationship of truck floor the surface temperature to transport time

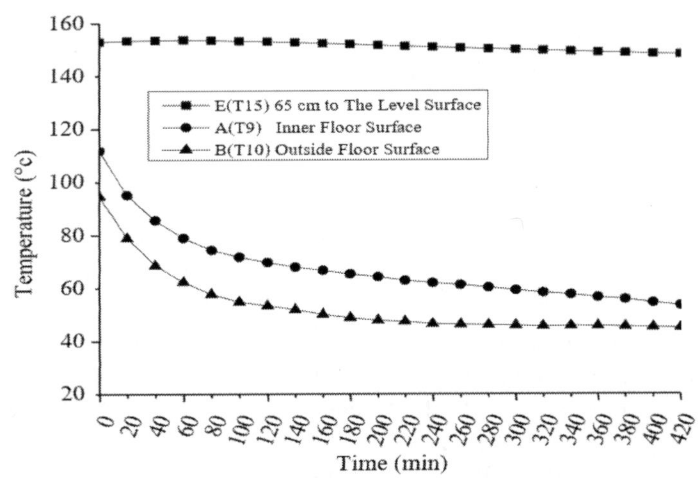

Figure 7. Relation of top surface temperature to transport time

The decrease in hot mix asphalt temperature, due to the temperature difference in the middle of the truck with the temperature around the surface of the truck wall inside and outside. At the start of transporting position temperature E (T15) 153,0°C, position A (T9) 111,9°C, and position B (T10) 94,7°C, the temperature difference (by 41,1°C and 17,2°C). Arriving at the location of positioning position E (T15) 148,3°C, position A (T9) 53,5°C, and position B (T10) 45,0°C, the temperature difference (by 94,8°C and 8,5°C). The rate of temperature loss from this hot mix asphalt is decreasing more with the increase of time of transportation.

3.6. Observation of Temperature Loss of Hot Mix Asphalt on Surface Upper Truck
The temperature observation data were obtained from Figure 1 position A(T11) and B(T12), for 420 minutes of transport shown in Figure 7. Figure 7 shows of temperature loss rate of hot mix asphalt on the truck upper surface area of the inner and outer, for 420 minutes of transport time. Based on Figure 1 of the initial temperature of position E (T15) 153,0^{0}C, to 148,3^{0}C, decrease (by 4,7^{0}C). Position A (T11) 92.6^{0}C, to 47.4^{0}C, decrease (by 45.2^{0}C). Position B (T12) 84.8°C, to 41.2^{0}C, decrease (by 43.6°C). The decrease in hot mix asphalt temperature, due to the temperature difference in the middle of the truck with the temperature around the surface of the truck wall inside and outside. At the start of transporting position temperature E (T15) 153,0^{0}C, position A (T11) 92,6^{0}C, and position B (T12) 84,8^{0}C, the temperature difference (by 60,4^{0}C, and 78,8^{0}C). Arriving at the location of positioning position E(T15) 148,3^{0}C, position A (T11) 47,4^{0}C, and position B (T12) 41,2^{0}C, the temperature difference (by 100,9^{0}C and 6,2^{0}C). As transport times increase, the rate of temperature loss increasingly. Differences in the hot mix asphalt temperature on the surface of the truck floor with the upper surface are is more due to different turbulence effect factors. On the floor surface turbulence outside the air that flows less, in other words, the flow is more laminar. While on the upper surface of air turbulence flowing higher.

4. Conclusion
From the experimental result got some conclusions:
1. Loss of average hot mix asphalt temperature to (six) side of the truck wall surface, floor surface and the top surface of the truck, for 420 minutes of transport is (45.2°C-65,7°C).
2. Highest temperature losses occur on the left and right truck wall surfaces, while the lowest on the surface of the rear truck's wall.
3. The results indicate that the maximum transport distance that meets the technical specifications is at the furthest distance of 165 km, with a minimum temperature of 130,1^{0}C, at a distance of 10 cm from the surface of the truck wall.
4. For the 6 (six) sides of the outer next truck surface, the average temperature loss is (43.6°C-53.2°C). The effect of the phenomenon causes hot mixed asphalt around the inside surface of the inner truck becomes hard (crusty).

5. Acknowledgments
The author would like to thank the Director of PT. Bohana Jaya Nusantara Regency of North Aceh as the leader of Asphalt Mixing Plant (AMP) company which has provided all facilities and permit sampling of hot mix Asphalt Concrete Binder Course (AC-BC).

6. References
[1] Aceh Regional Planning Agency 2015 http:// bappeda.acehprov.go.id
[2] Syaiful Bahri 2005 Analysis of Selection of Asphalt mixing plant (AMP) location at Padang-Sambutan Road Works *Proceedings of National Seminar* on Technology Management II, Surabaya (in Bahasa Indonesia)
[3] Bergman, Lavine, De Witt, Incropera 2007 *Fundamentals of Heat and Mass Transfer edition 6* (Wiley Publisher)
[4] Kementrian Pekerjaan Umum Direktorat Jenderal Bina Marga 2010 *General Specification Division 6 Asphalt Pavement Pp 6.1- 6.5* (in Bahasa Indonesia)
[5] Isaac L, Howard, L Gaylon, Baumgardner, S Walter, III Jordan, M Andrew, Menapace, S Walaa, Mogawer, H M James & In Jr 2014 *Journal of the Transportation Research Board* No **2408** pp 78–85
[6] Hashim. W, H R. Mohd, I N Norfarah, M I Nur, Yusoff, A E Mohd & H A Norhidayah 2015 *Journal of Teknologi* **78**: 4 pp 97–104
[7] Byzyka J, R Mujib, & C A Denis 2017 *Journal of traffic and transportation engineering* Volume **4** Issue 4 pp 360-371